Red Blood Cell Membranes

HEMATOLOGY

Series Editors

Kenneth M. Brinkhous, M.D.
Department of Pathology
University of North Carolina
School of Medicine
Chapel Hill, North Carolina

Sanford A. Stass, M.D.
Hematopathology Program
The University of Texas System Cancer Center
M. D. Anderson Hospital and Tumor Institute
Houston, Texas

Additional Volumes in Preparation

Red Blood Cell Membranes

Structure • Function • Clinical Implications

edited by

Peter Agre
The Johns Hopkins University
School of Medicine
Baltimore, Maryland

John C. Parker
University of North Carolina at Chapel Hill
School of Medicine
Chapel Hill, North Carolina

CRC Press
Taylor & Francis Group
Boca Raton London New York

CRC Press is an imprint of the
Taylor & Francis Group, an **informa** business

First published 1989 by Marcel Dekker, Inc.

Published 2019 by CRC Press
Taylor & Francis Group
6000 Broken Sound Parkway NW, Suite 300
Boca Raton, FL 33487-2742

© 1989 by Taylor & Francis Group, LLC
CRC Press is an imprint of Taylor & Francis Group, an Informa business

First issued in paperback 2019

No claim to original U.S. Government works

ISBN-13: 978-0-367-45114-1 (pbk)
ISBN-13: 978-0-8247-8022-7 (hbk)

Visit the Taylor & Francis Web site at
http://www.taylorandfrancis.com

and the CRC Press Web site at
http://www.crcpress.com

Library of Congress Cataloging-in-Publication Data

Red blood cell membranes : structure, function, clinical implications / edited
    by Peter Agre, John C. Parker.
        p.   cm. -- (Hematology ; v. 11)
    Includes bibliographies and index.
    ISBN 0-8247-8022-1
    1. Erythrocytes. 2. Cell membranes. 3. Membrane proteins.
4. Erythrocyte disorders. 5. Membrane disorders. I. Agre, Peter.
II. Parker, John C. III. Series: Hematology (New York, N.Y.) ; v. 11.
    [DNLM: 1. Erythrocyte Membrane--anatomy & histology.
2. Erythrocyte Membrane--physiology. W1 HE873 v. 11 / WH 150
R2883]
QP96.R29   1989
612'.111--dc19
DNLM/DLC
for Library of Congress   89-1076
                                                                     CIP

Robert L. Ney, M.D. 1933–1986

Clinician, Scholar, Scientist, Leader, and Friend

Bob Ney devoted the majority of his professional life to the
Departments of Medicine at the University of North Carolina
at Chapel Hill and The Johns Hopkins University, where he contributed
generously to the numerous needs of students, house officers,
and faculty, including the editors of this volume. Bob is sadly
missed, and this book is dedicated to his memory.

Series Introduction

For most of this century, hematology has followed a pattern of major scientific discoveries, improved understanding of disease, and rapid application of new knowledge in the clinic. The rate of advance continues at an accelerating pace, so that all but the most zealous have difficulty in keeping up with the literature in even a limited area of specialized interest. As the explosive development of knowledge continues apace, it is a continuing challenge to keep abreast of significant new developments as they impact on clinical and laboratory hematology. The Hematology series is designed to help in this respect, by providing up-to-date and expert presentations on important subject areas in our field. It is hoped that these works, both individually and collectively, will become important volumes for updating information and for reference for the clinician, investigator, teacher, and student, and in this manner contribute to the advancement of hematology.

This volume of the series is devoted to the red blood cell membrane, its structure and function, and abnormalities in disease states. The red blood cell provides a simple model to study basic cell function. From this has flowed an enormous amount of fundamental information, making the red blood cell the most thoroughly studied and best understood of any cell of the body. The erythroid cell system is an especially dynamic one, producing 2×10^6 red blood cells per second. The delineation of the composition of the red blood cell membrane, with its specialized proteins, and of the submembraneous cytoskeleton, with spectrin the major component, provides a clear picture of cellular architecture at the molecular level. The modern tools of protein purification and peptide synthesis, high-resolution electron microscopy, the new immunology, and molecular biology have all been required to reach the present level of understanding. The reader will find many intriguing concepts presented throughout the book, such as the mechanism of how the red blood cell in its differentiation loses its nucleus along with two of the three major filamentous systems of the cell—the microtubules and the intermediate filaments—retaining only the actin filaments. Another is the evolution of the Heinz body, with its attachment to the cytoplasmic aspect of the cell membrane, causing senescence and immunological removal of the cell from the circulation. The application of this new information to many disease states, especially the hereditary anemias, is the subject of several chapters.

The editors, Peter Agre and John C. Parker, and their coauthors have contributed significantly to the development of the present state of the art and are authorities in their respective subject areas. This book presents a well-documented and well-illustrated comprehensive picture of current knowledge of the field, from fundamental cell biology to the pathophysiology and clinical manifestations of red blood cell disorders. The field is an active one and the volume can also serve as an introduction to what lies ahead.

Kenneth M. Brinkhous
Series Coeditor

Preface

Transport physiologists have traditionally focused their efforts on mechanisms of solute and water movement through membranes, whereas biologists studying membrane structure have been concerned with protein–protein interactions. Despite sharing a fascination with the membrane of the red blood cell, members of the two disciplines have demonstrated little understanding of each other's work. As the level of science has matured in both fields, it has become apparent that students of transport and structure are both concerned with the same membrane constituents. For example, the most abundant integral membrane protein is known to the transport physiologist as the *anion transporter* and to the structural biologist as *band 3*, a principal site of attachment between the membrane skeleton and the lipid bilayer.

Recent developments in a cellular and molecular biology have contributed powerful insights into the biogenesis, structure, and organization of the red blood cell membrane. The entire nucleotide and amino acid sequences have recently been determined for several of the major polypeptides, permitting new approaches to understanding the physiological roles of these molecules. Clinical investigators, armed with the new technology, are discovering the molecular basis for congenital red blood cell membrane disorders that for decades had been the province of descriptive hematology. Such natural mutations have much to teach about structure-function relationships.

Investigators in basic red blood cell research now need to be fluent in the language of many disciplines. Furthermore, most lessons learned about the red blood cell membrane have proven generally applicable and have provided insights into the biology of complex nucleated cells. For this reason, students in many fields of biomedical science now need to be aware of important developments relating to the red blood cell and its membrane.

Despite a heyday of research spanning two decades, the red blood cell membrane continues to provide new avenues of inquiry. Even as this volume goes to press there are fresh ideas in the wind about membrane composition and function. It is the editors' hope that this volume may encourage its readers to indulge their curiosity about the workings of this elegant structure.

Peter Agre
John C. Parker

Contributors

Peter Agre, M.D. Associate Professor, Departments of Medicine and Cell Biology, The Johns Hopkins University School of Medicine, Baltimore, Maryland

Richard A. Anderson, Ph.D. Assistant Professor, Department of Pharmacology, University of Wisconsin Medical School, Madison, Wisconsin

George Vann Bennett, M.D., Ph.D. Professor and Investigator, Howard Hughes Medical Institute, Department of Biochemistry, Duke University Medical Center, Durham, North Carolina

Edward J. Benz, Jr., M.D. Professor of Internal Medicine and Human Genetics and Chief, Hematology Section, Departments of Internal Medicine and Human Genetics, Yale University School of Medicine, New Haven, Connecticut

David A. Berk, Ph.D.* Research Associate, Department of Mechanical Engineering and Materials Science, Duke University, Durham, North Carolina

Lee R. Berkowitz, M.D. Assistant Professor, Department of Medicine, University of North Carolina at Chapel Hill School of Medicine, Chapel Hill, North Carolina

Douglas Brailsford, Ph.D. Associate Professor, Department of Clinical Laboratory Science, School of Allied Health Professions, Loma Linda University, Loma Linda, California

---

**Current affiliation*: Postdoctoral Fellow, Department of Academic Pathology, The University of British Columbia, Vancouver, British Columbia, Canada

Brian S. Bull, M.D. Professor and Chairman, Department of Pathology and Laboratory Medicine, School of Medicine, Loma Linda University, Loma Linda, California

John G. Conboy, Ph.D. Assistant Research Biochemist, Cancer Research Institute, University of California, San Francisco, San Francisco, California

Philip B. Dunham, Ph.D. Professor, Department of Biology, Syracuse University, Syracuse, New York

Bernard G. Forget, M.D. Professor, Department of Medicine and Human Genetics, Yale University School of Medicine, New Haven, Connecticut

Otto Fröhlich, Ph.D. Assistant Professor, Department of Physiology, Emory University School of Medicine, Atlanta, Georgia

Kevin Gardner, M.D., Ph.D.* Research Fellow, Department of Cellular Biology and Anatomy, and Division of Hematology, Department of Medicine, The Johns Hopkins University School of Medicine, Baltimore, Maryland

Wendy P. Gati, Ph.D. Assistant Professor, McEachern Laboratory and Department of Pharmacology, University of Alberta, Edmonton, Alberta, Canada

David E. Golan, M.D., Ph.D. Assistant Professor, Departments of Medicine and of Biological Chemistry and Molecular Pharmacology, Harvard Medical School; and Hematology Division, Brigham and Women's Hospital, Boston, Massachusetts

Anne M. Griffin, M.D. Fellow in Hematology, Department of Medicine, University of North Carolina at Chapel Hill School of Medicine, Chapel Hill, North Carolina

Robert B. Gunn, M.D. Professor and Chairman, Department of Physiology, Emory University School of Medicine, Atlanta, Georgia

Mark Haas, M.D., Ph.D. Assistant Professor, Departments of Pathology and Cellular and Molecular Physiology, Yale University School of Medicine, New Haven, Connecticut

---

Current affiliation: Resident, Laboratory of Pathology, National Cancer Institute, National Institutes of Health, Bethesda, Maryland

Robert M. Hochmuth, Ph.D. Professor and Chairman, Department of Mechanical Engineering and Materials Science, Duke University, Durham, North Carolina

Rose M. Johnstone, Ph.D. Professor and Chairman, Department of Biochemistry, McGill University, Montreal, Quebec, Canada

Jack H. Kaplan, Ph.D. Professor, Department of Physiology, University of Pennsylvania, Philadelphia, Pennsylvania

Patricia A. King, Ph.D.* Research Associate, Department of Physiology, Emory University School of Medicine, Atlanta, Georgia

Elias Lazarides, Ph.D. Professor, Division of Biology, California Institute of Technology, Pasadena, California

Thomas L. Leto, Ph.D.† Associate Research Scientist, Department of Pathology, Yale University School of Medicine, New Haven, Connecticut

Philip S. Low, Ph.D. Professor, Department of Chemistry, Purdue University, West Lafayette, Indiana

Allan G. Lowe, Ph.D. Senior Lecturer in Biochemistry, Department of Biochemistry and Molecular Biology, School of Biological Sciences, University of Manchester, Manchester, England

Sally L. Marchesi, M.D., Ph.D. Associate Professor, Department of Laboratory Medicine and Pathology, Yale University School of Medicine, New Haven, Connecticut

Robert W. Mercer, Ph.D. Assistant Professor, Department of Cell Biology and Physiology, Washington University School of Medicine, St. Louis, Missouri

Narla Mohandas, D.Sc. Professor, Department of Laboratory Medicine and Cancer Research Institute, University of California, San Francisco, San Francisco, California

Mike M. Mueckler, Ph.D. Assistant Professor, Department of Cell Biology and Physiology, Washington University School of Medicine, St. Louis, Missouri

---

Current affiliations:
*Research Associate and Instructor, Department of Medicine, The University of Vermont College of Medicine, Burlington, Vermont
†Senior Staff Fellow, Laboratory of Clinical Investigation, National Institute of Allergy and Infectious Diseases, National Institutes of Health, Bethesda, Maryland

John Ngai, Ph.D.* Postdoctoral Research Fellow, Division of Biology, California Institute of Technology, Pasadena, California

John C. Parker, M.D. Professor, Department of Medicine, University of North Carolina at Chapel Hill School of Medicine, Chapel Hill, North Carolina

Alan R. P. Paterson, Ph.D. Professor, McEachern Laboratory and Departments of Biochemistry and Pharmacology, University of Alberta, Edmonton, Alberta, Canada

Wendell F. Rosse, M.D. Florence McAlister Professor of Medicine, Department of Medicine, Duke University Medical Center, Durham, North Carolina

Jay W. Schneider Department of Human Genetics, Yale University School of Medicine, New Haven, Connecticut

Betty W. Shen, Ph.D. Scientist, Biological, Environmental, and Medical Research Division, Argonne National Laboratory, Argonne, Illinois

David G. Shoemaker, Ph.D. Associate, Department of Physiology, Emory University School of Medicine, Atlanta, Georgia

Marilyn J. Telen, M.D. Assistant Professor, Division of Hematology/Oncology, Department of Medicine, and Associate Medical Director, Transfusion Service, Duke University Medical Center, Durham, North Carolina

Frank F. Vincenzi, Ph.D. Professor, Department of Pharmacology, University of Washington, Seattle, Washington

Adrian R. Walmsley, Ph.D. Research Associate, Department of Biochemistry, Leicester University, Leicester, England

Richard E. Waugh, Ph.D. Associate Professor, Department of Biophysics, University of Rochester School of Medicine and Dentistry, Rochester, New York

John C. Winkelmann, M.D.† Postdoctoral Fellow, Section of Hematology, Department of Medicine, Yale University School of Medicine, New Haven, Connecticut

---

Current affiliations:

*Postdoctoral Research Fellow, Howard Hughes Medical Institute, Columbia University College of Physicians and Surgeons, New York, New York

† Assistant Professor, Department of Medicine and Institute of Human Genetics, University of Minnesota Medical School, Minneapolis, Minnesota

Contents

Contents

Red Blood Cell Membranes

1

Recently Identified Erythrocyte Membrane-Skeletal Proteins and Interactions
Implications for Structure and Function

KEVIN GARDNER* *The Johns Hopkins University School of Medicine, Baltimore, Maryland*

GEORGE VANN BENNETT *Howard Hughes Medical Institute, Duke University Medical Center, Durham, North Carolina*

I. INTRODUCTION

A common approach in modern cell biology and medicine has been to elucidate the various processes and attributes of the living organism through an analysis of the behavior of its most basic constituents. Consequently, much of what is now known about the macromolecular assemblies responsible for cellular function and integrity has arisen from exhaustive in vitro characterization of the individual components that contribute to their structure. This approach, although limited to the test tube, has provided much of the guidance for the construction of conceptual frameworks from which the rules governing cell function and organization can be inferred. Such strategies have found great utility in the elucidation of the factors responsible for the strength and mechanical stability of the human erythrocyte membrane, properties it derives from an anastomosing network of proteins attached to its cytoplasmic surface. Referred to as the "membrane skeleton," this macromolecular assembly has been the focus of two decades of intense investigation. The cumulative result of these efforts has been the construction of a conceptual model of the membrane skeleton consistent with both the biomedical properties of its individual protein components and the

---

Current affiliation: National Cancer Institute, National Institutes of Health, Bethesda, Maryland

physiological characteristics of its macromolecular form. The operational impact of this construct has been as a schematic paradigm for detailed discussion of membrane skeletal structure in normal and abnormal erythrocytes and for initial studies of the more complex membranes in nonerythroid cells.

The purpose of this chapter will be to review what is known about the more recently identified proteins and interactions of the red blood cell membrane and membrane skeleton. Topics to be covered include red blood cell tropomyosin, myosin, protein 4.9 adducin, and protein 4.2. Finally, the discussion will be extended to include an update on the advances in the current thinking on how known elements of the membrane skeleton, such as ankyrin and protein 4.1, interact with the membrane and how such interactions may be modulated.

II. THE CURRENT CONCEPTUAL MODEL OF THE ERYTHROCYTE MEMBRANE SKELETON

A basic understanding of the anatomy of the red blood cell membrane skeleton is a prerequisite for a clear perception of how these newer elements and interactions may contribute to its structure and function. This chapter will therefore begin with a description of the fundamental aspects of the membrane skeleton (those aspects for which there is general agreement). For a more detailed discussion, the reader should refer to other chapters in this book as well as recent reviews [1-4]. A schematic drawing depicting the molecular organization of the membrane skeleton as well as some of the recently proposed interactions is shown in Figure 1.

The major component of the membrane skeleton is a rather large protein called spectrin [5]. Spectrin is a flexible rodlike molecule present in erythrocytes at ~200,000 copies per cell. It is composed of two similar but nonidentical subunits of 260,000 daltons (spectrin α subunit) and 225,000 daltons (spectrin β subunit) that are intertwined side-to-side to form a heterodimer ~100 nm in depth [6]. Spectrin heterodimers are polar molecules with binding sites for specific proteins along different aspects of its structure. A self-association between the head regions of two spectrin heterodimers creates a 200-nm-long tetramer [6,7]. The tetrameric species of spectrin is the predominant form of spectrin in the erythrocyte [8]. Hexamers and other oligomeric forms of spectrin have been observed in vitro and also may contribute to the structure of the membrane skeleton [9,10]. The tail portions of each spectrin heterodimer contain a binding site for actin filaments; spectrin tetramers and oligomers are thus multivalent for actin and can readily cross-link actin filaments [11-14].

Actin in the red blood cell is in the form of short oligomers composed of 12-20 monomers [15,16]. This property of actin is unique to the erythrocyte because in other cells actin is present as more extensively polymerized thin filaments containing hundreds of actin monomers. Calculations that compare the total mass of polymerized actin with the amount of filament ends in the red

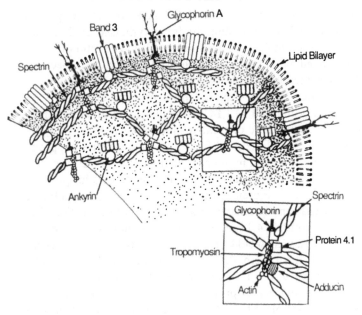

Figure 1 Schematic model for organization of proteins in the human erythrocyte membrane.

blood cell imply that there are ~25,000–30,000 actin oligomers per red blood cell [15]. The ratio of spectrin to actin oligomers estimated from this value indicates that each actin oligomer should associate with an average of about six spectrins. Such an arrangement predicts a basic morphology within the membrane skeleton where each actin oligomer is linked by spectrin tetramers to neighboring actin units, thus creating a continuous spectrin–actin lattice beneath the membrane (Figure 1). These predictions have been confirmed by high-resolution electron micrographs of isolated erythrocyte membrane skeletons [17–19]. The micrographs show a highly repeated and remarkably regular organization of the spectrin–actin complexes in which each complex is interlinked to adjacent complexes by multiple spectrin tetramers. In remarkable agreement with predictions based on relative amounts of spectrin and actin, five to eight spectrin tetramers are gathered about each actin core (Figure 1).

The major high-affinity attachment site between the membrane skeleton and the overlying lipid bilayer is provided by ankyrin [20–23;103]. Erythrocyte ankyrin is a slightly asymmetric globular protein with a molecular weight of 215,000 and it is present in the erythrocyte at 100,000 copies per cell [22]. It contains separate binding sites for both spectrin and the anion transporter (band 3) that is the major integral membrane protein of the red blood cell [24].

The membrane skeleton is thus tethered to the lipid bilayer by a stable spectrin–ankyrin–anion transporter linkage. A binding site for ankyrin is located on the beta subunit of spectrin ~20 nm from its head region [25]. The ratio of spectrin dimers to ankyrin in the red blood cell suggests that, on average, each spectrin tetramer is linked to the anion transporter by a single ankyrin.

Spectrin-actin interactions are very weak independently, but are greatly bilized by the presence of protein 4.1 [13,14]. Protein 4.1 and other proteins such as 4.9 and 4.2 were named based on relative mobility on SDS-polyacryl-amide electrophoresis gels [26]. Protein 4.1 has a molecular weight of approximately 78,000 daltons, and is present in red cells at about 200,000 copies per cell. The actin-binding tail regions of spectrin contain binding sites for protein 4.1 on each subunit [25–27,28]. Protein 4.1 increase the binding of spectrin to actin by an allosteric mechanism that increases spectrin affinity for actin. Alternately the participation of protein 4.1 in the formation of more stable ternary complexes with spectrin and actin would also be expected to enhance spectrin-actin interactions [29,30]. In any case, the actual molecular mechanism of protein 4.1 action remains obscure. The 1:1 ratio between the amount of spectrin dimers in the red blood cell and that of protein 4.1 implies that every spectrin in the red blood cell could have a protein 4.1 bound. However, because in vitro experiments suggest that each spectrin is capable of interacting with two molecules of protein 4.1 and that specific interactions exist between protein 4.1 and other proteins (see below), this impression is likely to be an oversimplification.

III. RECENTLY IDENTIFIED ERYTHROCYTE MEMBRANE–SKELETAL PROTEINS

A. Erythrocyte Tropomyosin

Tropomyosin is a well-conserved rodlike molecule present in many types of eukaryotic cells. Its functions are best understood in the contractile apparatus of muscle where, as a thin filament component, tropomyosin binds along both grooves of the actin filament helix to confer calcium sensitivity on the interaction between actin and myosin in the presence of troponin [31]. Tropomyosin is composed of two α helical subunits that are wrapped around each other in an elongated coiled-coil structure. In nonmuscle cells these subunits are slightly smaller than their counterpart in muscle, with molecular weights of 30,000 as compared with 35,000. Erythrocyte tropomyosin was first identified as polypeptides cross-reacting with chicken gizzard smooth-muscle tropomyosin that were present in red blood cell membranes isolated in the presence of millimolar concentrations of magnesium [32]. The erythroid form of tropomyosin is present as a 3:1 mixture of 29,000 dalton polypeptides and 27,000 polypeptides, respectively. These polypeptides have properties shared by other forms of tropomyosin, including: (a) Stokes radius, sedimentation coefficient, and high calculated frictional ratio; (b) stability to low pH and high temperature; (c) amino acid composition; (d) anomalous migration on sodium dodecyl sulfate (SDS) polyacrylamide gels in the presence of urea [32].

Erythrocyte tropomyosin is distinct from the other nonmuscle forms in that it binds well to actin at physiological concentrations of magnesium (2-3 mM), conditions where the nonmuscle forms bind weakly or not at all. In addition, the erythroid form of tropomyosin binds actin with a cooperativity that is not seen with its nonmuscle counterparts. This property suggests that red blood cell tropomyosin dimers may have the ability to bind each other in a tail-to-tail fashion, as has been proposed for the muscle forms of tropomyosin, which not only show cooperative binding to actin, but, like the erythroid form, bind well to actin filaments at lower concentrations of magnesium.

In vitro binding assays with erythrocyte tropomyosin indicate that a hetero-dimer of 27,000 and 29,000-dalton polypeptides may have both a higher affinity for actin and a lower magnesium requirement than the 29,000-dalton form [32]. It will be of interest in future work to separate the 27,000 and 29,000-dalton polypeptides and evaluate activities of homo- and heterodimers of these polypeptides.

Tropomyosin from other sources make actin filaments more rigid and resistant to depolymerization [33-36]. One function of erythrocyte tropomyosin may be to stabilize membrane-bound forms of actin. This property could have great importance in the biogenesis of the short actin filaments during erythroid differentiation. Because the intracellular concentrations of magnesium increase when the erythrocyte is deoxygenated [37,38], tropomyosin may provide a mechanism through which the stability of the membrane skeleton can be influenced by the oxygenation state of the red blood cell. Calmodulin, a protein homologous to troponin C, is present at micromolar concentrations in the erythrocyte [39,40] and evidence exists for an erythroid form of troponin I [41]. Considering these observations, the presence of tropomyosin in the red blood cell implies the possible existence of a troponin T-like molecule that may act with tropomyosin and the other analogues of the troponin complex to provide a calcium-sensitive means of regulating the interaction of actin with other actin-binding proteins within the membrane skeleton. This possibility is supported by the observation that tropomyosin is able to inhibit the association of red blood spectrin with actin [42].

B. Erythrocyte Myosin

Erythrocyte myosin was identified in erythrocytes as a polypeptide on SDS-polyacrylamide gels that cross-reacted with antibodies against platelet myosin [43]. Its purification from red blood cells confirms and extends earlier reports of myosinlike proteins and adenosine triphosphatase (ATPase) activity in erythrocyte-membrane-derived extracts [44,45]. The protein is remarkably similar to the nonmuscle myosins found in platelets and other tissues. Rotary shadowed micrographs of red blood cell myosin reveal a familiar 150-nm-long molecule with two globular heads and a rodlike tail. Other properties the erythrocyte myosin has in common with its nonmuscle counterparts include: (a) a calcium-activated and magnesium-inhibited ATPase activity; (b) high ATPase activity in the presence of 0.5 M KC1 and no divalent cations; (c) absent actin-activated activity in the presence of physiological concentration of KC1; (d) a dependence on myosin light-chain kinase phosphorylation for actin-activated ATPase activity [47]; and (e) the formation of 300–400-nm bipolar filaments at physiological concentrations of salt [43]. Red blood cell myosin is a hexamer composed of two identical heavy chains weighing 200,000 daltons each and two sets of light chains weighing 25,000 and 19,500 daltons. Two-dimensional peptide maps comparing the heavy chains of erythrocyte myosin with platelet myosin reveal that the heavy-chain polypeptides are nearly identical, suggesting a large degree of homology between erythrocyte myosin and the other nonmuscle forms. The major difference between the erythroid and the other nonmuscle forms resides in the light chains, because platelet myosin has light chains with molecular weights of 20,000 and 17,000, in contrast to the 25,000 and 19,500 forms of the erythroid species.

The bulk of erythrocyte myosin is not associated with membranes after hypotonic lysis. The amount of myosin retained with membranes is reported to double if the membranes are isolated in the presence of 2 mM magnesium, suggesting that binding sites for myosin may well exist on the membrane and/or membrane skeleton [43]. Currently there is no experimental evidence suggesting a function for myosin in the membrane skeleton, but a role for myosin in discocyte to echinocyte transformation and endocytosis has been hypothesized [44,45,48]. One possibility suggested by the role of myosin in skeletal muscle is that bipolar filaments of erythrocyte myosin could participate in sarcomere-like structures beneath the membrane. A significant difficulty with direct extension of the muscle paradigm to erythrocytes is that in erythrocytes actin filaments are quite small and already are crowded with spectrin and other molecules. An unorthodox possibility is that erythrocyte myosin does not require actin for movement, but instead interacts with another protein such as spectrin. In the absence of additional and experimental data concerning how myosin may interact with the membrane and membrane skeleton and how such an association might be regulated, the role of myosin in the mature erythrocyte remains a mystery.

C. Protein 4.9

Increasing interest in protein 4.9 has developed with the observation that it is actively phosphorylated by endogenous red cell cyclic-AMP-dependent protein kinases [49–52], as well as the calcium and phospholipid dependent protein kinase C [53–56]. Other investigators have noted higher levels of phosphorylation of protein 4.9 in sickled erythrocytes [57]. Because this pattern of phosphorylation could be mimicked in normal cells with calcium and ionophore, a link was suggested between the phosphorylation of protein 4.9 and increased intracellular calcium [58]. It now appears likely that this process requires the participation of protein kinase C.

Recently, protein 4.9 has been partially purified from erythrocyte membranes [59]. The procedure involved eluting the protein from detergent-insoluble membrane skeletons in low-ionic-strength alkaline buffers, followed by anion exchange chromatography. Protein 4.9 obtained in this fashion is a 145,000-dalton trimer that is thought to make up 1% of the total red blood cell membrane protein. When viewed under the electron microscope by low-angle rotary shadowing, partially pure preparations of this protein contain three-lobed structures with a diameter slightly greater than 10 nm.

Protein 4.9 induces the formation of F-actin bundles [59]. Evidence for this activity was twofold. First, the presence of protein 4.9 produced measurable and dose-dependent changes in the steady-state low-shear viscosity of actin solutions. Second, actin solutions polymerized in the presence of protein 4.9 contained F-actin bundles when examined under the electron microscope by negative staining. These filaments bundles are closely packed, up to 800 nm in diameter, and appear rigid and brittle. The bundles thus have properties that are indistinguishable from those of the actin bundles formed by other nonerythroid actin bundling proteins, such as intestinal microvillus fimbrin, villin, and sea urchin fascin [60–63].

Protein 4.9 also has an effect on the kinetic properties of actin polymerization [59]. The presence of protein 4.9 decreases the rate of actin filament elongation while also increasing the lag phase of actin polymerization when measured by either high-shear viscometry or fluorescence spectroscopy. Presently, it is not clear how protein 4.9, purified as described above, is capable of decreasing the rate of actin polymerization. Other proteins that decrease the rate of polymerization increase the concentration of actin necessary for polymerization (the critical concentration). Protein 4.9 does not share this property. It has been suggested that protein 4.9 may instead bind to the side of the actin filaments to alter its affinity for other monomers [59].

In view of what is currently known about protein 4.9, an understanding of its role in the mature erythrocyte membrane skeleton remains a dilemma because no actin bundles have been observed in red blood cells. One proposal is that protein 4.9 may be more active in the red blood cell prior to enucleation, where it might participate in the formation of actin bundles that later depolymerize up to spectrin-linked locations on the membrane. A logical conclusion of this reasoning would be that membrane skeletons should contain short "bundles" of actin filaments, yet the only observation in support of this possibility is the appearance of 15–100-nm diameter particles cross-linked by spectrin in thin-sectioned, negatively stained, and rotary-shadowed images of the cytoplasmic surface of red blood cell membranes [16–19,64–66]. Although these globular structures appear too large to accommodate a single actin filament, they can be more easily explained by the presence of several protein 4.1 molecules and other larger membrane skeletal proteins that are now known to bind to spectrin-actin complexes (see below).

Finally, there is provocative new evidence that the protein previously thought to be protein 4.9 is actually composed of two phosphoproteins that are clearly distinguished from one another by peptide maps and elution during anion exchange chromatography [67]. Interestingly, only one of these proteins cosediments with actin filaments. This observation is consistent in retrospect with the previous observation that the phosphorylation state of protein 4.9 depends strongly on the method of purification [59]. Protein 4.9 purified from membrane skeletons was more highly phosphorylated than that from spectrin-depleted inside-out vesicles. It may be that the two phosphorylation states of the protein 4.9 polypeptide observed in the prior study [59] actually represent two distinctly different proteins.

This new evidence indicates that the current thinking about protein 4.9 will have to undergo extensive reevaluation. Clearly, future studies of protein 4.9 will demand more complete physical characterization and purification of these proteins before a reliable assessment of their functional properties can be obtained. Until then the role of protein 4.9 in the membrane skeleton and the significance of its phosphorylation will remain an unanswered question.

D. Protein 4.2

Protein 4.2 is one of the more abundant components of the red blood cell membrane, accounting for 5% of the total membrane protein mass [25]. As a major component of the erythrocyte membrane, a possible role for protein 4.2 in membrane function seems likely, although this protein has only recently been isolated [68]. Cross-linking experiments suggested that protein 4.2 may exist in the red blood cell as a tetramer [69]. Evidence for a tight association between protein 4.2 and the anion transporter (band 3) is that detergent extracts of red blood cell membranes contain stable oligomers between protein 4.2 and band 3/band 6 or band 3/ankyrin complexes. Such complexes remain associated under a variety of nondenaturing conditions and could be co-isolated on sucrose gradients, by gel filtration and indirect immunoprecipitation with antiankyrin IgG [20–22,70].

Protein 4.2 has been purified by gel filtration from alkaline extracts of spectrin-depleted inside-out vesicles [68]. It is composed of 72,000-dalton polypeptide but exists in solution as a heterogeneous mixture of oligomers that appear to be in dimeric to trimeric states with an estimated molecular weight of 185,000. When viewed under the electron microscope by rotary shadowing or negative staining, protein 4.2 appears as a heterogeneous mixture of globular forms with diameters ranging between 8 and 15 nm. The hydrodynamic properties of this protein suggest that it is highly asymmetric, although this property is not readily appreciated in the electron micrographs [68].

Binding of purified protein 4.2 to inside-out vesicles stripped of endogenous protein 4.2 by extraction in alkaline buffers indicates that a binding site for protein 4.2 is present on the anion transporter. The interaction of purified protein 4.2 with depleted inside-out vesicles is specific and saturates at a monomeric protein 4.2 to monomeric transporter ratio of 1:1.5. Anion transporter fragment, a purified 43,000 peptide that contains most of the cytoplasmic domain [21,71], could completely inhibit the reassociation of protein 4.2 with the vesicles. This inhibition is reminiscent of the manner in which transporter fragment blocks ankyrin binding to inside-out vesicles [21]. An assessment of the concentration dependence of the inhibition suggests that the affinity of protein 4.2 for the anion transporter is in the range between 0.2 and 0.8 μM [68].

The actual stoichiometry of the protein 4.2/transporter complex in the red blood cell is uncertain. An understanding of this parameter is complicated further by the undefined heterogeneity of the oligomeric state of protein 4.2. Although the in vitro binding data suggest a stoichiometry of 1:1.5, the ratio of monomeric protein 4.2 to monomeric anion transporter in the cell is closer to 1:4. This discrepancy persists despite a careful consideration of possible stoichiometries of oligomeric protein 4.2 and transporter interactions, and suggests that the potential binding sites for protein 4.2 are incompletely occupied in the normal cell. Such a conclusion is supported further by the observation that unstripped inside-out vesicles (i.e., those containing their normal complement of endogenous protein 4.2) retain nearly half of the protein 4.2 binding capacity found on their stripped counterparts [68]. An equally plausible interpretation of this observation is that alternate binding sites for protein 4.2 distinct from the anion transporter may exist in the red blood cell.

The possible existence of a second binding site for protein 4.2 is supported by two lines of evidence. First, Scatchard plots of the binding of protein 4.2 to inside-out membrane vesicles are curvilinear, suggesting multiple nonidentical binding sites. Theoretically, this result could be explained by either the heterogeneity of the protein 4.2 ligand or negative cooperativity at the receptor sites. However, a second observation shows that endogenous protein 4.2 protects a class of protein 4.2 binding sites from mild chymotryptic proteolysis. These sites appear to be distinct from the anion transporter, which is otherwise degraded in spite of the presence of bound protein 4.2. Scatchard plots of the binding of exogenous protein 4.2 to these protected sites are linear, indicating binding to a single class of sites, notwithstanding the known heterogeneity of the ligand. Thus, protein 4.2 binds to the anion transporter in addition to an as yet uncharacterized receptor that is protected from mild proteolysis by bound protein 4.2 [68].

Other intersting aspects of protein 4.2 include a possible interaction with ankyrin and its heavy metal ion-induced phosphorylation. Co-isolation of ankyrin, anion transporter, and protein 4.2 has been observed in the past by a number of reports [20-22]. The recent purification of protein 4.2 demonstrates a co-isolation with ankyrin in the absence of anion transporter under a variety of conditions [68]. Although this apparent interaction could be an artifactual aggregation effect secondary to the harsh extraction conditions (pH 11) used for the purification of protein 4.2, the significance of this possible association warrants further investigation. A possible role for 4.2 may be to modulate the attachment of the anion transporter to the membrane skeleton by influencing the ankyrin-transporter linkage.

Protein 4.2 is phosphorylated preferentially in the presence of mercuric or cadmium ions at concentrations of 20 or 10 μM, respectively [72]. Although the overall phosphorylation of red blood cell membrane proteins is increased in the presence of these ions, the increase in protein 4.2 phosphorylation is nearly 25 to 50-fold greater than any of the other membrane components. The phosphorylation requires millimolar concentrations of magnesium ion and is inhibited by millimolar concentrations of calcium. Specific inhibitors of cyclic AMP-dependent kinase suggest that the phosphorylation may be catalyzed by membrane bound cyclic adenosine monophosphate (AMP-dependent kinases). It is thought that these heavy metal ions bind to the protein through covalent linkages with sulfhydryl groups to precipitate a conformational change that makes unique phosphorylation sites more accessible. This notion receives further support from the observation that protein 4.2 is the major Hg^{2+} binding site on the red blood cell and that specific sulfhydryl reactive agents such as p-chloromercuribenzenesulfonate and p-chloromercuribenzoate produce effects similar to that of Cd^{2+} and Hg^{2+} [72]. Such observations are consistent with previous findings that prior heat treatment of red blood cell membranes renders protein 4.2 more accessible to phosphorylation by purified erythrocyte cyclic AMP-dependent kinase [73]. Extracts of red blood cell membranes treated with Hg^{2+} contain a population of phosphorylated protein 4.2 that comigrates with spectrin complexes fractionated by gel filtration. Whether this observation represents an Hg^{2+}-induced interaction with spectrin complexes or a nonspecific aggregation (resulting from the oxidation of a protein with known tendencies for self-association) remains to be established. Until these matters can be resolved, the functional significance of this heavy metal ion–induced phosphorylation will remain quite obscure.

In summary, protein 4.2 is a major component of the red blood cell membrane that has only been recently purified. It is composed of a 72,000-dalton polypeptide present at >200,000 copies per cell. Under the current methods of purification the protein exists in solution as a heterogeneous mixture of dimeric to tetrameric globular oligomers with molecular diameters between 8 and 15 nm. Available data indicates that protein 4.2 binds to two different sites on the red blood cell membrane. One of these sites has been identified as the anion transporter, which confirms previous observations of the interaction. The identity of the second site remains to be determined. Phosphorylation of protein 4.2 can be induced specifically by exposure to heavy metal ions, although the significance of this phosphorylation is unclear. Future questions will center on the functional significance of the interaction of protein 4.2 with its sites on the membrane. A clearer definition of the molecular heterogeneity of preparations of purified protein 4.2 will be a prerequisite for a fuller understanding of the role of this protein in membrane function and stability.

E. Adducin

Adducin is a membrane skeletal protein that has been recently purified and characterized from human erythrocytes [74]. Present in the red blood cell at an estimated 30,000 copies per cell, it is not one of the major polypeptides of the membrane skeleton. Adducin exists in solution as a 200,000-dalton heterodimer formed from a single 103,000-dalton (α) subunit and a 97,000-dalton (β) subunit. Seen under the electron microscope adducin appears as a flattened globular protein with a molecular height of 5.4 nm and a molecular diameter of 12.4 nm. These slightly asymmetric dimensions indicate that adducin has the shape of an oblate ellipsoid, and are consistent with its hydrodynamic properties in solution [76]. Chymotryptic peptide map analysis of the α and β subunits of adducin indicate that these polypeptides are coded by separate genes, but share at least a 30% homology. These subunits therefore may have functions in common and are likely to have evolved from a common ancestral gene.

A striking feature of adducin is that it is a target for the calcium-dependent regulatory protein, calmodulin [74]. Calmodulin is a ubiquitous regulatory protein that confers micromolar calcium sensitivity to a variety of cellular process in many tissues [75]. The presence of a membrane-skeleton-associated target for calmodulin suggests the very interesting possibility that calcium may be acting through calmodulin to influence the mechanical properties of the erythrocyte membrane and membrane skeleton. This possibility is particularly relevant in view of the changes that occur in red blood cell mechanical stability and shape after ATP depletion and increased intracellular calcium [76].

Calmodulin photoaffinity labeling techniques have identified the 97,000-dalton β subunit of adducin as the binding site for calmodulin [74,77]. Quantitative assessment of the interaction of calmodulin with adducin reveals a K_d for binding of 0.2 μM. The interaction between adducin and calmoculin is physiologically meaningful because the concentration of calmodulin in the red blood cell is estimated to be 2.5 μM [39,40]. Therefore, in the normal red blood cell, the calcium-dependent occupation of adducin by calmodulin should approach 100%. Mammalian erythrocyte spectrin have also been reported to bind calmodulin under physiological conditions with a K_d = 6-20 μM [78-80]. The affinity of the calmodulin-spectrin interaction is far too low to be relevant to the erythrocyte in vivo because it is likely that <10% of the spectrin will be occupied by calmodulin in the normal functioning cell. Adducin is thus the predominant membrane-skeleton-associated calmodulin-binding protein in human erythrocytes.

Adducin is also a substrate for the calcium and phospholipid-dependent protein kinase C [81]. Protein kinase C is an enzyme that has been shown to be specifically activated by phorbol esters and is believed to have a major regulatory role in numerous cellular processes that modulate cell growth and differentiation [82,83]. In human erythrocytes, both subunits of adducin are phosphorylated by the endogenous protein kinase C. It is in fact one of the major substrates for the kinase, receiving as much as 3 mol of phosphate per mole of subunit. Adducin is also a substrate for the cyclic-AMP-dependent kinase within the red blood cell. At present the functional significance of phosphorylation of adducin by either enzyme remains to be determined.

Adducin binds specifically and with high affinity to spectrin–actin complexes in in-vitro assays [84,85]. Adducin binds with low affinity to actin filaments, and also associates weakly with spectrin alone. However, when spectrin is added to these mixtures, the binding dramatically increases in proportion to the amount of spectrin bound to the F-actin. Such results indicate that adducin either specifically recognizes preassembled spectrin–actin complexes, or binds weakly to either spectrin or actin alone to promote a ternary assembly of highly stable spectrin–actin–adducin oligomers. In either case, the obvious implication of these extrapolations is that adducin plays a significant role in stabilizing interactions within the spectrin–actin network.

Adducin also increases the amount of spectrin bound to F-actin in solution. At saturating concentrations of adducin the measured stoichiometry of spectrin to adducin in the spectrin–actin–adducin oligomers is at least two spectrins to one adducin molecule, suggesting that adducin may be multivalent for spectrin [84,85]. The ability of adducin to increase the assembly of spectrin onto F-actin could occur by three different mechanisms. One plausible mechanisms suggests that the increased spectrin binding could occur through an enhancement secondary to the greater thermodynamic stability of the spectrin–actin–adducin ternary complex. By a second possible mechanism, adducin may interact allosterically with either spectrin or actin to increase the affinity of one component for the other. A third and equally plausible mechanism is consistent with the high stoichiometry of spectrin to adducin in the spectrin–actin–adducin oligomers. By this mechanism, assembled spectrin–actin–adducin complexes may have the ability to bind additional spectrin molecules, thus increasing the amount of spectrin bound to F-actin in an adducin-dependent manner. Whatever the mechanism, the overall effect of adducin is to induce the assembly of spectrin onto F-actin. This activity is from whence adducin derives its name, which comes from the Latin "adducere," which means to "induce" or "bring together" [85].

Calmodulin at micromolar concentrations of calcium inhibits the ability of adducin to increase the binding of spectrin to F-actin [84,85]. This inhibition is specific for the adducin-dependent binding of spectrin because calmodulin neither inhibits the binding of adducin to spectrin–actin complexes nor the binding of spectrin to actin in the absence of adducin. Calmodulin has half-maximal effects at 1 μM and thus most likely interacts with the adducin component of the spectrin–actin–adducin oligomer to prevent the binding of additional spectrin molecules [85].

Adducin has features in common with protein 4.1 in that both proteins promote association of spectrin with actin. However, the properties of these proteins are distinct in many respects. Adducin binds very poorly or not at all to spectrin, requiring the presence of actin for the formation of stable adducin-spectrin–actin structural units. In contrast, protein 4.1 binds well to spectrin without requiring any other proteins [27,28,30,86,87]. The enhancement of spectrin binding in the presence of adducin occurs maximally at a spectrin-to-adducin stoichiometry of 2 to 1, whereas protein 4.1 enhances spectrin assembly by stabilizing spectrin–actin complexes with an average stoichiometry of spectrin to protein 4.1 of 1:1 or less [27,29]. These dissimilarities highlight the differences in the mechanism of action of adducin and protein 4.1 and may also account for the comparatively greater activity of protein 4.1 in vitro [13,29].

Negatively stained preparations of isolated membrane skeletons reveal a remarkably regular lattice that features a repeated structural unit containing five to eight spectrins gathered about a central actin core (Fig. 1). The putative ability of adducin to promote the interaction of multiple spectrins with spectrin-actin–adducin oligomers may have a role in the generation of this complex morphology.

Two preliminary observations indicate that protein 4:1 may block the interaction of adducin with spectrin–actin complexes [85]. First, protein 4.1 blocks the binding of adducin to isolated membrane skeletons. Moreover, protein 4.1 can increase the amount of spectrin bound to F-actin in the presence of adducin without causing comparable increases in the number of spectrin-actin-adducin complexes. It thus appears likely that protein 4.1 and adducin cannot share the same spectrin-actin complexes within the membrane skeleton. This exclusive arrangement suggests that there may be considerable molecular heterogeneity within the membrane skeleton. Protein 4.1 has binding sites on glycophorin and band 3 (see below), and a single spectrin dimer can bind as many as two protein 4.1 molecules [26,28]. The variety of potential binding sites for protein 4.1 provides an environment within the membrane skeleton that should leave enough unattended spectrin-actin junctions to accommodate the total population of adducin in each cell. Changes in the levels of calcium/calmodulin, protein phosphorylation, and phospholipid turnover (see below) potentially could influence this heterogeneity and consequently regulate membrane stability and elasticity.

It is tempting to speculate that adducin may be involved in the biogenesis of the membrane skeleton during erythroid development. Protein 4.1 is synthesized and assembled onto the membrane during the later stages of erythroid differentiation, after most of the spectrin has been incorporated [88]. It is conceivable that adducin may mediate an early assembly of an initially flexible and calcium-labile membrane skeleton that is later stabilized by protein 4.1. In avian erythrocyte systems, spectrin subunits are synthesized in numbers that exceed the possible binding sites on the membrane [89]. By its preferential recognition of spectrin-actin complexes, adducin could provide a sorting function that distinguishes between free spectrin or subunits synthesized in excess, and the actin-associated spectrin required for proper assembly of the membrane skeleton. This specificity is lacking in protein 4.1, because it binds well to both free spectrin and its individual α and β subunits despite the inability of the separate chains to bind actin independently [90]. Red blood cells completely deficient in protein 4.1 still manage to assemble a normal amount of spectrin and actin onto their membranes [91]. Factors such as adducin are likely to have a significant role in the early spectrin-actin assembly that occurs in the absence of protein 4.1.

There still remains much to be learned about adducin. The mechanism of its action will require further elucidation. Measured effects of calmodulin on adducin activity are quite subtle thus far. It is likely that the presence of other proteins such as tropomyosin and the "protein 4.9" proteins will be necessary to learn the full effect of calmodulin on adducin-dependent processes. Adducin binds actin, although with low affinity [85,86], and thus may be able to modulate aspects of actin polymerization or filament morphology. A membrane binding site for adducin has not been explored and the effects of protein phosphorylation on adducin activity are not yet known. These remain as tantalizing mysteries to be explored by future studies of this protein.

In summary, adducin is a new erythrocyte membrane-skeletal protein that modulates the assembly of spectrin onto actin. It is a 200,000-dalton heterodimer present in red blood cells at ~30,000 copies per cell and is a major target for calmodulin, and the calcium and phospholipid-dependent protein kinase C. Although the functional significance of protein kinase C phosphorylation of adducin is not known, micromolar concentration of calcium and calmodulin inhibits the ability of adducin to promote the binding of spectrin to F-actin. These properties indicate that adducin is likely to play a role in the processes through which calcium affects red blood cell membrane shape and stability.

IV. NEW INTERACTIONS AND MODALITIES OF REGULATION OF PROTEIN 4.1 AND ANKYRIN

A. Protein 4.1

Protein 4.1–Glycophorin Interactions and Its Modulation by Polyphosphoinositides

Recently protein 4.1 has been found to associate directly with a class of red blood cell integral membrane proteins referred to as the glycophorins [92–95]. The glycophorins are a class of sialoglycoproteins that include the MN and Ss blood group antigens [for review, see refs. 96,97]. Over the years, a variety of nomenclatures have developed to describe these proteins, but most commonly the three major elements of this group have been referred to as glycophorin A, B, and C [98]. Glycophorin A is the most abundant glycophorin, comprising nearly 75% of the total population of glycophorins. It is composed of 131 amino acids with 23 amino acids spanning the lipid bilayer and 35 amino acids contained in its cytoplasmic domain [99].

An interaction between protein 4.1 and the glycophorins was indicated by in vitro studies that measured the association of purified protein 4.1 with total isolates of glycophorin reconstituted into phosphatidylcholine lipid vesicles [93]. The protein specificity of this interaction was demonstrated by its lability to prior treatment of the vesicles with proteases such as chymotrypsin or trypsin. Moreover, the binding was saturable with an estimated affinity of 90 nM, and a capacity predicted by the fraction of liposome-bound glycophorins estimated to have their cytoplasmic domains accessible to the in solution [93].

A specific interaction with the glycophorins was demonstrated by the ability of antibodies directed against the cytoplasmic domain of glycophorin A to inhibit 70–80% of the protein 4.1 binding to total glycophorin vesicles. This level of inhibition was even greater when vesicles containing pure glycophorin A were used [93]. Previous reports have shown that protein 4.1 also has binding sites on glycophorin C [92,93]. The selectivity of the antibody inhibition suggested that glycophorins A, B, and C may each have unique binding sites for protein 4.1. This notion is substantiated by the finding that antibodies directed against the cytoplasmic domain of glycophorin B had no inhibitory effect on the binding of protein 4.1 to pure glycophorin A vesicles [93]. Furthermore, antibodies directed against the total glycophorins showed a 63% greater ability to inhibit the binding of protein 4.1 to erythrocyte inside-out vesicles (stripped of their endogenous ligand) than antibodies directed against glycophorin A alone [100].

The protein 4.1-glycophorin linkage is under strict control of the inositol triphosphate second messenger system [95]. Glycophorin reconstituted into vesicles containing phosphatidylinositol 4,5 biphosphate have greatly enhanced protein 4.1 binding activity. Phosphatidylinositol 4-phosphate had lesser effects and the action of both phosphoinositides was synergistic. The protein 4.1-glycophorin interaction can thus be added to the growing list of cellular processes found to be controlled by this novel transduction system. In the red blood cell, phosphoinositide levels could be influenced by the activity of myoinositolphosphate monoester kinases and phosphatases as well as the calcium-activated phospholipase C. Accordingly, decreases in intracellular ATP or increases in intracellular calcium could be expected to produce drops in polyphosphoinositides that would lead to a release of protein 4.1 from its binding site on the glycophorin, with a subsequent translocation to other locales thought to be alternate binding sites for protein 4.1 (see below).

Because protein 4.1 participates in the formation of a stable ternary complex with spectrin and actin, its additional interactions with the glycophorins suggests that it may serve a membrane linkage function similar to that of ankyrin. According to this scheme, spectrin would be tethered by its head portions to the lipid bilayer through a band 3-ankyrin-spectrin linkage with an additional coupling via its actin-binding tail region through a putative glycophorin-protein 4.1-spectrin linkage. However, there is currently no evidence to indicate that protein 4.1, glycophorin, spectrin, and actin can together form the stable oligomeric complexes necessary for function as a membrane linkage ensemble. Spectrin binding to erythrocyte inside-out vesicles is not affected by the presence of protein 4.1 [101,102]. No report describing a protein 4.1-mediated interaction between spectrin and glycophorin liposomes has been published at this time. An examination of Triton-solubilized red blood cell membranes indicate that very little glycophorin A remains associated with the membrane skeleton after solubilization of the lipid bilayer. Interestingly, red blood cells completely lacking glycophorin A are not deficient in protein 4.1 and remain normal in every other respect [103]. Moreover, human red blood cells in which protein 4.1 is either deficient or totally absent still maintain a full complement of spectrin on their membranes [91]. In contrast, glycophorin C is retained on membrane skeletons after Triton solubilization, and donors deficient in protein 4.1 have corresponding deficiencies in glycophorin C [92,104]. Thus, of the glycophorins, glycophorin C (glycoconnectin) is the most promising candidate for a second ankyrin-like membrane attachment site. A clear identification of stable complexes containing protein 4.1, glycophorin, and spectrin will be of primary importance in future evaluation of this possibility.

Protein 4.1-Anion Transporter Interactions

Protein 4.1 also binds to the cytoplasmic domain of the anion transporter close to but separate from the ankyrin binding site [100]. Binding was assessed by measuring the interaction of protein 4.1 with erythrocyte inside-out vesicles stripped of their endogenous ligand. The binding was saturable with a capacity that was far in excess of the amount of protein 4.1 in the normal cell, but was consistent with the endogenous amount of anion transporter. Quantitative assessment of this interaction provided evidence for two nonidentical receptors with estimated binding affinities of 20 and 500 nM [100]. Inhibition studies with either antibodies to the cytoplasmic domain of the anion transporter or the addition of pure preparations of the transporter 43,000-dalton cytoplasmic peptide could completely block the low-affinity binding sites. The portion of receptors unaffected by this treatment could be inhibited by antibodies against the glycophorins. It was concluded that protein 4.1 is capable of interacting with two separate binding sites on the red blood cell membrane: one a high-affinity site provided by glycophorin, and the other a low-affinity, high-capacity site provided by band 3 [100]. The red blood cell therefore contains a collection of potential binding sites for protein 4.1 that are in vast excess of the endogenous level of protein 4.1. A direct implication of this finding is that protein 4.1 is in a very dynamic state within the red blood cell (see below). Interestingly, experiments with transporter cytoplasmic domain and spectrin showed that this domain could displace the binding of protein 4.1 to spectrin, raising the possibility that the interaction of protein 4.1 with the anion transporter and spectrin may be mutually exclusive.

Protein 4.1 Phosphorylation

Protein 4.1 is a target for a number of protein kinases including cyclic AMP-dependent kinases, membrane associated casein kinases, and protein kinase C [52–56,105,106]. The functional significance of phosphorylation of 4.1 by cyclic AMP-dependent kinase as well as that of protein kinase C remain unknown at a cellular level. However, substantial effects can be elicited on phosphorylation of protein 4.1 by either the membrane-associated cyclic AMP-independent kinase or the cytosolic casein kinase of the erythrocyte [88]. Both enzymes incorporate 2 mol of phosphate per mole of protein 4.1 at apparently identical sites within its sequence. Phosphorylation of protein 4.1 by these enzymes leads to a fivefold decrease in its binding affinity for spectrin [86]. As determined from an analysis of the association of these proteins by rate zonal sedimentation of sucrose gradient, the affinity decreases from 2 μM to 9.4 μM [88]. However, it is unclear why the protein 4.1 binding affinity for spectrin reported in this study is nearly 10-fold lower than that of previous reports [27, 89]. Despite this discrepancy, the data nonetheless suggest that direct phosphorylation of protein 4.1 provides yet another means of modulating the dynamic distribution of protein 4.1 beneath the membrane. By this scheme, phosphoprotein 4.1 might be expected to translocate to other potential binding sites such as band 3 and the glycophorins.

Summary

Two classes of integral membrane proteins have been suggested to be receptors
for protein 4.1. The first class includes the glycophorin sialoglycoproteins. This
interaction occurs with high affinity and is positively regulated by the presence
of polyphosphoinositides. The second class is represented by the anion transporter. The protein 4.1-transporter interaction occurs with much lower affinity
than its interaction with glycophorin and has also been found to displace protein 4.1 from spectrin. There is currently no direct evidence to indicate that protein 4.1 can link the membrane skeleton to the lipid bilayer via these interactions because ternary complexes (functional units) containing either of these
membrane proteins coupled to spectrin via protein 4.1 have yet to be identified.
Direct phosphorylation of protein 4.1 also decreases its affinity for spectrin. In
view of this recent data, it now appears likely that the distribution of protein
4.1 within the membrane and membrane skeleton is highly dynamic, involving
translocations linked to the metabolic state of the cell through the actions of
specific protein and phospholipid kinases, phosphatases, and phosphodiesterases.

B. Ankyrin

Ankyrin Phosphorylation

Ankyrin, like protein 4.1, is an active substrate for several protein kinases [50-
52,105,107]. Incorporating nearly 7 mols of phosphate per mol of protein the
erythrocyte membrane-associated and cytosolic casein kinases are both particularly active cyclic AMP-independent ankyrin kinases [110]. Measurement of
the association of ankyrin with both red blood cell spectrin dimer and tetramer
show that ankyrin binds with much higher affinity to spectrin tetramer [108,
109], possibly through an allosterically coupled effect on the spectrin β subunit
[110]. Phosphorylation of ankyrin by these kinases [presumably with the
72,000 spectrin binding domains—see refs. 101,109], causes a four- to fivefold
decrease in its affinity for spectrin tetramer, although little effect was observed
on the interaction of ankyrin with spectrin dimer [108]. The results of these
studies as well as the studies on the phosphorylation of protein 4.1 (see above)
suggest a general role for protein phosphorylation in a molecular "relaxation"
of the membrane skeleton and its attachment.

Regulatory Domain of Ankyrin

Recent evidence suggests that ankyrin contains terminal regulatory domains at
opposite ends of the molecule that function to modulate its interaction with
spectrin and the anion transporter [111]. Erythrocytes contain a minor
186,000-dalton peptide (band 2.2) shown to be closely related to ankyrin by antibody cross-reactivity and peptide map analysis [20,101,112,113]. Although it is
likely that this peptide arises from proteolytic removal of a 29,000-dalton ter-

minal peptide, it is conceivable that this protein may also arise from an alternate splicing of mRNA transcribed from a single ankyrin gene [a process that has been described for other structural proteins; see ref. 114]. Band 2.2 has been isolated and binding studies with this protein show that it associates with spectrin with threefold higher affinity than ankyrin [111]. Band 2.2 also associates with twice the number of high-affinity anion transporter sites in membrane vesicles. Thus, through loss or deletion of a 29,000 terminal domain, protein 2.2 apparently becomes an activated form of ankyrin with membrane linkage capabilities that are enhanced both in affinity and capacity.

Erythrocytes contain a calcium-activated cytosolic protease, referred to as calpain I [115]. This protease reduces ankyrin to a 195,000-dalton form that is distinct from band 2.2 by virtue of its peptide map and physical properties [111]. The large calpain fragment is therefore thought to arise from removal of a domain on the opposite end of the molecule. Loss of this domain results in an eightfold reduction in affinity for the anion transporter and a twofold increase in the proportion of sites available for low-affinity binding. The molecular dimensions of the protein appear to be dramatically altered by this cleavage as hydrodynamic measurements of the cleaved protein show a measurable decrease in asymmetric projection or an extensive conformation rearrangement within the molecule secondary to the removal of the terminal domain.

These observations suggest that removal of terminal domains of ankyrin, either by site-directed proteolysis or possibly by alternative RNA splicing, may have a physiological role in erythrocytes. Such irreversible mechanisms are unlikely to occur routinely during the lifetime of a cell lacking protein synthesis. However, early and final stages in the life of an erythrocyte are points where an irreversible modification may be important. Although the relative proportion of band 2.2 does not change in the mature erythrocyte, its appearance in the later stages of erythropoiesis, possibly prior to the entrance of the cell into the circulation, could mark a point where the cell becomes committed to a predetermined composition of high-affinity linkages between the membrane skeletal and specific states of the anion transporter. Calpain I requires nonphysiological ($10 \mu M$) concentrations of calcium for activation, and red blood cells contain a potent calpain inhibitor (calpastatin) [115]. This protease is therefore not active in the normally functioning cells. However, prior to cell death, it is quite likely that agonal increases in intracellular calcium may occur secondary to membrane lipid damage. The resulting calpain-mediated decrease in ankyrin–band 3 interactions could function in the removal of senescent cells from the circulation. By this scheme, the slackened tethering of anion transporter to the membrane skeleton could lead to greater translational motion in the plane of the membrane and promote increased clustering of the anion transporter within the lipid bilayer. This clustering of the transporter could enhance their recognition by the low-affinity autoantibodies against the anion transporter that are thought to facilitate phagocytic removal of damaged cells during passage through the spleen [116,117].

Summary

Ankyrin is a large bipolar molecule with functional and regulatory domains localized at each terminus. Phosphorylation of ankyrin by membrane-bound and cytosolic protein kinases reduces its affinity for spectrin tetramer. A 29,000-dalton regulatory domain absent in the band 2.2 ankyrin-related peptide of the erythrocyte limits the interaction of ankyrin with band 3 and spectrin. The removal of this domain increases the accessibility of band 2.2 to high-affinity binding sites on the anion transporter, suggesting that processed forms of ankyrin may be able to influence the state of the transporter in the lipid bilayer. This processing may thus represent a novel mechanism whereby information encoded in peripheral proteins on the cytoplasmic surface can be communicated to the exterior of the cell through the modulation of transmembrane protein structure. Band 2.2 also has higher affinity for spectrin, indicating that the 29,000-dalton domain may exert allosteric influence over functional properties at both ends of the molecule. Calpain-catalyzed proteolytic cleavage of ankyrin (thought to be on the terminus opposite the 29,000-dalton ankyrin domain) produces a molecule with decreased binding affinity for the anion transporter and may have a role in "tagging" damaged or senescent cells for removal from the circulation.

V. FUTURE PERSPECTIVES

Our knowledge of the red blood cell membrane skeleton continues to grow. However, it is still far from complete because there are many unanswered questions and unexplored areas of importance. Although most of the major components of this structure have been fairly well characterized, there remains an undetermined number of proteins, present in small amounts, that despite their limited copies per cell may have significant influence on the organization of the membrane skeleton. The targets of these regulatory proteins are likely to be the specific interactions between the major structural components of the cell, e.g., spectrin–actin, spectrin–ankyrin, ankyrin–anion transporter, and spectrin–protein 4.1–actin linkages. Such effects may be far too subtle to be discovered by observation of simple binary interactions. Instead, it is clear that future in vitro analysis of membrane–skeletal protein interactions will have to be conducted under conditions that more closely approximate the in situ interactions in the cell. Thus, greater emphasis will have to be placed on the observation of ternary and higher order interactions.

A large percentage of the recently discovered elements of the membrane skeleton either bind actin or modulate its interaction with other proteins (e.g., tropomyosin, protein 4.9, and adducin). It may not be a coincidental finding that the three major substrates for protein kinase C fall into this category. Continued progress in the understanding of the role of protein phosphorylation and lipid turnover will depend on the ability to observe and assess the interactions of multicomponent systems.

These in vitro approaches are quite limited in providing extensive detail on the in situ organization of the membrane skeleton. The recent morphological studies on isolated membrane skeletons provide significant confirmatory data on the basic scheme of this organization; but these studies were conducted under conditions of ionic strength and pH that promoted spreading of the membrane skeletons. It is therefore likely that low-affinity interactions or those resulting from ternary or higher order linkages may be disrupted, damaged, or obscured. Nonetheless, these techniques do provide a potential means for identifying specific proteins within the membrane skeleton using antibody probes. Thus, binding sites for even the more minor components of the membrane skeleton may be accessible to identification and localization. Other morphological techniques that avoid or lessen the artifacts of tissue preparations (e.g., freeze fracture) may provide promising strategies for future elucidation of membrane skeletal organization.

Contributions from the field of molecular genetics offer great promise for providing information on protein structure and appearance during differentiation. Site-directed mutagenesis of genes encoding for membrane–skeletal elements may yield a wealth of information on the factors responsible for membrane stability and biogenesis.

The features of the membrane skeleton are most dynamic during its biogenesis. Functional evaluation of any new membrane components must therefore consider the possibility that some functions may be limited exclusively to the earlier stages in erythroid development. Careful analysis of naturally occurring mutants have shed light on this possibility. It has been demonstrated that the common underlying feature of spherocytic cells is a deficiency in spectrin (see Chapters 4 and 14). Important corollaries of this finding are that spectrin deficiency may arise from either a decrease in its gene product of a defect in the pathways that lead to its assembly on the membrane. The possible existence of factors that contribute to spectrin assembly on the membrane and their role in hemolytic disease will be an important area for future investigation.

ACKNOWLEDGMENT

Research from this laboratory was supported in part by grants from the National Institutes of Health: RO1 AM19808, RO1 GM33996, and Hematology Training Grant HL07535.

REFERENCES

1. Marchesi, V. T. (1985). Stabilizing infrastructure of cell membranes. *Ann. Rev. Cell Biol.* 1:531–461.
2. Goodman, S. R., and Zagon, I. S. (1986). The neural cell spectrin skeleton: A review. *Am. J. Physiol.*250:C347–360.
3. Bennett, V. (1985). The membrane skeleton of human erythrocytes and its implication for more complex cells. *Annu. Rev. Biochem.*54:273–304.
4. Zail, S. (1985). Clinical disorders of the red cell membrane skeleton. *CRC Crit. Rev. Oncol. Hematol.* 5:397–453.
5. Marchesi, V. T., and Steers, E. (1968). Selective solubilization of a protein component of the red cell membrane. *Science* 159:203.
6. Shotton, D. M., Burke, B. E., and Branton, D. (1979). The molecular structure of human erythrocyte spectrin: Biophysical and electron microscopic studies. *J. Mol. Biol.* 131:303–329.
7. Ungewickell, E., and Gratzer, W. (1978). Self-association of human spectrin. A thermodynamic and kinetic study. *Eur. J. Biochem.* 88:379–385.
8. Ji, T. H., Kiehm, D. J., and Middaugh, G. R. (1980). Presence of spectrin tetramer on the erythrocyte membrane. *J. Biol. Chem.* 255:2990–2993.
9. Morrow, J., and Marchesi, V. T. (1981). Self-assembly of spectrin oligomers in vitro: A basis for dynamic cytoskeleton. *J. Cell Biol.* 88:463–468.
10. Liu, S-C. P., Windisch, S. K., and Palek, J. (1984). Oligomeric states of spectrin in normal erythrocyte membranes: Biochemical and electron microscopic studies. *Cell* 37:587–594.
11. Brenner, S. L., and Korn, E. D. (1980). Spectrin/actin complex isolated from sheep erythrocyte accelerates actin polymerization by simple nucleation. *J. Biol. Chem.* 255:1670–1676.
12. Cohen, C. M., Tyler, J. M., and Branton, D. (1980). Spectrin-actin association studied by electron microscopy of shadowed preparations. *Cell* 21:875–883.
13. Ungewickell, E., Bennett, P. M., Calvert, R., Ohanian, V., and Gratzer, W. B. (1979). In vitro formation of a complex between cytoskeletal proteins of human erythrocytes. *Nature* 280:811–814.
14. Fowler, V. M., and Taylor, D. L. (1980). Spectrin plus band 4.1 cross-link actin. Regulation by micromolar calcium. *J. Cell Biol.* 85:361–376.
15. Pinder, J. C., and Gratzer, W. B. (1983). Structural and dynamic states of actin in the erythrocyte. *J. Cell Biol.* 96:768–775.
16. Shen, B. W., Joseph, R., and Steck, T. L. (1984). Ultrastructure of unit fragments of the skeleton of the human erythrocyte membrane. *J. Cell Biol.* 99:810–821.
17. Byers, T. J., and Branton, D. (1985). Visualization of the protein associations in the erythrocyte membrane skeleton. *Proc. Natl. Acad. Sci. U.S.A.* 2:6153–6157.
18. Shen, B. W., Josephs, R., and Steck, T. L. (1986). Ultrastructure of the intact skeleton of the human erythrocyte membrane. *J. Cell. Biol.* 102:997–1006.

19. Liu, S-C., Derick, L. H., and Palek, J. (1987). Visualization of the hexagonal lattice in the erythrocyte membrane skeleton. *J. Cell.Biol.* 104:527–536.
20. Bennett, V., and Stenbuck, P. J. (1979). The membrane attachment protein for spectrin is associated with band 3 in human erythrocyte membranes. *Nature* 280:468–473.
21. Bennett, V., and Stenbuck, P. J. (1980). Association between ankyrin and the cytoplasmic domain of band 3 isolated from human erythrocyte membranes. *J. Biol. Chem.* 255:6424–6432.
22. Bennett, V., and Stenbuck, P. J. (1980). Human erythrocyte ankyrin: Purification and properties. *J. Biol. Chem.* 255:2540–2548.
23. Hargreaves, W. R., Giedd, K. N., Verkleij, A., and Branton, D. (1980). Reassociation of ankyrin with band 3 in erythrocyte membranes and lipid vesicles. *J. Biol. Chem.* 255:11965–11972.
24. Jay, D., and Cantley, L. (1986). Structural aspects of the red cell anion exchange protein. *Ann. Rev. Biochem.* 55:511–538.
25. Tyler, J., Hargreaves, W., and Branton, D. (1979). Purification of two spectrin-binding proteins: Biochemical and electron microscopic evidence for site-specific reassociation between spectrin and band 2.1 and 4.1. *Proc. Natl. Acad. Sci. U.S.A.* 76:5192–5196.
26. Steck, T. L. (1974). Organization of protein in the human red blood cell membrane. *J. Cell Biol.* 62:1–19.
27. Tyler, J. M., Reinhardt, B. N., and Branton, D. (1980). Associations of erythrocyte membrane proteins: Binding of purified band 2.1 and 4.1 to spectrin. *J. Biol. Chem.* 255:7034–7039.
28. Cohen, C. M., and Langley, R. C. (1984). Functional characterization of human erythrocyte spectrin alpha and beta chains: Association with actin and erythrocyte protein 4.1. *Biochemistry* 23:4488–4495.
29. Cohen, C. M., and Foley, S. F. (1984). Biochemical characterization of complex formation by human erythrocyte spectrin, protein 4.1, and actin. *Biochemistry* 23:6091–6098.
30. Ohanian, V., Wolfe, L. C., John, K. M., Pinder, J. C., Lux, S. E., and Gratzer, W. B. (1984). Analysis of the ternary interaction of the red cell membrane skeletal proteins spectrin, actin and 4.1. *Biochemistry* 23:4416–4420.
31. Ebashi, S. (1974). Regulatory mechanism of muscle contraction with special reference to Ca^{2+}-troponin-tropomyosin system. *Essays Biochem.* 10:1–36.
32. Fowler, V. M., and Bennett, V. (1984). Erythrocyte membrane tropomyosin: Purification and properties. *J. Biol. Chem.* 259:5978–5989.
33. Maruyama, K., and Ohasi, K. (1978). Tropomyosin inhibits the interaction of F-actin and filamin. *J. Biochem. (Tokyo)* 84:1017–1019.
34. Bernstein, B. W., and Bamburg, J. R. (1982). Tropomyosin binding to F-actin protects the F-catin from disassembly by brain actin-depolymerizing factor (ADF). *Cell Motil.* 2:1–8.
35. Wegner, A. (1982). Kinetic analysis of actin assembly suggests that tropomyosin inhibits spontaneous fragmentation of actin filaments. *J. Mol. Biol.* 161:217–227.

36. Fattoum, A., Hartwig, J. H., and Stossel, T. P. (1983). Isolation and some structural and functional properties of macrophage tropomyosin. *Biochemistry* 22:1187–1193.
37. Bunn, H. F., Ransil, B. J., and Chao, A. (1971). The interaction between erythrocyte organic phosphates, magnesium ion, and hemoglobin. *J. Biol. Chem.* 246:5273–5279.
38. Gerber, G., Berger, H., Janig, G. R., and Rapoport, S. M. (1973). Interaction of haemoglobin with ions. Quantitative description of the state of magnesium, adenosine 5'triphosphate, 2,3-biphosphate, and human haemoglobin under stimulated intracellular conditions. *Eur. J. Biochem.* 38:562–571.
39. Jarrett, H., and Penniston, J. T. (1978). Purification of the Ca^{2+}-stimulated ATPase activator from human erythrocytes. *J. Biol. Chem.* 253:4676–4682.
40. Agre, P., Gardner, K., and Bennett, V. (1983). Association between human erythrocyte calmodulin and the cytoplasmic surface of human erythrocyte membranes. *J. Biol. Chem.* 258:6258–6265.
41. Maimon, J., and Puszkin, S. (1978). Erythrocyte troponin inhibitor-like proteins: Isolation and characterization. *J. Supramol. Struct.* 9:131–141.
42. Fowler, V. M., and Bennett, V. (1984). Tropomyosin: A new component of the erythrocyte membrane skeleton. In *Workshop on Erythrocyte Membranes*, G. J. Brewer (Ed.). Alan R. Liss, New York.
43. Fowler, V. M., Davis, J. Q., and Bennett, V. (1985). Human erythrocyte myosin: Identification and purification. *J. Cell Biol.* 100:47–55.
44. Nakao, M., Nakao, T., Yamazoe, S., and Yoshikawa, H. (1961). Adenosine triphosphate and the shape of erythrocytes. *J. Biochem.* 49:487–492.
45. Kirkpatrick, F. H., and Sweeney, M. L. (1980). Cytoplasmic and membrane-bound myosin(s). *Fed. Proc.* 39:2049a.
46. Schrier, S. L., Hardy, B., Junga, I., and Ma, L. (1981). Actin-activated ATPase in human red cell membranes. *Blood* 58:953–962.
47. Wong, A. J., Kiehart, D. P., and Pollard, T. D. (1985). Myosin from human erythrocytes. *J. Biol. Chem.* 260:46–49.
48. Fowler, V. M. (1986). An actomyosin contractile mechanism for erythrocyte shape transformations. *J. Cell. Biochem.* 31:1–9.
49. Guthrow, C. E. J., Allen, J. E., and Rasmussen, H. (1972). Phosphorylation of an endogenous membrane protein by an endogenous, membrane-associated cyclic adenosine 3', 5'-monophosphate-dependent protein kinase in human erythrocyte ghosts. *J. Biol. Chem.* 247:8145–8153.
50. Fairbanks, G., and Avruch, J. (1974). Phosphorylation of endogenous substrates by erythrocyte membrane protein kinases II. Cyclic adenosine monophosphate-stimulated reactions. *Biochemistry* 13:5514–5521.
51. Rubin, C. S. (1975). Adenosine 3', 5'-monophosphate-regulated phosphorylation of erythrocyte membrane proteins. *J. Biol. Chem.* 250:9044–9052.
52. Thomas, E. L., King, L. E., and Morrison, M. (1979). The uptake of cyclic-AMP by human erythrocytes and its effect on membrane phosphorylation. *Arch. Biochem. Biophys.* 196:459–464.

53. Ling, E., and Sapirstein, V. (1984). Phorbol ester stimulates the phosphorylation of rabbit erythrocyte band 4.1. *Biochem. Biophys. Res. Commun.* 120:291-298.
54. Horne, W. C., Leto, T. L., and Marchesi, V. T. (1985). Differential phosphorylation of multiple site in protein 4.1 and protein 4.9 by phorbol ester-activated and cyclic AMP-dependent kinases. *J. Biol. Chem.* 260:9073-9076.
55. Palfrey, H. C., and Waseem, A. (1985). Protein kinase C in the human erythrocyte. Translocation to the plasma membrane and phosphorylation of bands 4.1 and 4.9 and other membrane proteins. *J. Biol. Chem.* 260:16021-16029.
56. Faquin, W. C., Chahwala, S. B., Cantley, L. C., and Branton, D. (1986). Protein kinase C of human erythrocytes phosphorylates bands 4.1 and 4.9. *Biochim. Biophys. Acta* 887:142-149.
57. Dzandu, J. K., and Johnson, R. M. (1980). Membrane protein phosphorylation in intact normal and sickle cell erythrocytes. *J. Biol. Chem.* 255:6382-6386.
58. Johnson, R. M., and Dzandu, J. K. (1982). Calcium and ionophore A23187 induce the sickle cell membrane phosphorylation pattern in normal erythrocytes. *Biochim. Biophys. Acta* 692:218-222.
59. Siegel, D. L., and Branton, D. (1985). Partial purification and characterization of an actin-bundling protein, band 4.9, from human erythrocytes. *J. Cell. Biol.* 100:775-785.
60. Craig, S. W., and Powell, L. D. (1980). Regulation of actin polymerization by villin, a 95,000 dalton cytoskeletal component of intestinal brush border. *Cell* 22:739-746.
61. Bretscher, A., and Weber, K. (1980). Villin is a major protein of the microvillus cytoskeleton which binds both G and F actin in a calcium-dependent manner. *Cell* 20:839-847.
62. Glenney, J. F. Jr., Kaulfus, P., Matsudaira, P., and Weber, K. (1981). F-actin binding and bundling properties of fimbrin, a major cytoskeletal protein of microvillus core filaments. *J. Biol. Chem.* 256:9283-9288.
63. Otto, J. J., Kane, R. E., and Bryan, J. (1979). Formatin of filopodia in coelomocytes: Localization of fascin, a 58,000 dalton actin cross-linking protein. *Cell* 17:285-293.
64. Tsukita, S., Tsukita, S., and Ishikawa, H. (1980). Cytoskeletal network underlying the human erythrocyte membrane. Thin-section electron microscopy. *J. Cell Biol.* 85:567-576.
65. Nermut, M. (1981). Visualization of the "membrane skeleton" in human erythrocytes by freeze-etching. *Eur. J. Cell. Biol.* 25:265-271.
66. Timme, A. H. (1981). The ultrastructure of the erythrocyte cytoskeleton at neutral and reduced pH. *J. Ultrastruct. Res.* 77:199-209.
67. Horne, W. C., Miettinin, H., and Marchesi, V. T. (1986). Human erythrocyte band 4.9 is composed of two phosphoproteins. *J. Cell Biol.* 103:543a.

68. Korsgren, C., and Cohen, C. M. (1986). Purification and properties of human erythrocyte band 4.2. Association with the cytoplasmic domain of band 3. *J. Biol. Chem.* 261:5536–5434.
69. Steck, T. L. (1972). Crosslinking the major proteins of the isolated erythrocyte membrane. *J. Mol. Biol.* 66:295–305.
70. Yu, J., and Steck, T. L. (1975). Associations of band 3, the predominant polypeptide of the human erythrocyte. *J. Biol. Chem.* 250:9176–9184.
71. Steck, T. L., Ramos, B., and Strappazon, E. (1976). Proteolytic dissection of band 3, the predominant transmembrane polypeptide of the human erythrocyte membrane. *Biochemistry* 15:1153–1161.
72. Suzuki, K., Ikebuchi, H., and Terao, T. (1985). Mercuric and cadmium ions stimulate phosphorylation of band 4.2 protein on human erythrocyte membrane. *J. Biol. Chem.* 260:4526–4530.
73. Suzuki, K., Terao, T., and Osawa, T. (1981). Purification and characterization of a catalytic subunit of an adenosine $3':5'$-monophosphate-dependent protein kinase from human erythrocyte membranes. *J. Biochem. (Tokyo)* 89:1–11.
74. Gardner, K., and Bennett, V. (1986). A new erythrocyte membrane-associated protein with calmodulin binding activity. Identification and purification. *J. Biol. Chem.* 261:1139–1148.
75. Cheung, W. Y. (1980). Calmodulin plays a pivotal role in cellular regulation. *Science* 207:19–27.
76. Weed, R., LaCelle, P., and Merrill, E. (1969). Metabolic dependence of red cell deformability. *J. Clin. Invest.* 48:795–809.
77. Andreasen, T. J., Keller, C. H., LaPorte, D. C., Edelman, A. M., and Storm, D. R. (1981). Preparation of azidocalmodulin: A photoaffinity label for calmodulin-binding proteins. *Proc. Natl. Acad. Sci. U.S.A.* 78:2782–2785.
78. Husain, A., Howlett, G. J., and Sawyer, W. H. (1984). The interaction of calmodulin with human and avian spectrin. *Biochem. Biophys. Res. Commun.* 122:1194–1200.
79. Burns, N. R., and Gratzer, W. B. (1985). Interaction of calmodulin with the red cell and its membrane skeleton and with spectrin. *Biochemistry* 24:3070–3074.
80. Sears, D. E., Marchesi, V. T., and Morrow, J. S. (1986). A calmodulin and alpha-subunit binding domain in human erythrocyte spectrin. *Biochem. Biophys. Acta* 870:432–442.
81. Ling, E., Gardner, K., and Bennett, V. (1986). Protein kinase C phosphorylates a recently identified membrane skeleton-associated calmodulin-binding protein in human erythrocytes. *J. Biol. Chem.* 261:13875–13878.
82. Berridge, M. J., and Irvine, R. (1984). Inositol triphosphate, a novel second messenger in cellular signal transduction. *Nature* 312:315–321.
83. Nishizuka, Y. (1984). Turnover of inositol phospholipids and signal transduction. *Science* 225:1365–1370.
84. Gardner, K., and Bennett, V. (1987). Modulation of spectrin-actin assembly to erythrocyte adducin. *Nature* 328:359–362.

85. Gardner, K., and Bennett, V. (1987). Erythrocyte adducin: A new calmodulin regulated membrane-skeletal protein that modulates spectrin-actin assembly. *J. Cell. Biochem.* (in press).

86. Eder, P. S., Soong, C. J., and Tao, M. (1986). Phosphorylation reduces the affinity of protein 4.1 for spectrin. *Biochemistry* 25:1764–1770.

87. Podgorski, A., and Elbaum, D. (1985). Properties of red cell membrane proteins: mechanism of spectrin and band 4.1 interactions. *Biochemistry* 24:7871–7876.

88. Staufenbiel, M., and Lazarides, E. (1986). Assembly of protein 4.1 during chicken erythroid differentiation. *J. Cell. Biol.* 102:1157–1163.

89. Moon, R., and Lazarides, E. (1983). Beta-spectrin limits alpha-spectrin assembly on membranes following synthesis in a chicken erythroid cell lysate. *Nature* 305:62–65.

90. Calvert, R., Bennett, P., and Gratzer, W. B. (1980). Properties and structural role of the subunits of human spectrin. *Eur. J. Biochem.* 107:355–361.

91. Takakuwa, Y., Tchernia, G., Rossi, M., Benabadji, M., and Mohandas, N. (1986). Restoration of normal membrane stability to unstable protein 4.1-deficient erythrocyte membranes by incorporation of purified protein 4.1. *J. Clin. Invest.* 78:80–85.

92. Mueller, T. J., and Morrison, M. (1981). Glycoconnectin (PAS 2), a membrane attachment site for the human erythrocyte cytoskeleton. *Prog. Clin. Biol. Res.* 56:95–116.

93. Anderson, R. A., and Lovrien, R. E. (1984). *Nature* 307:655–658.

94. Shiffer, K. A., and Goodman, S. R. (1984). Protein 4.1: Its association with the human erythrocyte membrane. *Proc. Natl. Acad. Sci. U.S.A.* 81:4404–4408.

95. Anderson, R. A., and Marchesi, V. T. (1985). Regulation of the association of membrane skeletal protein 4.1 with glycophorin by a phosphoinositide. *Nature* 318:295–298.

96. Anstee, D. J. (1981). The blood group MNSs-active sialoglycoproteins. *Semin. Hematol.* 18:13–31.

97. Furthmayr, H., and Marchesi, V. T. (1984). Glycophorins: Isolation, orientation, and localization of specific domains. *Methods Enzymol.* 96:268–280.

98. Furthmayr, H. (1978). Glycophorins A, B, and C: A family of sialoglycoproteins. Isolation and preliminary characterization of trypsin-derived peptides. *J. Supramol. Struct.* 9:9.

99. Tomita, M., Furthmayr, H., and Marchesi, V. T. (1978). Primary structure of human erythrocyte glycophorin A. Isolation and characterization of peptides an complete amino acid sequence. *Biochemistry* 17:4756–4770.

100. Pasternack, G. R., Anderson, R. A., Leto, T. L., and Marchesi, V. T. (1985). Interactions between protein 4.1 and band 3. An alternative binding site for an element of the membrane skeleton. *J. Biol. Chem.* 260:3676–3683.

101. Bennett, V., and Stenbuck, P. J. (1979). Identification and partial purification of ankyrin, the high affinity membrane attachment site for human erythrocyte spectrin. *J. Biol. Chem.* 254:2533-2541.

102. Cohen, C. M., and Foley, S. F. (1982). The role of band 4.1 in the association of actin with erythrocyte membranes. *Biochim. Biophys. Acta* 688: 691-701.

103. Dahr, W., Uhlenbruck, G., Leikola, J., Wagstaff, W., and Landfried, K. (1976). Studies on the membrane glycoprotein defect of En (a-) erythrocytes. I. Biomedical aspects. *J. Immunogenet.* 3:329-346.

104. Alloisio, N., Morle, L., Bachir, D., Guetarni, D., Colonna, P., and Delauney, J. (1985). Red cell membrane sialoglycoprotein beta in homozygous and heterozygous 4.1 (-) hereditary elliptocytosis. *Biochim. Biophys. Acta* 816:57-62.

105. Hosey, M., and Tao, M. (1977). Protein kinases of rabbit and human erythrocyte membranes. Solubilization and characterization. *Biochim. Biophys. Acta* 482:348-357.

106. Leto, T. L., and Marchesi, V. T. (1984). A structural model of human erythrocyte protein 4.1. *J. Biol. Chem.* 259:4603-4608.

107. Bennett, V. (1977). Human erythrocyte spectrin: Phosphorylation in intact cells and purification of the ^{32}P-labelled protein in a non-aggregated state. *Life. Sci.* 21:433-440.

108. Lu, P. W., Soong, C. J., and Tao, M. (1985). Phosphorylation of ankyrin decreases its affinity for spectrin tetramer. *J. Biol. Chem.* 260:14958-14964.

109. Weaver, D. C., Pasternack, G. R., and Marchesi, V. T. (1984). The structural basis of ankyrin function II. Identification of two functional domains. *J. Biol. Chem.* 259:6170-6175.

110. Morrow, J. S., Giorgi, M., and Cianci, C. (1987). Spectrin oligomerization is allosterically coupled to ankyrin and protein 3. *J. Cell. Biochem.* *[Suppl.]* 11B:154. (abstract).

111. Hall, T. G., and Bennett, V. (1987). Regulatory domains of erythrocyte ankyrin. *J. Biol. Chem.* 262:10537-10545.

112. Luna, E. J., Kidd, G. H., and Branton, D. (1979). Identification by peptide analysis of the spectrin-binding protein in human erythrocytes. *J. Biol. Chem.* 254:2526-2532.

113. Yu, J., and Goodman, S. R. (1979). Syndeins: The spectrin binding protein(s) of the human erythrocyte membrane. *Proc. Natl. Acad. Sci. U.S.A.* 76:2340-2344.

114. Capetanaki, Y. G., Ngai, J., Flytzanis, C. N., and Lazarides, E. (1983). Tissue-specific expression of two mRNA species transcribed from a single vimentin gene. *Cell* 35:411-420.

115. Murakami, T., Hatanaka, M., and Murachi, T. (1981). The cytosol of human erythrocytes contain a highly calcium-sensitive thiol protease (Calpain I) and its specific inhibitor protein (calpastatin). *J. Biochem. (Tokyo)* 90:1809-1816.

116. Low, P. S., Waugh, S. M., Zinke, K., and Drenckhahn, D. (1985). The role of hemoglobin denaturation and band 3 clustering in red blood cell aging. *Science* 227:531–533.
117. Schluter, K., and Drenckhahn, D. (1986). Co-clustering of denatured hemoglobin with band 3: its role in binding of autoantibodies against band 3 to abnormal and aged erythrocytes. *Proc. Natl. Acad. Sci. U.S.A.* 83: 6137–6141.

2

Structure and Function of the Glucose Transporter

MIKE M. MUECKLER   *Washington University School of Medicine,*
St. Louis, Missouri

I. INTRODUCTION

The glucose transporter of human erythrocytes is one of the more extensively characterized transport systems of the facilitated diffusion type [for recent reviews, see 1-3]. This protein equilibrates glucose across the red blood cell membrane, providing the energy substrate required for anaerobic glycolysis. The metabolic energy requirement of the red blood cell is rather low, and transport is not rate-limiting for glucose utilization [4]. Interestingly, the red blood cells of adult nonprimates lack glucose transport activity, and apparently derive their metabolic energy primarily from plasma inosine, which is transported across the membrane via the nucleoside transporter [5].

Two kinetic characteristics of glucose transport in red blood cells are relevant to the structure and function of the protein. Asymmetry has been observed, in which glucose influx exhibits a higher V_{max} and lower K_m than efflux [6]. This presumably indicates a different conformation for the inward and outward-facing substrate binding sites of the protein. Exchange acceleration is manifested in a higher V_{max} for equilibrium exchange compared with net flux of sugar [7]. This observation suggests that the rate-limiting step in net transport is the conformational change of the unloaded carrier. Models concerning the structure and function of the glucose transporter must explain these important kinetic observations.

This review will summarize recent data concerning the structure and function of the glucose transporter. Emphasis will be placed on the application of the methods of molecular biology to this problem. The reader is referred to other recent reviews that emphasize different aspects of this topic [1-3].

II. PURIFICATION AND CHARACTERIZATION

Two developments were key to the purification of the glucose transporter:
liposome reconsitution methods that provide a functional assay for transport
activity [8], and the discovery that cytochalasin B acts as a relatively specific,
high-affinity inhibitor of glucose transport [9]. Kasahara and Hinkle [10]
achieved the first partial purification by Triton-X100 solubilization of red blood
cell membranes followed by diethylaminoethyl cellulose chromatography. Baldwin et al. [11] improved on the procedure by substituting octyl glucoside for
Triton-X100 and adding exogenous lipids, resulting in both higher yields and
specific activity (cytochalasin B binding). This procedure yields a preparation
that contains lipid and protein in a 3:1 weight ratio, the protein portion consisting of ~75% glucose transporter polypeptide [12]. Depending on the reconstitution method, the purified protein exhibits between 5 and 15% of the in vivo
transport activity [13,14].

The purified transporter migrates as a broad smear between 45 and 75 kD on
sodium dodecyl sulfate polyacrylamide gels [11]. This broad size distribution is
due to the presence of a single, N-linked oligosaccharide of the erythroglycan
type that is heterogeneous in size [15,16; also see below]. The oligosaccharide
is apparently unnecessary for transport function [13]. The protein contains
~15% carbohydrate by weight [16]. Treatment of the protein with endo-β-
galactosidase [15] or endoglycosidase F [17] before electrophoresis results in
a much sharper band migrating at 46 kD relative to soluble protein standards.
Amino acid composition analysis indicates that the protein is rich in hydrophobic residues [11].

The best evidence available suggests that the glucose transporter functions as
a monomer in the membrane. The transporter can be reconstituted into homogeneous unilamellar vesicles in which there is, on average, less than one cytochalasin B binding site per vesicle [14]. As the stoichiometry of cytochalasin B-
transporter polypeptide binding is ~1:1,* this indicates that the vesicles contain,
on average, less than one polypeptide. The number of functional transporter
units in the preparation can be determined by the value of the intravesicular volume that rapidly equilibrates with sugar divided by the known volume of the
vesicle. The results indicate that each glucose transporter polypeptide functions
in transport at 5% of the in vivo rate. The low efficiency observed is typical of
reconstituted transporters, and is likely due to the difference in the lipid and
protein environment in a reconstituted vesicle versus the native membrane, and
perhaps to the reconstitution procedure itself. Thus, the glucose transporter is

---

*Calculated from the value of 0.7:1 given in ref. 11 by correcting for the true M_r of the
transporter polypeptide (54 kD) [24] and the probable purity of the preparation (75%)
[12]. Thus, $0.7 \times 54/46 \times 1.25 = 1.03$.

capable of functioning as a monomer, and probably does so in its native state. Although this is in disagreement with the results of radiation inactivation analysis [18], it is difficult to conceive of the manner in which a polymeric channel-forming protein could function to any extent as a monomer. Additionally, the accuracy of radiation inactivation analysis as applied to membrane proteins is questionable [19]. However, it is conceivable that a polymeric state may modify the channel in such a way as to increase the efficiency of transport. Further evidence is clearly necessary to confirm the monomeric state of the native glucose transporter unit.

III. THE GLUCOSE TRANSPORTER IS A TRANSMEMBRANE PROTEIN

Both glucose transport function and the integrity of the polypeptide remain intact when erythrocytes are incubated with high levels of several different proteases [17]. This observation suggests that little of the protein is exposed on the extracytoplasmic face of the membrane. When the cytoplasmic domains of the protein are exposed, as in unsealed erythrocyte ghosts or "inverted" vesicles, trypsin digestion cleaves the protein into two large fragments [20,21]. The larger fragment (M_r 23–42 kD) runs as a diffuse band on polyacrylamide gels and bears the oligosaccharide chain, as judged by its conversion to a much sharper band of 23 kD after digestion with endo-β-galactosidase. The smaller band (18 kD) is insensitive to glycosidase treatment and probably contains the cytochalasin B binding site, as indicated by photoaffinity labeling of the protein with [^{3}H]cytochalasin B prior to protease digestion. Considering that the oligosaccharide chain lies on an extracytoplasmic domain of the protein [15], these results demonstrate that the larger tryptic fragment spans the membrane at least once. It can also be concluded that the cytochalasin B binding site and the glycosylated asparagine residue lie on the opposite halves of the transporter polypeptide. It has been argued on kinetic grounds that cytochalasin B binds to the inward-facing substrate binding site of the transporter [21]. Additionally, monoclonal antibodies directed against cytoplasmic domains of the transporter inhibit cytochalasin B binding [12,22]. However, neither of these observations constitutes proof that cytochalasin B binds to the inward-facing substrate-binding site, and this point will have to be established directly.

IV. STRUCTURE AND FUNCTION

Human glucose transporter cDNA was cloned from a HepG2 hepatoma λgt11 expression library [23]. Antiserum raised in rabbits against the human erythrocyte transporter was used to screen the library. The following pieces of evidence indicate that the isolated clone encodes a glucose transporter that is highly

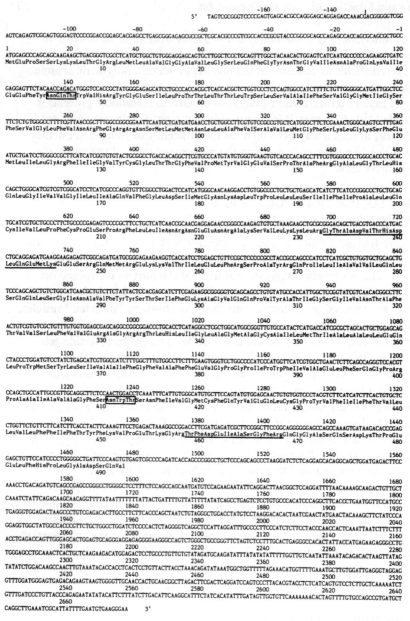

Figure 1 Nucleotide sequence of the human HepG2 hepatoma glucose transporter cDNA and deduced amino acid sequence. Numbering of nucleotides is above the sequence and the numbering of amino acid residues is below. The sequences of two peptides isolated after tryptic digestion of the erythrocyte transporter are underlined. The two possible sites of N-linked glycosylation are boxed. (From Ref. 24, with permission.)

homologous if not identical to the human erythrocyte protein: (a) The deduced amino acid composition closely matches that determined for the purified erythrocyte protein [11] ; (b) Twenty-six percent of the deduced HepG2 protein sequence was mapped by fast atom bombardment analysis of the trypsin-digested erythrocyte transporter; (c) The deduced N-terminal amino acid sequence agrees with that determined by Edman degradation sequence analysis of the purified erythrocyte transporter [24] in 14 of 18 positions. The discrepancies are likely due to problems with sequencing of the protein [23]. The deduced sequence matches perfectly the amino acid sequences of two internal tryptic peptides of the erythrocyte transporter [23] ; (d) Three distinct monoclonal antibodies directed against the erythrocyte protein [12] reacted with fusion proteins produced by in-phase homologus clones; and (3) Mouse fibroblast cell lines expressing high levels of protein encoded by the transfected cDNA exhibit increased basal levels of glucose transport [25].

The glucose transporter cDNA encodes a polypeptide of 492 residues (54,117 daltons) (Figure 1). This value is in reasonable agreement with the estimated size of the deglycosylated erythrocyte transporter (46 kD), because membrane proteins frequently exhibit increased mobility relative to soluble protein standards on sodium dodecyl sulfate polyacrylamide gels. This anomalous behavior was demonstrated directly by translation in an in vitro system of full-length synthetic glucose transporter mRNA derived from the cDNA [26]. The primary translation product exhibited an apparent M_r of 41 kD relative to soluble protein standards. The primary translation product immunoprecipitated after pulse-labeling of HepG2 cells migrated as a doublet of 46 and 38 kD that collapsed to a single band of 38 kD after boiling [27], indicating the likely propensity of this protein to form conformers of different mobility.

A hydropathy plot of the deduced HepG2 glucose transporter sequence is shown in Figure 2. Twelve membrane-spanning domains were predicted for the protein using the algorithm of Eisenberg et al. [28]. It is assumed that membrane-spanning domains comprise α-helical segments of about 21 amino acid residues. A postulated model for the two-dimensional structure of the protein in the membrane is also shown in Figure 2. This model is consistent with the results of the protease and glycosidase experiments described above [but see ref. 29].

There are two potential sites of N-linked glycosylation in the sequence, Asn^{45} and Asn^{411}. The following considerations indicate that only Asn^{45} bears an oligosaccharide chain: (a) Oligosaccharide is present only within a region corresponding to the N- or C-terminal third of the protein (not at both ends) [30] ; (b) Asn^{411} is predicted to lie within a membrane-spanning domain; (c) Experiments involving cleavage of the purified transporter at cysteine residues are consistent with a large fragment encompassing Asn^{411} that does not bear an oligosaccharide chain [30] ; (d) A polypeptide produced in vitro from a synthetic mRNA encoding a truncated HepG2 glucose transporter lacking Asn^{411}, but

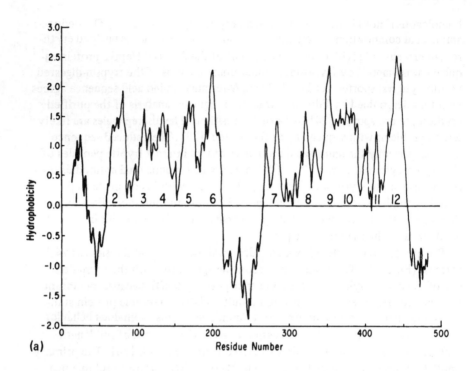

(a)

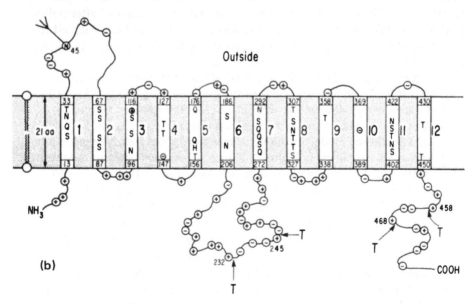

(b)

Figure 2

including Asn[45], was inserted into and fully glycosylated by rough microsomes [26]; (e) Antibodies raised against a synthetic peptide corresponding to residues 1-15 of the glucose transporter bind to the larger, carbohydrate-bearing tryptic fragment (M_r 23-42 kD; see above), but not to the 18 kD fragment lacking oligosaccharide [31].

Assuming the number of membrane-spanning domains is correct, the assignment of the oligosaccharide fixes the location of the N- and C-termini in the cytoplasm. This is consistent with the finding that digestion of red blood cells with high levels of several different proteases has no effect on the size of the transporter polypeptide [17]. The cytoplasmic orientation of the C-terminus has been verified directly by Davies et al. [31]. These investigators demonstrated that antibodies against a synthetic peptide corresponding to residues 447-492 of the transporter bind only to the cytoplasmic surface of the red blood cell membrane. The large, highly polar, central domain (residues 207-271) is localized in the cytoplasm as indicated by the isolation and sequencing of a tryptic fragment corresponding to residues 233-245 [23].

Analysis of the postulated extramembranous domains of the protein by the method of Chou and Fasman [32] revealed a clustering of β-turn propensity in the short linker regions, as expected [23]. The central polar domain (residues 207-271) has a very high α-helix propensity, and very little β-sheet propensity was detected overall in extramembranous segments of the protein [33]. Thus, the expected α-helical content of the glucose transporter based on analysis of the primary structure is somewhat greater than 64% (13% for the central polar region plus 51% fot the transmembrane segments). This compares favorably to the value of 70% α-helical content reportedly determined for the detergent-solubilized transporter by circular dichroism measurements [34]. A largely α-helical structure is consistent with the results of Fourier transformed infrared

---

Figure 2 (a) Hydropathy plot of the HeG2 glucose transporter sequence. Hydropathy values for a window of 21 amino acid residues were averaged, assigned to the middle residue of the span, and plotted with respect to position along the amino acid sequence. The numbers refer to putative membrane-spanning domains predicted by the algorithm of Eisenberg et al. [29]. (b) Proposed model for the orientation of the glucose transporter in the membrane. The 12 putative membrane-spanning domains are numbered and shown as rectangles. The relative positions of acidic (Glu, Asp) and basic (Lys, Arg) amino acid residues are indicated by circled (+) and (-) signs, respectively. Uncharged polar residues within the membrane-spanning domains are indicated by their single-letter abbreviations: S, serine; T, theonine; H, histidine; N, asparagine; Q, glutamine. The predicted position of the N-linked oligosaccharide as Asn[45] is shown. The arrows point to the positions of known tryptic cleavage sites in the native, membrane-bound erythrocyte transporter. (From Ref. 24, with permission.)

spectroscopy, which also indicated the presence of lesser degrees of β-structure and random coil [35].

According to the model, it would appear that the glucose transporter poly-peptide possesses a sufficient number of transmembrane segments to function as a monomer, but this is largely speculative. Five of the twelve putative membrane-spanning domains [3,5,7,8,11] (Figure 2) may form amphipathic helices [23]. These putative amphipathic helices are rich in hydroxyl and amide-containing side chains that would largely be localized to one face of the helix. Presumably, these side chains line the walls of an aqueous channel through which the sugar traverses the membrane. It is also possible that one or more of these groups may comprise part of the substrate-binding site.

Polarized Fourier transform infrared and ultraviolet circular dichroism measurements indicate that the α-helices in the glucose transporter are, on average, oriented roughly perpendicular to the plane of the membrane, showing tilt angles of <38° relative to the normal [34]. These measurements also indicate that D-glucose causes a slight reduction in the tilt angle. Thus, substrate binding may influence the conformation of a putative aqueous channel formed by α-helical domains of the protein. Hydrogen exchange studies [35,36] are consistent with the existence of an aqueous channel formed by the glucose transporter. These studies also detected changes in the conformation of the protein in the presence of various substrate analogues and inhibitors.

Wang et al. [37] have developed a proton nuclear magnetic resonance method for directly measuring the binding of D-glucose to both the outward and inward-facing substrate-binding sites of the transporter. These investigators demonstrated directly that, in the steady state, glucose-binding sites are present on both sides of the membrane, and that translocation of glucose as opposed to binding is the rate-limiting step in transport. If one accepts the premise that cytochalasin B does not directly affect the binding of glucose to the outward-facing site [21,22], their experiments also demonstrate that cytochalasin B recruits all binding sites to the intracellular compartment. This constitutes strong support for an alternating conformation mechanism of glucose transport [2]. Wang et al. [37] have proposed a sliding barrier model for the mechanism of transport. A single glucose binding site is proposed to exist in a cleft between two amphipathic helices comprising part of an aqueous channel. Free diffusion through the channel is limited by a helical plug at one end of the channel that is connected to segments of random coil. As the plug slides from one end of the channel to the other past the binding site (which may contain substrate), the binding site is alternately exposed to both sides of the membrane. Exchange acceleration is accounted for in this model by supposing that the binding of substrate to the cleft expands the diameter of the channel, allowing the plug to slide more easily through the channel. Asymmetry can also be explained if the position of the barrier or plug at either end of the channel influences the configuration of the binding site, or directly affects binding due to steric effects. Although highly speculative, models such as this one are useful in suggesting further avenues of experimentation.

V. THE GLUCOSE TRANSPORTER IN OTHER TISSUES

Glucose transporter cDNA has also been cloned from a rat brain expression library using antibodies against the human erythrocyte transporter [38]. The rat brain and human HepG2 transporters exhibit 97.6% sequence homology. Only 12 amino acid residues differ between the two 492-residue proteins, and most of these represent conservative substitutions. This high degree of conservation implies an unusual dependence of function on primary structure. The coding regions of the cDNAs are 89% homologous at the nucleotide level. This can be compared with the 87% homology observed between the nucleotide coding regions of the highly conserved histone H4 proteins of rats and humans [39]. Perhaps uniquely, this degree of homology also encompasses much of the 3' untranslated region as well as a portion of the 5' untranslated region of the mRNAs. This presumably reflects a functional role for a higher order structure of the glucose transporter mRNA. The structure of the glucose transporter gene(s) in rats and humans may give some insight into these tantalizing observations.

The glucose transporter cDNAs provide powerful tools for studying the tissue distribution of this protein. Messenger RNAs homologous to the 2.9 kb human HepG2/rat brain species have been detected in nearly every mammalian tissue examined [23,38,40]. Thus, it is likely that the same or a very similar glucose transporter is expressed in fat, skeletal muscle, brain microvessels, fibroblasts, kidney, colon, heart muscle, and lymphoid cells.

Interestingly, a possible exception is the liver. The rate of glucose transport across liver cells is of the same order of magnitude as in human erythrocytes [41], where the transporter constitutes ~5% of total membrane protein [12]. Although a 2.9 kb mRNA homologous to the HepG2 species can be detected in isolated rat hepatocytes, the intensity of the signal on Northern blots is nearly two orders of magnitude lower than for HepG2 cells, and at lower hybridization stringencies additional bands appear that are not seen in other tissues [40]. Thus, it is possible that a unique glucose transporter is expressed in liver. This interpretation is consistent with the lack of immunological cross-reactivity between the red blood cell and liver proteins [38], and the apparent differences observed in the affinity of cytochalasin B for the two species [42]. It is also consistent with the different physiological role played by the transporter in liver as compared with most tissues. Depending on the metabolic circumstances, glucose may be transported either into or out of hepatocytes, whereas under normal physiological conditions, in most tissues the glucose concentration within cells is always lower than that in the blood, and glucose is only transported into cells.

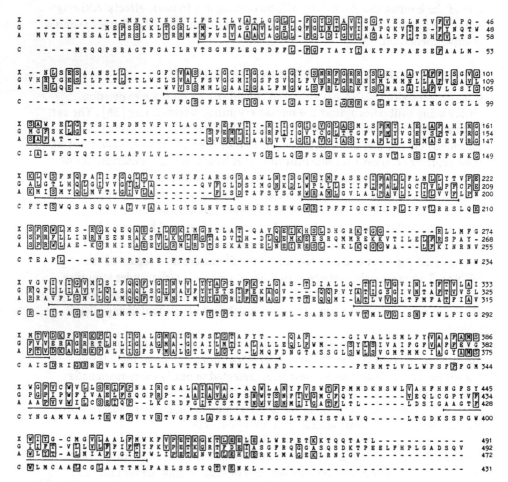

Figure 3 Aligned sequences of the *E. coli* xylose-H$^+$, arabinose-H$^+$, and citrate transporters with the human hepatoma glucose transporter. Residues are boxed in the xylose, glucose, and arabinose transporters if they occur more than once at an aligned position in the three sugar transporter sequences. They are only boxed in the citrate transporter if they also occur in the aligned positions of two or three of the sugar transporters. The predicted membrane-spanning regions of the glucose transporter are underlined. (From Ref. 43, with permission.)

VI. EVOLUTIONARY ASPECTS

Maiden et al. [43] have described intriguing homologies between the mammalian glucose transporter and two procaryotic sugar–proton symport systems. These investigators cloned and sequenced the genes encoding the xylose and arabinose transporters of *Escherichia coli*. The proteins contain 491 and 472 amino acid residues, respectively, and hydropathic analysis indicates that their predicted two-dimensional structures are strikingly similar to that of the glucose transporter [43]. All three transporters are predicted to possess two hydrophobic domains consisting of six transmembrane helices each, separated by a smaller, central polar domain (Figure 2). When aligned for optimal homology, there are 131 and 142 identities between the glucose transporter and the arabinose and xylose transporters, respectively (Figure 3). Many of the nonidentities represent conservative amino acid substitutions, so that when these are taken into account, the overall homology among the three proteins is nearly 40%. Limited homology was also observed between the *E. coli* citrate transporter and the three sugar transporters [43]. Analysis of the observed homologies should prove invaluable in designing mutagenesis experiments to investigate the role of specific regions of the proteins in transport function.

Maiden et al. [43] have noted the symmetrical two-dimensional structure of the proteins about the central polar domain, suggesting that all three proteins may be descended from a very ancient precursor whose gene underwent duplication. This is supported by the presence of a conserved pentameric motif (Arg-X-Gly-Arg-Arg, where Arg may be replaced by Lys) that is present in duplicate in all three proteins. There is one copy of this motif at an equivalent position of each hydrophobic domain (at positions 89–93 and 387–391 of the glucose transporter) consistent with the two domains having arisen from the duplication of an ancestral gene. Interestingly, the motif is also present at equivalent positions in the deduced amino acid sequences of the lac permease and the citrate and melibiose transporters [44]. Other than this motif, the lac permease and melibiose transporters show little, if any, sequence homology to the glucose, arabinose, or xylose transporters. However, the presence of this conserved motif combined with a similarity in their hydropathic profiles suggests that all six transporters may be descendents of a common ancestor.

VII. FUTURE STUDIES

Much progress has been made in the past 5 years in our understanding of the structure and function of the facilitated-diffusion glucose transporter. The elucidation of the primary structure and the availability of cDNA clones should allow the application of recombinant DNA techniques to structure-function studies.

Such studies may illuminate the role of specific regions and residues of the protein in transport function, but major progress in the elucidation of the mechanism of glucose transport most ultimately await the determination of the three-dimensional structure of the transporter. In the meantime, a combination of biophysical and molecular-biological approaches is needed for a further advance in our knowledge of the structure and function of this important protein.

ACKNOWLEDGMENT

I thank Drs. Martin C. J. Maiden, Peter J. F. Henderson, and Stephen A. Baldwin for sharing manuscripts prior to publication. The author was supported by a grant from the Diabetes Research and Training Center of Washington University School of Medicine.

REFERENCES

1. Wheeler, T. J., and Hinkle, P. C. (1985). The glucose transporter of mammalian cells. *Annu. Rev. Physiol.* 47:503.
2. Lienhard, G. E., Baldwin, J. M., Baldwin, S. A., and Gorga, F. R. (1983). The glucose transporter of human erythrocytes. In *Structure and Function of Membrane Proteins*, E. Quagliariello and F. Palmieri (Eds.). Elsevier Science Publishers, Amsterdam, p. 325.
3. Simpson, A., and Cushman, S. W. (1986). Hormonal regulation of mammalian glucose transport. *Annu. Rev. Biochem.* 55:1059.
4. Elbrink, J., and Bihler, I. (1975). Membrane transport: its relation to cellular metabolic rates. *Science* 188:1177.
5. Kwong, F. Y. P., Baldwin, S. A., Scudder, P. R., Jarvis, S. M., Choy, M. Y. M., and Young, J. D. (1986). Erythrocyte nucleoside and sugar transport. *Biochem. J.* 240:349.
6. Miller, D. M. (1971). The kinetics of selective biological transport. V. Further data on the erythrocyte monosaccharide transport system. *Biophys. J.* 11:915.
7. Eilam, Y., and Stein, W. D. (1972). A simple resolution of the kinetic anomaly in the exchange of different sugars across the membrane of the human red blood cell. *Biochim. Biophys. Acta* 266:161.
8. Kasahara, M., and Hinkle, P. C. (1976). Reconstitution of D-glucose transport catalyzed by a protein fraction from human erythrocytes in sonicated liposomes. *Proc. Natl. Acad. Sci. USA* 73:396.
9. Lin, S., and Spudich, J. A. (1974). Biochemical studies on the mode of action of cytochalasin B. Cytochalasin B binding to red cell membrane in relation to glucose transport. *J. Biol. Chem.* 249:5778.

10. Kasahara, M., and Hinkle, P. C. (1977). Reconstitution and purification of the D-glucose transport from human erythrocytes. *J. Biol. Chem.* 253:7384.

11. Baldwin, S. A., Baldwin, J. M., and Lienhard, G. E. (1982). Monosaccharide transporter of the human erythrocyte. Characterization of an improved preparation. *Biochemistry* 21:3836.

12. Allard, W. J., and Lienhard, G. E. (1985). Monoclonal antibodies to the glucose transporter from human erythrocytes. Identification of the transporter as a $M_r = 55,000$ protein. *J. Biol. Chem.* 260:8668.

13. Wheeler, T. J., and Hinkle, P. C. (1981). Kinetic properties of the reconstituted glucose transporter from human erythrocytes. *J. Biol. Chem.* 256:8907.

14. Baldwin, J. M., Gorga, J. C., and Lienhard, G. E. (1981). The monosaccharide transporter of the human erythrocyte. Transport activity upon reconstitution. *J. Biol. Chem.* 256:3685.

15. Gorga, F. R., Baldwin, S. A., and Lienhard, G. E. (1979). The monosaccharide transporter from human erythrocytes is heterogeneously glycosylated. *Biochem. Biophys. Res. Commun.* 91:955.

16. Sogin, D. C., and Hinkle, P. C. (1978). Characterization of the glucose transporter from human erythrocytes. *J. Supramol. Struct.* 8:447.

17. Lienhard, G. E., Crabb, J. H., and Ransome, K. J. (1984). Endoglycosidase F cleaves the oligosaccharides from the glucose transporter of the human erythrocyte. *Biochim. Biophys. Acta* 769:404.

18. Cuppoletti, J., Jung, C. Y., and Green, F. A. (1981). Glucose transport carrier of human erythrocytes. Radiation target size measurement based on flux inactivation. *J. Biol. Chem.* 256:1305.

19. Jorgensen, P. L. (1982). Mechanism of the Na^+, K^+ pump. Protein structure and conformations of the pure $(Na^+ + K^+)$-ATPase. *Biochim. Biophys. Acta* 694:27.

20. Deziel, M. R., and Rothstein, A. (1984). Proteolytic cleavages of cytochalasin B binding components of band 4.5 proteins of the human red blood cell membrane. *Biochim. Biophys. Acta* 776:10.

21. Deves, R., and Krupka, R. M. (1978). Cytochalasin B and the kinetics of inhibition of biological transport. A case of asymmetric binding to the glucose transporter. *Biochim. Biophys. Acta* 510:339.

22. Chen, C.-C., Kurokawa, T., Shaw, S.-Y., Tillotson, L. G., Kalled, S., and Isselbacher, K. J. (1986). Human erythrocyte glucose transporter: normal asymmetric orientation and function in liposomes. *Proc. Natl. Acad. Sci. USA* 83:2652.

23. Mueckler, M., Caruso, C., Baldwin, S. A., Panico, M., Blench, I., Morris, H. R., Allard, W., Lienhard, G. E., and Lodish, H. F. (1985). *Science* 229:941.

24. Lundahl, P., Greijer, E., Lindblom, H., and Fagerstam, L. G. (1984). Fractionation of human red cell membrane proteins by ion-exchange chromatography in detergent on mono Q with special reference to the glucose transporter. *J. Chromatogr.* 297:129.

25. M. Mueckler (1986), Unpublished data.
26. Mueckler, M., and Lodish, H. F. (1986). The human glucose transporter can insert posttranslationally into microsomes. *Cell* 44:629.
27. Haspel, H. C., Birnbaum, M. J., Wilk, E. W., and Rosen, O. M. (1986). Biosynthetic precursors and in vitro translation products of the glucose transporter of human hepatocarcinoma cells, human fibroblasts, and murine preadipocytes. *J. Biol. Chem.* 260:7219.
28. Eisenberg, D., Schwarz, E., Komaromy, M., and Wall, R. (1984). Analysis of membrane and surface protein sequences with the hydrophobic moment plot. *J. Mol. Biol.* 179:125.
29. Shanahan, M. F., and D'Artel-Ellis, J. (1984). Orientation of the glucose transporter in the human erythrocyte membrane. Investigation by in situ proteolytic dissection. *J. Biol. Chem.* 259:13878.
30. Cairns, M. T., Elliot, D. A., Scudder, P. R., and Baldwin, S. A. (1984). Proteolytic and chemical dissection of the human erythrocyte glucose transporter. *Biochem. J.* 221:179.
31. Davies, A., Meeran, K., Cairns, M. T., and Baldwin, S. A. (1987). Peptide-specific antibodies as probes of the orientation of the glucose transporter in the human erythrocyte membrane. *J. Biol. Chem* 262:9347.
32. Chou, P. Y., and Fasman, G. D. (1974). Prediction of protein conformation. *Biochemistry* 13: 222.
33. M. Mueckler (1986), Unpublished data.
34. Chin, J. J., Jung, E. K. Y., and Jung, C. Y. (1986). Structural basis of human erythrocyte glucose transporter function in reconstituted vesicles. Alpha-helix orientation. *J. Biol. Chem.* 261:7101.
35. Alvarez, J., Lee, D. C., Baldwin, S. A., and Chapman, D. (1987). Fourier transform infrared spectroscopic study of the structure and conformational changes of the human erythrocyte glucose transporter. *J. Biol. Chem.* 262:3502.
36. Jung, E. K. Y., Chin, J. J., and Jung, C. Y. (1986). Structural basis of human erythrocyte glucose transporter function in reconstituted system. Hydrogen exchange. *J. Biol. Chem.* 261:9155.
37. Wang, J.-F., Falke, J. J., and Chan, S. I. (1986). A proton NMR study of the mechanism of the erythrocyte glucose transporter. *Proc. Natl. Acad. Sci. USA* 83:3277.
38. Birnbaum, M. J., Haspel, H. C., and Rosen, O. M. (1986). Cloning and characterization of a cDNA encoding the rat brain glucose-transporter protein. *Proc. Natl. Acad. Sci. USA* 83:5784.
39. Wells, D. E. (1986). Compilation analysis of histones and histone genes. *Nuc. Acid. Res.* 14:119.
40. Flier, J. S., Mueckler, M., and Lodish, H. F. (1987). Distribution of glucose transporter mRNA transcripts in tissues of rat and man. *J. Clin. Invest.* 79: 657.
41. Craik, J. D., Elliott, K. R. F. (1979). Kinetics of 3-O methyl glucose transport in isolated rat hepatocytes. *Biochem. J.* 182:503.

42. Axelrod, J. D., and Pilch, P. F. (1983). Unique cytochalasin B binding characteristics of the hepatic glucose carrier. *Biochemistry* 22: 2222.
43. Maiden, C. J., Davis, E. O., Baldwin, S. A., Moore, D. C. M., and Henderson, P. J. F. (1987). Mammalian and bacterial sugar transport proteins are homologous. *Nature* 325:641.

3

Intermediate Filament Expression in Erythroid Differentiation and Morphogenesis

JOHN NGAI* and ELIAS LAZARIDES   *California Institute of Technology, Pasadena, California*

I. INTRODUCTION

In multicellular organisms, the maturation of distinct cell lineages yields differentiated cell and tissue types capable of performing specialized functions. The study of differentiating cell systems allows an opportunity to correlate changes in gene expression with changes in cell structure and function. Indeed, the characteristics of any given cell ultimately are a reflection and manifestation of its underlying structure and organization. The specification and development of cell shape and structure, a process referred to as morphogenesis, is therefore a critical feature of cellular differentiation. The vertebrate erythrocyte has provided an excellent model system for the study of gene expression and morphogenesis in the development of a specialized cell.

The events and stages of erythropoiesis have been well defined, and the structure of the mature red blood cell has been characterized in exhaustive detail. In this review we will consider the expression of one cytoskeletal protein, the intermediate filament subunit vimentin, during erythropoiesis in birds and mammals. We will discuss the regulation of vimentin gene expression as well as the known intracellular interactions in which vimentin filaments partake in the context of a comparison of avian and mammalian erythropoiesis. The rationale for

---

Current affiliation: Columbia University College of Physicians and Surgeons, New York, New York

comparing vimentin expression in these terms is the expectation that differences
or similarities in cytoskeletal protein expression in the maturation of the morphologically dissimilar nucleated avian erythrocyte and anuclear mammalian
erythrocyte may reveal the roles played by these proteins in morphogenesis. Furthermore, the molecular basis for any divergence in cytoskeletal protein expression in birds and mammals can be explored. Our objectives therefore are twofold. First, we wish to understand the factors regulating vimentin expression,
using erythropoiesis as a representative model system. Second, using the same
erythropoietic system, we wish to elucidate the role of vimentin filaments in
determining cell shape, with the ultimate goal of demonstrating function by the
experimental manipulation of expression.

II. INTERMEDIATE FILAMENTS: GENERAL STRUCTURE
AND DEVELOPMENTALLY REGULATED EXPRESSION

Intermediate filament proteins comprise a heterogeneous family of cytoskeletal
proteins that display a characteristic 10 nm-diameter filament morphology. By
immunological, biochemical, and protein and DNA sequence criteria, five classes
of cytoplasmic intermediate filament proteins have been defined in higher verte-
brates. The keratins (40–70 kD) are expressed in epithelial cells, desmin (50 kD)
in smooth and striated muscle, glial fibrillary acidic protein (GFAP, 50 kD) in
glial cells, neurofilaments (70 kD major subunit plus 160 kD and 200-kD components) in neuronal cells, and vimentin (52 kD) in a wide variety of cell types,
and often in immature stages of cell differentiation [reviewed in refs. 1 and 2].
All intermediate filament subunit types share a highly conserved α-helical central
rod domain structure of 310–340 amino acid residues, flanked by variable sub-
unit-specific nonhelical head- and tail-pieces [2–10]. The characteristic 10-nm
filament structure is believed to be due to the polymerization of subunits via a
tetrameric organization of the conserved central rod domains in coiled-coiled
dimers [3,4,11–13], which is influenced by the variable nonhelical domains
[3,14]. The presumed cell-type–specific functions of intermediate filament
subunits, inferred by their cell- and tissue-specific expression, most likely reside
in the variable nonhelical flanking domains, which probably are disposed radi-
ally at the peripheries of the filament [15–17]. In cases in which vimentin coexists with desmin, GFAP, or neurofilament protein, heteropolymeric filaments
are observed [18–25]. An exception is the coexpression of keratin with vimentin
or other filament subunits, wherein segregated homopolymeric filaments are observed [7,26–28]; the apparent inability of the keratins to copolymerize with
other intermediate filament subunits may reflect the greater degree of divergence
between the keratins and the other four classes of intermediate filament proteins.

A particular intermediate filament protein subunit may be induced during
the terminal differentiation of many cell lineages. In some cases, such as in fibro-

blasts, lens fibers, and avian erythrocytes, vimentin is the major intermediate filament subunit in the mature state [28–31]. In contrast, vimentin is completely replaced by neurofilament protein during terminal differentiation of chick spinal cord neurons [32–33]. Studies on chicken myogenesis in vitro have shown that vimentin is expressed in immature myoblasts, and desmin synthesis is induced after fusion of myoblasts into syncytial myotubes [19,34]. However, unlike the differentiation of spinal cord neurons, vimentin and desmin are coexpressed through the adult muscle stages in chickens [35]. Vimentin is also coexpressed with neurofilament protein in some mouse retinal neurons [36] and with GFAP in astrocytes [33,37,38]. An interesting situation is found in chicken erythrocytes, in which the major 70-kD neurofilament subunit is expressed at substoichiometric amounts relative to vimentin [25]. Hence, a cell-type-specific intermediate filament protein subunit often may partially or completely replace vimentin during terminal differentiation. In the mammalian erythropoietic and B-lymphocyte lineages, vimentin expression is repressed at late stages of differentiation, and in both cases no other intermediate filament protein is expressed or induced [39–41]. The tissue-specific and developmentally regulated expression of intermediate filament proteins therefore provides an interesting model for gene regulation. The diverse and variable nature of vimentin expression in several cell lineages makes the study of this intermediate filament type particularly interesting. In the next section we will describe the gross morphological characteristics of the mammalian and avian red blood cell with the intent of presenting these two cell types as model systems for studying the regulated expression of vimentin.

III. GENERAL FEATURES OF THE MAMMALIAN AND THE AVIAN RED BLOOD CELL

The mammalian erythrocyte is perhaps one of the most intensively studied cell in terms of its morphology and structural organization. The mature mammalian red cell is anuclear and devoid of other intracellular organelles, and assumes a biconcave disk-shaped configuration (Figure 1A). The integrity and deformability of the red blood cell, necessary for its survival of repeated passages through narrow capillaries, are functions of the properties of the cell membrane. The relative simplicity of this cell has allowed the components of its plasma membrane to be defined in great detail. A meshlike protein network lines the inner surface of the mammalian erythrocyte plasma membrane, and is believed to confer on this membrane its properties of strength and elasticity and to influence the lateral mobility of membrane proteins [reviewed in refs. 42–45; the reader is referred to these reviews for more extensive discussions—only the highlights of these studies will be considered here in brief]. The major proteins of this network, termed the membrane skeleton, are α- and β-spectrin, which exist as an

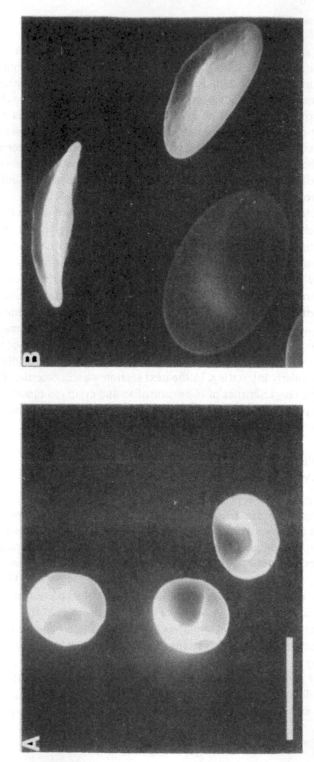

Figure 1 Scanning electron micrographs of human red blood cells (A) and chicken red blood cells (B). The cell at the top of panel B is seen from an edge-on perspective. Bar = 5 μm.

$(\alpha,\beta)_2$ tetramer and complex with actin to form a two-dimensional protein network. This spectrin–actin-based complex is bound to the plasma membrane primarily via a high-affinity interaction with ankyrin, which in turn is bound to a subset of transmembrane anion transporter molecules [46-51]. Protein 4.1 facilitates the interaction of spectrin and actin (and hence, the formation of the two-dimensional network) by forming a ternary complex with these two components [52-55]. Protein 4.1 also has been demonstrated to interact directly with two intrinsic membrane proteins, glycophorin and anion transporter [56, 57], as well as with membrane phospholipid [58]. Hence, the membrane skeleton may associate with the plasma membrane through a variety of different interactions. Intermediate filaments are not found in the mature mammalian red blood cell. As shown by immunofluorescence studies of human erythropoiesis in vivo, vimentin is expressed early in differentiation, but is lost at the early- to miderythroblastic stage, well prior to enucleation [41].

Although avian red blood cells perform a physiological function similar to the one performed by mammalian red blood cells, the physical characteristics of these two types of erythrocytes are quite different in several respects. The most striking and obvious difference between avian and mammalian erythrocytes is one of shape. The avian red blood cell retains its nucleus throughout its existence, and has the shape of a somewhat flattened biconvex ellipsoidal disk (Figure 1B). The major membrane skeletal components initially described in mammalian erythrocytes have been identified in chicken erythrocytes [59-64]. However, unlike the mammalian red blood cell, the chicken erythroid membrane skeleton interacts with two classes of cytoskeletal elements not present in mammalian red blood cells. The first is a structure known as the marginal band, which is an aggregation or bundle of microtubules that apparently attaches to the plasma membrane, and is disposed circumferentially at the periphery of the cell where the ellipsoidal faces meet [65]. Microtubule disruption experiments have suggested that the marginal band is important for the establishment of the ellipsoidal shape of the nucleated red blood cell [66], but its presence may not be necessary for the maintenance of this shape in the mature state [66,67]. However, the marginal band appears to play a role in providing resistance to deformation due to environmental mechanical stress [68,69], such as that which occurs in the circulation. Although the binding sites of the marginal band microtubules on the membrane skeleton have yet to be identified, it is apparent that such putative receptors for the structure have an anisotropic localization on the plasma membrane.

The second cytoskeletal system absent in mammalian erythrocytes and present in nucleated erythrocytes is a structure originally described in the latter as the transmarginal band material, a filamentous network that spans the cytoplasm and appears to enmesh the centrally located nucleus [68]. It has been suggested that the transmarginal band material may play a role in nuclear anchoring [68,70–72], and perhaps in the maintenance of the elliptical shape of the cell [68]. Initially, the filaments of the transmarginal band material were characterized as resembling intermediate filaments [68,70–72], and indeed later were demonstrated to be vimentin filaments [31]. The intermediate filament-associated protein, synemin, was also localized to these filaments at ~200 nm periodically and at points of filament junctions or branchings [73], suggesting a filament cross-linking role for synemin. Conventional electron microscopic images have shown vimentin filaments in close association with both the plasma membrane and the nucleus [68,70–72]. By freeze-fracture scanning electron microscopy and transmission electron microscopy of membrane replicas [73–75], it appears that the vimentin filament network in avian red blood cells attaches to a defined and segregated area of the plasma membrane; vimentin filaments appose the inner surface of the plasma membrane only on the two flattened faces of the cells, displaced and distinct from the region occupied by the marginal band. Hence, there must exist additional determinants on the plasma membrane or membrane skeleton that specify the anisotropic membrane interaction with vimentin filaments.

IV. INTRACELLULAR INTERACTIONS OF VIMENTIN FILAMENTS

Microinjection of living nonerythroid cells with antibodies specific for nonerythroid spectrin results in the precipitation of spectrin in the cell, with a concomitant partial collapse of the vimentin filament network [76]. These results suggest that vimentin may interact with the plasma membrane via an association with membrane skeletal components. In vitro binding studies have provided a first step in characterizing potential intracellular binding sites for vimentin. It has been demonstrated that spectrin can bind vimentin filaments in a saturable manner [77]. Furthermore, spectrin appears to mediate the binding of vimentin to erythrocyte plasma membranes via lateral associations with the filaments [77]. This type of interaction of plasma membranes with vimentin filaments observed in vitro is consistent with electron microscopic images of vimentin filaments looping out from the inner surface of the plasma membrane inner surface in situ [31,68–73]. In contrast, others have shown that vimentin binds to erythrocyte plasma membranes via the membrane skeletal component ankyrin [78]. This binding also appears to be saturable, with a K_D of 10^{-6} to 10^{-7} M, and can be inhibited by antiankyrin antibodies [78]. At present it is unclear which of these two binding activities (i.e., spectrin–vimentin or ankyrin–vimentin) predominates in vivo. The interaction of vimentin with ankyrin is mediated by the N-terminal head domain of the vimentin molecule [79]. Similar analyses have revealed that vimentin can interact via its C-terminal tail domain with nuclear lamin B in vitro, displaying K_D of ~10^{-7} M [80, 81]. The above observations have been interpreted to indicate that vimentin filaments are vectorially assembled from karyotic nucleation sites toward plasma membrane-bound ankyrin binding sites [80]. However, several topological considerations can be addressed by this interpretation only with difficulty. First, the proposed vectorial behavior of vimentin filaments is inconsistent with the generally accepted model of antiparallel (and therefore apolar) intermediate filament substructure [2-4,11-13]. Second, the association of vimentin with the plasma membrane most likely involves a more complicated scheme than simple ankyrin or spectrin binding. This becomes apparent when one considers the observation that vimentin filaments in chicken red blood cells interact with the plasma membrane only in segregated domains [73-75]. Because both ankyrin and spectrin are distributed throughout the membrane skeleton, other factors must exist to influence the site, and perhaps the extent of binding underneath the cell membrane. Third, binding of vimentin to the nuclear lamina in situ would require that vimentin filaments traverse the nuclear membrane, presumably through nuclear pores. Although seemingly unlikely, this type of configuration theoretically is possible. In support of this scheme are early electron microscopic observations of intermediate filament-nuclear interactions after extraction of the nuclear envelope, and close association of intermediate filaments with nuclear pore complexes [70-72,82]. Clearly the above-described in vitro binding experiments have raised several intriguing possibilities for intermediate filament localization in situ. It remains to be determined which of these proposed interactions occur in vivo, and the extent to which they may be modified by other factors in the cell.

V. FUNCTIONAL IMPLICATIONS OF DIVERGENT CYTOSKELETAL GENE EXPRESSION IN AVIAN AND MAMMALIAN RED BLOOD CELLS

In both avian and mammalian erythropoiesis, early progenitors and erythroblastic cells are relatively large, round cells; during terminal differentiation, the nuclei condense and overall cell volumes decrease with concomitant changes in cell morphology. In avian erythropoiesis, the cells attain their characteristic oblong shape, whereas in mammalian erythropoiesis, nuclei and other organelles are expelled and the reticulocytes mature further to form flattened biconcave disks. The contrast in the morphologies of nucleated and anuclear erythrocytes suggests that the divergent regulation of cytoskeletal protein expression in each lineage underlies these differences and is an important feature of the respective terminal differentiation programs. As discussed above, marginal band microtubules are believed to play a role in the establishment of the oblong shape of the nucleated red cell [66]. The mere retention of the nucleus in nonmammalian erythroid cells physically precludes the red blood cell from assuming a biconcave shape. However, although nuclear loss is necessary for the cell to adopt this configuration, it is not by itself sufficient, because the anuclear red blood cells of the *Camelidae* are biconvex and ellipsoid, and contain marginal bands [83,84]. What are the structural and mechanistic differences between mammalian and nonmammalian red blood cell lineages that are responsible for the loss of the nucleus in one case and the retention of the nucleus in the other?

Early ultrastructural studies have shown that enucleation of differentiating erythroblasts occurs by an extrusion process that grossly resembles an asymmetric cytokinesis [85-87], wherein the nucleus and a thin rim of cytoplasm are budded off from the part of the cell that later becomes the reticulocyte. However, the apparent similarity of the enucleation process to cell division may be a superficial one, because such critical components of cytokinesis as the contractile ring are absent [88]. Because intermediate filaments have been observed morphologically to span the cell and entrap the nucleus, it has been suggested that they play a structural role in nuclear anchoring and nuclear centration [68, 70-73]. It is possible that the presence of vimentin filaments during the late erythroblastic stages of mammalian erythropoiesis may physically prevent the expulsion of the nucleus. The loss of vimentin filaments therefore may be necessary to facilitate either the movement of the nucleus to its eccentric position before enucleation [85,87], or the actual process of enucleation itself.

During and just prior to enucleation in mouse erythroid cells, striking changes in the distribution and mobility of membrane proteins occur [89-93]. The mobilities of cell surface receptors for the lectins concanavalin A and wheat-germ agglutinin decrease as differentiation proceeds [90,91], and concanavalin A receptors become concentrated over budding nuclei in cells undergoing enucleation [89,90]. Conversely, the distribution of the major membrane skeleton protein spectrin, which initially is uniform over immature cells, becomes dramatically segregated in the incipient reticulocyte (i.e., away from the budding nucleus) during enucleation [90,92]. It has been proposed that the redistribution of spectrin is necessary for the developing red blood cell to accumulate sufficiently high levels of spectrin [94]. In avian erythropoiesis, this type of mechanism may not be necessary, because the reduction in cell volume takes place without enucleation, and spectrin synthesis remains high during late stages of differentiation [60]. The reorganization of the membrane skeleton therefore appears to be an important feature of mammalian erythropoiesis, and may in fact be either actively or passively involved in the enucleation process. If we again consider that vimentin interacts directly with the membrane skeleton [77-79], then it seems possible that vimentin filaments may need to be cleared from the cell not only to facilitate the enucleation process, but also to facilitate the requisite segregation of membrane skeletal components.

The observations discussed in this section suggest a structural and/or morphogenetic role for vimentin filaments in differentiating avian and mammalian red blood cells, although this suggestion can only be inferred from morphological data. Indeed, it has been difficult to demonstrate any role for intermediate filaments directly. For example, microinjection of anti-intermediate filament antibodies into cultured cells can result in a collapse of the intermediate filament network with no obvious effects on the cell [95-97]. However, as we will discuss in the following sections of this review, the differences in vimentin expression in avian and mammalian erythropoiesis can be exploited to determine not only the potential roles of vimentin in differentiation, but also the factors governing the divergent erythropoietic patterns of expression. We will summarize experimental results concerning the levels of regulation important for vimentin expression during erythropoiesis. The conclusions from these studies will then serve as a foundation for further investigations designed to experimentally manipulate the expression of vimentin during erythropoiesis as an initial attempt to discern any possible structural or morphogenetic consequences of aberrant vimentin expression.

VI.   ASSEMBLY OF VIMENTIN FILAMENTS FROM A SOLUBLE PRECURSOR SUBUNIT POOL

In studying the expression and functional potentials of intermediate filaments, it is important to understand the mechanisms involved in governing their regulation. Two obvious points of regulation could be the formation of filaments from precursor subunits and filament turnover. The kinetics of vimentin filament assembly have been studied in vivo using chicken embryonic erythroid cells [98, 99]. This is a convenient model system to study this process, because definitive chicken embryonic erythroid cells synthesize abundant amounts of vimentin and are relatively easy to isolate in large numbers. The approach for these experiments has been to monitor the synthesis of vimentin by the factionation of newly synthesized protein into different subcellular components. Metabolic labeling studies of these cells have indicated that newly synthesized vimentin enters a rapidly saturable ($\leq$ 30 min) soluble fraction, whereas the appearance of vimentin radioactivity in the cytoskeletal fraction shows a slight lag, and then increases linearly [98]. Pulse-chase studies have further shown that newly synthesized vimentin exits the soluble fraction rapidly (half-life < 20 min), and chases into the cytoskeletal compartment [98]. These results demonstrate that vimentin filaments are rapidly assembled posttranslationally from a soluble precursor pool of vimentin subunits, most likely by simple self-assembly [98]. The intermediate filament-associated protein synemin displays different assembly kinetics than those seen for vimentin [99]. Although synemin also is incorporated into the cytoskeletal fraction posttranslationally, synemin assembly exhibits a greater lag time than for vimentin, and saturation of the soluble pool occurs very slowly [99]. The kinetics of vimentin and synemin assembly suggest that the elongation of vimentin filaments produces synemin binding sites [99]. This interpretation is consistent with the observation of synemin localization at periodic intervals on vimentin filaments [31].

Although vimentin is a known substrate of cyclic adenosine monophosphate-dependent protein kinases [100], phosphorylation of newly synthesized vimentin is observed only after it is incorporated into the cytoskeleton [99], and the extent of phosphorylation appears to be substoichiometric [98]. These observations are seemingly paradoxical with recent studies of vimentin polymers in vitro showing that stoichiometric phosphorylation of vimentin filaments induces disaggregation [101]. It is possible that the observed substoichiometric phosphorylation of cytoskeletal vimentin in vivo may have little effect on filament stability, whereas hyperphosphorylation may induce filament instability.

Filamentous vimentin in chicken red blood cells is degraded only very slowly, and is stable for at least 4-6 h [98]. Studies of vimentin turnover in other cell systems have revealed that vimentin is extremely stable, showing a half-life comparable to the cell generation time [102]. The rapidity of vimentin assembly and the stability of vimentin filaments suggests that assembly and degradation do not serve as primary points of regulation in filament expression. Rather, it appears that the level of vimentin synthesis directly determines the amount and extent of vimentin polymerization, and hence filament accumulation.

VII. REGULATION OF VIMENTIN EXPRESSION IN ERYTHROPOIESIS OCCURS AT THE mRNA LEVEL

The results described in the previous section indicate that the expression of vimentin filaments is determined at the level of vimentin protein synthesis, i.e., translationally or by mRNA abundance. This has been examined in greater detail in the maturation of chicken embryonic red blood cells. Circulating erythroid cells from chicken embryos can be readily isolated at defined stages of differentiation [103], and changes in gene expression during erythropoiesis can be correlated with differences in protein or filament accumulation. Early and late polychromatophilic erythroblasts of the primitive series lineage are found in four-day-old embryos, and decline markedly in the circulation by approximately eight to 10 days of embryogenesis [103]. At 10 days of development, circulating cells are ~25% mature primitive cells, ~35% mid- to late-polychromatophilic definitive series erythroblasts, and ~35% mature definitive erythrocytes; at 15 days, 60-75% of the red blood cells are mature definitive erythrocytes, and the remaining fraction of cells are late polychromatophilic definitive erythroblasts [103]. Using cloned DNA probes specific for vimentin sequences, it was found that primitive series erythroid cells from four-day-old embryos express low amounts of vimentin mRNA, whereas definitive series cells possess high amounts of this message, increasing severalfold from 10 to 15 days of embryogenesis [104]. These results demonstrate the high level of expression and induction of steady state vimentin mRNA within the definitive series erythroid lineage. The differences in relative abundances of vimentin mRNA in these differentiating cells are similar to the changes observed at the protein level, suggesting that vimentin filament expression in avian erythropoiesis is determined primarily by the prevalence of vimentin mRNA [104].

The absence of vimentin filaments in mature mammalian red blood cells is the result of repressed expression in the erythroblastic stages of erythropoiesis [41]. Unfortunately, unlike the case with avian cells, it is difficult to isolate large numbers of differentiating mammalian erythroid cells at defined stages of differentiation for biochemical or molecular analysis. However, many of the events occurring in mammalian embryogenesis are faithfully recapitulated in the differentiation of Friend virus-transformed murine erythroleukemia (MEL) cells in vitro [105; reviewed in refs. 106-107]. Incubation of MEL cells in culture with agents such as dimethylsulfoxide (DMSO) or hexamethylene-bisacetamide results in the maturation of these cells from basophilic erythroblasts to orthochromatophilic normoblasts, cessation of cell proliferation, induction of heme synthetic enzyme activities, globin mRNA synthesis and accumulation, and hemoglobin synthesis [105-107]. In addition, the membrane skeleton components spectrin, anion transporter, and protein 4.1 are induced during MEL cell differentiation [108-110]. The differentiation of MEL cells therefore represents a useful experimental system for studying the molecular and physiological changes in mammalian erythropoiesis.

The loss of vimentin filaments during mammalian erythropoiesis in vivo indeed is faithfully reproduced in MEL cell differentiation in vitro [111]. By immunofluorescence microscopy, vimentin filaments are present in undifferentiated cells and diminish rapidly during inducer-mediated MEL cell maturation; these changes are preceded by an extensive decrease in vimentin protein synthesis [111]. The decline in vimentin synthesis is paralleled by a similarly rapid and extensive decrease in steady-state vimentin mRNA (~25-fold by four days of induction) [111], further suggesting that the level of vimentin synthesis is determined primarily by mRNA abundance. The removal of vimentin filaments from differentiating MEL cells after the cessation of de novo vimentin synthesis most likely occurs by simple dilution from cell division. Although it is possible that vimentin filament turnover and removal may be accelerated by, for example, a Ca^{2+}-dependent vimentin protease [112,113], the observed persistence of vimentin filaments expressed from transfected vimentin genes in differentiating MEL cells renders this type of mechanism unlikely (see Section XI) [114]. By comparing vimentin expression in avian and mammalian erythropoiesis, it becomes evident that the major point of regulation lies at the level of steady-state mRNA, although in the former case this regulation is positive, and in the latter case, it is negative.

VIII. THE VIMENTIN GENE: STRUCTURE AND
REGULATION

To better understand the mechanisms governing vimentin mRBA accumulation during erythroid differentiation, we must digress and consider the structure of the vimentin gene and mRNA. The structures of the chicken, hamster, and human vimentin genes have been elucidated by conventional molecular cloning techniques [6,10,115]. Vimentin genes exist as single copies in their respective haploid genomes and comprise ~8 kb of DNA. A comparison of chicken and hamster genes demonstrates a common organization of nine exons, with the positions of intron–exon borders precisely conserved [6,10]; protein-coding sequences show an overall 80% similarity at the nucleic acid level, yielding an amino acid sequence similarity of 87% [6,10]. The human and hamster vimentin genes exhibit 91 and 97% similarities at the nucleotide and amino acid levels, respectively [115]. In spite of these similarities, several notable features distinguish the chicken, hamster, and human vimentin genes. First, when comparing the chicken gene with the mammalian genes, both the 5' noncoding region and the 5' portion of the first exon corresponding to the N-terminal variable domain are markedly less conserved than the rest of the coding sequence (~35 and ~ 60% nucleotide identity, respectively) [6,10,115]. Second, the chicken vimentin gene encodes three mRNA species that differ in the lengths of their 3' noncoding regions [10,116]. Four consensus polyadenylation signals [5' AATAAA-3'] reside 250, 298, 533, and 554 nt downstream of the translation termination codon, and the second through fourth signals are used in chicken tissues [117], producing mRNAs ~2.0 and ~2.3 kb in length [116]. In contrast, mammalian vimentin genes encode only one mRNA species of 2.0–2.1 kb [6,111,115]. A comparison of hamster and chicken vimentin sequences at their 3' ends reveals that the hamster gene has retained only one functional signal, corresponding to the second chicken polyadenylation site [6]; 83% of the 3'-noncoding regions are identical in hamster and chicken vimentin mRNAs [6]. Based on sequence analysis of a human vimentin cDNA, it has been found that the human vimentin mRNA 3' untranslated region is significantly shorter [57 nt from translation termination codon to poly(A) tail] than that found in either the chicken or hamster mRNAs [115]. In the human sequence, there is no consensus 5'-AATAAA-3' polyadenylation signal, although the sequence 5'-GATAAA-3' is found ~14 nt upstream of the poly(A) tract [115]. However, it is formally possible that the observed human vimentin cDNA with the shorter 3' sequence may have derived from the use of a cryptic polyadenylation site, and that a functional and preferred site may well exist further downstream. This curious area of divergence of human and hamster vimentin genes remains to be confirmed and examined in greater detail by sequence analysis of human genomic DNA clones.

The 3' termini of chicken vimentin mRNAs are used in a tissue-specific pattern. Whereas the second through fourth polyadenylation sites are used with slightly variable but near-equivalent efficiencies in the majority of cells and tissues examined (e.g., lens, spinal cord, smooth and skeletal muscle, fibroblasts, and primitive embryonic erythroid cells), in definitive embryonic erythroid cells the second polyadenylation site is used preferentially, producing mainly the 2.0 kb RNA [104,114]. Although the biological effects of producing multiple mRNAs that differ only in the lengths of their 3' untranslated regions is not known, the developmentally regulated and tissue-specific expression of chicken vimentin mRNAs suggests that these differences are functionally significant. Sequences involved in mRNA stability have been localized to 3' untranslated regions of other mRNAs [118-120]. It is therefore possible that the differential use of chicken vimentin 3' termini may affect message stability or posttranscriptional processing. It is also possible that vimentin 3' untranslated sequences may differentially affect subcellular compartmentation, and hence the site of vimentin protein synthesis and polymerization. In situ hybridization studies have shown that vimentin RNA exhibits a preferential nuclear or perinuclear localization in chick embryonic fibroblasts [121]. However, in these experiments vimentin RNA was detected using a probe complementary to all three mRNA species, so that discrimination of the localization of individual mRNA species was not possible [121]. The contribution of chicken vimentin 3'-untranslated sequences to subcellular mRNA distribution remains an interesting issue to be explored. We should note that any explanation for the importance of tissue-specific expression of multiple vimentin mRNAs must be reconciled with the presence of only one vimentin mRNA in mammals, although any functional requirement for multiple vimentin mRNAs may have been lost or altered since the divergence of birds and mammals.

The diversity in vimentin gene expression, both in terms of the number of different cell lineages in which vimentin is expressed and the varying developmental regulation within each lineage (outlined in Section II), presents a complex problem of gene regulation. Perhaps there are multiple cis-linked regulatory sequences that are responsible for the positive or negative regulation in different cell lineages. For example, multiple tissue-specific regulatory elements have been demonstrated to be responsible for the differential expression of the mouse α-fetoprotein gene in yolk sac, visceral ectoderm, fetal liver, or gut [122,123]. An analysis of human vimentin 5'-flanking sequences in fact has revealed the presence of multiple transcriptional regulatory elements [124]. Successive deletions from the 5' end of the promoter region have defined a negative-regulatory element surrounded by two positive-regulatory elements [124].

The vimentin gene is growth-regulated in some cell types under certain culturing conditions. In human, mouse, and hamster fibroblasts and human lymphocytes, vimentin mRNA rises to peak levels in the G1 phase of the cell cycle

[125,126]. Quiescent mouse 3T3 cells stimulated to divide by serum or plate-let-derived growth factor (PDGF), but not by epidermal growth factor or insul-in, induce the levels of vimentin mRNA [115]. In contrast, vimentin expression is markedly elevated when human promyelocytic leukemia cells (HL-60) are induced by phorbol esters to cease proliferating and to undergo terminal differen-tiation [115,127]. The human vimentin 5'-flanking transcriptional elements de-scribed above are by themselves sufficient to confer growth regulation in PDGF- and serum-stimulated fibroblasts [124]. It will be interesting to see if these multiple elements are able to confer differential tissue-specific expression of the vimentin gene as well. As we will describe in the next section, the multiplicity of vimentin regulatory elements is manifested at both the transcriptional and post-transcriptional levels.

IX. TRANSCRIPTIONAL AND POSTTRANSCRIPTIONAL REGULATION OF VIMENTIN mRNA LEVELS IN DIFFERENTIATING MURINE ERYTHROLEUKEMIA CELLS

How are vimentin mRNA levels determined in erythroid differentiation? This question has been addressed by studying vimentin gene transcription rates in iso-lated nuclei from differentiating MEL cells [114]. In these types of studies, nu-clei are incubated in the presence of radioactive ribonucleotide precursors; radio-active incorporation into RNA represents elongation of nascent RNA chains, and hence polymerase density on the gene of interest [128,129]. By hybridizing the labeled RNA with filter-bound cloned DNA fragments, transcription of a given gene of interest can be estimated. It was found that vimentin gene trans-cription rates decrease maximally to ~40% of control levels at 96 h of DMSO-mediated MEL cell differentiation [114]. This relatively small decrease in trans-cription contrasts with a ~25-fold decrease in steady-state vimentin mRNA over a comparable period of differentiation [111]. The large disparity in gene trans-cription and steady-state mRNA levels suggest that posttranscriptional mecha-nisms are also involved in determining vimentin mRNA levels in mammalian erythropoiesis. It remains possible that in vitro nuclear transcription experiments systematically underestimate the changes in vimentin gene transcription in dif-ferentiating MEL cells. However, it should be noted that when in vitro trans-cription rates and steady-state mRNA are compared over a time course of dif-ferentiation [111,114], a constant difference in these two parameters is not ob-served. Specifically, vimentin gene transcription falls to ~56% of control levels at 24 h of differentiation and is maintained at ~40% from 72-96 h of differen-tiation, whereas vimentin mRNA abundances fall to ~30, ~10, and ~4% at 24, 72, and 96 h, respectively. These results suggest that the relatively limited changes in transcription are not due to a systematic and artifactual underesti-mate; rather, the accelerated decline in vimentin mRNA levels is mediated by a developmentally regulated posttranscriptional mechanism. A combination of transcriptional and posttranscriptional effects has been shown to influence the overall accumulation of hormonally responsive RNAs, cellular thymidine kinase RNA, histone RNA, and dihydrofolate reductase RNA [130-139]. Direct in vivo RNA labeling will be necessary to determine the contributions of transcription, posttranscriptional RNA processing and transport, and RNA turnover toward establishing the developmentally regulated levels of vimentin mRNA. Further in vitro nuclear transcription sutdies and in vivo labelings are also needed to determine the degree to which these mechanisms are used for controlling vimentin mRNA abundances in avian red blood cell development.

X. EVOLUTIONARY DIVERGENCE OF THE MECHANISMS GOVERNING VIMENTIN GENE EXPRESSION IN AVIAN AND MAMMALIAN ERYTHROPOIESIS

The positively regulated expression of vimentin in avian erythropoiesis and negatively regulated expression of vimentin in mammalian erythropoiesis presents an interesting evolutionary question. Namely, what factors or processes governing the regulation of vimentin expression have diverged to render these patterns? The results of the experiments discussed thus far indicate that the expression of vimentin filaments is regulated primarily by mRNA accumulation [104,111]. We will focus our attention on this level of regulation in an attempt to account for the observed divergence. Two simple alternative models can be entertained. In the first model (Model I of Figure 2), a change in *trans*-acting regulatory factors that recognize conserved *cis*-linked sequences is responsible for the observed divergence. In avian erythropoiesis, a positively acting factor interacts with a vimentin regulatory sequence in such a way as to activate or derepress vimentin expression, whereas in differentiating mammalian erythroid cells, a negatively acting factor either replaces or precludes the positive activity by interacting with the same sequence, resulting in the deactivation or repression of vimentin expression. In the second model (Model II of Figure 2), a difference in the *cis*-acting vimentin regulatory sequences results in the activation (or derepression) in avian erythropoiesis and repression (or deactivation) in mammalian erythropoiesis. This model predicts that if the erythroid-specific, positively acting factors are conserved from birds to mammals, they are prevented from acting on mammalian vimentin by a divergence in the target sequence itself. The repression of vimentin expression in mammalian erythropoiesis therefore would occur by the interaction of the putative divergent sequence with a negative factor, which may or may not be unique to the mammalian class. This model is particularly attractive if we presume that a limited number of *trans*-acting factors regulates a larger number of genes during erythropoiesis.

The two models for divergence of erythroid vimentin expression can be tested by studying the behavior of transfected chicken vimentin gene sequences in dif-

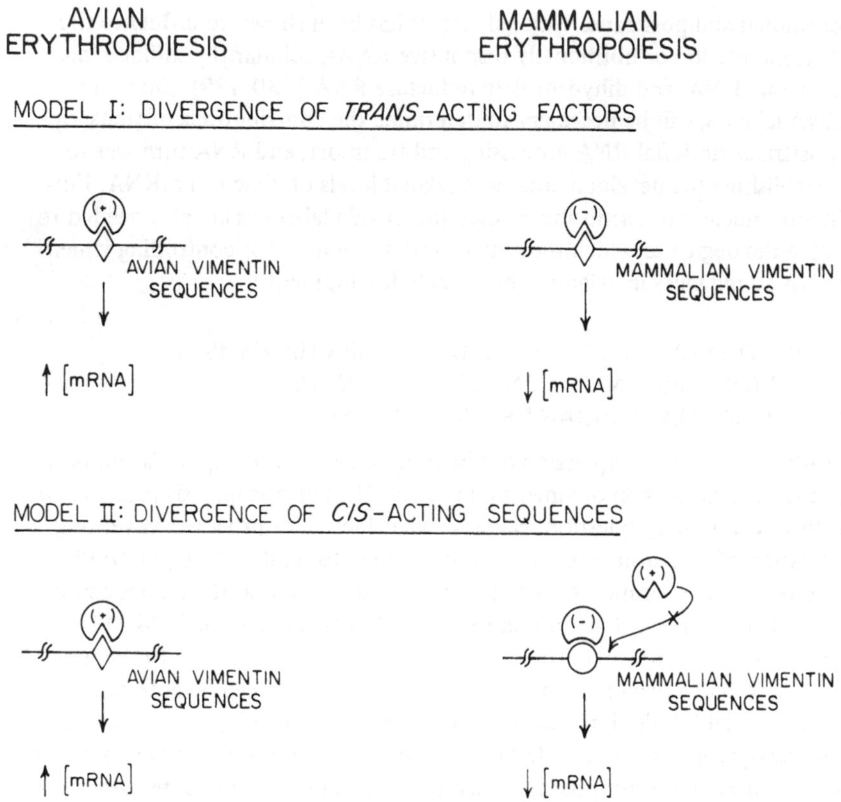

Figure 2 Two alternative models to account for the divergent regulation of vimentin gene expression in avian and mammalian erythropoiesis. For details and explanation, the reader is referred to Section X of the text.

ferentiating MEL cells. The expression of transfected globin genes in MEL cells indeed has proven fruitful in defining the cis-acting regulatory sequences responsible for the developmental regulation of these genes [140–145]. If a transfected chicken vimentin gene is downregulated during MEL cell differentiation in a manner similar to its endogenous counterpart, the scheme of Model I is favored. In this case, both chicken and mouse genes would be responding to the same factors via interactions with conserved sequences. If, on the other hand, a transfected chicken gene is upregulated as it is in its normal erythroid environment, then Model II is favored. This result would suggest that the negatively acting factor(s) in differentiating MEL cells is unable to interact with the chicken sequences. Moreover, an induction of chicken vimentin expression would also suggest that positively acting factors have been conserved from birds to mammals. When DNA containing the entire 8-kb chicken vimentin gene plus 2.5-kb 5' and 3' flanking sequences is transfected into MEL cells, the steady-state level of chicken vimentin and mRNA increase upon chemically mediated differentiation [114]. In contrast, a transfected hamster vimentin gene with a similar amount of flanking sequences is negatively regulated (as is the endogenous murine vimentin gene) [114], indicating that the observed expression of chicken vimentin sequences is not an artifact of gene transfection. In vitro nuclear run-on transcription analyses have shown that the increase in chicken vimentin mRNA during MEL cell differentiation is due primarily to a commensurate increase in transcription [114]. The results suggest that avian and mammalian vimentin *cis*-acting sequences have diverged at two levels: at one level the mammalian sequences have changed so that they respond to a transcriptional repression or deactivation, and at a second level the mammalian sequences are responsive to a negative posttranscriptional mechanism, whereas the avian sequences are not. The induction of chicken vimentin gene expression in differentiating MEL cells further suggests the presence of general erythroid factors that are functionally conserved to interact with the chicken sequences. Together these observations support Model II, as described above and outlined in Figure 2.

The expression of chicken vimentin genes in MEL cells allows us to ask if the mechanisms dictating the differential utilization of polyadenylation sites in definitive chicken erythroid cells (i.e., preferential use of site 2 relative to sites 3 and 4) have been maintained and conserved in mammalian erythroid cells. The polyadenylation sites of the chicken vimentin gene in fact are used with roughly equivalent efficiencies in both undifferentiated and differentiated MEL cells [114], suggesting that the mechanism for definitive chicken erythroid-specific polyadenylation site selection in mammalian erythroid cells is either absent or functionally incompatible with the chicken sequence. This apparent divergence in 3' terminus selection is juxtaposed with the apparent functional conservation of mammalian erythroid factors that are capable of inducing chicken vimentin expression.

XI. EXPRESSION OF VIMENTIN FILAMENTS FROM TRANSFECTED CHICKEN VIMENTIN GENES IN DIFFERENTIATING MURINE ERYTHROLEUKEMIA CELLS

Murine erythroleukemia cells expressing vimentin genes can be further studied for the possible consequences of inappropriate intermediate filament accumulation. We have already discussed the observations that chicken vimentin genes transfected in MEL cells are induced upon terminal differentiation. In these studies it was further found that chicken vimentin protein synthesis and assembly into stable filaments also increases during the maturation of these transfected cells [114]. The increase in cytoskeletal chicken vimentin during differentiation occurs while the levels of endogenous murine vimentin protein declines, suggesting that the normal decrease in vimentin filaments in mammalian erythropoiesis is determined primarily by the diminution of vimentin protein synthesis, and not by accelerated filament turnover. For example, if the removal of vimentin filaments from differentiating MEL cells is facilitated by the activation of a Ca^{2+}-dependent vimentin-specific protease [112,113] in response to the developmentally regulated Ca^{2+} ion influx [146,147], we would expect both chicken and murine vimentin to be degraded, because this protease has been shown to be functionally conserved from fish to mammals [148]. Because the results show a roughly equivalent increase in chicken vimentin mRNA levels, protein synthesis, and protein accumulation [114], these observations confirm earlier kinetics studies on vimentin synthesis and assembly showing that filament accumulation is determined primarily by subunit synthesis [98,99].

The inappropriately induced expression of vimentin filaments has no readily discernable effects on MEL cell differentiation [114]. For example, typical parameters of MEL cell maturation such as cessation of cell proliferation, reduction in cell volume, and induction of globin mRNA levels, are unaffected. In Section V we discussed the possible importance of removing vimentin filaments to facilitate the process of enucleation. In theory, this possibility can be tested by determining if any aberrations in enucleation occur in mammalian erythroid cells overexpressing vimentin filaments during terminal differentiation. Transfected MEL cells may not be ideally suited for these studies, however, because under most culturing conditions MEL cells rarely differentiate to the point of enucleation [149]. An examination of the effects of vimentin overexpression during the latest stages of erythropoiesis (i.e., enucleation and reticulocyte maturation) should be facilitated by the experimentally targeted expression of exogenous vimentin genes to red blood cells of transgenic mice. The MEL cell transfection studies described above have established the feasibility of such experiments, because we now know that the expression of vimentin RNA is by itself sufficient to determine the accumulation of vimentin filaments in erythropoiesis.

XII. CONCLUSION

In this review, we have attempted to address two major questions using avian and mammalian erythropoiesis as model systems. The first question concerns the mechanisms governing the developmentally regulated expression of vimentin. In definitive series avian erythroid cells, vimentin is expressed at high levels during terminal differentiation, and vimentin filaments are present in the mature state [31]. The regulation of this expression appears not to lie at the level of assembly or turnover, because filaments are formed by rapid self-assembly of soluble precursor subunits, and once formed, are stable [98,99]. The determining factor of filament assembly (and hence, filament accumulation) therefore is the rate or extent of vimentin subunit synthesis. In both avian and mammalian erythroid differentiation, the synthesis of vimentin protein is determined primarily by mRNA abundances [104,111]. However, unlike the high level of expression in avian red blood cells, differentiating mammalian red blood cells repress vimentin mRNA levels extensively, as demonstrated in MEL cell differentiation in vitro [111]. A comparison of vimentin expression in avian and mammalian erythropoiesis thus exemplifies the potential dynamic positive and negative regulation of the vimentin gene, and raises the issue of the molecular basis for the evolutionary divergence of vimentin expression in erythropoiesis. Studies on the behavior of transfected vimentin genes in differentiating MEL cells indicate that the observed differences in vimentin expression in avian and mammalian erythropoiesis are due to the divergence of *cis*-linked vimentin regulatory sequences [114]. This conclusion is consistent with the notion that *trans*-acting factors have been conserved between avian and mammalian erythroid cells, and a major difference in the two erythropoietic programs has evolved by a change in the target vimentin sequence itself. These studies further suggest that mammalian vimentin sequences have evolved in such a way as to respond to both transcriptional and posttranscriptional mechanisms. It will be of interest to determine the identity of the sequence elements responsible for this difference in regulation, and their possible relationship to other erythropoiesis-specific regulatory sequences.

The functional ramifications of divergent vimentin filament expression in avian and mammalian red blood cells raise the second major question addressed in this review: What are the potential structural and morphogenetic roles of vimentin filaments? By ultrastructural criteria, vimentin filaments appear to anchor the nucleus in nucleated red blood cells [68,70–72]. From this interpretation it follows that the process of enucleation or membrane reorganization in mammalian erythroid differentiation may require the prior elimination of vimentin filaments. These ideas can be tested by experimentally manipulating the expression of vimentin in differentiating erythroid systems. By inappropriately inducing vimentin synthesis and accumulation during mammalian erythropoiesis, it can be determined if any alterations in the normal developmental and morphogenetic programs are produced. Thus far it has been shown that vimentin filaments are stably expressed from transfected genes in differentiating MEL cells [114]. An understanding of the effects of aberrantly persistent vimentin expression on enucleation and other events characteristic of the latest stages of erythropoiesis (stages occurring beyond the capacity of most MEL cell systems) await further similar studies in transgenic animals. In the future it should be possible to continue comparing vimentin expression in avian and mammalian erythropoiesis, and using the observed differences as an approach to dissect the effects that vimentin filaments have on differentiation and morphogenesis.

ACKNOWLEDGMENT

We thank Ms. Stephanie Canada for her assistance in preparing this manuscript, and Mr. Pat Koen for his assistance with scanning electron microscopy. This work was supported by grants from the National Institute of Health, the National Science Foundation, and by a Developmental Biology Grant from the Lucille P. Markey Charitable Trust.

REFERENCES

1. Lazarides, E. (1982). Intermediate filaments: A chemically heterogeneous, developmentally regulated class of proteins. *Ann. Rev. Biochem.* 51:219–250.
2. Steinert, P. M., Steven, A. C., and Roop, D. R. (1985). The molecular biology of intermediate filaments. *Cell* 42:411–419.
3. Geisler, N., and Weber, K. (1982). The amino acid sequence of chicken muscle desmin provides a common structural model for intermediate filament proteins. *EMBO J.* 1:1649–1656.
4. Hanukoglu, I., and Fuchs, E. (1982). The cDNA sequence of a human epidermal keratin: divergence of sequence but conservation of structure among intermediate filament proteins. *Cell* 31:243–252.
5. Hanukoglu, I., and Fuchs, E. (1983). The cDNA sequence of a type II cytoskeletal keratin reveals constant and variable structural domains among keratins. *Cell* 33:915–924.
6. Quax, W., Egberts, W. V., Hendriks, W., Quax-Jeuken, Y., and Bloemendal, H. (1983). The structure of the vimentin gene. *Cell* 35:215–223.
7. Quax, W., van der Broek, L., Egberts, W. V., Ramaekers, F., and Bloemendal, H. (1985). Characterization of the hamster desmin gene: expression and formation of desmin filaments in nonmuscle cells after gene transfer. *Cell* 43:327–338.
8. Lewis, S. A., Balcarek, J. M., Krek, V., Shelanski, M., and Cowan, N. J. (1984). Sequence of a cDNA clone encoding mouse glial fibrillary acidic protein: Structural conservation of intermediate filaments. *Proc. Natl. Acad. Sci. USA* 81:2743–2746.
9. Lewis, S. A., and Cowan, N. J. (1986). Anomalous placement of introns in a member of the intermediate filament multigene family: an evolutionary conundrum. *Mol. Cell. Biol.* 6:1529–1534.
10. Zehner, Z. E., Li, Y., Roe, B. A., Paterson, B. M., and Sax, C. M. (1987). The chicken vimentin gene: Nucleotide sequence, regulatory elements, and comparison to the hamster gene. *J. Biol. Chem.* 262:8112–8120.
11. Geisler, N., Kaufmann, E., and Weber, K. (1982). Protein chemical characterization of three structurally distinct domains along the protofilament unit of desmin 10 nm filaments. *Cell* 30:277–286.
12. Aebi, U., Fowler, W. E., Rew, P., and Sun, T. T. (1983). The fibrillar substructure of keratin filaments unraveled. *J. Cell Biol.* 97:1131–1143.

13. Steinert, P. M., Rice, R. H., Roop, D. R., Trus, B. L., and Steven, A. C. (1983). Complete amino acid sequence of a mouse epidermal keratin subunit and implications for the structure of intermediate filaments. *Nature* 302:794–800.
14. Traub, P., and Vorgias, C. E. (1983). Involvement of the *N*-terminal polypeptide of vimentin in the formation of intermediate filaments. *J. Cell Sci.* 63:43–67.
15. Fraser, R. D. B., and MacRae, T. P. (1983). The structure of the α-keratin microfibril. *Biosci. Rep.* 3:517–525.
16. Steven, A. C., Hainfield, J. T., Trus, B. L., Wall, J. S., and Steinert, P. M. (1983). The distribution of mass in heteropolymer intermediate filaments assembled in vitro: STEM analysis of vimentin/desmin and bovine epidermal keratin. *J. Biol. Chem.* 258:8323–8329.
17. Steven, A. C., Hainfield, J. T., Trus, B. L., Wall, J. S., and Steinert, P. M. (1983). Epidermal keratin filaments assembled in vitro have masses-per-unit-length that scale according to average subunit mass: structural basis for homologous packing of subunits in intermediate filaments. *J. Cell Biol.* 97:1939–1944.
18. Gard, D. L., Bell, P. B., and Lazarides, E. (1979). Coexistence of desmin and the fibroblastic intermediate filament subunit in muscle and nonmuscle cells: identification and comparative peptide anslysis. *Proc. Natl. Acad. Sci. USA* 76:3894–3898.
19. Gard, D. L., and Lazarides, E. (1980). The synthesis and distribution of desmin and vimentin during myogenesis in vitro. *Cell* 19:263–275.
20. Steinert, P. M., Idler, W. W., Cabral, F., Gottesman, M. M., and Goldman, R. D. (1981). In vitro assembly of homopolymer and copolymer filaments from intermediate filament subunits of muscle and fibroblastic cells. *Proc. Natl. Acad. Sci. USA* 78:3692–3696.
21. Quinlan, R. A., and Franke, W. W. (1982). Heteropolymer filaments of vimentin and desmin in vascular smooth muscle tissue and cultured baby hamster kidney cells demonstrated by chemical crosslinking. *Proc. Natl. Acad. Sci. USA* 79:3452–3456.
22. Quinlan, R. A., and Franke, W. W. (1983). Molecular interactions in intermediate-sized filaments revealed by chemical cross-linking. Heteropolymers of vimentin and glial filament protein in cultured human glioma cells. *Eur. J. Biochem.* 132:477–484.
23. Sharp, G., Osborn, M., and Weber, K. (1982). Occurrence of two different intermediate filament proteins in the same filament in situ with a human glioma cell line. *Exp. Cell Res.* 141:385–395.
24. Ip, W., Danto, S. I., and Fischman, D. A. (1983). Detection of desmin-containing intermediate filaments in cultured muscle and nonmuscle cells by immunoelectron microscopy. *J. Cell Biol.* 96:401–408.
25. Granger, B. L., and Lazarides, E. (1983). Expression of the major neurofilament subunit in chicken erythrocytes. *Science* 221:553–556.
26. Kreis, T. E., Geiger, B., Schmid, E., Jorcano, J. L., and Franke, W. W. (1983). De novo synthesis and specific assembly of keratin filaments in non-

epithelial cells after microinjection of mRNA for epidermal keratin. *Cell* 32: 1125–1137.

27. Giudice, G. J., and Fuchs, E. (1987). The transfection of epidermal keratin genes into fibroblasts and simple epithelial cells: evidence for inducing a type I keratin by a type II gene. *Cell* 48:453–463.
28. Franke, W. W., Schmid, E., Osborn, M., and Weber, K. (1978). Different intermediate-sized filaments distinguished by immunofluorescence microscopy. *Proc. Natl. Acad. Sci. USA* 75:5034–5038.
29. Bradley, R. H., Ireland, M., and Maisel, H. (1979). The cytoskeleton of chick lens cells. *Exp. Eye Res.* 28:441–453.
30. Ramaekers, F. C. S., Osborn, M., Schmid, E., Weber, K., Bloemendal, H., and Franke, W. W. (1980). Identification of the cytoskeletal proteins in lens-forming cells, a special epithelioid cell type. *Exp. Cell Res.* 127:309–327.
31. Granger, B. L., Repasky, E. A., and Lazarides, E. (1982). Synemin and vimentin are components of intermediate filaments in avian erythrocytes. *J. Cell Biol.* 92:299–312.
32. Tapscott, S. J., Bennett, G. S., and Holtzer, H. (1981). Neuronal procursors in the chick neural tube express neurofilament proteins. *Nature* 292:836–838.
33. Tapscott, S. J., Bennett, G. S., Toyama, Y., Kleinbart, F., and Holtzer, H. (1981). Intermediate filament proteins in the developing chick spinal cord. *Dev. Biol.* 86:40–54.
34. Bennett, G. S., Fellini, S. A., Toyama, Y., and Holtzer, H. (1979). Redistribution of intermediate filament subunits during skeletal myogenesis and maturation in vitro. *J. Cell. Biol.* 82:577–584.
35. Granger, B. L., and Lazarides, E. (1979). Desmin and vimentin coexist at the periphery of the myofibril Z disc. *Cell* 18:1053–1063.
36. Dräger, U. C. (1983). Coexistence of neurofilaments and vimentin in a neurone of adult mouse retina. *Nature* 303:169–172.
37. Yen, S.-H., and Fields, K. L. (1981). Antibodies to neurofilament, glial filament, and fibroblast intermediate filament proteins bind to different cell types of the nervous system. *J. Cell Biol.* 88:115–126.
38. Schnitzer, J., Franke, W. W., and Schachner, M. (1981). Immunocytochemical demonstration of vimentin in astrocytes and ependymal cells of developing and adult mouse nervous system. *J. Cell Biol.* 90:435–447.
39. Traub, U. E., Nelson, W. J., and Traub, P. (1983). Polyacrylamide gel electrophoretic screening of mammalian cells cultured in vitro for the presence of the intermediate filament protein vimentin. *J. Cell Sci.* 62:129–147.
40. McTavish, C. F., Nelson, W. J., and Traub, P. (1983). Synthesis of vimentin in a reticulocyte cell-free system programmed by poly(A)-rich RNA from several cell lines and rat liver. *Eur. J. Biochem.* 130:211–221.
41. Dellagi, K., Vainchenker, W., Vinci, G., Paulin, D., and Brouet, J. C. (1983). Alteration of vimentin intermediate filament expression during differentiation of human hemopoietic cells. *EMBO J.* 2:1509–1514.

42. Branton, D., Cohen, C. M., and Tyler, J. (1981). Interaction of cytoskeletal proteins on the human erythrocyte membrane. *Cell* 24:24–32.
43. Bennett, V. (1985). The membrane skeleton of human erythrocytes and its implications for more complex cells. *Ann. Rev. Biochem.* 54:273–304.
44. Marchesi, V. T. (1985). Stabilizing intrastructure of cell membranes. *Annu. Rev. Cell Biol.* 1:531–561.
45. Lazarides, E. (1987). From genes to structural morphogenesis: the genesis and epigenesis of a red blood cell. *Cell* 51:345–356.
46. Bennett, V., and Stenbuck, P. J. (1979). Identification and partial purification of ankyrin, the high affinity membrane attachment site for human erythrocyte spectrin. *J. Biol. Chem.* 254:2533–2541.
47. Bennett, V., and Stenbuck, P. J. (1979). The membrane attachment protein for spectrin is associated with band 3 in human erythrocyte membranes. *Nature* 280:468–473.
48. Bennett, V., and Stenbuck, P. J. (1980). Association between ankyrin and the cytoplasmic domain of band 3 isolated from the human erythrocyte membrane. *J. Biol. Chem.* 255:6424–6432.
49. Luna, E. J., Kidd, G. H., and Branton, D. (1979). Identification by peptide analysis of the spectrin-binding protein in human erythrocytes. *J. Biol. Chem.* 254:2526–2532.
50. Yu, J., and Goodman, S. R. (1979). Syndeins: the spectrin binding protein(s) of the human erythrocyte membrane. *Proc. Natl. Acad. Sci. USA* 76:2340–2344.
51. Hargreaves, W. R., Giedd, K. N., Verkleij, A., and Branton, D. (1980). Reassociation of ankyrin with band 3 in erythrocyte membranes and in lipid vesicles. *J. Biol. Chem.* 255:11965–11972.
52. Ungewickell, E., Bennett, P. M., Calvert, R., Ohanian, V., and Gratzer, W. B. (1979). In vitro formation of a complex between cytoskeletal proteins of the human erythrocyte. *Nature* 280:811–814.
53. Fowler, V., and Taylor, D. L. (1980). Spectrin plus band 4.1 cross-link actin. Regulation by micromolar calcium. *J. Cell Biol.* 85:361–376.
54. Cohen, C. M., and Korsgren, C. (1980). Band 4.1 causes spectrin-actin gels to become thixotropic. *Biochem. Biophys. Res. Comm.* 97:1429–1435.
55. Ohanian, V., Wolfe, L. C., John, K. M., Pinder, J. C., Lux, S. E., and Gratzer, W. B. (1984). Analysis of the ternary interaction of the red cell membrane skeletal proteins spectrin, actin, and 4.1. *Biochemistry* 23:4416–4420.
56. Anderson, R. A., and Lovrien, R. E. (1984). Glycophorin is linked by band 4.1 protein to the human erythrocyte membrane skeleton. *Nature* 307:655–658.
57. Pasternack, G. R., Anderson, R. A., Leto, T., and Marchesi, V. T. (1985). Interaction between protein 4.1 and band 3. An alternative binding site for an element of the membrane-skeleton. *J. Biol. Chem.* 260:3676–3683.
58. Sato, S. B., and Ohnishi, S. (1983). Interaction of a peripheral protein of the erythrocyte membrane, protein 4.1, with phosphatidylserine-containing liposomes and erythrocyte inside-out vesicles. *Eur. J. Biochem.* 130:19–25.

59. Chan, L.-N. L. (1977). Changes in the composition of plasma membrane proteins during differentiation of embryonic chick erythroid cells. *Proc. Natl. Acad. Sci. USA* 74:1062–1066.
60. Weise, M. J., and Chan, L. L. (1978). Membrane protein synthesis in embryonic chick erythroid cells. *J. Biol. Chem.* 253:1892–1897.
61. Repasky, E. A., Granger, B. L., and Lazarides, E. (1982). Widespread evidence of avian spectrin in non-erythroid cells. *Cell* 29:821–833.
62. Jay, D. G. (1983). Characterization of the chicken erythrocyte anion exchange protein. *J. Biol. Chem.* 258:9431–9436.
63. Granger, B. L., and Lazarides, E. (1984). Membrane skeletal protein 4.1 of avian erythrocytes is composed of multiple variants that exhibit tissue-specific expression. *Cell* 37:595–607.
64. Nelson, W. J., and Lazarides, E. (1984). Goblin (ankyrin) in striated muscle: identification of the potential membrane receptor for erythroid spectrin in muscle cells. *Proc. Natl. Acad. Sci. USA* 81:3292–3296.
65. Fawcett, D. W. (1959). Electron microscopic observations on the marginal band of nucleated erythrocytes. *Anat. Rec.* 133:379.
66. Barrett, L. A., and Dawson, R. B. (1974). Avian erythrocyte development: microtubules and the formation of the disk shape. *Dev. Biol.* 36:72–81.
67. Behnke, O. (1970). A comparative study of microtubules of disk-shaped blood cells. *J. Ultrastruct. Res.* 31:61–75.
68. Cohen, W. D. (1978). Observations on the marginal band system of nucleated erythrocytes. *J. Cell. Biol.* 78:260–273.
69. Joseph-Silverstein, J., and Cohen, W. D. (1984). The cytoskeleton system of nucleated erythrocytes. III. Marginal band function in mature cells. *J. Cell Biol.* 98:2118–2125.
70. Harris, J. R., and Brown, J. N. (1971). Fractionation of the avian erythrocyte: an ultrastructural study. *J. Ultrastruct. Res.* 36:8–23.
71. Virtanen, I., Kurkinen, M., and Lehto, V.-P. (1979). Nucleus-anchoring cytoskeleton in chicken red blood cells. *Cell Biol. Int. Rep.* 3:157–162.
72. Woodcock, C. L. F. (1980). Nucleus-associated intermediate filaments from chicken erythrocytes. *J. Cell Biol.* 85:881–889.
73. Granger, B. L., and Lazarides, E. (1982). Structural associations of synemin and vimentin filaments in avian erythrocytes revealed by immunoelectron microscopy. *Cell* 30:263–275.
74. Haggis, G. H., and Phipps-Todd, B. (1977). Freeze-fracture for scanning electron microscopy. *J. Microsc.* 111:193–201.
75. Haggis, G. H., and Bond, E. F. (1979). Three-dimensional view of the chromatin in freeze-fractured chicken erythrocyte nuclei. *J. Microsc.* 115:225–234.
76. Mangeat, P. H., and Burridge, K. (1984). Immunoprecipitation of nonerythrocyte spectrin within live cells following microinjection of specific antibodies: relation to cytoskeletal structures. *J. Cell Biol.* 98:1363–1372.
77. Langley, R. C., and Cohen, C. M. (1987). Cell type-specific association between two types of spectrin and two types of intermediate filaments. *Cell Motil. Cytoskel.* 8:165–173.

78. Georgatos, S. D., and Marchesi, V. T. (1985). The binding of vimentin to human erythrocyte membranes: a model system for the study of intermediate filament-membrane interactions. *J. Cell Biol.* 100:1955–1961.
79. Georgatos, S. D., Weaver, D. C., and Marchesi, V. T. (1985). Site specificity in vimentin-membrane interactions: intermediate filament subunits associate with the plasma membrane via their head domains. *J. Cell Biol.* 100:1962–1967.
80. Georgatos, S. D., and Blobel, G. (1987). Two distinct attachment sites for vimentin along the plasma membrane and the nuclear envelope in avian erythrocytes: a basis for a vectorial assembly of intermediate filaments. *J. Cell Biol.* 105:105–115.
81. Georgatos, S. D., and Blobel, G. (1987). Lamin B constitutes an intermediate filament attachment site at the nuclear envelope. *J. Cell Biol.* 105: 117–125.
82. Lehto, V.-P., Virtanen, I., and Kurki, P. (1978). Intermediate filaments anchor the nuclei in nuclear monolayers of cultured human fibroblasts. *Nature* 272:175–177.
83. Andrew, W. (1965). *Comparative Hematology.* Grune & Stratton, New York.
84. Cohen, W. D., and Terwilliger, N. B. (1979). Marginal bands in camel erythrocytes. *J. Cell Sci.* 36:97–107.
85. Skutelsky, E., and Danon, D. (1967). An electron microscopic study of nuclear elimination from the late erythroblast. *J. Cell Biol.* 33:625–635.
86. Skutelsky, E., and Danon, D. (1970). Comparative study of nuclear expulsion from the late erythroblast and cytokinesis. *Exp. Cell Res.* 60:427–436.
87. Simpson, C. F., and Kling, J. M. (1967). The mechanism of denucleation in circulating erythroblasts. *J. Cell Biol.* 35:237–245.
88. Repasky, E. A., and Eckert, B. S. (1981). A reevaluation of the process of enucleation in mammalian erythroid cells. In: *The Red Cell: Fifth Ann Arbor Conference.* Alan R. Liss, New York, pp. 679–690.
89. Skutelsky, E., and Farquhar, M. G. (1976). Variations in distribution of Con A receptor sites and anionic groups during red blood cell differentiation in the rat. *J. Cell Biol.* 71:218–231.
90. Geiduschek, J. B., and Singer, S. J. (1979). Molecular changes in the membranes of mouse erythroid cells accompanying differentiation. *Cell* 16:149–163.
91. Hunt, R. C., and Marshall, L. M. (1979). The interaction of lectins with the surface of differentiating erythroleukemic cells. *J. Cell Sci.* 38:315–329.
92. Hunt, R. C., and Marshall, L. M. (1981). Membrane protein redistribution during differentiation of cultured human erythroleukemic cells. *Mol. Cell. Biol.* 1:1150–1162.
93. Marshall, L. M., Thureson-Klein, A., and Hunt, R. C. (1984). Exclusion of erythrocyte-specific membrane proteins from clathrin-coated pits during differentiation of human erythroleukemic cells. *J. Cell Biol.* 98:2055–2063.
94. Chang, H., Langer, P. J., and Lodish, H. F. (1976). Asynchronous synthesis of erythrocyte membrane proteins. *Proc. Natl. Acad. Sci. USA* 73:3206–3210.

95. Klymkowsky, M. W. (1981). Intermediate filaments in 3T3 cells collapse after intracellular injection of a monoclonal anti-intermediate filament antibody. *Nature* 291:249–251.
96. Gawlitta, W., Osborn, M., and Weber, K. (1981). Coiling of intermediate filaments induced by microinjection of a vimentin-specific antibody does not interfere with locomotion and mitosis. *Eur. J. Cell Biol.* 26:83–90.
97. Lin, J. J.-C., and Feramisco, J. R. (1981). Disruption of the in vivo distribution of the intermediate filaments in fibroblasts through the microinjection of a specific monoclonal antibody. *Cell* 24:185–193.
98. Blikstad, I., and Lazarides, E. (1983). Vimentin filaments are assembled from a soluble precursor in avian erythroid cells. *J. Cell Biol.* 96:1803–1808.
99. Moon, R. T., and Lazarides, E. (1983). Synthesis and posttranslational assembly of intermediate filaments in avian erythroid cells: vimentin assembly limits the rate of synemin assembly. *Proc. Natl. Acad. Sci. USA* 80:5495–5499.
100. O'Connor, C. M., Gard, D. L., and Lazarides, E. (1981). Phosphorylation of intermediate filament proteins by cAMP-dependent protein kinases. *Cell* 23:135–143.
101. Inagaki, M., Nishi, Y., Nishizawa, K., Matsuyama, M., and Sato, C. (1987). Site-specific phosphorylation induces disassembly of vimentin filaments in vitro. *Nature* 328:649–652.
102. McTavish, C. F., Nelson, W. J., and Traub, P. (1983). The turnover of vimentin in Ehrlich ascites tumor cells. *FEBS Lett.* 154:251–256.
103. Bruns, G. A. P., and Ingram, V. M. (1973). The erythroid cells and haemoglobins of the chick embryo. *Philos. Trans. R. Soc. Lond. B.* 266:225–305.
104. Capetanaki, Y. G., Ngai, J., Flytzanis, C. N., and Lazarides, E. (1983). Tissue-specific expression of two mRNA species transcribed from a single vimentin gene. *Cell* 35:411–420.
105. Friend, C., Scher, W., Holland, J. G., and Sato, T. (1971). Hemoglobin synthesis in murine virus-induced leukemic cells in vitro: stimulation of erythroid differentiation by dimethyl sulfoxide. *Proc. Natl. Acad. Sci. USA* 68:378–382.
106. Marks, P. A., and Rifkind, R. A. (1978). Erythroleukemic differentiation. *Ann. Rev. Biochem.* 47:419–448.
107. Marks, P. A., Sheffery, M., and Rifkind, R. A. (1987). Induction of transformed cells to terminal differentiation and the modulation of gene expression. *Canc. Res.* 47:659–666.
108. Eisen, H., Bach, R., and Emery, R. (1977). Induction of spectrin in erythroleukemic cells transformed by Friend virus. *Proc. Natl. Acad. Sci. USA* 74:3898–3902.
109. Sabban, E. L., Sabatini, D. D., Marchesi, V. T., and Adesnik, M. (1980). Biosynthesis of erythrocyte membrane protein band 3 in DMSO-induced Friend erythroleukemia cells. *J. Cell. Physiol.* 104:261–268.
110. Granger, B. L., and Lazarides, E. (1985). Appearance of new variants of membrane skeletal protein 4.1 during terminal differentiation of avian erythroid and lenticular cells. *Nature* 313:238–241.

111. Ngai, J., Capetanaki, Y. G., and Lazarides, E. (1984). Differentiation of murine erythroleukemia cells results in the rapid repression of vimentin gene expression. *J. Cell Biol.* 99:306–314.
112. Nelson, W. J., and Traub, P. (1981). Properties of a Ca^{2+}-activated protease specific for the intermediate-sized filament protein vimentin in Ehrlich-ascites-tumor cells. *Eur. J. Biochem.* 16:51–57.
113. Nelson, W. J., and Traub, P. (1982). Purification and further characterization of the Ca^{+}-activated proteinase specific for the intermediate filament proteins vimentin and desmin. *J. Biol. Chem.* 257:5544–5553.
114. Ngai, J., Bond, V. C., Wold, B. J., and Lazarides, E. (1987). Expression of transfected vimentin genes in differentiating murine erythroleukemia cells reveals divergent cis-acting regulation of avian and mammalian vimentin sequences. *Mol. Cell. Biol.* 7:3955–3970.
115. Ferrari, S., Battini, R., Kaczmarek, L., Rittling, S., Calabretta, B., de Riel, J. K., Philiponis, V., Wei, J.-F., and Baserga, R. (1986). Coding sequence and growth regulation of the human vimentin gene. *Mol. Cell. Biol.* 6: 3614–3620.
116. Zehner, Z. E., and Paterson, B. M. (1983). Characterization of the chicken vimentin gene: single copy gene producing multiple mRNAs. *Proc. Natl. Acad. Sci. USA* 80:911–915.
117. Zehner, Z. E., and Paterson, B. M. (1983). Vimentin gene expression during myogenesis: two functional transcripts from a single copy gene. *Nucleic Acids Res.* 11:8317–8332.
118. Lüscher, B., Stauber, C., Schindler, R., and Schümperli, D. (1985). Faithful cell-cycle regulation of a recombinant mouse histone H4 gene is controlled by sequences in the 3'-terminal part of the gene. *Proc. Natl. Acad. Sci. USA* 82:4389–4393.
119. Meijlink, F., Curran, T., Miller, A. D., and Verma, I. M. (1985). Removal of a 67-base pair sequence in the noncoding region of protooncogene *fos* converts it to a transforming gene. *Proc. Natl. Acad. Sci. USA* 82:4987–4991.
120. Shaw, G., and Kamen, R. (1986). A conserved AU sequence from the 3' untranslated region of GM-CSF mRNA mediates selective mRNA degradation. *Cell* 46:659–667.
121. Lawrence, J. B., and Singer, R. H. (1986). Intracellular location of messenger RNAs for cytoskeletal proteins. *Cell* 45:407–415.
122. Godbout, R., Ingram, R., and Tilghman, S. M. (1986). Multiple regulatory elements in the intergenic region between the α-fetoprotein and albumin genes. *Mol. Cell. Biol.* 6:477–487.
123. Hammer, R. E., Krumlauf, R., Camper, S. A., Brinster, R. L., and Tilghman, S. M. (1987). Diversity of alpha-fetoprotein gene expression in mice is generated by a combination of separate enhancer elements. *Science* 235: 53–58.
124. Rittling, S. R., and Baserga, R. (1987). Functional analysis and growth factor regulation of the human vimentin promoter. *Mol. Cell. Biol.* 7: 3908–3915.

125. Hirschhorn, R. R., Aller, P., Yuan, Z.-A., Gibson, C. W., and Baserga, R. (1984). Cell-cycle-specific cDNAs from mammalian cells temperature sensitive for growth. *Proc. Natl. Acad. Sci. USA* 81:6004–6008.
126. Kaczmarek, L., Calabretta, B., and Baserga, R. (1985). Expression of cell-cycle-dependent genes in phytohemagglutinin-stimulated human lymphocytes. *Proc. Natl. Acad. Sci. USA* 82:5375–5379.
127. Bernal, S. D., and Chen, L. B. (1982). Induction of cytoskeleton-associated proteins during differentiation of human myeloid leukemic cell lines. *Canc. Res.* 42:5106–5116.
128. McKnight, G. S., and Palmiter, R. D. (1979). Transcriptional regulation of the ovalbumin and conalbumin genes by steroid hormones in chick oviduct. *J. Biol. Chem.* 254:9050–9058.
129. Groudine, M., Peretz, M., and Weintraub, H. (1981). Transcriptional regulation of hemoglobin switching in chicken embryos. *Mol. Cell. Biol.* 1: 281–288.
130. Brock, M. L., and Shapiro, D. J. (1983). Estrogen stabilizes vitellogenin mRNA against cytoplasmic degradation. *Cell* 34:207–214.
131. Robins, D. M., Paek, I., Seeburg, P. H., and Axel, R. (1982). Regulated expression of human growth hormone genes in mouse cells. *Cell* 29:623–631.
132. Paek, I., and Axel, R. (1987). Glucocorticoids enhance stability of human growth hormone mRNA. *Mol. Cell. Biol.* 7:1496–1507.
133. Merrill, G. F., Hauschka, S. D., and McKnight, S. L. (1984). tk enzyme expression in differentiating muscle cells is regulated through an internal segment of the cellular tk gene. *Mol. Cell. Biol.* 4:1777–1784.
134. Lewis, J. A., and Matkovich, D. A. (1986). Genetic determinants of growth phase-dependent and adenovirus 5-responsive expression of the Chinese hamster thymidine kinase gene are contained within thymidine kinase mRNA sequences. *Mol. Cell. Biol.* 6:2262–2266.
135. Stewart, C. J., Ito, M., and Conrad, S. E. (1987). Evidence for transcriptional and posttranscriptional control of the cellular thymidine kinase gene. *Mol. Cell. Biol.* 7:1156–1163.
136. Schümperli, D. (1986). Cell cycle regulation of histone gene expression. *Cell* 45:471–472.
137. Santiago, C., Collins, M., and Johnson, L. F. (1984). In vitro and in vivo analysis of the control of dihydrofolate reductase gene transcription in serum-stimulated mouse fibroblasts. *J. Cell Physiol.* 118:79–86.
138. Farnam, P. J., and Schimke, R. (1985). Transcriptional regulation of mouse dihydrofolate reductase in the cell cycle. *J. Biol. Chem.* 260:7675–7680.
139. Leys, E. J., Crouse, G. F., and Kellems, R. E. (1984). Dihydrofolate reductase gene expression in cultured mouse cells is regulated by transcript stabilization in the nucleus. *J. Cell Biol.* 99:180–187.
140. Chao, M. V., Mellon, P., Charnay, P., Maniatis, T., and Axel, R. (1983). The regulated expression of β-globin genes introduced into mouse erythroleukemia cells. *Cell* 32:483–493.

141. Wright, S., de Boer, E., Grosveld, F. G., and Flavell, R. A. (1983). Regulated expression of the human β-globin gene family in murine erythroleukemia cells. *Nature* 305:333–336.
142. Charnay, P., Mellon, P., and Maniatis, T. (1985). Linker scanning mutagenesis of the 5′-flanking region of the mouse β-major-globin gene: sequence requirements for transcription in erythroid and non-erythroid cells. *Mol. Cell. Biol.* 5:1498–1151.
143. Charnay, P., Treisman, R., Mellon, P., Chao, M., Axel, R., and Maniatis, T. (1984). Differences in human α- and β-globin gene expression in mouse erythroleukemia cells: the role of intragenic sequences. *Cell* 38:251–263.
144. Wright, S., Rosenthal, A., Flavell, R., and Grosveld, F. (1984). DNA sequences required for regulated expression of β-globin genes in murine erythroleukemia cells. *Cell* 38:265–273.
145. Chao, M. V. (1986). Expression of transfected genes. In *Gene Transfer*, R. Kucherlapati (Ed.). Plenum Press, New York, pp. 223–241.
146. Chapman, L. F. (1980). Effect of calcium on differentiation of Friend leukemia cells. *Dev. Biol.* 79:243–246.
147. Levenson, R., Houseman, D., and Cantley, L. (1980). Amiloride inhibits murine erythroleukemia cell differentiation: evidence for a Ca^{2+} requirement for commitment. *Proc. Natl. Acad. Sci. USA* 77:5948–5952.
148. Nelson, W. J., and Traub, P. (1982). Intermediate (10 nm) filament proteins and the Ca^{2+}-activated proteinase specific for vimentin and desmin in the cells from fish to man: an example of evolutionary conservation. *J. Cell Sci.* 57:25–49.
149. Volloch, V., and Houseman, D. (1982). Terminal differentiation of murine erythroleukemia cells: physical stabilization of end stage cells. *J. Cell Biol.* 93:390–394.

4

The Erythrocyte Cytoskeleton in Hereditary Elliptocytosis and Spherocytosis

SALLY L. MARCHESI   *Yale University School of Medicine, New Haven, Connecticut*

I. INTRODUCTION

Shape change in red blood cells is a common finding in a variety of inherited and acquired disease states. Hemoglobinopathies, abnormal cytoplasmic enzymes, metabolic depletion states, intravascular trauma, and the normal aging process itself may produce irreversible alterations in membrane structure accompanied by changes in deformability and in other rheologic properties of the red blood cell.

The inherited hemolytic anemias, which are characterized by abnormal erythrocyte shape without evidence of enzymopathy or hemoglobinopathy, have traditionally been attributed to intrinsic membrane defects. Although an attractive idea, this notion has had little experimental support until the early 1980s, when methods for study of normal membrane structure became sufficiently advanced to allow real progress in the understanding of red blood cell membrane pathophysiology. This chapter will focus on the quantitative and qualitative changes in red blood cell cytoskeletal proteins that have been described in hereditary elliptocytosis and spherocytosis.

II. HEREDITARY ELLIPTOCYTOSIS

Hereditary elliptocytosis (HE) has been estimated to occur in hospitalized American and French patients at a rate of 25-50 per 100,000 population [1,2]. However, there is no firm data on the true incidence of HE in the general population and within specific ethnic groups, in part because HE is frequently asymptomatic. Inheritance of HE is typically autosomal with high penetrance,

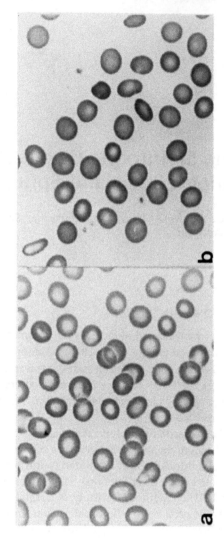

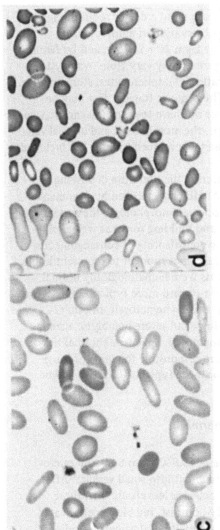

Figure 1 Peripheral blood smears (Wright stained) (a) Normal, (b) Mild HE, without anemia; note ovalocytes, elliptocytes, spherocytes, (c) HE with compensated hemolysis, (d) HPP; hemolytic anemia, s/p splenectomy. Smears in (b), (c), and (d) are all from patients with mutant spectrin αI domains (see Fig. 4).

but with variable expression; hemolysis and anemia to a degree requiring splenectomy are uncommon. Diagnosis is made by observation of elliptocytes on peripheral smear, which may range in number from 20% to 75%, and by family studies. The red blood cell morphology (Figure 1) may vary from ovalocytic (axial ratio > 1.5:1) to cigar-shaped (axial ratio 3:1); spherocytes, stomatocytes, and fragmented cells may also be present. The osmotic fragility test may be normal, decreased, or increased and thus does not have the same diagnostic value as it does in hereditary spherocytosis. Indeed, the morphologic and clinical variability of HE is such that a classification scheme developed by Lux [3] included nine categories.

Since 1955, scattered reports have appeared in the literature describing neonates or young children with variant forms of HE characterized by fulminant hemolysis and markedly poikilocytic red blood cell morphology; the smears typically include microspherocytes, fragments, and budding forms as well as elliptocytes [4-10]. In some cases, the children were offspring of consanguineous families with HE, and the disorder in these cases was thought to represent homozygous HE. In other cases, one or both parents were hematologically normal. This "poikilocytic HE of infancy" may evolve toward more typical elliptocytosis during the first year of life, accompanied by a rising hematocrit. In other children, fragmented red blood cell morphology and severe hemolytic anemia persist, necessitating transfusion and/or splenectomy. Specific abnormalities of cytoskeletal proteins have been found that are responsible for some of these poikilocytic hemolytic anemias, as well as for more typical HE. These will be discussed in detail below.

A. Red Blood Cell Thermolability and Abnormal Spectrin in Hereditary Elliptocytosis

Normal erythrocytes heated to temperatures > 48.8°C form buds and microspherocytes; at temperatures > 50°C they fragment into small vesicles. After cooling to room temperature, the red blood cells are less elastic than those maintained at room temperature [11,12]. By contrast, red blood cells heated to 47°C do not fragment, and retain normal elasticity. Calorimetric studies of isolated red blood cell ghosts show four structural transitions as determined by changes in circular dichroism over the temperature range 45-80°C [13]. One of these, designated the "A" transition, occurs at 49°C not only in ghosts, but in the spectrin extracted from them. These studies suggested that heat-induced loss of elasticity in the red blood cell is due to denaturation of spectrin, and implied that spectrin is an important determinant of red blood cell deformability.

In 1975, Zarkowsky and co-workers [9] studied three children from two kindreds with poikilocytic hemolytic elliptocytosis whose peripheral smears resembled those seen in thermal injury. Heat-induced fragmentation of red blood cells from these patients was found to occur at 45°C instead of the expected fragmentation temperature of 49°C. Thermal denaturation of spectrin extracted from these ghosts occurred at 44°C in contrast to the normal denaturation temperature of 49°C [14]. These observations implicated spectrin in the pathogenesis of hemolytic anemia in these children and provided the basis for calling this variant of elliptocytosis "hereditary pyropoikilocytosis" or "HPP." Although the occurrence of HPP was not and still is not predictable from family studies, it usually occurs in families with typical dominant HE.

Zarkowsky [10] subsequently studied two infants with poikilocytic elliptocytosis whose red blood cells fragmented at 45°C in the neonatal period. When studied again at five to six months of age, both had more typical elliptocytosis, and their red blood cells now resisted fragmentation until 47°C. Red blood cells of four kindreds with typical HE (without red blood cell fragmentation) were also found to have a mild increase in thermolability with fragmentation and/or spectrin denaturation occurring at 48°C [15], intermediate between the normal temperature of 49°C and the 45°C fragmentation temperature of red blood cells in HPP.

B. Abnormal Spectrin in Hereditary Elliptocytosis

Electrophoresis of red blood cell ghosts in most HE/HPP families shows a normal complement of membrane proteins, although spectrin content may be moderately decreased in HPP membranes [16]. A search for qualitative abnormalities in HE spectrin appeared initially to be a formidable task, because the combined molecular weight of the spectrin α and β subunits is 560,000. The problem was approached by Marchesi and co-workers [17-19] by analysis of intermediate-size tryptic peptides of intact spectrin and of isolated α and β subunits subjected to electrophoresis on two-dimensional [isoelectric focusing/sodium dodecyl sulfate (SDS)] polyacrylamide gels [20]. The resulting peptide maps were further analyzed by ^{125}I limit chymotryptic mapping and by monoclonal antibody immunoblotting, from which a chemical domain structure was established for both α and β subunits [18] as shown in Figure 2. The parent domains of both subunits are labeled and some of the daughter peptides are also identified. These maps are remarkably reproducible from one normal subject to another, and on repeated samples from the same subject, whether normal or abnormal. Although there is some variability in the number and intensity of tryptic fragments, particularly those with molecular weights less than 30 kD, the location of parent domains and of the larger daughter peptides are invariant.

The sensitivity of this technique to minor changes in the spectrin molecule is illustrated by the recognition of structural polymorphisms in the αII and αIII domains [21,22]. The αII polymorphisms have been found to occur commonly in the black population, and family studies show the expected autosomal dom-

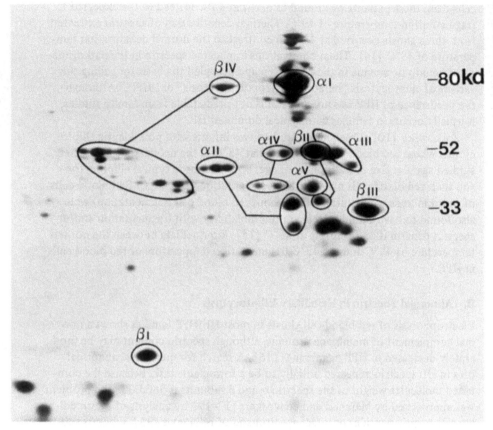

Figure 2 Two-dimensional (IEF/SDS) polyacrylamide gel electrophoresis of intermediate sized tryptic peptides of spectrin extracted from normal red cells. 200 µg of spectrin in 20 mM, Tris, pH 8.0, was digested with trypsin at 1:20 w/w ratio at 2°C for 90 minutes, electrophoresed in two dimensions (IEF/SDS PAGE), and stained with Coomassie blue. The major domains of the α and β subunits, as defined by tryptic digestion, are shown; in some cases, peptides derived from further tryptic cleavage of a parent domain are also shown.

inant inheritance. In our experience, the αIII mutations occur much less commonly and are also restricted to the black population. Four types of αII domain peptides are recognized on the tryptic maps: type 1 is the wild type shown in Fig. 2; in type 2, all αII peptides show an increase in molecular weight of ~4000 D and a basic shift in isoelectric point (pI); in type 3, the αII peptides have the 4000 D increase in molecular weight without change in pI; in type 4, the αII peptides show the basic shift in pI at the wild-type molecular weight.

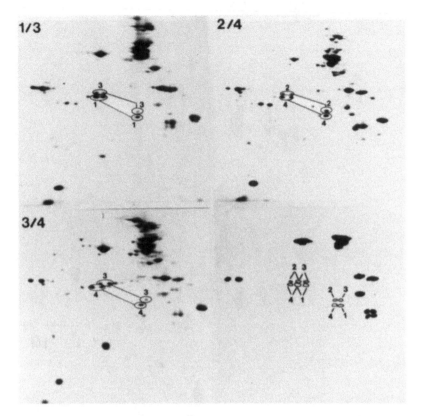

Figure 3 Two-dimensional IEF/SDS-PAGE of spectrin from individuals doubly heterozygous for αII variants 1,2,3, and 4. A schematic representation of the relative positions of Types 1,2,3, and 4 αII peptides is shown in the lower right panel; solid spots in this drawing are for reference to other spectrin peptides seen on two-dimensional electrophoresis. Reproduced from Knowles, et al. *J. Clin. Invest.*, 1984, Vol. 73, pp. 973–979 by copyright permission of the authors and the American Society for Clinical Investigation.

Figure 3 shows spectrins from hematologically normal individuals that are heterozygous for two αII polymorphic types.

The only αIII polymorphism recognized to date is characterized by a ∼4000 D increase in molecular weight without a change in pI (see Fig. 4b). Polymorphic αII and αIII domains have not as yet been shown to have any pathophysiologic effect on normal or abnormal red blood cell cytoskeletons.

Two-dimensional electrophoresis of tryptic peptides from spectrin of patients with HE and HPP has revealed three discrete mutations in the *N*-terminal 80 kD

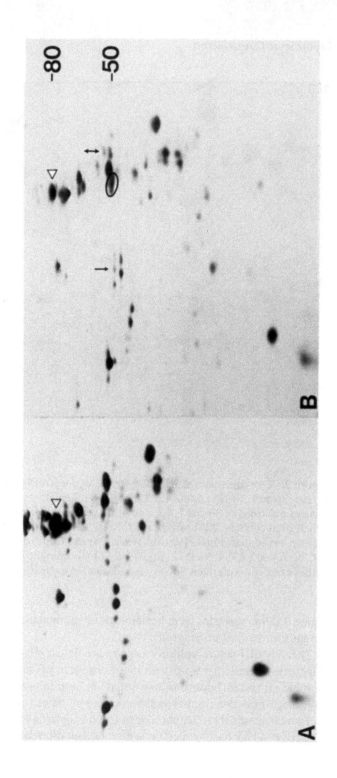

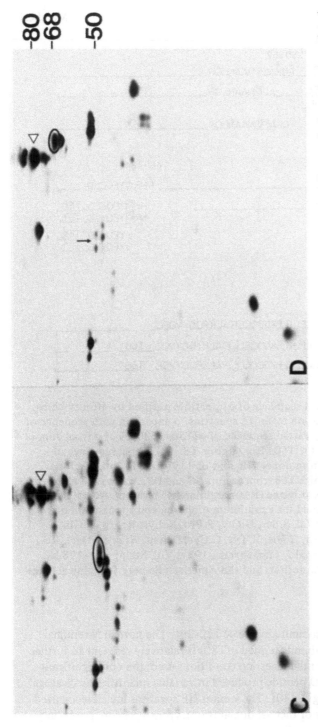

Figure 4 Intermediate tryptic digests of normal and HE spectrin. A, Normal spectrin; B, C, and D spectrin from three unrelated individuals with HE. The parent αI (αIT80) is marked in each digest by an open triangle. New peptides in panel B (T50a) in C (T50b), and in D (T68) are circled. The arrows in B and D indicate polymorphic αII domains type 1/2 as described in Fig. 3. The double ended arrow in B shows a high molecular weight polymorphism of the αIII domain (22). Reproduced from Marchesi, et al *J. Clin. Invest.*, 1987, Vol. 80 pp 191–198 by copyright permission of the authors and the American Society for Clinical Investigation.

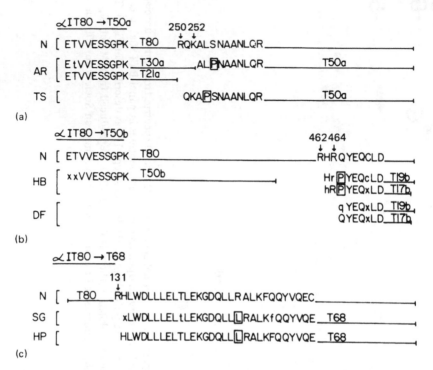

Figure 5 Partial amino acid sequence of αI peptides purified by affinity chromatography and electroelution from HE spectrins: comparison with sequence of the normal αI domain. (a) Partial sequence of αIT50a, T30a, T21 a from donor AR and partial sequence of αIT50a from donor TS. (b) Partial sequence of αIT50b, T19b and T17b from donor HB, and of T19b and T17b from donor DF. (c) Partial sequence of αIT68 from donors SG and HP. Arrows and numbers mark tryptic cleavage sites at lysine (K) or arginine (R) residues. Actual sequencing of peptides was continued for a minimum of eleven and a maximum of 44 residues. E:Glu, T:Thr, V:Val, S:Ser, G:Gly, P:Pro, K:Lys, R:Arg, Q:Gln, A:Ala, L:Leu, P:Pro, N:Asn, H:His, Y:Tyr, C:Cys, D:Asp, W:Try, F:Phe. Reproduced from Marchesi, et al *J. Clin. Invest.*, 1987, Vol. 80, pp 191–198 by copyright permission of the authors and the American Society for Clinical Investigation.

domain of the spectrin α subunit (Figure 4) [22–29]. The normal N-terminal tryptic domain of the α subunit (termed αIT80) is relatively resistant to further tryptic cleavage under the mild digestion conditions used; the only significant cleavage product is a 74-kD peptide produced in variable quantities by cleavage near the N-terminus at Arg[39] [30]. The variant HE spectrins are all recognized by increased susceptibility of their αI domains to tryptic digestion, producing

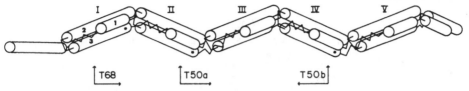

Figure 6 Model of the spectrin αI domain (αIT80) according to Speicher (31) ⟨━━━⟩: predicted helical segments, ⊏ : nonhelical connectors ∧∧∧∨ : beta sheet. Arrows mark sites of tryptic cleavage and the locations of new tryptic peptides in HE spectrins; asterisks mark sites of amino acid substitution or insertion described in the text for spectrins αIT80 → T68, T50a, and T50b. Reproduced from Marchesi et al *J. Clin. Invest.*, 1987, Vol. 80, pp 191–198 by copyright permission of the authors and the American Society for Clinical Investigation.

αI fragments at 46–50 kD or at 65–68 kD that are not present (or present in trace quantities) in normal spectrin digests. The molecular weights of these peptides are estimated from two-dimensional gels, and thus reported sizes of these fragments vary from one laboratory to another. For the purposes of this chapter, the three αI fragments generated by tryptic digestion of HE spectrins will be designated αIT50a, αIT50b, and αIT68. The mutant spectrins will be designated αIT80 → T50a, αIT80 → T50b, and αIT80 → T68, respectively.*

A fourth HE spectrin variant distinguished by Palek and co-workers undergoes increased cleavage of αIT80 to αIT74 [24]. Because this cleavage occurs in normal spectrin to varying degrees, its significance with regard to expression of HE is difficult to evaluate, at least in our experience.

The sites of cleavage producing T50a, T50b, and T68 have been identified in six unrelated patients, two with each of the three mutations [29]; these are shown in Figure 5. It can be seen that the three pathways of spectrin αI domain cleavage are unique and that they do not share a common final pathway. Figure 5 also shows that spectrin from the two patients with αIT80 → T50a has a proline substituted for Ser^{255} or Leu^{254}, three and four residues distant, respectively, from the tryptic cleavage site that produces T50a. Spectrin from one of the two subjects studied with αIT80 → T50b has a proline substituted for glutamine at residue 465, three residues from the cleavage that produces T50b and two small C-terminal peptides (T19b and T17b). Cleavage of αIT80 to T68 at Arg^{131} was associated with insertion of a leucine at residue 149 in both HE spectrins of this type studied. According to the molecular model of spectrin devel-

---

*αT50a and αT68 are $Sp\alpha^{I/46}$ and $Sp\alpha^{I/65}$ in the terminology of Palek and co-workers [26].

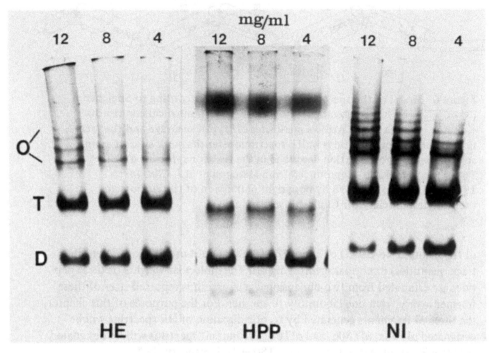

Figure 7 Non-denaturing gel electrophoresis of mutant spectrin αIT80 → T50a from members of a family with HE and HPP. Spectrin extracted from red cell ghosts at low ionic strength was concentrated and incubated for 3 hrs at 30°C prior to electrophoresis on 2–4% acrylamide gels. The subjects with HE and HPP are siblings. D:dimer, T:tetramer, O:higher oligomers. The diffuse band at the top of the HPP sample is hemoglobin. Reproduced from Knowles, et al *J. Clin. Invest.*, 1983, Vol. 71, pp 1867–1877 by copyright permission of the authors and the American Society for Clinical Investigation.

oped by Speicher (Figure 6) [31], all of the substitutions or insertions in the α subunit described above occur within helical segments of the repeat units and are theoretically capable of altering tertiary structure sufficiently to allow for the new tryptic cleavages observed.

   A functional consequence of the mutations described in the αI domain is to limit spectrin self-association, a process thought to be important in the assembly and stability of the membrane cytoskeleton [22-29,32-34]. This is not surprising, because the oligomerization process is known to depend on an intact αI domain [36,36]. The impairment in spectrin self-association in patients with HPP may be very severe; in some cases almost all spectrin remains in dimer form

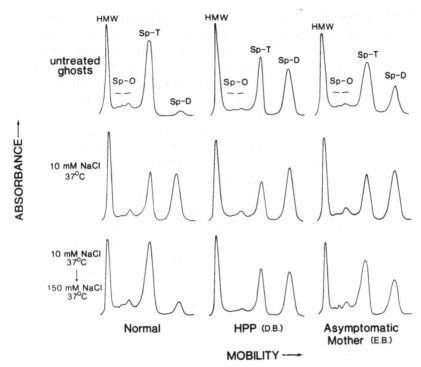

Figure 8 Spectrin tetramer → dimer transformation in the membrane. Ghosts were incubated in 5mM NaPO₄ buffer (pH 7.4) containing 10 mM NaCl at 37° C for 30 min. to initiate Sp-T → Sp-D transformation. Subsequently, a portion of the ghosts was reincubated in isotonic buffer containing 150 mM NaCl at 37° C for 30 min. to reverse the transformation. After the incubation, ghosts were washed and extracted overnight with low ionic-strength buffer at 0–4°C. Spectrin species in the extracts from untreated ghosts (top panels), hypotonic buffer incubated ghosts (middle panels), and isotonic buffer reincubated ghosts (bottom panels) were analyzed by non denaturing gel electrophoresis and densitometry. Reproduced from Liu, et al *J. Clin. Invest.*, 1981, Vol. 68, pp 597–605 by copyright permission of the authors and the American Society for Clinical Investigations.

even after incubation at high concentration that converts normal spectrin into a series of tetrameric and higher oligomers. Two illustrations of impaired oligomer formation in HE and HPP are shown in Figures 7 and 8.

Spectrin αI abnormalities in HE are not uncommon. More than 25 families have been recognized to date with HE/HPP and unstable spectrin αI domains of types αIT80 → T50a, → T50b, and → T68 [22-29,32-34]. Most of these families have been black or North African. Although the inheritance pattern is generally autosomal dominant, expression of the disease among family members may range from mild HE without significant hemolysis to striking red blood cell fragmentation and severe hemolysis requiring transfusion and eventually splenectomy. Studies of families with any of the three αI mutants described above indicate that the clinical severity of disease is roughly proportional to the fraction of unstable αI T80 and to the degree of impairment of spectrin self-association [22], but do not explain the variability of expression among members of a single family. A few patients with HPP have been described [23,37] whose spectrin digests show no intact αI domain and who are presumed homozygotes for the αI mutant or are double heterozygotes for a mutant αI domain and another abnormality, possibly abnormal α chain synthesis.

Mentzer and co-workers [38] have described a "carrier state" for HPP in a hematologically normal father whose red blood cell ghosts demonstrated increased mechanical fragility and whose spectrin showed impaired self-association. The same investigators have also suggested that the expression of HE may be modulated by 2,3-diphosphoglycerate levels. They found that incorporation of 2.55 mM 2,3-DPG into HE ghosts decreased their mechanical stability to that of HPP ghosts, whereas 2,3-DPG concentrations of 6–9 mM were required to produce the same effect in normal ghosts [39]. The suggestion was made that the high 2,3-DPG levels of neonatal red blood cells (resulting from reduced binding by hemoglobin F) may explain the transient poikilocytosis and fragmentation of red blood cells observed in many infants with HE.

Two groups of investigators have reported the presence of shortened spectrin β subunits in affected members of three families with hereditary elliptocytosis (Figure 9) [40–42]. The apparent molecular weights of the variant β subunits are 214,000–216,000 (normal 220,000–230,000). The missing piece in each case appears to be near the C-terminus because none of the mutant β subunits are phosphorylated, and all have defective spectrin self-association.

A shortened spectrin α subunit has also been associated with HE [43]. A new band was seen to migrate between the α and β spectrin subunits on SDS polyacrylamide gel electrophoresis (PAGE) of ghosts from six affected members of one kindred. The new band (αM) reacted with polyclonal anti-α spectrin antiserum on immunoblots and was not phosphorylated by [32]P. Spectrin self-association (measured by dimer content) was impaired in proportion to the αM content, which varied from 10–45% of the total α subunit in affected family members.

C. Abnormal Spectrin:Ankyrin:Band 3 Interactions in Hereditary Elliptocytosis

A family with recessively inherited elliptocytosis demonstrated decreased spectrin content (70% of normal) in red blood cells of affected members [44]. The

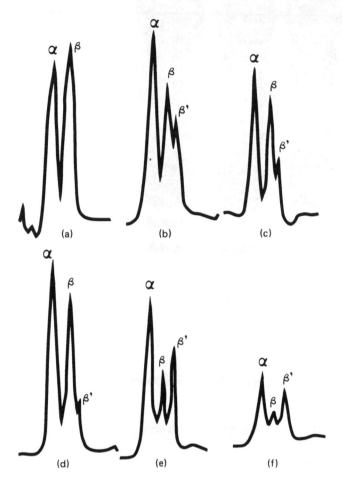

Figure 9 SDS PAGE densitometric tracings of spectrin from a subject with elliptocytosis and an abnormal β subunit (Mr = 214,000) coexisting with a decrease in the normal β subunit (Mr = 220,000). (a) Control 4°C extract, (b) proband's 4°C extracts; (c), (d), and (e) spectrin oligomer peak, spectrin tetramer peak, and spectrin dimer peak, respectively, each isolated by gel filtration of the proband's 4°C extract. (f) Pure spectrin dimer further isolated from (e) by sucrose gradient velocity centrifugation. Reproduced from Dhermy et al *J. Clin. Invest.*, 1982, Vol 70, pp 707–715 by copyright permission of the authors and the American Society for Clinical Investigation.

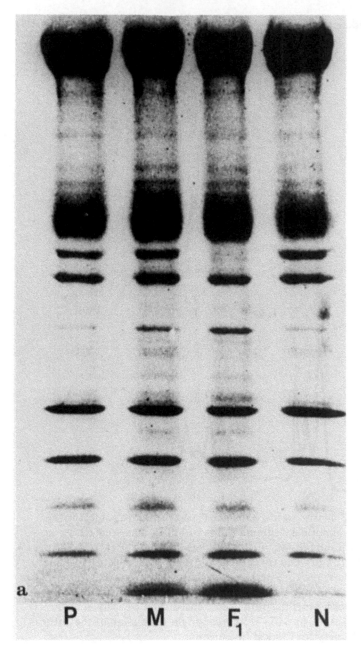

Figure 10 (a) Membranes of erythrocytes of a father (P), mother (M), and child (F$_1$), and of a normal subject (N). Note in F$_1$ the nearly complete absence of band 4.1 (4$_1$). Two very faint bands are visible in the position of 4.1 indicated by two arrows in (b). All other proteins appear quantitatively normal in the three patients (compare with gel labelled N). Reproduced from Feo, et al.[46] with permission of the authors and the editors of the *Nouvelle Revue Française D'Hematologie*.

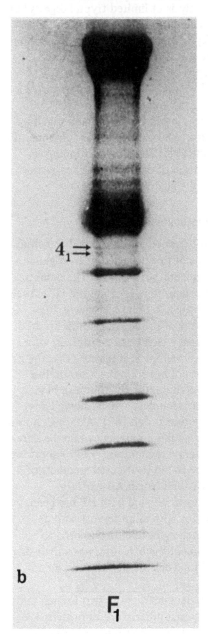

Figure 10b

spectrin appeared qualitatively normal by analysis of limited tryptic digests but demonstrated abnormal self-association and abnormal binding to ankyrin, suggesting an abnormality at the proposed ankyrin binding site near the C-terminus of the β subunit. The decrease in membrane spectrin was considered to be secondary to its abnormal binding to ankyrin.

Investigation of two other HE kindreds by Agre et al. [45] demonstrated a decrease in binding sites for ankyrin on spectrin-depleted inside-out vesicles (IOVs) of affected family members, which was interpreted as an abnormality in band 3 receptor sites. One of these patients (who had HPP) was subsequently found to have mutant spectrin αIT80 → T68 [29]. The ankyrin binding abnormality in this case may have been a reflection of damage to, or rearrangement of, band 3 receptor sites secondary to the hemolytic anemia.

D. Abnormal Protein 4.1 in Hereditary Elliptocytosis

Deficiency of red blood cell cytoskeletal protein 4.1 has been linked to hereditary elliptocytosis. The most striking example of this occurred in an Algerian family studied by Tchernia and co-workers (Figure 10) [46,47]. The parents, who were cousins, had decreased red blood cell protein 4.1 and mild HE. Three of their five children had no red blood cell protein 4.1 and had severe, life-threatening hemolytic elliptocytosis. Deficiency of protein 4.1 in this kindred resulted in marked instability of the membrane, which was reversible in vitro by exchange hemolysis with purified normal protein 4.1, suggesting that the defect in this kindred is in protein 4.1 itself and not in its binding sites on the cytoskeleton [48]. These conclusions were corroborated by Southern blot analysis of genomic DNA from affected family members, which showed a rearrangement of DNA upstream from the initiation codon [49]. Although homozygous 4.1 deficiency is uncommon, heterozygous 4.1 deficiency may be a frequent cause of HE in some populations. Moderate reductions (~30%) in protein 4.1 content of red blood cell membranes were found in affected members of 4 of 10 kindreds with mild HE studied in France and North Africa [50].

A model for study of protein 4.1 deficiency associated with HE has been described in a canine pedigree [51]. A male dog with elliptocytosis and decreased protein 4.1 content of his red blood cell cytoskeleton was bred to his daughter, also with protein 4.1 deficiency. One of the offspring was a male dog with total absence of protein 4.1 in his red blood cell membranes and hemolytic HE.

Shortened forms of protein 4.1 have been demonstrated in two French families [52-54]. In one family [52,53] heterozygous inheritance of protein 4.1 deficiency (4.1-) and of a shortened protein 4.1 (4.1Presles) occurred independently. Family members heterozygous for 4.1- had mild HE; those with 4.1Presles were hematologically normal; furthermore, 4.1Presles did not affect

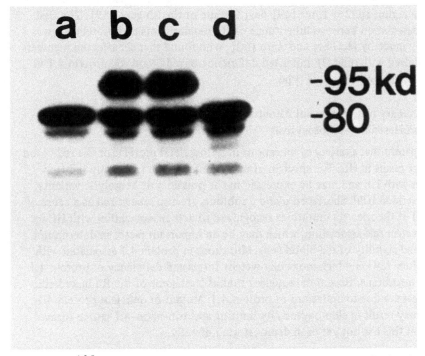

Figure 11 I^{125} immunoblots of erythrocyte ghosts using rabbit polyclonal anti 4.1 and I^{125} stap A protein. a, normal subject; b, father with HE; c, son with HE; d, normal son.

expression of 4.1– HE. In the other family [54], heterozygosity for the shortened protein 4.1 itself was associated with HE.

A high-molecular-weight form of protein 4.1 in a family from the southeastern United States is associated with mild HE (Figure 11) [55]. Studies of the high-molecular-weight protein 4.1 (4.1$^{Hurdle Mills}$) indicate that the site of insertion of amino acid sequence is within or adjacent to the central 8–10 kD domain of normal protein 4.1, which is the binding site for spectrin/actin [56]. Conboy et al. [57] found two species of 4.1 mRNA in family members heterozygous for protein 4.1$^{Hurdle Mills}$, one 5.6 kb in length and the other 5.8 kb (Chapter 7).

It has been recognized for many years that there are at least two genetically distinct types of HE, those linked to the Rh locus on chromosome 1 and those that are not. Phenotypic differences between Rh and non-Rh-linked HE were suggested by Geerdink [58], who found that non-Rh-linked HE was expressed by more hemolysis and anemia than Rh-linked HE. These observations have taken on renewed interest, because protein 4.1 has been mapped to chromosome 1 in the region 1p32 → 1pter [49] near the site of the Rh gene [59]. The first direct association between inheritance of Rh determinants and protein 4.1 was recently made by McGuire and Agre [60], who found that six affected members of a kindred with mild HE inherited deficiency of red blood cell protein 4.1 in linkage with their Rh blood type.

E. Summary of Cytoskeietal Abnormalities Associated with Hereditary Elliptocytosis

It is apparent that a variety of mutations in cytoskeletal proteins of the red blood cell may result in HE. The spectrin αI mutations are seen frequently in black families with HE and may be expressed in the poikilocytic hemolytic variants of HE termed HPP. Shortened α and β subunits are also recognized as a cause of HE. All of the spectrin mutations recognized to date in association with HE impair spectrin self-association, which may be an important factor in determining shape and stability of red blood cells. Mutations in protein 4.1 associated with HE include low- and high-molecular-weight forms and deficiency of protein 4.1 on the membrane. Recent data suggest that at least some of the Rh-linked elliptocytoses are due to mutations of protein 4.1. Mutant or deficient protein 4.1 states may result in elliptocytosis by limiting spectrin–actin–4.1 lattice formation, but this has not yet been demonstrated directly.

F. Southeast Asian Ovalocytosis

Hereditary ovalocytosis (or elliptocytosis) without clinical symptoms or anemia is common in Southeast Asia; it occurs in > 20% of the population in some areas of New Guinea [61] and is estimated at 30% in the Temuan [62]. The high frequency of ovalocytosis in a region where malaria is endemic suggested that it might confer resistance to the parasite, as is the case for red blood cells containing hemoglobin S and C and in G6PD deficiency. This was confirmed by Kidson et al. [63], who showed that ovalocytes from Melanesians in Papua New Guinea are resistant to invasion in culture by both *Plasmodium falciparum* and *P. knowlesi*. These observations indicate that the mechanism of resistance involves a feature common to the invasion process for both organisms, and not to membrane receptors specific for each *Plasmodium* species.

It has been shown that ovalocytes of Papua New Guineans and of Malayan aborigines have reduced deformability when studied either by the micropipette technique [64] or by ektacytometry [61]. By the latter method, Malayan ovalocytes have been found to require an 8 to 10-fold increase in applied shear stress to deform to the same extent as normal cells [61]. Furthermore, successful parasite invasion was inversely related to membrane deformability, suggesting that rigidity of the ovalocyte is in some way related to resistance to infection.

III. HEREDITARY SPHEROCYTOSIS

Hereditary spherocytosis (HS) has been estimated to occur with a frequency of
~25 per 100,000 population in the northern United States and England [65],
similar to the frequency of HE. MacKinney [65] found in a study of 26 families with HS that more than one-half of the affected patients gave no history
suggestive of the disease, so that the actual incidence of HS may be significantly
higher. Hereditary spherocytosis is usually inherited in an autosomal dominant
manner with an estimated penetrance of 90% [65]; however, in a study of 100
children with HS [66], 25 lacked a family history. Of the 25, family history was
incomplete in six: the remaining 19 children presumably had new mutant HS
or had one of the nondominant forms of spherocytosis recently characterized by
Agre et al. [67–69].

A diagnosis of hereditary spherocytosis is made in the clinical laboratory by
observing an increase in red blood cell osmotic fragility of HS red blood cells
compared with normal red blood cells after an overnight incubation of heparin-
ized whole blood at 37°C. Under these conditions, normal cells do not hemolyze
at concentrations of NaCl above 0.65% and < 50% of normal red blood cells
hemolyze at 0.55% NaCl. The increase in osmotic fragility seen in HS cells varies
widely, but 50% hemolysis occurs typically at ~.65% NaCl and there is frequently a population of extremely fragile cells that hemolyzes in isotonic
(0.85%) sodium chloride.

The moderately shortened life span of red blood cells in typical HS is usually
well compensated by increased bone marrow production of erythrocytes. In
some cases, however, there is mild anemia that may be exacerbated by intercur-
rent illness or surgery, under which conditions bone marrow compensatory mechanisms may be relatively suppressed. The role played by the spleen in shorten-
ing the survival of circulating HS erythrocytes is demonstrated dramatically by
the excellent clinical response of patients of any age to splenectomy; hematocrits usually return to normal and reticulocyte counts decrease. Spherocytes are
still demonstrable in the peripheral blood after splenectomy, sometimes in increased numbers, and the osmotic fragility remains abnormal.

A. Genetic Studies in HS

Chromosome analysis of kindreds with HS suggests that there is a determinant
gene for dominant spherocytosis on chromosome 8. A large family with HS has
been described with a balanced translocation of the short arms of chromosomes
8 and 12 t(8;12) (p11;p13) [72]; in another family, a mother and son with
spherocytosis were shown to have a balanced translocation between the short
arms of chromosomes 3 and 8 t(3;8) (p21;p11) [71]. In both families the break
point on chromosome 8 was at p11. Two siblings with severe hemolytic spherocytosis and congenital anomalies were found to have a deletion of a portion of

the short arm of chromosome 8 [72]. We have observed a similar child with multiple congenital abnormalities including Hirschsprung's disease and growth retardation who has deletion of the short arm of chromosome 8 at p11-p21 [73]. Reports of seven other patients with 8p⁻ syndromes [74,75] do not mention the presence of spherocytosis, but at least five of these deletions began at p21 or 22. The specific molecular explanation for chromosome-8-linked HS is not known.

B. Membrane Protein Studies in Dominant
Hereditary Spherocytosis

The composition of red blood cell membranes from patients with dominant HS has generally appeared normal. Agre and his colleagues, however, note small to moderate reductions in red blood cell spectrin content (63–81% of normal) in patients with dominant HS (69); the decrease in spectrin content in these patients correlates inversely with increases in osmotic fragility and with the fractional reduction in membrane shear elasticity [76]. Spectrin extracted from individuals with dominant HS has appeared normal by one-dimensional gel analysis, and there have been no consistent abnormalities reported in HS spectrin tryptic peptides analyzed by two-dimensional (IEF/SDS) PAGE.

Two studies from Australia suggest that in some kindreds with HS, spectrin is bound unusually tightly to the membrane. Sheehy and Ralston [77] reported that spectrin from 1 of 12 HS patients was inextractable from membrane ghosts at low ionic strength, conditions under which release of spectrin from normal and other HS ghosts is complete. In a study by Hill et al. [78], spectrin dissociated from HS ghosts and from Triton shells more slowly than normal when the rate of extraction was slowed so that kinetics could be followed accurately (Figure 12); however, the total amount of extractable spectrin was normal. Paradoxically, the ability of the spectrin-depleted HS vesicles to rebind spectrin (either from normal or HS cells) was impaired, suggesting that the defect in these HS cells resides in abnormal binding sites for spectrin on the membrane. By contrast, a study by Goodman et al. [79] of nine patients from three kindreds with HS showed normal binding of spectrin to IOVs. The differences between these studies could reflect the heterogeneous nature of HS, but it should be noted that there were significant differences between the two studies in methodology and in the absolute amount of spectrin that bound to IOVs.

Two groups of investigators [80,81] have looked directly at binding of HS spectrin to protein 4.1 by a variety of methods including immunoprecipitation of spectrin-4.1 complexes, sedimentation of spectrin-4.1-actin complexes through a sucrose cushion or density gradient, and binding of spectrin to a 4.1 affinity column (Figure 13). Both studies reported that spectrin from some kindreds with HS (one of six [81], and two of four [80]) is defective in its

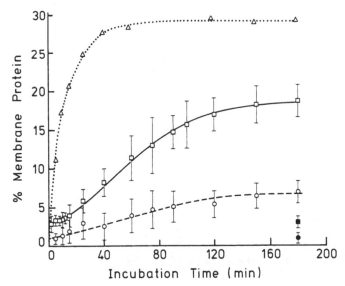

Figure 12 Kinetics of the dissociation of protein from erythrocyte cell membranes at low ionic strength. The percentage of total protein solubilized represents the fractions that remains in the supernatants after centrifugation at 150,000 g for 30 s (total centrifugation time including acceleration and deceleration, 4 min.). Results for normal ghosts in 0.1mM-imidazole+0.1mM EDTA, pH 7.6 at 37°C are indicated by △. Results for normal ghosts (□) and HS ghosts (○) in 1.2 mM KCl/ 0.1 mM EDTA/0.2mM. imdazole (pH 7.6) at 23°C also shown. The error bars represent means (±1 S.D.) for membranes obtained from seven different HS patients and from paired normal subjects. The filled symbols (■ and ●) represent control experiments in which membranes were maintained in a solution containing 6mM-KCl, 0.1mM-imidazole (pH7.6) at 23°C. Reproduced from Hill, et al. [78] with permission of the authors and the editors of the *Biochemical Journal*.

capacity to bind to protein 4.1. Analysis of spectrin from one of these families [82] suggests that a structural defect in the βIV domain causes oxidant injury, which in turn reduces the capacity of the HS spectrin to interact with protein 4.1. Sulhydryl reduction of the HS spectrin improved protein 4.1 binding, while mild oxidation of normal spectrin impaired its protein 4.1 binding capacity.

Finally, there have been two reports from Japan of marked reduction in protein 4.2 in red blood cell membranes of five patients from three kindreds with hereditary spherocytosis [83,84]. Because little is known of the nature and function of protein 4.2, the significance of this finding in terms of membrane stability is unknown.

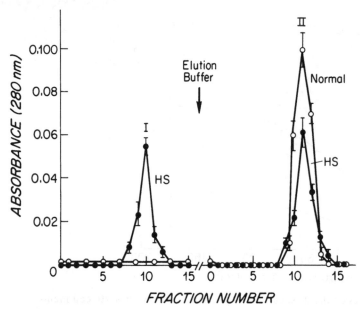

Figure 13 Separation of hereditary spherocytosis spectrin dimer into normal
and defective subpopulations. Spectrin dimer from patient 3 (HS) and from a
normal control were chromatographed on a column of immobilized normal
protein 4.1 under conditions (physiologic ionic strength, 4°C) in which spectrin-protein 4.1 binding was favored and spectrin-spectrin self-association
was suppressed. Note that all normal spectrin was detained, but that 41±2 percent (mean ± S.D.) of the hereditary spherocytosis spectrin failed to bind
and emerged in the void volume of the column (peak I). Chromatography
was then switched (arrow) to conditions (low ionic strength, 23°C) that released all the normal spectrin and the remaining hereditary-spherocytosis
spectrin (peak II). The total column recovery averaged 93 percent. Reproduced from Wolfe, et al.[81] with permission of the authors and the editors
of the *New England Journal of Medicine*.

C. Phosphorylation of Membrane Proteins in Hereditary Spherocytosis

Incubation of intact red blood cells with inorganic ^{32}P or of membranes with
adenosine triphosphate-^{32}P–labeled results in phosphorylation of a number of
membrane proteins, which include the spectrin β subunit, ankyrin, proteins 4.1,
4.9, and band 6 (G3PD). A number of membrane research groups have looked
for effects of phosphorylation on protein-protein interactions and for a link
between the phosphorylation state of membrane proteins and abnormal shape
or deformability of the cell. Several groups have reported abnormal phosphory-

lation of membrane proteins [85-87] in intact HS cells and in isolated HS membranes. Others [88-90] have not confirmed these findings but have instead pointed out pitfalls in interpretation of these complex studies, particularly when phosphorylation of isolated membranes is the method used. Furthermore, incubation of normal, intact red blood cells with ^{32}P [91] revealed no correlation between the state of phosphorylation of spectrin and its extractability from the membrane, its oligomeric state, nor in its ability to rebind to spectrin-depleted IOVs. In the same study, metabolic depletion and crenation of cells was not accompanied by dephosphorylation of spectrin. Thus, at the present time we do not know how the extent of phosphorylation of membrane proteins relates to any known pathophysiologic state.

D. Recessive Spherocytosis

Six spontaneous single gene mutations in the house mouse have been identified, all of which cause severe hemolytic anemia [92]. These mutations are inherited in an autosomal recessive manner and are termed nb, ja, and sph (four types). All mutant red blood cells have spherocytic morphology and all are characterized by low spectrin content ranging from 0-55% of normal. The severity of anemia produced by each mutant correlates with the degree of spectrin deficiency.

Studies of in vitro synthesis and assembly of cytoskeletal proteins by reticulocytes of the nb mutant mice [92] suggest that this form of spherocytosis is due to a primary deficiency of ankyrin, and that secondary spectrin deficiency develops during assembly of the cytoskeleton. Reticulocytes from mice homozygous for sph, sph2bc, and sph2j do not synthesize detectable amounts of α spectrin, whereas the ja/ja mutant is deficient in β spectrin production. The sph locus was found to be on chromosome 1, and is presumed to be the structural gene for α spectrin. The α spectrin gene has since been cloned in both mice and humans and has been assigned in both species to chromosome 1 [93]. The human α spectrin gene has been localized further to chromosome 1q22-1q25 by in situ hybridization to human metaphase chromosomes [93].

Nondominant forms of spherocytosis in humans have been described that resemble the murine spherocytosis described above in that spectrin content on red blood cell membranes of affected individuals may be as low as 30% of normal compared with spectrin contents of 63-81% in dominant HS [67-69] (Chapter 24). In contrast to dominant HS, recessively inherited spherocytosis may present early in childhood as a severe hemolytic anemia requiring transfusions and early splenectomy. The parents of affected individuals are normal hematologically, but their red blood cells have mildly decreased spectrin content and mildly increased osmotic fragility [69].

Tryptic peptides of spectrin from seven of the families with recessive HS reported by Agre et al. [67,69] have been analyzed in our laboratory by two-

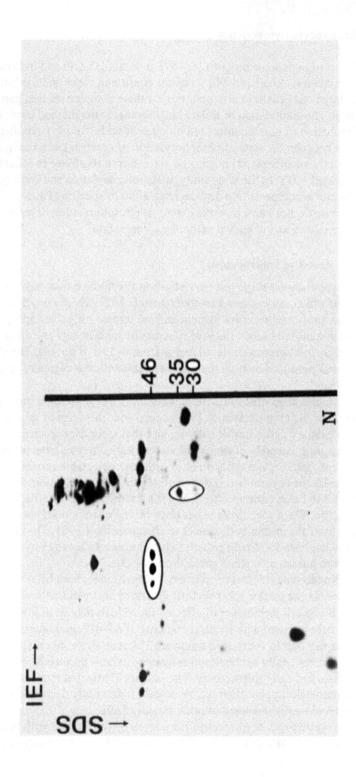

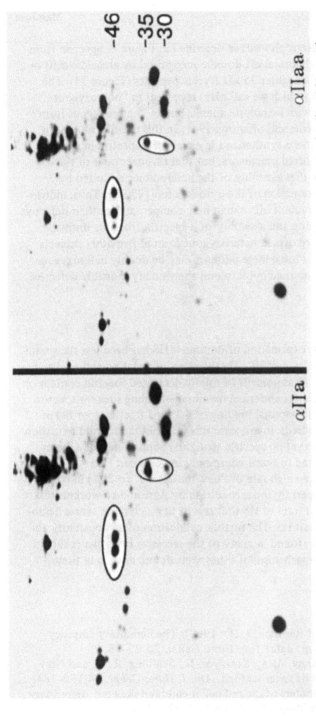

Figure 14 Two-dimensional IEF/SDS gelectrophoresis of intermediate tryptic peptides of spectrin from a family with recessive HS. The αII domain peptides are circled in all three panels. The normal αII domain peptides at 46, 35, and 30 kD are circled on the digest of normal spectrin (N). On the bottom right panel the circled αII domain peptide lack the most basic (left) spot seen in normal spectrin, and include a new, more acidic spot at 46 kD. In addition, the 35 kD spot is slightly more acidic than the normal 35 kD peptide. This represents a mutant αII domain in homozygous form (αIIaa). The digest on the bottom left contains both normal and acidic αII peptides; this individual is heterozygous for the normal and mutant αII domain (αIIa).

103

dimensional (IEF/SDS) electrophoresis as described in Figure 2. Spectrin from
four families contains an abnormal αII domain recognized by an acidic shift in
pI of its parent 46 kD and daughter 35 kD tryptic peptides (Figure 14). The
acidic shift in αII peptides (which we call αIIa) is present in "homozygous
form" in family members with hemolytic anemia and in "heterozygous form"
in their asymptomatic parents and offspring [94]. In this regard, it has been
demonstrated that α spectrin is synthesized in excess of β spectrin in both
avian and mammalian erythroid precursors, but that turnover rates of the soluble species are similar, so that assembly of the heterodimer is limited by β
spectrin, at least early in formation of the cytoskeleton [95,96]. Thus, individuals heterozygous for the mutant αII doman may compensate for their defective
allele by adequate production and assembly of α spectrin from the normal
allele. We have also observed αIIa in heterozygous form in five other subjects
with nondominant spherocytosis; these patients may be doubly heterozygous
for αIIa and a second unidentified cytoskeletal abnormality, possibly defective
α subunit synthesis.

IV. CONCLUSION

The search for a molecular explanation of dominant HS has been less successful
than a similar search in the case of HE. The positive information gathered so far
can be summarized as: (a) measurement of mildly decreased spectrin content on
the membrane; (b) evidence for abnormal membrane binding sites for spectrin
in several kindreds; and (c) abnormal binding of 4.1 by a fraction (~40%) of
spectrin from three HS kindreds; in one case this is attributed to mild oxidation
of the affected spectrin. As yet no specific structural abnormality of a membrane protein has been found to occur commonly in dominant HS.

The previously recognized high rate of "new mutations" (~20%) in HS has
been explained at least in part by the recognition by Agre and co-workers of a
recessive (or nondominant) form of HS that results in much more severe hemolysis than occurs in dominant HS. The further delineation of an apparently abnormal spectrin αII domain found in many of the recessive HS patients should
help in understanding the mechanism of spherocytosis and anemia in these
families.

REFERENCES

1. Bannerman, R. M., and Renwick, J. H. (1962). The hereditary elliptocytoses: clinical and linkage data. *Ann. Hum. Genet.* 26:23–38.
2. Morton, N. E., Mackinney, A. A., Kosower, N., Schilling, R. F., and Gray,
   M. P. (1962). Genetics of spherocytosis. *Am. J. Hum. Genet.* 14:170–184.
3. Lux, S. E. (1983). Disorders of the red cell membrane skeleton: Hereditary
   spherocytosis and hereditary elliptocytosis. In *The Metabolic Basis of*

Inherited Disease, Fifth ed. J. B. Stanley, J. B. Wyngaarden, D. S. Fredrickson, J. L. Goldstein, and M. S. Brown (Eds.). McGraw Hill, New York, pp 1573-1605.

4. Pryor, D. S., and Pitney, W. R. (1967). Hereditary elliptocytosis: A report of two families from New Guinea. *Br. J. Haematol.* 13:126-134.
5. Lipton, E. L. (1955). Elliptocytosis with hemolytic anemia: The effects of splenectomy. *Pediatrics* 15:67-81.
6. Austin, R. F., and Desforges, J. F. (1969). Hereditary elliptocytosis: An unusual presentation of hemolysis in the newborn associated with transient morphologic abnormalities. *Pediatrics* 44:196-200.
7. Grech, J. L., Cachia, E. A., Calleja, F., and Pullicino, F. (1961). Hereditary elliptocytosis in two Maltese families. *J. Clin. Pathol.* 14:365-373.
8. Evans, J. P. M., Baines, A. J., Hann, I. M., Al-Hakim, I., Knowles, S. M., and Hoffbrand, A. V. (1983). Defective spectrin dimer-dimer association in a family with transfusion dependent homozygous hereditary elliptocytosis. *Br. J. Haematol.* 54:163-172.
9. Zarkowsky, H. S., Mohandas, N., Speaker, C. B., and Shohet, S. B. (1975). A congenital hemolytic anaemia with thermal sensitivity of the erythrocyte membrane. *Br. J. Haematol.* 29:537-543.
10. Zarkowsky, H. S. (1979). Heat-induced erythrocyte fragmentation in neonatal elliptocytosis. *Br. J. Haematol.* 41:515-518.
11. Williamson, J. R., Shanahan, M. O., and Hochmuth, R. M. (1975). The influence of temperature on red cell deformability. *Blood* 46:611-624.
12. Rakow, A. L., and Hochmuth, R. M. (1975). Thermal transition in the human erythrocyte membrane: Effect on elasticity. *Biorheology* 12:1-3.
13. Brandts, J. F., Erickson, L., Lysko, K., Schwartz, A. T., and Taverna, R. D. (1977). Calorimetric studies of the structural transitions of the human erythrocyte membrane. The involvement of spectrin in the A transition. *Biochemistry* 16:3450-3454.
14. Chang, K., Williamson, J. R., and Zarkowsky, H. S. (1979). Effect of heat on the circular dichroism of spectrin in hereditary pyropoikilocytosis. *J. Clin. Invest.* 64:326-328.
15. Tomaselli, M. B., John, K. M., and Lux, S. E. (1981). Elliptical erythrocyte membrane skeleton and heat-sensitive spectrin in hereditary elliptocytosis. *Proc. Natl. Acad. Sci. USA* 78:1911-1915.
16. Coetzer, T. L., and Palek, J. (1986). Partial spectrin deficiency in hereditary pyropoikilocytosis. *Blood* 67:919-924.
17. Speicher, D. W., Morrow, J. S., Knowles, W. J., and Marchesi, V. T. (1980). Identification of proteolytically resistant domains of human erythrocyte spectrin. *Proc. Natl. Acad. Sci. USA* 77:5673-5677.
18. Speicher, D. W., Morrow, J. S., Knowles, W. J., and Marchesi, V. T. (1982). A structural model of human erythrocyte spectrin. *J. Biol. Chem.* 257:9093-9101.
19. Yurchenco, P. D., Speicher, D. W., Morrow, J. S., Knowles, W. J., and Marchesi, V. T. (1982). Monoclonal antibodies as probes of domain structure of the spectrin α subunit. *J. Biol. Chem.* 257:9102-9107.

20. O'Farrell, P. H. (1975). High resolution two-dimensional electrophoresis of proteins. *J. Biol. Chem.* 250:4007–4021.
21. Knowles, W. J., Bologna, M. L., Chasis, J. A., Marchesi, S. L., and Marchesi, V. T. (1984). Common structural polymorphisms in human erythrocyte spectrin. *J. Clin. Invest.* 73:973–979.
22. Marchesi, S. L., Knowles, W. J., Morrow, J. S., Bologna, M., and Marchesi, V. T. (1986). Abnormal spectrin in hereditary elliptocytosis. *Blood* 67:141–151.
23. Knowles, W. J., Morrow, J. S., Speicher, D. W., Zarkowsky, H. S., Mohandas, N., Mentzer, W. C., Shohet, S. B., and Marchesi, V. T. (1983). Molecular and functional changes in spectrin from patients with hereditary pyropoikilocytosis. *J. Clin. Invest.* 71:1867–1877.
24. Lawler, J., Liu, S. C., and Palek, J. (1984). A molecular defect of spectrin in a subset of patients with hereditary elliptocytosis: Alterations in the α-subunit domain involved in spectrin self-association. *J. Clin. Invest.* 73: 1688–1695.
25. Lecomte, M. C., Dhermy, D., Solis, C., Ester, A., Feo, C., Gautero, H., Bournier, O., and Boivin, P. (1985). A new abnormal variant of spectrin in black patients with hereditary elliptocytosis. *Blood* 65:1208–1217.
26. Lawler, J., Coetzer, T. L., Palek, J., Jacob, H. S., and Luban, N. (1985). Spα$^{I/65}$: A new variant of the α subunit of spectrin in hereditary elliptocytosis. *Blood* 66:706–709.
27. Alloisio, N., Guetarni, D., Morle, L., Pothier, B., Ducluzeau, M. T., Soun, A., Colonna, P., Clerc, M., Philippe, N., and Delaunay, J. (1986). Spα$^{I/65}$ hereditary elliptocytosis in North Africa. *Am. J. Hematol.* 23:113–122.
28. Lambert, S., and Zail, S. (1987). A new variant of the α subunit of spectrin in hereditary elliptocytosis. *Blood* 69:473–478.
29. Marchesi, S. L., Letsinger, J. T., Speicher, D. W., Marchesi, V. T., Agre, P., Hyun, B., and Gulati, G. (1987). Mutant forms of spectrin α-subunits in hereditary elliptocytosis. *J. Clin. Invest.* 80:191–198.
30. Speicher, D. W., Davis, G., and Marchesi, V. T. (1983). Structure of human erythrocyte spectrin II. The sequence of the αI domain. *J. Biol. Chem.* 258:14938–14947.
31. Speicher, D. W., and Marchesi, V. T. (1984). Erythrocyte spectrin is comprised of many homologous triple helicle segments. *Nature* 311:177–180.
32. Lawler, J., Liu, S. C., and Palek, J. (1982). Molecular defect of spectrin in hereditary pyropoikilocytosis. *J. Clin. Invest.* 70:1019–1030.
33. Coetzer, T., and Zail, S. (1982). Spectrin tetramer-dimer equilibrium in hereditary elliptocytosis. *Blood* 59:900–906.
34. Liu, S. C., Palek, J., Prchal, J., and Castleberry, R. P. (1981). Altered spectrin dimer-dimer association and instability of erythrocyte membrane skeletons in hereditary pyropoikilocytosis. *J. Clin. Invest.* 68:597–605.
35. Morrow, J. S., Speicher, D. W., Knowles, W. J., Hsu, C. J., and Marchesi, V. T. (1980). Identification of functional domains of human erythrocyte spectrin. *Proc. Natl. Acad. Sci. USA* 77:6592–6596.

36. Morrow, J. S., and Marchesi, V. T. (1981). Self-assembly of spectrin oligomers in vitro: A basis for a dynamic cytoskeleton. *J. Cell Biol.* 88: 463–468.
37. Lawler, J., Palek, J., Liu, S.-C., Prchal, J., and Butler, W. (1983). Molecular heterogeneity of hereditary pyropoikilocytosis: Identification of a second variant of the spectrin of α subunit. *Blood* 62:1182–1189.
38. Mentzer, W. C., Turetsky, T., Mohan, N., Schrier, S., Wu, C. C., and Koenig, H. (1984). Identification of the hereditary pyropoikilocytosis carrier state. *Blood* 63:1439–1446.
39. Mentzer, W. C., Jr., Iarocci, T. A., Mohandas, N., Lane, P. A., Smith, B., Lazerson, J., and Hays, T. (1987). Modulation of erythrocyte membrane mechanical stability by 2,3-diphosphoglycerate in the neonatal poikilocytosis/elliptocytosis syndrome. *J. Clin. Invest.* 79:943–949.
40. Dhermy, D., Lecomte, M. C., Garbarz, M., Bournier, O., Galand, C., Gautero, H., Feo, C., Alloisio, N., Delaunay, J., and Boivin, P. (1982). Spectrin β-chain variant associated with hereditary elliptocytosis. *J. Clin. Invest.* 70:707–715.
41. Ohanian, V., Evans, J. P., and Gratzer, W. B. (1985). A case of elliptocytosis associated with a truncated spectrin chain. *Br. J. Haematol.* 61:31–39.
42. Pothier, B., Morle, L., Alloisio, N., Ducluzeau, M. T., Caldani, C., Feo, C., Garbarz, M., Chaveroche, I., Dhermy, D., Lecomte, M. C., Boivin, P., and Delaunay, J. (1987). Spectrin Nice ($\beta^{220/216}$): A shortened β-chain variant associated with an increase of the $\alpha^{I/74}$ fragment in a case of elliptocytosis. *Blood* 69:1759–1765.
43. Lane, P. A., Shew, R. L., Iarocci, T. A., Mohandas, N., Hays, T., and Mentzer, W. C. (1987). Unique α-spectrin mutant in a kindred with common hereditary elliptocytosis. *J. Clin. Invest.* 79:989–996.
44. Zail, S. S., and Coetzer, T. L. (1984). Defective binding of spectrin to ankyrin in a kindred with recessively inherited hereditary elliptocytosis. *J. Clin. Invest.* 74:753–762.
45. Agre, P., Orringer, E. P., Chui, D. H., and Bennett, V. (1981). A molecular defect in two families with hemolytic poikilocytic anemia. *J. Clin. Invest.* 68:1566–1576.
46. Feo, C. J., Fischer, S., Piau, J. P., Grange, M. J., and Tchernia, C. (1980). Premiere observation de l'absence d'une proteine de la membrane erythrocytaire (bande 4.1) dans un cas d'anemia elliptocytaire familiale. *Nouv. Rev. Fr. Hematol.* 22:315–325.
47. Tchernia, G., Mohandas, N., and Shohet, S. B. (1981). Deficiency of skeletal membrane protein band 4.1 in homozygous hereditary elliptocytosis. *J. Clin. Invest.* 68:454–460.
48. Takakuwa, Y., Tchernia, G., Rossi, M., Benabadji, M., and Mohandas, N. (1986). Restoration of normal membrane stability to unstable protein 4.1-deficient erythrocyte membranes by incorporation of purified protein 4.1. *J. Clin. Invest.* 78:80–85.
49. Conboy, J., Mohandas, N., Tchernia, G., and Kan, Y. W. (1986). Molecular basis of hereditary elliptocytosis due to protein 4.1 deficiency. *N. Engl. J. Med.* 315:680–684.

50. Alloisio, N., Morlé, L., Dorléac, E., Gentilhomme, O., Bachir, D., Guetarni, D., Colonna, P., Bost, M., Zouaoui, Z., Roda, L., Roussel, D., and Delaunay, J. (1985). The heterozygous form of 4.1 (x) hereditary elliptocytosis [the 4.1 (x) trait]. *Blood* 65:46–51.

51. Smith, J. E., Moore, K., Arens, M., Rinderknecht, G. A., and Ledet, A. (1983). Hereditary elliptocytosis with protein band 4.1 deficiency in the dog. *Blood* 61:373–377.

52. Alloisio, N., Dorléac, E., Girot, R., andDelaunay, J. (1981). Analysis of the red cell membrane in a family with hereditary elliptocytosis: Total or partial decrease of protein 4.1. *Hum. Genet.* 58:68–71.

53. Morlé, L., Garbarz, M., Alloisio, N., Girot, R., Chaveroche, I., Boivin, P., and Delaunay, J. (1985). The characterization of protein 4.1 Presles, a shortened variant of RBC membrane protein 4.1. *Blood* 65:1511–1517.

54. Garbarz, M., Dhermy, D., Lecomte, M. C., Féo, C., Chaveroche, I., Galand, C., Bournier, O., Bertrand, O., and Boivin, P. (1984). A variant of erythrocyte membrane skeletal protein band 4.1 associated with hereditary elliptocytosis. *Blood* 64:1006–1015.

55. Letsinger, J. T., Agre, P., Marchesi, S. L. (1986). High molecular weight protein 4.1 in the cytoskeletons of hereditary elliptocytes (abstract). *Am. Soc. Hematol.* (suppl 1): 38a.

56. Correas, I., Speicher, D. W., and Marchesi, V. T. (1986). Structure of the spectrin-actin binding site of erythrocyte protein 4.1. *J. Biol. Chem.* 261: 13362–13366.

57. Conboy, J., Kan, Y. W., Agre, P., and Mohandas, N. (1986). Molecular characterization of hereditary elliptocytosis due to an elongated protein 4.1 (abstract). *Blood* 68 (suppl. 1): 34a.

58. Geerdink, R. A., Nijenhuis, L. E., and Huizinga, J. (1967). Hereditary elliptocytosis: linkage data in man. *Ann. Hum. Genet.* 30: 363–378.

59. McKusick, V. A. (1986). In *Mendelian Inheritance in Man*, Seventh Ed. Johns Hopkins University Press, Baltimore, Maryland, p. 103.

60. McGuire, M., Smith, B. L., and Agre, P. (1987). Distinct variants of erythrocyte 4.1 inherited in linkage with elliptocytosis and Rh type in three white families. *Blood* 72:287–293.

61. Mohandas, N., Lie-Injo, L. E., Friedman, M., and Mak, J. W. (1984). Rigid membranes of malayan ovalocytes: A likely genetic barrier against malaria. *Blood* 63:1385–1392.

62. Baer, A., Injo, L. E., Welch, Q. B., and Lewis, A. N. (1976). Genetic factors and malaria in the Temuan. *Am. J. Hum. Genet.* 28:179–188.

63. Kidson, C., Lamont, G., Saul, A., and Nurse, G. T. (1981). Ovalocytic erythrocytes from Melanesians are resistant to invasion by malaria parasites in culture. *Proc. Natl. Acad. Sci. USA* 78:5829–5832.

64. Saul, A., Lamont, G., Sawyer, W. H., and Kidson, C. (1984). Decreased membrane deformability in Melanesian ovalocytes from Papua New Guinea. *J. Cell Biol.* 98:1348–1354.

65. MacKinney, A. A., Jr. (1965). Hereditary spherocytosis: Clinical family studies. *Arch. Intern. Med.* 116:257–262.

66. Krueger, H. C., and Burgert, E. O., Jr. (1966). Hereditary spherocytosis in 100 children. *Mayo Clin. Proc.* 41:821–830.
67. Agre, P., Orringer, E. P., and Bennett, V. (1982). Deficient red cell spectrin in severe, recessively inherited spherocytosis. *N. Engl. J. Med.* 306:1155–1161.
68. Agre, P., Casella, J. F., Zinkham, W. H., McMillan, C., and Bennett, V. (1985). Partial deficiency of erythrocyte spectrin in hereditary spherocytosis. *Nature* 314:380–383.
69. Agre, P., Asimos, A., James, B. S., Casella, F., and McMillan, C. (1986). Inheritance pattern and clinical response to splenectomy as a reflection of erythrocyte spectrin deficiency in hereditary spherocytosis. *N. Engl. J. Med.* 315:1579–1583.
70. Kimberling, W. J., Fulbeck, T., Dixon, L., and Lubs, H. A. (1975). Localization of spherocytosis to chromosome 8 or 12 and report of a family with spherocytosis and a reciprocal translocation. *Am. J. Hum. Genet.* 27:586–594.
71. Bass, E. B., Smith, S. W., Jr., Stevenson, R. E., and Rosse, W. F. (1983). Further evidence for location of the spherocytosis gene on chromosome 8. *Ann. Intern. Med.* 99:192–194.
72. Chilcote, R. R., Jones, B., Dampier, C., LeBeau, M., Rowley, J., Verlinsky, Y., and Pergament, E. (1984). Dysmorphic syndrome, spherocytosis and partial depletion of the short arm of chromosome 8. *Pediatr. Res.* 18:220A.
73. Marchesi, S., McIntosh, S., Collins, F., and Francke, U. (1985). Unpublished data.
74. Reiss, J. A., Brenes, P. M., Chamberlin, J., Magenis, R. E., and Louvrien, E. W. The 8p-syndrome. *Hum. Genet.* 47:135–140.
75. Beighle, C., Karp, L. E., Hanson, J. W., Hall, J. G., and Hoehn, H. (1977). Small structural changes of chromosome 8: Two cases with evidence for deletion. *Hum. Genet.* 38:113–121.
76. Waugh, R. E., and Agre, (1988). Reductions of erythrocyte membrane viscoelastic coefficients reflect spectrin deficiencies in hereditary spherocytosis. *J. Clin. Invest.* 81:133–141.
77. Sheehy, R., and Ralston, G. B. (1978). Abnormal binding of spectrin to the membrane of erythrocytes in some cases of hereditary spherocytosis. *Blut* 36:145–148.
78. Hill, J. S., Sawyer, W. H., Howlett, G. J., and Wiley, J. S. (1981). Hereditary spherocytosis of man: Altered binding of cytoskeletal components to the erythrocyte membrane. *Biochem. J.* 201:259–266.
79. Goodman, S. R., Weidner, S. A., Eyster, M. E., and Kesselring, J. J. (1982). Binding of spectrin to hereditary spherocyte membranes. *J. Mol. Cell Cardiol.* 14:91–97.
80. Goodman, S. R., Shiffer, K. A., Casoria, L. A., and Eyster, M. E. (1982). Identification of the molecular defect in the erythrocyte membrane skeleton of some kindreds with hereditary spherocytosis. *Blood* 60:772–784.
81. Wolfe, L. C., John, K. M., Falcone, J. C., Byrne, A. M., and Lux, S. E. (1982). A genetic defect in the binding of protein 4.1 to spectrin in a kindred with hereditary spherocytosis. *N. Engl. J. Med.* 307:1367–1374.

82. Becker, P. S., Morrow, J. S., and Lux, S. E. (1987). Abnormal oxidant sensitivity and β chain structure of spectrin in hereditary spherocytosis associated with defective spectrin-protein 4.1 binding. *J. Clin. Invest.* 80: 557-565.

83. Nozawa, Y., Noguchi, T., Iida, H., Fukushima, H., Sekiya, T., and Ito, Y. (1974). Erythrocyte membrane of hereditary spherocytosis: Alteration in surface ultrastructure and membrane proteins, as inferred by scanning electron microscopy and SDS-Disc gel electrophoresis. *Clin. Chim. Acta.* 55: 81-85.

84. Hayashi, S., Koomoto, R., Yano, A., Ishigami, S., Tsujino, G., Saeki, S., and Tanaka, T. (1974). Abnormality in a spectrin protein of the erythrocyte membrane in hereditary spherocytosis. *Biochem. Biophys. Res. Commun.* 57:1038-1044.

85. Greenquist, A. C., and Shohet, S. B. (1976). Phosphorylation in erythrocyte membranes from abnormally shaped cells. *Blood* 48:877-885.

86. Thompson, S., and Maddy, A. H. (1981). The abnormal phosphorylation of spectrin in human hereditary spherocytosis. *Biochim. Biophys. Acta* 649:31-37.

87. Matsumoto, N., Yawata, Y., and Jacobs, H. S. (1977). Association of decreased membrane protein phosphorylation in HS. *Blood* 49:223-239.

88. Wolfe, L. C., and Lux, S. E. (1978). Membrane protein phosphorylation of intact normal and hereditary spherocytic erythrocytes. *J. Biol. Chem.* 253:3336-3342.

89. Zail, S. S., and Van Den Hoek, A. K. (1975). Studies on protein kinase activity and the binding of adenosine 3'5-monophosphate by membranes of hereditary spherocytosis erythrocyte. *Biochem. Biophys. Res. Commun.* 66:1078-1086.

90. Beutler, E., Guinto, E., and Johnson, C. (1976). Human red cell protein kinase in normal subjects and patients with hereditary spherocytosis, sickle cell disease, and autoimmune hemolytic anemia. *Blood* 48:887-898.

91. Anderson, J. M., and Tyler, J. M. (1980). State of spectrin phosphorylation does not affect erythrocyte shape or spectrin binding to erythrocyte membranes. *J. Biol. Chem.* 255:1259-1265.

92. Bodine, D. M., Birkenmeier, C. S., and Barker, J. E. (1984). Spectrin deficient inherited hemolytic anemias in the mouse: Characterization by spectrin synthesis and mRNA activity in reticulocytes. *Cell* 37:721-729.

93. Huebner, K., Palumbo, A. P., Isobe, M., Kozak, C. A., Mona, S., Rovers, G., Croce, C. M., and Curtis, P. J. (1985). The α-spectrin gene is on chromosome 1 in mouse and man. *Proc. Natl. Acad. Sci. USA* 82:3790-3793.

94. Winkelmann, J. C., Marchesi, S. L., Watkins, P., Linnenbach, A. J., Agre, P., and Forget, B. G. (1986). Recessive hereditary spherocytosis is associated with an abnormal α spectrin subunit. *Clin. Res.* 34:474A.

95. Blikstad, I., Nelson, W. J., Moon, R. T., and Lazarides, E. (1986). Synthesis and assembly of spectrin during avian erythropoiesis: Stoichiometric assembly but unequal synthesis of α and β spectrin. *Cell* 68 (suppl 1): 35a.

96. Hanspal, M., Dainiak, N., and Palek, J. (1986). Synthesis and assembly of spectrin in human erythroblasts. *Blood* 68(5)Suppl. 1:p. 35a.

5

Spectrin Genes

JOHN C. WINKELMANN,* THOMAS L. LETO,† and BERNARD G. FORGET
Yale University School of Medicine, New Haven, Connecticut

I. INTRODUCTION

The red blood cell cytoskeleton, a lattice of extrinsic proteins on the cell membrane's inner surface, endows the mature erythrocyte with its unique shape. It also provides resilience and durability while allowing remarkable deformability. Spectrin is the most abundant protein in the membrane skeleton, comprising ~75% of its mass [1]. Its singular molecular properties and interactions with other cytoskeleton proteins are vital to the function and in vivo survival of the erythrocyte.

Spectrin is a heterodimer, consisting of α and β subunits of 240 and 220 kD, respectively. These subunits associate in antiparallel fashion to form a 1000 Å flexible rod that serves as the basic structural element in the erythrocyte membrane skeleton [2]. Extensive analysis of the primary structure of human red blood cell spectrin reveals that both subunits possess a unique 106 amino acid homologous repeating motif throughout their entire lengths [3]. Speicher and Marchesi hypothesize that each homologous repeating segment folds into a bundle of three α helices. They estimate a total of 20 segments in the α subunit and 18 in the β subunit.

Spectrin is not confined to the erythrocyte. Antisera raised against red blood cell spectrin detect spectrinlike proteins in many, if not all, cell types [4] and in a wide variety of animal species. The precise number and function of the

---

Current affiliations:
*University of Minnesota Medical School, Minneapolis, Minnesota
†National Institute of Allergy and Infectious Diseases, National Institutes of Health, Bethesda, Maryland

various nonerythroid spectrins are unknown. These will be discussed in greater detail below.

Molecular analysis of the spectrin genes that encode this important group of proteins is in a formative stage. Our goal in this chapter is to review the current state of this evolving field. Where possible, we focus on the human spectrin genes, because these have the most direct relevance to clinical disorders; we also discuss work on other species when it provides further insight.

II. NORMAL SPECTRIN GENES

A. Red Blood Cell α Spectrin Gene

Molecular Cloning of α-Spectrin Genomic DNA

A variety of genetic abnormalities of the amino-terminal 80 kd or "αI" tryptic domain of human erythroid α spectrin have been described [2,5,6]. These usually result in impaired spectrin dimer self-association and clinical elliptocytosis and/or pyropoikilocytosis. To directly obtain a molecular probe for this important region of the α spectrin gene, Linnenbach et al. designed a 90 nucleotide synthetic gene fragment (90-mer) derived from known amino acid sequence [7]. This fragment was cloned into the single-stranded bacteriophage M13. Primed DNA synthesis with radioactive deoxyribonucleotides followed by excision of the 90-mer by restriction endonuclease digestion yielded a very high specific-activity probe. A bacteriophage λ library of total human genomic DNA was screened with this probe by hybridization. By this means, a phage clone with a 16.8 kb DNA insert, denoted λ3021, was isolated (Figure 1). Nucleotide sequence analysis of the region found to hybridize with the synthetic 90-mer confirmed its identity as part of the chromosomal α-spectrin gene based on an exact correspondence of translated coding DNA sequence with known amino acid sequence. Hybridization of a portion of the gene to a Southern blot of total human genomic DNA was consistent with a single gene copy for α spectrin; similar results were also obtained using an α spectrin cDNA probe [8].

Further sequence analysis of the gene 10 kb upstream (5′) from the 90-mer hybridization site has uncovered the coding DNA for the N-terminus of α spectrin (B. G. Forget, 1986, unpublished data). These data revealed an additional 6 amino acids at the amino terminus that were previously unknown. Adjacent to the 5′ coding DNA the promoter region of the gene was also found within the λ3021 clone. The promoter and surrounding DNA very likely contain control sequences responsible for erythroid-specific expression of the human α spectrin gene. Detailed structural and functional analysis of this important region may yield insight into the mechanism by which expression is controlled.

Molecular Cloning of α Spectrin cDNA

The first mammalian erythroid α spectrin molecular clone was reported by Cioe and Curtis [9]. They isolated a 750 base pair (bp) cDNA clone by antibody-

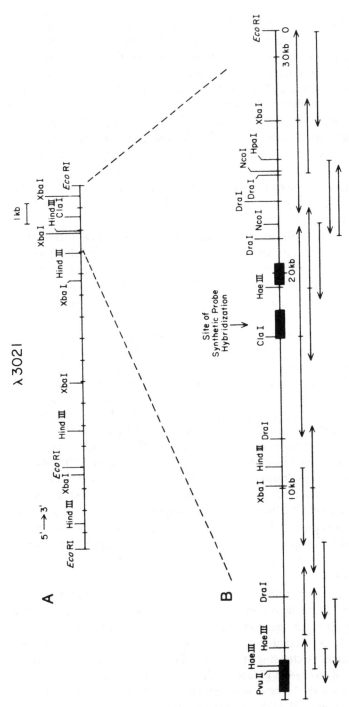

Figure 1 *Structural analysis of α-spectrin genomic clone λ3021*. A. Restriction map of the 16.8 kb insert found to hybridize to the synthetic α-spectrin gene fragment [7]. B. Blow-up of the region hybridizing to the probe. A fine restriction map is shown. Arrows illustrate the nucleotide sequencing strategy. Dark boxes denote exon (coding) sequences. Reprinted with permission of Linnenbach et al. [7].

screening [10] of an expression library, constructed in a pUC plasmid vector, using antibodies reactive against total red blood cell membranes. The source of mRNA was anemic mouse spleen. This murine cDNA was then used to screen a human K562 cell cDNA library by colony hybridization [8]. By this means a 713 bp human α spectrin cDNA was isolated, and its identity confirmed by correspondence to stretches of known amino acid sequence. Comparison between human and murine cDNA sequences in the overlapping 360 bp revealed 82% nucleotide sequence homology and 83% homology at the amino acid sequence level. Alignment of this derived peptide sequence with peptide sequence previously reported by Speicher and Marchesi [3] shows that this clone encodes a sequence starting near the end (residue 87 of 106) of repeat 14 and extending to the beginning (residue 5) or repeat 17. These 237 amino acids represent ~12% of the entire spectrin α subunit.

To analyze the entire primary structure of α spectrin and allow nucleotide coding sequence comparisons between various normal and abnormal spectrins, an important goal has been to obtain a full-length cDNA. The large size of the α-spectrin mRNA (8 kb) has made this a formidable task. Synthesis and cloning of cDNA by current techniques does not usually allow cloning a cDNA of this size as a single event. Therefore, our strategy has been to isolate multiple, large, overlapping clones that together span the entire coding and flanking sequences of the α spectrin mRNA. To this end, we have made use of cDNA libraries synthesized from human fetal liver and adult erythroid bone marrow poly (A)+ mRNA. First-strand synthesis was primed with oligodeoxythymidylic acid (dT). which biases cloning toward the 3' (C-terminal) end, or random hexamers, which allow internal priming and increased cloning of 5' (N-terminal) cDNA. These libraries were constructed in the bacteriophage expression vector λgt11, developed by Young and Davis [11]. Screening was performed in different ways. First, a labeled exon probe derived from the 5' α-spectrin genomic clone λ3021 was hybridized to the random-primed cDNA libraries. This approach yielded 5' cDNA clones. Next, taking advantage of the expression feature of λgt11, we screened the dT-primed fetal liver library with polyclonal antispectrin antisera. Of 10⁶ recombinants probed, 37 gave positive signals. To determine which of these might contain 3' α spectrin cDNA, these antibody-positive phage clones were counterscreened with a synthetic 20 base oligonucleotide mixture designed to match known C-terminal amino acid sequence.

The first 5' and 3' α spectrin cDNA clones isolated by the above methods neither overlapped internally nor extended to their respective ends. Therefore, we performed repeated rounds of cDNA library screening, with DNA fragments derived from the ends of existing cDNA clones, in an effort to "walk," or obtain overlapping clones extending in the desired directions. The results of this experiment are summarized in Figure 2, which aligns homologous repeating segments of α spectrin with the various molecular clones that encode portions of this protein. Identities of all cDNA clones have been confirmed by restriction

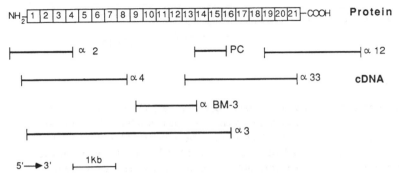

Figure 2 *Alignment of human α-spectrin cDNA clones.* The α-spectrin protein is schematically represented according to the 106 amino acid homologous model of Speicher and Marchesi [3]. The identity of cDNA clones has been verified by hybridization to Southern blots, restriction mapping and/or nucleotide sequencing. The names of the various clones and their alignment with the protein are shown. The first human α-spectrin cDNA to be isolated [8] is designated PC. Many internally overlapping clones are omitted.

endonuclease mapping and nucleotide sequencing. Internally overlapping clones are not shown, except the first human α spectrin cDNA, isolated by Curtis et al. [8], which is identified as "PC." An entire α spectrin cDNA appears to be in hand, albeit in pieces. Preliminary sequence data suggest that there are at least 21 homologous 106 amino acid repeats in α spectrin, rather than 20 predicted by Speicher and Marchesi (P. Curtis and B. G. Forget, 1988, unpublished data).

Huebner and co-workers [12] assigned the human (and murine) α spectrin gene to chromosome 1 by hybridization of an α spectrin cDNA to Southern blots containing genomic DNA derived from human X rodent somatic hybrid cell lines. In situ hybridization resulted in further sublocalization to 1q22–1q25.

B. Red Blood Cell β Spectrin Gene

The first molecular clones for human erythroid β spectrin have been obtained from cDNA [13,14]. An avian red blood cell β spectrin cDNA has also been reported [15]. The approach taken in our laboratory was similar to that described for α spectrin: The 37 antispectrin antibody-positive λgt11 recombinants from the dT-primed human fetal liver library were counterscreened with a 17 base oligonucleotide mixture designed to match a known C-terminal β spectrin amino acid sequence. A 2.7 kb cDNA insert, designated β28 in Figure 3, was confirmed to encode β spectrin by exact correspondence of deduced amino acid sequence to known amino acid sequences near the C-terminus of β-spectrin. A 3.3 kb cDNA, β8, extends further in the 5′ direction. In Figure 3 these clones are aligned with the 18-repeats of β spectrin originally defined by Speicher and

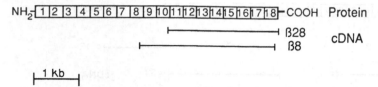

Figure 3 *Alignment of human β-spectrin cDNA clones.* The β-spectrin protein is schematically represented according to the 18 repeat model of Speicher and Marchesi [3]. The identity of cDNA clones has been verified by hybridization, restriction mapping and/or nucleotide sequencing [16]. The alignment of the cDNA and protein is illustrated. Internally overlapping clones are omitted.

Marchesi [3]. Together they comprise >50% of the estimated β spectrin coding sequence. Prchal and co-workers also reported the isolation of β-spectrin cDNA clones from a human reticulocyte cDNA expression library [14]. Together these encode a sequence extending from repeat 10 to repeat 12. The C-terminal end of β spectrin includes sites for self-association, ankyrin binding, and phosphorylation [2]. Therefore, the primary structure of the described cDNA clones is of considerable interest. Efforts are currently under way to obtain a full-length cDNA and, ultimately, to determine the genomic DNA structure of the β spectrin gene and its promoter. To this end a large cDNA clone has recently been isolated that extends from repeat 4 through the C-terminus, when aligned with the β spectrin peptide [16a].

The human β spectrin gene has been assigned to chromosome 14 by cDNA hybridization to human chromosomes separated by a fluorescence-activated cell sorter [14] and by hybridization to Southern blots of somatic hybrid cell DNA [16]. More recently, the gene has been sublocalized to bands 14q23–14q24.2 [16a]. The mouse β spectrin gene [16b] has been localized to chromosome 12 [16c], which, like human chromosome 14, also carries the gene for the immunoglobulin heavy-chain locus.

C. Nonerythroid Spectrin Genes

Occurrence of Nonerythroid Spectrins

Within a relatively short time, workers in several independent laboratories became aware of the existence of spectrinlike molecules in a variety of nonerythroid tissues [4,17–22]. These conclusions were based on observations that these molecules shared a number of structural and functional features similar to erythrocyte spectrin. Now that several of the nonerythroid spectrins have been characterized in some detail the following common features have been used to define this family of proteins: these proteins are composed of two large (>200 kD) nonidentical subunits that are isolated in forms that closely resemble the extended rodlike appearance of erythrocyte spectrin tetramers; all of these proteins exhibit some degree of immunochemical cross-reactivity with erythrocyte spectrin; and they share a number of common functional properties, which include the ability to bind to calmodulin and actin filaments.

Based both on structural and immunochemical criteria at least two α and several β spectrins have been identified in mammals. A number of different β spectrin subunits have been identified, but only one type of alpha chain has been detected in nonerythroid tissues [23,24]. The nonerythroid α subunit appears to be similar in size to the erythroid α chain, but these polypeptides are chemically distinct and exhibit only weak immunochemical cross-reactivity [22–25]. Avian spectrins, however, are composed of the same α chain in both erythroid and nonerythroid tissues [22,26]. β spectrin subunits exhibit marked variability in size and antigenicity [26,27].

Several tissues are known to produce more than one type of spectrin. These isoforms can be differentially expressed during development and segregated into distinct cytoskeletal domains. The β chains appear to endow tissue-specific functions responsible for this segregation. Thus, for example, the β subunit from a unique spectrin found in chicken intestinal epithelial cells associates with actin filament bundles in the terminal web cytoskeleton (TW 240/260). This spectrin lacks the ability to associate with the membrane-linking proteins ankyrin [28] and band 4.1 [29]. These cells also coexpress another spectrin, which is abundant in brain (240/235 kD; sometimes called fodrin) and localizes to the inner plasma membrane surface [25]. Similarly, neurons can express immunochemically distinct spectrins which segregate into different cytoskeletal compartments [24,30,31]. The fodrin β subunit exhibits altered affinities for erythrocyte ankyrin [32] and protein 4.1 [29] and it has been proposed that these modified interactions could allow for the assembly of different spectrin isoforms into distinct membrane skeletal domains.

In light of these observations an understanding of the genetic basis for these multiple isoforms has become an issue of considerable interest. Although the analysis of nonerythroid α spectrin genes has defined, to some extent, the relationship between erythroid and nonerythroid α chains, comparatively little genetic information is available to account for the diversity of β subunits.

cDNA Cloning

Nonerythroid α spectrin cDNA clones have been isolated from chicken [33,34], rat, and human sources [35,35a]. These clones were selected from expression libraries using antibodies reactive with the nonerythroid α chains. The deduced polypeptide sequences derived from these clones have allowed alignment of these genes with carboxyl-terminal portions of the α chain as illustrated in the schematic in Figure 4. All of these cDNAs gave deduced peptide sequences that were homologous to the internal repeat structure characteristic of both α and β

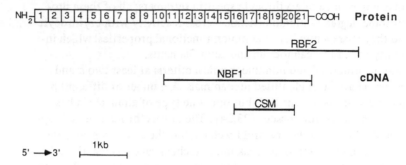

Figure 4 *Alignment of nonerythroid α-spectrin cDNA clones.* The α-spectrin protein is schematically represented according to the 106 amino acid homologous repeat model of Speicher and Marchesi [3], which also holds true in the nonerythroid α subunit [33–35]. The alignment is shown of the protein with rat brain (RBF2), human neuroblastoma (NBF1), and chicken smooth muscle (CSM) cDNAs [33–35], which encode proteins that are greater than 95% homologous across species. Based on comparison to homologous repeats of erythroid α spectrin the non-erythroid α subunit is thought to contain 21 homologous repeat segments.

erythrocyte spectrin. The chicken sequence contained one long open reading frame encoding three complete and two incomplete spectrin repeats, whereas the rat sequence contained a 3′ polyadenylated terminus preceded by ~1000 bp of untranslated sequence. Although most nonerythroid α spectrin repeats were like erythrocyte spectrin, in that they contained 106 amino acids with the invariant tryptophan and other hydrophobic and charged residues in the same conserved locations, the last two repeats showed less homology to other repeats [35]. This departure from the normal repetitive motif was suggested to represent alterations for a specializing function, such as actin binding, which has been localized to this terminal region.

Homology Among Species

The analysis of nonerythroid α spectrin sequences across species revealed a remarkable conservation of protein structure in comparison to the erythroid α spectrin. Thus, cDNA sequences aligned in Figure 4 from chicken and humans were 82% homologous, but the predicted amino acid sequences of the proteins are >95% homologous [35]. In contrast, analogous portions of erythroid α spectrin cDNAs from mouse and humans drifted to a comparable extent, with 82% nucleotide sequence homology, but the predicted amino acid sequences are only 83% homologous [10]. These findings, demonstrating an extraordinary degree of conservation among the nonerythroid α spectrins, suggested that considerable structural and functional constraints are imposed in the nonerythroid tissues, whereas a rapid divergence in α spectrin structure has occurred within the recent evolution of mammalian erythrocytes.

Evidence confirming that the mammalian erythroid and nonerythroid α spectrins have recently diverged from a common ancestral gene has been obtained by comparison of the peptide sequences deduced from these cDNAs [35]. This comparison showed 50–66% amino acid identity between analogous repeats in the two spectrins, whereas the homologies between other nonanalogous repeats showed significantly lower homology (<36%). The abrupt divergence between erythroid and nonerythroid α spectrins is thought to have occurred since the evolution of birds, because these species express a common α spectrin in both erythroid and nonerythroid tissues that appears to arise from a single gene [36]. The specialization of erythrocytes in mammals has resulted in the loss of the nucleus and cytoskeletal elements including microtubules and intermediate filaments. These events have perhaps affected the functional requirements imposed on other remaining cytoskeletal proteins, including spectrin.

The chromosomal location of the human nonerythroid α spectrin gene was determined by hybridization to Southern blots containing DNA from Chinese hamster X human somatic hybrid cell lines using both the rat and human cDNAs as probes [35]. Characteristic human restriction fragments were detected only in cell lines that had retained copies of human chromosome 9, whereas all other human chromosomes were ruled out by at least three discordant hybrids. These results were confirmed and further refined by in situ hybridization to human metaphase chromosomes, where significant signals were only observed on the distal band of the long arm of chromosome 9. Taken together, these results are consistent with a single gene for the nonerythroid α spectrin, designated NESA1, on chromosome 9, bands q33–q34 [35].

Nonerythroid β Spectrins

The diversity of nonerythroid β spectrins has for some time been considered the product of multiple genes both in mammals and birds, based both on high resolution peptide mapping and immunochemical studies [26,30–32]. These predictions were studies in chickens through analysis of β spectrin transcripts using an erythroid β spectrin cDNA as a probe [15]. This clone was selected from an erythroid cDNA expression library and its identity confirmed by in vitro translation of hybrid-selected mRNA. Hybridization to RNA blots using this cDNA has detected different sized transcripts in at least some chicken tissues that express polypeptides immunoreactive with anti-β spectrin antibody. In particular, the transcript detected in myotubes was larger than the erythroid mRNA, although the differences between these gene products, designated β and β', has not been defined. Whether these transcripts are the product of the same or different genes is unknown. In other tissues (fibroblasts and intestinal epithelia), which express β spectrin variants that are less closely related, mRNAs were not detected by hybridization using the erythroid β spectrin probe, suggesting that other less homologous β spectrin genes are expressed in these tissues.

Recently, human skeletal muscle β spectrin cDNA clones have been isolated [36a]. Analysis has revelaed these to be the product of the single-copy "erythroid" β spectrin gene. A structural difference between erythroid and muscle isoforms appears pre-mRNA. Another recent report describes the isolation of a human genomic clone encoding a homologous, unique nonerythroid β spectrin [16a].

III. ABNORMAL SPECTRIN GENES

A. General Considerations

One would expect abnormalities of spectrin structure, function, or quantity to adversely affect the erythrocyte. In fact, several human hereditary hemolytic anemias manifest such alterations [2,5,6,37,38]. In no instance has the ultimate genetic lesion been pinpointed. The elucidation of these abnormalities at the gene level is certain to enhance understanding of the structure–function relationships within the spectrin molecule and the overall assembly of the membrane skeleton. It will also allow improved detection of the responsible genes in humans.

The newly isolated molecular clones described in the preceding section have allowed initial clinical studies. The methodology employed thus far is limited to Southern blotting, whereby patient genomic DNA is digested with restriction endonucleases, separated by agarose gel electrophoresis, transferred to nitrocellulose or nylon matrix and hybridized to one or more molecular probes [39]. This technique easily detects gross gene defects (e.g., large deletions/insertions, gene rearrangements) within the region probed. Minor structural changes (e.g., point mutations, small deletions/insertions) may go undetected, unless they happen to alter or create a site for a restriction enzyme used in the experiment. When a single restriction enzyme site is altered in a patient, it may represent a random variation in DNA sequence called restriction fragment length polymorphisms (RFLP) that is unrelated to the genetic defect, or it may be due to the actual mutation causing the defect. Ultimately the analysis of these hereditary hemolytic anemias will employ molecular cloning, nucleotide sequencing and in vitro expression of patients' spectrin genes.

B. Abnormal α Spectrin Genes

α Spectrin Gene RFLPS

A fortuitous development in the clinical study of α spectrin genes has been the discovery of frequently occurring RFLPs for three restriction enzyme sites in normal human DNA [40]. When a portion of the cloned chromosomal α spectrin gene (probe 3021 E1) (Figure 5a) corresponding to the αI spectrin domain is used as a hybridization probe in Southern blots of genomic DNA digested with

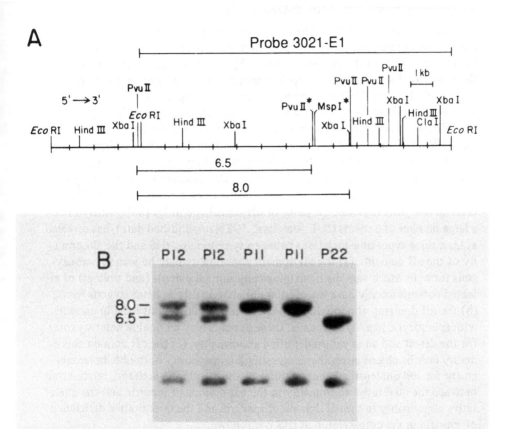

Figure 5 *The α-spectrin gene Pvu II restriction fragment length polymorphism (RFLP).* (A) The probe used to detect RFLPs [40], designated 3021 E1, is a 13 kb *Eco*R I subclone of λ3021 (restriction map shown in Fig. 1) [7]. Asterisks indicate polymorphic restriction sites. (B) The inheritance of the *Pvu* II RFLP in a family. Genomic DNA is analyzed by Southern blotting [39] and hybridization to probe 3021 Eco1. The parents (lanes 1–2) are heterozygotes (P1, 2) for the polymorphism. Their offspring (lanes 3–5) are either homozygous for one form (P1, 1) or the other (P2, 2) of the RFLP.

Pvu II, Xba I, or Msp I, polymorphisms are detected with high frequency in normal individuals. Figure 5b provides an example of the inheritance of the Pvu II polymorphism in one family. The parents are both heterozygotes (P1, 2); two offspring are homozygous for one form (P1, 1); and one is homozygous for the other (P2, 2). In this manner these polymorphisms can be followed in families with inherited hemolytic anemias. Presence of absence of coinheritance of the polymorphism with the disorder can support or refute linkage of the disease to the α-spectrin gene.

Autosomal Recessive Hereditary Spherocytosis

Autosomal recessive hereditary spherocytosis (rHS) has been associated with a marked quantitative deficiency of spectrin in mice [41] and humans [42]. Experimental studies of the murine models [43] have uncovered defects in: (a) the α spectrin locus, leading to impaired spectrin incorporation into the cytoskeleton (sph^{ha}/sph^{ha}); (b) the α spectrin locus, leading to undetectable levels of α spectrin mRNA by in vitro translation (sph/sph); (c) the synthesis of β spectrin (ja/ja); and (d) the accumulation of ankyrin (nb/nb). By inference, human rHS, a rare disorder, might similarly result from diverse mechanisms.

It has recently been demonstrated that many rHS patients have a unique abnormality of the α spectrin protein [44; see also Chapter 4]. Partial tryptic cleavage of spectrin, followed by two-dimensional gell electrophoresis [45], reveals an acid shift in isoelectric point of αII domain peptides [44]. Analysis of a large number of patients (S. L. Marchesi, 1987, unpublished data), has revealed at least three types of associations between symptomatic rHS and the abnorma - ity of the αII domain: (a) the αII domain abnormality may be seen in homozygous form, in which case the hematologically normal parents (and siblings) of affected patients usually have the αII domain abnormality in heterozygous form; (b) the αII domain: abnormality may be seen in heterozygous form in patients with spherocytic hemolytic anemia; these patients may be double heterozygotes for the defect and an as yet unidentified abnormality; (c) the αII domain abnormality may be absent altogether, suggesting homozygosity or double heterozygosity for still unidentified abnormalities of the membrane skeleton, relationship between the qualitative abnormality in the αII domain of spectrin and the qualitative abnormality in the αII domain of spectrin and the quantitative deficiency of spectrin in the cytoskeleton in rHS is unknown.

To determine whether rHS in patients with no identifiable abnormality in the α subunit of spectrin is also associated with the α spectrin gene, α spectrin RFLP inheritance was studied in two such families. One kindred was uniformative (all individuals were heterozygous for the RFLPs) but the other kindred demonstrated an inheritance pattern consistent with tracking of the disorder with the α spectrin gene. However, much larger numbers of patients would be necessary to establish linkage. It is also noteworthy that restriction endonuclease digestion of DNA from rHS patients and hybridization to either α or β spectrin probes revealed no gross gene deletions or rearrangements.

Autosomal Dominant Hereditary Spherocytosis

The primary defect of the membrane skeleton in autosomal dominant hereditary spherocytosis (dHS), the most common inherited hemolytic anemia in Caucasians, is largely unknown. A mild quantitative deficiency of spectrin has been demonstrated in the majority of patients [38,46]. Rare patients demonstrate abnormal spectrin–protein 4.1 interactions [47,48]. To investigate the relation-

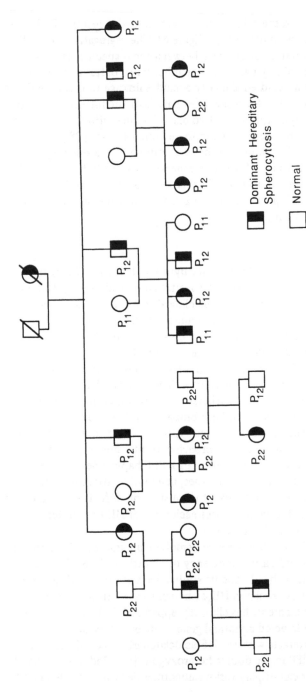

Figure 6 *Inheritance of the Pvu II α-spectrin gene RFLP in a caucasian kindred with autosomal dominant hereditary spherocytosis (dHS).* Peripheral blood leukocyte DNA was isolated, digested with *Pvu* II, blotted [39] and hybridized to probe 3021 Eco1 [49]. Heterozygotes (P1, 2), homozygotes (P1, 1) and homozygotes (P2, 2) are as indicated. Hereditary spherocytosis (diagnosed by osmotic fragility) is as shown. Segregation of the α-spectrin genes and dHS in this family is inconsistent with the disorder being linked to this locus.

ship between dHS and the α spectrin gene, we have studied α spectrin RFLPs in
a large Caucasian kindred with dHS [49] (Figure 6). The segregation of the *Pvu*
II polymorphism and dHS rules out linkage between the α spectrin gene and
dHS in this kindred. Preliminary studies of the inheritance pattern of a protein
4.1 gene RFLP in the same kindred similarly excludes linkage to this gene (Tang
Tang, 1987, unpublished data). Retrospective analysis of the inheritance pattern
of αII domain protein polymorphisms in two black families with dHS also rules
out linkage of dHS to the α spectrin gene in these families. Agre [38] specu-
lates that, because β spectrin subunit synthesis is limiting in the assembly of
avian erythrocyte spectrin dimers [50], this gene would seem to be a more likely
candidate to be involved in human dHS. This hypothesis remains to be tested directly. A newly described β spectrin gene, RFLP, will allow linkage analysis to
be performed in families with dHS [16a].

Other Disorders

Hereditary elliptocytosis (HE) and hereditary pyropoikilocytosis (HPP) are
additional genetic disorders with links to α spectrin (reviewed in [6]). Hereditary elliptocytosis is a disorder with variable clinical severity that is associated
with a number of different defects of membrane skeleton proteins. A subset
of affected individuals exhibit structurally abnormal spectrin. At least three different types of abnormal proteolytic cleavage of the 80 kD αI tryptic domain
are seen in such patients. Recent investigations have demonstrated an amino
acid substitution to proline in two forms and an insertion of a leucine in a third
[51]. These molecular alterations are sufficient to cause impairment of spectrin
self-association and elliptocytosis. Southern blot analysis has been performed on
DNA from a number of such patients and, as expected, failed to detect any gross
defect of α spectrin gene structure. An approach currently being applied to the
analysis of such patients is the determination of RFLP haplotypes that are
associated with different peptide abnormalities, in an initial attempt to determine the degree of genetic heterogeneity of the α spectrin defects prior to direct
studies of gene structure. One would expect each form of HE to manifest a
subtle genetic mutation at the DNA level.

Hereditary pyropoikilocytosis is a severe hemolytic anemia with in vitro red
blood cell thermolability and partial deficiency of spectrin [52,53]. This disorder is related to HE: virtually all patients studied have abnormal αI domain
tryptic cleavages of spectrin as seen in HE, although to a relatively greater extent. Typically, one of the parents has clinically apparent HE. Investigators have
hypothesized than an undetected abnormal gene contributed by one (usually
asymptomatic) parent interacts with the HE-associated αI defect of the other
parent, thus leading to HPP in the double heterozygote [6]. Red blood cells of
the asymptomatic carrier can display various abnormalities when studied in vitro

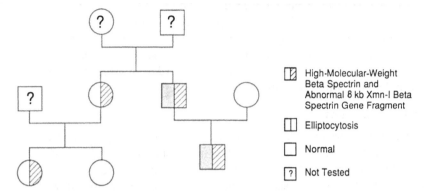

Figure 7 *Inheritance of high molecular weight β spectrin, the 8 kb Xmn I β-spectrin gene fragment and elliptocytosis in a family.* Peripheral blood leukocyte DNA was isolated, digested with *Xmn* I, blotted [39] and hybridized to β-spectrin cDNA. A unique 8 kb *Xmn* I restriction fragment is coinherited with the large β-spectrin variant. The elliptocytosis in this family is thought to be due to an α-spectrin structural defect. (Johnson et al., personal communication.)

[52]. To date, no direct molecular analysis of spectrin genes of HPP patients has been reported.

C. Abnormal β Spectrin Genes

High Molecular Weight β Spectrin Variant

Johnson and Ravindranath have found a large (260 kD) β spectrin that is inherited in the heterozygous state in a family with HE [54]. Subsequent analysis has shown that patients with this variant have no clinical abnormality; HE in this family is thought to be due, rather, to an αI domain lesion of spectrin (R. M. Johnson et al., personal communication). It is challenging to envision the genetic mechanism that has given rise to this very large β spectrin subunit that apparently functions normally in the cytoskeleton. We probed Southern blots of DNA from members of this family with β spectrin cDNA. Digestion with *Xmn* I gave an 8 kb β spectrin gene fragment in patients with the 260 kD β spectrin that was not seen in DNA of unaffected individuals [54a]. Digestion with other enzymes also revealed abnormal fragments in DNA of affected individuals. The mapping data were consistent with the presence of a deletion or rearrangement at the 3' end of the β spectrin gene [54a]. A family study (Figure 7) confirmed the close association of the high molecular weight (HMW) β variant and the 8 kb *Xmn* I fragment. Every individual with one also has the other, whereas individuals lacking the abnormal *Xmn* I fragment also lack the HMW β variant. Several possible genetic mechanisms could result in this large β variant. One attractive hypothesis is that the mutation event disrupts a normal alternative splicing mechanism, thereby allowing inclusion, in the erythroid β spectrin mRNA, of one or more exon-encoding sequences that may normally be expressed only in certain non-erythroid β spectrins. Several precedents exist for such tissue-specific alternative splicing schemes (reviewed in [55]). A newly discovered example is found in erythroid and nonerythroid isoforms of protein 4.1 [56].

Hereditary Elliptocytosis

Several cases of HE have been reported that exhibit truncated β spectrin chains that lack C-terminal phosphorylation sites and manifest impaired spectrin self-association [57,58,58a]. DNA from patients with two forms of truncated β spectrin (provided by Profs. J. Delaunay and P. Boivin) have been probed with C-terminal β spectrin cDNA after Southern blotting. Their β spectrin genes have no discernable abnormality by this method (unpublished results). Therefore, the gene defect is subtle, perhaps a nonsense mutation causing premature chain termination.

IV. CONCLUSION

A. General Considerations

The properties of spectrin proteins and the molecular clone thus far isolated allow several predictions regarding the size, number, expression, and evolution of the spectrin genes. First, like their protein products, the spectrin genes are very large. The coding sequence alone within each spectrin gene must be in excess of 6 kb to account for the known size of each subunit. The size of mouse α spectrin mRNA, detected by cDNA hybridization to blots of gel-fractionated RNA, is ~8 kb [9]. Human β spectrin and nonerythroid α spectrin mRNAs are 7.5 and 8 kb, respectively [16,35]. Therefore, the chromosomal spectrin genes may each be 80 kb or more in length, depending on the length of intron DNA sequences. Their actual total size is, as yet, unknown.

Second, the number of human spectrin genes, based on protein analysis, must be at least four, depending on the number of nonerythroid β subunit genes that exist to encode the various tissue-specific isoforms. It is possible that the spectrin gene family is considerably larger than this minimum figure. As was discussed previously, molecular clones have been isolated for human erythrocyte α [7,8] and β [13,14,16] spectrin. The human nonerythroid α spectrin has also been cloned [35]. Recent reports describe the molecular cloning of human nonerythroid β spectrins [16a,36a]. The human spectrin genes studied thus far are each unique and dispersed throughout the genome: erythroid α and β genes are on chromosomes 1 [12] and 14 [14,16], respectively; the nonerythroid α gene is on chromosome 9 [35].

The regulation of spectrin gene expression in erythroid maturation has yet to be directly investigated. However, spectrin protein biosynthesis, turnover,

subunit assembly, and cytoskeletal binding have been intensively studied in nucleated red blood cells of the chicken [49,59–61]. Early reports of similar
studies in mammalian model systems are also beginning to emerge [62–65]. In
the chicken, both α and β spectrin synthesis (as well as ankyrin and protein 4.1
synthesis) is ongoing at the stage of the most immature erythroblast. Unassembled α subunits accumulate in the cytoplasm in a threefold excess over β
subunits, primarily as a result of differential rates of turnover. Woods and
Lazarides hypothesize that the equilibrium between β spectrin degradation and
assembly into dimers is a limiting, regulatory event in the cascade of cytoskeleton biogenesis [50]. Binding of spectrin dimers to the membrane skeleton depends on the presence of band 3, the integral membrane anion channel protein
which, via ankyrin, provides an attachment site for spectrin. Band 3, like globin,
accumulates as the red blood cell terminally differentiates, thus enabling the
cytoskeleton to assemble. In mammalian erythroid maturation one might predict that early red blood cell precursors, like other nucleated cells, synthesize
nonerythroid α subunits and that a switch occurs to the erythroid form with
progressive differentiation. The observation that erythroid and nonerythroid
spectrins are coproduced in Friend erythroleukemia cells [66] supports this
prediction. Red blood cell spectrin gene expression is, therefore, likely to be
an early epigenetic marker of erythroid differentiation. Informative differences
might exist between the developmental regulation of spectrin and globin gene
expression that could cast light on early events in erythropoiesis. The tissue-
specific regulation of β spectrin gene expression remains unexplored, awaiting
molecular cloning of unique nonerythroid β spectrin genes.

Finally, the existence of homologous repeating segments within the spectrin
genes raises interesting questions concerning the evolution of this gene family.
Did these genes evolve from multiple duplications and fusions of a small, pri-
mordial gene with its own functional attributes? Did the various spectrin genes
arise by divergence from a single, common ancestor? Do instances exist in
which a single gene encodes more than one spectrin protein, accomplished by
alternative mRNA splicing, or differential use of transcription initiation sites?
These queries remain unanswered, but molecular tools are now in hand to allow
their direct study.

B. Molecular Clones Derived from Spectrin Genes

As described in detail above, cDNA encoding the full length of human erythrocyte α spectrin has been isolated. A molecular clone for a portion of the
chromosomal gene and the gene promoter has also been described. The human
β spectrin cDNA thus far isolated encodes ~80% of the C-terminal end of the
erythroid β subunit.

cDNA clones encoding portions of nonerythroid α spectrin have been obtained from three species. Comparison reveals strict sequence conservation of

nonerythroid α spectrin genes and proteins between these species. The somewhat more distant homology between human nonerythroid α spectrin and erythrocyte α spectrin suggests that the former gave rise to the latter and that an accelerated divergence subsequently occurred in the erythroid α spectrin gene. The finding of only a single ("nonerythroid") α spectrin in birds also supports this theory and is consistent with the evolution of the erythroid spectrin gene from the nonerythroid gene after the origin of mammals.

C. Early Clinical Studies Using Spectrin Gene Probes

The red cell α spectrin gene has, within the region encoding for the N-terminal αI tryptic domain, frequently occurring RFLPs for three restriction enzymes. These have been useful in excluding a close association between dominant HS and the α spectrin gene. Restriction fragment length polymorphism inheritance fails to exclude a relationship between α spectrin and recessive HS in a subset of families in which there is deficient but apparently structurally normal α spectrin protein.

In several inherited hemolytic anemias that were studied directly with spectrin gene probes, no gross gene deletions or rearrangements have been detected. However, a unique 8 kb β spectrin gene *Xmn* I restriction fragments has been found in association with a HMW variant of β spectrin.

D. Prospects for the Future

The available molecular probes should make it possible to dissect the complete structure of the spectrin genes and analyze their expression. They should facilitate the molecular cloning of a variety of abnormal spectrin genes and the pinpointing of their molecular lesions. In the process, a great deal more fundamental information about the role of spectrin in the cytoskeleton, and the tissue-specific regulation of its synthesis, may be learned.

ACKNOWLEDGMENTS

We thank D. Dillman and K. Haggerty for excellent secretarial assistance in the preparation of this manuscript. The authors' work described in this manuscript was supported in part by grants from the National Institutes of Health.

REFERENCES

1. Cohen, C. M. (1983). The molecular organization of the red cell membrane skeleton. *Semin. Hematol.* 20:141–158.
2. Knowles, W., Marchesi, S. L., and Marchesi, V. T. (1983). Spectrin: structure, function, and abnormalities. *Semin. Hematol.* 20:159–174.
3. Speicher, D. W., and Marchesi, V. T. (1984). Erythrocyte spectrin is comprised of many homologous triple helical segments. *Nature* 311:177–180.

4. Goodman, S. R., Zagon, I. S., and Kulikowski, R. R. (1981). Identification of a spectrin like protein in nonerythroid cells. *Proc. Natl. Acad. Sci. USA* 78:7570–7574.
5. Marchesi, S. L., Knowles, W. J., Morrow, J. S., Bologna, M., and Marchesi, V. T. (1986). Abnormal spectrin in hereditary elliptocytosis. *Blood* 67:141–151.
6. Palek, J. (1985). Hereditary elliptocytosis and related disorders. *Clin. Haematol.* 14:45–87.
7. Linnenbach, A. J., Speicher, D. W., Marchesi, V. T., and Forget, B. G. (1986). Cloning of a portion of the chromosomal gene for human erythrocyte α spectrin by using a synthetic gene fragment. *Proc. Natl. Acad. Sci. USA* 83:2397–2401.
8. Curtis, P. J., Palumbo, A., Ming, J., Fraser, P., Cioe, L., Pacifico, P., Shane, S., and Rovera, G. (1985). Sequence comparison of human and murine erythrocyte alpha-spectrin cDNA. *Gene* 36:357–362.
9. Cioe, L., and Curtis, P. (1985). Detection and characterization of a mouse α spectrin cDNA clone by its expression in *Escherichia coli. Proc. Natl. Acad. Sci. USA* 82:1367–1371.
10. Helfman, D. M., Feramisco, J. R., Fiddes, J. C., Thomas, G. F., and Hughes, S. H. (1983). Identification of clones that encode chicken tropomyosin by direct immunological screening of a cDNA expression library. *Proc. Natl. Acad. Sci. USA* 80:31–35.
11. Young, R. A., and Davis, R. W. (1983). Efficient isolation of genes by using antibody probes. *Proc. Natl. Acad. Sci. USA* 80:1194–1198.
12. Huebner, K., Palumbo, A. P., Isobe, M., Kozak, C. A., Monaco, S., Rovera, G., Croce, C. M., and Curtis, P. J. (1985). The α-spectrin gene is on chromosome 1 in mouse and man. *Proc. Natl. Acad. Sci. USA* 82:3790–3793.
13. Winkelmann, J. C., Leto, T. L., Linnenbach, A. J., Speicher, D. W., Marchesi, V. T., and Forget, B. G. (1985). Molecular cloning of human erythrocyte beta spectrin cDNA. *Blood* 66:41a.
14. Prchal, J. T., Morley, B. J., Yoon, S.-H., Coetzer, T. L., Palek, J., Conboy, J. G., and Kan, Y. W. (1987). Isolation and characterization of cDNA clones for human erythrocyte β spectrin. *Proc. Natl. Acad. Sci. USA* 84:7468–7472.
15. Moon, R. T., Ngai, J., Wold, B. J., and Lazarides, E. (1985). Tissue-specific expression of distinct spectrin and ankyrin transcripts in erythroid and nonerythroid cells. *J. Cell. Biol.* 100:152–160.
16. Winkelmann, J. C., Leto, T. L., Watkins, P. C., Eddy, R., Shows, T. B., Linnenbach, A. J., Sahr, K. E., Kathuria, N., Marchesi, V. T., and Forget, B. G. (1988). Molecular cloning of the cDNA for human erythrocyte beta spectrin *Blood* 72:328–334.
16a. Forget, B. G., Chang, J. G., Coupal, E., Fukushima, Y., Stanislovits, P., Costa, F., Byers, M., Winkelmann, J., Agre, P., Marchesi, V. T., Shows, T. B., and Watkins, P. (1988). Molecular genetics of the human β-spectrin gene. *Clin. Res.* 36:612A.

16b. Cioe, L., Laurila, P., Meo, P., Krebs, K., Goodman, S., and Curtis, P. J. (1987). Cloning and nucleotide sequence of a mouse erythrocyte β-spectrin cDNA. *Blood* 70:915–920.

16c. Laurila, P., Cioe, L., Kozak, C. A., and Curtis, P. J. (1987). Assignment of mouse beta-spectrin gene to chromosome 12. *Somatic Cell Mol. Genet.* 13:93–98.

17. Levin, J., and Willard, M. (1983). Redistribution of fodrin (a component of the cortical cytoplasm) accompanying capping of cell surface molecules. *Proc. Natl. Acad. Sci. USA* 80:191–195.

18. Glenney, J. R., Jr., Glenney, P., Osborne, M., and Weber, K. (1982). An f-actin and calmodulin binding protein from isolated intestinal brush borders has a morphology related to spectrin. *Cell* 28:843–854.

19. Bennett, V., Davis, J., and Fowler, W. S. (1982). Brain spectrin, a membrane associated protein related in structure and function to erythrocyte spectrin. *Nature* 299:126–131.

20. Glenney, J. R., Glenney, P., and Weber, K. (1982). F-actin binding and cross-linking properties of porcine brain fodrin, a spectrin-related molecule. *J. Biol. Chem.* 95:478–486.

21. Burridge, K., Kelly, T., and Mangeat, P. (1982). Nonerythrocyte spectrins: actin-membrane attachment proteins occurring in many cell types. *J. Biol. Chem.* 95:478–486.

22. Repasky, S. A., Granges, B. L., and Lazarides, E. (1982). Widespread occurrence of avian spectrin in nonerythroid cells. *Cell* 29:821–833.

23. Glenney, J. R., and Glenney, P. (1983). Fodrin is the general spectrin-like protein found in most cells whereas spectrin and the TW protein have a restricted distribution. *Cell* 34:503–512.

24. Glenney, J. R., and Glenney, P. (1984). Comparison of spectrin isolated from erythroid and nonerythroid sources. *Eur. J. Biochem.* 144:529–539.

25. Harris, A. S., Greene, L. A. D., Ainger, K. J., and Morrow, J. S. (1985). Mechanism of cytoskeletal regulation (I): Functional differences correlate with antigenic dissimilarity in human brain and erythrocyte spectrin. *Biochim. Biophys. Acta* 830:147–158.

26. Nelson, W. J., and Lazarides, E. (1983). Switching of subunit composition of muscle spectrin during myogenesis in vitro. *Nature* 304:364–368.

27. Riederer, B. M., Zagon, I. S., and Goodman, S. R. (1986). Brain spectrin (240/235) and brain spectrin (240/235E): Two distinct spectrin subtypes with different locations within mammalian neural cells. *J. Cell. Biol.* 102:2088–2097.

28. Howe, C. L., Sacramore, L. M., Mooseker, M. S., and Morrow, J. W. (1986). Mechanisms of cytoskeletal regulation: Modulation of membrane affinity in avian brush border and erythrocyte spectrins. *J. Cell Biol.* 101:1379–1385.

29. Coleman, T. R., Harris, A. S., Mische, S. M., Mooseker, M. S., and Morrow, J. S. (1987). Beta spectrin bestows protein 4.1 sensitivity on spectrin-actin interactions. *J. Cell Biol.* 104:519–526.

30. Lazarides, E., Nelson, W. J., and Kasamatsu, T. (1984). Segregation of two spectrin forms in the chicken optic system: a mechanism for establishing restricted membrane-cytoskeletal domains in neurons. *Cell* 36:269–278.

31. Siman, R., Adhoot, M., and Lynch, G. (1987). Ontogeny, compartmentation, and turnover of spectrin isoforms in rat central neurons. *J. Neurosci.* 7:55–64.

32. Harris, A. S., Anderson, J. P., Greene, L. A. D., Ainger, K. J., and Morrow, J. S. (1986). Mechanisms of cytoskeletal regulation: functional and antigenic diversity in human erythrocyte and brain beta spectrin. *J. Cell Biochem.* 30:51–69.

33. Birkenmeier, C. S., Bodine, D. M., Repasky, E. A., Helfman, D. M., Hughes, S. H., and Barker, J. E. (1985). Remarkable homology among the internal repeats of erythroid and nonerythroid spectrin. *Proc. Natl. Acad. Sci. USA* 82:5671–5675.

34. Wasernius, V. M., Saraste, M., Knowles, J., Virtanen, I., and Lehto, V. P. (1985). Sequencing of the chicken nonerythroid spectrin cDNA reveals an internal repetitive structure homologous to the human erythrocyte spectrin. *EMBO J.* 4:1325–1430.

35. Leto, T. L., Fortugno-Erikson, D., Barton, B. E., Yang-Feng, T. L., Francke, U., Morrow, J. S., Marchesi, V. T., and Benz, E. J., Jr. (1988). Comparison of nonerythroid alpha spectrin genes reveals strict homology among diverse species. *Mol. Cell Biol.* 8:1–9.

35a. McMahon, A. P., Gievelhaus, D. H., Champion, J. E., Bailes, J. A., Lacey, S., Carritt, B., Henchman, S. K., and Moon, R. T. (1987). cDNA cloning, sequencing and chromosome mapping of a non-erythroid spectrin, human α fodrin. *Differentiation* 34:68–78.

36. Birkenmeier, C. S., Bodine, D. M., and Barker, J. E. (1986). Evolution of the spectrin gene family: Conservation of internal repeats among spectrin isoforms. In *Membrane Skeletons and Cytoskeleton–Membrane Associations*, V. Bennett, C. M. Cohen, S. E. Lux, and J. Palek (Eds.), pp. 187–194.

36a. Winkelmann, J. C., and Forget, B. G. (1988). Isolation of human skeletal muscle clones which hybridize to human erythrocyte β spectrin cDNA. *Clin. Res.* 36:422A.

37. Becker, P. S., and Lux, S. E. (1985). Hereditary spherocytosis and related disorders. *Clin. Haematol.* 14:15–43.

38. Agre, P., Asimos, A., Casella, J. F., and McMillan, C. (1986). Inheritance pattern and clinical response to splenectomy as a reflection of erythrocyte spectrin deficiency in hereditary spherocytosis. *N. Engl. J. Med.* 35:1579–1583.

39. Southern, E. J. (1975). Detection of specific sequences among DNA fragments separated by gel electrophoresis. *J. Mol. Biol.* 98:503–517.

40. Hoffman, N., Stanislovitis, P., Watkins, P. C., Klinger, K. W., Linnenbach, A. J., and Forget, B. G. (1987). Three RFLPs are detected by an alpha spectrin genomic clone. *Nucl. Acids Res.* 15:4696.

41. Greenquist, A., Shohet, S. B., and Bernstein, S. E. (1978). A marked reduction of spectrin in hereditary spherocytosis in the common house mouse. *Blood* 51:1149–1155.
42. Agre, P., Orringer, E. P., and Bennett, V. (1982). Deficient red-cell spectrin in severe, recessively inherited spherocytosis. *N. Engl. J. Med.* 306:1155–1161.
43. Bodine, D. M., Birkenmeier, C. S., and Barker, J. E. (1984). Spectrin deficient inherited hemolytic anemias in the mouse: characterization by spectrin synthesis and mRNA activity in reticulocytes. *Cell* 37:721–729.
44. Winkelmann, J. C., Marchesi, S. L., Watkins, P., Linnenbach, A. J., Agre, P., and Forget, B. G. (1986). Recessive hereditary spherocytosis is associated with an abnormal alpha spectrin subunit. *Clin. Res.* 34:474A.
45. Speicher, D. W., Knowles, W. J., and Marchesi, V. T. (1980). Identification of proteolytically resistant domains of human erythrocyte spectrin. *Proc. Natl. Acad. Sci. USA* 77:5673–5677.
46. Agre, P., Casella, J. F., Zinkham, W. H., McMillan, C., and Bennett, V. (1985). Partial deficiency of erythrocyte spectrin in hereditary spherocytosis. *Nature* 314:380–383.
47. Goodman, S. R., Shiffer, K. A., Caseria, L. A., and Eyster, M. E. (1982). Identification of the molecular defects in the erythrocyte membrane skeleton of some kindreds with hereditary spherocytosis. *Blood* 60:772–784.
48. Wolfe, L. C., John, K. M., Falcone, J. C., Byrne, A. M., and Lux, S. E. (1982). A genetic defect in the binding of protein 4.1 to spectrin in a kindred with hereditary spherocytosis. *N. Engl. J. Med.* 307:1367–1374.
49. Winkelmann, J. C., Marchesi, S. L., Gillespie, F. P., Agre, P., and Forget, B. G. (1986). Dominant hereditary spherocytosis is not closely linked to the gene for alpha spectrin. *Blood* 68:50a.
50. Woods, C. M., and Lazarides, E. (1985). Degradation of unassembled α- and β-spectrin by distinct intracellular pathways: regulation of spectrin topogenesis by β-spectrin degradation. *Cell* 40:959–969.
51. Marchesi, S. L., Letsinger, J. T., Speicher, D. W., Marchesi, V. T., Agre, P., Hyun, B., and Gulati, G. (1987). Mutant forms of spectrin α-subunits in hereditary elliptocytosis. *J. Clin. Invest.* 80:191–198.
52. Mentzer, W. C., Turetsky, T., Mohandas, N., Schrier, S., Wu, C.-S. C., and Koenig, H. (1984). Identification of the hereditary pyropoikilocytosis carrier state. *Blood* 63:1439–1446.
53. Coetzer, T. L., and Palek, J. (1986). Partial spectrin deficiency in hereditary pyropoikilocytosis. *Blood* 67:919–924.
54. Johnson, R. M., and Ravindranath, Y. (1985). A new variant of spectrin with a large beta-spectrin chain associated with hereditary elliptocytosis. *Blood* 66:33a.
54a. Winkelmann, J. C., Ravindranath, Y., Johnson, R. M., and Forget, B. G. (1988). High molecular weight β-spectrin variant may result from a two kilobase 3′ β-spectrin gene deletion. *Clin. Res.* 36:423A.

55. Leff, S. E., Rosenfeld, M. G., and Evans, R. M. (1986). Complex transcriptional units: Diversity in gene expression by alternative RNA processing. *Annu. Rev. Biochem.* 55:1091–1117.

56. Tang, T., Leto, T. L., Correas, I., Alonso, M. A., Marchesi, V. T., and Benz, E. J. (1986). Identification by molecular cloning of a lymphoid isoform of protein 4.1 with a deletion in the spectrin-actin binding domain. *Blood* 68:41a.

57. Dhermy, D., LeComte, M. C., Garbarz, M., Bournier, O., Galand, C., Gautero, H., Feo, C., Alloisio, N., Delaunay, J., and Boivin, P. (1982). Spectrin β-chain variant associated with hereditary elliptocytosis. *J. Clin. Invest.* 70: 707–715.

58. Ohanian, V., Evans, J. P., and Gratzer, W. B. (1985). A case of elliptocytosis associated with a truncated spectrin chain. *Br. J. Haematol.* 61: 31–39.

58a. Pothier, B., Morle, L., Alloisio, N., Ducluzeau, M. T., Caldani, C., Feo, C., Garbarz, M., Chaveroche, I., Dhermy, D., Lecomte, M. C., Boivin, P., and Delaunay, J. (1987). Spectrin Nice ($\beta^{220/216}$): a shortened β-chain variant associated with an increase of the $\alpha^{I/74}$ fragment in a case of elliptocytosis. *Blood* 69:1759–1765.

59. Moon, R. T., and Lazarides, E. (1984). Biogenesis of the avian erythroid membrane skeleton: receptor-mediated assembly and stabilization of ankyrin and spectrin. *J. Cell Biol.* 98:1899–1904.

60. Lazarides, E., and Moon, R. T. (1984). Assembly and topogenesis of the spectrin-based membrane skeleton in erythroid development. *Cell* 37: 354–356.

61. Woods, C. M., and Lazarides, E. (1986). Spectrin assembly in avian erythroid development is determined by competing reactions of subunit homo- and hetero-oligomerization. *Nature* 321:85–89.

62. Hanspal, M., and Palek, J. (1985). Synthesis and assembly of spectrin in mammalian erythroid cells: α spectrin is synthesized in excess of β spectrin. *Blood* 66:35a.

63. Hanspal, M., Dainiak, N., and Palek, J. (1986). Synthesis and assembly of spectrin in human erythroblasts. *Blood* 68:35a.

64. Koury, M. J., Bondurant, M. C., and Mueller, T. J. (1986). The role of erythropoietin in the production of principle erythrocyte proteins other than hemoglobin during terminal erythroid differentiation. *J. Cell Physiol.* 126:259–265.

65. Pfeffer, S. R., Huima, T., and Redman, C. M. (1986). Biosynthesis of spectrin and its assembly into the cytoskeletal system of Friend erythroleukemic cells. *J. Cell Biol.* 103:103–113.

66. Glenney, J., and Glenney, P. (1984). Co-expression of spectrin and fodrin in Friend erythroleukemic cells treated with DMSO. *Exp. Cell. Res.* 152: 15–21.

6

Na,K-ATPase Structure

ROBERT W. MERCER   *Washington University School of Medicine, St. Louis, Missouri*

JAY W. SCHNEIDER and EDWARD J. BENZ, Jr.   *Yale University School of Medicine, New Haven, Connecticut*

I. INTRODUCTION

The Na,K-adenosine triphosphatase (Na,K-ATPase), also known as the sodium pump or sodium-potassium pump, is a membrane-associated enzyme responsible for maintaining the high internal K concentration and low internal Na concentration characteristic of most animal cells. It couples the hydrolysis of ATP to the transport of Na and K across the plasma membrane against their respective electrochemical gradients. For each ATP hydrolyzed the Na,K-ATPase normally expels three Na ions and takes in two K ions. As outlined in Figure 1, the Na,K-ATPase is fundamental to several diverse cellular functions. These functions include: (a) regulation of cellular volume; (b) maintenance of the high internal K concentration required for several intracellular enzymes; (c) generation of the transmembrane Na gradient necessary for the uphill transport of sugars, amino acids and ions; (d) maintenance of the ion gradients essential for the membrane potential and the excitability of the membrane; and (e) the transport of Na across the epithelia. Generally up to one-third of an animal cell's energy requirement is consumed in fueling the Na,K-ATPase. In electrically active tissues or tissues involved in salt transport up to 70% of the cell's total energy requirement may be used by the pump.

Since Skou's first description in 1957 of a Na- and K-dependent ATPase in crab nerve [1] and the subsequent studies demonstrating its identity with the Na,K pump, much information about the structure and function of the Na,K-

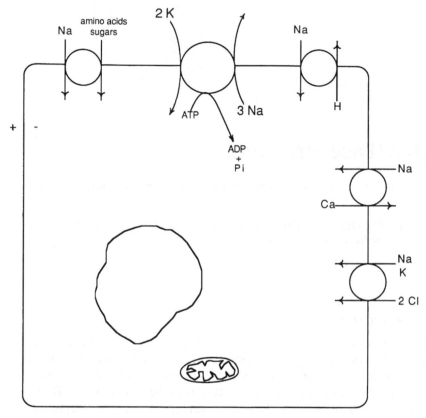

Figure 1 Schematic or a representative cell showing the influence of the Na,K-ATPase on transport.

ATPase has been acquired (reviewed in [2,3]). Although these studies have provided many insights into the structure of the Na,K-ATPase, relatively little has been known concerning the amino acid sequence of the enzyme. However, an important advance in the understanding of the structure of the Na,K-ATPase was recently obtained when the complete amino acid sequences for the Na,K-ATPase subunits were determined [4–6]. This chapter will focus on these recent advances in defining the structure of the Na,K-ATPase.

The Na,K-ATPase consists of two noncovalently linked polypeptides, a catalytic α subunit, with a molecular weight of ∼100,000 and a smaller glycosylated β subunit with a molecular weight of ∼55,000 [2,3]. A small peptide with a molecular weight of ∼10,000, termed the γ subunit, has also been identified in purified preparations of the enzyme [7,8]. This polypeptide was considered to be a contaminant of purification until it was shown that the γ-polypeptide, along with the α subunit, could be covalently labeled from the extracellular surface by a photoaffinity labeled deriviative of the specific Na,K-ATPase inhibitor, ouabain [8-11]. The specific labeling of the γ subunit by photolabeled ouabain strongly suggests that it is associated with the Na,K-ATPase.

Most functions of the Na,K-ATPase have been localized to the α subunit. The α subunit contains the binding sites for ATP and ouabain; it is phosphorylated by ATP and undergoes ligand dependent conformational changes accompanying the binding, occlusion, and translocation of ions [3]. The exact functions of the β and γ subunits remain unknown. The γ subunit does not appear to be essential for ATPase activity [12], but recent evidence suggests that the β subunit has an important role in Na,K-ATPase function. In ouabain-resistant HeLa cells, both α and β subunits are amplified, suggesting that both subunits are required for the acquisition of ouabain resistance [13,14]. Moreover, the high degree of homology between β subunits from several diverse species [6,14-16] supports the view that the β subunit is important for Na,K-ATPase structure and function.

The Na,K-ATPase belongs to a widely distributed class of cation translocating enzymes that includes the sarcoplasm reticulum and plasma membrane Ca-ATPase, the H,K-ATPase from gastric mucosa, the fungal H-ATPases, and the bacterial K-ATPase. These enzymes directly convert the metabolic energy derived from the hydrolysis of ATP into the work required to move cations against their concentration gradients. Similarities in the structure and the reaction mechanism of these widely divergent proteins suggests that these enzymes share a common gene origin. Therefore, one can hope that conclusions drawn from one of these transport proteins may provide information relevant to the others.

It is unclear how cation translocating enzymes couple the hydrolysis of ATP to the transport of cations across the membrane. A thorough knowledge of the structure of these enzymes should facilitate understanding of the molecular mechanism of transport. For the Na,K-ATPase, knowledge of the kinetic mechanism has come largely from studies with human red blood cells [1,17], (Chapter 16). However, red blood cells have an extremely low density of Na,K-ATPase in their membranes, making them an inadequate source for the purification and isolation of the enzyme. Therefore, the structure of the Na,K-ATPase has usually been studied using preparations with high specific activities of the enzyme. Generally the starting material for the isolation of the Na,K-ATPase is from a tissue specialized in Na transport such as the avian salt gland [18], the shark rectal gland [7,19], or the outer medulla of the mammalian kidney [20-22], or from electrically active tissue such as the mammalian brain [23] or the electric organ of the eel [19,24]. By studying the enzyme isolated from these tissues it is hoped that the general observations of Na,K-ATPase structure and function will be applicable to the Na,K-ATPase in the other tissues.

```
                                                                                                              -1
Human     MGKGV
Pig       MGKGV
Sheep     MGKGV
Rat       MGKGV
Torpedo   MGKGA

                                                                                                              100
Human     GRDKYEPAAVSEQGDKKGKK  GKKDRDMD*ELKKEVSMDDH  KLSLDELHRKYGTDLSRGLT  SARAAEILARDGPNALTPPP  TTPEWIKFCRQLFGGFSMLL
Pig       GRDKYEPAAVSEHGDK**KK  ÅKKERDMD*ELKKEVSMDDH  KLSLDELHRKYGTDLSRGLT  PARAAEILARDGPNALTPPP  TTPEWQKFCRQLFGGFSMLL
Sheep     GRDKYEPAAVSEHGDK**KK  ÅKKERDMD*ELKKEVSMDDH  KLSLDELHRKYGTDLSRGLT  TARAAEILARDGPNALTPPP  TTPEWQKFCRQLFGGFSMLL
Rat       GRDKYEPAAVSEHGDKKSKK  ÅKKERDMD*ELKKEVSMDDH  KLSLDELHRKYGTDLSRGLT  PARQAEILARDGPNALTPPP  TTPEWQKFCRQLFGGFSMLL
Torpedo   ASEKYQPAATSENA*KNSKK  SKSKTIDIDELKKEVSLDDH  KINLDELHQKYGTDLIQGLT  PARAKEILARDGPNALTPPP  TTPEWIKFCRQLFGGFSILL

                                                                                                              200
Human     WIGAILCFLAYSIQAATEEE  PQNDNLYLGVVLSAVVIITG  CFSYYQEAKSSKIMESFKNM  VPQQALVIRNGERKMSINAEE  VVVGDLVEVKGGDRIPADLR
Pig       WIGAILCFLAYGIQAATEEE  PQNDNLYLGVVLSAVVIITG  CFSYYQEAKSSKIMESFKNM  VPQQALVIRNGERKMSINAEE  VVVGDLVEVKGGDRIPADLR
Sheep     WIGAVLCFLAYGIQAATEEE  PQNDNLYLGVVLSAVVIITG  CFSYYQEAKSSKIMESFKNM  VPQQALVIRNGERKMSINAEE  VVVGDLVEVKGGDRIPADLR
Rat       WIGAILCFLAYGIRSAATEE  PQNDLYLGVVLSAVVIITG   CFSYYQEAKSSKIMESFKNM  VPQQALVIRNGERKMSINAED  VVVGDLVEVKGGDRIPADLR
Torpedo   WIGAILCFLAYGIQVATYDN  PANDNLYLGVVLSTVVIITG  CFSYYQEAKSSKIMDSFKNM  VPQQALVIRDGEKSSINAEQ   VVVGDLVEVKGGDRIPADLR

                                                                                                              300
Human     IISANGCKVDNSSLTGESEP  QTRSPDFTNENPLETRNIAF  FSTNCVEGTARGIVVTGDR  TVMGRIATIASGLEGGQTPI  AAEIEHFIHIITGVAVFLGV
Pig       IISANGCKVDNSSLTGESEP  QTRSPDFTNENPLETRNIAF  FSTNCVEGTARGIVVTGDR  TVMGRIATIASGLEGGQTPI  AAEIEHFIHIITGVAVFLGV
Sheep     IISANGCKVDNSSLTGESEP  QTRSPDFTNENPLETRNIAF  FSTNCVEGTARGIVVTGDR  TVMGRIATIASGLEGGQTPI  AAEIEHFIHIITGVAVFLGV
Rat       IISANGCKVDNSSLTGESEP  QTRSPDFTNENPLETRNIAF  FSTNCVEGTARGIVVTGDR  TVMGRIATIASGLEGGQTPI  AEIEHFIHLITGVAVFLGV
Torpedo   IISACSCKVDNSSLTGESEP  QSRSPEYSSENPLETKNIAF  FSTNCVEGTARGIVINTGDH  TVMGRIATIASGLEVGQTPI  AAEIEHFIHIITGVAVFLGV

                                                                                                              400
Human     SFFILSLILEYTWLEAVIFL  IGIIVANVPEGLLATVTVCL  TLTAKRMARKNCLVKNLEAV  ETLGSTSTICSDKTGTLTQN  RMTVAHMWFDNQIHEADTTE
Pig       SFFILSLILEYTWLEAVIFL  IGIIVANVPEGLLATVTVCL  TLTAKRMARKNCLVKNLEAV  ETLGSTSTICSDKTGTLTQN  RMTVAHMWFDNQIHEADTTE
Sheep     SFFILSLILEYTWLEAVIFL  IGIIVANVPEGLLATVTVCL  TLTAKRMARKNCLVKNLEAV  ETLGSTSTICSDKTGTLTQN  RMTVAHMWFDNQIHEADTTE
Rat       SFFILSLILEYTWLEAVIFL  IGIIVANVPEGLLATVTVCL  TLTAKRMARKNCLVKNLEAV  ETLGSTSTICSDKTGTLTQN  RMTVAHMWFDNQIHEADTTE
Torpedo   SFFILSLILGYTWLEAVIFL  IGIIVANVPEGLLATVTVCL  TLTAKRMARKNCLVKNLEAV  ETLGSTSTICSDKTGTLTQN  RMTVAHMWFDNQIHEADTTE

                                                                                                              500
Human     NQSGVSFDKTSATWLALSRI  AGLCNRAVFQANQENLPILK  RAVAGDASESALLKCIELCC  GSVKEMRERYAKIVEIPFNS  TNKYQLSIHKNPNTSEPQHL
Pig       NQSGVSFDKTSATWLALSRI  AGLCNRAVFQANQENLPILK  RAVAGDASESALLKCIELCC  GSVKEMRERYIKIVEIPFNS  TNKYQLSIHKNPNTAEPBHL
Sheep     NQSGVSFDKTSATWLALSRI  AGLCNRAVFQANQDNLPILK  RAVAGDASESALLKCIEVCC  GSVKEMRERYAKIVEIPFNS  TNKYQLSIHKNANAGEPBHL
Rat       NQSGVSFDKTSATWLALSRI  AGLCNRAVFQANQENLPILK  RAVAGDASESALLKCIEVCC  GSVMEMREKYIKIVEIPFNS  TNKYQLSIHKNPNASEPKHL
Torpedo   NQSGISFDKTSLSWNALSRI  AALCNRAVFQAGQDSVPILK  RSVAGDASESALLKCIELCC  GSVSOMRDRNEKIVEIPFNS  TNKYQLSIHENDKADS*RYL
```

```
                                                                        600
Human    LVMKGAPERIIDRCSSILLH  GKEQPLDEELKDAFQNAYLE  LGGLGERVLGFCHLFLPDEQ  FPEGFQFDTDVNFPIDNLC  FVGLISMIDPPRAAVPDAVG
Pig      LVMKGAPERIIDRCSSILLH  GKEQPLDEELKDAFQNAYLE  LGGLGERVLGFCHLFLPDEQ  FPEGFQFDTDVNFPLDNLC  FVGLISMIDPPRAAVPDAVG
Sheep    LVMKGAPERIIDRCSSILLH  GKEQPLDEELKDAFQNAYLE  LGGLGERVLGFCHIMLPDEQ  FPEGFQFDTDVNFPVDNLC  FVGLISMIDPPRAAVPDAVG
Rat      LVMKGAPERIIDRCSSILLH  GKEQPLDEELKDAFQNAYLE  LGGLGERVLGFCHLLLPDEQ  FPEGFQFDTDEVNFPVDNLC FVGLISMIDPPRAAVPDAVG
Torpedo  LVMKGAPERIIDRCSTILLN  GEDKPLNEEMKEAFQNAYLE  LGGLGERVLGFCHLKLSTSK  FPEGYPFDVEERNFPITDLC FVGIMSMIDPPRAAVPDAVG

                                                                        700
Human    KCRSAGIKVIMVTGDHPITA  KAIAKGVGIISEGNETVEDI  AARLNIPVSQVNPRDAKACV  VHGSDLKDMTSEELDDILKY  HTEIVFARTSPQQKLIIVEG
Pig      KCRSAGIKVIMVTGDHPITA  KAIAKGVGIISEGNETVEDI  AARLNIPVSQVNPRDAKACV  VHGSDLKDMTSEQLDDILKY  HTEIVFARTSPQQKLIIVEG
Sheep    KCRSAGIKVIMVTGDHPITA  KAIAKGVGIISEGNETVEDI  AARLNIPVSQVNPRDARACV  VHGSDLKDMTSEELDDILKY  HTEIVFARTSPQQKLIIVEG
Rat      KCRSAGIKVIMVTGDHPITA  KAIAKGVGIISEGNETVEDI  AARLNIPVNQVNPRDAKACV  VHGSDLKDMTSEELDDILRY  HTEIVFARTSPQQKLIIVEG
Torpedo  KCRSAGIKVIMVTGDHPITA  KAIAKGVGIISEGNETVEDI  AARLNIPVNQVNPRDAKACV  VHGTDLKDLSHENLDDILHY  HTEIPFLVFIIANVLPLGT

                                                                        800
Human    CQRQGAIVAVTGDGVNDSPA  LKKADIGVAMGIAGSDVSKQ  AADMILLDDNFASIVTGVEE  GRLIFDNLKKSIAYTLTSNI  PEITPFLIFIIANIPLPLGT
Pig      CQRQGAIVAVTGDGVNDSPA  LKKADIGVAMGIAGSDVSKQ  AADMILLDDNFASIVTGVEE  GRLIFDNLKKSIAYTLTSNI  PEITPFLIFIIANIPLPLGT
Sheep    CQRQGAIVAVTGDGVNDSPA  LKKADIGVAMGIAGSDVSKQ  AADMILLDDNFASIVTGVEE  GRLIFDNLKKSIAYTLTSNI  PEITPFLIFIIANIPLPLGT
Rat      CQRQGAIVAVTGDGVNDSPA  LKKADIGVAMGIYGSDVSKQ  AADMILLDDNFASIVTGVEE  GRLIFDNLKKSIAYTLTSNI  PEITPFLIFIIANIPLPLGT
Torpedo  CQRQGAIVAVTGDGVNDSPA  LKKADIGVAMGIAGSDVSKQ  AADMILLDDNFASIVTGVEE  GRLIFDNLKKSIAYTLTSNI  PEITPFLVFIIANVLPLGT

                                                                        900
Human    VTILCIDLGTDMVPAISLAY  EQAESDIMKRQPRNPKTDKL  VNERLISMAYGQIGMIQALG  GFFTYFVILAENGFLPIHLL  GLRVDWDDRWINDVEDSYGQ
Pig      VTILCIDLGTDMVPAISLAY  EQAESDIMKRQPRNPKTDKL  VNEQLISMAYGQIGMIQALG  GFFTYFVILAENGFLPIHLL  GLRVNWDDRWINDVEDSYGQ
Sheep    VTILCIDLGTDMVPAISLAY  EQAESDIMKRQPRNPQTDKL  VNERLISMAYGQIGMIQALG  GFFTYFVMAENGFLPNHLL   GIRVLWDDRWINDVEDSYGQ
Rat      VTILCIDLGTDMVPAISLAY  EQAESDIMKRQPRNPKTDKL  VNERLISMAYGQIGMIQALG  GFFTYFVILAENGFLPEHLL  GIRETWDDRWINDVEDSYGQ
Torpedo  VTILCIDLGTDMVPAISLAY  ERAESDIMKRQPRNPKTDKL  VNERLISMAYGQIGMIQALG  GFFSYFVILAENGFLPIDLI  GIREKWDELWINDLEDSYGQ

                                                                       1000
Human    SIVVQWADLVICKTRRNSV   FQQGMKNKILIFGLFEETAL  AAFLSYCPGMGVALRMYPLK  PTWWFCAFPYSLLIFVYDEV
Pig      SIVVQWADLVICKTRRNSV   FQQGMKNKILIFGLFEETAL  AAFLSYCPGMGVALRMYPLK  PTWWFCAFPYSLLIFVYDEV
Sheep    SIVVQWADLVICKTRRNSV   FQQGMKNKILIFGLFEETAL  AAFLSYCPGMGVALRMYPLK  PTWWFCAFPYSLLIFVYDEV
Rat      SIVVQWADLVICKTRRNSV   FQQGMKNKILIFGLFEETAL  AAFLSYCPGMGAALRMYPLK  PTWWFCAFPYSLLIFVYDEV
Torpedo  SIVLVQWADLLICKTRRNSI  FQQGMKNKILIFGLFEETAL  AAFLSYTPGTDIALRMYPLK  PSWWFCAFPYSLILFLYDEA

Human    RKLIIRRRPGGWVEKETYY
Pig      RKLIIRRRPGGWVEKETYY
Sheep    RKLIIRRRPGGWVEKETYY
Rat      RKLIIRRRPGGWVEKETYY
Torpedo  RRFILRRNPGGWVEQETYY
```

Figure 2 Amino acid sequence of the Na,K-ATPase α subunits. Differences from the human sequence are underlined.

Since the observation of Kyte in 1971 [20] that the Na,K-ATPase consists of two subunits, the α and β subunits have been extensively studied. A large number of studies have been directed at identifying reactive groups, determining the molecular weight and amino acid composition, and characterizing the membrane topography of the subunits (reviewed in [1,2]). Recently the complete amino acid sequence for the α and β subunits from several species has been determined from their complementary DNA (cDNA). These results, along with the earlier biochemical studies, have helped to elucidate some aspects of the structure of the Na,K-ATPase subunits.

II. STRUCTURE OF THE α SUBUNIT

The complete amino acid sequences for the human [25], pig [26], sheep [4], rat [27], and *Torpedo* [5] α subunits have been recently determined. The alignment of the amino acid sequences from these species is shown in Figure 2. The deduced primary structure of the α subunits reveals a polypeptide consisting of $\sim$1000 amino acids with a molecular weight of $\sim$112,500. The initiation methionine is located five amino acids before the amino terminus of the mature polypeptide. It appears that an amino terminal pentapeptide is cleaved from the primary translation product during processing. The α subunit of the Na,K-ATPase lacks a cleaved hydrophobic signal peptide [28]. The absence of a signal peptide may be characteristic of transport proteins because the Ca and H,K-ATPases [29,30], the murine anion exchange protein [31], and the human glucose transporter [32] also lack a hydrophobic amino terminal signal sequence.

The degree of sequence homology between the human and *Torpedo* α subunits is 87% whereas the homology between the human and the other mammalian proteins is $\sim$98%. The Na,K-ATPase exhibits a high degree of homology with several other cation-transporting ATPases. As shown in Figure 3, there are several regions of high homology between the H-ATPase from *Neurospora* [33] and the mammalian Ca and H,K-ATPase [29,30]. These regions include the phosphorylation site, and several regions thought to be involved in ATP binding (see below). Consequently, regions of the ATPases that are homologous may be involved in the basic energy transduction process common to these proteins while the nonhomologous regions may be involved in cation selectivity, which is different for each enzyme.

The amino acid sequence, site-specific labeling [8,34-37], immunological [38-40], and proteolytic digestion studies [38,41-43] have provided some insight into the possible transmembrane orientation of the α subunit. Under defined conditions digestion of the native α subunit by chymotrypsin and trypsin results in a specific digestion pattern that has been used to study the arrangement of the polypeptide in the membrane [41-43]. For example, trypsin digestion of the α subunit, in the presence of Na, releases into the cytoplasm 20

```
Na,K-ATPase  176   INAEEVVVGDLVEVKGGDRIPADLRIISAN**GCKVDNSSLTGESEPQTRS
H,K-ATPase   189   INADELVVGDLVEMKGGDRVPADIRILSAE**GCKVDNSSLTGESEPQTRS
Ca-ATPase    140   IKARDIVPGDIVEVAVGDKVPADIRILSIKSTTLRVDQSILTGESVSVIKH
H-ATPase     191   IEAPEVVPGDILQVEEGTIIPADGRIVTDD*AFLQVDQSALTGESLAVDKH


Na,K-ATPase  238   AFFSTNCVEGTARGIVVYTGDRTVMGRIATLASGLEGGQTP
H,K-ATPase   252   AFFSTMCLEGTAQGLVVSTGDRTIIGRIASLASGVENEKTP
Ca-ATPase    208   LFSGTNIAAGKALGIVATTGVSTEIGKIRDQMAATEQDKTP
H-ATPase     245   VFASSAVKRGEAFVVITATGDNTFVGRAAALVNAASGGSGH


                                                                    *
Na,K-ATPase  320   IGIIVANVPEGLLATVTVCLTLTAKRMARKNCLVKNLEAVETLGSTSTICSDKTGTLTQN
H,K-ATPase   333   MAIVVAYVPEGLLATVTVCLSLTAKRLASKNCVVKNLEAVETLGSTSVICSDKTGTLTQN
Ca-ATPase    300   VALAVAAIPEGLPAVITTCLALGTRRMAKKNAIVRSLPSVETLGCTSVICSDKTGTLTTN
H-ATPase     327   LAITIIGVPVGLPAVVTTTMAVGAAYLAKKKAIVQKLSAIESLAGVEILCSDKTGTLTKN


                         ↓
Na,K-ATPase  500   LVMKGAPERILDRCSSILLHGKEQPLDEELKDAFQ
H,K-ATPase   514   LVMKGAPERVLERCSSILIKGQELPLDEQWREAFQ
Ca-ATPase    512   MFVKGAPEGVIDRCNYVRVGTTRVPMTGPVKEKIL
H-ATPase     471   TCVKGAPLFVLKTVEEDHPIPEEVDQAYKNKVAEF


Na,K-ATPase  588   DPPRAAVPDAVGKCRSAGIKVIMVTGDHPITAK
H,K-ATPase   602   DPPRATVPDAVLKCRTAGIRVIMVTGDHPITAK
Ca-ATPase    601   DPRRKEVMGSIQLCRDAGIRVIMITGDNKGTAI
H-ATPase     534   DPPRHDTYKTVCEAKTLGLSIKMLTGDAVGIAR


                   ─────────────────────
Na,K-ATPase  682   EIVFARTSPQQKLIIVEGCQRQGAIVAVTGDGVNDSPALKKADIGVAMGIAGSDVSKQAAD
H,K-ATPase   696   EMVFARTSPQQKLVIVESCQRLGAIVAVTGDGVNDSPALKKADIGVAMGIAGSDAAKNAAD
Ca-ATPase    673   ACCFARVEPSHKSKIVEYLQSYDEITAMTGDGVNDAPALKKAEIGIAMGS*GTAVAKTASE
H-ATPase     604   ADGFAEVFPQHKYNVVEILQQRGYLVAMTGDGVNDAPSLKKADTGIAVEG*SSDAARSAAD
```

Figure 3 Regions of homology between the human α subunit of the Na,K-ATPase and the rabbit Ca-ATPase, rat H,K-ATPase and *Neurospora* H-ATPase. An asterisk identifies the phosphorylated aspartic acid, an arrow the lysine labeled by FITC, and a bar the tryptic peptide labeled by FSBA. Numbering for the α subunit is based on Figure 2.

residues from the amino-terminal end (amino acids 1-20 in Figure 2) and results in an "invalid" Na,K-ATPase with several catalytic defects [42-45]. Another cytoplasmic site, cleaved more slowly in the presence of Na, results in a 78-kD fragment that remains associated with the membrane [41]. Carboxy-terminal analysis of the 78 kD fragment indicates that it represents the carboxy-terminal region of the intact polypeptide [43]. The amino-terminal residue of the 77-kD fragment is Ile, which would most likely place this site at Ile^{266} (Figure 2).

Specific labeling of the ATP binding site has also confined several segments of the α subunit to the cytoplasmic side of the membrane. The ATP analogue 5'-(p-fluorsulfonyl)benzoyladenosine (FSBA) both labels and inhibits the Na,K-ATPase [36,46,47]. The inactivation and labeling of the Na,K-ATPase by FSBA is specifically prevented by ATP [36]. Studies using radioactive FSBA have shown that FSBA is incorporated into two tryptic peptides [36] (amino acids 658-667 and 704-722 in Figure 2). Another probe of the ATP binding site is fluorescein isothiocyanate (FITC). Fluorescein isothiocyanate reacts covalently to inhibit the Na,K-ATPase and this inhibition can be prevented by ATP [48, 49]. Fluorescein isothiocyanate labels a Lys residue (504) in the tryptic peptide corresponding to residues 499-509 [34,35]. The sites reactive to FSBA and FITC are presumably involved in ATP binding and are therefore at the cytoplasmic surface of the enzyme.

During the catalytic cycle an aspartate residue of the α subunit is phosphorylated by ATP [50,51]. Determination of the amino acid sequence surrounding this site indicates that Asp^{372} is phosphorylated [52]. This location is also consistent with the phosphorylation site of the Ca-ATPase [53] and with the analysis of α subunit peptide fragments [43]. Thus, several segments of the α subunit can be localized to the cytoplasmic surface of the membrane. Several regions of the α subunit have also been shown to be extramembranous. To identify regions of the enzyme exposed to the aqueous environment, several soluble peptide fragments released after extensive proteolysis have been isolated and characterized [26,36,54]. These fragments should represent regions of the α subunit not embedded or tightly associated with the membrane. The location of the extramembranous peptides of the α subunit are shown in a linear map in Figure 4a (open bars).

Using reactive reagents that partition in the bilayer several portions of the α subunit presumably within or associated with the membrane have been labeled [55,56]. On exposure to light, the lipophilic carbene precursor, 1-tritiospiro [adamantane-4,3'-diazirine], labels several fragments of the Na,K-ATPase α subunit [56]. Determination of the amino-terminal sequence of these peptides has identified these putative membrane domains to the regions shown in Figure 4a (hatched bars). Thus, the membrane orientation and membrane association of several regions of the Na,K-ATPase α subunit have been tentatively identified.

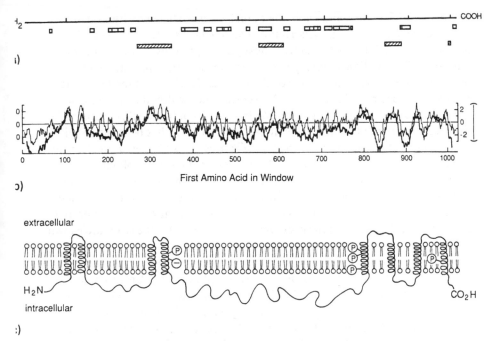

Figure 4 The Na,K-ATPase α subunit. (a) The location of extramembranous peptides of the α-subunit subunit are identified as open bars and putative membrane domains as hatched bars. Peptide fragments determined from references 26, 36, 54, and 56. See text for details. (b) Hydropathy plot of the human α subunit generated by the methods of Engleman et al. [102] and Kyte and Doolittle [103]. Hydrophobic regions are above the x-axis and hydrophilic regions below. C. Model of the transmembrane orientation of the human Na,K-ATPase α subunit. The putative membrane buried proline residues are shown. As proposed by Brandl and Deber [104] the redirection of the peptide chain caused by the proline peptide bond may provide the reversible conformational change requisite for the regulation of an ion channel.

Figure 4b shows the hydropathy profile of the human Na,K-ATPase α subunit. The α subunit from other species, the rabbit cardiac Ca-ATPase, and the yeast and *Neurospora* H-ATPases have similar profiles [4,5,25,29,33,57]. Hydrophobic regions are plotted above the x-axis whereas hydrophilic regions are below it. The hydropathy analysis, when combined with the other results, can be used to construct a model for the path of the Na,K-ATPase α subunit in the membrane. As shown in Figure 4c, these studies predict a cytoplasmic amino-terminus followed by four transmembrane spanning regions, which are followed by a large cytoplasmic domain consisting of roughly one-third of the polypeptide. The four transmembrane domains are predicted by hydropathy. and secondary structure analysis assuming an α-helical stretch of at least 20 hydrophobic residues. As mentioned above, the large cytosolic domain contains the FITC and FSBA binding regions and the aspartate residue that is phosphorylated. Interestingly, one of the peptides labeled with the lipophilic reagent 1-tritiospiro [adamantane-4,3'-diazirine] lies within the cytoplasmic domain. The isolation of soluble peptide fragments containing this region makes its involvement with the membrane unlikely. However, it has been suggested that the corresponding region of the Ca-ATPase consists of alternating β strands and α helices, which fold to give parallel β sheets that are hydrophobic on both faces [29]. Therefore, although this region is extramembranous, it may have some hydrophobic characteristics. The localization of several antibody binding sites, all known to be at the cytoplasmic surface of the enzyme, to tryptic and chymotryptic fragments of the α subunit is consistent with the structure shown [40].

Analysis of the carboxy-terminal portion of the enzyme suggests the presence of from two to five membrane-spanning regions [4,5,25,26]. Unfortunately, little information from site-specific labeling or proteolytic digestion exists for this portion of the protein. However, two regions have been labeled with the lipophilic probe 1-tritiospiro [adamantane-4,3'-diazirine], and two soluble peptide fragments (Figure 4a) have been isolated. These results are consistent with the presence of four or five transmembrane segments. The existence of five membrane-spanning domains would place the carboxy-terminus at the extracellular surface. However, the hydrophobic region corresponding to amino acid residues 980–996 may only be associated with the membrane and not traverse it. Because there are few external tryptic sites [58,59], the isolation of a soluble peptide fragment corresponding to the carboxy-terminus suggests that this region of the protein is intracellular. However, the position of the carboxy terminus of the α subunit with respect to the membrane will have to be resolved experimentally.

Other transmembrane orientations for the carboxy-terminal portion of the α subunit have also been suggested [26]. If the first membrane-spanning domain (M5) is extended, its length could be sufficient to cross the membrane twice, as was proposed for the analogous portion of the Ca-ATPase [29]. This would place the junction between M6 and M7 at the extracellular surface, and with four subsequent membrane-spanning regions the carboxy-terminus at the extracellular surface. This arrangement also places a potential phosphorylation site (R-R-N-S-V, 936–940 in Figure 2) for a cAMP-dependent protein kinase at the cytoplasmic surface [60]. It is apparent that the present data are consistent with several possible models describing the arrangement of the α subunit in the membrane. The determination of which model accurately describes the arrangement of the α subunit in the membrane must await further study.

Table 1 Alignment of Homologous Sequences in Adenine Nucleotide Binding Proteins

Protein	Residues	Sequences																		
Rat Na,K-ATPase	544–561	G	E	R	V	–	L	G	F	C	H	L	L	L	P	D	E	Q	F
Rabbit Ca-ATPase	612–628	L	C	R	Q	–	A	G	I	R	V	I	M	I	T	G	D	N	K
Rat H,K-ATPase	614–630	K	C	R	T	–	A	G	I	R	V	I	M	V	T	G	D	H	P
Bovine ATPase β	244–261	Y	F	R	D	Q	E	G	Q	D	V	L	L	F	I	D	N	I	F
Escherichia coli ATPase β	230–246	K	F	R	D	–	E	G	R	D	V	L	L	F	V	D	N	I	Y
E. coli ATPase α	268–284	Y	F	R	D	–	R	G	E	D	A	L	I	I	Y	D	D	L	S
ATP/ADP translocase	277–294	V	L	R	G	M	G	G	A	F	V	L	V	L	Y	D	E	I	K
Adenylate kinase	105–121	F	E	R	K	–	I	G	Q	P	T	L	L	L	Y	V	D	A	G
Phosphofructokinase	88–104	Q	L	K	K	–	H	G	I	Q	G	L	V	V	V	G	G	D	G

Residues enclosed in boxes are identical or conserved substitutions. Sequences for the nontransporting ATPases are compiled in Walker et al. [63].

As mentioned above, several regions presumably involved in ATP binding and hydrolysis have been identified. These regions include the phosphorylation site (Asp^{372}), the two tryptic peptides labeled with FSBA, and the FITC binding site (Lys^{504}). Like the phosphorylated aspartate residue, the amino acids that are labeled by FSBA are presumed to be close to the terminal phosphate of ATP [61]. On the other hand, FITC is thought to bind to the portion of the Na,K-ATPase that is normally occupied by the adenine ring [62]. Comparison of the primary sequence of the Na,K-ATPase α subunit with enzymes involved in ATP binding may identify other portions of the enzyme responsible for adenine nucleotide binding. Table 1 shows a homologous region from several divergent proteins, all of which are involved in the binding of ATP. Notably, these proteins have a conserved sequence of R or K, X-X or X-X-X, G-X-X-X-L, followed by two or three hydrophobic residues [63,64]. As shown in Table 1, the amino acid sequence of the rat α subunit corresponding to residues 546–555 is homologous to this conserved region of the adenine nucleotide binding proteins. It has been suggested that this sequence forms a hydrophobic β sheet structure important in nucleotide binding. Also, studies with phosphofructokinase have suggested that the conserved aspartic acid residue (boxed in Table 1) is important for the binding of magnesium in the magnesium-ATP substrate complex [63]. Thus, there appear to be several regions of the Na,K-ATPase that interact with ATP. Not surprisingly, most of these regions are conserved in the other ion-transporting ATPases. As shown in Figure 3, the phosphorylation site, the FITC binding region, and one of the tryptic peptides labeled with FSBA are in regions conserved in the ATPases. These regions, although far apart in the primary sequence, may provide clues about the arrangement of the polypeptide chain around the ATP binding site. Presumably the folding of the polypeptide is such that these regions must be in close proximity to form the tertiary structure necessary for the binding of ATP.

Comparison of the amino acid sequences from the various α subunits and the other transport ATPases has provided other insights into Na,K-ATPase structure. As previously mentioned, when the enzyme is in the E_1 form, the amino-terminus of the α subunit is rapidly cleaved with trypsin. Cleavage of the α subunit at this site appears to alter the equilibrium between the E_1 and E_2 conformations of the enzyme. This tryptic site (K, 20) is in a conserved lys-rich region, which surprisingly, is also found in the H,K-ATPase [30]. It has been proposed that this region may be involved in the conformational shift that occurs during cation occlusion and may function as a movable ion-selective gate that controls the passage of ions to and from their binding site [27].

The determination of the complete amino acid sequences of the α subunit from ouabain-sensitive and insensitive species makes it possible to identify amino acid residues that may be involved in the binding of ouabain. The enzyme from the human, pig, and sheep is more than 1000 times more sensitive to ouabain than the enzyme from rodents [64,65]. Also, the rat α(+) isoform of the

Table 2 Amino Acid Differences Between the Rat (Ouabain-insensitive) and Rat α(+), Human, Pig, and Sheep (Ouabain-sensivite) Na,K-ATPase α Subunits

	Residue					
	114	122	125	469	555	877
Rat	R	P	D	K	L	F
Rat α+	L	S	N	R	N	S
Human	Q	Q	N	R	F	I
Pig	Q	Q	N	R	F	I
Sheep	Q	Q	N	R	M	N

Na,K-ATPase (discussed below) is much more sensitive to ouabain than the α form. As shown in Table 2, there are only six amino acid differences in which the rat differs from the rat α(+) isoform and the other mammalian species. Of the six differences only three of these are in potential extracellular domains. If the differences in ouabain sensitivity are caused by amino acid changes at the ouabain binding site, and not by differences at some location distant from ouabain binding, then presumably one of these amino acids should be involved in ouabain binding. Site-directed mutagenesis of these amino acids will undoubtedly provide information about their involvement in ouabain resistance.

III. Na,K-ATPase α SUBUNIT ISOFORMS

As first shown by Sweadner, two molecular forms of the α subunit exist in mammals [23]. In addition to the renal form, designated α, another Na,K-ATPase α subunit termed α(+) has been identified. As mentioned before, the α(+) isoform (so named because of its larger molecular weight by sodium dodecylsulfate gel electrophoresis) can be distinguished from the renal form by its higher affinity for ouabain [23,66]. The α(+) isoform, when compared with α, also has an increased sensitivity to N-ethylmaleimide and pyrithiamin, and is relatively resistant to digestion by trypsin [23,67]. Both isoforms are present in the brain; however, it appears that the renal form is expressed in the glia cells whereas both forms are expressed in the neurons [23,68,69]. In adipocytes and skeletal muscle both isoforms are present, but it appears that the α(+) form of the enzyme is selectively sensitive to stimulation by insulin [70]. Determination of the amino-terminal sequences of the two isoforms has shown that they have distinct, although homologous, amino-terminal ends [71]. Recently, the cDNAs coding for the α(+) isoform and a previously unidentified α subunit, designated α3, have been isolated and sequenced [27]. Figure 5 shows the complete amino acid sequences for the rat α, α(+) and α3, α subunit isoforms. The

```
a     MGKGV  -1
a(+)  MGBGA
a3    *****


a     GRDKYEPAAVSEHGD**KKS KKAK*KERDMDELKKEVSMD DHKLSLDELHRKYGTDLSRG LTPARPAEILARDGPNALTP PPTTPEWKFCRQLFGGFSM
a(+)  GRE*YSPAATTAENGGGKK* *KQK**EKELDELKKEVAMD DHKLSLDELGRKYNVDLSKG LTNQRAQDILARDGPNALTP PPTTPEWKFCRQLFGGFSI
a3    ****MGDKKDDKSSP**KKS *KAK*ERRDLDDLKKEVAMT EHKMSVEEVCRKYNTDCVQG LTHSKAQEILARDGPNALTP PPTTPEWKFCRQLFGGFSI


a     LLWIGAILCFLAYGIRSATE EEPPNDDLYLGVVLSAVVII TGCFSYYQEAKSSKIMESFK NMVPQQALVIRNGERMSINA EDVVVGDLVEVKGGDRIPAD
a(+)  LLWIGAILCFLAYGILAAME DEPSNDNLYLGIVLAAVVIV TGCFSYYQEAKSSKIMDSFK NMVPQQALVIREGERMQINA EEVVVGDLVEVKGGDRVPAD
a3    LLWIGAILCFLAYGIQAGTE DDPSGDNLYLGIVLAAVVII TGCFSYYQEAKSSKIMESFK NMVPQQALVIREGERMQVNA EEVVVGDLVELKGGDRVPAD


a     LRIISANGCKVDNSSLTGESEP QTRSPDFTNENPLETRNIAF FSTNCVEGTARGIVVYTGDR TVMGRIATLASGLEGGQTPI AEEIEHFIHLITGVAVFL
a(+)  LRIISSHGCKVDNSSLTGESEP QTRSPEFTHENPLETRNICF FSTNCVEGTARGIVIATGDR TVMGRIATLASGLEVGQTPI AMEIEHFIQLITGVAVFL
a3    LRIISAHGCKVDNSSLTGESEP QTRSPDCTHDNPLETRNIIF FSTNCVEGTARGVVVATGDR TVMGRIATLASGLEVGKTPI ALEIEHFIQLITGVAVFL


a     GVSFFILSLILEYTWLEAVI FLIGIIVANVPEGLLATVTV CLTLTAKRMARKNCLVKNLE AVETLGSTSTICSDKTGTLT QNRMTVAHMWFDNQIHEADT
a(+)  GVSFFVLSLIIGYSWLEAVI FLIGIIVANVPEGLLATVTV CLTLTAKRMARKNCLVKNLE AVETLGSTSTICSDKTGTLT QNRMTVAHMWFDNQIHEADT
a3    GVSFFILSLILGYTWLEAVI FLIGIIVANVPEGLLATVTV CLTLTAKRMARKNCLVKNLE AVETLGSTSTICSDKTGTLT QNRMTVAHMWFDNQIHEADT


a     TENQSGVSFDKTSATWF*AL SRIAGLCNRAVFQANQENLP ILKRAVAGDASESALLKCIE VCCGSVMEMREKYTKIVEIP FNSTNKYQLSIHKNPNASEP
a(+)  TEDQSGATFDKRSPTWT*AL SRIAGLCNRAVFKAGQENLS VSKRDTAGDASESALLKCIE LSCGSVRKWRDRNPKVAEIP FNSTNKYQLSIHEREDSPOS
a3    TEDQSGTSFDK*SHTWVSAL SHIAGLCNRAVFKGGQDNIP VLKRDVAGDASESALLKCIE LSSGSVKLMRERNKKVAEIP FNSTNKYQLSIHETEDPNDN
```

```
a     KHLLVMKGAPERILDRCSSI  LLHGKEQLDEELKDAFQNA  YLELGGLGERVLGFCHLLLP  DEQFPEGFQFDTDEVNFPVD  NLCFVGLISMIDPPRAAVPD
a(+)  *HVLVMKGAPERILDRCSTI  LVQGKELPLDKFMQDAFQNA  YMELGGLGERVLGFCQLNLP  SGKFPBGFKFDTDELNFPTE  KLCFVGLMSMIDPPRAAVPD
a3    RYLLVMKGAPERILDRCATI  LLQGKEQPLDEEMKEAFQNA  YLELGGLGERVLGFCHYYLP  EEOFPKGFAFDCDQVNFTID  NLCFVGLMSMIDPPRAAVPD

a     AVGKCRSAGIKVIMVTGDHP  ITAKAIAKGVGIISEGNETV  EDIAARLNIPVNQVNPRDAK  ACVVHGSDLKDMTSEELDDI  LRYHTEIVFARTSPQQKLII
a(+)  AVGKCRSAGIKVIMVTGDHP  ITAKAIAKGVGIISEGNETV  EDIAARLNIPVSQVNPREAK  ACVVHGSDLKDMTSEQLDEI  LRQHTEIVFARTSPQQKLII
a3    AVGKCRSAGIKVIMVTGDHP  ITAKAIAKGVGIISEGNETV  EDIAARLNIPVSQVNPRDAK  ACVIHGTDLKDETSQQIDEI  LQNHTEIVFARTSPQQKLII

a     VEGCQRQGAIVAVTGDGVND  SPALKKADIGVAMGIVGSDV  SKQAADMILLDDNFASIVTG  VEEGRLIFDNLKKSIAYTLT  SNIPEITPFLIFIIANIPLP
a(+)  VEGCQRQGAIVAVTGDGVND  SPALKKADIGIAMGISGSDV  SKQAADMILLDDNFASIVTG  VEEGRLIFDNLKKSIAYTLT  SNIPEITPFLLFIIANIPLP
a3    VEGCQRQGAIVAVTGDGVND  SPALKKADIGVAMGIVGSDV  SKQAADMILLDDNFASIVTG  VEEGRLIFDNLKKSIAYTLT  SNIPEITPFLLFIMANIPLP

a     LGTVTILCIDLGTDMVPAIS  LAYEQAESDIMKRQPRNPKT  DKLVNERLISMAYGQIGMIQ  ALGFFFTYFVILAENGFLPF  HLLGIRETWDDRWINDVEDS
a(+)  LGTVTILCIDLGTDMVPAIS  LAYEQAESDIMKRQPRNSQT  DKLVNERLISMAYGQIGMIQ  ALGFFSYFVILAENGFLPS   RLLGIRLDWDDRTNDLEDS
a3    LGTITILCIDLGTDMVPAIS  LAYEQAESDIMKRQPRNPKT  DKLVNERLISMAYGQIGMIQ  ALGFFSYFVILAENGFLPG   NLLGIRLNWDDRTVNDLEDS

a     YGQQWTYEQRKIVEFTCHTA  FFVSIVVQWADLIVICKTRR  NSVFQQGMKNKILIFGLFEE  TALAAFLSYCPGMGAALRMY  PLKPTWWFCAFPYSLLIFVY
a(+)  YGQEWTYEQRKVVEFTCHTA  FFASIVVQWADLLICKTRR   NSVFQQGMKNKILIFGLLEE  TALAAFLSYCPGMGVALRMY  PLKVTWWFCAFPYSLLIFIY
a3    YGQQWTYEQRKVVEFTEHTA  FFVSIVVQWADLLICKTRR   NSVFQQGMKNKILIFGLFEE  TALAAFLSYCPGMDVALRMY  PLKPSWWFCAFPYSLLIFVY

a     DEVRKLIIRRRPGGWVEKETYY
a(+)  DEVRKLILRRYPGGWVEKETYY
a3    DEVRKLIIRRNPGGWVEKETYY
```

Figure 5 The amino acid sequence of the Na,K-ATPase α subunit isoforms. Differences from the α sequence are underlined.

degree of sequence homology based on alignment is: α and $\alpha(+)$, 86%; α and $\alpha3$, 85%; and $\alpha(+)$ and $\alpha3$, 86%. As can be seen in Figure 5, the most amino acid divergence occurs in the amino-terminus and in the cytoplasmic region from amino acids 403 to 503. The greatest similarities occur around the phosphorylation site, the major hydrophobic regions, and the cytoplasmic region that includes the FSBA reactive peptides (amino acids 589-785). Interestingly, the amount of amino acid sequence homology between the rat α subunit isoforms is approximately the same as between the α isoforms from the distantly related rat and *Torpedo*. If the degree of homology can be considered to be related to the time of divergence, then the divergence in the genes coding for the α subunit isoforms must have been a distant evolutionary event.

The determination of the nucleotide sequence [27] and hybridization analysis of rat genomic DNA [72] indicate that each α subunit isoform is the product of a different gene. Chromosomal mapping localizes the gene coding for the α isoform to chromosome 1p in the human and chromosome 3 in the mouse, whereas the genes coding for human $\alpha(+)$ and $\alpha3$ are on chromosomes 1q and 19, respectively [73]. The gene for the β subunit resides on chromosome 1q [73].

Hybridization analysis of total cellular RNA and in situ hybridization of whole rat embryos has revealed several interesting and surprising results concerning the expression of α subunit isoforms. Figure 6 shows the RNA hybridization analysis of α subunit isoform expression in rat tissues. The cDNA from the isoforms was used to probe the RNA products of α subunit gene expression. Figure 6a shows that in all tissues examined, a single ~ 4.5 kb, α mRNA is expressed. Although α RNA was found in all tissues examined, the highest levels of α transcription are in the kidney. Characterization of $\alpha(+)$ transcription (Figure 6b) reveals $\alpha(+)$ mRNA in the brain, diaphragm muscle, heart muscle, and spinal cord. Also, in these tissues two distinct mRNAs are present, one mRNA with a molecular weight corresponding to the α isoform and the other with a slightly larger molecular weight of ~6.5 kb. The relative abundance of the two mRNA species appears to be tissue-specific, with the larger transcript predominating in the neural tissues and the smaller in the muscle. As shown in Figure 6c, the previously unidentified isoform, $\alpha3$, could only be detected in the brain and spinal cord.

Hybridization analysis of RNA isolated from brain, heart, and kidney of 18-day-old embryonic rats indicated that α and $\alpha(+)$ expression was similar to that found in the adult tissues. However, although $\alpha3$ mRNA cannot be detected in adult heart, it is present in relatively high concentrations in the fetal heart. Thus, in addition to its expression in brain, $\alpha3$ represents a fetal-specific form of rat heart Na,K-ATPase.

In situ hybridization of rat embryos and brain and kidney sections indicate that the highest level of α isoform expression is in the transport epithelia, especially the kidney. The cranial and dorsal root ganglia also exhibit high degrees

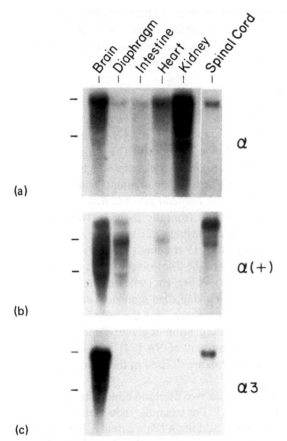

(a)

(b)

(c)

Figure 6   RNA blot analysis of α subunit mRBAs in adult rat tissues. Twenty-five micrograms of total cellular RNA was hybridized to α (a), α(+) (b), or α3 (c) cDNA probes. The two horizontal bars at the left of each panel indicate the positions of 28S and 18S ribosomal RNAs.

of α expression. Characterization of α3 expression demonstrates high levels of expression in the brain and spinal cord. Expression of α3 in the brain appears to be neuron specific with high levels of expression in the large neurons of the cerebellum, cortex, and hippocampus, the Purkinje cells of the cerebellum, and the pyramidal cells of the cortex and hippocampus [72].

Thus, it seems clear that α subunit expression is regulated in a tissue-specific fashion. The α isoform is found mainly in the kidney and transport epithelia, α(+) is expressed largely in the heart and brain, and α3 mainly in neuronal tissue. It is easy to speculate that the α subunit isoforms may have functional differences that are important in the regulation of Na,K-ATPase activity in the different tissues. However, the exact physiological functions of these different α subunit isoforms have not been determined yet. Undoubtedly, an understanding of the significance of the α subunit isoforms will provide insights to the function of the Na,K-ATPase.

IV. STRUCTURE OF THE β SUBUNIT

The complete amino acid sequences for the human [15], pig [26], sheep [16], rat [14], and *Torpedo* [6] β subunits are shown in Figure 7. The subunits consist of 302–305 amino acids and have an approximate molecular weight of 35,000. The mammalian β subunits exhibit a high degree of amino acid homology (~95%) and are 62% homologous with the *Torpedo* protein. When comparing the mammalian β subunits with the β subunit from the *Torpedo*, two regions are highly conserved, a hydrophobic region consisting of 28 amino acids (35–63, 89% homology) and an uncharged region of 19 residues (233–250, ~89% homology). As indicated by arrows in Figure 7 the *Torpedo* β subunit has four potential sites for *N*-linked glycosylation (Asn-X-Ser/Thr) whereas the mammalian subunits have three. A substitution in the mammalian sequences at amino acid 114 converts an asparagine residue to Lys, eliminating the most amino terminal glycosylation site. Isolation and characterization of glycosylated peptide fragments [26] and in vitro expression of β subunit cDNA [74] have clearly demonstrated that all of the glycosylation sites are utilized in the mammalian β subunits.

It also appears that the β subunit has one or two disulfide bonds that may be important for activity of the enzyme [75,76]. For example, reduction of the β subunit with 2-mercaptoethanol completely inhibits ATPase activity, suggesting that a change in the structure of the β subunit can lead to a loss in activity of the enzyme [76]. Isolation and characterization of β subunit tryptic fragments indicates that Cys^{159} and Cys^{175} are cross-linked by a sulfide bond [54]. It is not yet clear whether this is the disulfide bond important for ATPase activity.

The β subunit has at least one transmembrane segment because it possesses antigenic determinants at the cytoplasmic surface [77], and can be covalently labeled from the extracellular surface [78]. Isolation of several soluble tryptic fragments corresponding to most of the exposed polypeptide demonstrate that a majority of the β subunit is extramembranous [26,54], and labeling studies using Bolton-Hunter reagent indicate that most of it is extracellular [79]. The hydropathy profile of the human β subunit is shown in Figure 8. The hydropathy profiles and the predicted secondary structures for all the β subunits are nearly identical, implying that the amino acid substitutions have not changed the overall structural organization of the peptides. The hydropathy profile suggests the presence of a single transmembrane domain corresponding to the highly conserved region consisting of residues 35–62. The presence of a single transmem-

```
Human    MARGKAKEEG*SWKKFIWNS  EKKEFLGRTGGSWFKILLFY  VIFYGCLAGIFIGTIQVMLL  TISEFKPTYQDRVAPPGLTQ  IPQIQKTEISFRNDPKSYE
Pig      MARGKAKEEG*SWKKFIWNS  EKKEFLGRTGGSWFKILLFY  VIFYGCLAGIFIGTIQVMLL  TISEFKPTYQDRVAPPGLTQ  IPQSQKTEISFRNDPQSYE
Sheep    MARGKAKEEG*SWKKFIWNS  EKKEFLGRTGGSWFKILLFY  VIFYGCLAGIFIGTIQVMLL  TISEFKPTYQDRVAPPGLTQ  IPQIQKTEIAFRNDPKSYM
Rat      MARGKAKEEG*SWKKFIWNS  EKKEFLGRTGGSWFKILLFY  VIFYGCLAGIFIGTIQVMLL  TISELKPTYQDRVAPPGLTQ  IPQIQKTEISFRNDPKSYE
Torpedo  MAREKSTDDGGGWKKFLWDS  EKKQVLGRTGISWFKIFVFY  LIFYGCLAGIFIGTIQVMLL  TISDFEPKYQDRVAPPGLSH  SPYAVKTEISFSVSNPNSYE


Human    AYVLNIVRFLEKYKDSAQRD  DMIFEDCGDVPSEPKERGDF  NHERGERKVCRFKLEWLGNC  SGLNDETYGYKEGKPCIIIK  LNRVLGFKPKPPKNES.LET
Pig      SYVVSIVRFLEKYKDLAQKD  DMIFEDCGNVPSELKERGEY  NNERGERKVCRFRLEWLGNC  SGLNDETYGYKQGKPCVIIK  LNRVLGFKPKPPKNES.LET
Sheep    TYVDNIDNFIKKYRDSAQKD  DMIFEDCGNVPSELKDRGEF  NNEQGERKVCRFKLEWLGNC  SGLNDETYGYKEGKPCVIIK  LNRVLGFKPKPPKNES.LET
Rat      AYVLNIIRFLEKYKDSAQKD  DMIFEDCGSMPSEPKERGEF  NHERGERLVCRFKLDWLGNC  SGLNDESYGYKEGKPCIIIK  LNRVLGFKPKPPKNES.LET
Torpedo  NEVNGLKELLKNYNESKQDG  NTPFEDCGVIPADYITRGPI  EESQGQKRVCRFLLQWLKNC  SGIDDESYGYSEGKPCIIAK  INRILGFIPKPPKNGTDIEE


Human    YPV*MKYNPNVLPVQCTGKR  DEDKDKVGNVEYFGLGNSPGF  PLQYYPYYGKLLQPKYLQP  LLAVQFTNLTMDTEIRIECK  AYGENIGYSEKDRFQGRFDV
Pig      YPV*MKYNPYVLPVHCTGKR  DEDKEKVGTMEYFGLGGYPGF  PLQYYPYYGKLLQPKYLQP  LMAVQFTNLTMDTEIRIECK  AYGENIGYSEKDRFQGRFDV
Sheep    YPV*MKYNPYVLPVQCTGKR  DEDKEKVGSIEYFGLGGYPGF  PLQYYPYYGKLLQPKYLQP  LLAVQFTNLTMDTEIRIECK  AYGENIGYSEKDRFQGRFDV
Rat      YPLTMKYNPNVLPVQCTGKR  DEDKDKVGNIEYFGMGGEYGF  PLQYYPYYGKLLQPKYLQP  LLAVQFTNLTLDTEIRIECK  AYGENIGYSEKDRFQGRFDV
Torpedo  ALQ*ANYNQVIPIHCQAKK   EEDKVRLGTIEYFGMGGVGGF  PLQYYPYYGKBLQNYLQP   LVGIQFTNLTHNVLRVECK   VEGDNIAYSEKDRSLGRFEV


Human    KIEVKS
Pig      KIEVKS
Sheep    KIEVKS
Rat      KIEVKS
Torpedo  KIEVKS
```

Figure 7 Amino acid sequence of the Na,K-ATPase β subunits. Differences from the human sequence are underlined. Potential glycosylation sites are noted by arrows.

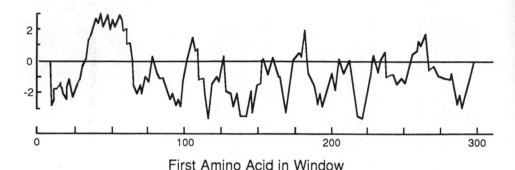

First Amino Acid in Window

Figure 8 Hydropathy profile of the human Na,K-ATPase β subunit. Hydropathy plot of the β subunit was generated by the method of Kyte and Doolittle [103]. Hydrophobic regions are above the x-axis and hydrophilic regions below.

brane domain would predict a β subunit with a highly charged cytoplasmic amino-terminus, followed by a single transmembrane region and a large extracellular carboxy-terminal domain. A model of the β subunit showing the transmembrane orientation, location of the glycosylation sites, and position of the disulfide bond is shown in Figure 9. The conserved region consisting of residues 233–250 and comprising only uncharged amino acids is shown to be associated with the membrane. Although this region probably does not span the membrane it may be in contact or embedded in the membrane or involved in hydrophobic interactions with the α subunit.

In contrast to the α subunit, the rat and human β subunit gene encodes four distinct mRNA species that are expressed in a tissue-specific fashion. The β subunit mRNA levels also vary considerably from tissue to tissue, suggesting that transcriptional control mechanisms may in part account for the differences in Na,K-ATPase activity. In the rat the relative levels of β subunit mRNAs generally mimic the levels of mRNA coding for the α subunit, with one notable exception: no hybridizable β subunit mRNAs can be detected in liver cells even though mRNA coding for the α subunit is present [14]. These results imply that the steady-state levels of β subunit mRNA in liver cells may be extremely low or that liver cells have a structurally divergent form of the β subunit. An alternative explanation is that the β subunit may not be present in rat hepatocytes [80]. Also, different tissues exhibit distinctive preferences for individual transcripts. For example, rat kidney accumulates the largest β subunit mRNA in preference to the other three, whereas different mRNAs are preferred in the bladder and brain. The presence of multiple polyadenylation sites suggests that the different mRNA transcripts may be a result of the utilization of alternate polyadenylation signals. In this regard it is interesting that the 3′ noncoding regions of

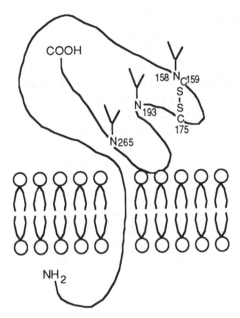

Figure 9 Model of the β subunit of the Na,K-ATPase. The three glycosylation sites and the position of the disulfide bridge are shown. Modified from Ohta, et al., [54]. See text for details.

human, rat, and *Torpedo* β subunit mRNA exhibit a remarkable degree of homology. For example, a region of > 140 nucleotides in the 3′ untranslated region of the rat mRNA differs by only one nucleotide when compared with the human sequence. This high degree of nucleotide sequence conservation suggests that the 3′-untranslated region plays an important role in β subunit biogenesis.

V. STRUCTURE OF THE γ SUBUNIT
The γ subunit is a hydrophobic peptide with a molecular weight of 10,000–12,000. As mentioned before, the strongest evidence that the γ subunit is associated with the Na,K-ATPase is that it can be photoaffinity labeled with ouabain derivatives [8-11]. Quantitation of γ subunit from lamb and shark Na,K-ATPase indicates ∼1 γ subunit per α subunit, suggesting that the γ subunit is present in the enzyme in stoichiometric amounts [81]. When the lamb kidney γ peptides are chromatographed in nonpolar solvents two distinct proteolipid fractions, designated γ1 and γ2 can be identified. Determination of the amino acid compositions of the γ peptides demonstrates that γ1 and γ2 have similar amino acid

compositions, although γ1 has more hydrophobic and fewer charged residues
[81]. Although little is known about the role of the γ subunit, the γ subunits
from lamb kidney and shark rectal gland have similar amino acid compositions,
implying that the amino acid sequence is conserved and therefore important for
Na,K-ATPase structure or function [12]. However, detergent treatment of avian
and shark Na,K-ATPase under nondenaturing conditions separates the γ subunit
from the α and β subunits and the ATPase activity, indicating that the γ subunit
is not essential for the hydrolysis of ATP [12]. Other evidence suggesting that
the γ subunit is not essential for activity comes from analogy with the Ca-ATP-
ase, which when highly purified contains a small hydrophobic peptide similar to
the γ subunit. MacLennan and colleagues have been able to reconstitute the
transport activity of the Ca-ATPase into phospholipid vesicles after removing
all the small hydrophobic peptides from the enzyme [82]. Although it appears
that the γ subunit is not required for catalytic activity of the Na,K-ATPase, the
subunit may have a regulatory role. However, speculation concerning the exact
nature and function of the γ peptide is best constrained until there is at least
a rudimentary understanding of the subunit.

VI. STRUCTURE OF THE Na,K-ATPase

Although much information is known about the structure of the α and β subunits, relatively little is known concerning the subunit interactions of the Na,K-
ATPase. The exact subunit structure of the enzyme and the minimum functional
unit for the active transport of cations has not yet been determined. Uncertainty
in the molecular weights of the α and β subunits has caused some ambiguity in
the determination of the molar α/β ratio. Assuming molecular weights for the α
and β subunits of 112,000 and 45,000, respectively, the quantitation of the α/β
mass ratio by several different methods generates an α/β ratio of approximately
one [22,83–86]. Crosslinking studies [85], high performance gel chromatography [87], and measurements of the ratio of the amino-terminal residues for
the subunits [88] also support a 1:1 ratio. Thus, several lines of evidence seem
to support a model for the Na,K-ATPase in which the number of α and β subunits are equal.

Although there is general agreement for a 1:1 α/β ratio, there is much less
agreement about whether the αβ unit alone represents the active form of the enzyme (reviewed in [1,2]). However, there is some evidence suggesting that the
αβ unit has enzymatic activity. In 1981, Brotherus et al. demonstrated that
solubilization of pig kidney Na,K-ATPase in the n-dodecyl octaethylene glycol
monoether ($C_{12}E_8$) results in predominantly αβ units that retain ATPase activity [89]. More recently it has been shown that each soluble αβ unit can bind
one molecule of ATP [90], undergo transitions between the E_1 and E_2 conformations of the α subunit [91], and when reconstituted into phospholipid

vesicles can transport Na and K [92]. However, it is not clear whether the $\alpha\beta$ protomer by itself can form the pathway for the transport of cations across the membrane or if active transport requires an association between $\alpha\beta$ units in the membrane. Indeed, there are several lines of evidence suggesting that the active unit of the Na,K-ATPase involves an aggregation of the $\alpha\beta$ protomers. Molecular weight determinations using analytical ultracentrifugation [93,94], low-angle laser light scattering [87], and radiation inactivation [86,95,96], and evidence from cross-linking [97-99] and electron microscopy [100,101] support oligo- metric models for Na,K-ATPase-mediated transport. Thus, although it seems certain that the $\alpha\beta$ protomer has independent catalytic activity, it cannot be excluded that the transport of Na and K across the cellular membrane requires the formation of oligomeric complexes.

VII. CONCLUSION

Tremendous strides have been made in the understanding of the structure of the Na,K-ATPase. The recent isolation of the DNA coding for the α and β subunits will undoubtedly provide opportunities for further understanding this important enzyme. In view of the recent advances it may be possible to correlate some of the functional studies of the Na,K-ATPase with the structural aspects of the en- zyme to obtain some fundamental observations on the molecular mechanisms of transport.

REFERENCES

1. Skou, J. C. (1957). The influence of some cations on an adenosine triphos- phate from peripheral nerves. *Biochim. Biophys. Acta* 23:394–401.
2. Glynn, I. M. (1985). The Na^+, K^+ transporting adenosine triphosphatase. In *The Enzymes of Biological Membrane*, Vol. 3, 2nd ed. A. Martonosi (Ed.). Plenum, New York, pp. 35–114.
3. Jorgensen, P. L.(1982). Mechanism of the Na^+, K^+ pump. Protein structure and conformation of the pure ($Na^+ + K^+$)-ATPase. *Biochim. Biophys. Acta* 694:27–68.
4. Shull, G. E., Schwartz, A., and Lingrel, J. B. (1985). Amino-acid sequence of the catalytic subunit of the ($Na^+ + K^+$)ATPase deduced from a comple- mentary DNA. *Nature* 316:691–695.
5. Kawakami, K., Noguchi, S., Noda, M., Takahashi, H., Ohta, T., Kawamura, M., Nojima, H., Nagano, K., Hirose, T., Inayama, S., Hayashida, H., Miyata, T., and Numa, S. (1985). Primary sequence of the α-subunit of *Torpedo californica* ($Na^+ + K^+$)-ATPase deduced from the cDNA sequence. *Nature* 316:733–736.
6. Noguchi, S., Noda, M., Takahashi, H., Kawakami, K., Ohta, T., Nagano, K., Hirose, T., Inayama, S., Kawamura, M., and Numa, S. (1986). Primary struc- ture of the β-subunit of *Torpedo californica* ($Na^+ + K^+$)-ATPase deduced from the cDNA sequence. *FEBS Lett.* 196:315–319.

7. Hokin, L. E., Dahl, J. L., Duupree, J. D., Dixon, J. F., Hackney, J. F., and Perdue, J. F. (1973). Studies on the characterization of the sodium-potassium transport of adenosine triphosphatase. X. Purification of the enzyme from rectal gland of *Squalus acanthias. J. Biol. Chem.* 248:2593–2605.

8. Forbush, B., Kaplan, J. H., and Hoffman, J. F. (1978). Characterization of a new photoaffinity derivative of ouabain: labeling of the large polypeptide and of a proteolipid component of the Na,K-ATPase. *Biochemistry* 17: 3667–3676.

9. Rogers, T. B., and Lazdunski, M. (1979). Photoaffinity labeling of the digitalis receptor in the (sodium + potassium)-activated adenosinetriphosphatase. *Biochemistry* 18:135–140.

10. Rogers, T. B., and Lazdunski, M. (1979). Photoaffinity labelling of a small protein component of a purified (Na$^+$+K$^+$)ATPase. *FEBS Lett.* 98:373–376.

11. Collins, J. H., Forbush, B., Lane, L. K., Ling, E., Schwarz, A., and Zot, A. (1982). Purification and characterization of an (Na$^+$ + K$^+$)-ATPase proteolipid labeled with a photoaffinity derivative of ouabain. *Biochim. Biophys. Acta* 686:7–12.

12. Hardwicke, P. M. D., and Freytag, J. W. (1982). A proteolipid associated with Na,K-ATPase is not essential for ATPase activity. *Biochem. Biophys. Res. Commun.* 102:250–257.

13. Schneider, J. W., Mercer, R. W., Caplan, M., Emanuel, J. R., Sweadner, K. J., Benz, E. J., Jr., and Levenson, R. (1985). Molecular cloning of rat brain Na,K-ATPase α-subunit cDNA. *Proc. Natl. Acad. Sci. USA* 82:6357–6361.

14. Mercer, R. W., Schneider, J. W., Savitz, A., Emanuel, J., Benz, E. J., Jr., and Levenson, R.(1986). Rat-brain Na,K-ATPase β-chain gene: Primary structure, tissue-specific expression, and amplification in ouabain-resistant HeLa C$^+$ cells. *Mol. Cell. Biol.* 6:3884–3890.

15. Kawakami, K., Nojima, H., Ohta, T., and Nagano, K. (1986). Molecular cloning and sequence analysis of human Na,K-ATPase β-subunit. *Nucleic Acids Res.* 14:2833–2844.

16. Schull, G. E., Lane, L. K., and Lingrel, J. B. (1986). Amino-acid sequence of the β-subunit of the (Na$^+$ + K$^+$)ATPase deduced from a cDNA. *Nature* 321: 429–431.

17. Hoffman, J. F. (1986). Active transport of Na$^+$ and K$^+$ by red blood cells. In *Physiology of Membrane Disorders*, T. E. Andreoli, J. F. Hoffman, D. D. Fanestil, and S. G. Schultz, (Eds.). Plenum, New York, pp. 221–234.

18. Hopkins, B. E., Wagner, H., and Smith, T. W. (1976). Sodium- and potassium-activated adenosine triphosphatase of the nasal salt gland of the duck (*Anas platyrhynchos*). Purification, characterization, and NH$_2$-terminal amino acid sequence of the phosphorylating polypeptide. *J. Biol. Chem.* 251:4365–4371.

19. Perrone, J. R., Hackney, J. F., Dixon, J. F., and Hokin, L. E. (1975). Molecular properties of purified (sodium and potassium)-activated adenosine triphosphatases and their subunits from the rectal gland of *Squalus acanthias*

and the electric organ of *Electrophorus electricus. J. Biol. Chem.* 250:4178–4184.

20. Kyte, J. (1971). Purification of the sodium- and potassium-dependent adenosine triphosphatase from canine renal medulla. *J. Biol. Chem.* 246:4157–4165.

21. Jorgensen, P. L., and Skou, J. C. (1971). Purification and characterization of (Na$^+$ + K$^+$)-ATPase. I. Influence of detergents on the activity of (Na$^+$ + K$^+$)-ATPase in preparations from the outer medulla of rabbit kidney. *Biochim. Biophys. Acta* 233:366–380.

22. Jorgensen, P. L. (1974). Purification and characterization of (Na$^+$ + K$^+$)-ATPase. III. Purification from the outer medulla of mammalian kidney after selective removal of membrane components by SDS. *Biochim. Biophys. Acta* 356:36–52.

23. Sweadner, K. J. (1979). Two molecular forms of (Na$^+$ + K$^+$)-stimulated ATPase in brain. Separation and difference in affinity for strophanthidin. *J. Biol. Chem.* 254:6060–6067.

24. Dixon, J. F., and Hokin, L. E. (1978). A simple procedure for the preparation of highly purified (sodium + potassium) adenosine triphosphatase from the rectal gland of *Squalus acanthias* and the electric organ of *Electrophorus electricus. Arch. Biochem.* 86:378–385.

25. Kawakami, K., Ohta, T., Nojima, H., and Nagano, K. (1986). Primary structure of the α-subunit of human Na,K-ATPase deduced from cDNA sequence. *J. Biochem.* 100:389–397.

26. Ovchinnikov, Y. A., Modyanov, N. N., Broude, N. E., Petrukhin, K. E., Grishin, A. V., Arzamazova, N. M., Aldanova, N. A., Monastyrskaya, G. S., and Sverdlov, E. D. (1986). Pig kidney (Na$^+$ + K$^+$)-ATPase. *FEBS Lett.* 201:237–245.

27. Schull, G. E., Greeb, J., and Lingrel, J. B. (1986). Molecular cloning of three distinct forms of the Na$^+$, K$^+$-ATPase α-subunit from rat brain. *Biochemistry* 25:8125–8132.

28. Blobel, G., and Dobberstein, B. (1975). Transfer of proteins across membranes. I. Presence of proteolytically processed and unprocessed nascent immunoglobulin light chains on membrane-bound ribosomes of murine myeloma. *J. Cell Biol.* 67:835–851.

29. MacLennan, D. H., Brandl, C. J., Korczak, B., and Green, N. M. (1985). Amino-acid sequence of a Ca^{2+} + Mg^{2+}-dependent ATPase from rabbit muscle sarcoplasmic reticulum, deduced from its complementary DNA sequence. *Nature* 316:696–700.

30. Schull, G. E., and Lingrel, J. B. (1986). Molecular cloning of the rat stomach (H$^+$ + K$^+$)-ATPase. *J. Biol. Chem.* 261:16788–16791.

31. Kopito, R. R., and Lodish, H. F. (1985). Primary structure and transmembrane orientation of the murine anion exchange protein. *Nature* 316:234–238.

32. Mueckler, M., Caruso, C., Baldwin, S. A., Panico, M., Blench, I., Morris, H. R., Allard, W. J., Lienhard, G. E., and Lodish, H. F. (1985). Sequence and structure of a human glucose transporter. *Science* 229:941–945.

33. Hager, K. M., Mandala, S. M., Davenport, J. W., Speicher, D. W., Benz, E. J., Jr., and Slayman, C. W. (1986). Amino acid sequence of the plasma membrane ATPase of *Neurospora crassa*:Deduction from the genomic and cDNA sequences. *Proc. Natl. Acad. Sci. USA* 83:7693–7697.

34. Farley, R. A., Tran, C. H., Carilli, C. T., Hawke, D., and Shrively, J. E. (1984). The amino acid sequence of a fluorescein-labeled peptide from the active site of (Na,K)-ATPase. *J. Biol. Chem.* 259:9532–9535.

35. Kirley, T. L., Wallick, E. T., and Lane, L. K. (1984). The amino acid sequence of the fluorescein isothiocyanate reactive site of lamb and rat kidney Na^+ and K^+-dependent ATPase. *Biochem. Biophys. Res. Comm.* 125:767–773.

36. Ohta, T., Nagano, K., and Yoshida, M. (1986). The active site structure of Na^+/K^+-transporting ATPase: Location of the $5'(p$-fluorosulfonyl) benzoyladenosine binding site and soluble peptides released by trypsin. *Proc. Natl. Acad. Sci. USA* 83:2071–2075.

37. Le, D. T. (1986). Purification of labeled cyanogen bromide peptides of the α polypeptide from sodium ion and potassium ion activated adenosine-triphosphatase modified with N-[^{3}-H]ethylmaleimide. *Biochemistry* 25:2379–2386.

38. McDonough, A. A., Hiatt, A., and Edelman, I. S. (1982). Characteristics of antibodies to guinea pig $(Na^+ + K^+)$-adenosine triphosphatase and their use in cell-free synthesis studies. *J. Membr. Biol.* 69:13–22.

39. Collins, J. H., Zot, A. S., Ball, W. J., Jr., Lane, L. K., and Schwartz, A. (1983). Tryptic digest of the α subunit of lamb kidney $(Na^+ + K^+)$-ATPase. *Biochim. Biophys. Acta* 742:358–365.

40. Farley, R. A., Ochoa, G. T., and Kudrow, Y. (1986). Location of major antibody binding domains on α-subunit of dog $Na^+ + K^+$-ATPase. *Am. J. Physiol.* 250:C896–906.

41. Jorgensen, P. L. (1975). Purification and characterization of $(Na^+ + K^+)$-ATPase. V. Conformational changes in the enzyme. Transitions between the Na-form and the K-form studied with tryptic digestion as a tool. *Biochim. Biophys. Acta* 401:399–415.

42. Jorgensen, P. L. (1977). Purification and characterization of $(Na^+ + K^+)$-ATPase. VI. Differential tryptic modification of catalytic functions of the purified enzyme in the presence of NaCl and KCl. *Biochim. Biophys. Acta* 466:97–108.

43. Castro, J., and Farley, R. A. (1979). Proteolytic fragmentation of the catalytic subunit of the sodium and potassium adenosine triphosphatase. Alignment of tryptic and chymotryptic fragments and the location of sites labelled with ATP and iodoacetate. *J. Biol. Chem.* 254:2221–2228.

44. Jorgensen, P. L., and Klodos, I. (1978). Purification and characterization of $(Na^+ + K^+)$-ATPase. VII. Tryptic degradation of the Na-form of the enzyme resulting in selective modification of dephosphorylation reactions of the $(Na^+ + K^+)$-ATPase. *Biochim. Biophys. Acta* 507:8–16.

45. Jorgensen, P. L., and Anner, B. M. (1979). Purification and chracterization of $(Na^+ + K^+)$-ATPase. VII. Altered Na^+:K^+ transport ratio in vesicles

reconstituted with purified $(Na^+ + K^+)$-ATPase that has been selectively modified with trypsin in the presence of NaCl. *Biochim. Biophys. Acta* 555:485–492.

46. Cooper, J. B., and Winter, C. G. (1980). 5'-p-fluorosulfonylbenzoyladenosine as an ATP site affinity probe for $Na^+ + K^+$-ATPase. *J. Supramol. Struct.* 13:165–174.

47. Cooper, J. B., Johnson, C., and Winter, C. G. (1983). Affinity labeling studies of the ATP binding site of canine kidney Na,K-ATPase. In *Current Topics in Membranes and Transport*, Vol. 19, J. F. Hoffman and B. Forbush (Eds.). Academic Press, New York, pp. 367–370.

48. Karlish, S. J. D. (1979). *Na,K-ATPase Structure and Kinetics*, J. C. Skou and J. G. Norby (Eds.). Academic Press, London, pp. 115–128.

49. Karlish, S. J. D. (1980). Characterization of conformational changes in (Na,K)ATPase labeled with fluorescein at the active site. *J. Bioenerg. Biomembr.* 12:111–136.

50. Albers, R. W., Fahn, S., and Koval, G. J. (1963). The role of sodium ions in the activation of *Electrophorus* electric organ adenosine triphosphatase. *Proc. Natl. Acad. Sci. USA* 50:474–481.

51. Post, R. L., Sen, A. K., and Rosenthal, A. S. (1965). A phosphorylated intermediate in adenosine triphosphate-dependent sodium and potassium transport across kidney membranes. *J. Biol. Chem.* 240:1437–1445.

52. Bastide, F., Meissner, G., Fleischer, S., and Post, R. L. (1973). Similarity of the active site of phosphorylation of the adenosine triphosphatase for transport of sodium and potassium in kidney to that for transport of calcium ions in the sarcoplasmic reticulum of muscle. *J. Biol. Chem.* 248:8385–8391.

53. Allen, G., and Green, N. M. (1976). A 31-residue tryptic peptide from the active site of the $[Ca^{++}]$-transporting adenosine triphosphatase of rabbit sarcoplasmic reticulum. *FEBS Lett.* 63:188–192.

54. Ohta, T., Yoshida, M., Hirano, K., and Kawamura, M. (1986). Structure of the extra-membranous domain of the β-subunit of (Na,K)-ATPase revealed by the sequences of its tryptic peptides. *FEBS Lett.* 204:297–301.

55. Farley, R. A., Goldman, D. W., and Bayley, H. (1980). Identification of regions of the catalytic subunit of (Na-K)ATPase embedded within the cell membrane. *J. Biol. Chem.* 255:860–864.

56. Nicholas, R. A. (1984). Purification of the membrane-spanning tryptic peptides of the α polypeptide from sodium and potassium ion activated adenosine-triphosphatase labeled with 1-tritiospiro[adamantane-4,3'-diazirine]. *Biochemistry* 23:888–898.

57. Serrano, R., Kielland-Brandt, M. C., and Fink, G. R. (1986). Yeast plasma membrane ATPase is essential for growth and has homology with $(Na^+ + K^+)$, K^+- and Ca^{2+}-ATPases. *Nature* 319:689–693.

58. Giotta, G. J. (1975). Native $(Na^+ + K^+)$-dependent adenosine triphosphatase has two trypsin-sensitive sites. *J. Biol. Chem.* 250:5159–5164.

59. Karlish, S. J. D., and Pick, U. (1981). Sidedness of the effects of sodium and potassium on the conformational state of the sodium-potassium pump. *J. Physiol.* 312:505–529.

60. Krebs, E. G., and Beavo, J. A. (1979). Phosphorylation-dephosphorylation of enzymes. *Annu. Rev. Biochem.* 48:923–959.
61. Colman, R. F. (1983). Affinity labeling of purine nucleotide sites in proteins. *Annu. Rev. Biochem.* 52:67–91.
62. Cantley, L. C., Carilli, C. T., Farley, R. A., and Perlman, D. M. (1982). Location of binding sites on the (Na,K)-ATPase for fluorescein-5'-isothiocyanate and ouabain. *Ann. N.Y. Acad. Sci.* 402:289–291.
63. Walker, J. E., Saraste, M., Runswick, M. J., and Gay, N. J. (1982). Distantly related sequences in the α- and β-subunits of ATP synthase, myosin, kinases and other ATP-requiring enzymes and a common nucleotide binding fold. *EMBO J.* 1:945–951.
64. Wallick, E. T., Pitts, B. J. R., Lane, L. K., and Schwartz, A. (1980). A kinetic comparison of cardiac glycoside interactions with the Na$^+$ + K$^+$-ATPases from skeletal muscle and from the kidney. *Arch. Biochem. Biophys.* 202:442–449.
65. Ahmed, K., Rohrer, D. C., Fullerton, D. S., Detto, T., Kitatsuji, E., and From, A. H. L. (1983). Interaction of (Na$^+$ + K$^+$)-ATPases and digitalis genins. *J. Biol. Chem.* 258:8092–8097.
66. Marks, M. J., and Seeds, N. W. (1978). A heterogeneous ouabain-ATPase interaction in mouse brain. *Life Sci.* 23:2735–2744.
67. Matsuda, T., Iwata, H., and Cooper, J. R. (1984). Specific inactivation of α(+) molecular form of (Na$^+$ + K$^+$)-ATPase by pyrithiamin. *J. Biol. Chem.* 259:35858–35863.
68. Specht, S. C., and Sweadner, K. J. (1984). Two different Na,K-ATPases in the optic nerve: Cells of origin and axonal transport. *Proc. Natl. Acad. Sci. USA* 81:1234–1238.
69. Atterwill, C. K., Cunningham, V. J., and Balazs, R. (1984). Characterization of Na$^+$, K$^+$-ATPase in cultured and separated neuronal and glial cells from rat cerebellum. *J. Neurochem.* 43:8–18.
70. Lytton, J., Lin, J. C., and Guidotti, G. (1985). Identification of two molecular forms of (Na$^+$ + K$^+$)-ATPase in rat adipocytes. *J. Biol. Chem.* 260:1177–1184.
71. Lytton, J. (1985). The catalytic subunits of the (Na$^+$ + K$^+$)-ATPase α and α(+) isozymes are the products of different genes. *Biochem. Biophys. Res. Comm.* 132:764–769.
72. Schneider, J. W., Mercer, R. W., Gilmore-Herbert, M., Utset, M. F., Lai, C., Green, A., and Benz, E. J., Jr. (1988). Tissue specificity, localization in brain, and cell-free translation of mRNA encoding the α3 isoform of Na,K-ATPase. *Proc. Natl. Acad. Sci. USA* 85:284–288.
73. Schneider, J. W., Yang-Feng, T. L., Lindgren, V., Mercer, R. W., Green, A., Benz, E. J., Jr., and Francke, U. (1987). Mapping of the genes for Na,K-ATPase subunits identifies candidate for myotonic dystrophy. *Am. J. Hum. Genet.* 41:A183.
74. Gilmore-Herbert, M., Mercer, R. W., Schneider, J. W., and Benz, E. J., Jr. (1988). In vitro expression of the α and β subunits of the Na,K-ATPase, J. C. Skou and J. G. Nørby (Eds.). Alan R. Liss, New York, pp. 71–76.

75. Esmann, M. (1982). Sulphydryl groups of (Na$^+$ + K$^+$)-ATPase from rectal glands of *Squalus acanthias*. *Biochim. Biophys. Acta* 668:251–259.

76. Kawamura, M., and Nagano, K. (1984). Evidence for essential disulfide bonds in the β-subunit of (Na$^+$ + K$^+$)-ATPase. *Biochim. Biophys. Acta* 774:188–192.

77. Girardet, M., Geering, K., Frantes, J. M., Geser, D., Rossier, B. C., Kraehen-buhl, J.-P., and Bron, C. (1981). Immunochemical evidence for a trans-membrane orientation of both (Na$^+$, K$^+$)-ATPase subunits. *Biochemistry* 20:6684–6691.

78. Hall, C., and Ruoho, A. (1980). Ouabain-binding-site photoaffinity probes that label both subunits of Na$^+$, K$^+$-ATPase. *Proc. Natl. Acad. Sci. USA* 77:4529–4533.

79. Dzhandzhugazyan, K. N., and Jorgensen, P. L. (1985). Asymmetric orienta-tion of amino groups in the α-subunit and the β-subunit of (Na$^+$ + K$^+$)-ATPase in tight right-side-out vesicles of basolateral membranes from outer medulla. *Biochim. Biophys. Acta* 817:165–173.

80. Hubert, J. J., Schenk, D. B., Skelly, H., and Leffert, H. L. (1986). Rat hepatic Na$^+$ + K$^+$-ATPase: alpha subunit isolation by immunoaffinity chromatography and structural analysis by peptide mapping. *Biochemistry* 25:4163–4167.

81. Reeves, A. S., Collins, J. H., and Schwartz, A. (1980). Isolation and charac-terization of (Na,K)-ATPase proteolipid. *Biochem. Biophys. Res. Comm.* 95:1591–1598.

82. MacLennan, D. H., Reithmeier, R. A. F., Shshan, V., Campbell, K. P., Le-Bel, D., Herrman, T. R., and Shamoo, A. E. (1980). Ion pathways in pro-teins of the sarcoplasmic reticulum. *Ann. N.Y. Acad. Sci.* 358:138–148.

83. Peterson, G. L., and Hokin, L. E. (1980). Improved purification of brine-shrimp (*Artemia saline*) (Na$^+$ + K$^+$)-activated adenosine triphosphatase and amino-acid and carbohydrate analogues of the isolated subunits. *Biochem. J.* 192:107–118.

84. Peters, W. H. M., De Pont, J. J. H. H. M., Koppers, A., and Bonding, S. L. (1981). Studies on the (Na$^+$ + K$^+$)-ATPase. XLVII. Chemical composition, molecular weight and molar ratio of the subunits of the enzyme from rabbit kidney outer medulla. *Biochim. Biophys. Acta* 641:55–70.

85. Craig, W. S., and Kyte, J. (1980). Stoichiometry and molecular weight of the minimum asymmetric unit of canine renal sodium and potassium ion-activated adenosine triphosphatase. *J. Biol. Chem.* 255:6262–6269.

86. Bonting, S. L., Swarts, H. G. P., Peters, W. H. M., Schuurmans Stekhoven, F. M. A. H., and De Pont, J. J. H. H. M. (1983). Magnesium-induced and conformational changes in Na,K-ATPase. In *Current Topics in Membranes and Transport*, Vol. 19. J. F. Hoffman and B. Forbush (Eds.). Academic Press, New York, pp. 403–424.

87. Hayashi, Y., Takagi, T., Maezawa, S., and Matsui, H. (1983). Molecular weights of αβ-protomeric and oligomeric units of soluble (Na$^+$ + K$^+$)-ATP-ase determined by low-angle laser light scattering after high-performance gel chromatography. *Biochim. Biophys. Acta* 748:153–167.

88. Cantley, L. C. (1981). Structure and mechanism of the (Na$^+$ + K$^+$)-ATPase. *Curr. Top. Bioenerg.* 11:201–237.

89. Brotherus, J. R., Moller, J. V., and Jorgensen, P. L. (1981). Soluble and active renal Na,K-ATPase with maximum protein molecular mass 170,000 ± 9000 daltons; formation of larger subunits by secondary aggregation. *Biochem. Biophys. Res. Commun.* 100:146–154.

90. Jensen, J., and Ottolenghi, P. (1983). ATP binding to solubilized (Na$^+$ + K$^+$)-ATPase. The abolition of subunit-subunit interaction and the maximum weight of the nucleotide-binding unit. *Biochim. Biophys. Acta* 731:282–289.

91. Jorgensen, P. L., and Andersen, J. P. (1986). Thermoinactivation and aggregation of $\alpha\beta$ units in soluble and membrane-bound (Na,K)-ATPase. *Biochemistry* 25:2889–2897.

92. Brotherus, J. R., Jacobsen, L., and Jorgensen, P. L. (1983). Soluble and enzymatically stable (Na$^+$ + K$^+$)-ATPase from mammalian kidney consisting predominantly of protomer $\alpha\beta$-units. Preparation, assay and reconstitution of active Na$^+$,K$^+$ transport. *Biochim. Biophys. Acta* 731:290–303.

93. Esmann, M., Skou, J. C., and Christiansen, C. (1979). Solubilization and molecular weight determination of Na,K-ATPase from rectal glands of *Squalus acanthias. Biochim. Biophys. Acta* 567:410–420.

94. Esmann, M., Christiansen, C., Karlsson, K. A., Hansson, G. C., and Shou, J. C. (1980). Hydrodynamic properties of solubilized (Na$^+$ + K$^+$)-ATPase from rectal glands of *Squalus acanthias. Biochim. Biophys. Acta* 603:1–12.

95. Ellory, J. C., Green, J. R., Jarvis, S. M., and Young, J. D. (1979). Measurement of the apparent molecular volume of membrane-bound transport systems by radiation inactivation. *J. Physiol.* 295:10–11P.

96. Ottolenghi, P., Ellory, J. C., and Klein, R. (1983). Radiation inactivation analysis of NaK-ATPase. In *Current Topics in Membranes and Transport,* Vol. 19. J. F. Hoffman and B. Forbush (Eds.). Academic Press, New York, pp. 139–143.

97. Askari, A., and Huang, W-H. (1980). Na$^+$,K$^+$-ATPase: Half of the subunits cross-linking reactivity suggests an oligomeric structure containing a minimum of four catalytic subunits. *Biochem. Biophys. Res. Commun.* 93:448–453.

98. Askari, A., Huang, W-H., and Antieau, J.M. (1980). Na$^+$, K$^+$-ATPase: Ligand-induced conformational transitions and alterations in subunit interactions evidenced by cross-linking studies. *Biochemistry* 19:1132–1140.

99. Huang, W-H., and Ascari, A. (1981). Phosphorylation-dependent cross-linking of the α-subunits in the presence of Ca^{2+} and o-phenanthroline. *Biochim. Biophys. Acta* 645:54–58.

100. Skriver, E., Maunsbach, A. B., and Jorgensen, P. L. (1981). Formation of two-dimensional crystals in pure membrane-bound Na$^+$,K$^+$-ATPase. *FEBS Lett.* 131:219–222.

101. Hebert, H., Jorgensen, P., Skriver, E., and Maunsbach, A. B. (1982). Crystallization patterns of membrane-bound (Na$^+$ + K$^+$)-ATPase. *Biochim. Biophys. Acta* 689:571–574.

102. Engelman, D. M., Steitz, T. A., and Goldman, A. (1986). Identifying non-polar transbilayer helices in amino acid sequences of membrane proteins. *Annu. Rev. Biophys. Chem.* 15:321–353.
103. Kyte, J., and Doolittle, R. F. (1982). A simple method for displaying the hydrophobic character of a protein. *J. Mol. Biol.* 157:105–132.
104. Brandl, C. J., and Deber, C. M. (1986). Hypothesis about the function of membrane-buried proline residues in transport proteins. *Proc. Natl. Acad. Sci. USA* 83:917–921.

7

Characterization of the Gene Coding for Human Erythrocyte Protein 4.1
Implications for Understanding Hereditary Elliptocytosis

JOHN G. CONBOY and NARLA MOHANDAS   *Cancer Research Institute, University of California, San Francisco, San Francisco, California*

I. INTRODUCTION

The red blood cell, as it continuously circulates, must be able to undergo extensive deformation and to resist fragmentation. These two essential qualities require a highly deformable yet remarkably stable membrane. Membrane deformability and mechanical stability appear to be regulated by an extensive network of structural proteins that underlies the lipid bilayer and is associated with it by protein–protein and protein–lipid interactions [1–3]. When the integrity of the membrane skeleton is disrupted by structural abnormalities or deficiencies of its protein components, the membrane becomes susceptible to fragmentation and the red blood cells in turn lose their characteristic discoid shape. Evidence to support this contention has been derived from membrane stability studies of red blood cells from patients with defined quantitative or qualitative skeletal protein abnormalities [2,4–12].

The red blood cell membrane has been well characterized biochemically and is composed of a lipid bilayer, integral proteins band 3 and glycophorins, and a skeletal protein network of spectrin, actin, ankyrin, tropomyosin, adducin, and proteins 4.1 and 4.9 [13–15]. A simple model depicting the most abundant constituents of the membrane skeleton, and the major binding interactions that link these proteins together, is shown in Figure 1. Spectrin heterodimers of α and β chains self-associate by head-to-head binding to form tetramers, and are

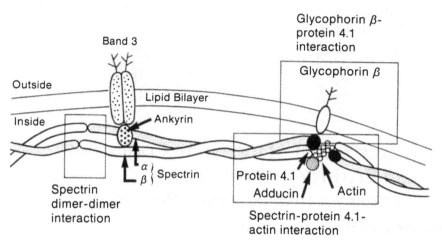

Figure 1 A schematic diagram of the erythrocyte membrane organization based
on our current understanding of various protein associations. The three key protein associations identified to date to be important in regulating membrane properties are shown in enclosed rectangles. These include spectrin dimer-dimer interaction, glycophorin β-protein 4.1 interaction, and spectrin-protein 4.1-actin interaction. Adducin, a recently identified membrane skeleton-associated calmodulin binding protein, promotes binding of spectrin to actin filaments.

further cross-linked at their tails with short actin oligomers through complexes
with protein 4.1 and adducin. This higher-order structure is attached to the
membrane by at least two interactions: a linkage of β spectrin to integral membrane protein band 3 by ankyrin; and a direct binding of protein 4.1 to glycophorin C (sialoglycoprotein β), another integral membrane protein [15-17].

Protein 4.1 is a major component of this membrane skeleton with ~200,000
copies of it present in each cell [18,19]. The designation of "4.1" is based on
the electrophoretic mobility of this protein in sodium dodecylsulfate (SDS)
polyacrylamide gels to a position between band 3 (a principal anion transport
protein in the erythrocyte membrane) and band 5 (actin) [20]. In vitro binding
studies have indicated that protein 4.1 forms a complex with short actin filaments to cross-link spectrin heterodimers at their tail ends [13,21,22]. It also
appears to play a role in anchoring the membrane skeleton to the lipid bilayer
through its interaction with glycophorin C (sialoglycoprotein β) [16,17].
Through these interactions, protein 4.1 appears to play a key role in regulating
erythrocyte membrane physical properties of deformability and mechanical
stability. Confirmation of this important functional role has been obtained in
studies showing that protein 4.1 deficiency results in marked mechanical instability of the membrane, and that normal membrane stability can be restored

to unstable protein 4.1-deficient membranes by reincorporation of the purified protein 4.1 into these membranes [7]. A number of recent studies suggest that both qualitative and quantitative defects in protein 4.1 result in hereditary elliptocytosis [7,23,24].

We have chosen a molecular biological approach to study the biochemistry and genetics of erythrocyte protein 4.1. In a effort to convey the strengths and suitability of this approach to study of the erythrocyte membrane, we have chosen to emphasize the strategies and methods used as well as the results obtained. The major topics considered in this chapter are the following: (a) isolation of cloned DNA sequences encoding protein 4.1; (b) analysis of the primary structure of the protein 4.1 polypeptide deduced from the nucleotide sequence of these clones; (c) structural organization of the protein 4.1 gene in the human genome, including its mapping to a specific chromosome; and (d) study of mutant protein 4.1 genes in patients with hereditary elliptocytosis due to qualitative and quantitative defects in protein 4.1.

II. PROTEIN 4.1 cDNA CLONING

The particular strategy used to clone the erythroid protein 4.1 gene was as follows: A DNA library representing only the subset of genes actively expressed in peripheral blood reticulocytes was constructed using standard enzymatic techniques to synthesize a complementary DNA (cDNA) copy of the complex reticulocyte messenger RNA (mRNA) population [25]. We were confident that this reticulocyte cDNA library would contain 4.1 cDNA sequences, because it was possible to demonstrate the presence of protein 4.1 mRNA in reticulocytes by the functional assay shown in Figure 2. Human reticulocyte mRNA was translated into protein using a standard cell-free protein synthesizing system (a rabbit reticulocyte lysate) [26] in the presence of $[^{35}S]$ methionine, and the labeled proteins were visualized by SDS-polyacrylamide gel electrophoresis (SDS-PAGE) and autoradiography (lane 1). To show that one of the newly-made proteins was protein 4.1, the translation mixture was immunoprecipitated with a monospecific anti-protein 4.1 antibody. A single polypeptide protein 4.1 was detected (lane 3). The polypeptide was not precipitated by nonimmune rabbit serum (lane 2).

The vector into which the reticulocyte cDNA was cloned is a specially modified version of the bacteriophage lambda (λ) designated as λgt 11 [27]. This cloning vehicle is termed an "expression vector" because the inserted cDNAs are inserted into an active gene (β galactosidase) within the vector, allowing them to be transcribed and translated. In our library, bacteria harboring reticulocyte cDNAs in λgt11 should synthesize reticulocyte proteins and be amenable to screening with specific antibodies. When affinity-purified anti-protein 4.1 IgG was used to screen the library for clones producing protein 4.1 antigen(s), immunoreactive clones were detected at a frequency of 1 in 50,000. Five such

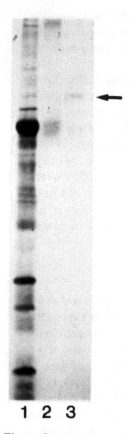

Figure 2 Detection of protein 4.1 mRNA in human reticulocytes by in vitro translation and immunoprecipitation. Shown is an autoradiograph of SDS-PAGE gel (6–15%:). Lanes: (1) Total [^{35}S] methionine-labeled proteins synthesized; (2) proteins immunoprecipitated by using nonimmune rabbit serum; (3) proteins immunoprecipitated by using rabbit anti-human protein 4.1 antibody. (Taken from reference 28).

clones were propagated using standard bacteriological techniques and vector DNA was isolated for further characterization. The putative protein 4.1 cDNA inserts were released from the vector cloning site by cleavage with restriction endonuclease EcoRI, and size determination showed that the cDNA fragments ranged in size from 0.6–1.2 kilobases (kb). This first generation of candidate 4.1 cDNA probes was then used to isolate additional overlapping clones, which together extend over 4 kb of sequence. Figure 3 shows a composite map of these overlapping clones, including as reference points the cleavage sites for several

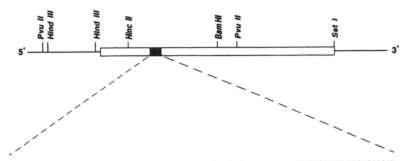

TCATACAGGTCCATGACTCCAGCTCAGGCTGACTTGGAGTTTCTTGAGAATGCCAAAAAGTTGTCTATGTATGGAGTTGAT
ThrProAla Gln Ala AspLeuGluPhe LeuGluAsn Ala Lys LysLeu

Figure 3 Restriction map constructed by analysis of the isolated λgt 11 cDNA clones for protein 4.1. Box represents the coding domain. The amino acid sequence predicted by nucleotides 1303–1350 was identical to a peptide of authentic protein 4.1 sequenced and indicated below the nucleotide sequence.

restriction endonucleases. The nucleotide sequence of these clones has been determined and analyzed to reveal the following information [28]. A 1764-nucleotide "open reading frame," beginning with the translation start codon for methionine (AUG) and ending with the translation termination codon UGA, encodes a 588-amino acid protein. Confirmation that this sequence represents bona fide protein 4.1 was obtained when the amino acid sequence predicted by nucleotides 1303–1350 was found to be identical to a peptide of authentic protein 4.1 we had sequenced earlier over a stretch of 16 consecutive amino acids.

III. CHARACTERIZATION OF ERYTHROCYTE PROTEIN 4.1 cDNA

Nucleotide sequence and predicted amino acid sequence of erythrocyte protein 4.1 derived from this study are shown in Figure 4. A single long open reading frame at positions 799–2562 in cDNA encodes a 588-amino acid protein 4.1 molecule, which has a calculated molecular mass of 66,303 daltons. Amino acid residues 1–42 of the predicted sequence are identical to the amino-terminus of erythrocyte protein 4.1 determined by direct protein sequencing (T. L. Leto, D. W. Speicher, and V. T. Marchesi, personal communication), confirming the assignment of the initiation codon to the AUG signal at positions 799–801. In addition, amino acids 405–471 of the predicted sequence are identical to another peptide of protein 4.1 (the "8k" spectrin-actin binding domain) [41].

The unusually long 5' untranslated region contains four potential AUG translation initiation sites at nucleotides 261–263, 279–281, 461–463, and 611–613.

```
TCC CGT GGA GCA GAG  GGG CAA AGT GGC AGG  AAC CTC TTA AAG GGC  GAG ACG GGC GGC GAC  CGA GAA CGC GGT CGG  CCC GGT CCC CGC CGC  ACC CAG CCC AGA GAA  GAG TTT AGT GAC TGA  120
GGC CGA AAA TTC ACA  GCA CCA ACA GAA GGA  AGA GGG TGA GGA AGC  CAT AAA CTC AGG CCA  ACA AGA ACC TCA GCA  GGA GGA ATC TTG TCA  AAC AGC AGC TGA AGG  AGA TAA TTG GTG TGA  240
ACA CAA GGT GAA AGC  TTC TAA TGG AGA CAC  TCC TAC ACA TGA AGA  CTT GAC CAA GAA CAA  GGA GCG GAC ATC AGA  AAG CAG AGG ACT TTC  ACG ACT ATT CTC CTC  GTT TCT CAA AAG GCC  360
CAA ATC TCA GGT GTC  CGA GGA AGA AGG CAA  AGA AGT AGA GTC AGA  TAA AGA AAA AGG TGA  AGG AGG TCA GAA AGA  GAT AGA ATT TGG AAC  CAG TCT TGA TGA AGA  GAT CAT TTT AAA GGC  480
CCC AAT TGC AGC TCC  TGA ACC GGA ACT CAA  AAC AGA CCC ATC TTT  GGA TCT TCA TTC ATT  AAG CAG TGC AGA AAC  ACA GCC TCA CCA TTA  CAA TTA AGA ATT ATT  TTT AGA GTC TTC TTA  600
TTT CTG AAG CAT GTG  AAT ATT ATC TCG ATC  GTT AAA AGT CCT GCT  CAG GAA GAA CTC AGA  GAA GAT CCA GAT TCT  GAA ATT AAG GAA GGA  GAA GGA CTT GAA GAG  TGC TCC AAA ATA GAA  720
                                                                                                                                                   10
                                                                                                             Met His Cys Lys  Val Ser Leu Leu Asp  Asp Thr Val Tyr Glu
GTA AAA GAA GAA AGC  CCT CAA TCA AAA GCA  GAA ACA GAA TTA AAA  GCT TCC CAA AAA CCA  ATC AGA AAA CAC AGG  AAC ATG CAC TGC AAG  GTT TCT TTG TTG GAT  GAC ACA GTT TAT GAA  840
                     20                                        30                                        40                                        50
Cys Val Val Glu Lys  His Ala Lys Gly Gln  Asp Leu Leu Lys Arg  Val Cys Glu His Leu  Asn Leu Leu Glu Glu  Asp Tyr Phe Gly Leu  Ala Ile Trp Asp Asn  Ala Thr Ser Lys Thr
TGT GTT GTG GAG AAA  CAT GCT AAG GGA CAA  GAT TTG CTT AAA CGA  GTA TGT GAG CAT CTC  AAT CTT TTG GAA GAA  GAC TAT TTT GGT CTA  GCC ATT TGG GAT AAC  GCA ACC TCT AAG ACA  960
                     60                                        70                                        80                                        90
Trp Leu Asp Ser Ala  Lys Glu Ile Lys Lys  Gln Val Arg Gly Val  Pro Trp Asn Phe Thr  Phe Asn Val Lys Phe  Tyr Pro Pro Asp Pro  Ala Gln Leu Thr Glu  Asp Ile Thr Arg Tyr
TGG CTG GAT TCC GCC  AAA GAA ATA AAA AAG  CAG GTT CGT GGT GTC  CCT TGG AAT TTT ACA  TTT AAT GTA AAG TTT  TAT CCA CCT GAC CCA  GCA CAG TTA ACA GAA  GAC ATA ACA AGA TAT  1080
                     100                                       110                                       120                                       130
Tyr Leu Cys Leu Gln  Leu Arg Gln Asp Ile  Val Ala Gly Arg Leu  Pro Cys Ser Phe Ala  Thr Leu Ala Leu Leu  Gly Ser Tyr Thr Ile  Gln Ser Glu Leu Gly  Asp Tyr Asp Pro Glu
TAT TTA TGT CTT CAG  CTT CGG CAG GAC ATA  GTT GCA GGA CGT CTG  CCC TGT TCC TTT GCA  ACC TTA GCA TTA TTA  GGT TCT TAC ACC ATC  CAG TCT GAA CTG GGA  GAC TAC GAC CCA GAA  1200
                     140                                       150                                       160                                       170
Leu His Gly Val Asp  Tyr Val Ser Asp Phe  Lys Leu Ala Pro Asn  Gln Thr Lys Glu Leu  Glu Glu Lys Val Met  Glu Leu His Lys Ser  Tyr Arg Ser Met |Thr  Pro Ala Gln Ala Asp
CTC CAT GGC GTG GAT  TAT GTT AGT GAT TTT  AAA CTG GCC CCG AAT  CAG ACC AAG GAA CTT  GAA GAG AAG GTC ATG  GAA CTG CAT AAG TCA  TAC AGG TCC ATG|ACT  CCA GCT CAG GCT GAC  1320
                     180                                       190                                       200                                       210
Leu Glu Phe Leu Glu  Asn Ala Lys Lys Leu | Ser Met Tyr Gly Val  Asp Leu His Lys Ala  Lys Asp Leu Glu Gly  Val Asp Ile Ile Leu  Gly Val Cys Ser Ser  Gly Leu Leu Val Tyr
TTG GAG TTT CTT GAG  AAT GCC AAA AAG TTG | TCT ATG TAT GGA GTT  GAT CTT CAT AAA GCA  AAG GAC TTG GAA GGA  GTA GAT ATC ATC CTA  GGT GTC TGC TCT AGT  GGC CTT CTG GTT TAC  1440
                     220                                       230                                       240                                       250
Lys Asp Lys Leu Arg  Ile Asn Arg Phe Pro  Trp Pro Lys Val Leu  Lys Ile Ser Tyr Lys  Arg Ser Ser Phe Phe  Ile Lys Ile Arg Pro  Gly Glu Gln Glu Gln  Tyr Glu Ser Thr Ile
AAA GAT AAG CTG AGA  ATT AAC CGC TTC CCT  TGG CCC AAA GTG CTG  AAG ATT TCT TAT AAA  CGT AGT AGC TTT TTC  ATC AAG ATT CGG CCT  GGA GAG CAA GAG CAG  TAT GAA AGT ACC ATC  1560
                     260                                       270                                       280                                       290
Gly Phe Lys Leu Pro  Ser Tyr Arg Ala Ala  Lys Lys Leu Trp Lys  Val Cys Val Glu His  His Thr Phe Phe Arg  Leu Thr Ser Thr Asp  Thr Ile Pro Lys Ser  Lys Phe Leu Ala Leu
GGA TTC AAA CTT CCC  AGT TAC CGA GCA GCT  AAG AAA TTA TGG AAA  GTC TGT GTA GAA CAT  CAC ACG TTT TTC AGA  TTG ACA TCT ACA GAC  ACC ATT CCC AAA AGC  AAA TTT CTT GCG CTA  1680
                     300                                       310                                       320                                       330
Gly Ser Lys Phe Arg  Tyr Ser Gly Arg Thr  Gln Ala Gln Thr Arg  Gln Ala Ser Ala Leu  Ile Asp Arg Pro Ala  Pro His Phe Glu Arg  Thr Ala Ser Lys Arg  Ala Ser Arg Ser Leu
GGA TCC AAA TTT CGA  TAC AGT GGC CGG ACT  CAA GCT CAG ACC AGG  CAA GCT AGT GCT CTA  ATT GAC AGG CCT GCC  CCA CAC TTC GAG CGT  ACA GCA AGT AAA CGG  GCG TCC CGG AGC CTC  1800
                     340                                       350                                       360                                       370
Asp Gly Ala Ala Ala  Val Asp Ser Ala Asp  Arg Ser Pro Arg Pro  Thr Ser Ala Pro Ala  Ile Thr Gln Gly Gln  Val Ala Glu Gly Gly  Val Leu Asp Ala Ser  Ala Lys Lys Thr Val
GAT GGA GCA GCA GCT  GTC GAT TCG GCA GAC  CGA AGT CCT CGG CCC  ACT TCT GCA CCT GCC  ATT ACT CAG GGT CAG  GTT GCA GAA GGT GGC  GTC CTA GAT GCC TCT  GCT AAA AAA ACA GTG  1920
                     380                                       390                                       400                                       410
Val Pro Lys Ala Gln  Lys Glu Thr Val Lys  Ala Glu Val Lys Lys  Glu Asp Glu Pro Pro  Glu Gln Ala Glu Pro  Glu Pro Thr Glu Ala  Trp Lys Lys Lys Arg  Glu Arg Leu Asp Gly
GTC CCT AAA GCA CAG  AAG GAA ACA GTG AAG  GCT GAA GTG AAA AAG  GAA GAC GAG CCA CCT  GAG CAA GCT GAG CCA  GAG CCC ACA GAA GCA  TGG AAG AAA AAG AGA  GAA AGA CTA GAT GGT  2040
                     420                                       430                                       440                                       450
Glu Asn Ile Tyr Ile  Arg His Ser Asn Leu  Met Leu Glu Asp Leu  Asp Lys Ser Gln Glu  Glu Ile Lys Lys His  His Ala Ser Ile Ser  Glu Leu Lys Lys Asn  Phe Met Glu Ser Val
GAA AAC ATT TAT ATC  AGA CAT AGC AAT TTA  ATG TTG GAG GAT TTA  GAC AAG AGT CAA GAG  GAG ATC AAA AAA CAT  CAT GCC AGC ATC AGT  GAG CTG AAA AAG AAC  TTC ATG GAG TCT GTA  2160
                     460                                       470                                       480                                       490
Pro Glu Pro Arg Pro  Ser Glu Trp Asp Lys  Arg Leu Ser Thr His  Ser Pro Phe Arg Thr  Leu Asn Ile Asn Gly  Gln Ile Pro Thr Gly  Glu Gly Pro Pro Leu  Val Lys Thr Gln Thr
CCA GAA CCA CGG CCT  AGT GAA TGG GAT AAA  CGC TTA TCC ACT CAC  TCA CCC TTC CGA ACT  CTT AAC ATC AAT GGG  CAA ATC CCC ACA GGA  GAA GGA CCT CCC CTG  GTG AAG ACA CAA ACT  2280
                     500                                       510                                       520                                       530
Val Thr Ile Ser Asp  Asn Ala Asn Ala Val  Lys Ser Glu Ile Pro  Thr Lys Asp Val Pro  Ile Val His Thr Glu  Thr Lys Thr Ile Thr  Tyr Glu Ala Ala Gln  Thr Val Lys Gly Gly
GTC ACC ATC TCA GAT  AAT GCC AAT GCT GTG  AAA AGT GAA ATC CCA  ACC AAA GAC GTC CCT  ATT GTC CAC ACT GAG  ACC AAG ACC ATC ACT  TAT GAG GCT GCC CAG  ACT GTA AAA GGT GGG  2400
                     540                                       550                                       560                                       570
Ile Ser Glu Thr Arg  Ile Glu Lys Arg Ile  Val Ile Thr Gly Asp  Ala Asp Ile Asp His  Asp Gln Val Leu Val  Gln Ala Ile Lys Glu  Ala Lys Glu Gln His  Pro Asp Met Ser Val
ATT TCA GAG ACA CGT  ATT GAA AAG AGA ATT  GTG ATC ACA GGA GAT  GCT GAT ATT GAC CAT  GAT CAG GTC CTT GTA  CAA GCC ATC AAG GAG  GCA AAG GAG CAG CAC  CCA GAC ATG TCA GTG  2520
                     580
Thr Lys Val Val Val  His Gln Glu Thr Glu  Ile Ala Asp Glu END
ACC AAG GTG GTC GTC  CAC CAG GAG ACC GAG  ATT GCT GAT GAG TGA  GCT CAG GAA CTA ACC  TAC CCC AAC TCT GCC  CTT CTC CCA TCC AAG  AGA AAC CAC GAA AAT  GAT AAA GAA GCT AAC  2640
CTG CCA TAG TCA GAC  TTC AGA CTT TCA AGA  TTA TTC TAA ATC ACC  AGA AAA TTA ATT TCA  GTT TCT ATT GGG AGT  TTA TAC CAA GAG ATT  CTT CTA GAT CTC ATT  GAT CCT TTT GAA GAG  2760
CTT TTT CTA TAT TAG  GAT ATC AGA ATT GTT  CAA CTT TTC ACT CTA  TAG ACT GTT TTA AGA  GTT TTG GGG GGT TTT  TAA TTG GGT GGT TTG  TAA CCC CTT CAG CCT  AG   2867
```

Figure 4 Nucleotide sequence of protein 4.1 cDNA and predicted amino acid sequence of erythrocyte protein 4.1. Boxed area indicates amino acid sequence determined by analysis of a cyanogen bromide-cleaved tryptic peptide of purified protein 4.1. (Taken from reference 28.)

These upstream sites are unlikely to function as translation signals for the following reasons: All four lie in a different reading frame from the functional signal at positions 799–801, all four are followed immediately by in-phase termination codons, and only one contains the required purine nucleotide at position –3 of the consensus sequence surrounding functional AUG translation initiation sites. In contrast, the AUG at positions 799–801 does conform to the consensus sequence and is followed by a long open reading frame. To verify that the long 5' untranslated region represents true protein 4.1 cDNA sequences and not a cloning artifact, two fragments of cDNA extending from nucleotides 1–254 and 254–766 were subcloned separately and radiolabeled for use in RNA blotting experiments. Both 5' probes hybridized to a 5.6 kb mRNA from reticulocytes, indistinguishable from that recognized by coding-region probes.

IV. PRIMARY AND SECONDARY STRUCTURE OF PROTEIN 4.1

The amino acid sequence derived from protein 4.1 cDNA was used to predict the major secondary structural features of the protein, as well as to characterize the biochemical properties of the various functional domains of the protein (Figure 5). These domains, defined originally by limited chymotryptic digestion, serve distinct functional roles critical to the integrity of the membrane skeleton [29]. The amino-terminal 30-kD "membrane-binding" domain, which provides one type of membrane-skeleton-lipid bilayer linkage, is very hydrophobic (36% aromatic plus Ile, Leu, Met, and Val residues). Among these hydrophobic residues are 33 potential chymotryptic cleavage sites that are not digested unless high enzyme/substrate ratios are used. In addition to its overall hydrophobicity, this domain contains all seven Cys residues present in the molecule. Two potential cAMP-dependent phosphorylation sites at Ser^{185} and Ser^{237}, and three potential O-linked glycosylation sites at Thr^{51}, Thr^{74}, and Thr^{151} are also present in this domain. Secondary structure prediction indicates that this domain contains a substantial β sheet structure, in addition to a number of amphipathic α helical structures.

The 16-kD domain, which resides between the 30-kD membrane-binding domain and the 8-kD spectrin-binding domain, is quite hydrophilic (only 14% hydrophobic residues) and has an unusually high proline content (9.3%). α Helical structure is the principal feature of the secondary structure of this domain.

The "8-kD" spectrin-binding domain is highly charged (46% Asp, Glu, Arg, Lys, and His). Secondary structure predictions indicate that this region contains a long α helix as its most prominent feature. A potential cAMP-dependent phosphorylation site is located at Ser^{467}. The carboxy-terminal domain (22–24 kD), as reported earlier, is quite acidic. The most remarkable feature of this domain

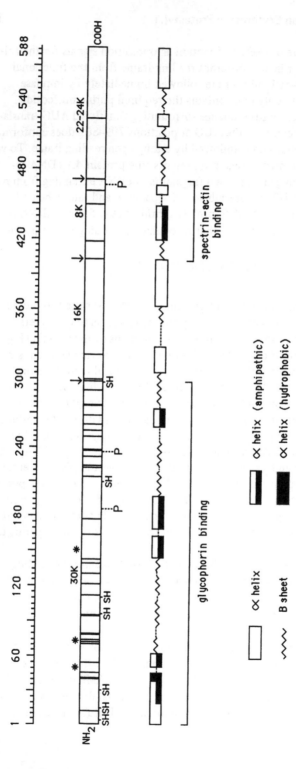

Figure 5 Secondary structural features of protein 4.1. Scale indicates amino acid number. Arrows indicate boundaries of the four domains of protein 4.1 molecule identified by limited chymotryptic cleavage. Potential chymotryptic sites are indicated by vertical lines within the closed box. Potential glycosylation sites (Asn-Xaa-Ser) are indicated by asterisks (*), potential cAMP-dependent phosphorylation sites (basic-basic-Xaa-Ser) are indicated by "p"; "SH" denotes cystein residues. (Taken from reference 28.)

is its short length: it contains 117 amino acids with a calculated size of 12,636 D, far below its apparent size of 22-24 kD in the SDS-PAGE system.

V.   ERYTHROCYTE PROTEIN 4.1 ISOFORMS

The initial characterization of protein 4.1 cDNA enabled us to deduce the complete amino acid sequence from three overlapping cDNA clones of what was presumed at the time to be a unique polypeptide. As expected, the predicted protein sequence was identical to that of several subdomains determined by direct protein sequencing. Unexpectedly, however, recent studies provide evidence for heterogeneity in erythrocyte protein 4.1 structure, i.e., several structural isoforms may exist. From the original reticulocyte cDNA library, we have now characterized clones that yield different amino acid sequences affecting the structure of the 8-kD spectrin–actin binding domain, the 22/24-kD carboxy-terminal domain, as well as the 30-kD membrane-binding domain of the protein 4.1 molecule [30]. The cDNAs encoding these protein 4.1 isoforms are characterized by extensive regions of perfect homology, interrupted at defined sites by insertion or deletion of sequence domains that appear to represent exons of a single protein 4.1 gene. The predicted structure of two of the isoforms involving the 8-kD spectrin–actin binding domain are characterized by deletion of 21 amino acids from one isoform and not the other. In the carboxy-terminal 22-24-kD domain, three structural isoforms are created by alternate expression of a 43- or 34-amino acid peptide, or both, into an otherwise identical constituitively expressed sequence. Analysis of the genomic sequence encoding the 34-amino acid addition to the 22/24 kD carboxy-terminal domain showed that the coding sequence is flanked by intronlike sequences that exhibit homology to 3' and 5' splice sites, implying that the addition or deletion of coding sequences in protein 4.1 is the result of alternative mRNA splicing. These data imply that the erythroid cells are capable of synthesizing multiple isoforms of protein 4.1 and suggest that these different isoforms may account for the multiple functions of this structural protein in membrane assembly and dynamic regulation of membrane properties. However, it should be noted that although these studies have helped define the molecular basis for generation of various isoforms, the relative expression of these isoforms at different stages of erythroid development and the possible differences in the functional characteristics of the various isoforms is yet to be defined.

VI.   ASSIGMENT OF THE PROTEIN 4.1 GENE
TO CHROMOSOME 1

Erythroid protein 4.1 cDNA was used as a probe to determine the location of the protein 4.1 gene in the human genome. Human metaphase chromosomes prepared from cultured lymphocytes were stained with fluorescent dyes and

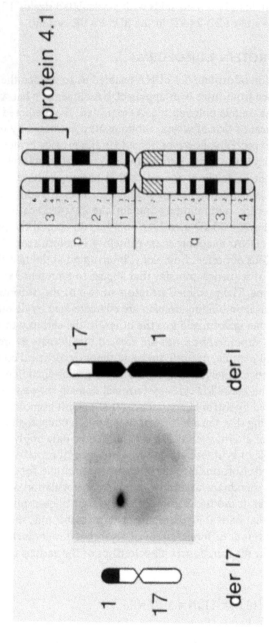

Figure 6 Chromosomal assignment of the erythrocyte protein 4.1 gene. Left panel shows hybridization of the 2.5-kb probe to derivative chromosomes 1 (der 1) and 17 (der 17) sorted from cell line GM201, which contains a chromosome 1 : 17 translocation. An idiogram of chromosome 1 illustrating the segment in which the protein 4.1 gene is licated is shown in the right panel. (Taken from Ref. 31.)

sorted into 20 fractions using a fluorescence-activated cell sorter. DNA from each chromosomal fraction was fixed on nitrocellulose filter disks and hybridized to radiolabeled protein 4.1 cDNA. When bound probe was visualized by autoradiography, only chromosome 1 gave a positive signal, thus enabling the assignment of protein 4.1 to human chromosome 1 [31].

The gene was sublocalized to a smaller region within chromosome 1 by hybridization of the cDNA probe to chromosomes sorted from a cell line (GM-201) carrying a balanced translocation between chromosome 1 and 17. Figure 6 shows the derivative 17 chromosome (der 17) hybridized strongly to the protein 4.1 probe, whereas derivative chromosome 1 (der 1) did not. Because the derivative 17 chromosome carries the portion of the short arm (p) of chromosome 1 from p32 to the terminus, the protein 4.1 gene must be localized in this region of the short arm of chromosome 1.

The assignment of the erythrocyte protein 4.1 gene to chromosome 1p32 → 1pter provides a clue to the pathogenesis of one form of hereditary elliptocytosis. Previously, some elliptocytic defects have been linked to the locus for the Rh factor [32,33], which has been assigned to chromosome 1 in the region p32-p36. The proximity of the protein 4.1 gene to the Rh locus suggests that these cases of elliptocytosis may be due to a defect in the structure of protein 4.1.

VII. MOLECULAR BASIS OF HEREDITARY ELLIPTOCYTOSIS

Hereditary elliptocytosis is a heterogeneous red blood cell disorder that is characterized phenotypically by abnormally shaped erythrocytes and a variable degree of hemolytic anemia [34]. In the most severe form, it can cause neonatal jaundice and transfusion-dependent hemolytic anemia, but in the majority of individuals the condition is clinically mild. The biochemical basis of the hereditary elliptocytosis is in large part related to abnormalities of red blood cell membrane skeletal proteins spectrin and protein 4.1 [34]. Spectrin-related hereditary elliptocytosis has been described in association with at least four spectrin α-subunit variants and with two β-subunit variants [5,8,9,12,35,36]. Protein 4.1-related elliptocytosis was first reported in a consanguineous Algerian family to be the result of a quantitative deficiency of this protein [37, 38]. Complete deficiency of protein 4.1 in red blood cells from this family is characterized by elliptocytes and fragmented cells with marked anemia, whereas partial deficiency of protein 4.1 results in elliptocytes with very mild anemia. In an in vitro assay, red blood cell membranes from these patients exhibited markedly decreased membrane stability, and reconstitution of these membranes with an exogenous source of purified protein 4.1 resulted in restoration of membrane stability to near-normal levels [7]. Recently, two qualitative defects in protein 4.1 have also been identified in hereditary elliptocytosis [23,24,39]. One mutant form of protein 4.1 is characterized by a high-

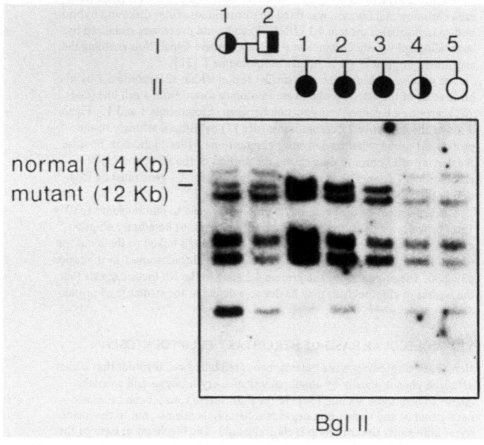

Figure 7 Southern blot analysis of Bgl II-digested genomic DNA. The pedigree shows the protein 4.1 phenotypes of the persons whose DNA is analyzed directly below. Subject II-5 had normal levels of protein 4.1, I-1, I-2 and II-4 had fifty percent of the normal levels of protein 4.1, and II-1, II-2, and II-3 had no detectable levels of protein 4.1. A 2.5-kb nick-translated cDNA probe was used in this analysis. (Taken from reference 31.)

molecular-weight protein 4.1 variant (100 kD versus 80 kD for normal protein 4.1), whereas the other mutant form is characterized by a low-molecular weight 4.1 variant (65–68 kD). Both these mutant forms of protein 4.1 were inherited in linkage with Rh phenotypes consistent with the localization of the protein 4.1 gene on chromosome 1 in close proximity to the Rh locus. Using the protein 4.1 cDNA probes, we have begun to investigate the molecular basis of hereditary elliptocytosis due to protein 4.1 deficiency and due to the mutant high-molecular-weight protein 4.1 variant.

The pedigree of the Algerian family with hereditary elliptocytosis due to protein 4.1 deficiency is shown in Figure 7. The clinical and laboratory findings in this family have been described in detail earlier [38]. The three homozygously affected daughters (II-1, II-2 and II-3) had marked elliptocytosis and complete deficiency of protein 4.1 in their red blood cell membranes. The heterozygous parents and the one heterozygous daughter (II-4) had a milder elliptocytosis and 50% of the normal levels of protein 4.1 in their red blood cell membranes. A fifth daughter (II-5) was hematologically normal and the membrane content of protein 4.1 was normal. The gene structure of protein 4.1 in this family was examined by Southern blot analysis using genomic DNA extracted from peripheral blood leukocytes [31]. The restriction fragments generated by digestion with endonuclease Bg1 II and visualized by hybridization to a 2.5-kb cDNA probe are shown in Figure 7. A variation in the pattern of genomic DNA correlated precisely with the presence or absence of protein 4.1 in the various family members. The three homozygously affected daughters had a 12-kb band, whereas the hematologically normal daughter had a 14-kb band. The heterozygous parents and the heterozygously affected daughter had both the normal 14-kb band and the mutant 12-kb band.

An abnormal restriction fragment can result from either a point mutation affecting a Bg1 II site or a more extensive gene rearrangement. To assess these two possibilities, Southern blot analysis was repeated with several other restriction enzymes including Eco RI and Pvu II. In analyses with all these enzymes, DNA from homozygously affected family members yielded a pattern different from that resulting from normal DNA, whereas DNA from heterozygous family members yielded a composite pattern that included both normal and mutant fragments. This finding of an abnormal restriction fragment associated with protein 4.1 deficiency with a number of different enzymes suggests the possibility of extensive gene rearrangement.

The boundaries of the gene arrangements were delineated by performing Southern blot analysis using probes derived from different domains of the protein 4.1 cDNA. As shown in Figure 8, when Pst I-digested genomic DNA was probed with three different segments of the cDNA clones: a 254 base pair (bp) fragment from the 5' untranslated portion of mRNA (probe A), a 502-bp fragment immediately preceding the initiation codon (probe B), and a 756-bp fragment from the protein-coding region (probe C), distinct differences were noted. Although probes A and C yielded identical patterns in normal and affected family members, probe B showed abnormal patterns in the analysis with the mutant gene, allowing the localization of the DNA rearrangement to region B of the protein 4.1 gene.

The consequences of this gene rearrangement on processing of mRNA was investigated by performing Northern blot analysis on reticulocyte RNA isolated from blood obtained from a normal donor and a protein 4.1-deficient in-

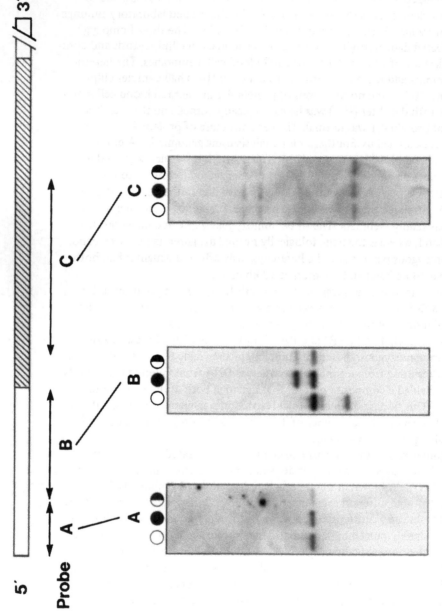

Figure 8 Southern blot analysis of Pst-I-digested DNA from normal and protein 4.1-deficient subjects, using three segments of the protein 4.1 cDNA as a probe. A partial restriction map of protein 4.1 cDNA (top) shows the probes used and their position with respect to untranslated sequences (open box) and protein-coding sequences (hatched box). (Taken from reference 31.)

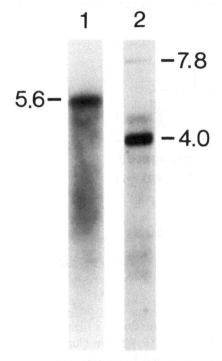

Figure 9 Analysis of reticulocyte RNA from a normal control and a protein 4.1-deficient family member. Lane 1 shows the analysis of 0.5 μg of poly A+ RNA from a normal control; lane 2 shows analysis of total RNA from 10 ml of blood from a homozygous patient with protein 4.1 deficiency. The probe used was a 2.5-kb cDNA fragment. The weakly hybridizing band at 4.8 kb in the sample deficient in protein 4.1 resulted from cross-hybridization with 28S ribosomal DNA, since there is homology between the 5' end of the mRNA and this ribosomal RNA. (Taken from reference 31.)

dividual. Polyadenylated (poly A+) reticulocyte RNA from a normal control and total reticulocyte RNA from a homozygously affected family member were analyzed by agarose gel electrophoresis and hybridization with the 2.5 kb cDNA probe (Figure 9). Protein 4.1 mRNA from the normal control was 5.6 kb in length, whereas RNA from the homozygous family member showed a major aberrant protein 4.1 mRNA species at 4.0 kb and a minor aberrant species at 7.8 kb. Thus, the principal protein 4.1 mRNA produced from the mutant gene is ~1.6 kb shorter than the normal 5.6-kb mRNA. It is likely that because the rearrangement occurs near the initiation codon for the translation, this altered message would probably be incapable of normal translation into protein 4.1, thus explaining the complete absence of protein 4.1 in the red blood cell membranes of these patients.

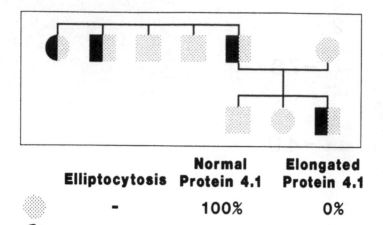

	Elliptocytosis	Normal Protein 4.1	Elongated Protein 4.1
	−	100%	0%
	+	∼50%	∼50%

Figure 10 Pedigree and phenotype of family with hereditary elliptocytosis due to an elongated protein 4.1.

The pedigree of a family with a hereditary elliptocytosis due to a high-molecular-weight variant of protein 4.1 is illustrated in Figure 10. Western blot analysis of red blood cell membrane proteins with anti-protein 4.1 antibody showed that, in addition to the normal 80-kD protein, a higher-molecular-weight 100-kD protein 4.1 variant is present in red blood cells from all members with elliptocytosis. The five normal family members have only the 80-kD species of protein 4.1. Radioimmunoassay showed no reduction in total protein 4.1 content in red blood cells from the elliptocytic members, suggesting that the reduction in the normal variant is compensated by the presence of higher-molecular-weight variant of protein 4.1 [24]. To characterize the mutation in the protein 4.1 gene in this family, we performed Northern blot analysis of reticulocyte RNA and Southern blot analysis of genomic DNA using cloned protein 4.1 cDNA as a molecular hybridization probe. The heterozygous HE individuals had two species of mRNA, 5.6 and 5.8 kb in length instead of the single 5.6 kb mRNA found in normal individuals [40]. The novel 5.8-kb mRNA is most likely lengthened by rearrangement and/or insertion of sequences within its protein coding region, thereby generating the high-molecular-weight variant protein. The recent finding of insertion of a new peptide fragment between the 8-kD spectrin–actin binding domain and the 16-kD domain support this suggestion [39]. In contrast to the Algerian family with protein 4.1 deficiency, Southern blot analysis did not reveal gross rearrangement of the mutant gene. These findings suggest that a mutation in the protein 4.1 gene results in an altered processing of its mRNA, producing the novel larger mRNA and the larger protein species observed in this family.

VIII. SUMMARY

The application of molecular biology to the study of the human erythrocyte membrane skeleton is outlined in this chapter. After describing the strategy employed to clone DNA sequences encoding the skeletal protein, protein 4.1, we have shown how these sequences may be used to deduce detailed structural and functional information about the 4.1 polypeptide. The use of cloned 4.1 DNA sequences as probes to delineate the molecular basis for hereditary elliptocytosis in two families is also discussed. Extension of these approaches permit the generation of molecular description of the structural proteins that constitute the red blood cell skeleton as well as the genetic mechanisms regulating the expression of these proteins in erythroid and nonerythroid cells.

ACKNOWLEDGMENT

The authors thank Dr. Y. W. Kan for his invaluable help and constant encouragement. We also thank Mr. James Harris for his expert assistance in the preparation of this chapter. The work described has been supported in part by NIH grant DK 32094.

REFERENCES

1. Evans, E. A., and Hochmuth, R. M. (1977). A solid–liquid composite model of the red cell membrane. *J. Membr. Biol.* 30:351.
2. Chasis, J. A., and Mohandas, N. (1986). Erythrocyte membrane deformability and stability: Two distinct membrane properties are independently regulated by skeletal protein associations. *J. Cell Biol.* 103:343.
3. Waugh, R. E. (1982). Temperature dependence of the yield shear resultant and the plastic viscosity coefficient of erythrocyte membrane. *Biophys. J.* 39:273.
4. Liu, S. C., and Palek, J. (1980). Spectrin tetramer-dimer equilibrium and the stability of erythrocyte membrane skeletons. *Nature* 285:586.
5. Liu, S. C., Palek, J., Prschal, J., and Castleberry, R. P. (1981). Altered spectrin dimer–dimer association and instability of erythrocyte membrane skeletons in hereditary pyropoikilocytosis. *J. Clin. Invest.* 68:597.
6. Mohandas, N., Clark, M. R., Heath, B. P., Rossi, M., Wolfe, L. C., Lux, S. E., and Shohet, S. B. (1982). A technique to detect reduced mechanical stability of red cell membranes: relevance to elliptocyte disorders. *Blood* 59:768.
7. Takakuwa, Y., Tchernia, G., Rossi, M., Benabadji, M., and Mohandas, N. (1986). Restoration of normal membrane stability to unstable protein 4.1-deficient erythrocyte membranes by incorporation of purified protein 4.1. *J. Clin. Invest.* 78:80.
8. Knowles, W. J., Morrow, J. S., Speicher, D. W., Zarkowsky, H. S., Mohandas, N., Shohet, S. B., and Marchesi, V. T. (1982). Molecular and functional changes in spectrin from patients with hereditary pyropoikilocytosis. *J. Clin. Invest.* 71:1867.

9. Lane, P. A., Shew, R. L., Ioarocci, T. A., Mohandas, N., Hays, T., and Mentzer, W. (1987). A unique alpha spectrin mutant in a kindred with common hereditary elliptocytosis. *J. Clin. Invest.* 79:989.

10. Waugh, R. E., and Agre, P. (1988). Reductions of erythrocyte membrane viscoelastic coefficients reflect spectrin deficiencies in hereditary spherocytosis. *J. Clin. Invest.* 81:133.

11. Zail, S. S., and Coetzer, T. L. (1984). Defective binding of spectrin to ankyrin in a kindred with recessively inherited hereditary elliptocytosis. *J. Clin. Invest.* 74:753.

12. Dhermy, D., Lecomte, M. C., Garbarz, M., Bournier, O., Galand, C., Goutero, H., Feo, C., Alloisio, N., Delauney, J., and Boivin, P. (1982). Spectrin β-chain variant associated with hereditary elliptocytosis. *J. Clin. Invest.* 70: 707.

13. Branton, D., Cohen, C. M., and Tyler, J. (1981). Interaction of cytoskeletal proteins on the human erythrocyte membrane. *Cell* 24:24.

14. Marchesi, V. T. (1983). The red cell membrane skeleton: recent progress. *Blood* 61:1.

15. Bennett, V. (1985). The membrane skeleton of human erythrocytes and its implication for more complex cells. *Ann. Rev. Biochem.* 54:273.

16. Reid, M. E., Takakuwa, Y., Tchernia, G., Jensen, R. H., and Mohandas, N. (1987). Protein 4.1 binds to sialoglycoprotein β and regulates its membrane contents in human erythrocytes. *Blood* 70(suppl 1):42A.

17. Mueller, T. J., and Morrison, M. (1981). Glycoconnectin (PAS2), a membrane attachment site for the human erythrocyte cytoskeleton. In *Erythrocyte Membrane 2: Recent Clinical and Experimental Advances*, W. C. Kruckeberg, J. W. Eaton, and G. J. Brewer (Eds.). Alan R. Liss, New York, p. 95.

18. Tyler, J. M., Reinhardt, B. M., and Branton, D. (1980). Association of erythrocyte membrane proteins: binding of purified bands 2.1 and 4.1 to spectrin. *J. Biol. Chem.* 255:7034.

19. Goodman, S. R., Yu, J., Whitfield, C. F., Culp, E. N., and Posnak, E. J. (1982). Erythrocyte membrane skeletal protein bands 4.1a and b are sequence-related phosphoproteins. *J. Biol. Chem.* 257:4564.

20. Fairbanks, G., Steck, T. L., and Wallach, D. F. H. (1971). Electrophoretic analysis of the major polypeptides of the human erythrocyte membrane. *Biochemistry* 10:2606.

21. Fowler, V., and Taylor, D. L. (1980). Spectrin plus band 4.1 cross-link actin: regulation by micromolar calcium. *J. Cell Biol.* 85:361.

22. Cohen, C. M., and Korsgren, C. (1980). Band 4.1 causes spectrin–actin gels to become thixiotropic. *Biochem. Biophys. Res. Commun.* 97:1429.

23. Morle, L., Garbarz, M., Alloisio, N., Girot, R., Chaveroche, I., Boivin, P., and Delauney, J. (1985). The characterization of protein 4.1 Presles, a shortened variant of rbc membrane protein 4.1. *Blood* 65:1511.

24. McGuire, M., and Agre, P. (1987). Three distinct variants of protein 4.1 in Caucasian hereditary elliptocytosis. *Clin. Res.* 35:428A.

25. Maniatis, T., Fritsh, E. F., and Sambrook, J. (1982). *Molecular Cloning: A Laboratory Manual.* Cold Spring Harbor, New York.

26. Pelham, H. R. B., and Jackson, R. J. (1976). An efficient mRNA-dependent translation system from reticulocyte lysates. *Eur. J. Biochem.* 67:247.

27. Young, R. A., and Davis, R. W. (1983). Efficient isolation of genes by using antibody probes. *Proc. Natl. Acad. Sci. (USA)* 80:1194.

28. Conboy, J., Kan, Y. W., Shohet, S. B., and Mohandas, N. (1986). Molecular cloning of protein 4.1, a major structural element of the human erythrocyte membrane skeleton. *Proc. Natl. Acad. Sci. (USA)* 83:9512.

29. Leto, T. L., and Marchesi, V. T. (1984). A structural model of human erythrocyte protein 4.1. *J. Biol. Chem.* 259:4603.

30. Conboy, J., Chan, J., Mohandas, N., and Kan, Y. W. (1987). Unexpected diversity of erythroid protein 4.1 isoforms generated by alternative mRNA splicing. *Blood* 70:(suppl 1):38A.

31. Conboy, J., Mohandas, N., Tchernia, G., and Kan, Y. W. (1986). Molecular basis of hereditary elliptocytosis due to protein 4.1 deficiency. *N. Engl. J. Med.* 315:680.

32. Morton, N. E. (1956). The detection and estimation of the linkage between the genes for elliptocytosis and the Rh blood type. *Am. J. Hum. Genet.* 8: 80.

33. Lovric, V. A., Walsh, R. J., and Bradley, M. A. (1965). Hereditary elliptocytosis: genetic linkage with the Rh chromosome. *Aust. Ann. Med.* 14:162.

34. Palek, J. (1985). Hereditary elliptocytosis and related disorders. *Clin. Haematol.* 14:45.

35. Marchesi, S. L., Letsinger, J. T., Speicher, D. W., Marchesi, V. T., Agre, P., Hyun, B., and Glutai, G. (1987). Mutant forms of spectrin α-subunits in hereditary elliptocytosis. *J. Clin. Invest.* 80:191.

36. Johnson, R. M., and Ravindranath, Y. (1985). A new variant of spectrin with a large beta-spectrin chain associated with hereditary elliptocytosis. *Blood* 66(suppl 1):33A.

37. Feo, C., Fischer, S., Piau, J. P., Grange, M. J., and Tchernia, G. (1980). Première observation de l'absence d'une proteine de la membrane erythrocytaire (bande 4.1) dans un cas d'ançmie elliptocytaire familiale. *Nouv. Rev. Fr. Hematol.* 22:315.

38. Tchernia, G., Mohandas, N., and Shohet, S. B. (1981). Deficiency of skeletal membrane protein band 4.1 in homozygous hereditary elliptocytosis. *J. Clin. Invest.* 68:1454.

39. Letsinger, J. T., Agre, P., and Marchesi, S. L. (1986). High molecular weight protein 4.1 in the cytoskeletons of hereditary elliptocytosis. *Blood* 68(suppl 1);38A.

40. Conboy, J., Kan, Y. W., Agre, P., and Mohandas, N. (1986). Molecular characterization of hereditary elliptocytosis due to an elongated protein 4.1. *Blood* 68(suppl 1):61A.

41. Correas, I., Speicher, D. W., and Marchesi, V. T. (1986). Structure of the spectrin–actin binding site of erythrocyte protein 4.1. *J. Biol. Chem.* 261: 13362.

8

Regulation of Protein 4.1-Membrane Associations by a Phosphoinositide

RICHARD A. ANDERSON *University of Wisconsin Medical School, Madison, Wisconsin*

I. INTRODUCTION

The cortical cytoskeleton of the vertebrate erythrocyte is an interconnecting meshwork formed by oligomerization of the major component, spectrin, and by associations between spectrin and actin [1-3]. The rodlike spectrin dimer end-to-end associates to form tetramers and by further addition of dimers, higher oligomers [4]. At the end of spectrin opposite to the oligomerization site, spectrin associates with actin [5-9]. In the erythrocyte, spectrin and actin are tightly bound to the plasma membrane. This association is mediated by proteins that serve to link spectrin to transmembrane protein receptors [1-3]. Spectrin, in the absence of actin, binds with high affinity to the membrane through an interaction with ankyrin, which in turn associates with the anion channel band 3 [1].

Erythrocyte actin is in the form of a protofilament composed of 16-18 actin monomers [10-16]. The mechanism of stabilizing actin protofilaments is not known, but involves interactions with actin binding proteins and other cytoskeletal proteins [12]. These include an interaction with tropomyosin, proteins 4.9, spectrin-protein 4.1, and possibly a calmodulin-binding protein [10-17]. Spectrin alone has a weak affinity for actin protofilaments. However, in the presence of protein 4.1 the spectrin-actin association is dramatically increased, indicating that protein 4.1 acts as a cofactor for the spectrin-actin assembly [5-7, 18]. Each actin protofilament has the capacity to bind as many as eight spectrin-protein 4.1 complexes, but in the cell there appear to be on average five to six spectrin-protein 4.1 complexes associated with each actin protofilament [12-

187

14]. The resulting oligomeric structure is a junction point or "junctional complex" that has the potential of regulation by phosphorylation of its component proteins, and as discussed here, by the metabolism of the phosphoinositides [19–22].

The cytoskeleton in all cells must have associations with the plasma membrane. In turn, these associations must be regulated in such a fashion that the interaction can be turned "off and on" in response to changes in the cellular requirements [23–26]. This is particularly important for motile cells, but is also required for intracellular transport of components, for phagocytosis, and for other functions [1,23–26]. Potential mechanisms for cytoskeletal-membrane on-on–off switches include protein phosphorylation, protease cleavage, other post-translational modifications, or a phospholipid cofactor that is metabolically active and is an obligate cofactor for cytoskeletal assembly at membrane receptors [19–22,27]. Although the latter mechanism initially seems the most complex, it may fit into a larger scheme of cellular regulatory events, many of which are regulated by the metabolism of the phosphoinositides [28–30].

II. GLYCOPHORIN MAY REGULATE CELL SHAPE BY INTERACTING WITH THE CYTOSKELETON

The impetus in looking for an interaction between the glycophorins and the cytoskeleton came from early work with ligands that bind to the erythrocyte and inhibit the discocyte–echinocyte erythrocyte shape change [31,32]. The ligands are lectins (carbohydrate-binding proteins) or antibodies, and both must be specific for glycophorin [31–33]. The effect of the lectins on cell shape change was so dramatic as to suggest that the lectin receptor was fundamentally involved in the underlying mechanism driving the shape conversion. Another interesting observation was that the lectins blocked cell shape change in the forward, discocyte–echinocyte, and reverse, echinocyte–discocyte direction. Lectin binding to the cell membrane alone did not induce a shape change, except at very high amounts of lectin bound, close to binding saturation [32,34]. The shape change induced at these concentrations is not the echinocyte, but a boat-shaped elliptical cell [34], similar to shapes observed in a number of hemolytic states [35,36]. The discocyte–echinocyte shape conversion is induced by three methods: (a) the binding of amphipathic ligands such as fatty acids, anionic detergents, lyso-phospholipids, and others that bind to the outer lipid leaflet but are not membrane permeable; (b) metabolic depletion of adenosine triphosphate (ATP), and (c) increased cytoplasmic Ca^{2+} carried across the membrane by ionophore A23187 [36–39]. The latter methods induce shape change by modifying components in the cell cytoplasm, the first by modifying the ectoplasmic side of the cell membrane, possibly by expanding the outer lipid leaflet of the bilayer [39]. Lectins inhibit the shape change induced by all of the above

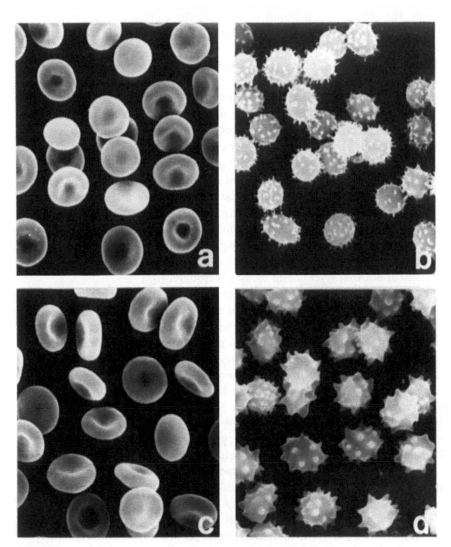

Figure 1 Scanning electron micrographs of red cells demonstrating lectin effectiveness as inhibitors of red cell shape change. a, Red cells pretreated with wheat germ lectin followed by addition of Ca^{++} and ionophore A23187; b, red cells treated only with Ca^{++} and ionophore A23187; c, desialated red cells pretreated with osage orange lectin followed by addition of sodium laurate, an echionocytic ligand; d, desialated red cells treated only with sodium laurate.

methods. The efficacy of the lectins at inhibiting shape change induced by Ca^{2+}-A23187 (a and b) or the fatty acid sodium laurate (c and d) is demonstrated in Figure 1.

There are two kinds of oligosaccharides attached to glycophorin, a Ser/Thr-linked tetrasaccharide of which there are 15 mol per glycophorin monomer, and one Asn-linked oligosaccharide [40]. To inhibit cell shape change, the lectin must bind to glycophorin's Ser/Thr-linked oligosaccarides; lectin binding to the Asn-linked oligosaccharides, as in the case of *Phasoulus vulgarus* lectin, has no effect on cell-shape change. Interestingly, however, this lectin can detect erythrocyte shape change by its interaction with glycophorin, suggesting that the Asn-linked oligosaccharide is more exposed in the echinocyte than in the discocyte shape [41]. Further, the lectin must be multivalent; although the monovalent lectin, ricin, binds to Ser/Thr-linked oligosaccharides, it does not block cell shape change, likely because it is monovalent [31-34,50]. Lectins that specifically bind to glycolipids or other membrane glycoproteins, such as band 3, have no effect on cell shape change, indicating that mere binding to the membrane is not sufficient but that glycophorin binding is required [32].

The effect of lectins on cell-shape change was quantitated by titrating the bound lectin:cell stoichiometry required to inhibit shape conversion. Figure 2 shows the effect of increasing amounts of bound lectin on the rate of cell shape change for both laurate and Ca^{2+}-A23187 induced shape change. The striking aspect of these curves are twofold: First, inhibition of the rate of cell-shape change for both Ca^{2+} and laurate-induced conversion required the same number of lectin molecules bound per cell: $\sim3.6 \times 10^5$. Second, although increasing the concentration of Ca^{2+} or laurate increases the rate of shape change, the number of lectin molecules required to inhibit shape change remains the same. One interpretation of these results is that increased cytoplasmic Ca^{2+} and laurate induce cell-shape change by acting on the same lectin receptor, and that a specific number of lectin receptors are involved in the mechanism of cell-shape change. As a result, when the stoichiometry of lectin binding reaches 3.6×10^5 lectin molecules per cell, further binding has only a minor effect on the inhibition of cell-shape change, indicating saturation of lectin inhibition.

The stoichiometry of lectin molecules bound per cell required for inhibition of cell-shape change corresponds to the number of putative glycophorin dimers per cell. This suggests that each glycophorin dimer requires a bound lectin before inhibition of shape change is complete. Lectins inhibit the shape changes initiated by amphipathic ligand binding to the outside surface (outer bilayer leaflet) of the membrane, by decreased ATP, or by increased cytoplasmic Ca^{2+} Adenosine triphosphate depletion and increased Ca^{2+} both act on the cytoplasmic side of the membrane. This suggests that glycophorin plays a role both in ligand and metabolic-induced shape change. A mechanistic explanation of these results was proposed to involve an association between glycophorin and a

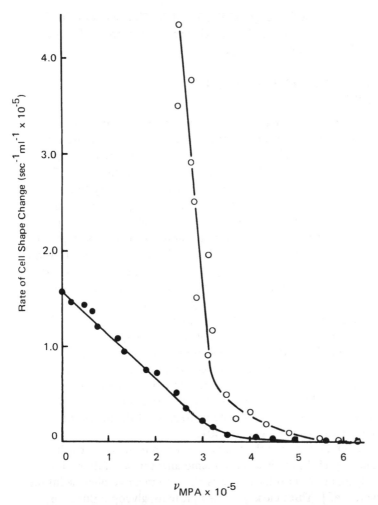

Figure 2 A plot of the rate of red cell shape change as a function of the number of lectin molecules bound per cell (ν_{MPA}). The rate or velocity of cell shape change was determined by fitting a plot of fraction echinocyte versus time with a linear regression. Plots of fraction echinocyte versus time are quite linear for the fraction echinocyte ~0.6; above this, the curves became nonlinear (32). The rate of shape change was determined by multiplying the slope of the line by the cell concentration (10^7 cells/ml). Lectin bound was determined using affinity purified and ^{125}I-labeled osage orange lectin [32,50].

cytoplasmic component, possibly a cytoskeletal protein. Such an association
could play a role in lectin inhibition of shape change and perhaps in the underlying mechanism of shape change [31,32].

III. ASSOCIATION OF PROTEIN 4.1 WITH
GLYCOPHORIN

Glycophorin is a generic name for a group of transmembrane proteins that are
heavily glycosylated on the extracellular domain, with a particularly large content of sialic acid [40,43–45]. These glycoproteins have a peptide core that
spans the membrane with a single helical intramembranous segment of ~22
hydrophobic amino acid residues. The glycophorins have the ability to form
dimers in detergent solution that are quite stable, dissociating only at elevated
temperatures [42]. Subunit association involves the transmembrane region; a
methionine residue (residue 81) within this region appears to be important for
assembly of glycophorin A dimers. The glycophorins also have the ability to form
heterodimers. At the bilayer interface, the cytoplasmic segment of glycophorins each have a cluster of positively charged amino acid residues that are particularly well placed to interact with negatively charged phospholipids. This is followed by a polar segment that extends into the cell cytoplasm and is rich in
negatively charged amino acid residues [40,43–45]. Because glycophorin is a
transmembrane protein, the phospholipid bilayer must be disrupted with detergents to extract it from the membrane. However, after extraction, glycophorin
is unusual in that it is very soluble in aqueous solution in the absence of detergents. These unique solubility properties result from formation of a protein-
phospholipid micellar structure, which is soluble because of the high content of
glycosylation on the extracellular segment [40].

Glycophorin was extracted from the human erythrocyte membrane using
lithium diiodosalicylate (LIS) to disrupt the membrane; this was followed by
partitioning the glycophorins into the aqueous phase from a two-phase solution
of phenol and water [45]. Three closely related proteins, glycophorins A, B,
and C, are isolated by this method; glycophorin A comprises 75% of the total
weight, with glycophorins B and C comprising equal amounts of the remainder
[40,44]. The glycophorins can be further fractionated by chromatography on a
Bio Gel A 0.5 column in the presence of 0.1% Ammonyx-LO (N,N-dimethyl-
laurylamine oxide) or purified to homogeneity by preparative sodium dodecyl-
sulfate-polyacrylamide gel electrophoresis (SDS-PAGE) and elution from the
gel [40,44]. Glycophorin, isolated using LIS or other methods, retains bound
phospholipids that are largely negatively charged; these are removed by extraction with chloroform:methanol:12:1 N HCl (200:100:1) or by chromatography
in the presence of 0.2% SDS [46]. After complete lipid extraction, glycophorin
remains soluble in water, in the absence of detergents as a 7–8S protein micelle
[19].

To study the association of glycophorin with other membrane proteins, purified glycophorins were reconstituted into large phosphatidylcholine liposomes using a detergent-dialysis method [47,48]. Glycophorin reconstituted into liposomes for these studies was extracted with cold (−10°C) ethanol. This treatment extracts most neutral lipids, but not the bound negatively charged phospholipids, such as the phosphoinositides or phosphatidylserine. Before reconstitution into liposomes, glycophorin was desialated with *Clostridium perfringens* neuraminidase to remove the sialic acid residues that contribute a large negative charge to glycophorin that may bind proteins by electrostatic interactions [48]. The detergent octyl-β-D-glucoside was used to solubilize the phospholipid and membrane protein, forming a detergent–phospholipid–protein mixed micellar solution. Dialysis of the octyl-glucoside, facilitated by its high critical micelle concentration (CMC), results in the formation of phosphatidylcholine liposomes containing reconstituted glycophorin [47,48]. These liposomes were separated from micellar glycophorin not reconstituted by chromatography on Sepharose CL-4B; liposomes elute in the void volume whereas micellar glycophorin is retained in the included volume. In later experiments [19], liposomes and glycophorin-phospholipid micelles were separated by sucrose density gradient sedimentation. In this case, liposomes float at 7% sucrose, whereas glycophorin-phospholipid micelles sediment into the gradient [19]. In either case, large liposomes were then selected for by sedimentation at 100,000 g for 30 min. The size of the resulting liposomes facilitated studying the association of membrane skeletal proteins with reconstituted glycophorin because liposomes with or without bound protein could be easily sedimented and separated from proteins that did not bind [48].

Determining selective binding of cytoskeletal proteins to liposomes required that the cytoskeleton be disassembled into soluble monomeric components before addition to liposomes. This was accomplished by selectively solubilizing membrane skeletal proteins by sequential extraction of membranes with low-ionic-strength buffer followed by high-ionic-strength buffer [2,49]. Extraction with 0.1 mM ethylenediamine tetraacetic acid and 0.1 mM diisopropylfluorophosphate at 37°C solubilizes the spectrin, actin, and a small amount of protein 4.1 as well as other components in lower concentration [1–3]. The membranes, on loss of spectrin and actin, endovesiculate, forming inside-out membrane vesicles (IOVs) that were further extracted with buffered 1 M KCl, solubilizing proteins 2.1 (ankyrin), 4.1, 4.9, and 6 from the membrane (Fig. 3) [48,49]. To determine if any membrane components specifically bound to glycophorin, the soluble protein extracts were combined with liposomes containing reconstituted glycophorin. As a control, protein extracts were combined with an equivalent amount of liposomes lacking glycophorin. The solutions after incubation were centrifuged, sedimenting liposomes and bound proteins. The pelleted liposomes, when examined by SDS-PAGE, bound a number of proteins nonspecifically, but

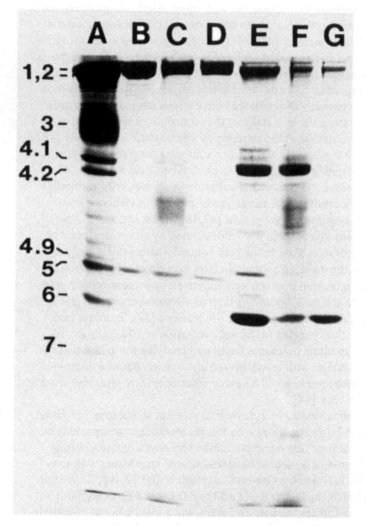

Figure 3 The interaction of membrane skeletal proteins with phosphatidylcholine liposomes containing desialated glycophorin or lacking glycophorin.
SDS-PAGE shows membrane skeletal protein extracts and the membrane skeletal proteins that bound to liposomes containing glycophorin or control liposomes lacking glycophorin. Samples are: A, total membrane proteins, 25 μg; B,
low-ionic-strength extracted protein, 7 μg; C, proteins from the low-ionic-
strength extract bound to liposomes containing glycophorin; D, as in C except
liposomes lack glycophorin; E, 1 M KCl extracted protein, 14 μg; F, protein
from the 1 M KCl extract bound to liposomes containing glycophorin; G, same
as in F except liposomes lack glycophorin. The buffer was (in mM); 130 KCl, 20
NaCl, 10 Tris, 1 $MgCl_2$, 0.5 β-mercapoethanol, 0.1 diisopropylfluorophosphate
(DFP), pH 7.6 (iso-KCl buffer). From Anderson and Lovrien (48).

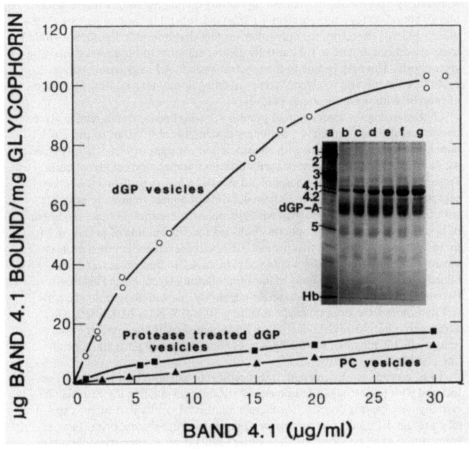

Figure 4 Binding of protein 4.1 to phosphatidylcholine liposomes containing glycophorin (o), protease-treated liposomes containing glycophorin (■), and liposomes that lack glycophorin (▲). SDS-PAGE of protein 4.1 bound to liposomes containing glycophorin is shown. Samples are: a, 30 μg of total membrane protein; b–g, protein 4.1 bound to liposomes (the protein 4.1 concentration is varied from 2 to 30 μg/ml); 70 μg of desialoglycophorin was loaded per lane. The buffer is iso-KCl. From Anderson and Lovrien (48).

liposomes containing glycophorin selectively bound protein 4.1 (Figure 3). In a similar experiment, liposomes were combined with membrane skeletal proteins extracted with 1 M KCl, in this case containing some spectrin, but mainly proteins 2.1, 4.1, 4.9, and band 6. This solution was then applied to a Sepharose CL-4B column that the liposomes elute from in the column void volume and were completely separated from proteins not bound. These liposomes, when analyzed by SDS-PAGE, showed that liposomes containing reconstituted glyco-phorin bound 40-fold more protein 4.1 than did control liposomes lacking glyco-phorin [48,50]. Spectrin, the only other protein that bound to liposomes within these conditions, bound in substantially greater amounts to liposomes containing glycophorin. This may be due to the spectrin–protein 4.1 association, which could link the spectrin to glycophorin-containing liposomes, resulting in elution of spectrin with these liposomes [48,50].

Characterizing the association of protein 4.1 with liposomes containing glyco-phorin revealed that protein 4.1 binding was saturable at ~120 μg of protein 4.1 bound per mg of glycophorin with an association constant of ~10^7 M^{-1} (Figure 4). The association constant varied with different preparations of glycophorin. Preparations retaining more phospholipid showed a higher protein 4.1 binding capacity [19,32,50]. However, protein 4.1 did not appear to directly associate with the phospholipids co-isolated with glycophorin, because protease treatment of liposomes containing glycophorin abolished specific binding of protein 4.1 to these liposomes, indicating that protein 4.1 binding is dependent on a protease-sensitive peptide. Nonspecific binding of protein 4.1 to liposomes was deter-mined using equivalent amounts of liposomes lacking glycophorin (Figure 4). Because glycophorin is known to retain negatively charged phospholipids, con-trol liposomes were also prepared containing 30 mol % phosphatidylserine in addition to phosphatidylcholine. These liposomes showed the same or lower nonspecific binding than did the liposomes containing only phosphatidylcho-line [19,32,50].

To be physiologically relevant, it was critical to establish that protein 4.1 as-sociated with the cytoplasmic segment of glycophorin and not the extracellular carbohydrate-bearing domain. There are a number of results that support spe-cific protein 4.1 binding to cytoplasmic segments of the glycophorins. First, re-constitution of glycophorin into liposomes results in an asymmetric orientation of the glycophorin across the bilayer. The majority, ~70–80%, of the glycophor-ins are oriented with the extracellular carbohydrate-bearing domains located at the outside surface of the liposome [47,48,51-53]. Liposomes, with glycophor-in so asymmetrically oriented, would result in a low protein 4.1/glycophorin-binding stoichiometry at saturation if protein 4.1 associated with the cytoplas-mic domain of glycophorin. The observed stoichiometry at saturation, of 120 μg of protein 4.1/mg of glycophorin (1 mol protein 4.1/20 mol glycophorin), cor-responds well with an association of protein 4.1 with the cytoplasmic domain of glycophorin.

To gain direct evidence that protein 4.1 associates with the cytoplasmic seg-ment of glycophorin, a soluble peptide derived from tryptic cleavage of glyco-phorin, the T4 peptide, was tested for inhibition of protein 4.1 binding to mem-branes. The results demonstrated that the cytoplasmic T4 peptide, even at high concentrations (500 μg/ml), did not inhibit the binding of protein 4.1 to lipo-

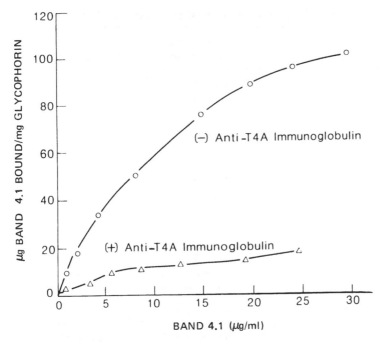

Figure 5 Binding of protein 4.1 to liposomes containing glycophorin (o) and liposomes containing glycophorin that have been pretreated with sheep antibody specific for glycophorin's cytoplasmic domain (△).

somes containing glycophorin or to IOVs previously stripped of extrinsic proteins with high salt and urea. These data suggest that a more structured cytoplasmic segment, perhaps requiring the transmembrane domain, may be required for the association of glycophorin with protein 4.1.

Direct support for protein 4.1 binding to the cytoplasmic domain of glycophorin was provided by antibodies specific for the cytoplasmic domain of glycophorin A [54,55]. Antiserum was raised in sheep by immunizing against the cyanogen bromide fragment-2 (CB2) containing the cytoplasmic domain of glycophorin A [54]. Pure antibody specific for the cytoplasmic domain of glycophorin was then isolated from antiserum by affinity chromatography using the T4 fragment as the affinity ligand attached to Sepharose 4B. The T4 fragment is the major portion of the cytoplasmic domain of glycophorin A and is prepared by tryptic digestion of glycophorin A [43]. To inhibit protein 4.1 binding, anti-T4 antibody was preincubated with liposomes containing glycophorin. Subsequent protein 4.1 binding showed that liposomes with saturating amounts of bound anti-T4 antibody had 70–80% fewer protein 4.1 binding sites

Table 1   Inhibition of Protein 4.1 Binding to Vesicles by Anti-T4 Antibody[a]

Treatment	Vesicle type	Protein 4.1 bound/ vesicle protein	Binding inhibition
		µg/mg	%
None	PC only	21 ± 2	–
None	GP-A,B,C[b]	107 ± 5	–
Nonimmune IgG	GP-A,B,C	112 ± 6	–
Anti-T4	GP-A,B,C,	33 ± 2	69
None	GP-A	85 ± 3	–
Anti-T4	GP-A	4 ± 1	95
None	GP-B,C[c]	120 ± 7	–
Anti-T4	GP-B,C	89 ± 6	26
None	s-IOV	120 ± 4	–
Anti-T4	s-IOV	69 ± 8	43

[a]^{125}I-labeled protein 4.1 (25 µg) was added to phosphatidylcholine liposomes containing desialated glycophorin (150 µg) or as a control no glycophorin, which was pretreated with anti-T4 (300 µg) or nonimmune I_gG (300 µg). Inverted membranes (150 µg) were treated identically. The final volume was 500 µl in iso-KCl buffer. After incubation, the liposomes or IOVs were centrifuged at 150,000 X g for 40 min, and the supernatant and pellet analyzed for protein 4.1.
[b]Nonspecific binding has been subtracted.
[c]Contains glycophorin A, ~20%.
PC = phosphatidylcholine; GP = glycophorin; s-IOV = stripped inside-out membrane vesicles.

(Figure 5). Purified sheep IgG from nonimmune serum showed no inhibition of protein 4.1 binding to liposomes containing glycophorin. Liposomes that contained pure glycophorin A bound protein 4.1, and anti-T4 antibody completely inhibited protein 4.1 binding (>90%) to these liposomes (Table 1). Moreover, when glycophorin B and C, separated from glycophorin A by gel filtration [40, 44], were reconstituted into phosphatidylcholine liposomes, protein 4.1 specifically bound to these liposomes, but the binding was only marginally inhibited by anti-T4 antibody [50]. Such behavior is expected because the anti-T4 antibody does not cross-react with glycophorins B or C [54]. The anti-T4A antibody also inhibited protein 4.1 binding to urea-KCl-stripped IOVs (s-IOVs), adding further evidence that the glycophorins provide membrane binding sites for protein 4.1 (Table 1) [50]. However, antibodies specific for the glycophorins only inhibit a fraction of the protein 4.1 binding to the membrane, suggesting that other sites are involved [48,50,56].

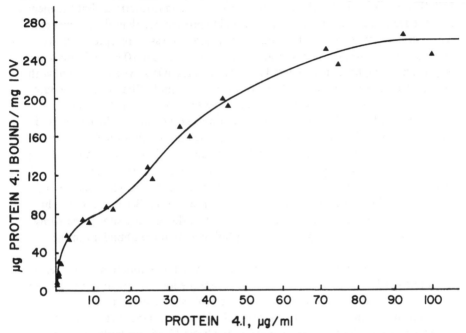

Figure 6 Binding of ^{32}P-protein 4.1 to inverted membrane vesicles. Protein 4.1 labeled with ^{32}P was added to s-IOVs (60 μg) and incubated for 30 min at 37°C or for 2 hr at 0°C in 200 μl of isotonic KCl buffer. Samples were layered over a 10% sucrose cushion and the IOVs were sedimented through the sucrose; proteins which are not bound remain in the supernatant; both the pellet and the supernatant were analyzed by liquid scintillation counting. From Pasternack et al. [56].

IV. THERE ARE TWO PROTEIN 4.1 BINDING
   RECEPTORS ON THE MEMBRANES

To determine if protein 4.1 indeed does bind to multiple receptor sites on the membrane, the binding isotherm of protein 4.1 to inverted erythrocytes was determined. As a probe to follow binding, membrane-bound protein 4.1 was labeled on IOVs using γ-^{32}P-ATP and the membrane-bound cyclic adenosine monophosphate (cAMP)-activated kinase to increase labeling [57,58]. The IOVs, washed free of γ-^{32}P-ATP, were extracted with 1 M KCl and the ^{32}P-protein 4.1 was purified by anion exchange chromatography [19,49]. The binding isotherm of ^{32}P-labeled protein 4.1 to s-IOVs is saturable and composed of high- and low-affinity sites, the high-affinity sites with an approximate K_a of 6×10^7 M^{-1}, and the low-affinity sites with an approximate K_a of 2×10^6

M^{-1} (Figure 6) [56]. A striking feature of the binding isotherm is that the membranes have a much larger capacity for binding protein 4.1 than the normal content found in native membranes, indicating a large excess of receptors. Native erythrocyte membranes have a ratio of one protein 4.1 per 10 band 3 monomers or per five glycophorin monomers [1,3]. However, s-IOVs, when saturated with protein 4.1, have a ratio of ~ 1.4 mol of protein 4.1/mol of band 3. These data suggest that there are at least two binding receptors for protein 4.1. Data discussed above (Table 1) demonstrate that antibodies specific for the cytoplasmic segment of glycophorin A block a fraction of the binding sites on S-IOVs, but these experiments do not identify which of the binding sites, low- or high-affinity sites, correspond to the protein receptor on the membrane. In the red blood cell membrane, there are two major classes of transmembrane proteins, band 3 and the glycophorins, which are available as protein receptors. Because the stoichiometry of protein 4.1 binding to s-IOVs was high, the binding sites presumably correspond to protein receptors (or phospholipids) that are abundant in the membrane.

Several experiments were used to determine which membrane proteins served as protein 4.1 binding receptors and which were the low- and high-affinity receptors on the membrane. A number of specific probes were used to identify the receptors, including: the soluble cytoplasmic domain of band 3 (the 43 K fragment) derived by chymotryptic digestion of s-IOVs [1,56], soluble micellar glycophorin, and antibodies specific for the cytoplasmic domains of band 3 and glycophorin as well as antibodies specific for all of the glycophorins, A, B, and C. These probes were used as potential specific inhibitors of protein 4.1 binding to determine if protein 4.1 could be competitively displaced from its membrane binding sites. In these experiments, ^{32}P-protein 4.1 was added to s-IOVs, treated as discussed in the legend for Table 2, and the samples were centrifuged, separating the membranes (pellet fraction) from the supernatant fractions, enabling bound and free protein to be analyzed by SDS-PAGE or scintillation counting. Without additional treatments, the majority of the protein 4.1 bound to the s-IOVs and was found in the pellet fraction. When s-IOVs, reconstituted with close to saturating amounts of protein 4.1, were exposed to a large excess of 43K fragment with respect to protein 4.1, the distribution of protein 4.1 changed dramatically, with the majority now appearing in the supernatant (Table 2). Similarly, pretreatment of the vesicles with affinity-purified antibody specific for the cytoplasmic domain of band 3 also reduced the binding of protein 4.1 to membrane vesicles. These data suggest that band 3 is also a protein 4.1 binding site on the membrane. As demonstrated in Table 2, anti-43K pretreatment inhibits protein 4.1 rebinding by ~65%. Antibody to glycophorin A inhibits by 30%, and antibody to all glycophorins inhibits by 49%, suggesting that glycophorins B and C may also serve as binding sites, as previously proposed [48,59]. When these antibodies were combined, a significant increase in inhibition was observed, sug-

Table 2 Inhibition of Protein 4.1 Binding to Membrane Vesicles[a]

Treatment	Protein 4.1 bound/ s-IOV protein	Inhibition
	µg/mg	%
None	237 ± 12	—
Anti-43K	83 ± 2	65
Anti-CB2	167 ± 13	30
Anti-glycophorins	121 ± 5	49
Anti-43K + anti-CB2	63 ± 3	74
Anti-43K + anti-glycophorins	63 ± 4	74
None	152	—
43K fragment[b]	49	68
Micellar-GP[c]	5	97
None	8 ± 0.5	—
Micellar-GP	0.4 ± 0.05	96

[a]Urea-KCl stripped inverted membranes [19] were combined with antibodies (100 µg) and incubated at 37°C followed by addition of ^{32}P-protein 4.1 [50]. When testing for inhibition by 43K (100 µg) or micellar glycophorin (10 µg), all components were combined at the same time and then incubated. The final volume of samples was 100 µl. Anti-spectrin antibodies had no effect on protein 4.1 binding, indicating that antibody inhibition is specific.
[b]The 43K fragment and micellar glycophorin were added after protein 4.1 was bound to IOVs.
[c]Micellar glycophorin contains bound intrinsic phospholipid.
IOV = Stripped inside-out membrane vesicles; GP = glycophorins.

gesting that two classes of receptors were affected by this treatment. Inhibition by antiglycophorin and by anti-43K was not additive, possibly reflecting binding to phospholipids or nonspecific binding. Antispectrin had a minimal effect, thereby excluding a nonspecific inhibition by immunoglobulin. Micellar glycophorin also inhibited binding to s-IOVs within these conditions, again demonstrating that glycophorin is a binding site. Indeed, micellar glycophorin inhibited binding at low levels of protein 4.1 bound to s-IOVs, in the high-affinity region of the binding isotherm, suggesting that micellar glycophorin inhibits binding to the high-affinity sites. In all instances, the distribution of the ^{32}P-labeled protein 4.1, demonstrated by autoradiography of the SDS-PAGEs, corresponded to the distribution of protein 4.1 stained with Coomassie blue, indicating that protein 4.1 labeled with ^{32}P by cAMP-activated protein kinase retained normal mem-

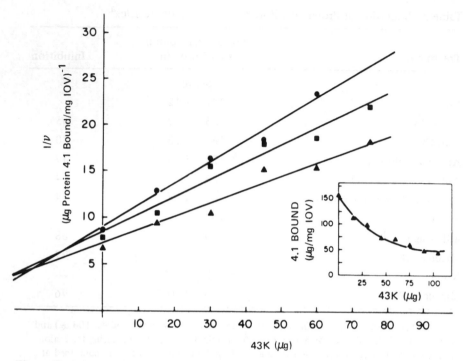

Figure 7 Inhibition of protein 4.1 binding to membrane vesicles by the soluble cytoplasmic domain of band 3. Stripped membrane vesicles (30 μg) were reconstituted with 5, 6.5, or 8 μg of ^{32}P-protein 4.1, and then incubated with increasing amounts of the 43K fragment of band 3 for 30 min at 37°C [50]. The final volume was 100 μl. The data were plotted by the method of Dixon [60]. From Pasternack et al. [56].

brane binding activity and could be used in quantitating the association of protein 4.1 with the membrane.

The amount of protein 4.1 bound to the reconstituted membrane vesicles at saturation was significantly greater than the amount of protein 4.1 normally present in erythrocyte membranes. In contrast, the amount of protein 4.1 remaining on the vesicles after competition by 43K or inhibition by anti-43K approximated the level of protein 4.1 found in native membranes. This result was examined quantitatively by measuring the amount of ^{32}P-labeled protein 4.1 remaining bound when increasing concentrations of 43K were added to inhibit binding. This curve shows that a maximum of approximately 65% of the bound protein 4.1 could be displaced from the vesicles when a large molar excess of 43K was added (Figure 7). When additional data was plotted according to the method of Dixon [60], the inhibition of protein 4.1 binding to s-IOVs

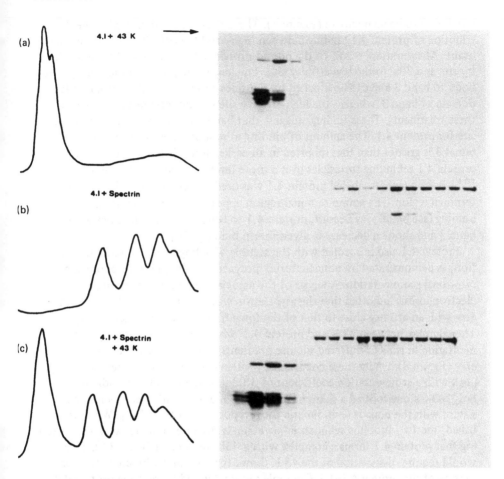

Figure 8 Protein 4.1, when associated with band 3, does not bind spectrin. Protein 4.1 (96 μg) (a), spectrin (312 μg) (b), and/or 43K fragment (316 μg) (c) were incubated for 15 min at 4 C in 25 mM KCl, 10 mM Tris, 2 mM MgCl₂, 1 mM 2-mercaptoethanol at pH 7.5. The sample (250 μl) was layered on a 5–20% sucrose gradient in the same buffer and centrifuged for 16 hr at 200,000 X g at 2°C. The left-hand panels show tracings of the absorbance at 280 nm during fractionation; the right-hand panels show analysis of the corresponding gradient fractions by SDS-PAGE. The arrow indicates direction of sedimentation. From Pasternack et al. [56].

was found to be competitive (Figure 7). These data are consistent with the distribution of protein 4.1 binding between high- and low-affinity sites on the membrane. At saturation, ~30% of the bound protein 4.1 is bound to the high-affinity site and 70% to the low-affinity site. The low-affinity site is inhibited by antibody to band 3 and can be blocked by peptides derived from the cytoplasmic domain of band 3, whereas the high-affinity site was not affected by either of these treatments. These findings suggest that band 3 is the low-affinity binding site for protein 4.1. The amount of binding at saturation with ^{32}P-labeled protein 4.1 is greater than that reported in an earlier study [61], which measured protein 4.1 rebinding to vesicles over a more limited range. In this study [61], ^{125}I-Bolton Hunter-labeled protein 4.1 was used, which modifies the amino-terminal region. The amino-terminal region appears to be involved in membrane binding (see below); as a result, protein 4.1 so labeled may not interact with band 3 and shows a decrease in glycophorin binding [50].

Protein 4.1 also interacted with the soluble 43K fragment of band 3 in solution, as demonstrated by nondenaturing polyacrylamide gel electrophoresis and rate-zonal sedimentation. Analysis of the association by nondenaturing gel electrophoresis indicated that the association was saturable at a 1:1 stoichiometry with an affinity close to that of the low-affinity membrane association. The complex between 43 K and protein 4.1, when resolved by rate-zonal sedimentation in iso-KCl-buffered sucrose gradients, also suggests a 1:1 stoichiometry (Figure 8). Under these conditions, protein 4.1 alone sedimented as a sharp peak with a sedimentation coefficient of 3.0S near the top of the gradient.Similarly, 43K alone formed a sharp peak at 4.2S, a sedimentation coefficient consistent with the noncovalent dimeric form reported by others [62]. When combined, the two proteins sedimented as a single discrete complex at 4.4S, suggesting that protein 4.1 forms a complex with a 43K monomer (Figure 8A). This would require dissociation of the 43 K dimer for protein 4.1 binding. The formation of the protein 4.1–4.3 K complex was inhibited by reacting protein 4.1 with N-ethylmaleimide, indicating that native protein 4.1 was required, and implying that the association is specific [56].

It has been proposed that protein 4.1 links the spectrin–actin complex to the lipid bilayer through interactions with spectrin and integral membrane proteins [3,5–7,63,64]. The role of band 3 as a possible linkage for the cytoskeleton was investigated using rate-zonal sedimentation to determine if a ternary complex could be formed between protein 4.1–spectrin and the 43 K fragment of band 3. Protein 4.1 forms a complex with spectrin in low-ionic-strength buffer, cosedimenting when analyzed by sucrose density gradient sedimentation (Figure 8B) [49,56]. The addition of excess 43 K to preformed spectrin–protein 4.1 complexes did not result in formation of a ternary complex between protein 4.1–spectrin and the 43 K fragment; instead, 43 K displaced the majority of protein 4.1 bound to spectrin (Figure 8C). This result suggested

that 43 K inhibits the interaction between protein 4.1 and spectrin, possibly
by competition for a common binding site on protein 4.1 or through an allosteric mechanism. However, when spectrin and actin are added to preformed
complexes of 43 K fragment and protein 4.1, the spectrin–protein 4.1–actin
complex was formed and did not appear to be affected, even by high concentrations of the 43 K fragment. This demonstrates that the protein 4.1–4.3 K
association is of substantially lower affinity than that of the spectrin–protein
4.1–actin association. As a result, protein 4.1 does not appear to link the spectrin–actin complex to the membrane through an association with the cytoplasmic domain of band 3; instead this may be a regulatory binding site. This suggests that another protein 4.1 binding site is involved in the linkage between
spectrin–actin–protein 4.1 and the membrane.

V. A PHOSPHOINOSITIDE IS A COFACTOR FOR
GLYCOPHORIN-PROTEIN 4.1 ASSOCIATIONS

Glycophorin isolated from human erythrocyte membranes using the LIS-phenol
method is soluble in aqueous solution as a protein–phospholipid mixed micelle
containing 14–18 glycoprotein monomers with a sedimentation coefficient in
the 7–8S range [19,44]. Preparations of glycophorin vary considerably in their
content of residual intrinsic lipid, depending on isolation conditions [46,62–64].
Glycophorin isolated by most methods, including the LIS-phenol method, contains acidic phospholipids that remain bound to glycophorin, even after extraction with neutral chloroform:methanol, and are efficiently removed only by extraction with acidified chloroform:methanol or by chromatography in SDS [46,
62–64].

Conceptually, a transmembrane protein in the form of a soluble micelle-like
aggregate is in an environment quite different from that of the membrane. However, glycophorin in the micellar form does retain intrinsic phospholipid, and
its cytoplasmic domain is exposed to the solvent because antibodies specific for
the cytoplasmic domain bind to glycophorin micelles [54]. Moreover, when micellar glycophorin and protein 4.1 were combined, micellar glycophorin bound
protein 4.1, resulting in a increase in the sedimentation coefficient of both components when analyzed by sucrose density gradient sedimentation. To determine
whether this interaction was specific, micellar glycophorin was covalently attached to Affi-Gel 10, and the immobilized glycophorin Affi-Gel beads were
used as an affinity matrix (Figure 9). Membrane skeletal proteins, which were
extracted from IOVs with 1 M KCl buffer, were applied to a column containing
glycophorin–Affi-Gel beads in isotonic KCl buffer. After washing with the same
buffer, the bound protein(s) was eluted with 1 M KCl buffer. Analysis of the
flow through and bound protein by SDS-PAGE showed that protein 4.1 bound
to the glycophorin-Affi-Gel beads (Figure 9). Further, protein 4.1 bound to im-

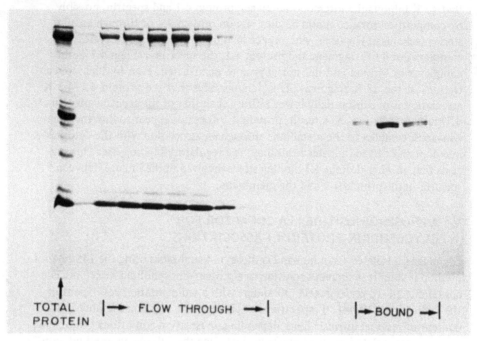

Figure 9 Affinity purification of protein 4.1 on desialated glycophorin immobilized in Affi-Gel 10. Elution profile of extrinsic membrane proteins extracted with 1 M KCl from red cell inside-out membrane vesicles (IOVs) and
then applied to a column, in iso-KCl buffer, containing desialoglycophorin (with
bound intrinsic phospholipid) covalently attached to Affi-Gel 10. The protein
bound to the glycophorin-Affi-Gel 10 column was eluted with 1 M KCl, 10 mM
Tris, 2 mM EDTA, at pH 7.5. From Anderson [32].

mobilized glycophorin within physiologic conditions of ionic strength and pH
and was eluted from the column with high ionic strength, conditions identical to
those used to extract protein 4.1 from erythrocyte membranes. These studies indicated that protein 4.1 associates specifically with micellar glycophorin even
though the glycophorin may be in an environment different from that of a phospholipid membrane.

It became apparent on repeating these experiments that different preparations of glycophorin had widely variable protein 4.1 binding capacities. The
variability between glycophorin preparations was found to be related to the
amount of residual intrinsic phospholipid bound to the glycophorin [19,50].
When glycophorin was stripped of all phospholipid, it retained no protein 4.1
binding activity [50]. However, if the stripped glycophorin was then recon-

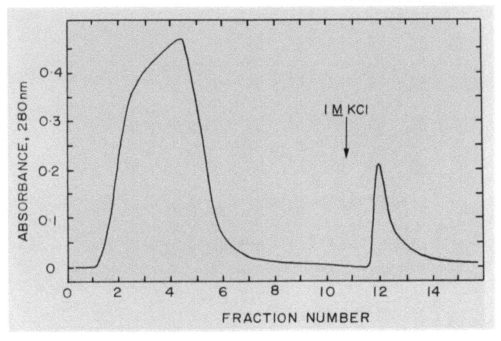

Figure 9 (continued)

stituted with the extracted phospholipid, it regained its capacity to bind protein 4.1. These observations demonstrated that the capacity of glycophorin to bind protein 4.1 is dependent on a phospholipid component that co-isolated with glycophorin [19,32,50].

A number of different phospholipids remain bound to glycophorin isolated by the LIS-phenol method. The majority of these are acidic phospholipids, such as phosphatidylserine [PS], phosphatidic acid [PA], and the polyphosphoinositides [46,52,53,65–67]. To determine which of these phospholipids is involved in regulating glycophorin's protein 4.1 binding capacity, glycophorin previously stripped of phospholipid was reconstituted with pure phospholipids and assayed for protein 4.1 binding capacity. Phospholipid was reconstituted into the glycophorin micelles by dissolving the phospholipid and glycophorin with the detergent octyl-β-D-glucoside. The detergent-phospholipid-protein mixed micelles, on removal of the detergent by dialysis, form glycophorin–phospholipid mixed micelles that are very stable. Incorporated ^{32}P-labeled phospholipids completely co-migrate with glycophorin micelles when analyzed by sucrose density gradient centrifugation [19]. In addition to phospholipids mentioned above, phosphatidylcholine (PC), phosphatidylinositol (PI), and phosphatidylethanolamine (PE)

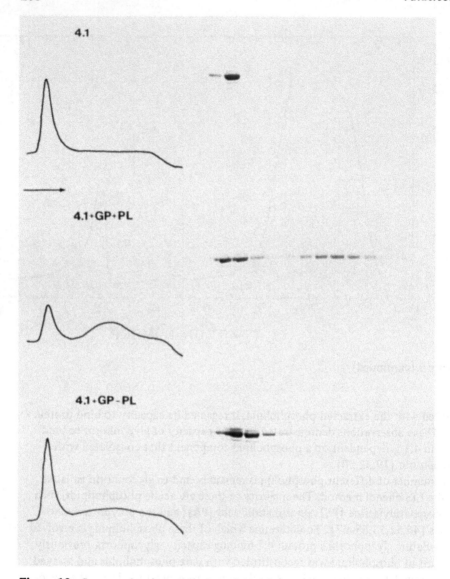

Figure 10 Sucrose density gradient analysis of the association of protein 4.1 with glycophorin-phospholipid micelles. Shown are absorbance profiles at 280 nm and corresponding fractions analyzed by SDS-PAGE. Samples are: Protein 4.1 alone, protein 4.1 plus glycophorin containing intrinsic phospholipid, protein 4.1 plus glycophorin stripped of phospholipid, protein 4.1 plus glycophorin reconstituted with PI, protein 4.1 plus glycophorin reconstituted with PIP, protein 4.1 plus glycophorin reconstituted with PIP$_2$. The phospholipid stripped glycophorin was treated as others, but without added phospholipid. Purified protein 4.1 and/or glycophorin at a 1:2 molar stoichiometry was applied to a 5–20% sucrose density. Gradient. The gradients were centrifuged at 200,000 X g for 18 hr in a Beckman SW-40 rotor. From Anderson and Marchesi [19].

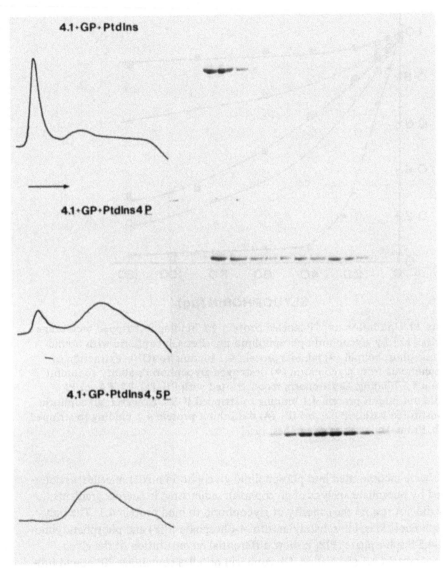

Figure 10 (continued)

were reconstituted into glycophorin micelles. The capacity of glycophorin-phospholipid micelles to associate with protein 4.1 was analyzed by density gradient sedimentation, which separates free protein 4.1 from protein 4.1 bound to glycophorin (Figure 10).

The only phospholipids that reconstituted the ability of glycophorin to bind protein 4.1 were the polyphosphoinositides. Although all of the other phospho-

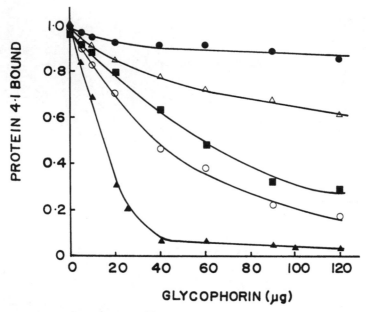

Figure 11 Inhibition of ^{32}P-labeled protein 4.1 binding to stripped membrane vesicles [17] by glycophorin-phospholipid micelles. Glycophorin with bound intrinsic phospholipid (■) inhibits protein 4.1 binding to IOVs; extraction of phospholipids from glycophorin (●) destroyed glycophorin's ability to inhibit protein 4.1 binding. Glycophorin reconstituted with PE, PC, PS, PA, or PI (△) did not inhibit protein 4.1 binding to stripped IOVs. However, glycophorin reconstituted with PIP (○) or PIP₂ (▲) did inhibit protein 4.1 binding to stripped IOVs. From Anderson and Marchesi [19].

lipids were incorporated into phospholipid-glycophorin mixed micelles, as determined by phosphate analysis of glycophorin sedimented in sucrose gradients, they did not restore the capacity of glycophorin to bind protein 4.1. The polyphosphoinositides, phosphatidylinositol-4-phosphate (PIP) and phosphatidylinositol-4,5-bisphosphate (PIP_2), show differential reconstitution of the glycophorin–protein 4.1 association. Glycophorin micelles containing PIP bound protein 4.1 with relatively low affinity compared with micelles containing an equivalent amount of PIP_2, which bound protein 4.1 with a high apparent affinity (Figure 10).

Micellar glycophorin reconstituted with polyphosphoinositides contains a specific and high-affinity binding site for protein 4.1. If glycophorin is a membrane binding site, then micellar glycophorin should be a competitive inhibitor of protein 4.1 binding to s-IOVs (Figure 11). Indeed, when micellar glycophorin was added in increasing concentration to s-IOVs previously reconstituted with [32]P-

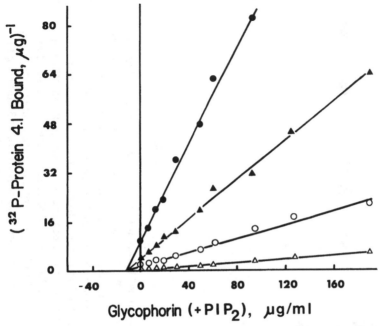

Figure 12 Inhibition of protein 4.1 binding to membrane vesicle by glyco-phorin-PIP_2 mixed micelles. Stripped IOVs (60 μg) were reconstituted with 0.1, 0.2, 0.5 and 2.0 μg of ^{32}P-protein 4.1 as in Fig. 6, and glycophorin-PIP_2 mixed micelles were added and the solution incubated for 30 min further. Data was plotted according to the method of Dixon [60].

labeled protein 4.1, glycophorin micelles retaining intrinsic phospholipid, mainly PS and PIP_2, inhibited protein 4.1 binding. Glycophorin-PIP_2 showed the greatest inhibition of protein 4.1 binding to red blood cell membranes (Figure 11) and its inhibition of binding was competitive (Figure 12), demonstrated by plotting the data according to the method of Dixon [60]. Not only did glycophorin-PIP_2 micelles completely inhibit protein 4.1 binding, but this occurred at a ratio of protein 4.1/s-IOV protein of 1.7 μg/mg, which is in the high-affinity region of the protein 4.1 binding isotherm. In comparison to inhibition by the 43K fragment, inhibition by glycophorin-PIP_2 is complete, suggesting that glyco-phorin-PIP_2 is the high-affinity binding site for protein 4.1 on the membrane.

Although glycophorin-PIP_2 competitively inhibits protein 4.1 binding to membranes, there are two possible intrinsic artifacts in these experiments that cannot be overlooked. First, glycophorin in a protein–phospholipid micelle is in an environment very different from that surrounding glycophorin and the polyphosphoinositides in the native lipid bilayer. Second, the data collected

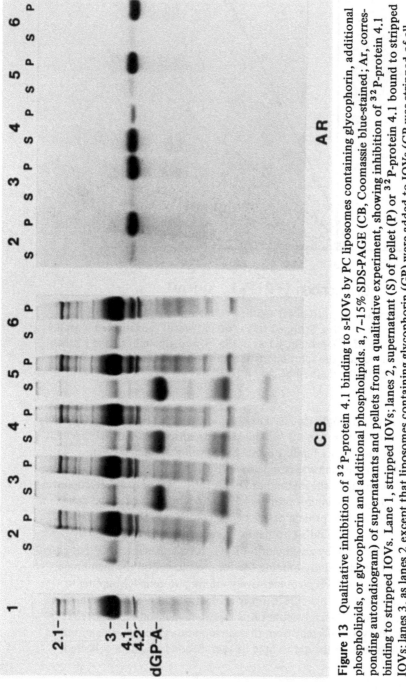

Figure 13 Qualitative inhibition of ^{32}P-protein 4.1 binding to s-IOVs by PC liposomes containing glycophorin, additional phospholipids, or glycophorin and additional phospholipids. a, 7–15% SDS-PAGE (CB, Coomassie blue-stained; Ar, corresponding autoradiogram) of supernatants and pellets from a qualitative experiment, showing inhibition of ^{32}P-protein 4.1 binding to stripped IOVs. Lane 1, stripped IOVs; lanes 2, supernatant (S) of pellet (P) or ^{32}P-protein 4.1 bound to stripped IOVs; lanes 3, as lanes 2 except that liposomes containing glycophorin (GP) were added to IOVs (GP was stripped of all phospholipid before reconstitution into liposomes); lanes 4, as lanes 2 except that liposomes containing GP and PIP$_2$ were added to IOVs; lanes 5, as lanes 2 except that liposomes containing GP and PIP were added to IOVs; lanes 6 as lanes 2 except that liposomes containing PIP$_2$ alone were added to IOVs. Stripped IOVs (35 µg of protein) reconstituted with 1.2 µg of ^{32}P-protein 4.1 were combined with liposomes, layered over a 10% sucrose cushion, and centrifuged at 38,000 × g for 30 min. Within these conditions IOVs sediment through, whereas liposomes float at the interface of. the sucrose cushion.

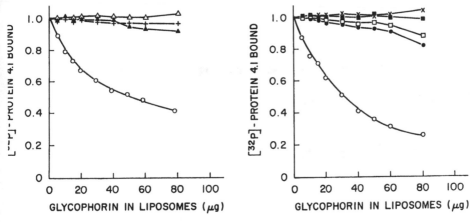

Figure 14 Quantitative inhibition of ^{32}P-protein 4.1 binding to stripped IOVs. Conditions are as in Fig. 13. Phosphatidylcholine liposomes contain: left panel, GP alone (+), GP and PIP (▲), GP and PIP$_2$ (o), and PIP$_2$ alone (△); right panel, PIP$_2$ and PIP (X); GP and PI (■); GP and PS (□); GP and PA (●); GP, PIP$_2$ and PIP (o). The ratio of added phospholipid to total phospholipid in the liposome were kept close to physiological. For example, phosphatidylserine was used in concentrations up to 30 mole %, whereas phosphatidic acid was used up to 5 mole %. The amount of PIP$_2$ used in this case was a 2:1 PIP$_2$ to glycophorin molar stoichiometry, or 0.3 mole percent of phospholipid; at higher molar concentrations of PIP$_2$, more inhibition of protein 4.1 bindng to IOVs was observed.

using the glycophorin-polyphosphoinositide micelle does not rule out a direct interaction between protein 4.1 and the polyphosphoinositides. To address these concerns, glycophorin stripped of phospholipid was reconstituted into PC liposomes. The phospholipid composition of these liposomes was varied by addition of other phospholipids during liposome preparation. The effect of liposome phospholipid composition on the glycophorin–protein 4.1 interaction was studied by direct binding of protein 4.1 to liposomes and by competition of protein 4.1 binding to IOVs by liposomes.

Direct binding of protein 4.1 to liposomes demonstrated that PC liposomes containing glycophorin and PIP$_2$ bind ^{32}P-labeled protein 4.1 saturable and with high affinity ($K_a > 10^7 \ M^{-1}$). In contrast, liposomes containing glycophorin and PS, PE, PA, PI, or PIP showed only nonspecific protein 4.1 binding. Liposomes containing only PIP$_2$ bound protein 4.1 with low affinity ($K_a < 10^5 \ M^{-1}$) but greater than nonspecific binding [19,32]. The association of protein 4.1 with liposomes containing both glycophorin and PIP$_2$ indicates that the glycophorin-PIP$_2$ complex is a binding site on membranes. However, if glycophorin-PIP$_2$ is an erythrocyte membrane binding site for protein 4.1, then liposomes containing

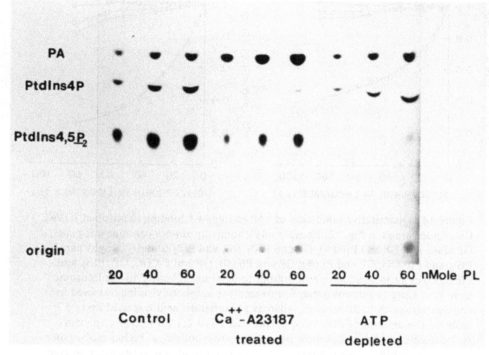

Figure 15 Autoradiogram of thin-layer chromatogram (TLC) onto which have been applied s-IOVs prepared from ^{32}P-labeled control cells, cells treated with Ca[++] and ionophore A23187 (100% echinocytes), or ATP-depleted cells (100% echinocytes). The major ^{32}P-labeled components are the phosphoinositides and Pa; ^{32}P-labeled protein, seen at the origin, represents a minor amount of label.

glycophorin-PIP$_2$ should compete with erythrocyte membranes for protein 4.1 binding.

To test this hypothesis, s-IOVs were reconstituted with ^{32}P-labeled protein 4.1 at a protein 4.1/membrane ratio in which the majority of the protein 4.1 would be bound to high-affinity sites. Liposomes were then combined with these IOVs, and the partitioning of ^{32}P-protein 4.1 between liposomes and IOVs was quantitated, after separation of liposomes and IOVs (Figure 13). Liposomes were separated from IOVs by taking advantage of the low density of liposomes compared with IOVs. A mixture of liposomes and IOVs was layered over a 10% sucrose cushion. The IOVs, which are quite dense, sediment through the cushion and were separated from the liposomes that float at the interface of the sucrose cushion. When the supernate and pellet of such an experiment were analyzed by SDS-PAGE, it was found that liposomes and IOVs were completely separated.

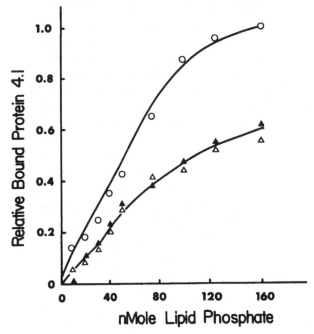

Figure 16   The binding of ^{32}P-protein 4.1 to membrane vesicles from control (o) and phosphoinositide-depleted erythrocytes, by ATP depletion (△) or by Ca^{++}-A23187 treatment (▲).

Further, only liposomes containing both glycophorin and PIP$_2$ effectively competed with IOVs for protein 4.1 binding (Figure 13, lanes 4). Quantitation of the competition of ^{32}P-labeled protein 4.1 binding to IOVs by liposomes is shown in Figure 14. Clearly, liposomes containing glycophorin and PIP$_2$ compete much more effectively than do liposomes containing glycophorin and other phospholipids. These data demonstrate that in a phospholipid membrane the association between glycophorin and protein 4.1 is regulated only by the phospholipid phosphatidylinositol-4,5-bisphosphate, indicating that in a membrane the specificity for a phosphoinositide is much higher than in the glycophorin micelle. Although the data in Figure 14 show only inhibition with liposomes containing glycophorin and 0.3 mol % PIP$_2$, increasing the mole fraction of PIP$_2$ in the liposomes increases their ability to inhibit protein 4.1 binding to membranes; this demonstrates that the concentration of PIP$_2$ in the membrane modulates the affinity of protein 4.1 binding to glycophorin.

  In the red blood cell, as in all other eukaryotic cells, PI is sequentially phosphorylated on the fourth hydroxyl of the *myo*-inositol ring to yield phosphatidylinositol-4-phosphate, which is further phosphorylated by another kinase on the fifth hydroxyl, yielding phosphatidylinositol-4,5-diphosphate [28-30,68,69]. The PIs are also sequentially dephosphorylated by phosphatases that are highly specific for each monoester phosphate. The cycle of phosphorylation-dephosphorylation in the erythrocyte, as well as in other cells, expends much metabolic energy and thus is likely to be tightly regulated. The erythrocyte also contains a membrane-bound Ca^{2+}-activated phospholipase C that specifically cleaves the polyphosphoinositides when the intracellular Ca^{2+} concentrations is $> 10\ \mu M$ [70]. In the red blood cell, the membrane concentration of PIP_2 decreases when the cellular ATP concentration drops or when the Ca^{2+} is increased [70-74]. Under both conditions, the cell is transformed from the normal biconcave disk to a spiculated sphere, the echinocyte [70-74]. The shape transformation appears to correspond with the metabolic changes of the phosphoinositides [74].

Because protein 4.1 appears to bind with high affinity only to glycophorin complexed with PIP_2, a decrease in the membrane concentration of PIP_2 should lead to a decrease in the binding capacity of the membranes for protein 4.1. To test this hypothesis, membranes depleted of PIP_2 by increased cellular Ca^{2+} or ATP depletion (Figure 15) were isolated and stripped of membrane skeletal components. On quantitating the binding of protein 4.1 to these membranes, a significant decrease in the protein 4.1 binding capacity of polyphosphoinositide-depleted membranes was found (Figure 16), lending additional support to the hypothesis that polyphosphoinositide metabolism is important in regulation of protein 4.1 associations with the membrane. This leaves the question: Is the spectrin–protein 4.1–actin lattice attached to the membrane through the association with glycophorin-PIP_2?

Spectrin and actin, when combined, form only a low-affinity interaction; however, when protein 4.1 is added, a high-molecular-weight complex is formed that sediments to the bottom of a sucrose density gradient on centrifugation (Figure 17) [5-9]. The association of glycophorin with this complex was measured using ^{125}I-labeled glycophorin reconstituted into glycophorin–phospholipid micelles. Glycophorin micelles containing any phospholipid other than PIP or PIP_2 when combined with spectrin, protein 4.1, actin, or any single or joint combination of these proteins, did not form an association with these proteins. However, glycophorin micelles reconstituted with PIP_2, when combined with spectrin and analyzed by rate zonal sedimentation, appear to form a low-affinity association, enough to skew the normally sharp glycophorin micelle peak (Figure 17b). Addition of micellar glycophorin-PIP_2 to spectrin and protein 4.1 results in a high-molecular-weight complex that sediments to the bottom of the gradient (Figure 17b). When spectrin–protein 4.1–actin complexes are formed and glycophorin-PIP_2 micelles are added, the glycophorin–PIP_2 micelles plummet to the bottom of the density gradient along with the spectrin–protein 4.1–actin complex, indicating that glycophorin–PIP_2 forms an association with spectrin-protein 4.1-actin complexes. This association requires PIP_2, and the association of glycophorin-PIP_2 with the spectrin-protein 4.1-actin complex is inhibited by excess protein 4.1, suggesting that the association requires protein 4.1 (not shown). The addition of excess protein 4.1 does not hinder, and indeed promotes, the assembly of the spectrin-protein 4.1-actin complex. These results suggest that the spectrin-protein 4.1-actin complex and possibly the native junctional complexes are linked to the membrane by an association with glycophorin-PIP_2 and perhaps are regulated by this interaction.

A considerable amount is known about protein 4.1 structure, both alignment of proteolytic fragments and the location of function domains involved in spectrin-actin binding and membrane binding [8,9,57,75]. The amino-terminal domain of protein 4.1 appears to be involved in the association of protein 4.1 with glycophorin-PIP_2. This domain is very resistant to proteolysis, is quite basic (pI > 8), contains all of the cysteine residues present in the protein, and has a molecular weight by SDS-PAGE of ~30 K (Figure 19) [57,66]. When protein 4.1 is mildly digested by α-chymotrypsin and then the digest is applied to a column containing glycophorin linked to Affi-Gel 10, the 30 K fragment and fragments containing 30 K are the only peptides retained by the column and eluted with 1 M KCl (Figure 18). The association of the 30 K domain with glycophorin specifically requires glycophorin-PIP_2, indicating that this selectivity lies within this domain. Although the 30 K domain specifically binds glycophorin, this peptide alone does not have the ability to compete with protein 4.1 binding to membranes, suggesting that other parts of the molecule are required for high-affinity binding. However, antibodies and Fab fragments of antibodies specific for the 30 K domain of protein 4.1 do inhibit binding to membranes. Further, modification of the cysteine residues completely inhibited association with glycophorin-PIP_2. Taken together these data suggest that the 30 K fragment contains a major portion of the site involved in membrane binding. Interestingly, band 3 also appears to interact with the 30 K domain, although likely at a different site than glycophorin, because band 3-protein 4.1 interactions and protein 4.1-spectrin interaction cannot occur simultaneously. Perhaps the sites involved in spectrin-actin and band 3 binding are close in the folded protein, even though they are distant in the sequence (Figure 19).

VI. DO NONERYTHROID PROTEIN 4.1 ANALOGUES REQUIRE A PHOSPHOINOSITIDE FOR MEMBRANE BINDING?

Immunocrossreactive analogues of protein 4.1 have been identified in all tissues studied so far [76-83]. In surveying a number of tissues and cells by Western blotting techniques using antibodies raised against erythroid protein 4.1, a large number of immunologically reactive forms were detected. These isoforms vary

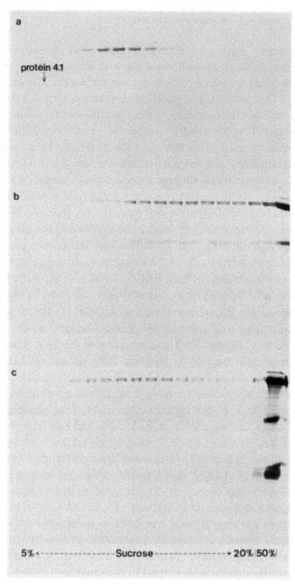

Figure 17 Spectrin-protein 4.1-actin associates with glycophorin-PIP_2. G-actin isolated from rabbit muscle [9] was dialyzed against 2 mM Tris, 0.2 mM Mg-ATP, 10 mM 2-mercaptoethanol, 0.5 mM NaN_3, pH 8.0, until addition of 213 μg to protein 4.1 (100 μg), spectrin (300 μg), and/or glycophorin-phospholipid mixed micelles (100 μg). The components after mixing were in Iso-KCl buffer (400 μl total vol) with 0.1 mM ATP and 2 mM $MgCl_2$, pH 7.5. The samples were loaded onto a 5-20% sucrose gradient and centrifuged for 12 hr at 200,000 g. The left-hand panels show analysis of the gradient fractions by SDS-PAGE; the direction of sedimentation is from left to right. Right hand panels are sedi-

218

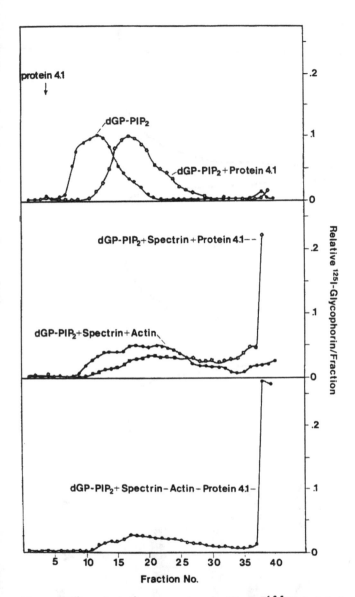

Figure 17 (continued) mentation profiles of ^{125}I-labeled glycophorin in glyco-phorin-phospholipid mixed micelles when the indicated mixtures of components are added. Left hand panels are: a, sedimentation of protein 4.1 bound to glyco-phorin-PIP$_2$ micelles; the position of protein 4.1 alone is indicated; b, sedimenta-tion of spectrin, protein 4.1, and glycophorin-PIP$_2$ micelles; see right-hand panel for glycophorin profile; c, sedimentation of spectrin, protein 4.1, actin, and glycophorin-PIP$_2$ micelles; the right-hand panel shows that glycophorin-PIP$_2$ micelles sediment with the spectrin-protein 4.1-actin complex. Glycophorin-phospholipid micelles with phospholipids other than PIP$_2$ do not interact with spectrin-protein 4.1-actin.

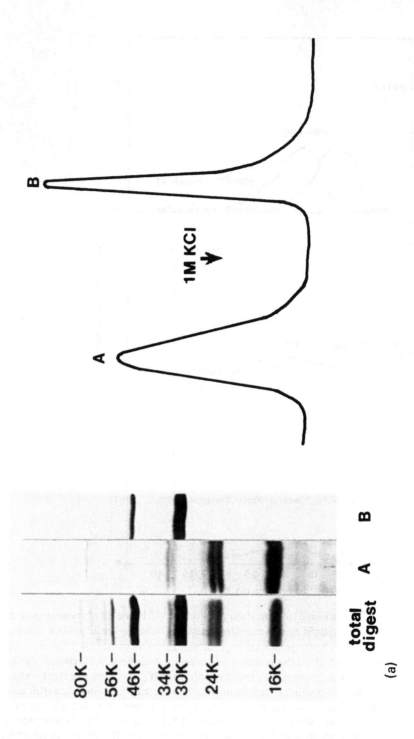

(a)

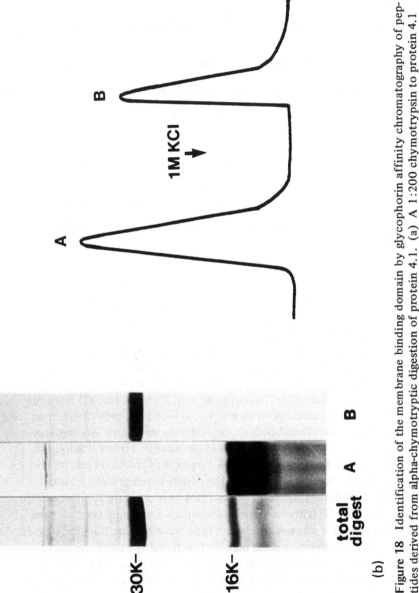

(b)

Figure 18 Identification of the membrane binding domain by glycophorin affinity chromatography of peptides derived from alpha-chymotryptic digestion of protein 4.1. (a) A 1:200 chymotrypsin to protein 4.1 digest; (b) a 1:20 chymotrypsin to protein 4.1 (30 min at 0 C). In both cases, only peptides that contain the 30 K aminoterminal domain bind to the glycophorin-Affi-Gel column [57,75]. Lane A is flow-through; B is 1 M KCl elution.

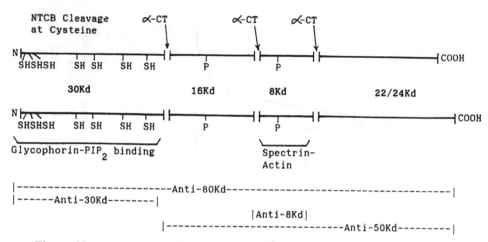

Figure 19 Structural and functional model of protein 4.1 showing location of antibody reactivities.

in molecular weight from 175–30 K and are expressed in cells in a tissue-specific fashion [82]. Immunolocalization and cell fractionation of some of these isoforms suggest that nonerythroid protein 4.1 is not restricted to the plasma membrane [83]. Indeed, these data suggest that there are isoforms located at the nuclear membrane, along stress fibers possibly associated with small vesicles, as well as on the plasma membrane [76–83]. If all of these isoforms are associated with membrane structures, then perhaps analogues of glycophorin with bound PIP_2 are also required for these associations. If the protein 4.1 isoforms bind to membranes by a mechanism similar to red blood cell protein 4.1, then it is likely that these nonerythroid isoforms would have membrane binding sites homologous to the 30 K domain of erythroid protein 4.1. To test this hypothesis, antibodies that were raised to the 30 K fragment of red blood cell protein 4.1 (Figure 19) were used to determine if the nonerythroid protein 4.1 analogues contain homologous regions. The results shown in Figure 20 demonstrate that isoforms in T and B lymphocytes and the spleen and lymph node all have regions that are immunocrossreactive to the membrane binding site of red blood cell protein 4.1, suggesting that there may be functional as well as structural relatedness. An interesting observation is that the 175 K protein 4.1 analogue in these tissues was detected by the anti-8 kD and anti-30 K antibodies but not by the polyclonal antibody toward protein 4.1. This suggests that the 175 K nonerythroid analogue is quite different from erythroid protein 4.1, but retains similarity in the membrane binding and the spectrin–actin binding domain [82].

More direct evidence for PIP_2 regulation of a protein 4.1 isoform's interaction with the membrane comes from the interaction of synapsin I with glycophorin micelles. Synapsin I is a phosphoprotein found in the synaptic region of

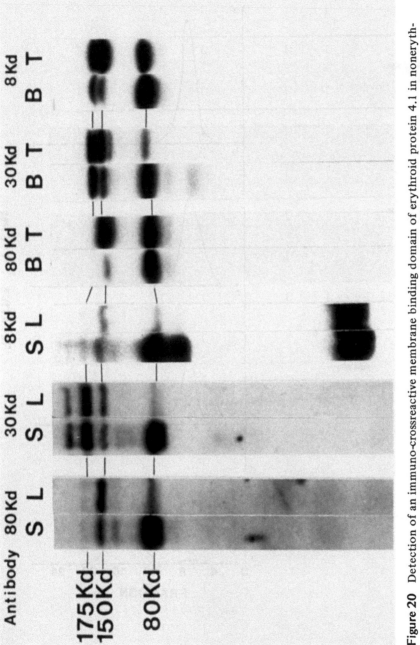

Figure 20 Detection of an immuno-crossreactive membrane binding domain of erythroid protein 4.1 in nonerythroid protein 4.1 analogues. The three panels on the left show antibody reactivity by Western Blotting of homogenates of rat spleen (S) and the lymph node (L). The three panels on the right show immunoreactivity toward B and T human lymphocyte membranes. All protein 4.1 analogues contain the 30K membrane binding domain and the 8K spectrin–actin binding domain.

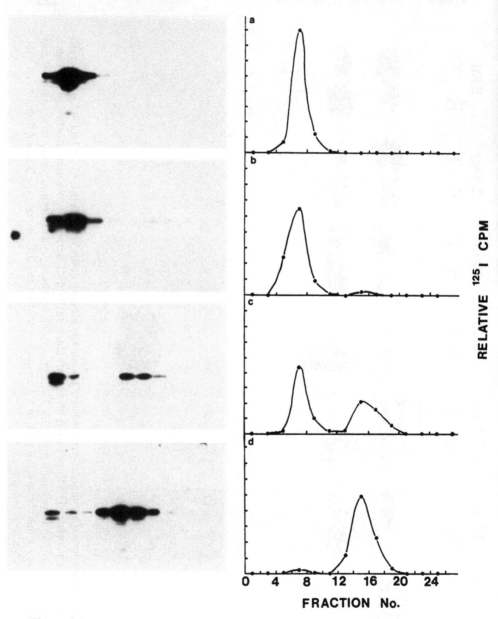

Figure 21

neuronal cells [84]. This protein is thought to associate specifically with clear synaptic vesicles and to be involved in neuronal transmission [84]. Recent evidence suggests that porcine synapsin I and porcine erythrocyte protein 4.1 are immunohomologous proteins [78]. Although antibodies prepared against human protein 4.1 did not immunocrossreact with porcine synapsin1 (our unpublished results), suggesting less homology than originally proposed, synapsin I did appear to have functional homology with protein 4.1. When a crude preparation of synapsin I was combined with micellar glycophorin–phospholipid combinations and analyzed by sucrose density gradient sedimentation, as in Figure 10, synapsin I bound to glycophorin micelles containing PIP or PIP_2, but not glycophorin micelles containing other phospholipids (Figure 21). Like the protein 4.1-glycophorin association, synapsin I also bound to glycophorin micelles containing PIP_2 with higher affinity than to micelles containing PIP. The association between synapsin I and glycophorin appeared to be specific because other proteins in the mixture did not bind to glycophorin-PIP_2 micelles. The exciting possibility of phosphoinositide regulation of synapsin I–synaptic vesicle binding becomes even more exciting with a recent observation that some exocytotic vesicules have both bound fodrin (a spectrin analog) and a phospholipase C that is specific for phosphoinositides, both of which may play a part in secretion [85-88]. Activation of this phospholipase C directly by Ca^{2+} or by Ca^{2+}-calmodulin could result in dissociation of synapsin I, or in nonneuronal cells, other protein 4.1 analogues from the secretory vesicles. At the same time, a fusogenic lipid, 1,2-diacylglycerol, is formed. This could be the signal for release of exocytotic vesicles from cytoskeletal interactions, allowing transport through the actin-based cytoskeleton to the plasma membrane where fusion and release of vesicle contents could take place.

Figure 21   Synapsin I, a putative protein 4.1 analogue, binds to glycophorin and requires PIP_2 for binding. A preparation bovine of synapsin I [78] containing about 5 μg of synapsin I (65 μg of total protein) was combined with 20 μg of glycophorin-phospholipid micelles, layered on a 5–20% sucrose gradient and centrifuged for 16 hr at 200,000 X g. The gradients were fractionated and analyzed by SDS-PAGE and Western Blotting using an anti-synapsin I antibody. The left-hand panels show Western blots of the SDS-PAGE of the sedimentation gradients. The panels are: a, glycophorin-PI micelles + synapsin I, the sedimentation of synapsin I here is identical to synapsin I alone; b, glycophorin-PIP + synapsin I; c, glycophorin-PIP_2 + synapsin I, greater than 90% of the synapsin I is bound to glycophorin. Synapsin I was the only protein in the mixture which bound to glycophorin-PIP_2.

VII. IMPLICATIONS

The characteristic biconcave disk shape of the red blood cell is maintained by the cell's active metabolism. When the cytoplasmic pool of ATP is depleted, the red blood cell undergoes a dramatic shape transformation to a spiculated sphere, the echinocyte [38,39,73,74]. The function of ATP in maintaining the discoid shape is not directly related to exclusion of Ca^{2+} ion or preservation of ionic gradients [39]. Instead, ATP was thought to maintain cell shape by phosphorylation of membrane proteins; in particular, spectrin [89]. However, recent results have shown that the transformation to the echinocyte shape on ATP depletion was complete before substantial dephosphorylation of spectrin or any other membrane protein had occurred, suggesting that another phosphorylated substrate is involved [73].

Possible candidates for phosphorylated substrates are the polyphosphoinositides, because phosphorylation of this class of phospholipids likely consumes more high energy phosphate than the combined phosphorylation of all the membrane proteins [69,90]. The polyphosphoinositides, which are located entirely on the cytoplasmic leaflet of the bilayer, are sequentially phosphorylated from PI to PIP and then to PIP_2. In turn, PIP_2 is dephosphorylated sequentially to PIP and then PI. The resulting futile cycle occurs naturally in the cell, and more complete dephosphorylation occurs on ATP depletion. The depletion of the phosphoinositides, dephosphorylation or cleavage of the head group from PIP_2, recently has been shown to coincide with the diskocyte-to-echinocyte transformation induced by ATP depletion or by increased cytoplasmic Ca^{2+}. The rephosphorylation of the polyphosphoinositides, on ATP repletion, also corresponds with the reversal of cell shape change back to the diskocyte [74]. The hydrolysis of the polyphosphoinositides induced by increased cytoplasmic Ca^{2+} is irreversible because Ca^{2+} activates a phospholipase C that cleaves polyphosphoinositides, producing myo-inositol polyphosphate and 1,2-diacylglycerol [70-72]. The de novo synthesis of PI by enzymatic condensation of phosphatidic acid and myo-inositol does not take place in the red blood cell, nor does the red blood cell pick up PI from plasma lipoproteins. As a result, activation of phospholipase C results in a permanent loss of inositol phospholipid from the red blood cell membrane [70-72]. However, red blood cell shape change induced by Ca^{2+} is initially reversible because there is a pool of PI in the membrane, which can be rephosphorylated, although repeated activation of the phospholipase C would eventually deplete the pool of phosphoinositides. Depletion of the polyphosphoinositides in the erythrocyte may be important in erythrocyte senescence, clearance from the circulation, and also in some hemolytic diseases such as sickle cell disease and hereditary pyropoikilocytosis [32, 36].

Many hypotheses have been proposed to account for the molecular events that take place when red blood cells change shape [36,39,73,74]. The results

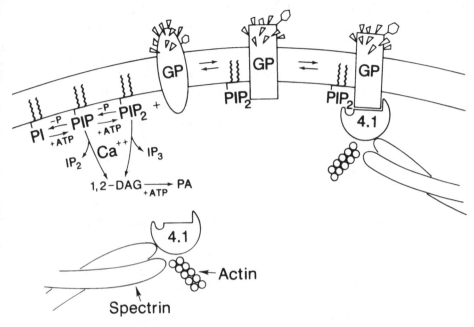

Figure 22 A model showing the regulation, by polyphosphoinositide metabolism, of the association of protein 4.1 (and spectrin-actin) with the cytoplasmic domain of glycophorin. The interaction of PIP_2 with the cytoplasmic domain of glycophorin and in the subsequent binding of protein 4.1 (and spectrin-actin) may be important in maintenance and regulation of red cell shape. Similar regulatory mechanisms in nonerythroid cells may be involved in diverse functions.

presented here provide yet another possible mechanism (Figure 22). The membrane content of the polyphosphoinositides dramatically decreases in response to metabolic depletion of ATP or increased cytoplasmic Ca^{2+}. In both cases, the membrane content of PIP_2 fluctuates the most. As a result, on ATP depletion or increased cytoplasmic Ca^{2+}, the membrane content of PIP_2 available for glycophorin binding would be decreased, resulting in the glycophorin-protein 4.1 association being turned off. This, in turn, would significantly alter the amount of protein 4.1 bound to the membrane through an association with glycophorin, possibly by shifting the association to band 3. Coincident with the unbinding of PIP_2 and protein 4.1, glycophorin could undergo a reversible refolding that is transmitted across the membrane. Conceptually, such a transmembrane event could be blocked by lectin binding to the extracellular domain of glycophorin. This molecular mechanism could tie together the observations that lectins that bind glycophorin also block cell shape change, with the fluctuations in the membrane content of the phosphoinositides that coincide with cell shape change. In turn, echinocytic ligands, which bind to the outside surface of the membrane, might initiate a change in cell shape by inducing a transmembrane refolding of glycophorin that alters the glycophorin–protein 4.1 association on the inside surface of the membrane. This could explain lectin inhibition of shape change induced by this class of echinocytic ligands. Although such a model is hypothetical, it appears to be supported by the data collected on lectin inhibition of shape change and polyphosphoinositide regulation of glycophorin–protein 4.1-mediated attachment of spectrin and actin to the membrane. Indeed, direct evidence for the involvement of cytoskeletal-membrane linkages in regulation of cell shape comes from the opposite end of spectrin, where ankyrin links spectrin to the membrane. If the spectrin–ankyrin–band 3 linkage is broken by selective proteolysis of ankyrin, the cell shape is irreversibly converted to the echinocyte and is unresponsive to chloropromazine or Mg-ATP [91]. Therefore, as has been proposed [19,31–36,48,91], it appears that a tight coupling between the membrane, the integral membrane proteins, and the associated cytoskeleton is necessary to maintain the normal red blood cell shape. This phospholipid bilayer–integral membrane protein–cytoskeletal coupling may be necessary for the molecular events by which echinocytic and stomatogenic agents induce cell shape change, suggesting that the membrane acts as a "trilayer couple" when the cell undergoes shape change [36].

Appreciation of the functional significance of two binding sites for protein 4.1 necessitates a dynamic view of the membrane skeleton. The deformability and shape of the erythrocyte, and likely other cells, are related to metabolic events. For example, changes in the levels of 2,3-diphosphoglycerate and PI-4,5-diphosphate may cause alterations in the membrane skeleton, in response to the cellular environment [39,92,93]. Although the molecular details whereby these metabolites exert their effects are incomplete, the findings described here suggest a mechanism by which metabolic changes in the erythrocyte may influence membrane skeletal assembly. There are approximately three times as many potential binding sites for protein 4.1 on band 3 than there are on the glycophorins. Based simply on the relative amounts of the two proteins in the membrane, it is likely that most of the protein 4.1 would be bound to band 3. However, glycophorin with bound PIP_2 forms a much higher affinity binding site for protein 4.1, putatively bringing along spectrin and actin, because their association with the membrane requires protein 4.1 [5,63,64]. Factors that modify the states of phosphorylation of this phospholipid might alter this balance, allowing protein 4.1 to shift from glycophorin to band 3 sites, but this would only occur if protein 4.1 is not associated with spectrin and actin. The binding of specific ligands to the external segments of glycophorin could provide a stimulus for the shift of 4.1 from one site, and this could have pronounced effects on the membrane skeleton. There is evidence for a functional linkage between glycophorin and band 3. When glycophorin was cross-linked in erythrocyte ghosts by antibody to its extracellular segment, the rotational diffusion of band 3 decreased [94]. This finding was taken to demonstrate a direct association of glycophorin with band 3 in the membrane. Treatment of leaky ghosts with antispectrin antibody failed to affect this process, suggesting that mediation of this phenomenon by the membrane skeleton was unlikely. If there is some form of association between glycophorin and band 3, the functional significance of two membrane binding sites for protein 4.1 may be quite important because protein 4.1 could translocate the short distance between the two sites. Such a flip-flop between adjacent membrane binding sites, metabolically regulated by PIP_2, could result in a very dynamic molecular ensemble that is potentially responsive to intracellular and extracellular stimuli.

In the erythrocyte the actin protofilament (~16 monomers) is thought to have five to six spectrin and protein 4.1 molecules associated with it, as well as tropomyosin, protein 4.9, and the calmodulin binding 105/100 kD dimer [10-17]. This oligomeric unit is thought to bind to the membrane through an association mediated by protein 4.1 [5,12-16,63,64]. Such a junctional complex has five to six copies of protein 4.1, each having the potential to bind one glycophorin receptor. If this indeed happens, then the affinity of the junctional complex for the membrane would be multiplicatively increased with each glycophorin bound, analogous to a multivalent antibody whose affinity (avidity) is increased toward a polymeric antigen, compared with a monovalent antibody. This effect would bestow on the junctional complex not only a high affinity for the membrane, but also a high potential for regulation by even small changes in the concentration of PIP_2. This is consequently predicted because each protein 4.1 that interacts with a glycophorin receptor requires a PIP_2 molecule to be bound to glycophorin. As a result, a junctional complex with five molecules of protein 4.1, if it were to interact with five glycophorin receptors, would require at least an equal number of PIP_2 molecules. This would give junctional complex-membrane binding a factor of $[PIP_2]^{-5}$ in the equation describing this association. This requirement in the concentration of PIP_2 could make the association between the junctional complex and the membrane highly cooperative and very dependent on membrane concentration and metabolism of PIP_2.

The interaction of protein 4.1 in an oligomeric complex with glycophorin may also explain why the loss of glycophorin A, in $E_n(a-)$ cells, may not drastically affect the physical properties of these cells [95]. In the membrane, there are about ~200,000 molecules of protein 4.1, ~600,000 molecules of glycophorin A, and ~150,000-200,000 molecules of glycophorins B and C combined [1,3,55]. In the membrane, there would be between 30,000 and 40,000 copies of each junctional complex; as a result, simple attachment of the junctional complex to the membrane could be fulfilled by substantially fewer glycophorin molecules than are present in the normal cell. However, because the nature of the interaction between the junctional complex and glycophorin-PIP$_2$ remains vague, the role this plays in regulation of cytoskeletal assembly is unknown.

Recent studies have demonstrated that a variety of nonerythroid cells contain analogs of the proteins that comprise the erythrocyte membrane skeleton. Protein 4.1 [76–83], band 3 (1), and spectrin [1,2,96–99] all have their counterparts in a growing list of functionally diverse cells. The cytoskeletal–membrane associations discussed here may reflect the potential complexities characteristic of interactions between the membrane skeleton and integral proteins imbedded in the lipid bilayer. These interactions are likely to be regulated by metabolic events within the cells, some of which may be triggered by stimulation of surface receptors. The simplified membrane of the erythrocyte may, in this instance, highlight the functions as well as the structures that are to be found in other cells.

REFERENCES

1. Bennett, V. (1985). The membrane skeleton of human erythrocytes and its implication for more complex cells. *Annu. Rev. Biochem.* 54:273–304.
2. Marchesi, V. T. (1985). Stabilizing infrastructure of cell membranes. *Annu. Rev. Cell Biol.* 1:531–561.
3. Cohen, C. M., Tyler, J. M., and Branton, D. (1980). Interaction of cytoskeletal proteins of the human erythrocyte membrane. *Cell* 21:875–883.
4. Morrow, J. S., and Marchesi, V. T. (1981). Self assembly of spectrin oligomers in vitro: A basis for a dynamic cytoskeleton. *J. Cell Biol.* 88:463–468.
5. Cohen, C. M., and Korsgren, C. (1980). The role of band 4.1 in the association of actin with erythrocyte membranes. *Biochim. Biophys. Acta* 688:691–701.
6. Fowler, V., and Taylor, D. L. (1980). Spectrin plus band 4.1 cross-link actin. Regulation by micromolar calcium. *J. Cell Biol.* 85:361–376.
7. Ungewickell, E., Bennett, P. M., Calvert, R., Ohanian, V., and Gratzer, W. B. (1979). In vitro formation of a complex between cytoskeletal proteins of the human erythrocyte. *Nature* 280:811–814.
8. Correas, I., Leto, T. E., Speicher, D. W., and Marchesi, V. T. (1986). Identification of the functional site of erythrocyte protein 4.1 involved in spectrin-actin associations. *J. Biol. Chem.* 261:3310–3315.
9. Correas, I., Speicher, D. W., and Marchesi, V. T. (1986). Structure of the spectrin-actin binding site of erythrocyte protein. *J. Biol. Chem.* 261:3310–3315.
10. Atkinson, M. A., Morrow, J. S., and Marchesi, V. T. (1982). The polymeric state of actin in the human erythrocyte cytoskeleton. *J. Cell Biochem.* 18:493–505.
11. Pinder, J. C., and Gratzer, W. B. (1983). Structural and dynamic states of actin in the erythrocyte. *J. Cell Biol.* 96:768–775.

12. Fowler, V. M. (1986). New views of the red cell network. *Nature* 322:777–778.
13. Matsuzaki, F., Suto, K., and Ikai, A. (1985). Structural unit of the erythrocyte cytoskeleton. Isolation and electron microscopic examination. *Eur. J. Cell Biol.* 39:153–160.
14. Beaven, G. H., Jean-Baptiste, L., Ungewickell, E., Baines, A. J., Shahbakhti, F., Pinder, J. C., Lux, S. E., and Gratzer, W. B. (1985). An examination of the soluble oligomeric complexes extracted from the red cell membrane and their relation to the membrane cytoskeleton. *Eur. J. Cell Biol.* 36:299–306.
15. Shen, B. W., Josephs, R., and Steck, T. L. (1986). Ultrastructure of the intact skeleton of the human erythrocyte membrane. *J. Cell Biol.* 102:997–1006.
16. Byers, T. J., and Branton, D. (1985). Visualization of the protein associations in the erythrocyte membrane skeleton. *Proc. Natl. Acad. Sci. USA* 82:6153–6157.
17. Gardner, K., and Bennett, V. (1986). A new erythrocyte membrane-associated protein with calmodulin binding activity. *J. Biol. Chem.* 261:1339–1348.
18. Ohanian, V., Wolfe, L. C., John, K. M., Pinder, J. C., Lux, S. E., and Gratzer, W. B. (1984). Analysis of the ternary interaction of the red cell membrane skeletal proteins spectrin, actin, and 4.1. *Biochemistry* 23:4416–4420.
19. Anderson, R. A., and Marchesi, V. T. (1985). Regulation of the association of membrane skeletal protein 4.1 with glycophorin by a polyphosphoinositide. *Nature* 318:295–298.
20. Eder, P. S., Soong, C., and Tao, M. (1986). Phosphorylation reduces the affinity of protein 4.1 for spectrin. *Biochemistry* 25:1764–1770.
21. Tao, M., Conway, R., and Cheta, S. (1980). Purification and characterization of a membrane-bound protein kinase from human erythrocytes. *J. Biol. Chem.* 266:2563–2568.
22. Lu, P., Soong, C., and Tao, M. (1985). Phosphorylation of ankyrin decreases its affinity for spectrin tetramer. *J. Biol. Chem.* 260:14958–14964.
23. Weber, K., and Osborn, M. (1982). *The Cytoskeleton.* National Cancer Institute, Bethesda, Monograph 60, pp. 31–46.
24. Shay, J. W. (Ed.) (1984). *Cell and Muscle Motility, Vol. 5, The Cytoskeleton.* Plenum, New York.
25. Stossel, T. P., Chaponnier, C., Ezzell, R. M., Hartwig, J. H., Janmey, P. A., Kwiatkowski, D. J., Lind, S. E., Smith, D. B., Southwick, F. S., Yin, H. L., and Zaner, K. S. (1985). Nonmuscle actin-binding proteins. *Annu. Rev. Cell Biol.* 1:353–402.
26. Geiger, B. (1979). Membrane-cytoskeletal interactions. *Cell* 18:193–205.
27. Lynch, G., and Baudry, M. (1984). The biochemistry of memory: A new and specific hypothesis. *Science* 224:1057–1063.
28. Nishizuka, Y. (1984). The role of protein kinase C in cell surface signal transduction and tumour promotion. *Nature* 308:693–698.
29. Berridge, M. J., and Irvine, R. F. (1984). Inositol triphosphate, a novel second messenger in cellular signal transduction. *Nature* 312:315–321.

30. Hokin, L. E. (1985). Receptors and phosphoinositide-generated second messengers. *Annu. Rev. Biochem.* 54:205–235.
31. Anderson, R. A., and Lovrien, R. E. (1984). Erythrocyte membrane sidedness in lectin control of the Ca^{++}-A23187-mediated diskocyte-echinocyte conversion. *Nature* 292:158–160.
32. Anderson, R. A. (1986). Glycophorin-protein 4.1: a transmembrane protein complex that may regulate the erythrocyte membrane skeleton. In *Membrane Skeletons and Cytoskeletal Membrane Associations*, V. Bennett, C. M. Cohen, S. Lux, and J. Palek (Eds.). Alan R. Liss, New York, pp. 223–241.
33. Chassis, J. A., Mohandas, N., Mentzer, W., Walker, P., and Shohet, S. B. (1983). Glycophorin A interacts with the membrane skeleton to influence membrane deformability. *Blood* 62:30a.
34. Lovrien, R. E., and Anderson, R. A. (1980). Stoichiometry of wheat germ agglutinin as a morphology controlling agent and as a morphology protective agent for the human erythrocyte. *J. Cell Biol.* 80:534–548.
35. Feo, C. J., Fischer, S., Pisu, J. P., Grange, M. J., and Tchernia, G. (1980). First observation of a membrane protein deficiency (band 4.1) in a case of hereditary elliptocytosis (Fr). *Nouv. Rev. Fr. Hematol.* 22:315.
36. Morrow, J. S., and Anderson, R. A. (1986). Editorial: Shaping the too fluid bilayer. *Lab. Invest.* 54:237–240.
37. Bessis, M. (1973). Normal and abnormal red cell shape. In *Red Cell Shape*, M. Bessis, R. I. Weed, and P. F. Leblond (Eds.). Springer-Verlag, New York, pp. 1–24.
38. Nakao, M., Nakao, T., and Yamazoe, S. (1960). Adenosine triphosphate and maintenance of the shape of human red blood cells. *Nature* 187:945–947.
39. Sheetz, M. P. (1983). Membrane skeletal dynamics: Role in modulation of red cell deformability, mobility of transmembrane protein. *Semin. Hematol.* 20:175–188.
40. Furthmayr, H. (1981). Glycophorin A: A model membrane protein. In: *Biology of Complex Carbohydrates*, V. Ginsburg and P. Roggins (Eds.). Wiley, New York, pp. 123–197.
41. Singer, J. A., and Morrison, M. (1975). Effect of metabolic state on phytohemagglutinin-P agglutination of normal human erythrocytes. *Biochim. Biophys. Acta* 406:553–563.
42. Schulte, T. H., and Marchesi, V. T. (1979). Conformation of human erythrocyte glycophorin A and its constituent peptides. *Biochemistry* 18:275–280.
43. Tomita, M., Furthmayr, H., and Marchesi, V. T. (1978). Primary structure of human erythrocyte glycophorin A. Isolation and characterization of peptides and complete amino acid sequence. *Biochemistry* 17:4756–4770.
44. Furthmayr, H., and Marchesi, V. T. (1984). Glycophorins: Isolation, orientation, and localization of specific domains. *Methods Enzymol.* 96:268–280.
45. Marchesi, V. T., and Andrews, E. P. (1971). Glycoprotein: Isolation from cell membranes with lithium diiodosalicylate. *Science* 147:1247–1248.

46. Armitage, I., Shapiro, D. L., Furthmayr, H., and Marchesi, V. T. (1977). ^{31}P nuclear magnetic resonance evidence for polyphosphoinositide associated with one hydrophobic segment of glycophorin A. *Biochemistry* 16: 1317–1320.

47. Mimms, L. T., Zampighi, G., Nozaki, Y., Tanford, C., and Reynold, J. A. (1981). Reconstitution of glycophorin into large phosphatidylcholine vesicles using the detergent octyl-β-D-glucoside. *Biochemistry* 20:833–840.

48. Anderson, R. A., and Lovrien, R. E. (1984). Glycophorin is linked by band 4.1 to the human erythrocyte membrane skeleton. *Nature* 307:655–658.

49. Tyler, J. M., Reinhardt, B. N., and Branton, D. (1980). Associations of erythrocyte membrane proteins. Binding of purified bands 2.1 and 4.1 to spectrin. *J. Biol. Chem.* 255:7034–7039.

50. Anderson, R. A. (1982). Glycophorin-protein 4.1 is a transmembrane protein complex which is linked to the human erythrocyte membrane skeleton and may be an avenue for lectin control of morphology [Dissertation]. University of Minnesota, Minneapolis.

51. Ong, R. L., Marchesi, V. T., and Prestegard, J. H. (1981). Small unilamellar vesicles containing glycophorin A. Chemical characterization of proton nuclear resonance studies. *Biochemistry* 20:4283–4292.

52. Van Zoelen, E. J. J., Verkleij, A. J., Zwaal, R. F. A., and Van Deenen, L. L. M. (1978). Incorporation and asymmetric orientation of glycophorin in reconstituted protein containing vesicles. *Eur. J. Biochem.* 86:539–546.

53. Van Zoelen, E. J. J., Van Dijok, P. W. M., DeKruijff, B., and Van Deenen, L. L. M. (1978). Effect of glycophorin incorporation on the physico-chemical properties of phospholipid bilayers. *Biochim. Biophys. Acta* 514:9–24.

54. Cotmore, S. F., Furthmayr, H., and Marchesi, V. T. (1977). Immunochemical evidence for the transmembrane orientation of glycophorin A localization of ferritin-antibody conjugates in intact cells. *J. Mol. Biol.* 113:539–553.

55. Furthmayr, H. (1978). Structural comparison of glycophorins and immunochemical analysis of genetic variants. *Nature* 271:519–524.

56. Pasternack, G. R., Anderson, R. A., Leto, T. L., and Marchesi, V. T. (1985). Interaction between protein 4.1 and band 3. An alternative binding site for an element of the membrane skeleton. *J. Biol. Chem.* 260:3676–3683.

57. Leto, T. L., Correas, I., Tobe, T., Anderson, R. A., and Horne, W. C. (1986). Structure and function of erythrocyte protein 4.1. In *Membrane Skeletons and Cytoskeletal Membrane Associations*, V. Bennett, C. M. Cohen, S. Lux, and J. Palek (Eds.). Alan R. Liss, New York, pp. 201–209.

58. Wolfe, L. C., and Lux, S. E. (1978). Membrane protein phosphorylation of intact normal and hereditary spherocytic erythrocytes. *J. Biol. Chem.* 253: 3336–3342.

59. Mueller, T., and Morrison, M. (1981). Glycoconnectin (PAS 2), a membrane attachment site for the human erythrocyte cytoskeleton. In *Erythrocyte Membranes 2: Recent Clinical and Experimental Advances*. W. Kruckeberg, J. Eaton, and J. Brewer (Eds.). Alan R. Liss, New York, pp. 95–112.

60. Dixon, M. (1953). The determination of enzyme inhibitor constants. *Biochem. J.* 55:170–171.
61. Shiffer, K. A., and Goodman, S. R. (1984). Protein 4.1: Its association with the human erythrocyte membrane. *Proc. Natl. Acad. Sci. U.S.A.* 81:4404–4408.
62. Appell, K. C., and Low, P. S. (1981). Partial structural characterization of the cytoplasmic domain of the erythrocyte membrane protein, band 3. *J. Biol. Chem.* 256:11104–11111.
63. Cohen, C. M., and Foley, S. F. (1980). Spectrin dependent and -independent association of F-actin with the erythrocyte membrane. *J. Cell Biol.* 86:694–698.
64. Fowler, V. M., Luna, E. J., Hargreaves, W. R., Taylor, D. L., and Branton, D. (1981). Spectrin promotes the association of F-actin with the cytoplasmic surface of the human erythrocyte membrane. *J. Cell Biol.* 88:388–395.
65. Shukla, B. D., Coleman, R., Finean, J. B., and Michell, R. H. (1979). Are polyphosphoinositides associated with glycophorin in human erythrocyte membranes. *Biochem. J.* 179:441–444.
66. Buckly, J. K. (1978). Coisolation of glycophorin A and polyphosphoinositides from human erythrocyte membranes. *Can. J. Biochem.* 56:349–351.
67. Van Zoelen, E. J. J., Zwaal, R. F. A., Reuvers, F. A. M., Demel, R. A., and Van Deenen, L. L. M. (1977). Evidence for the preferential interaction of glycophorin with negatively charged phospholipids. *Biochim. Biophys. Acta* 464:482–492.
68. Michell, R. H. (1975). Inositol phospholipids and cell surface receptor function. *Biochim. Biophys. Acta* 415:81–147.
69. Hawthorne, T. N., and Kai, M. (1980). Metabolism of phosphoinositides. In *Handbook of Neurochemistry*, A. Lajtha (Ed.). Plenum Press, New York, pp. 491.
70. Allan, D., and Michell, R. H. (1978). A calcium-activated polyphosphoinositide phosphodiesterase in the plasma membrane of human and rabbit erythrocytes. *Biochim. Biophys. Acta* 508:277–286.
71. Downes, P., and Michell, R. H. (1982). Phosphatidylinositol-4-phosphate and phosphatidylinositol-4,5-bisphosphate: Lipids in search of a function. *Cell Calcium* 3:467–502.
72. Raval, P. J., and Allan, D. (1985). The effects of phorbol ester, diacylglycerol, phospholipase C and Ca^{++} ionophore on protein phosphorylation in human and sheep erythrocytes. *Biochem. J.* 232:43–47.
73. Patel, V. P., and Fairbanks, G. (1986). Relationship of major phosphorylation reactions and Mg-ATPase activities to ATP-dependent shape change of the human erythrocyte membranes. *J. Biol. Chem.* 261:3170–3177.
74. Ferrell, J. E., and Huestis, W. H. (1984). Phosphoinositide metabolism and the morphology of human erythrocytes. *J. Cell Biol.* 98:1992–1998.
75. Leto, T. E., and Marchesi, V. T. (1984). A structural model of human erythrocyte protein 4.1. *J. Biol. Chem.* 259:4603–4608.
76. Aster, J. C., Brewer, G. J., and Maisel, H. (1986). The 4.1-like proteins of the bovine lens: Spectrin-binding proteins closely related in structure to red blood cell protein 4.1. *J. Cell Biol.* 103:115–122.

77. Cohen, C. M., Foley, S. F., and Korsgren, C. (1982). A protein immuno-
    logically related to erythrocyte band 4.1 is found on stress fibres of non-
    erythroid cells. *Nature* 299:648–650.
78. Baines, A. J., and Bennett, V. (1985). Synapsin I is a spectrin-binding pro-
    tein immunologically related to erythrocyte protein 4.1. *Nature* 315:410–
    413.
79. Baines, A. J., and Bennett, V. (1986). Synapsin I is a tubulin binding pro-
    tein. *Nature* 316:670–675.
80. Granger, B. L., and Lazarides, E. (1984). Membrane skeletal protein 4.1
    of avian erythrocytes is composed of multiple variants that exhibit tissue-
    specific expression. *Cell* 37:595–607.
81. Staufenbiel, M., and Lazarides, E. (1986). Assembly of protein 4.1 during
    chicken erythroid differentiation. *J. Cell Biol.* 102:1157–1163.
82. Anderson, R. A., Correas, I., Mazzucco, C. E., Castle, D. E., and Marchesi,
    V. T. (1988). Tissue specific analogues of erythrocyte protein 4.1 retain
    functional domains. *J. Cell Biochem.* 37:269–284.
83. Correas, I., Anderson, R. A., Mazzucco, C. E., and Marchesi, V. T. (1988).
    Immunoreactive forms of protein 4.1 are associated with the nucleus and
    mitotic apparatus. (in press).
84. De Camilli, P., Cameron, R., and Greengard, P. (1983). Synapsin I (protein
    I), a nerve terminal-specific phosphoprotein. I. Its general distribution in
    synapses of the central and peripheral nervous system demonstrated by im-
    munofluorescence in frozen and plastic sections. *J. Cell Biol.* 96:1337–
    1354.
85. Perrin, D., and Aunis, D. (1986). Reorganization of α-fodrin induced by
    stimulation in secretory cells. *Nature* 315:589–582.
86. Perrin, D., Langly, O. K., and Aunis, D. (1987). Anti-α-fodrin inhibits secre-
    tion from permeabilized chromaffin cells. *Nature* 326:498–501.
87. Burgoyne, R. D., and Cheek, T. R. (1987). Role of fodrin in secretion. *Na-
    ture* 326:488.
88. Creutz, C. E., Dowling, L. G., Kyger, E. M., and Franson, R. C. (1985).
    Phosphatidylinositol-specific phospholipase C activity of chromaffin
    granule-binding proteins. *J. Biol. Chem.* 260:7171–7173.
89. Sheetz, M. P., and Singer, S. J. (1977). On the mechanism of ATP-induced
    shape change in human erythrocyte membranes. *J. Cell Biol.* 73:638–646.
90. Muller, E., Hegewald, H., Jaroszewicz, K., Cumme, G. A., Hoppe, H., and
    Frunder, H. (1986). Turnover of phosphomonoester groups and compart-
    mentation of polyphosphoinositides in human erythrocytes. *Biochem. J.*
    235:775–783.
91. Jinbu, Y., Sato, S., Nakao, M., Tsukita, S., and Ishikawa, H. (1984). The
    role of ankyrin in shape and deformability change of human erythrocyte
    ghosts. *Biochim. Biophys. Acta* 773:237–242.
92. Sheetz, M. P., and Casaly, J. (1980). 2,3-Diphosphoglycerate and ATP dis-
    sociate erythrocyte membrane skeletons. *J. Biol. Chem.* 255:9955–9960.
93. Wolfe, L. C., Lux, S. E., and Ohanian, V. (1981). Spectrin-actin binding
    in vitro: Effect of Protein 4.1 and polyphosphates. *Supramol. Struct.*
    5(suppl):123.

94. Nigg, E. A., Bron, C., Girardet, M., and Cherry, R. A. (1980). Band 3-glycophorin A association in erythrocyte membranes demonstrated by combining protein diffusion measurements with antibody-induced cross-linking. *Biochemistry* 19:1887–1893.
95. Dahr, W., Uhlenbruch, G., Wagstaff, W., and Leikola, J. (1976). Studies on the membrane glycoprotein of En(a–) erythrocytes. II MN antigenic properties of En(a–) erythrocytes. *J. Immunogenet.* 3:383–393.
96. Levine, J., and Willard, M. (1983). Redistribution of fodrin (a component of the cortical cytoplasm) accompanying capping of cell surface molecules. *Proc. Natl. Acad. Sci. USA* 80:191–195.
97. Mangeat, P. H., and Burridge, K. (1984). Immunoprecipitation of non-erythrocyte spectrin within live cells following microinjection of specific antibodies: Relation to cytoskeletal structures. *J. Cell Biol.* 98:1363–1377.
98. Repasky, E. A., Symer, D. E., and Bankert, R. B. (1984). Spectrin immunofluorescence distinguishes a population of naturally capped lymphocytes in situ. *J. Cell Biol.* 99:350–355.
99. Simian, R., Baudry, M., and Lynch, G. (1986). Regulation of glutamate receptor binding by the cytoskeletal protein fodrin. *Nature* 313:225–228.

9

Interaction of Native and Denatured Hemoglobins with Band 3
Consequences for Erythrocyte Structure and Function

PHILIP S. LOW   *Purdue University, West Lafayette, Indiana*

Because of its abundance, ease of isolation, and required function, hemoglobin (Hb) has been one of the most heavily studied proteins in the literature. However, most previous investigations of Hb have dealt solely with the isolated macromolecule in aqueous solutions and only recently have attempts been made to study the interaction of Hb with its surrounding shell, the erythrocyte membrane. The purpose of this review is first to summarize what is currently understood concerning the interaction of Hb with its major binding site on the erythrocyte membrane, the cytoplasmic domain of band 3, and then to discuss the structure and consequences of the interaction of denatured Hb with the same membrane site. The former topic has been briefly reviewed [1]; however, the latter theme has never been summarized in literature.

I. INTERACTION OF NATIVE HEMOGLOBIN WITH BAND 3

A. Introduction

Hemoglobin is predominantly a tetramer $(\alpha_2\beta_2)$ containing two pairs of homologous heterodimeric $(\alpha\beta)$ subunits. In its relaxed (oxygenated) state, several important intersubunit interactions are broken, permitting a fraction of the tetramers to dissociate. At usual Hb concentrations within the red blood cell (5 mM tetramer), the oxygenated dimer can be present at as high as 0.1 mM concentration [2,3]. The same oxygen-induced changes in Hb can influence

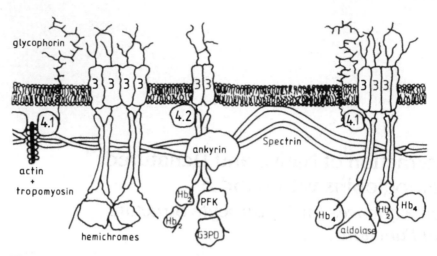

Figure 1 Hypothetical model of the human erythrocyte membrane showing a possible arrangement of band 3 and the various peripheral proteins with which it interacts.

the protein's affinity for small ligands such as diphosphoglycerate and H$^+$ [2], and, as will be pointed out below, this relaxed $\rightleftharpoons$ tense state equilibrium in Hb can also modulate its interaction with the cytoplasmic domain of band 3.

The N-terminal cytoplasmic domain of band 3 (cdb3) is the totally exposed, water-soluble cytoplasmic pole of the major anion transport protein of the membrane, band 3 (for a review see [4]). The 43,000-dalton cytoplasmic domain is structurally and functionally independent of the complementary 55,000-dalton membrane-spanning (transport) domain [4–9]. The cdb3 can be cleaved from the membrane without significantly modifying any of its major structural or functional properties [4,5], and in its isolated dimeric form it behaves as a highly elongated flexible molecule [10,11]. The primary function of cdb3 appears to be to provide an anchor for the linkage of peripheral proteins to the membrane. Those proteins reported to reside at least partly on cdb3 include ankyrin (band 2.1 or syndein, i.e., the connection to the cytoskeleton), band 4.1, band 4.2, glyceraldehyde-3-phosphate dehydrogenase, aldolase, phosphofructokinase, Hb, and hemichromes, a denatured form of Hb [1,4,5,12–28]. A hypothetical model of the disposition of band 3 and its associated proteins in the membrane is depicted in Figure 1.

B. Interaction of Hemoglobin with the Red Blood Cell Membrane at Low pH and Ionic Strength

When erythrocyte membranes are titrated with hemoglobin in low ionic strength buffers near pH 6, two distinct classes of binding sites are observed [23-26]. The low-affinity sites have been attributed both to the cytoplasmic pole of gly-cophorin and to anionic inner leaflet lipids. Evidence for these assignments includes (a) the resistance of the low-affinity sites to proteolysis with chymotrypsin, (b) the reduction of these sites in rabbit erythrocytes (which lack glyco-phorin), (c) the presence of similar low-affinity sites in synthetic vesicles reconstituted with anionic lipids and/or glycophorin, and (d) the partial loss of these sites by treatment with phospholipase C. The high-affinity sites, on the other hand, have been demonstrated to reside on the cytoplasmic domain of band 3 [23-28]. Support for this contention comes from (a) the loss of these sites on selective chymotryptic removal of the cdb3, (b) the presence of similar high-affinity binding sites on the isolated cdb3, (c) the displacement of Hb from these sites by glyceraldehyde-3-phosphate dehydrogenase, a known ligand of band 3, and (d) the perturbation of these sites by reaction of the cells with DIDS, a specific covalent inhibitor of the membrane-spanning domain of band 3. The binding of Hb to cdb3 occurs rapidly [29], is of high affinity [24,30], and depends strongly on pH and ionic strength (vide infra). At low Hb concentrations there is evidence for cooperativity in band 3 binding [26], although this is not universally accepted [25]. At saturating Hb concentrations, binding to all copies of band 3 is complete, yielding 10^6 high-affinity sites per cell [24,25]. The subsite on band 3 involved in the Hb interaction is at the extreme N-terminus of the cytoplasmic domain [1,27,31], where Hb competes for binding with glyceraldehyde-3-phosphate of dehydrogenase and aldolase [1,24,25,31]. This site on cdb3 is extraordinarily acidic, containing no basic amino acids, a blocked N-terminus, and more than 50% acidic residues (18 of the last 33 amino acids) [32], (see sequence in Figure 2). Furthermore, at least one of the tyrosines in this region can be phosphorylated by endogenous and exogenous tyrosine kinases [33-36], adding further possible negative charges to this region.

Two lines of evidence suggest that the interaction between Hb and cdb3 is structurally specific and not a trivial consequence of random electrostatic interactions. First, myoglobin, a monomeric homologue of similar charge and protomeric structure, displays little or no affinity for band 3 [29,31,37]. Likewise, cytochrome c and other cationic proteins exhibit no tendency to compete with Hb for band 3. This suggests binding based solely on electrostatic interactions does not achieve the same affinity displayed by Hb for cdb3. Second, phosphorylation of band 3 on a tyrosine residue within the Hb binding sequence completely blocks Hb binding [38], even though this additional negative charge would be expected to enhance any nonspecific electrostatic interaction. Therefore, steric or conformational factors, in addition to electrostatic interactions, must contribute to the stability of the Hb–cdb3 complex.

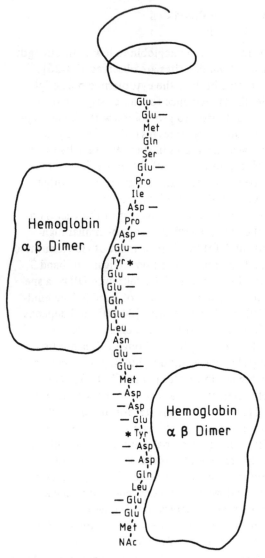

Figure 2 Sketch of the N-terminal 33 amino acids of band 3 in possible association with two αβ dimers of oxyHb. In an extended conformation the length of
the 33 residues stretch could reach nearly 120 Å [45], clearly long enough to
accommodate two dimers of Hb simultaneously. The predominant interacting
form of Hb, the tetramer in its deoxygenated or tense state, probably binds in
the region occupied by the lower hemoglobin dimer, as evidenced by its competition with glyceraldehyde-3-phosphate dehydrogenase and its displacement by
phosphorylation of tyrosine 8. Acidic amino acids (–) and possible phosphorylation sites (*) are labeled.

C. Hemoglobin–cdb3 Interactions Under Intracellular Conditions

As noted above, both pH and ionic strength profoundly influence the strength of the Hb–cdb3 interaction. Although dissociation constants in the submicromolar range are commonly reported for dilute solutions of the complex at low ionic strength near pH 6 [24,29,30], as these solution variables are elevated into the physiological range the complex rapidly dissolves [24–26,31,39,40]. As justification for studying this apparently nonphysiological interaction, researchers have noted that hemoglobin concentrations in the cell are much higher than can be achieved in vitro [27,28], implying that even under weak, physiological binding conditions an interaction should occur in vivo. Evidence for this contention has, in fact, arisen from several sources. First, Eisinger and colleagues have employed Förster energy transfer methods to evaluate the distance of Hb from the membrane in intact cells [41,42]. Their data suggest a close approach of the protein to the membrane in situ with a pH dependence of binding qualitatively similar to the behavior seen in vitro, i.e., more Hb appears to contact the membrane as the cellular pH is lowered. Second, there is much evidence to indicate that Hb in the intact cell alters the properties of band 3 and the surrounding membrane. Those characteristics that are reported to change in the presence of elevated Hb include the mobility of band 3 and the morphology and flexibility of the membrane (vide infra). Third, studies of the behavior of Hb in concentrated solutions [43] demonstrate that the protein's activity is not a simple linear function of its concentration. Indeed, at high intracellular Hb concentrations, the activity of hemoglobin (i.e., its effective reactive concentration) appears to be at least 50-fold larger than its true concentration. This nonideal behavior at high intracellular concentrations not only suggests that weak associations will be much stronger than trivial extrapolations would predict, but it also argues that due to linked equilibria the pH dependence of binding will be shifted toward the physiological pH range in vivo (Low, P. S., unpublished observations).

The most convincing evidence that Hb associates with cdb3 in vivo was recently obtained by Chêtrite and Cassoly [28] in binding experiments conducted under physiological conditions. Using an immobilized preparation of cdb3 suspended in 10 mM-bis-Tris (pH 7.2) containing 120 mM NaCl, they measured a K_D of 4×10^{-4} M for deoxyhemoglobin. In a slightly different buffer of similar ionic strength and pH, Walder et al. [27] obtained a related K_D of 3.1×10^{-4} M for the interaction of Hb with a synthetic peptide corresponding to the binding site on band 3. Using these dissociation constants and taking into account the antagonistic effect of diphosphoglycerate on the interaction [27,28], the former group was able to estimate that ~50% of band 3 should be ligated by Hb in vivo.

D. Structure of the Complex Between Hemoglobin and Cytoplasmic Domain of Band 3

As mentioned previously, the Hb binding site on cdb3 has been localized to the extreme N-terminus, within 56 (and probably within 23) residues of the N-

terminal N-acetylated methionine. In an effort to visualize the structure of this complex, Walder et al. [27] cocrystallized deoxyhemoglobin with the first 11 residues of band 3 and solved its X-ray diffraction pattern to 5 Å resolution. The difference map showed that the N-terminal peptide extends deep (~ 18 Å) into the central cavity between the β chains, along the dyad symmetry axis of Hb. The acidic cdb3 peptide was found to have an extended conformation with only 5 to 7 of its 11 residues in contact with Hb. However, the Hb residues forming this contact are largely cationic, thus providing an explanation of the pH and salt sensitivity of the complex, i.e., the strongly anionic band 3 residues were juxtaposed to the predominantly cationic residues of the Hb cleft. Likewise, the antagonistic effect of diphosphoglycerate on the Hb–cdb3 complex [26–28] could be explained, because the same site in the central cavity of Hb is known to bind diphosphoglycerate. The observed preference of cdb3 for deoxy over oxyhemoglobin [27,28] was also obvious from the crystallographic data because the central cavity becomes occluded in oxyhemoglobin, i.e., the rotation of the oxygenated $\alpha\beta$ dimers closes off the diphosphoglycerate cleft. Thus, for oxyhemoglobin to bind similarly it would either have to remain in the "tense conformation" or dissociate into $\alpha\beta$ dimers, thereby exposing the cationic residues of the intersubunit interface.

The above crystallographic picture of the cdb3–Hb complex is well supported by biochemical data. Obstruction of the entrance to the central cavity in deoxyhemoglobin by cross-linking its β subunits from $Lys^{82}\ \beta_1$ to $Lys^{82}\ \beta_2$ blocks the binding of cdb3 as well as its N-terminal peptide [27]. Likewise, addition of diphosphoglycerate to a Hb–band 3 reaction mixture strongly inhibits complex formation [26,28]. Evaluation of the O_2 binding curve of Hb in the presence of either cdb3 or the undecameric peptide shows significantly reduced O_2 affinity [27], confirming the preference of cdb3 for the deoxy form. When oxyhemoglobin does bind cdb3, it is found to associate at least initially as the $\alpha\beta$ "openface" dimer [44,45], consistent with the preference of band 3 for the cationic sites in the intersubunit binding cleft. Finally, the tendency for cross-linking reagents to connect primarily the β chain of Hb to cdb3 [46,47] would be predicted from the residues forming the contacts in the crystallographic structure.

Since the concentration of fully deoxygenated Hb in a normal circulating erythrocyte is low, attention must also be paid to the nature of the partially oxygenated Hb–cdb3 complexes. Both kinetic and sedimentation data suggest, as stated above, that oxyhemoglobin binds primarily to cdb3 as a dimer with a stoichiometry of two oxyhemoglobin dimers per cdb3 monomer [44,45]. Once bound, the two adjacent oxyhemoglobin dimers have the ability to isomerize to form a single tetramer. Whether this tetramer is in the "tense" or "relaxed" state or whether it forms a complex structurally analogous to the deoxyhemoglobin tetramer cannot be determined from the data. Significantly, sequence data on band 3 also points to the presence of two potential Hb binding sites, because residues 12-23 are an almost direct repeat of residues 1–11 [32]. Thus, if this 23 residue stretch of cdb3 were in an extended conformation (at 3.6 Å per residue [48] the total length calculates to 83 Å), there would be more than enough length along the peptide for two $\alpha\beta$ Hb dimers to bind (Figure 2). This presence of a potential second Hb site on band 3 could explain why only partial competition is observed between Hb and several glycolytic enzymes for this region of band 3 [4].

E. Influence of Hemoglobin Binding on Red Blood Cell Membrane Properties

Although band 3 binding shifts the Hb–O_2 binding curve to the right, there is clearly too little band 3 (1.2×10^6 copies per cell) to exert a measurable impact on the oxygen-carrying capacity of the erythrocyte [27,28]. However, by the converse argument, in cases in which Hb modulates the properties of band 3 or its associated proteins, the interaction in vivo has good potential to alter membrane behavior. Thus, the question arises, does Hb modulate any band 3 or erythrocyte membrane properties?

Although the data are insufficient to allow any molecular interpretations, numerous observations suggest that band 3 and its surrounding membrane behave differently in the presence of Hb. Thus, Hb binding changes the reactivity of the two sulfhydryl groups of cdb3 [30], even though they are well removed from the Hb binding site. Similarly, cooperativity is observed in Hb binding at low Hb concentrations [26], suggesting that the interaction with one subunit of band 3 can be communicated to its pair. Band 3 is found to rotate five to eight times more rapidly in ghosts than in the intact cell [49], implying that Hb removal in some way releases important motional constraints on band 3 (however, other explanations are also possible) [49]. Hemoglobin, but not myoglobin, is found to distort the shape of erythrocyte ghosts [25] and to modulate the Mg-ATP–induced echinocyte–diskocyte transition [50]. Addition of Hb to ghosts also promotes membrane resealing, enhances endocytotic inside-out vesicle formation in the presence of Mg-ATP, and increases the internal volume of isolated membranes more than threefold [50,51]. In the absence of Mg-ATP, Hb actually interferes with inside-out vesicle formation during spectrin depletion [51a]. Although none of the above observations has a clear molecular explanation, because cdb3 provides the high affinity point of interaction of Hb with the membrane and because Hb can cross-link between adjacent band 3 molecules [52], it is not unlikely that cdb3–Hb complex formation can explain much of the above data.

To this point in the review, no attempt has been made to distinguish among the common variants of Hb (i.e., HbA, HbS, HbC) in their interactions with band 3. At least in the well-documented case of HbS, a difference in behavior

does exist. Thus, HbS has a higher affinity for cdb3 than does HbA [39,40,53],
perhaps accounting in part for its increased adherence to the membrane. Hemoglobin S may also nucleate on the membrane during polymerization; however,
this hypothesis is still disputed [54,55]. Still, in view of the proposal that extending the delay time preceeding HbS polymerization during deoxygenation
might be of therapeutic value, further studies of this phenomenon seem war-
ranted. Probably the most dramatic impact of Hb variants on band 3 and erythrocyte properties occurs as these less stable variants denature. As will be described below, the binding of denatured Hbs to the membrane exerts a dramatic
effect on both erythrocyte membrane structure and function.

II. INTERACTION OF DENATURED HEMOGLOBIN
WITH BAND 3

A. Formation of Denatured Hemoglobin In Vivo

Not all Hb remains in its native state until the erythrocyte is retired by the
spleen or other components of the reticuloendothelial system. On the contrary, denatured Hb can often be found associated with the cytoplasmic membrane surface, either dispersed uniformly on the inner leaflet or aggregated into
dense particles termed Heinz bodies. Although Hb denaturation must occur to
some extent even in young, healthy cells, the denatured Hb aggregates are most
commonly seen under four stressful conditions. First, erythrocytes containing
unstable Hbs, especially Hbs with mutations in the β chain near the heme, show
elevated levels of Heinz bodies [56-61]. Presumably the Heinz bodies form
faster in these hemoglobinopathies than they can be removed by the reticuloendothelial system. Second, in the thalassemias, where either the α or β chain
is produced in excess, the unincorporated chains often precipitate on the membrane [62,63]. Third, Heinz bodies are seen in cells containing normal Hb under
conditions of oxidant stress. This stress can either arise due to ingestion of oxidant generating drugs such as phenylhydrazine [64-67] or as a result of a
genetic inability to maintain an adequate reducing environment in the cell, e.g.,
as in the case of glucose-6-phosphate dehydrogenase or glutathione peroxidase
deficiency diseases [68-70]. Finally, aggregates of denatured Hb often appear
in normal cells as they age [71-72]. This senescence-related increase in Heinz
bodies is especially pronounced in splenectomized individuals, in whom a major
organ of altered or aged cell removal has been excised [73].

Whether natural or drug-promoted, Heinz body formation is believed to follow a similar pathway. Methemoglobin that is not reduced back to its functional
state can gradually convert to a reversible hemichrome and then to an irreversible hemichrome, which can aggregate to form a Heinz body [58,68,74]. The
hemichromes are not totally unfolded, because they retain most of their heme
and exist in dimeric and tetrameric forms, probably not too different from

native Hb [16,61,68]. It has, in fact, been suggested that a major distinction
between hemichrome and Hb lies in the oxidation state of the iron and in the
occupation of its sixth coordination position by the distal His or some other
ligand endogenous to the protein [74]. Regardless of the difference, hemichrome has properties not shared with native Hb that permit it to interact much
more avidly with band 3, causing the aggregation of band 3 in the membrane
and a disruption of normal membrane structure (vide infra).

B. The Interaction of Hemichromes with Band 3

The avid association of hemichromes with band 3 was first discovered when a
colorless solution of cdb3 was added to a clear, reddish-brown solution of hemichromes. Rather than obtaining the anticipated transparent mixture of the two,
a heavy precipitate formed virtually instantaneously and settled to the bottom
of the test tube. This precipitate was later shown to be a regular copolymer of
the two components, containing 2.5 hemichrome tetramers (or five hemichrome
dimers) per cdb3 dimer [16,17]. Curiously, this subunit stoichiometry remained invariant, regardless of the ratio of the two proteins in the reaction solution or the length of time of polymerization. This invariant stoichiometry was
taken as evidence that the copolymerization of cdb3 and hemichromes is driven
by the molecular specificity of interacting sites on the multivalent proteins and
not by a random aggregation of "sticky" polypeptides. Further evidence for this
molecular specificity is the observation that tyrosine phosphorylation of the N-
terminus of band 3 inhibits this interaction (Low, P. S., unpublished observations), much the way it blocks the binding of Hb and the glycolytic enzymes [38].

The binding of hemichromes to cdb3 appears to be of much higher affinity
than the interaction of native Hb. On addition of cdb3 to a solution containing
a 20-fold molar excess of Hb over hemichrome, only the later is detected in the
copolymer [16]. Further, native Hb is virtually ineffective as an inhibitor of
the copolymerization process, even though it is known to interact at the same
site on band 3. It is, therefore, anticipated that when hemichromes exist in the
cell, their binding to band 3 will take precedence over the Hb interaction.

Although the oxidative cross-linking of hemichromes facilitates their copolymerization, disulfide bonding was found not be be essential, i.e., the macroscopic copolymer formed adequately in the presence of excess reducing agents.
Curiously, the site on the hemichrome that interacts with band 3 was shown to
be outside the central cavity, because cross-linking reagents that occluded this
cavity did not inhibit polymerization [17]. This feature turns out to be important, because the binding of the cdb3 dimer off the dyad axis of hemichrome
assures that there are at least two equivalent binding sites, one on each $\alpha\beta$
dimer of the hemichrome tetramer. Such a duality of sites is an essential feature
for regular copolymerization. Although low-affinity binding sites were also de-

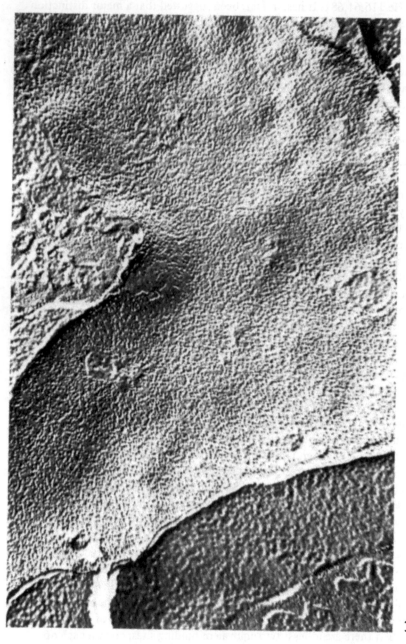

(a)

(b)

Figure 3 Freeze-fracture electron micrographs of the protoplasmic face of human erythrocyte membranes (a) before and (b) after treatment with 15 mM phenylhydrazine. Note the even distribution of particles on the P-face prior to modification with phenylhydrazine and the clustered distribution after this treatment. Micrographs provided by K. Platt-Aloia, W. W. Thompson, and I. W. Sherman, University of California, Riverside.

tected in binding studies on intact ghosts, the high-affinity hemichrome sites were shown to be on cdb3, because these sites were specifically blocked by addition of excess glyceraldehyde-3-phosphate dehydrogenase or by selective cleavage of cdb3 from the membrane [16].

C. Copolymerization of Band 3 and Hemichromes In Vivo Causes Clustering of Band 3 and Other Proteins in the Membrane

There are several lines of evidence that the copolymerization of band 3 with hemichromes observed and characterized in vitro manifests itself in situ by forcing a clustering of band 3 in the membrane. For example, cells treated with phenylhydrazine (used to generate hemichromes in situ) display a patchy distribution of protoplasmic face intramembrane particles (Figure 3b), whereas unmodified cells show an even particle distribution (Figure 3a). Because these intramembrane particles are thought to consist predominantly of band 3 oligomers, the observed particle aggregation suggests that hemichromes promote a redistribution of band 3 in situ. This same phenylhydrazine-induced band 3 aggregation in intact erythrocytes has also been documented by immunofluorescence microscopy [75], where antibodies to the cytoplasmic domain of band 3 reveal a clustered distribution in the modified cells and a smooth distribution in the untreated erythrocytes. Importantly, the locations of the band 3 clusters invariably coincide with the sites of membrane-bound hemichrome aggregates (visualized by phase contact microscopy in the same field of cells) [75]. This colocalization of band 3 aggregates with hemichrome aggregates further demonstrates that the copolymerization reaction documented above also occurs in situ.

Erythrocytes containing naturally occurring Heinz bodies and other hemichrome aggregates also show the expected coclustering of hemichromes with band 3. For example, in α thalassemia, where excess β chains precipitate on the membrane, intramembrane particle aggregation is observed at sites of hemichrome binding [62,76]. Similarly, in sickle cells, where Heinz bodies are commonly seen attached to the membrane, band 3 is frequently found clustered over the Heinz body sites [77,78]. Figure 4 shows two micrographs of a single field of sickle cells examined both by phase contrast microscopy, to visualize the membrane-associated Heinz bodies (Figure 4a), and by immunofluorescence microscopy to observe the distribution of band 3 (Figure 4b). In most cases, a correlation of Heinz body binding with band 3 aggregation is observed.

Because band 3 is associated in situ with several major membrane proteins (Figure 1), it is logical to inquire whether the hemicrhome-induced band 3 clustering might also extend to these proteins, i.e., will band 3-associated components be dragged along during clustering? In an investigation of a large sampling of cells from five different donors, it was indeed found that an aggregation of other polypeptides does occur [77]. Thus, when sickle cells were stained with antibodies to ankyrin, ~37% of the Heinz bodies were found associated with a

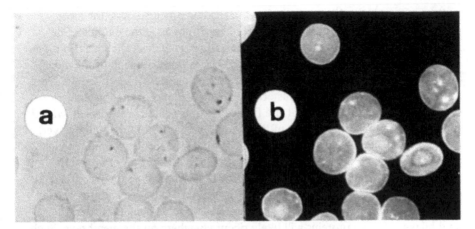

Figure 4 Colocalization of Heinz bodies with clusters of band 3 in blood from sickle hemoglobinopathy patients. The erythrocytes were spread on glass slides, stained with a rabbit antibody to cdb3 and visualized with rhodamine-conjugated anti-rabbit I_gG. (a) Phase contrast microscopy. (b) Immunofluorescence microscopy. Reproduced from [77] with permission from *J. Clin. Invest.*

corresponding ankyrin aggregate (the remaining Heinz bodies showed no corresponding ankyrin cluster). Likewise, 22% of the hemichrome aggregates showed a colocalized density of glycophorin. The frequency of coclustering of band 3 with Heinz bodies in these samples was 64%. This observation that the frequency of codistribution is highest for band 3 confirms, in fact, that band 3 is the primary receptor for hemichrome binding. Whether the other membrane components are aggregated directly by hemichromes or as a consequence of band 3 clustering cannot be answered from the data.

D. Speculations on the Consequences of Hemichrome-Induced Membrane Reorganization

Erythrocytes containing denatured Hb exhibit a number of pathological properties not characteristic of cells devoid of hemichromes. These properties include elevated nonspecific ion and sugar permeabilities [62,79-82], reduced cell filterability/deformability [83-85], shortened cell life span [62,64,65,68,86,87], enhanced hemolysis [16,65,66,84], increased adherence to vascular endothelium [85-94], elevated surface-associated antibodies [75,86,95,112], altered cytoskeletal components [84,96], modified phospholipid organization [97,104], and elevated oxidant stress [65,98,99]. Because most of the above lesions have either been observed in nonsickle cells (e.g., in thalassemias and other hemoglobinopathies) or in oxygenated sickle cells of normal morphology, the abnormalities cannot be readily attributed to the sickling process. Thus, the most reasonable commonality among all affected erythrocytes is the binding of denatured Hb to the membrane. As will be hypothesized below, the redistribution of band 3 and concomitant microscopic reorganization/distortion of the membrane at these sites can account for many of the above defects.

On a microscopic scale, the clustering of band 3 and associated proteins should have important structural and functional consequences at the following 3 sites: (a) the cytoplasmic surface of the membrane, (b) within the lipid bilayer, and (c) on the external cell surface. At the cytoplasmic surface, because band 3 provides the major linkage to the cytoskeleton, a redistribution of interacting cytoskeletal elements should naturally occur. If enrichment of the membrane in ankyrin [77], band 4.1 (Low, P. S., unpublished observations), and presumably actin can occur at Heinz body sites, then a corresponding depletion from the lipid bilayer of these proteins will likely occur elsewhere on the membrane. Such a redistribution of cytoskeletal elements could account, at least in part, for the altered fragility of these cells. Add to this the rigidifying influence of cross-linked hemichrome patches on the inner membrane surface and much of the loss of cell deformability/filterability might also be explained [83]. Further, as pointed out by many others [100-104], the weakening or disruption of the cytoskeleton can lead to enhanced phospholipid flip-flop, thus providing a possible mechanism for the observed abnormal phospholipid distribution in such cells [97,104].

Within the lipid bilayer, at sites of band 3/glycophorin clustering, imperfections in the permeability barrier of the bilayer are predicted to develop. It has been shown, for example, that when band 3 is clustered by exogenous cross-linking reagents, nonspecific leaks of ions and polar nonelectrolytes such as sugars arise [82]. By a similar mechanism, the observed aggregation of integral membrane components could cause part of the leakiness characteristic of hemichrome-containing cells.

The tethering of band 3 and glycophorin, etc., at the cytoplasmic membrane surface will also influence properties determined at the exoplasmic membrane surface. In fact, cells containing denatured Hb have been known for some time to display an uneven distribution of both intramembrane particles and cell surface charges [62,76,77,93,105]. As has been pointed out by others [89,93], this clustered distribution of cell surface charges (primarily carried on glycophorins) may be responsible for the tendency of these cells to adhere to vascular endothelium. Because this adhesive potential seems to correlate closely with the severity of sickle cell disease [88,89,93], the role of Hb denaturation in causing the major symptoms of sickle cell disease may deserve greater scrutiny.

It has also been observed that the spleen can "clean" Heinz body–containing cells of their hemichrome aggregates in a process termed pitting [64,77,106]. The surviving, pitted cell apparently continues to circulate, but with a reduced

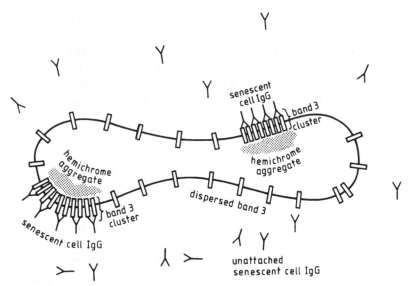

Figure 5 Sketch of a proposed mechanism for the selective antibody mediated removal of aged cells from a mixed population of erythrocytes. Due to prior clustering of band 3 upon hemichrome binding, senescent cell antibodies which interact too weakly to bind monovalently, attach to the membrane at sites of band 3 aggregation. The autologous antibody-coated erythrocyte is then removed from circulation by macrophages, much like antibody coated virus or bacterium. Since Hb denaturation is the trigger for cell removal, the mechanism assures that the red cell is eliminated as soon as it begins to falter in its primary function of O_2 transport.

band 3/glycophorin-to-spectrin ratio due to the selective loss of the band-3-glycophorin–hemichrome complex [107]. Although the rigidity of the hemichrome cross-linked patch may provide one mechanism whereby the Heinz body site is selectively recognized, the clustering of the external poles of band 3 and glycophorin might also generate this identification marker for the spleen [77].

Finally, we have proposed elsewhere [75] that the clustering of band 3 at the external membrane surface provides a recognition site for binding of senescent cell autologous antibodies. It has been suggested for some time that band 3 is the senescent cell antigen recognized on the surfaces of aged cells by autoantibodies responsible for their removal [108–111]; however, the molecular distinction between band 3 in old and young erythrocytes has never been clearly defined. To promote the selective removal of aged cells from the blood (3.6 × 10^{11} senescent erythrocytes are removed daily), this antigenic marker must not only be distinct, but it must also be strongly age-dependent [110]. We have suggested that the age-related appearance of hemichromes and their subsequent clustering of band 3 generates this marker [75]. As graphically depicted in Figure 5, the binding of a bivalent IgG (or pentavalent IgM) to an antigenic receptor will be many orders of magnitude stronger if the receptor has been previously clustered in the membrane. The actual enhancement in binding affinity, E, upon clustering of band 3 can be quantitatively evaluated with the expression:

$$ E = \frac{I}{V} \left[\frac{2}{X_0} \right]^{V-1} $$

where v is the valence of the antibody and X_0 is the mole fraction of band 3 in the membrane. Assuming a normal concentration of 600,000 band 3 dimers per cell, the enhancement factor for antibody recognition calculates to 10^3 and 10^{12} for IgG and IgM binding, respectively. If affinity of the senescent cell antibody for band 3 is normally very weak, recognition of band 3 and hence red blood cell removal will only occur on hemichrome-induced clustering of the protein, an event that presumably takes place solely in aged or abnormal cells. Much evidence exists to demonstrate that clustering of band 3 promotes autologous antibody binding [75,111]. Therefore, the hemichrome-induced lesion may also be responsible for the elevated antibody content [86,95,111] and shortened life span [62,64,65,68,86,87] of hemichrome-containing cells.

Although the above speculations have concentrated on the consequences of hemichrome binding and the resulting membrane reorganization, the absence of any discussion of other possible sources of erythrocyte lesions was not meant to imply they need not be considered. Indeed, the oxidation of lipids, modification of cell surface carbohydrate residues, cross-linking and/or oxidation of cytoskeletal proteins, and diminution of certain enzyme activities must all be carefully evaluated in a total picture of the abnormal erythrocyte's pathology [98,106, 113,114]. However, where Hb denaturation is a prominent event, its binding to the membrane and the corresponding structural and functional consequences must not be overlooked [115].

NOTE ADDED IN PROOF

A study has recently appeared [116] which lends strong biochemical support to our contention that hemichrome denaturation leads to band 3 clustering in the membrane and that this clustering in turn generate a major site for autologous IgG binding. Basically, we have isolated the hemichrome-membrane protein aggregates from sickle cells using two independent procedures. Compositional analysis of these aggregates by SDS polyacrylamide gel electrophoresis demonstrated that they are 67% by weight hemichromes; 13% band 3; and 5%

or less ankyrin, band 4.9, actin, and band 4.1. Most important, although the aggregates represented only 1.3% of the total membrane protein, the isolated aggregates contained 75% of the total cell surface bound autologous IgG. Thus, the clustered site clearly represents the predominant locus of autologous IgG binding to the membrane surface. It was also demonstrated that the stability of the aggregate was inextricably linked to the presence of the hemichromes. Taken together, these observations strongly implicate the role of Hb denaturation and band 3 clustering in triggering aged/altered cell removal.

ACKNOWLEDGMENT

The above review was written while the author was on sabbatical leave at the Biozentrum der Universität Basel and supported by NIH grant GM24417. The author thanks Regula Niederhauser for typing the manuscript.

REFERENCES

1. Kaul, R. K., and Köhler, H. (1983). Interaction of hemoglobin with band 3: A review. *Klin. Wochenschr.* 61:831.
2. Perutz, M. F. (1978). Hemoglobin structure and respiratory transport. *Sci. Am.* 239:68.
3. Ip, S. H. C., and Ackers, G. K. (1977). Thermodynamic studies on subunit assembly in human hemoglobin. *J. Biol. Chem.* 252:82.
4. Low, P. S. (1986). Structure and function of the cytoplasmic domain of band 3: center of erythrocyte membrane-peripheral protein interactions. *Biochim. Biophys. Acta* 864:145.
5. Bennett, V., and Stenbuck, P. J. (1980). Association between ankyrin and the cytoplasmic domain of band 3 isolated from the human erythrocyte membrane. *J. Biol. Chem.* 255:6424.
6. Appell, K. C., and Low, P. S. (1982). Evaluation of the structural interdependence of membrane-spanning and cytoplasmic domains of band 3. *Biochemistry* 21:2151.
7. Lepke, S., and Passow, H. (1976). Effects of incorporated trypsin on anion exchange and membrane proteins in human red blood cell ghosts. *Biochim. Biophys. Acta* 455:353.
8. Grinstein, S., Ship, S., and Rothstein, A. (1978). Anion transport in relation to proteolytic dissection of band 3 protein. *Biochim. Biophys. Acta* 507:294.
9. Beth, A. H., Balasubramanian, K., Wilder, R. T., Venkataramu, S. D., Robinson, B. H., Dalton, L. R., Pearson, D. E., and Park, J. H. (1981). Structural and motional changes in glyceraldehyde-3-phosphate dehydrogenase upon binding to the band 3 protein of the erythrocyte membrane examined with $[^{15}N, ^{2}H]$ maleimide spin label and electron paramagnetic resonance. *Proc. Natl. Acad. Sci. USA* 78:4955.

10. Low, P. S., Westfall, M. A., Allen, D. P., and Appell, K. C. (1984). Characterization of the reversible conformational equilibrium of the cytoplasmic domain of erythrocyte membrane band 3. *J. Biol. Chem.* 259:13070.
11. Appell, K. C., and Low, P. S. (1981). Partial structural characterization of the cytoplasmic domain of the erythrocyte membrane protein, band 3. *J. Biol. Chem.* 256:11104.
12. Hargreaves, W. R., Giedd, K. N., Verkleij, A., and Branton, D. (1980). Reassociation of ankyrin with band 3 in erythrocyte membranes and in lipid vesicles. *J. Biol. Chem.* 255:11965.
13. Pasternack, G. R., Anderson, R. A., Leto, T. L., and Marchesi, V. T. (1985). Interactions between protein 4.1 and band 3: An alternative binding site for an element of the membrane skeleton. *J. Biol. Chem.* 260:3676.
14. Murthy, S. N. P., Liu, T., Kaul, R. K., Köhler, H., and Steck, T. L. (1981). The aldolase binding site of the human erythrocyte membrane is at the NH_2 terminus of band 3. *J. Biol. Chem.* 256:11203.
15. Tsai, I. H., Murthy, S. N. P., and Steck, T. L. (1982). Effect of red cell membrane binding on the catalytic activity of glycerladehyde-3-phosphate dehydrogenase. *J. Biol. Chem.* 257:1438.
16. Waugh, S. M., and Low, P. S. (1985). Hemichrome binding to band 3: Nucleation of Heinz bodies on the erythrocyte membrane. *Biochemistry* 24:34.
17. Waugh, S. M., Walder, J. A., and Low, P. S. (1987). Partial characterization of the copolymerization reaction of erythrocyte membrane band 3 with hemicrhomes. *Biochemistry* 26:1777.
18. Higashi, T., Richards, C. S., and Uyeda, K. (1979). The interaction of phosphofructokinase with erythrocyte membranes. *J. Biol. Chem.* 254:9542.
19. Jenkins, J. D., Kezdy, F. J., and Steck, T. L. (1985). Mode of interaction of phosphofructokinase with the erythrocyte membrane. *J. Biol. Chem.* 260:10426.
20. Korsgren, C., and Cohen, C. M. (1986). Purification and properties of human erythrocyte band 4.2: Association with the cytoplasmic domain of band 3. *J. Biol. Chem.* 261:5536.
21. Yu, J., and Steck, T. L. (1975). Isolation and characterization of band 3, the predominant polypeptide of the human erythrocyte membrane. *J. Biol. Chem.* 250:9170.
22. Bennett, V. (1982). Isolation of an ankyrin-band 3 oligomer from human erythrocyte membranes. *Biochim. Biophys. Acta* 689:475.
23. Rauenbuehler, P. B., Cordes, K. A., and Salhany, J. M. (1982). Identification of the hemoglobin binding sites on the inner surface of the erythrocyte membrane. *Biochim. Biophys. Acta* 692:361.
24. Shaklai, N., Yguerabide, J., and Ranney, H. M. (1977). Classification and localization of hemoglobin binding sites on the red blood cell membrane. *Biochemistry* 16:5593.
25. Salhany, J. M., Cordes, K. A., and Gaines, E. D. (1980). Light scattering measurements of hemoglobin binding to the erythrocyte membrane. Evidence for transmembrane effects related to a disulfonic stilbene binding to band 3. *Biochemistry* 19:1447.

26. Premachandra, B. R. (1986). Interaction of hemoglobin and its component α and β chains with band 3 protein. *Biochemistry* 25:3455.
27. Walder, J. A., Chatterjee, R., Steck, T. L., Low, P. S., Musso, G. F., Kaiser, E. T., Rogers, P. H., and Arnone, A. (1984). The interaction of hemoglobin with the cytoplasmic domain of band 3 of the human erythrocyte membrane. *J. Biol. Chem.* 259:10238.
28. Chêtrite, G., and Cassoly, R. (1985). Affinity of hemoglobin for the cytoplasmic fragment of human erythrocyte membrane band 3. *J. Mol. Biol.* 185:639.
29. Kirschner-Zilber, I., and Shaklai, N. (1982). The specificity of hemoglobin for band 3 membrane sites. *Biochem. Int.* 5:309.
30. Salhany, J. M. (1983). Binding of cytosolic proteins to the erythrocyte membrane. *J. Cell. Biochem.* 23:211.
31. Murthy, S. N. P., Kaul, R. K., and Köhler, H. (1984). Hemoglobin binds to the amino-terminal 23-residue fragment of human erythrocyte band 3 protein. *Hoppe-Seylers Z. Physiol. Chem.* 365:9.
32. Kaul, R. K., Murthy, S. N. P., Reddy, A. G., Steck, T. L., and Köhler, H. (1983). Amino acid sequence of the N-terminal 201 residues of human erythrocyte membrane, band 3. *J. Biol. Chem.* 258:7981.
33. Dekowski, S. A., Rybicki, A., and Drickamer, K. (1983). A tyrosine kinase associated with the red cell membrane phosphorylates band 3. *J. Biol. Chem.* 258:2750.
34. Phan-Dinh-Tuy, F., Henry, J., and Kahn, A. (1985). Characterization of human red blood cell tyrosine kinase. *Biochem. Biophys. Res. Commun.* 126:304.
35. Harrison, M. L., Low, P. S., and Geahlen, R. L. (1984). T and B lymphocytes express distinct tyrosine protein kinases. *J. Biol. Chem.* 259:9348.
36. Foulkes, J. G., Mathey-Pevot, B., Guild, B. C., Prywes, R., and Baltimore, D. (1985). A comparison of the protein-tyrosine kinases encoded by Abelson murine leukemia virus and Rous sarcoma virus. *Cancer Cells* 3:329.
37. Shaklai, N., Yguerabide, J., and Ranney, H. (1977). Interaction of hemoglobin with red blood cell membranes as shown by a fluorescent chromophore. *Biochemistry* 16:5585.
38. Low, P. S., Allen, D. P., Zioncheck, T. F., Chari, P., Willardson, B. M., Geahlen, R. L., and Harrison, M. L. (1987). Tyrosine phosphorylation of band 3 inhibits peripheral protein binding. *J. Biol. Chem.* 262:4592.
39. Fischer, S., Nagel, R. L., Bookchin, R. M., Roth, Jr., E. F., and Nagel, I. T. (1975). The binding of hemoglobin to membranes of normal and sickle erythrocytes. *Biochim. Biophys. Acta* 375:422.
40. Fung, L. W.-M., Litvin, S. D., and Reid, T. R. (1983). Spin-label detection of sickle hemoglobin-membrane interaction of physiological pH. *Biochemistry* 22:864.
41. Eisinger, J., Flores, J., and Salhany, J. M. (1982). Association of cytosol hemoglobin with the membrane in intact erythrocytes. *Proc. Natl. Acad. Sci. USA* 79:408.

42. Eisinger, J., and Flores, J. (1983). Cytosol-membrane interface of human erythrocytes: A resonance energy transfer study. *Biophys. J.* 41:367.

43. Ross, P. D., and Minton, A. P. (1977). Analysis of non-ideal behavior in concentrated hemoglobin solutions. *J. Mol. Biol.* 112:437.

44. Cassoly, R. (1983). Quantitative anslysis of the association of human hemoglobin with the cytoplasmic fragment of band 3 protein. *J. Biol. Chem.* 258: 3859.

45. Salhany, J. M., and Shaklai, N. (1979). Functional properties of human hemoglobin bound to the erythrocyte membrane. *Biochemistry* 18:893.

46. Sayare, M., and Fikiet, M. (1981). Cross-linking of hemoglobin to the cytoplasmic surface of human erythrocyte membranes. *J. Biol. Chem.* 256: 13152.

47. Bank, A., Mears, G., Weiss, R., O'Donnell, J. V., and Natta, C. (1974). Preferential binding of β-S-globin chains associated with stroma in sickle cell disorders. *J. Clin. Invest.* 54:805.

48. Cary, R. B., and Pauling, L. (1953). Fundamental dimensions of polypeptide chains. *Proc. R. Soc. Lond. (Biol.)* B141:10.

49. Beth, A. H., Conturo, T. E., Venkataramu, S. D., and Staros, J. V. (1986). Dynamics and interactions of the anion channel in intact human erythrocytes: An ESR spectroscopic study employing a new membrane-impermeant bifunctional spin-label. *Biochemistry* 25:3824.

50. Wiedenmann, B., and Elbaum, D. (1983). Effect of hemoglobin A and S on human erythrocyte ghosts. *J. Biol. Chem.* 258:5483.

51. Premachandra, B. R., and Bajaj, V. (1986). Band 3 is involved in hemoglobin-induced resealing of erythrocyte membranes. *Fed. Proc.* 45:1832.

51a. Cordes, K. A., and Salhany, J. M. (1982). Cytosolic protein binding to band 3 protein inhibits endocytosis of isolated human erythrocyte membranes. *Biochem. J.* 207:595.

52. Cassoly, R., and Salhany, J. M. (1983). Spectral and oxygen-release kinetic properties of human hemoglobin bound to the cytoplasmic fragment of band 3 protein in solution. *Biochim. Biophys. Acta* 745:134.

53. Shaklai, N., Sharma, V. S., and Ranney, H. M. (1981). Interaction of sickle cell hemoglobin with erythrocyte membranes. *Proc. Natl. Acad. Sci. USA* 78:65.

54. Goldberg, M. A., Lalos, A. T., and Bunn, H. F. (1981). The effect of erythrocyte membrane preparations on the polymerization of sickle hemoglobin. *J. Biol. Chem.* 256:193.

55. Shibata, K., Cattam, G. L., and Waterman, W. R. (1980). Acceleration of the rate of deoxyhemoglobin S polymerization by the erythrocyte membrane. *FEBS Lett.* 110:107.

56. Jacob, H. S., and Winterhalter, K. H. (1972). Mechanisms of formation of Heinz bodies and their attachments to red blood cell membranes in the unstable hemoglobinopathies. In *VI Intl. Symp. Strukt. Funkt. Erythrozyten*, S. Rapoport and F. Jung (Eds.). Akademie-Verlag: Berlin, pp. 93-96.

57. Carrell, R. W., and Lemann, H. (1969). The unstable hemoglobin hemolytic anemias. *Semin. Hematol.* 6:116.

58. Winterbourn, C. C., and Carrell, R. W. (1974). Studies of hemoglobin de-
    naturation and Heinz body formation in the unstable hemoglobins. *J. Clin.
    Invest.* 54:678.
59. Rifkind, R. A., and Dannon, D. (1965). Heinz body anemia: An ultrastruc-
    tural study. I. Heinz body formation. *Blood* 25:885.
60. Eisinger, J., Flores, J., Tyson, J. A., and Shohet, S. B. (1985). Fluorescent
    cytoplasm and Heinz body hemoglobin Köln erythrocytes: Evidence for
    intracellular heme catabolism. *Blood* 65:886.
61. Sears, D. A., and Luthra, M. G. (1983). Membrane-bound hemoglobin in
    the erythrocytes of sickle cell anemia. *J. Lab. Clin. Med.* 102:694.
62. Rachmilewitz, E. A., Shinar, E., Shalev, O., Galili, U., and Schrier, S. L.
    (1985). Erythrocyte membrane alterations in β-thalassemia. *Clin. Haematol.*
    14:163.
63. Nathan, D. G., and Shohet, S. B. (1970). Erythrocyte ion transport defects
    and hemolytic anemia: "hydrocytosis" and "desiccytosis." *Semin. Hematol.*
    7:381.
64. Rifkind, R. A. (1965). Heinz body anemia: An ultrastructural study. II.
    Red cell sequestration and destruction. *Blood* 26:433.
65. Bates, D. A., and Winterbourn, C. C. (1984). Haemoglobin denaturation,
    lipid peroxidation and hemolysis in phenylhydrazine-induced anaemia.
    Biochim. Biophys. Acta 798:84.
66. Itano, H. A., Hirota, K., Vedvick, T. S. (1977). Ligands and oxidants in
    ferrihemochrome formation and oxidative hemolysis. *Proc. Natl. Acad.
    Sci. USA* 74:2556.
67. Snyder, L. M., Fortier, N. L., Trainor, J., Jacobs, J., Leb, L., Lubin, B.,
    Chiu, D., Shohet, S., and Mohandas, M. (1985). Effect of hydrogen perox-
    ide exposure on normal human erythrocyte deformability, morphology,
    surface characteristics, and spectrin-hemoglobin cross-linking. *J. Clin. In-
    vest.* 76:1971.
68. Jandl, J. H., Engle, L. K., and Allen, D. W. (1960). Oxidative hemolysis and
    precipitation of hemoglobin. I. Heinz body anemias as an acceleration of
    red cell aging. *J. Clin. Invest.* 39:1818.
69. Hopkins, J., and Tudhope, G. R. (1974). Glutathione peroxidase deficiency
    with increased susceptibility to erythrocyte Heinz body formation. *Clin.
    Sci. Molec. Med.* 47:643.
70. Allen, D. W., Groat, J. D., Finkel, B., Rank, B. H., Wood, P. A., and Eaton,
    J. W. (1983). Increased adsorption of cytoplasmic proteins to the erythro-
    cyte membrane in ATP-depleted normal and pyruvate kinase-deficient ma-
    ture cells and reticulocytes. *Am. J. Hematol.* 14:11.
71. Sears, D. A., Friedman, J. M., and White, D. R. (1975). Binding of intra-
    cellular protein to the erythrocyte membrane during incubation: The pro-
    duction of Heinz bodies. *J. Lab. Clin. Invest.* 86:722.
72. Campwala, H. Q., and Desforges, J. F. (1982). Membrane-bound hemi-
    chrome in density separated ghosts of normal (AA) and sickled (SS) cells.
    J. Lab. Clin. Med. 99:25.
73. Selwyn, J. G. (1955). Heinz bodies in red cells after splenectomy and after
    phenacetin administration. *Br. J. Haematol.* 1:173.

74. Peisach, J., Blumberg, W. E., and Rachmilewitz, E. A. (1975). The demonstration of ferrihemochrome intermediates in Heinz body formation following the reduction of oxyhemiglobin A by acetylphenylhydrazine. *Biochim. Biophys. Acta* 393:404.

75. Low, P. S., Waugh, S. M., Zinke, K., and Drenckhahn, D. (1985). The role of hemoglobin denaturation and band 3 clustering in red blood cell aging. *Science* 227:531.

76. Lessin, L. S., Jensen, W., and Klug, P. (1972). Ultrastructure of the normal and hemoglobinopathic RBC membrane. Freeze-etching and stereoscan electron microscope studies. *Arch. Int. Med.* 129:306.

77. Waugh, S. M., Willardson, B. M., Kannon, R., Labotka, R. J., and Low, P. S. (1986). Heinz bodies induce clustering of band 3, glycophorin, and ankyrin in sickle cell erythrocytes. *J. Clin. Invest.* 78:1155.

78. Schlüter, K., and Drenckhahn, D. (1986). Co-clustering of denatured hemoglobin with band 3: Its role in binding of autoantibodies against band 3 to abnormal and aged erythrocytes. *Proc. Natl. Acad. Sci. USA* 83:6137.

79. Wiley, J. S. (1981). Increased erythrocyte cation permeability in thalassemia and conditions of marrow stress. *J. Clin. Invest.* 67:917.

80. Shalev, O., Mogilner, S., Shinar, E., Rachmilewitz, E. A., and Schrier, L. S. (1984). Impaired erythrocyte calcium homeostasis in β-thalassemia. *Blood* 64:564.

81. Orringer, E. P., and Parker, J. C. (1977). Selective increase of potassium permeability in red blood cells exposed to acetylphenylhydrazine. *Blood* 50:1013.

82. Deuticke, B., Poser, B., Lütkemeier, P., and Haest, C. W. M. (1983). Formation of aqueous pores in the human erythrocyte membrane after oxidative cross-linking of spectrin by diamide. *Biophys. Biochim. Acta* 731:196.

83. Reinhart, W. H., Sung, L. A., and Chien, S. (1986). Quantitative relationship between Heinz body formation and red blood cell deformability. *Blood* 68:1376.

84. Vilsen, B., and Nielsen, H. (1984). Reaction of phenylhydrazine with erythrocytes. Cross-linking of spectrin by disulfide exchange with oxidized hemoglobin. *Biochem. Pharmacol.* 33:2739.

85. Evans, E., Mohandas, N., and Leung, A. (1984). Static and dynamic rigidities of normal and sickle erythrocytes. Major influence of cell hemoglobin concentration. *J. Clin. Invest.* 73:477.

86. Green, G. A., Rehn, M. M., and Kalra, V. K. (1985). Cell-bound autologous immunoglobulin in erythrocyte subpopulations from patients with sickle cell disease. *Blood* 65:1127.

87. Knyszynski, A., Danon, D., Kahane, I., and Rachmilewitz, E. A. (1979). Phagocytosis of nucleated and mature β-thalassemic red blood cells by mouse macrophages *in vitro*. *Br. J. Haematol.* 43:251.

88. Hebbel, R. P., Boogaerts, M. A. B., Eaton, J. W., and Steinberg, M. H. (1980). Erythrocyte adherence to endothelium in sickle-cell anemia: A possible determinant of disease severity. *N. Engl. J. Med.* 302:992.

89. Hebbel, R. P., Schwartz, R. S., and Mohandas, N. (1985). The adhesive sickle erythrocyte: Cause and consequence of abnormal interactions with endothelium, monocytes/macrophages and model membranes. *Clin. Haematol.* 14:141.

90. Mohandas, N., and Evans, E. (1984). Adherence of sickle erythrocytes to vascular endothelial cells: Requirement for both cell membrane changes and plasma factors. *Blood* 64:282.

91. Mohandas, N., and Evans, E. (1985). Sickle erythrocyte adherence of vascular endothelium. Morphologic correlates and the requirement for divalent cations and collagen-binding plasma proteins. *J. Clin. Invest.* 76:1605.

92. Smith, B. D., and La Celle, P. L. (1986). Erythrocyte-endothelial cell adherence in sickle cell disorders. *Blood* 68:1050.

93. Hebbel, R. P., Yamada, O., Moldow, C. F., Jacob, H. S., White, J. G., and Eaton, J. W. (1980). Abnormal adherence of sickle erythrocytes to cultured vascular endothelium. Possible mechanism for microvascular occlusion in sickle cell disease. *J. Clin. Invest.* 65:154.

94. Kaul, D. K., Fabry, M. E., and Nagel, R. L. (1986). Vaso-occlusion by sickle cells: Evidence for selective trapping of dense red cells. *Blood* 68:1162.

95. Petz, L. D., Yam, P., Wilkinson, L., Garratty, G., Lubin, B., and Mentzer, W. (1984). Increased IgG molecules bound to the surface of red blood cells of patients with sickle cell anemia. *Blood* 64:301.

96. Platt, O. S., Falcone, J. F., and Lux, S. E. (1986). Molecular defect in the sickle erythrocyte skeleton. Abnormal spectrin binding to sickle inside-out vessicles. *J. Clin. Invest.* 75:266.

97. Lubin, B., Chiu, D., Bastasky, J., Roelofsen, B., and Van Deenen, L. L. M. (1981). Abnormalities in membrane phospholipid organization in sickled erythrocytes. *J. Clin. Invest.* 67:1643.

98. Hebbel, R. P. (1985). Auto-oxidation and a membrane-associated Fenton reagent: A possible mechanism for development of membrane lesions in sickle erythrocytes. *Clin. Haematol.* 14:129.

99. Hebbel, R. P., Eaton, J. W., Balasingann, M., and Steinberg, M. H. (1982). Spontaneous oxygen radical generation by sickle erythrocytes. *J. Clin. Invest.* 70:1253.

100. Bergmann, W. L., Dressler, V., Haest, C. W. M., and Deuticke, B. (1984). Cross-linking of SH-groups in erythrocyte membrane enhances transbilayer reorientation of phospholipids. Evidence for a limited access of phospholipids to the reorientation sites. *Biochim. Biophys. Acta* 769:390.

101. Haest, C. W. M., Plasa, G., Kamp, D., and Deuticke, B. (1978). Spectrin as a stabilizer of the phospholipid asymmetry in the human erythrocyte membrane. *Biochim. Biophys. Acta* 509:21.

102. Lubin, B., and Chiu, D. (1982). Membrane phospholipid organization in pathologic human erythrocytes. *Prog. Clin. Biol. Res.* 96:137.

103. Mohandas, N., Wyatt, J., Mel, S. F., Rossi, M. E., and Shohet, S. B. (1982). Lipid translocation across the human erythrocyte membrane: Regulatory factors. *J. Biol. Chem.* 257:6537.

104. Franck, P. F. H., Bevers, E. M., Lubin, B. H., Comfurius, P., Chiu, D. T. Y., Op den Kamp, J. A. F., Zwaal, R. F. A., van Deenen, L. L. M., and Roelofsen, B. (1985). Uncoupling of the membrane skeleton from the lipid bilayer. The cause of accelerated phospholipid flip-flop leading to an enhanced procoagulant activity of sickled cells. *J. Clin. Invest.* 75:183.

105. Kahane, I., Polliack, A., Rachmilewitz, E. A., Bayer, E. A., and Skutelsky, E. (1978). Distribution of sialic acids on the red blood cell membrane in β-thalassemia. *Nature* 271:674.

106. Wolfe, L. (1985). The red cell membrane and the storage lesion. *Clin. Haematol.* 14:259.

107. Miyahara, K., and Spiro, M. J. (1984). Nonuniform loss of membrane glycoconjugates during in vivo aging of human erythrocytes: Studies of normal and diabetic red cell saccharides. *Arch. Biochem. Biophys.* 232: 310.

108. Lutz, H. U., and Stringaro-Wipf, G. (1983). Identification of a cell-age-specific antigen from human red blood cells. *Biomed. Biochim. Acta* 42: 117.

109. Kay, M. M. B., Goodman, S. R., Sorensen, K., Whitfield, C. F., Wong, P., Zaki, L., and Rudloff, V. (1983). Senescent cell antigen is immunologically related to band 3. *Proc. Natl. Acad. Sci. USA* 80:1631.

110. Kay, M. M. B. (1983). Appearance of a terminal differentiation antigen on senescent and damaged cells and its implications for physiologic auto-antibodies. *Biomembr. Patholog. Membr.* 11:119.

111. Lutz, H. U. (1981). Elimination alter Erythrocyten aus der Zirkulation: Freilegung eines zellalterspezifischen Antigens auf alternden Erythro-zyten. *Schweiz. Med. Wochenschr.* 111:1507.

112. Galili, U., Clark, M. R., and Shohet, S. B. (1986). Excessive binding of natural anti-alpha-galactosyl immunoglobulin G to sickle erythrocytes may contribute to extravascular cell destruction. *J. Clin. Invest.* 77:1.

113. Lux, S. E., John, K. M., and Karnovsky, M. J. (1976). Irreversible deformation of spectrin-actin lattice in irreversibly sickled cells. *J. Clin. Invest.* 58:955.

114. Fukuda, M., Fukuda, M. N., Hakomori, S., and Papyannopoulou, T. (1981). Anomalous cell surface structure of sickle cell anemia erythrocytes as demonstrated by cell surface labeling and endo-β-galactosidase treatment. *J. Supramol. Struct. Cell. Biochem.* 17:289.

115. Claster, S., Clark, M. R., Wagner, G. M., Chiu, D. T. Y., Mentzer, W., and Lubin, B. (1986). Association of membrane bound hemichrome with clinical severity in sickle cell disease. *Blood* 68:60a.

116. Kannan, R., Labotka, R., and Low, P. S. (1988). Isolation and Characterization of the Hemichrome-stabilized Membrane Protein Aggregates from Sickle Erythrocytes: Major Site of Autologous Antibody Binding. *J. Biol. Chem.* 263:13766.

10

Ultrastructure and Function of Membrane Skeleton

BETTY W. SHEN   *Argonne National Laboratory, Argonne, Illinois*

I. INTRODUCTION

A. General Background

The membrane of the human erythrocyte is reinforced along its entire cytoplasmic surface by a two-dimensional network of peripheral proteins that closely adhere to the membrane proper through specific protein-protein interactions [1-9]. This durable, flexible, and elastic network functions to stabilize the membrane bilayer without compromising its deformability [10-14], thus enabling the red blood cell to withstand the shear stress during its turbulent passage through the vasculature. The erythrocytes respond to the mechanical pressure by transient deformation while traversing through microcapillaries and subsequently return to the biconcave disk shape. This rapid, reversible deformation of the erythrocytes allows for maximum contact between the erythrocyte membrane and the wall of the microcapillaries, thus facilitating the exchange of oxygen and electrolytes while maintaining minimum viscosity in the circulatory system. Perturbations of the skeleton have been shown to cause irreversible alterations in the permeability, integrity, deformation, and shape of cells, leading to red blood cell pathophysiology [15-18].

B. Basic Organization of the Red Blood Cell Skeleton

The proteins essential to the integrity of the skeleton are band 1 plus band 2 (α and β subunits of spectrin, respectively), band 4.1, and actin [19]. Spectrin exists in situ as heterodimers, tetramers, and higher oligomers, with tetramers as the predominant form [20]. The actin is thought to be associated into homooligomers or protofilaments [21–26]. Spectrin has been shown to bind to and link rabbit muscle actin filaments in reconstituted systems [25–27]. Such complexes are stabilized by the presence of band 4.1 proteins, which associate in vitro with the ends of spectrin tetramers [28–35].

The skeleton is linked to the membrane proper both by a second association of band 4.1 protein, either with an integral glycoprotein, glycophorin [36,37], or band 3 [38], and by the association of band 2.1 protein (ankyrin) with both spectrin and band 3 [39–46]. At least two additional proteins are implicated in the organization and function of the skeletal network. Membrane tropomyosin is believed to play an important role in regulating the length and stability of the actin protofilament by a mechanism similar to that of its muscle analogue [47]. Band 4.9 functions as an actin-"bundling" protein [48] to regulate the distribution of actin protofilaments. Recently, myosin has been purified from erythrocytes, and a question has been raised regarding the possible role of an actomyosin contractile system in the properties of the red blood cells [49,50]. Spectrin, actin, band 2.1, and band 4.1 have been detected in cytoskeleton–plasma membrane complexes in various nucleated cells (for a review, see [2]). Thus, the red blood cell can serve as a general model for the analysis of the molecular interactions between the cytoskeleton and plasma membrane, as well as the structure and function of cytoskeletons in all cells.

The structure, function, and biogenesis of individual proteins are covered in detail in separate chapters of this volume; here we will concentrate on the organizational aspects of these proteins and their role in the structure and function of the membrane skeleton.

The skeleton is a shell that retains the approximate size and shape of the intact erythrocyte [7,8,19,51]. Its location is shown by spectrin-specific ferritin-conjugated antibodies to be closely associated with the cell membrane [52]. The skeleton can be isolated by extraction with Triton X-100, which dissolves away the lipid constituents and leaves the intact skeleton with a number of integral membrane proteins that serve as the attachment sites for the skeleton [7,8,19]. These accessory proteins can be extracted by high-salt treatments, leaving a basic skeleton containing mainly spectrin, actin, and band 4.1. Direct visualization and detailed definition of the molecular architecture of the skeleton have been hampered in the past by the high surface concentration of proteins at the plane of the membrane and by the lack of recognizable structural periodicity among the proteins. Therefore, general consensus concerning the organization of the red blood cell skeleton has been derived predominantly from various biochemical and biophysical studies of purified components and their reconstituted ensembles. Thus, this chapter will first review the isolation and properties of individual components and their limited ensembles.

II. STRUCTURAL COMPONENTS AND SMALL RECONSTITUTED ENSEMBLES

The spectrin and actin of the skeleton can be dissociated from the membrane into soluble components by extraction with low-ionic-strength alkaline buffers, which leads to fragmentation of the membrane bilayer and formation of the inside-out vesicles (IOVs) [53-55]. Bands 2.1 and 4.1 remain tightly associated with the outer surface of the IOVs and can be extracted together by high-salt treatment [32,56] or separately by sodium cholate elution of band 4.1 followed by 1.0 M KCl dissociation of band 2.1 [39,42]. The soluble components can be fractionated by gel permeation or anion-exchange column chromatography into purified components and analyzed by chemical and physical methods.

A. Isolation and Properties of Spectrin

Spectrin, the major constituent of the membrane skeleton, accounts for ~75% of the skeletal protein mass. It consists of two homologous polypeptides of 240,000 and 220,000 daltons, designated α and β subunits, respectively [2-6, 57,58]. The spectrin exists in the low-salt extract as heterodimers ($\alpha\beta$), tetramers ($\alpha\beta)_2$, and higher oligomers [20, 59–62]. These different forms of spectrin can be separated by electrophoresis on agarose-polyacrylamide composite gels under nondenaturing conditions and resolved by sepharose 4B or agarose column chromatography [63–65]. Spectrin extracted at $0°C$ consists mainly of tetramers, whereas spectrin extracted at $37°C$ is predominantly dimeric [64]. At ambient temperature, the spectrin dimers and tetramers are in a thermodynamic equilibrium and interconvert readily. At $0°C$, however, the rate of interconversion between dimers and tetramers is so slow that their respective forms can be trapped for long intervals [57]. This slow interconversion at low temperature, together with the fact that the spectrin tetramer is the predominant species isolated from membrane ghosts that have been kept strictly at or below $4°C$, suggests that spectrin exists in situ as tetramers.

The α and β subunits of spectrin form stable antiparallel double-strand heterodimers that can only be separated under denaturing conditions [58,66, 67]. The isolated subunits (α and β) reassociate rapidly under nondenaturing conditions. The kinetics of reassociation exhibit positive cooporativity, suggesting multiple domains of interactions [58]. Conformational analysis based on partial amino acid sequence of tryptic fragments of spectrin indicates that both the α and β subunits are composed of homologous repeating segments that are 106 amino acids long [68–70] and that these repeating segments have a high probability of forming triple-helical domains interconnected by flexible, short, random coil hinges. The amino terminus (N-terminus) of the α chain and the carboxylic acid terminus of the β chain both participate in the head-to-head interactions that are important for the formation of tetramers [71]. An 80,000-

dalton tryptic fragment from the N-terminus of the α chain is shown to bind to the β subunit of the second heterodimer in the formation of tetramers. (For a topical review on spectrin before 1980, see [72].)

Spectrin is phosphorylated at multiple sites close to the N-terminus of the β subunits by a cyclic adenosine monophosphate (cAMP)-independent protein kinase and possibly by cAMP- and calmodulin-dependent protein kinases as well [73-75]. However, phosphorylation and dephosphorylation of spectrins have no effect on the dimer–dimer interaction or on their affinity for other skeletal proteins.

B. Erythrocyte Actin

Actin in the erythrocyte skeleton is thought to be associated into homooligomers or protofilaments that can nucleate the formation of F-actin [21-27]. Purified erythrocyte actin is capable of forming 7-nm-diameter filaments with the double helical morphology [23]. It activates myosin adenosine triphosphatase (ATPase) and shares many other common properties with striated muscle actin [76,77]. However, despite these similarities, early attempts to visualize actin filaments in membrane or membrane skeletons were generally unsuccessful. Much of the evidence for the existence of F-actin in the skeleton is based on the ability of red blood cell membranes and oligomeric spectrin–actin–band 4.1 complexes derived from membrane skeletons to nucleate the polymerization of globular actin monomers (G-actin) under conditions in which spontaneous F-actin formation would be slow or nonexistent [78-80]. It has been shown that G-actin binds to the cytoplasmic but not the extracellular surface of red blood cell ghosts and that IOVs that have been stripped of endogenous actin and spectrin by low-ionic-strength extraction bind little G-actin [24]. However, when a crude spectrin extract containing primarily spectrin, actin, and band 4.1 protein was added back to stripped vesicles, binding of G-actin was restored. Data based on cytochalasin B binding and cellular actin content suggested that the actin filaments in the membrane are short and contain $<$ 10-13 monomers each [21, 26]. Electron microscopic measurement of actin filaments extracted from phalloiden-treated membranes, however, suggests that some of the filaments may be as long as 100 nm or more and may contain 30-60 actin monomers [22].

C. Band 4.1 Protein and Its Interaction with Spectrin-Actin Complexes

Band 4.1 comprises ~5% of the red blood cell skeletal mass. It consists of two homologous proteins of 78,000 and 82,000 daltons [36,80-82]. They migrate as a closely spaced doublet designated 4.1a and 4.1b [81] in sodium dodecyl sulfate polyacrylamide gel electrophoresis (SDS-PAGE) with the discontinuous buffer system of Laemmli [83]. Bands 4.1a and 4.1b are phosphorylated to similar degrees in erythrocytes by both cAMP-dependent and cAMP-independent protein kinases [73,84–88] and by a phorbol ester-activated protein kinase C [88–90]. The phosphorylation of band 4.1, in vitro, significantly reduces its affinity but not its binding stoichiometry for spectrin [74]. Spectrin and band 4.1 act together to cause the formation of short actin filaments from G- and F-actin [91]. Purified spectrin or band 4.1 alone, however, lacks the capability of enhancing the nucleation of G-actin or the severing of F-actin. It was suggested that spectrin and band 4.1 constitute a filament-severing and capping system that acts by binding the pointed (slow-growing) end of the F-actin.

The interaction of spectrin with actin and band 4.1 has been studied extensively by a variety of techniques, including viscometry [28,30], electron microscopy [27,32], sedimentation analysis [29], and binding studies of radio-isotope-labeled skeletal components [27,31,35]. Current evidence indicates that both spectrin dimers and tetramers interact weakly with F-actin (but not with G-actin). However, only tetramers are capable of linking rabbit muscle actin filaments in reconstituted systems because they possess two actin binding sites, one on each heterodimer close to the distal ends of the tetramer [27,32]. Binding of spectrin to actin requires both α and β subunits because isolated subunits were inactive [92]. The weak interactions between spectrin and actin are markedly enhanced and stabilized by the presence of band 4.1 proteins [35,93], which were shown by electron microscopy of platinum-shadowed specimens to be associated, in vitro, with the ends of spectrin tetramers close to the actin binding sites [33,34].

D. Role of Ankyrin

Band 2.1, ankyrin, is located at the cytoplasmic surface of the red blood cell membrane ghost and is responsible for the high-affinity, saturable binding of spectrin to the membrane bilayer [94]. Ankyrin is highly susceptible to proteolytic degradation in the erythrocyte, forming a group of partially digested polypeptides that migrate slightly ahead of the intact protein in the SDS gels [95, 96]. Strictly speaking, ankyrin is not one of the structural components of the skeleton because it can be extracted from the Triton shells of the membranes by a high concentration of salts without compromising the integrity of the network. Nonetheless, this protein plays an important role in the functioning of the skeleton and warrants a brief discussion in this context. Purified ankyrin and its proteolytic fragments compete for and inhibit spectrin binding to the surface of the IOVs that were formed during the low-salt extraction of the red blood cell membrane ghosts [41–43,97]. Removal of ankyrin (and band 4.1 protein) from the surface of IOVs by high-salt treatment abolishes their spectrin-binding ability, which can be restored on the addition of purified ankyrin, whereas selective removal of band 4.1 protein by acetic acid and sodium cholate has no effect on the reassociation of spectrin with the membrane vesicle [41].

Recently, binding studies have indicated that purified ankyrin exhibits different affinities for spectrin dimers and tetramers [73,98], and that the affinities of ankyrin for spectrin are not appreciably affected by the phosphorylation of the spectrin. On the other hand, phosphorylation of ankyrin by cAMP-independent protein kinases significantly reduces its affinity for spectrin tetramers but not spectrin dimers [73]. It is not surprising that phosphorylation of ankyrin affects its affinity for spectrin because the phosphorylation sites are localized in the spectrin-binding domain of this asymmetrical macromolecule [41,42,98]. However, the reason for the difference in the effect of phosphorylation between dimers and tetramers is unknown.

Electron microscopic [32–34] amd biochemical evidence [42,67,71,100] indicates that the location of the ankyrin-binding site is ~20 nm from the proximal end (to the dimer–dimer interaction site) of the filamentous spectrin dimer on a specific region of the β subunit and that each spectrin tetramer is capable of binding two molecules of ankyrin. However, the number of ankyrin molecules is only half of the total number of spectrin dimers in the red blood cells [101]. Interestingly, band 3 protein, the attachment site for ankyrin to the membrane bilayer, is also present in large excess [42–43,99]. The physiological significance of this excess of binding sites for ankyrin on both the skeletal network and membrane bilayer is not clear. An intelligent guess is that the spectrin–ankyrin–band 3 complex is one of the sites involved in the ATP-dependent shape changes of erythrocytes [75].

Ankyrin links spectrin to the membrane by direct association with the cytoplasmic domain of band 3 protein, the anion transporter [42,99]. Ankyrin contains a high-affinity binding site for both the cytoplasmic domain of band 3 and the spectrin (for review see [2–4]). These binding sites are structurally distinct and can be isolated in nonoverlapping tryptic fragments of ~82 and 55 kDa, respectively [98,100].

III. ULTRASTRUCTURE OF COMPLEXES DERIVED FROM INTACT SKELETONS

A. Spectrin and High-Molecular-Weight Complexes Isolated by Low-Ionic-Strength Extraction

The integrity of the red blood cell skeleton and its attachment to the membrane proper relies on a small number of well-established specific interactions: spectrin dimer–spectrin dimer, spectrin–actin–band 4.1, spectrin–band 4.1–glycophorin (or band 3), spectrin–ankyrin–band 3, and actin–actin interactions. These interactions are sensitive to temperature, ionic strength, and pH to varying degrees and can be selectively dissociated to cause the release of various types of fragments. (See [2–6] for a general review.) A few other minor proteins, i.e., bands 4.2 [102], 4.9 [48], tropomyosin [47], and myosin [49,50], may also regulate the organization and stability of the skeleton as well as its linkage to the membrane bilayer; however, the interactions involved are not well defined.

Of all the experimental parameters, ionic strength and temperature have been used most extensively and effectively in the dissociation of the intact skeletons. In the presence of low-ionic-strength medium containing ethylenediamine tetraacetic acid (EDTA), the cytoplasmic network of spectrin and actin dissociates completely into soluble components, leaving band 4.1 and ankyrin largely attached to the outer (cytoplasmic) surface of the inverted membrane vesicles [53]. The types of protein interactions disrupted by low-ionic-strength extraction are spectrin–actin, spectrin–band 4.1, actin–actin, and spectrin dimer–dimer interactions (the last type of dissociation occurs significantly only at elevated temperature). The soluble proteins, mainly spectrin and actin, can be separated from membrane vesicles by centrifugation and fractionated by gel permeation chromatography on sepharose 4B or agarose columns [57,64,65]. The major proteins isolated from the supernatant of the low-salt extract at 37°C are spectrin dimers that constitute the smallest native complexes of the skeleton. They retain the ability to reassociate into tetramers, interact with actin, band 4.1, and ankyrin, and reassemble with band 4.1 and actin into ternary complexes that resemble the native oligomeric complexes. These spectrin dimers also reassociate with ankyrin and band 4.1 at the cytoplasmic surface of the IOVs [103] and incorporate into the membranes of erythrocytes from spectrin-deficient mice, where they increase the mechanical stability of such reconstituted ghosts [104].

The spectrin dimers isolated by low-salt extraction appear to be long, thin, flexible filaments ~100 nm in length when visualized by platinum shadowing of specimens sprayed with a nebulizer from a volatile buffer containing a high percentage of glycerol [57]. Tetramers formed by head-to-head reassociation of these filamentous dimers are structurally and functionally indistinguishable from tetramers isolated by low-ionic-strength extraction of membrane ghosts at 0°C. A consistently small but variable amount of the spectrin extracted at 35°C from red blood cell membranes with low-ionic-strength buffer exists in the form of a stable, high-molecular-weight oligomeric complex consisting of actin, band 4.1, and band 4.9 [105]. The molar ratio of actin to spectrin dimers in these high-molecular-weight complexes varies between 2.5 and 4, whereas the molar ratio of band 4.1 to spectrin dimer is close to unity. The number of spectrin dimer molecules per cytocholasin binding site (i.e., actin protofilament) varied somewhat from preparation to preparation, with a mean value of 15 ± 2. However, electrophoresis under nondenaturing condition showed that free spectrin is present in these high-molecular-weight fractions. It is undoubtedly formed by dissociation of the complex with time, although the bulk remains undissociated for long periods. Those high-molecular-weight complexes appear to be "spiderlike" particles containing 6–10 convoluted spectrin filaments radiating from a central core of oligomeric actin [105].

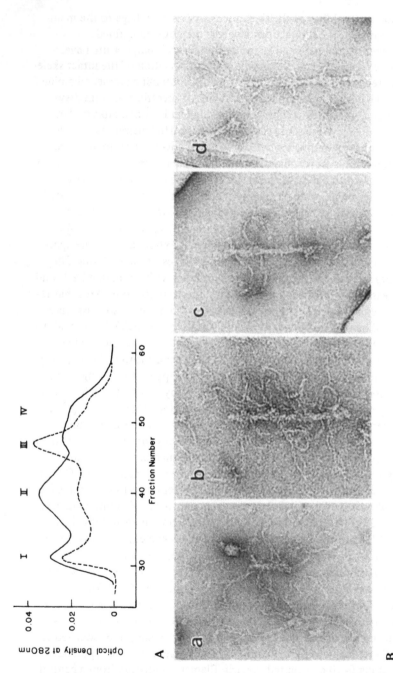

Figure 1 (A) Chromatographic profile of skeletal fragments. Skeletons from ghosts were incubated in 2 mM NaPi (pH7) for 30 min (solid line) and 45 min (dash line), and chromatographed on a Fractogel TSK HW-75 column (2.6 × 100 cm) in the same buffer containing 0.5 mM dithiothreitol in the cold room at a flow rate of 40 ml/hour. The effluent was monitored at 280 nm and collected into 6.5 ml fractions. The amount of protein in peak I as indicated by UV

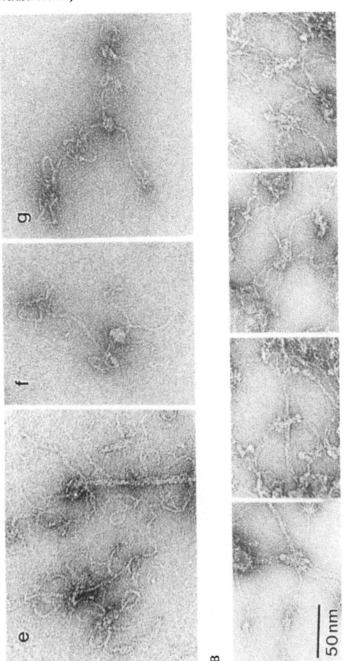

Figure 1 (continued) (B) Tropomyosin-depleted unit complexes of heat-dissociated skeleton. Materials eluted under the bulk and at the leading edge of peak II, and under peak III are shown in panels a–d, e, and in f and g, respectively. (C) Actin-spectrin complexes in tropomyosin-containing intact skeletons that have been expanded by incubation in low ionic strength buffer. (Reprinted from Ref. 106.)

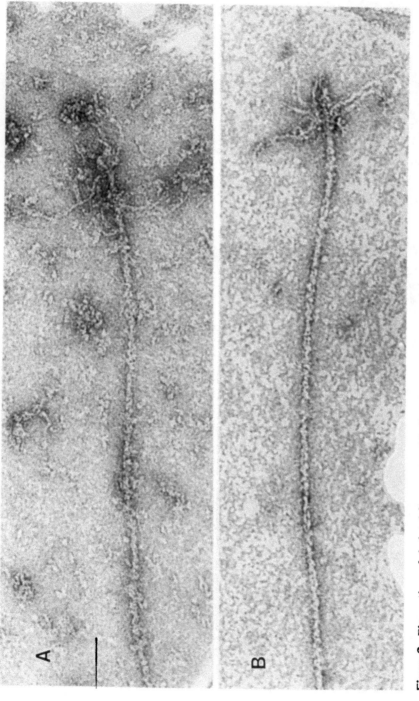

Figure 2 Elongation of skeletal fragments with G-actin, panels A and B, and the decoration of elongated F-actin with S1 fragments of heavy meromyosin, panels C and D. Preparation of peak II fragments (see Fig. 1) were incubated at room temperature with G-actin (~64 μg/ml final concentration) in 50 mM NaCl, 3 mM NaPi (pH 7). The reactions were stopped by the addition of an equal volume of ice-cold phosphate buffered saline (PBS), and the samples were adsorbed onto

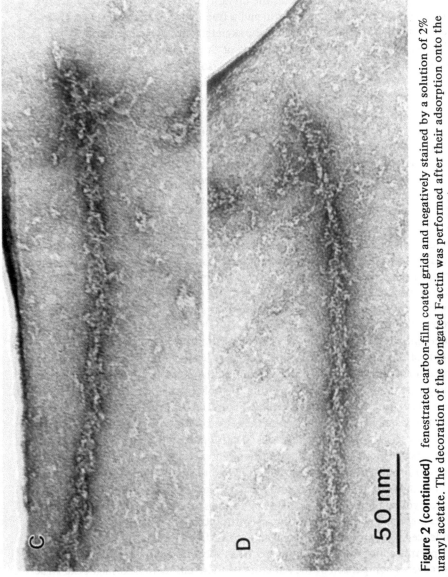

Figure 2 (continued) fenestrated carbon-film coated grids and negatively stained by a solution of 2% uranyl acetate. The decoration of the elongated F-actin was performed after their adsorption onto the grid prior to staining with uranyl acetate. (Reprinted from Ref. 106.)

B. Unit Complexes Derived From Heat-Dissociated Intact Skeletons

Unit complexes that retain native spectrin–actin–band 4.1 interactions can also be generated by heat-induced dissociation of intact skeletons that have been released from the membrane by nonionic detergents [106]. When the lipid bilayer is dissolved away by Triton X-100, band 4.1 remains strongly associated with spectrin (and actin) and maintains the integrity of the spectrin–actin networks. The accessory proteins, i.e., ankyrin and a fraction of band 3, as well as glycophorin, that remain associated with the skeletal network can be removed by high-salt extraction of the Triton-residue of the membrane, leaving a basic network of spectrin, actin, and band 4.1. These isolated skeletons break down in a reproducible manner when exposed to low-ionic-strength buffer at elevated temperature and produce fragments that can be fractionated into four major fractions by gel-filtration column chromatography [106]. The majority of the proteins originally in the skeleton fall under two peaks eluted between the residual lipid vesicles at the void volume and free proteins. The size of the heat-dissociated fragments decreases continuously during the first hour of incubation as evidenced by the decrease in the area under the second peak and the increase of that under the third peak. Afterward, the size of the complexes remains unchanged even after prolonged incubation at 37°C.

Basically, two types of unit complexes, both containing spectrin, actin, and band 4.1, were isolated by heat-induced dissociation of stripped and tropomyosin-depleted skeletons in 2 mM sodium phosphate (NaPi), pH 7.0. The larger unit complexes consist of a long and rigid central core of short F-actin with clusters of spectrin emanating from discrete sites along the actin filament. The smaller complexes typically consist of three to four strands of spectrin dimers radiating from a globular core of oligomeric actin. The structure of the smaller complexes sometimes resembles that of the Triskelionlike forms noted recently in fragments of the erythrocyte skeletons [20]. The leading edge of the second peak at earlier time points contains larger ensembles that consist of two or more unit complexes interconnected by long, thin, flexible filaments ~200 nM in length. (For illustration of unit complexes, see Figure 1.) The photomicrographs of the negatively stained unit complexes over the perforation of fenestrated carbon film show clear images of filamentous spectrin radiating from short, double helical cores of F-actin. The identity of these central cores is confirmed by their ability to nucleate the polymerization of G-actin and by the arrowhead decoration of those short, rigid cores with the S-1 fragments of rabbit muscle heavy meromyosin (Figure 2) [106]. The key to the success of the above method lies in its presumed ability to dissociate spectrin tetramers exclusively. However, as evidenced by the presence of small amount of free proteins, limited dissociation of spectrin–actin, and actin–actin, as well as spectrin–band 4.1 has occurred. Moreover, a comparison of the structure of these unit complexes with the structure of the intact, least-perturbed, and tropomyosin-containing skeletons suggests that substantial G-actin migration and annealing of actin protofilament have occurred during the preparation procedures (see next section). The release of spectrin, and perhaps band 4.1, is most likely caused by the erosion of G-actin from actin protofilaments that became unstable due to the removal of its stabilizing element—tropomyosin. A reinvestigation of the heat-dissociated unit complexes using unstripped skeletons and under conditions that prevent the dissociation of membrane tropomyosin is of critical importance to our understanding of the structural organization of the native skeletons.

Concurrent with the above study, Matsuzaki et al. [107] reported the isolation of a structural unit of bovine erythrocyte cytoskeleton that sedimented with a Sweberg coefficient of 26S. This 26S unit fragment is generated by heat-dissociation of Triton shells that were stripped of accessory proteins by mild heat treatment at 25°C for 20 min in 5 mM NaPi. Because equilibrium between spectrin dimers and tetramers at 25°C might cause, to a certain extent, the dissociation and release of skeletal components and because portions of the skeletons were removed as large aggregates after incubation at 37°C, only 50% of the protein originally present in the Triton shells was recovered in this 26S fraction by sedimentation in a linear density gradient of glycerol. A portion of the proteins in this 26S fraction is shown by electrophoresis in nondenaturing gels and electron microscopy to be free spectrin. Electron microscopy of low-angle rotary-platinum-shadowed preparations revealed that the structure of the complexes in this 26S fraction is similar to that of the high-molecular-weight complexes reported by Beaver et al., which sediment at a velocity of 32 ± 2S [105]. The stoichiometry of the spectrin dimer–actin–band 4.1 in these 26S complexes was estimated to be 5–10–10. The actin-to-spectrin ratio of 2 is slightly lower than the value reported in previous studies, whereas the ratio of band 4.1 to spectrin is twice the literature value; however, it agrees with the value obtained in the binding study of purified spectrin and band 4.1 (for a review, see [2]). This different stoichiometry and the recovery of only half of the skeletal proteins in the 26S fraction suggest that some loss of actin and spectrin might have occurred during the differential heat treatment and extensive washing of the Triton shells.

IV. ULTRASTRUCTURE OF INTACT SKELETONS

A. Thin Section of Embedded Membrane Ghost

Although individual spectrin molecules and limited combinations of cytoskeleton components have been clearly visualized by electron microscopy of platinum-shadowed or negatively stained preparations, the overall organization of the cytoskeleton has been more difficult to discern. Thin sections of membrane fixed with a tannic acid-glutaraldehyde mixture have provided crucial views of the red blood cell skeleton in situ [108].

Figure 3 Platinum-carbon (Pt-C) replica of the protoplasmic surface of an RBC ghost freeze-fractured and freeze-dried after prefixation with formaldehyde. Washed red cells were resuspended in PBS and applied to Alcian blue coated mica. The attached cells were lysed in hypotonic PBS, then fixed with formaldehyde. The fixed cells were next sandwiched between the mica support and a copper plate, dipped rapidly into liquid nitrogen for 20 sec and fractured. After freeze drying, the cells were shadowed with Pt-C. (For details, see reference 109.) Photomicrograph: courtesy of Dr. M. V. Nermut, National Institute for Medical Research, Mill Hill, London, England.

Obliquely cut areas of the membrane have an evenly distributed but disorganized meshwork of filaments that are 9 nm in cross-sectional diameter, interconnected by scattered round nodules ~25 nm in diameter. In areas cut close to the bilayer, a dense layer of granular particles 10-13 nm in diameter appeared. In sections cut perpendicular to the plane of the membrane bilayer, these granular particles appeared to attach vertically to the bilayer and extend through the trilamellar structure of membrane bilayer [108]. The 9-nm filaments were identified as spectrin on the basis of their morphologic similarity to purified spectrin and to spectrin reassociated with the surface of the spectrin-depleted membrane vesicles [108].

The composition of the granular particles that serve as the cross-bridge between the membrane proper and the filamentous network was not definitively identified at the time of investigation; undoubtedly they must contain a mixture of band 3-ankyrin and band 4.1-glycophorin complexes. The actual identity of those granular particles needs to be confirmed by specific antibody decoration. The scattered round spots to which several filamentous components appear to attach are not as easily discernible as other structural features.

B. Skeleton in Freeze-Fractured Membrane Disk

Carbon replicas of the freeze-fractured, freeze-dried, and platinum–carbon shadowed membrane revealed an extensive anastomosing network covering the entire surface of the fractured membrane disk (Figure 3) [109]. This network consists of groups of particles ~10 nm in diameter interconnected by filamentous elements. Treatment of the protoplasmic surface of the red blood cell ghost in situ with EDTA removed the filamentous elements. Both the length and number of filaments were substantially smaller than those of spectrin filaments obtained by other methods.

The lengths of the filaments, ranging from 40-140 nm with more than 50% between 70 and 90 nm, are consistent with the argument that the spectrin dimer has two sites of interaction with globular proteins. However, the number of filaments per erythrocyte is nearly 10 times less than the number of spectrin molecules derived from stain intensity of protein bands after SDS-PAGE. It was argued that some of the spectrin could have been removed during freeze-fracture so that the number of filaments was underestimated in the freeze-fracture study whereas it was overestimated by gel analysis. However, it is difficult to reconcile a 10-fold difference. One plausible explanation is that the spectrins in the membrane ghosts in situ are in a condensed rather than a filamentous state and that only a very small fraction of the spectrin molecules were extended due to the adsorption of the ghosts to the positively charged mica and/or the freezing of the specimens.

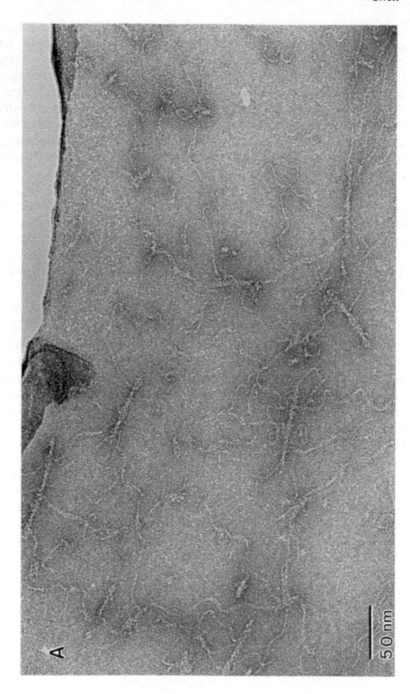

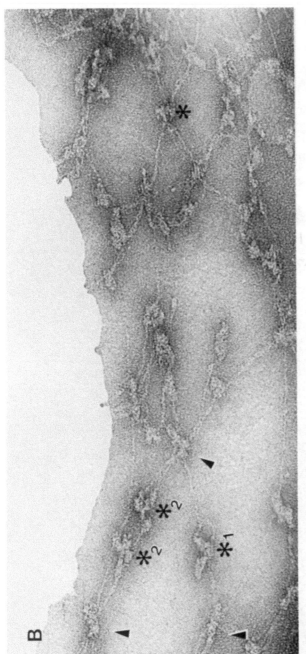

Figure 4 Ultrastructure and composition of human red cell membrane skeleton. Electron micrograph of (A) stripped, (B) unstripped, and (C) partially stripped membrane skeletons that have been expanded electrostatically, and sodium dodecyl sulfate gel electrophoretic pattern of ghosts (lane A) and skeletons extracted with various concentrations of NaCl in 0.25 M increments from 0 to 1.5 (lanes B-H). Arrow heads indicate actin protofilaments while asterisks and stars represent band 3-ankyrin complexes and ankyrin, respectively. The subscripts 1 and 2 correspond to the number of globular complexes on the spectrin tetramers. Results in panels B and C suggest that the spectrin tetramers can have either one or both of its two ankyrin binding sites occupied, depending on whether it is closely associated with another spectrin tetramer.

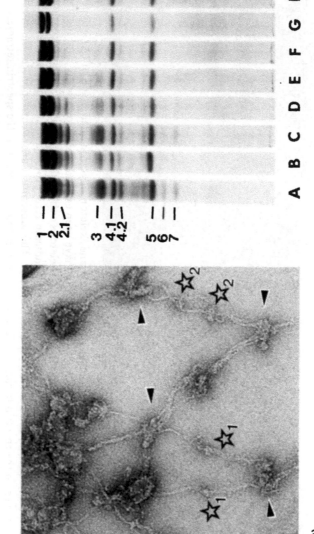

Figure 4 (continued)

C. Molecular Architecture of the Expanded Skeleton

Early studies of negatively stained specimens of intact skeleton revealed a fluffy, disorganized network of nodules and filaments joined together to form an irregular network the size of the ghost. The gross morphology of the cytoskeleton appears to be affected by conditions that alter the net negative charge of spectrin. Regions of the network contract in the presence of polylysine or in low-pH media, leading to the artifactual expansion of areas interconnecting those condensed patches. This low-pH-induced organizational change is at least partially reversible [110].

A series of recent studies with negatively stained specimens has provided the first ultrastructural definition of intact skeleton both with and without accessory proteins. The success of these studies lies in the spreading of the network either by low-pH-induced contraction of regions that are strongly attached to the carbon support, leading to the expansion of neighboring areas [111] or by an electrostatically driven, uniform expansion of the network at low temperature [112,113].

Electron microscopy of the negatively stained specimens shows that the intact skeletons of erythrocytes are organized into short actin protofilaments of homogeneous length interconnected by multiple strands of spectrin tetramers. The spectrin tetramer originates from one actin protofilament, extends through head-to-head interdimer interaction, and terminates at a neighboring actin protofilament. The spectrin filaments do not cross or extend beyond the most proximal actins. The difference between the stripped and unstripped (i.e., replated with accessory proteins) skeletons is evident. In the micrographs of the replated skeletons, the major accessory proteins (band 3–ankyrin complexes) can be easily localized, whereas they are absent in the micrographs of the stripped networks [112]. In skeletons that are partially stripped, both large (band 3–ankyrin complex) and small (ankyrin) globular masses are detected close to the middle of the spectrin tetramers (for details see Figure 4). The actin-spectrin junctions in the skeletal networks are easily depicted. Approximately five to eight strands of spectrin tetramers radiate from a common actin protofilament toward the neighboring actin filaments [111,112] that are, in turn, interconnected by spectrin tetramers forming a polygon with actin protofilaments and spectrin tetramers as the vertices and edges, respectively [111].

This pattern of irregular polygons of spectrin tetramers has recently been shown to extend over the entire network of the membrane cytoskeleton with hexagons as the most predominant feature [113] (Figure 5). Moreover, the number of spectrin tetramers radiating from the actin vertex seems to reflect the stability of the skeletal network. Spectrin oligomers at higher orders, i.e., hexamers and octamers, are also detected among the polygonal lattice of tetramers [113]. However, it is not clear whether these hexameric and octameric spectrins are intrinsic entities of the skeletal network in situ or simply artifacts arising

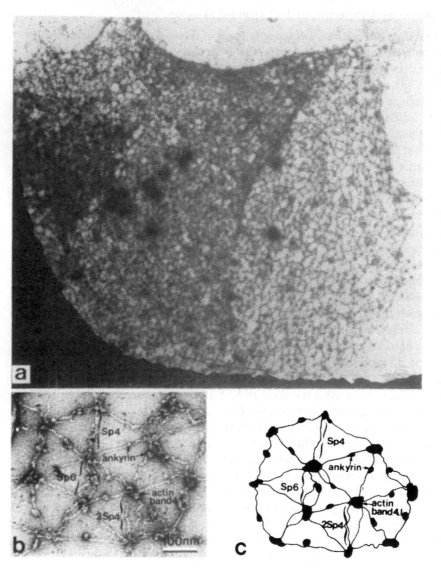

Figure 5 (a) Low magnification view of an intact membrane skelton visualized by negative staining with uranyl acetate showing an extensive polygonal protein network. (b) Portion of a single-leaflet membrane skeleton showing two overlapped hexagons and a pentagon with filamentous spectrin tetramers (Sp4) and hexamers (Sp6) as the sides of the polygon radiating from vertices of short actin protofilaments.

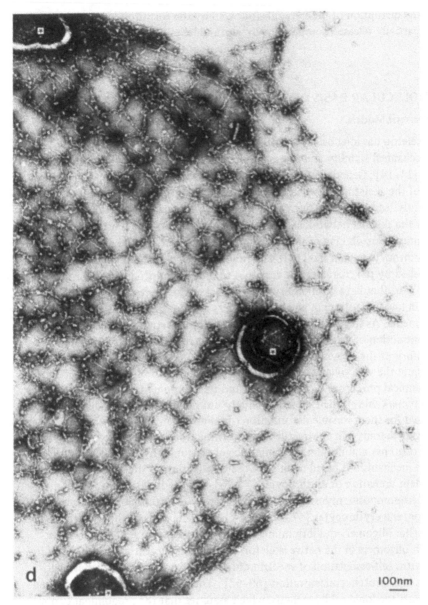

100nm

Figure 5 (continued)   (c) Schematic drawing of image in (b), and (d). Higher magnification view of a large area of a single-leaflet membrane skeleton. Asterisks indicate residual membrane vesicles. (For details, see ref. 113.) Photomicrographs: courtesy of S. C. Liu, L. H. Dericks, and J. Palek.

from the disruption of the network followed by the fortuitous cross-linking of these partially released components by glutaraldehyde.

V. MOLECULAR BASIS OF ERYTHROCYTE ELASTICITY

A. Current Models

The skeleton has long been recognized as the structural element that underlies the mechanical stability and reversible deformability of the erythrocyte membrane [11–14]. Genetic deficiencies and malfunctions in the structural components of the skeleton, as well as the proteins linking it to the membrane proper, frequently lead to varying degrees of hemolytic anemia [16–18]. In vitro disruption of the skeletal network causes the fragmentation of the lipid bilayer, whereas chemical cross-linking of the spectrin drastically reduces the deformability of the erythrocyte membrane [13]. The functional role of the skeleton is further established by the characteristic response to changes in pH and concentration of electrolytes that have been observed for intact erythrocytes; the skeleton expands in low-ionic-strength and alkali buffers and contracts in high-salt and acidic media. As for the intact erythrocyte, this ionic-strength-induced dilation and contraction of the skeleton is preceded by a lag period that is dependent on the history of the network [11,12].

Despite the vast volume of information existing on the structure, function, and chemical properties of the skeleton and its components, the molecular basis for its remarkable elasticity remains largely unknown. Various models have been proposed for the reversible deformation (or shape change) of the erythrocyte including an actomyosin contractile mechanism, a reversible spectrin tetramer–higher oligomer equilibrium, and an ionic gel elastomer hypothesis. The actomyosin mechanism is based mainly on the observation of the ATP- and calcium-dependent crenation of erythrocytes [75,76] and the isolation of a small quantity of tropomyosin, myosin, and skeleton-associated calmodulin-binding protein from the erythrocytes [47,49,50,114,115]. The basis of the spectrin tetramer–higher oligomer equilibrium mechanism is the detection of higher-order spectrin oligomers in the native skeleton of abnormal erythrocytes [63] and in the in vitro self-association of spectrin dimers and tetramers into higher oligomers at high protein concentration [60–62]. The latter observation is also used to support the ionic gel hypothesis, which assumes that the predominant component of the skeleton, spectrin, is an elastomer without specific structure that behaves like ionic gels that expand and contract as ionic strength changes [116]. Phosphorylation and dephosphorylation of ankyrin and band 4.1 protein, as well as phosphoinositol, is also presumed to regulate the shape and deformability of the erythrocyte [73,74,84–90,93].

B. Folded Spectrin Hypothesis

Recently, the dilation of the skeleton induced at low ionic strength was shown to be associated with a progressive extension of the spectrin molecules in the skeletal network (Figures 4 and 6) [112]. Although electron micrographs of negatively stained specimens of expanded intact skeleton and unit complexes show irregularly elongated and apparently flexible molecules that resemble spectrin isolated by low-ionic strength extraction and visualized by low-angle platinum shadowing [106,111,112], the least perturbed skeletons are condensed, and their spectrin appears compact. The conversion of compact spectrins in the condensed skeleton to the long, thin, flexible components observed in the expanded skeleton, under the influence of low-ionic-strength buffer, is a slow process that takes minutes for completion [112]. The above observation suggests that native spectrin associated with the membrane of the erythrocyte at static state may assume a specifically folded configuration stabilized by weak interactions among different domains of the spectrin dimers, and that the reversible conversion between this compact configuration and its long, thin filamentous counterpart may be the fundamental basis of the remarkable elasticity of the skeleton.

This folded-spectrin hypothesis has many appealing features: (a) It offers a rationale for the extensive conservation of secondary structure in spectrin molecules, (b) It is compatible with regulation of membrane elasticity and other spectrin-based mechanical properties through specific conformational changes, and (c) It provides a rationale for the observation of dense packing of peripheral proteins at the cytoplasmic surface of the red blood cell membrane. A corollary of this hypothesis is that spectrin exists in more than one conformation and can undergo reversible interconversion mediated by some yet-unidentified factor or factors acting alone or in combination. This hypothesis does not preclude the extended filament as one of many possible states of the spectrin molecule, and the concept of reversible conformational changes is compatible with the restoration of binding sites for band 4.1 protein and ankyrin on the low-ionic-strength–extracted spectrin on reversion of the ionic strength [67].

Up to now, ultrastructural analysis has been conducted with spectrin molecules isolated by exposure of ghost membranes to low-ionic-strength medium that contains EDTA [57]. Although these preparations provide a good starting point, it is not clear whether the long, thin, flexible, ($\sim$2000 Å) filament observed in the preparations represents the native conformation of spectrin in situ. It is not unlikely that the structure of native molecules have been altered because the spectrin molecules are free from all types of physical constraints and may be subject to alterations resulting from sonication, pH, and drying effects when prepared in low-ionic-strength buffers.

The discrepancy between the structure of spectrin of intact cells and that of purified spectrin may arise because spectrin has been universally isolated by

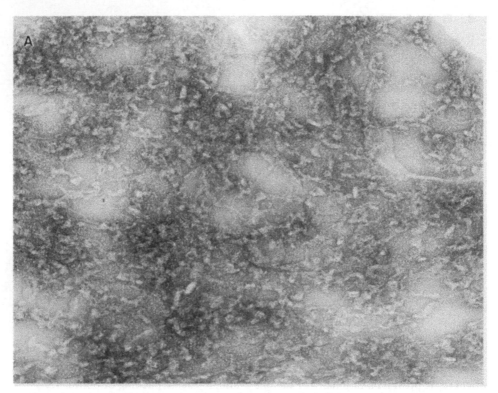

Figure 6 Electrostatic expansion of the unstripped membrane skeletons of human erythrocyte. The unstripped skeletons (i.e., replaced with intrinsic membrane proteins) of human erythrocyte membranes prepared under conditions that prevent the dissociation of membrane tropomyosin were expanded in 5 mM NaPi with the following additives: (A) 2 mM $MgCl_2$, (B) none, and (C-E) 0.5 mM dithiothreitol. Structure of skeleton in panel A, with dense patches of oblate features, is closest to that of the native network at static state as it is the least perturbed. Elimination of $MgCl_2$ from the dilating buffer causes substantial expansion of the compact features and produces images with patches of globular forms interconnected by segments of fibrous structures (panel B). Addition of dithiothreitol to the buffer results in further expansion of the network (panels C-E) revealing filaments approximately 2000 Å long; actin protofilaments of uniform length (arrow heads) with multiple strands of filaments emanate from their centers; and globular complexes (asterisks) at close to the midpoint of the filaments (also see Figures 4B, 4C and ref. 112).

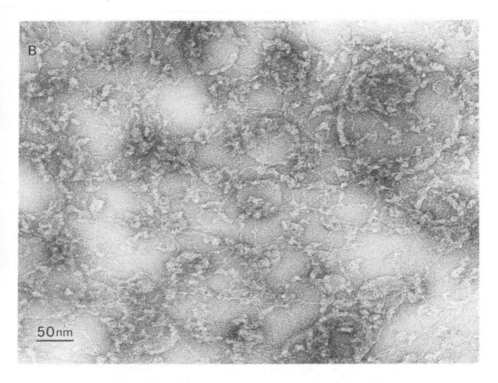

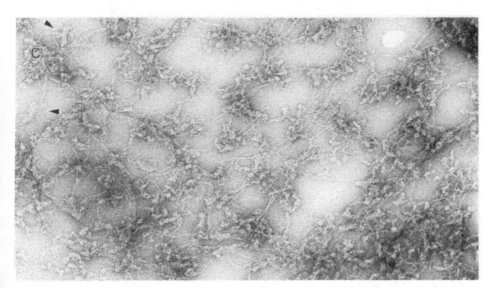

Figure 6 (continued)

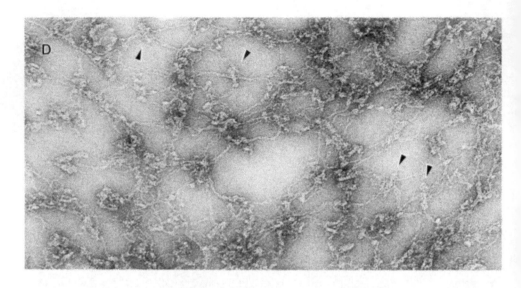

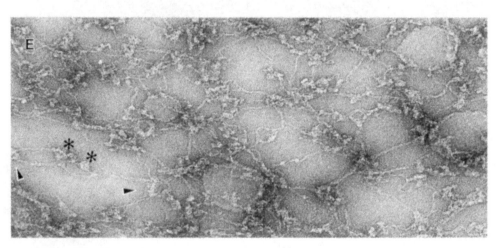

Figure 6 (continued)

treatments that promote its extension as they drive the dissociation of spectrin from actin, ankyrin, and band 4.1 protein. Previously, the elution volume of the spectrin isolated by low-salt extraction from sepharose 4B or agarose columns was shown to increase with increasing ionic strength of the effluent [64,65]. This salt-induced increase in elution volume can be interpreted as a salt-induced conformational change of the spectrin from an extended to the compact state. This observation is entirely consistent with the folded spectrin hypothesis.

The folded spectrin hypothesis deemphasizes molecular rearrangement as the fundamental basis for red blood cell deformability. However, because the individual components of the skeleton, especially spectrin and actin, are capable of dynamic interactions, even though these interactions are much slower than the rapid folding and unfolding of the spectrin molecules, rearrangements at the level of both the skeleton and integral membrane can occur. These rearrangements may account for changes in deformability and permeability of aged red blood cells and may contribute to their removal from the circulation.

Recently, the concept of compact spectrin oligomers was invoked to interpret the electron microscopic images of axially elongated membrane skeletons in which short, stubby, globular features are interconnected by long, thin filaments that seem to run parallel to each other over great length [117]. The authors suggested that the short, stubby globular forms are oligomers of spectrin and that the long, thin, linear filaments are F-actins that run continuously throughout the dimension of the skeleton. This interpretation represents an alternative model for the organization of red blood cell membrane skeleton that features lengthy, continuous actin filaments interconnected by short, nearly globular spectrin decamers and dodecamers. This alternative view is at variance with the conventional model proposed by Lux [118] and the folded spectrin hypothesis presented here. The Lux model, based on biochemical, biophysical, and electron microscopic evidence, features short actin filaments interconnected by filamentous spectrin tetramers. The folded spectrin hypothesis, even though it presents a different viewpoint on the structures of the spectrin tetramers and proposes an entirely different fundamental basis for the viscoelasticity of the skeleton, is itself a refinement of the current Lux model that is in need of experimental verification.

C. Length of Actin Filaments

Another important structural feature of the skeleton is actin. The actin in the erythrocyte plays a novel organizational and functional role. It is dispersed as a random but uniform array of stable, short protofilaments, each of which serves as an anchoring point for multiple strands of spectrin and at the same time, through the spectrin–actin–band 4.1–glycophorin interactions, serves as one type of cross-bridge between the network and the membrane bilayer. The other type of cross-bridge arises from spectrin–ankyrin–band 3 interaction. It is conceivable that the mechanical force that causes the deformation of the membrane bilayer is transmitted to the skeletal network through the cross-bridge, in turn causing the deformation of the skeletal network. It has been shown that erythrocytes containing aggregates of band 3 protein and ankyrin exhibit an enhanced tendency to release membrane vesicles and/or to hemolyze [119]. However, there is virtually no information on the correlation between the deformability and stability of the skeleton and the length, spatial distribution, and dynamic state of the actin protofilaments.

Although compelling evidence suggests that erythrocyte actin is oligomeric, direct detection of actin filaments in the skeleton by electron microscopic examination has not been successful, partly due to the high concentration of proteins at the plane of the membrane and partly due to the limited length of actin filaments. Recently, the entire length of erythrocyte F-actin has been measured in photomicrographs of negatively stained specimens, initially in unit complexes over the perforation of fenestrated grids and later in intact skeletons that have been expanded by electrostatic repulsion induced by buffers with low ionic strength or low pH [106,111,112]. In addition, G-actin accretion has been shown to occur at both the pointed and barbed ends of the actin protofilaments in the isolated unit complexes and in the intact skeletons prepared under conditions that promote the dissociation of tropomyosin [106,107,112]. The actin filaments in both cases can anneal into longer actin filaments while maintaining most, if not all, of their actin–spectrin–band 4.1 interactions.

The length of actin protofilaments depends strongly on the integrity of the skeletons and their method of preparation. In the intact and uniformly expanded skeletons that are replated with accessory proteins and presumably contain tropomyosin, the length of the actin protofilaments follows a very narrow Gaussian distribution with a mean value of 32 ± 5 nm [112], whereas the actin filaments in the heat-dissociated unit complexes of tropomyosin-depleted skeleton are longer and more dispersed with an average value of 53 ± 17 nm [106]. Moreover, most actin filaments in the stripped skeletons [112] and their fragments [106] contain more than one cluster of spectrin multimers, whereas the filaments in the least-perturbed skeletons invariably contain only one spectrin cluster [112]. This finding suggests that annealing of actin protofilament can occur in the stripped, tropomyosin-depleted skeletons through the accretion of G-actin at both ends of the actin filaments. Whether annealing of actin protofilaments and accretion of G-actin occurs in vivo as the result of aging (or maturation) of red blood cells or membrane damage remains to be investigated.

Because the intact and uniformly expanded skeletons are far less manipulated, their actin length is likely to be closer to that of the native form than that in stripped skeletons. The lengthy manipulations required for obtaining stripped skeletons and their fragments may both promote the dissociation of actin-stabilizing proteins, i.e. tropomyosin, and stimulate the migration of actin monomers from some filaments to others [98]. The average length of 32 nm for actin protofilament in the uniformly expanded, unstripped skeleton [112] compares favorably with the value of 33 nm reported for length of actin in erythrocyte skeletons that are partially stretched by low pH [97], and both values are equivalent to the axial repeat in paracrystals of nonmuscle tropomyosin.

These observations suggest that tropomyosin may impose a quantum length on actin protofilaments, perhaps by binding to the grooves, one on each side of the double helix, of the actin filaments. However, the content of tropomyosin in the uniformly expanded skeleton is uncertain because the last step in the rapid preparation of the expanded skeleton is dilution of the tropomyosin-containing skeleton by a buffer without Mg^{+2}, a condition that has been shown to promote the dissociation of tropomyosin [47]. The removal of magnesium ion seems to be necessary, but not sufficient, for the expansion of the skeletons. To confirm the stabilizing action of tropomyosin, the spatial disposition and content of erythrocyte tropomyosin in the expanded skeletons must be evaluated by immunoelectron microscopy and immunoblotting.

VI. CONCLUSION AND FUTURE PERSPECTIVES

Our understanding of the molecular architecture of the protein network responsible for the mechanical stability and deformability of the mammalian erythrocyte is increasing at a brisk pace. The work discussed in this review shows that ultrastructural studies of the intact skeleton and of the unit complexes derived from the native network not only reveal the precise spatial disposition of the major skeletal proteins, i.e., spectrin and actin, but also substantiate the types of specific protein-protein interactions inferred from biochemical and physical studies of the isolated skeletal components. The most intriguing nature of the work is the indication that the correlation of structure with function may be possible by using advanced techniques for electron microscopy.

Photomicrographs of negatively stained preparations of intact skeletons have already revealed the organizational role of the short actin filaments and the electrostatically driven extension of compact spectrin in the intact skeleton. It is, therefore, of interest to study the mechanism and reversibility of the unfolding of the native spectrin and the role of lipid bilayer in the refolding of the extended spectrins.

Another important aspect of the organization of the skeleton is the structure and dynamics of band 3, the anion transporter. This integral membrane protein not only regulates the anionic potential of the erythrocyte, but also, together with ankyrin and perhaps band 4.1, forms the major crossbridge between the membrane bilayer and the skeleton. The position of ankyrin–band 3 complexes relative to spectrin has been inferred from reconstitution studies with purified ankyrin and spectrin dimer and tetramers. However, direct evidence regarding the location of ankyrin–band 3 complexes in the native network is still lacking.

It has been shown that one-third of the band 3 proteins and all of the ankyrin molecules remain associated with the skeleton after solubilization of the membrane bilayer and that these accessory proteins can be extracted from the network by a molar concentration of sodium chloride [112]. Therefore, it is possible to determine the location of these ankyrin–band 3 complexes by comparing the structural difference between the stripped and nonstripped networks [111,112] (Figure 3). However, the isolation of a skeleton-associated calmodulin-binding protein with molecular weight and ionic strength dependencies similar to those of band 3 raises the question regarding the structural and functional heterogeneity of the broad protein band found by SDS-PAGE, which had been collectively called band 3. The question of interactions between band 3 and ankyrin as well as between ankyrin and calmodulin-binding protein and between band 3 and calmodulin must be investigated to differentiate band 3 from calmodulin and to determine the position of those proteins in the skeletal network.

Other structural elements that might play crucial regulatory and stabilizational roles and need to be investigated in detail include the length and distribution of actin filaments in normal and pathological erythrocyte membranes, and the localization of minor skeleton proteins such as band 4.1, 4.2, 4.9, myosin, and tropomyosin.

With recent advances in the electron microscopic study of biological specimens, the availability of specific antibodies against minor membrane components of the erythrocyte, and the accumulation of vast genetic, biochemical, biophysical, and clinical information, a full understanding of the structure–function relationship of the erythrocyte membrane is within reach. This undertaking can be extended to the structure and function relationship of the cytoskeleton in the nucleated cell, and to the mechanism underlying the interactions between the cytoskeleton and membrane in general.

ACKNOWLEDGMENTS

Research by the author has been funded in part by grants from the National Institute of Health (R01 HL 33254 to the author with T. L. Steck as Co-principal Investigator and P01 HL 30121, Biology of Sickle Cell Disease Program Project, E. Goldwasser, Program Director), and the American Heart Association (Grant 81-664). B. W. Shen was an Established Investigator of the American Heart Association (Grant 79-131). Valuable editorial assistance by Susan Barr is graciously acknowledged. The manuscript was prepared by LeAnna Westfall.

REFERENCES

1. Fowler, V. M. (1986). New views of the red cell network. *Nature* 322:777–778.
2. Bennett, V. (1985). The membrane skeleton of human erythrocytes and its implications for more complex cells. *Annu. Rev. Biochem.* 54:273–304.

3. Marchesi, V. T. (1985). Stabilizing infrastructure of cell membranes. *Annu. Rev. Cell Biol.* 1:531–561.
4. Cohen, C. (1983). The molecular organization of the red cell membrane skeleton. *Semin. Hematol.* 20:141–158.
5. Goodman, S. R., and Shiffer, K. (1983). The spectrin membrane skeleton of normal and abnormal human erythrocyte: a review. *Am. J. Physiol.* 244: C121–C141.
6. Branton, D., Cohen, C., and Tyler, J. (1981). Interaction of cytoskeletal proteins on the human erythrocyte membrane. *Cell* 24:24–32.
7. Hainfeld, J., and Steck, T. L. (1977). The submembrane reticulum of the human erythrocyte: a scanning electron microscopic study. *J. Supramol. Struct.* 6:301–307.
8. Sheetz, M. P., and Sawyer, D. (1978). Triton shells of intact erythrocytes. *J. Supramol. Struct.* 88:399–412.
9. Sheetz, M. P. (1979). Integral membrane protein interaction with Triton cytoskeletons of erythrocytes. *Biochim. Biophys. Acta.* 577:122–134.
10. Waugh, R., and Evans, E. (1979). Thermoelasticity of red blood cell membranes. *Biophys. J.* 26:115–132.
11. Lange, Y., Hodesman, R. A., and Steck, T. L. (1982). Role of the reticulum in the stability and shape of the isolated human erythrocyte membrane. *J. Cell Biol.* 92:714–721.
12. Johnson, R. M., Tylor, G., and Meyer, D. B. (1980). Shape and volume changes in erythrocyte ghosts and spectrin-actin networks. *J. Cell Biol.* 86: 371–376.
13. Mohandas, N., Chasis, J. A., and Shohet, S.B. (1983). The influence of membrane skeletons on red cell deformability, membrane material properties, and shape. *Semin. Hematol.* 20:225–242.
14. Sheetz, M. P. (1983). Membrane skeletal dynamics: role in modulation of red cell deformability, mobility of transmembrane proteins, and shape. *Semin. Hematol.* 20:175–188.
15. Tchernia, G., Mohandas, N., and Shohet, S. (1981). Deficiency of skeletal membrane protein band 4.1 in homozygous hereditary elliptocytosis. *J. Clin. Invest.* 68:454–460.
16. Palek, J., and Lux, S. E. (1983). Red cell membrane skeletal defects in hereditary and acquired hemolytic anemia. *Semin. Hematol.* 20:189–224.
17. Palek, J. (1985). Hereditary elliptocytosis and related disorders. *Clin. Haematol.* 14:45–87.
18. Shohet, S., and Lux, S. E. (1984). The erythrocyte membrane skeleton: Pathophysiology. *Hosp. Pract.* November:89–108.
19. Yu, J., Fischman, D. A., and Steck, T. L. (1973). Selective solubilization of proteins and phospholipids from red blood cell membrane by nonionic detergent. *J. Supramol. Struct.* 1:233–248.
20. Liu, S. C., Windisch, P., Kim, S., and Palek, J. (1984). Oligomeric states of spectrin in normal erythrocyte membranes: Biochemical and electron microscopic studies. *Cell* 37:587–594.

21. Brenner, S. L., and Korn, E. (1980). Spectrin-actin complex isolated from sheep erythrocytes accelerates actin polymerization by simple nucleation: Evidence for oligomeric actin in the erythrocyte cytoskeleton. *J. Biol. Chem.* 255:1670–1676.
22. Atkinson, M. A. L., Morrow, J. S., and Marchesi, V. T. (1982). The polymeric state of actin in the human erythrocyte cytoskeleton. *J. Cell Biochem.* 18:493–505.
23. Pinder, J. C., and Gratzer, W. B. (1983). Structural and dynamic state of actin in the erythrocyte. *J. Cell Biol.* 96:768–775.
24. Cohen, C., Jackson, P. T., and Branton, D. (1978). Actin-membrane interactions: Association of G-actin with the red cell membrane. *J. Supramol. Struct.* 9:113–124.
25. Lin, D. C., and Lin, S. (1979). Actin polymerization induced by a motility related high affinity binding complexes from human erythrocyte membranes. *Proc. Natl. Acad. Sci. USA* 76:2345–2349.
26. Lin, D. C. (1981). Spectrin-4.1-actin complex of the human erythrocyte: Molecular basis of its ability to bind cytochalasins with high affinity and to accelerate actin polymerization in vitro. *J. Supramol. Struct.* 15:129–138.
27. Cohen, C. M., Tyler, J. M., and Branton, D. (1980). Spectrin-actin associations studies by electron microscopy of shadowed preparations. *Cell* 21: 875–883.
28. Cohen, C., and Korsgren, C. (1980). Band 4.1 causes spectrin-actin gels to become thixotropic. *Biochem. Biophys. Res. Comm.* 27:1429–1435.
29. Ungewickell, E., Bennett, P. M., Calvert, R., Ohamair, V., and Gratzer, W. B. (1979). In vitro formation of a complex between cytoskeletal proteins of the human erythrocyte. *Nature* 280:811–814.
30. Fowler, V., and Taylor, D. L. (1980). Spectrin plus band 4.1 cross-link actin: Regulation by micromolecular calcium. *J. Cell Biol.* 85:361–376.
31. Cohen, C., and Foley, S. (1982). The role of band 4.1 in the association of actin with erythrocyte membrane. *Biochim. Biophys. Acta.* 688:691–701.
32. Cohen, C., Branton, D., and Tyler, J. (1982). Mapping functional sites on biological macromolecules. *Ultramicroscopy* 8:185–190.
33. Ohanian, V., et al. (1984). Analysis of the ternary interaction of the red cell membrane skeletal proteins spectrin, actin, and band 4.1. *Biochemistry* 23: 4416–4420.
34. Leto, T. L., and Marchesi, V. T. (1984). A structure model of human erythrocyte membrane skeletal protein 4.1. *J. Biol. Chem.* 259:4603–4608.
35. Correas, I., Leto, T. L., Speicher, D. W., and Marchesi, V. T. (1986). Identification of the functional site of erythrocyte protein 4.1 involved in spectrin-actin association. *J. Biol. Chem.* 261:3310–3315.
36. Mueller, T. J., and Morrison, M. (1981). Glycoconnectin (PAS-2) membrane attachment site for the human erythrocyte cytoskeleton. In *Erythrocyte Membranes 2: Recent Clinical and Experimental Advances*, W. Kruckberg, J. Easton, and G. Brewer (Eds.). Alan R. Liss, New York, pp. 95–112.
37. Anderson, R. A., and Lovrein, R. E. (1984). Glycophorin is linked by band 4.1 protein to the human erythrocyte skeleton. *Nature* 307:655–658.

38. Pasternack, G. R., Anderson, R. A., Leto, T. L., and Marchesi, V. T. (1985). Interaction between protein 4.1 and band 3. *J. Biol. Chem.* 260:3676–3683.
39. Bennett, V., and Stenbuck, P. J. (1979). Identification and partial purification of ankyrin, the high affinity membrane attachment site for human erythrocyte spectrin. *J. Biol. Chem.* 254:2533–2541.
40a. Tyler, J. M., Hargreaves, W. R., and Branton, D. (1979). Purification of two spectrin-binding protein. Biochemical and electron microscopic evidence for site-specific reassociation between spectrin and band 2.1 and 4.1. *Proc. Natl. Acad. Sci. USA* 76:5192–5196.
40b. Tyler, J. M., Reinhardt, B. N., and Branton, D. (1980). Association of rocyte membrane proteins: binding of purified band 2.1 and 4.1 to spectrin. *J. Biol. Chem.* 255:7034–7039.
41. Bennett, V., and Stenbuck, P. J. (1980). Human erythrocyte ankyrin: purification and properties. *J. Biol. Chem.* 255:2540–2548.
42. Bennett, V., and Stenbuck, P. J. (1979). The membrane attachment protein for spectrin is associated with band 3 in human erythrocyte membranes. *Nature* 280:468–473.
43. Hargraves, W. R., Giedd, K. N., Verkleijz, A., and Branton, D. (1980). Reassociation of ankyrin with band 3 in erythrocyte membranes and in lipid vesicles. *J. Biol. Chem.* 255:11965–11972.
44. Low, P. S. (1986). Structure and function of the cytoplasmic domain of band 3: center of erythrocyte membrane-peripheral protein interactions. *Biochim. Biophys. Acta.* 864:145–167.
45. Bennett, V., and Stenbuck, P. (1980). Association of ankyrin and cytoplasmic domain of band 3 isolated from the human erythrocyte membrane. *J. Biol. Chem.* 255:6421–6432.
46. Bennett, V. (1982). The molecular basis for membrane-cytoskeleton association in human erythrocyte. *J. Cell Biochem.* 18:49–65.
47. Fowler, V. M., and Bennett, V. (1984). Erythrocyte membrane tropomyosin: purification and properties. *J. Biol. Chem.* 259:5978–5988.
48. Siegel, D. L., and Branton, D. (1985). Partial purification and characterization of an actin-bundling protein, band 4.9, from human erythrocytes. *J. Cell Biol.* 100:775–785.
49. Fowler, V., Davis, J. Q., and Bennett, V. (1985). Human erythrocyte myosin: Identification and purification. *J. Cell Biol.* 100:47–55.
50. Wong, A. J., Kiehart, D. P., and Pollard, T. (1985). Myosin from human erythrocytes. *J. Biol. Chem.* 260:46–49.
51. Lux, S. E., John, K. M., and Karnovsky, M. J. (1976). Irreversible deformation of the spectrin-actin lattice in irreversibly sickled cells. *J. Clin. Invest.* 58:955–963.
52. Nicolson, G. L., Marchesi, V. T., and Singer, S. J. (1971). Localization of spectrin on the inner surface of human red blood cell membranes by ferritin conjugated antibodies. *J. Cell Biol.* 51:265–272.
53. Marchesi, V. T., and Steers, Jr., E. (1968). Selective solubilization of a protein components of the red cell membrane. *Science* 159:203–204.

54. Hoogeveen, J. T., Juliano, R., Coloman, R., and Rothstein, A. (1970). Water-soluble proteins of the red cell membrane. *J. Memb. Biol.* 3:156–172.

55. Fairbanks, G., Steck, T. L., and Wallach, D. F. H. (1971). Electrophoretic analysis of the major polypeptides of the human erythrocyte membrane. *Biochemistry* 10:2606–2617.

56. Weaver, D. C., and Marchesi, V. T. (1984). The structural basis of ankyrin function. I. Identification of two structural domains. *J. Biol. Chem.* 259: 6165–6169.

57. Shotton, M. D., Burk, B. E., and Branton, D. (1979). The molecular structure of human erythrocyte spectrin. Biophysical and electron microscopic studies. *J. Mol. Biol.* 131:303–329.

58. Yoshino, H., and Marchesi, V. T. (1984). Isolation of spectrin subunits and reassociation *in vitro*: Analysis by fluorescence polarization. *J. Biol. Chem.* 259:4496–4500.

59. Ungewickell, E., and Gratzer, W. (1978). Self-association of human spectrin: A thermodynamic and kinetic study. *Eur. J. Biochem.* 88:379–385.

60. Liu, S. C., Windisch, P., Kim, S., and Palek, J. (1984). Oligomeric states of spectrin in normal erythrocyte membranes: biochemical and electron microscopic studies. *Cell* 37:587–594.

61. Morrow, J., Haigh, W., and Marchesi, V. T. (1981). Spectrin oligomers: A structural feature of the erythrocyte cytoskeleton. *J. Supramol. Struct.* 17: 275–287.

62. Morrow, J., and Marchesi, V. T. (1981). Self-assembly of spectrin oligomers in vitro: a basis for a dynamic cytoskeleton. *J. Cell Biol.* 88:463–468.

63. Liu, S. C., Palek, J., Prchal, J., and Castlebury, R. P. (1981). Altered spectrin-dimer-dimer association and instability of erythrocyte membrane skeletons in hereditary pyropoikilocytosis. *J. Clin. Invest.* 68:597–605.

64. Ralston, G. B. (1975). The isolation of aggregates of spectrin from bovine erythrocyte membranes. *Aust. J. Biol. Sci.* 28:259–266.

65. Ralston, G. B. (1976). The influence of salt on the aggregation state of spectrin from bovine erythrocyte membranes. *Biochim. Biophys. Acta.* 443: 387–393.

66. Calvert, R., Bennett, P., and Gratzer, W. (1980). Properties and structural role of the subunit of human spectrin. *Eur. J. Biochem.* 107:355–361.

67, Litman, D., Hsu, C.-J., and Marchesi, V. T. (1980). Evidence that spectrin binds to macromolecular complexes on the inner surface of the red cell membrane. *J. Cell Sci.* 42:1–22.

68. Speicher, D. W., Davis, G., and Marchesi, V. T. (1983). Structure of human erythrocyte spectrin: II. The sequence of the α-I domain. *J. Biol. Chem.* 258:14938–14947.

69. Speicher, D. W., and Marchesi, V. T. (1984). Erythrocyte spectrin is comprised of many homologous triple helical segments. *Nature* 311:177–180.

70. Speicher, D. W., Davis, G., Yurchenco, P. D., and Marchesi, V. T. (1983). Structure of human erythrocyte spectrin I. Isolation of the α-I domain and its cyanogen bromide peptides. *J. Biol. Chem.* 258:14931–14937.

71. Morrow, J. S., Speicher, D. W., Knowles, W. J., Hsu, C. J., and Marchesi, V. T. (1980). Identification of functional domains of human erythrocyte spectrin. *Proc. Natl. Acad. Sci. USA* 77:6592–6596.
72. Marchesi, V. (1979). Spectrin: present status of a putative cytoskeletal protein of the red cell membrane. *J. Membrane Biol.* 51:101–131.
73. Lu, P.-W., Soong, C.-J,, Tao, M. (1985). Phosphorylation of ankyrin decreases its affinity for spectrin tetramer. *J. Biol. Chem.* 260:14958–14964.
74. Eder, P. S., Soong, C.-J., and Tao, M. (1985). Phosphorylation reduces the affinity of protein 4.1 for spectrin. *Biochemistry* 25:1764–1770.
75. Patel, V. P., and Fairbanks, G. (1986). Relationship of major phosphorylation reactions and MgATPase activities to ATP-dependent shape change of human erythrocyte membranes. *J. Biol. Chem.* 261:3170–3177.
76. Schrier, S., Hardy, B., and Junga, I. (1981). Actin-activated ATPase in human red cell membranes. *Blood* 58:953–961.
77. Nakashima, K., and Beutter, E. (1979). Comparison of structure and function of human erythrocyte and human muscle actin. *Proc. Natl. Acad. Sci. USA* 76:935–938.
78. Brenner, S., and Korn, E. (1979). Spectrin-actin interaction: phosphorylated and dephosphorylated spectrin tetramer cross-link F-actin. *J. Biol. Chem.* 254:8620–8627.
79. Pinder, J., Ungewickell, E., and Calvert, R. (1979). Polymerization of G-actin by spectrin preparations: Identification of the active constituent. *FEBS Lett.* 104:396–400.
80. Pinder, J., Bray, D., and Gratzer, W. (1975). Actin polymerization induced by spectrin. *Nature* 258:765–766.
81. Goodman, S. R., Yu, J., Whitfield, C. F., Culp, E. N., and Pasnak, E. J. (1982). Erythrocyte membrane skeletal protein band 4.1 a and b are sequence-related phosphoproteins. *J. Biol. Chem.* 257:4564–4569.
82. Ohanian, V., and Gratzer, W. (1984). Preparation of red-cell-membrane cytoskeletal constituents and characterization of protein 4.1. *Eur. J. Biochem.* 144:375–379.
83. Laemmli, U. K. (1970). Cleavage of structural proteins during the assembly of bacteriophage T4. *Nature* 227:680–685.
84. Hosey, M., and Tao, M. (1977). Phosphorylation of rabbit and human erythrocyte membrane by soluble adenosine 3′,5′-monophosphate-dependent and -independent protein kinase. *J. Biol. Chem.* 252:102–109.
85. Hosey, M., and Tao, M. (1976). An analysis of the autophosphorylation of rabbit and human erythrocyte membrane. *Biochemistry* 15:1561–1568.
86. Plut, D., Hosey, M., and Tao, M. (1978). Evidence for the participation of cytosolic protein kinases in membrane phosphorylation in intact erythrocytes. *Eur. J. Biochem.* 82:333–337.
87. Thomas, E., King, L., and Morrison, M. (1979). The uptake of cyclic AMP by human erythrocytes and its effect on membrane phosphorylation. *Arch. Biochem. Biophys.* 196:459–464.
88. Horne, A. C., Leto, T. L., and Marchesi, V. T. (1983). Differential phosphorylation of multiple sites in protein 4.1 and protein 4.9 by phobol ester-activated and c-AMP-dependent protein kinases. *J. Biol. Chem.* 260:9073–9076.

89. Ling, E., and Sapirstein, V. (1984). Phorbol ester stimulates the phosphory-lation of rabbit erythrocyte band 4.1. *Biochem. Biophys. Res. Comm.* 120: 291–298.

90. Palfrey, H. C., and Waseem, A. (1985). Protein kinase C in the human eryth-rocyte: Translocation to the plasma membrane and phosphorylation of band 4.1 and 4.9 and other membrane proteins. *J. Biol. Chem.* 260:16021–16029.

91. Pinder, J. C., Ohanian, V., and Gratzer, W. B. (1984). Spectrin and protein 4.1 as an actin filament capping complex. *FEBS Lett.* 169:161–164.

92. Cohen, C., and Langley, Jr., R. C. (1984). Functional characterization of human erythrocyte spectrin a and b chains: association with actin and erythrocyte protein 4.1. *Biochemistry* 23:4488–4495.

93. Wolfe, L., Lux, S., and Ohanian, V. (1980). Regulation of spectin-actin binding by protein 4.1 and polyphosphates. *J. Cell Biol.* 87:203a.

94. Bennett, V., and Branton, D. (1977). Selective association of spectrin with the cytoplasmic surface of human erythrocyte plasma membranes. *J. Biol. Chem.* 252:2753–2763.

95. Yu, J., and Goodman, S. (1979). Syndein: the spectrin-binding protein(s) of the human erythrocyte membrane. *Proc. Natl. Acad. Sci. USA* 76:2340–2344.

96. Luna, E. J., Kidd, G., and Branton, D. (1979). Identification by peptide analysis of the spectrin-binding proteins in human erythrocytes. *J. Biol. Chem.* 254:2526–2532.

97. Bennett, V. (1978). Purification of an active proteolytic fragment of the membrane attachment site for human erythrocyte spectrin. *J. Biol. Chem.* 253:2292–2299.

98. Weaver, D. C., Pasternack, G. R., and Marchesi, V. T. (1984). The structural basis of ankyrin function II. Identification of two functional domains. *J. Biol. Chem.* 259:6170–6175.

99. Bennett, V., and Stenbuck, P. (1979). The membrane attachment protein for spectrin is associated with band 3 in human erythrocyte membrane. *Nature* 280:468–473.

100. Wallin, R., Culp, E. N., Coleman, B. D., and Goodman, S. R. (1984). A structural model of human erythrocyte band 2.1: Alignment of chemical and functional domains. *Proc. Natl. Acad. Sci. USA* 81:4095–4099.

101. Bennett, V. (1979). Immunoreactive forms of human erythrocyte ankyrin are present in diverse cells and tissues. *Nature* 281:597–599.

102. Korsgren, G., and Cohen, C. M. (1986). Purification and properties of hu-man erythrocyte band 4.2: Association with the cytoplasmic domain of band 3. *J. Biol. Chem.* 261:5536–5543.

103. Tsukita, S., Tsukita, S., Ishikawa, H., Sato, S., and Nakao, M. (1981). Elec-tron microscopic study of reassociation of spectrin and actin with the hu-man erythrocyte membrane. *J. Cell Biol.* 90:70–77.

104. Shohet, S. B. (1979). Reconstitution of spectrin-deficient spherocyte mem-brane. *J. Clin. Invest.* 64:483–494.

105. Beaver, G. H., Jean-Baptiste, L., Ungewickell, E., Baines, A. J., Shah-bakhti, F., Pinder, J. C., Lux, S. E., and Gratzer, W. B. (1985). An examination of the soluble oligomeric complexes extracted from the red cell membrane and their relation to the membrane cytoskeleton. *Eur. J. Cell Biol.* 36:299–306.

106. Shen, B. W., Josephs, R., and Steck, T. L. (1984). Ultrastructure of unit fragments of the skeleton of the human erythrocyte membrane. *J. Cell Biol.* 99:810–821.

107. Matsuzaki, S., Sutoch, F. K., and Ikai, A. (1985). Structural unit of the erythrocyte cytoskeleton. Isolation and electron microscopic examination. *Eur. J. Cell. Biol.* 39:153–160.

108. Tsukita, S., Tsukita, S., and Ishikawa, H. (1980). Cytoskeletal network underlying the human erythrocyte membrane: Thin-section electron microscopy. *J. Cell Biol.* 85:567–576.

109. Nermut, M. V. (1981). Visualization of the "membrane skeleton" in human erythrocytes by freeze-etching. *Eur. J. Cell. Biol.* 25:265–271.

110. Timme, H. H. (1981). The ultrastructure of the cytoskeleton at neutral and reduced pH. *J. Ultrastruct. Res.* 77:199–209.

111. Byers, T. J., and Branton, D. (1985). Visualization of the protein associations in the erythrocyte membrane skeleton. *Proc. Natl. Acad. Sci. USA* 82:6151–6157.

112. Shen, B. W., Josephs, R., and Steck, T. L. (1986). Ultrastructure of the intact skeleton of the human erythrocyte membrane. *J. Cell Biol.* 102:997–1006.

113. Liu, S. C., Derick, L. H., and Palek, J. (1987). Visualization of the hexagonal lattice in the erythrocyte membrane skeleton. *J. Cell Biol.* 104:527–536.

114. Gardner, K., and Bennett, V. (1986). A new erythrocyte membrane-associated protein with calmodulin binding activity. Identification and purification. *J. Biol. Chem.* 261:1339–1348.

115. Ling, E., Gardner, K., and Bennett, V. (1986). Protein kinase C phosphorylates a recently identified membrane skeleton-associated calmodulin-binding protein in human erythrocytes. *J. Biol. Chem.* 261:13875–78.

116. Elgsaeter, A., Stokke, B. T., Mikkelsen, A., and Branton, D. (1986). The molecular basis of erythrocyte shape. *Science* 234:1217–1223.

117. Weinstein, R. S., Tazelaar, H. D., and Loew, J. M. (1986). Red cell comets: Ultrastructure of axial elongation of the membrane skeleton. *Blood Cells* 11:343–357. Commentary by C. M. Cohen, pp. 359–361, response to commentary, pp. 363–366.

118. Lux, S. E. (1979). Dissecting the erythrocyte membrane cytoskeleton. *Nature* 281:426–499.

119. Waugh, S. M., Willardson, B. M., Kannan, R., Lobotka, R. L., and Low, P. S. (1986). Heinz bodies induces clustering of band 3, glycophorin, and ankyrin in sickle cell erythrocytes. *J. Clin. Invest.* 78:1155–1160.

11

The Biochemistry of the Antigens of the Red Blood Cell Membrane

WENDELL F. ROSSE and MARILYN J. TELEN *Duke University Medical Center, Durham, North Carolina*

I. INTRODUCTION

Antigens are molecules that elicit the production of, are recognized by, and react with antibodies. Molecules of all sorts—proteins, polysaccharides, glycoproteins, glycolipids, lipoproteins, lipids, polynucleic acids, etc.—have these characteristics, so that the variety of antigens is almost endless.

Antigens are defined as such by their ability to react with antibodies. This reaction occurs at the antigen-binding pocket of the antibody molecule; this pocket has very restricted dimensions so that only three to five polysaccharide units or five to ten amino acid residues are able to interact within it. Thus, antigens are usually contained within larger molecules.

The antigens of the red blood cell are better known immunologically than those of almost any other cell because of their importance in blood transfusion. More than 600 different antigens on the red blood cell surface have been identified by their reactions with specific alloantibodies. The number of antigenic structures on the surface of the red blood cell is undoubtedly much greater, because of most of the known antigens are defined by alloantibodies (that is, antibodies made by persons lacking the specific antigens). Many surface structures are shared by all humans, and thus do not elicit production of alloantibodies, although autoantibodies or heteroantibodies (antibodies made in other species) may identify some of these structures.

For many years, the antigens of the red blood cell surface were identified solely by their immunological interactions. Specificity was assigned to an antibody by the identification of populations of individuals whose cells possessed the antigen and populations whose cells did not. Thus, an unknown antibody could be tested in these populations to determine if it reacted with the identified antigen. This technique was useful in identifying the occurrence of the antigen and permitted analysis of its genetic transmission and relationship to other antigens, thus leading to the description of families of antigens known as blood groups.

In recent years, there has been an attempt to determine the biochemical nature of the antigens of the red blood cell surface. These studies have led to insights not otherwise available by serological techniques into the origin of the antigens, their relationship to one another, and the biochemical basis of their interactions with antibody and with other antigens. This chapter reviews some of what is known about the biochemical nature of red blood cell antigens and relates this knowledge to that derived from serological studies. Where possible, the importance of antigen-bearing molecules in the structure and function of the cell is described.

II. THE POLYSACCHARIDE ANTIGENS

Many antigens of the red blood cell surface are composed of polysaccharide units. These consist of three or more sugars linked together and attached to either lipid or protein molecules. Only six sugars take part in antigens recognized by blood group alloantibodies [1,2,3]: glucose, galactose, L-fructose (a six-carbon sugar in which one carbon does not have an -OH group), N-acetylglucosamine (N-glu) (glucose with an -N-CO-CH$_3$ group attached on the second carbon), N-acetylgalactosamine (N-gal) (galactose with the same side chain at the second carbon) and N-acetylneuraminic acid or sialic acid (NANA), which has a carboxyl group on carbon 1, an N-acetyl group on carbon 4, and an additional -CHOH-CH$_2$OH group on carbon 6. Three other sugars—L-arabinose, D-mannose, and D-xylose—may be found in polysaccharide structures of the red blood cell surface but are not part of antigens recognized by alloantibodies.

These sugars are linked to one another by enzymes that are often very specific as to substrate (the molecule being linked) and acceptor molecule (the molecule to which it is being linked). These enzymes are genetically determined and their presence or absence determines the phenotype of the cell with respect to the polysaccharide antigens. The specific interaction with an antibody may depend on a single sugar or several sugars. Further, it may depend on the way in which the sugars are attached to one another. The connecting bond may need to be α (above the plane of the sugar) or β (below it). The sugars may need to be "branched"—that is, more than two sugars attached to a single sugar.

The complex polysaccharides are present either on glycoproteins or, most commonly, on glycosphingolipids. These glycolipid molecules are affixed in the membrane so that the lipid-soluble sphingosine portion of the molecule is inserted into the lipid bilayer and the polysaccharide side chain is spread along the

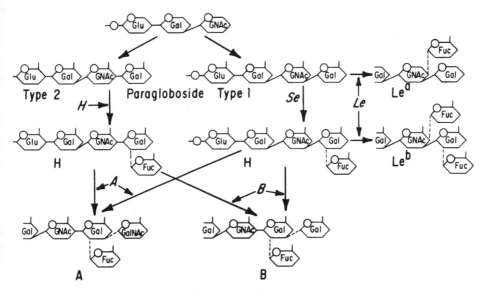

Figure 1 Modifications of the paragloboside molecules to form the H, A, B, Le^a, and Le^b antigens. The following abbreviations are used to denote saccharide residues: Glu - glucose; GNAc - N-acetylglucosamine; Gal - galactose; GalNac - N-acetylgalactosamine; Fuc - L-fucose. α glycosidic linkages between residues are indicated by interrupted lines, β linkages by solid lines. The symbols designating genes are italicized. The antigens are indicated in upright letters.

external surface of the bilayer. Thus, the entire surface of the cell may have a polysaccharide layer coating it. These molecules, like other nonpolymorphic glycosphingolipids, are probably important in cell recognition systems (see below).

A. The ABH Antigens

The fundamental unit of many polysaccharide blood group antigens consists of one of two tetrasaccharide units (called "paragloboside") that differ only in a single bond [2–4] (Figure 1). These are molecules of the so-called "lacto" series of glycosphingolipids. Alloantibodies react equally with antigens derived from either, but hybridoma-derived monoclonal antibodies can distinguish those derived from one chain or the other [5,6]. The red blood cell itself is thought to be able to make only the type 2 chain [7,8]. Other cells in the body make antigens with the type 1 chain; these chains and the antigens that they contain are secreted in saliva in a water-soluble form (a glycoprotein) or may occur in the serum as glycolipids associated with lipoproteins [9,10]. The antigens on these glycolipid molecules may associate passively with the red blood cell [11].

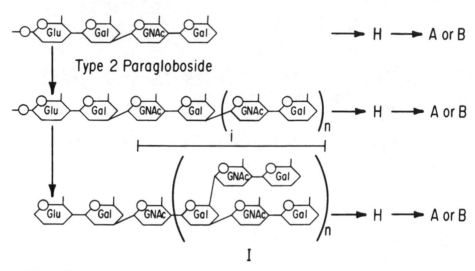

Figure 2   The structure of the extended and branched chains which characterize the I and i antigens. The symbols are the same as used in Figure 1.

Type 2 but not type 1 chains can be elongated by the addition of further
N-acetylglucosamine β1,4-galactose   units either linearly or in a branched form
[12,13] (Figure 2). The branching may be simple or more complex; however,
the terminal sugar is frequently a galactose-residue-linked β1–4 to the subtermin-
al N-acetylglucosamine.

The different forms of the fundamental chain are present in different
amounts on the red blood cells of newborn infants and adults. The red blood
cells of infants bear predominantly the straight chain elongated forms of the
backbone, whereas the red blood cells of adults bear the branched forms. Cold
agglutinin autoantibodies to both the elongated form and the branched form are
frequent; those reacting with the elongated form are called anti-i, those to the
branched form anti-I [14].

B.   The H Antigen

The first modification of the fundamental chain structure that leads to a specifically recognized alloantigen is the addition by a specific fucosyl transferase
of an L-fucose molecule to the terminal galactose residue by an α1–2 linkage
(Figure 1). The antigen that is formed is the H antigen [1–4]. All individuals
have the enzyme producing this antigen except those with the "Bombay" phenotype; such persons have red blood cells that lack the antigen [15] and serum
that does not contain the enzyme specifying this addition [16]. This phenotype

is rare, even among the Marathi-speaking people of Bombay in whom it was first found [17].

It has been suggested that the fucosyl transferase that forms the H antigen is able to act only on type 2 chains. The fucosylation of type 1 chains is carried out by a similar enzyme, which is produced by a gene misnamed the "secretor" or Se gene [18]. Because type 1 chains are produced in a soluble form by cells other than the red blood cell, the antigens that appear on them are present in secretions such as saliva.

The two genes, according to this model, are probably reduplications and are thus located next to one another. Most "Bombay" individuals lack both genes but some may possess the Se gene. Most normal individuals possess both genes but ~22% lack the Se gene; these individuals lack the H antigen (and A and B antigens) in their secretions. Because A and B do not appear in their saliva, they are called "nonsecretors."

C. The A and B Antigens

The A and B antigens are made from the H antigen on both type 1 and type 2 backbones by the addition of an N-gal (for A) or a gal (for B) to the terminal gal [2,4] (Figure 1). The enzymes that accomplish this (an N-acetylgalactosyl transferase for the A antigen [19] and a galactosyl transferase for B [20,21]) are apparently coded for by allelic genes; this suggests that a slight modification in the structure of the enzyme accounts for the specificity for N-acetylgalactose or galactose. Because both types of H substance can be modified by these enzymes, the A and B antigens occur both on the red blood cells and in the secretions of those with the secretor gene and the specific blood group gene. The degree of conversion of H to A or B is determined in part by the amount of the enzyme available for the reaction [22,23]; this may account for some of the variation seen in the strength of expression of these antigens although, in other cases, the variation in strength may be due to abnormal enzymes [24].

The immunodeterminant sugars can be removed by enzymatic action. Conversion of type B cells to type "O" by a specific galactosidase permits them to survive normally in a person with anti-B in the serum [25,26].

The red blood cells of some group A individuals are weakly reactive with antibody to the A antigen, and their serum may contain an antibody reacting with the cells of the more strongly reacting phenotype (A_1) [27]. These cells are said to have the A_2 phenotype. This type arises because the gene locus that codes for the enzymes responsible for the specific A and B antigens also may code for a third variant, an N-acetylgalactosaminyl transferase like that responsible for the A antigen but different in the acceptor molecule to which it is able to affix the sugar [28]. The usual enzyme (A_1) is able to glycosylate "globo-H" and an extended version of H substance as well as the simpler forms, whereas the variant enzyme is only able to glycosylate the simpler forms of H [29–31]. If only the

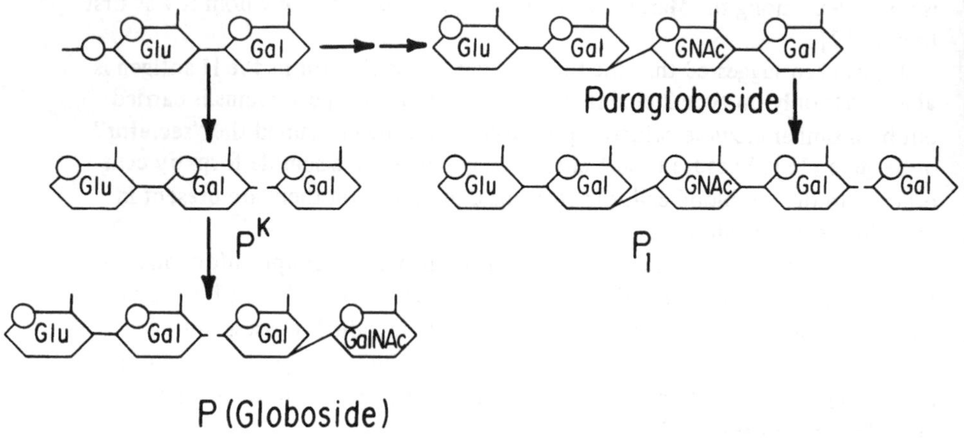

P (Globoside)

Figure 3 The biochemistry of the P antigens. The symbols are the same as used in Figure 1.

variant form is present, the expression of A antigen is weak and the cells have the phenotype A_2.

D. The Lewis Antigens

Type 1 paragloboside chains can be modified by a further enzyme that affixes an L-fucose molecule to the penultimate N-acetylgalactosamine by an $\alpha 1,4$ linkage [1,2]; this modification cannot occur on type 2 chains because the site is occupied by the 1-4 link with the terminal galactose (Figure 1). The enzyme that catalyzes this reaction is coded for by the Lewis gene [32,33]. If only the penultimate N-acetylgalactosamine is fucosylated, the antigen is called Lewis a; if the final galactose is also fucosylated (by the enzyme determined by the Se gene), the antigen is called Lewis b.

E. The P Antigens

The P antigens are also on glycosphingolipids [34,35,36] (Figure 3). Lactosyl ceramide is acted on by an enzyme that adds galactose in an $\alpha 1,4$ linkage; rarely, this enzyme is missing in persons who do not make antigens of the P system and whose red blood cells are therefore said to have the p phenotype [37,38]. The antigen that results from this modification of lactosyl ceramide is called the p^k antigen. Most adults have another enzyme that converts this to the P antigen by the addition of an N-acetylgalactosamine in a $\beta 1-3$ linkage. If this enzyme is lacking, the red blood cells have the rare p^k phenotype [36].

A second, seeming unrelated antigen of the P system is formed by the addition of a galactose in an $\alpha 1-4$ linkage to paragloboside; this antigen is called the

P_1 antigen [35,36]. This antigen appears to be genetically linked to the other P antigens by the fact that the individuals who do not make P^k (and thus cannot make P) also do not make P_1. The explanation for this is not apparent but may be due to sequential (reduplicated) genes that code for enzyme having the same substrate specificity but using different acceptors; if this is true, persons of the p phenotype lack both whereas persons of the P_1-negative phenotype lack only one.

The precise function of these polysaccharide antigens on red blood cells is not clear, but similar structures on other cells are probably involved in cell–cell recognition. Related antigens, called stage-specific embryonic antigens, occur and disappear at certain early stages of embryogenesis [39]. Some of these polysaccharide structures may be related to receptors for growth factors, as monoclonal antibodies to epidermal growth factor reacts with ALe^b and ALe^y structures [40,41]. Other related glycosphingolipids may act as receptors for microbial organisms [42]. These antigens and unusual related antigens may appear on malignant tissues, suggesting that the glycolipid coat is important in normal cellular development and may be altered in carcinogenesis [43,44].

III. THE GLYCOPROTEIN ANTIGENS

Although most proteins exposed on the external surface of the red blood cell bear some oligosaccharide moieties, certain classes of proteins bear a large burden of these; the majority of these glycoproteins belong to a class of molecules called glycophorins. The best characterized of these is glycophorin A (also known as sialoglycoprotein α or MN glycophorin), the glycophorin present in the greatest copy number [45]. This protein consists of three parts [46]: (a) A long external N-terminal segment of 70 amino acids that is highly glycosylated; 15 of the last 50 residues are Ser or Thr, each of which bears a tetrasaccharide with two sialic acid groups; a single Asn at residue 26 bears a complex sugar moiety, (b) A segment of 22 hydrophobic amino acids that are situated within the membrane lipid bilayer, (c) An intracytoplasmic segment of 39 amino acids that are hydrophilic, not glycosylated, and interact with the cytoskeleton [47]. Because of the large number of sialic acid residues, these proteins are responsible for much of the negative charge of the red blood cell membrane surface, as sialic acid is the only acidic component that is ionized at the pH of blood.

A number of antigens reactive with allo- or autoantibodies are present on glycophorin A; some of these antigens are formed by protein alone, some by carbohydrates alone, and some by the combination of both. The blood group antigens M and N are found at the amino terminus. They are due to a difference in the first and fifth amino acids resulting from allelic genes encoding glycophorin A [48] (Table 1). The three intervening amino acids are normally glycosylated [49], and most antibodies to M or N will not react with their antigens

Table 1 Variations in the Amino Terminus of
Glycophorin A Which Result in Specific Antigens

Antigen	Amino acid residues from amino terminus					
	1	2	3	4	5	6
		*	*	*		
M	*Ser*—Ser—Thr—Thr—*Gly*—Val—					
		*	*	*		
N	*Leu*—Ser—Thr—Thr—*Glu*—Val—					
		*	*	*		
M^c	*Ser*—Ser—Thr—Thr—*Glu*—Val—					
		(*)	(*)			
M^g	Leu—Ser—Thr—*Asn*—Glu—Val—					

*: O-linked tetrasaccharide present. (*): O-linked tetrasaccharide altered by presence of Asn residue. Adapted
from Salmon, Cartron and Rouger.

if this glycosylation is altered [50]. Other variations in the first five amino acids
account for several rare variants of the MN antigens (Table 1) [51-53].

The nonpolymorphic portion of the protein is also antigenic, but because
nearly everyone has the protein, antibodies to it are usually autoantibodies [54].
The rare individuals who lack the protein (presumably due to deletion of the
gene coding for it) are said to be En(a-) and are able, if transfused, to make alloantibodies that resemble these autoantibodies [55-57].

En(a-) red blood cells have been studied extensively. Because they are less
sialylated than normal cells [56], they are less negatively charged. This results in
increased agglutination by antibodies usually not capable of bringing about agglutination. Further, these cells are less easily invaded by *Plasmodium falciparum*
than are normal cells [58]. Nevertheless, they seem to survive normally in the
circulation and are not abnormal in shape.

A second species of glycoprotein, glycophorin B (also called sialoglycoprotein δ or Ss glycophorin), closely resembles glycophorin A. The first 26 amino
acids are virtually identical; in glycophorin B, the first five amino acids are in
the form constituting the N antigen regardless of the phenotype on glycophorin
A [49]. Residues 27-58 in glycophorin A are not present in glycophorin B, but
the homology is again evident if this insert is ignored. A cytoplasmic domain of
glycophorin B has not been identified and, if present, is certainly shorter than
that of glycophorin A.

The glycophorin B molecule has a polymorphism at residue 29; the presence of methionine at this point yields the S antigen, whereas a Thr gives rise to the s antigen [59]. Presumably, a change of a single nucleic acid has resulted in allelic genes responsible for these antigens. Additional polymorphisms lead to the expression of other rare antigens by glycophorin B [60]. In some persons, the gene for the entire Ss glycophorin may be deleted; when it is, both S and s are naturally missing. On such cells, a further antigen, called U, is also missing, suggesting that it resides in the nonpolymorphic portion of the molecule [61-63]. A partial deletion of the gene apparently also occurs, resulting in a protein lacking S and s but not U. Such deletions do not alter the physiology of the cell but may interfere with invasion by malarial parasites, because the S-s phenotype is more common in blacks than in whites.

Glycophorins A and B are probably the products of reduplicated genes close to one another. This is suggested by the great degree of homology between the two proteins and the fact that the genes for MN and Ss have been shown by classical genetic studies to be linked; thus, they must occur near one another on the same chromosome. Additional evidence regarding the relationship between glycophorins A and B comes from descriptions of a series of proteins that could only have arisen by unequal crossing over at the gene level; these proteins have an amino terminus like one of the glycophorins and a carboxy terminus like the other [64-66]. Such crossing over only occurs in genes that are near one another and that are homologous in large measure.

The gene encoding glycophorin A has been sequenced and been shown to comprise a single unit without introns [67]. The gene encoding glycophorin B may be derived from a reduplication of the glycophorin A gene from which the codons for amino acids 27-58 have been deleted. Work with erythroleukemia cells has shown that the genes for glycophorins A and B are coordinately regulated and give rise to distinct mRNAs [68].

Two less plentiful and less well-characterized glycoproteins are known as glycophorins C and D (or sialoglycoproteins β and γ, respectively). These proteins are apparently genetically unrelated or only distantly related to glycophorin A but are closely related to one another. Recent evidence suggests that both are coded by the same gene but that the smaller (glycophorin D) may be a truncated version of the larger.

The Gerbich (Ge) antigens are associated with both of these glycoproteins. Gerbich-negative cells are of several phenotypes: (a) The Leach phenotype lacks both β and γ sialoglycoproteins [69] and is common in certain parts of Papua-New Guinea [70]. The cells are elliptocytic [71], a point of great interest because the β and perhaps the γ sialoglycoproteins are anchored to the cytoskeleton by protein 4.1 [72]. The occurrence in the Melanesian population along with some direct evidence [73] suggests that cells of this phenotype probably resist malarial invasion. This phenotype is probably due to the deletion of the gene that codes for both glycophorins C and D, (b) Other persons are Ge-negative but have an abnormal sialoglycoprotein immunologically related to glycophorin C [58,74]. Their red blood cells are not elliptocytic, suggesting that the abnormal molecule is able to interact normally with the cytoskeleton and maintain normal cell shape. This molecule appears to result from deletions within the glycophorin molecule that have left the cytoplasmic portion intact.

The carbohydrate moieties of the glycophorins contribute enormously to the negative charge of the cell and to the complexity of the antigenic structure of the red blood cell surface. All glycophorins have multiple tetrasaccharides composed of an N-acetylgalactosamine linked to a Ser or a Thr residue; a galactose molecule is linked to it, and an N-acetylneuraminic acid molecule is linked to each of these sugars. When sialic acid is removed, an antigen known as the T or Thomsen-Freidenreich antigen is exposed [75]; most adults have "naturally-occurring" antibodies in their serum to this antigen. If the galactose molecule is not added, the Tn antigen results [76,77]; this may occur as the result of dysplasia of the cells in bone marrow diseases [78]. In some instances, the tetrasaccharide is altered and becomes a pentasaccharide, resulting in the Cad antigen [79,80]. This antigen occurs as an inherited characteristic; cells with this antigen are less easily invaded by some species of malaria [81].

A great number of other antigenic variants are associated with the glycophorins, particularly glycophorins A and B. Although they are not all biochemically characterized, many of these probably represent changes in single amino acids or peculiar fusion molecules.

IV. THE LIPOPROTEIN ANTIGENS

Certain proteins appear to be more intimately interactive with the lipid portion of the bilayer than the glycophorins. Examples are known in which the protein molecule crosses the lipid bilayer 12 times, suggesting that a large part of its structure is hydrophobic and thus able to interact with the lipid. The structure of some proteins is so hydrophobic that lipid must be present for the protein to assume its functional form. Such proteins are often involved in the transport of materials across the membrane.

A. The Rh Antigens

The Rh antigens are an extremely complex group composed of >30 variations. Although the structure of the individual Rh antigens is not known, it has been shown that one of the principle antigens (Rh_O or D) resides on a protein that depends on its interaction with the membrane for maintenance of conformation, and thus, antigenicity. Green first suggested that lipid was necessary for the expression of Rh antigens [82]. It is now clear that the presence of lipid serves to influence the orientation of the Rh protein, thus increasing the antigenicity of the molecule [83]. Numerous investigators have now shown the D antigen to be expressed by a membrane protein of ~32,000 daltons [84-86]. This protein is strongly associated with the membrane cytoskeleton [87,88]. It is unclear whether other Rh antigens, including CcEe, are carried by the same protein or by one or more similar but distinct molecules [89].

The protein carrying the D antigen has been partially characterized; it appears to lack carbohydrate and to require the presence of a free sulfhydryl group for antigen activity [84]. A form of the protein is present in both D(+) and D(-) cells [90]. In D(-) cells, it is not labeled by external surface labels, suggesting that in these cells it does not extend out of the external leaflet of the lipid membrane.

The Rh antigens are serologically complex and have led to the use of several different systems of nomenclature. This discussion will use the CDE nomenclature of Race and Fisher without necessarily implying agreement with all its genetic implications. There appear to be three antigenic loci that are regularly allomorphic—one expressing E or e, one expressing C or c, and one expressing D or nothing (at least no antigen has been identified serologically). These may be considered theoretically to be different parts of a molecule coded for by a single complex gene.

The sequences responsible for the expression of the E or e antigens are most likely near the external terminus. Those expressing the C or c antigens are apparently nearby whereas those expressing the D antigen may be at some distance and perhaps even on a different protein. This sequence is hypothesized because of the following facts: (a) The antigens of the C and E locus may form "compound antigens" so that an antibody appears to react with portions of both antigens (e.g., with c and e, the so-called anti-ce or anti-f) [91-94]. Compound antigens are not found between the D locus and either of the other two, (b) Expression at one or another locus may be missing; e.g., neither E nor e may be expressed. This may be due to a deletion of a portion of the protein. These deletions are progressive; the E locus alone may be deleted (giving the phenotype DC-) [95]. Both the C and the E locus may be deleted, giving the phenotype D- - [96]. However, the C locus is never deleted when the E locus is expressed.

It is not at all certain that there is an allelic expression at the D locus because, when D (the original Rh antigen) is not expressed, no antibody has been found that identifies an allelic antigen. It has been suggested that in Rh(D)-negative cells the D antigen is present but is not able to interact with the antibody; perhaps it is submerged in the membrane by a hydrophobic substitution nearby.

The D (Rh$_0$) antigen itself appears to be complex. Parts of it may be missing, resulting in diminished expression of D; this is the D-mosaic form of the so-called D^u (weak D) phenotype [97,98]. In other cases, the D^u phenotype arises in the setting of a gene encoding C on the paired chromosome, giving rise to a so-called *trans* effect of weakened D expression [99].

The other Rh loci may also have variant expressions that are detected by antibodies. Variations in the E locus, particularly of the e antigen, are more common among blacks [100]. Others, such as C^w, appear to be minor changes in the C antigen [101]. These may represent amino acid changes in these regions.

The remainder of the Rh protein (or proteins), which is relatively nonpolymorphic, is also antigenic. Usually the antibodies are auto- or heteroimmune. Many autoimmune antibodies appear to react with this portion of the molecule, as evidenced by the fact that they do not react with red blood cells that lack Rh antigens [102].

The phenotype lacking expression of the Rh antigens, called "Rh_null," may arise in two ways: (a) It may appear as a simple deletion of the protein, in which case both parents also appear to lack the antigens associated with that gene product; this phenotype is consistent with a deletion of the gene coding for the protein on which the Rh antigens reside [103]. (b) It may appear to be due to a gene that is not part of the gene for the Rh protein itself and that segregates independently of Rh antigens [104]. In this case, the parents appear to have normal Rh phenotypes (both Rh genes appear to be intact) but each is heterozygous for the responsible "suppressor" or "regulator" gene. The Rh_null individual has the genes for the Rh antigens, as evidenced by his ability to pass them to his offspring, but Rh antigens are not expressed on his cells. In fact, all Rh-associated antigens, and thus the Rh protein, appear to be missing. The explanation for this phenotype is not clear. Perhaps some posttranslational alteration of the Rh protein or perhaps another protein is needed to stabilize the Rh protein in the membrane.

Lack of the Rh protein(s), but not of the D antigen alone, has a deleterious effect on the organization of the red blood cell membrane. Cells lacking all Rh antigens are stomatocytic [105], leaky to cations, and have an altered phospholipid distribution. Further, they do not survive normally so that these persons have a compensated hemolytic anemia, suggesting that the Rh protein is essential to the organization of the membrane.

The Rh protein is bound to the membrane cytoskeleton [87,88]. This attachment presumably prevents movement of Rh protein molecules laterally in the plane of the membrane, even when reacted with antibody, thus preventing Rh antibodies from fixing complement. Even when all antigen sites are occupied, the antibody molecules are too far away to permit the formation of the doublet of complement-fixing sites necessary to fix Clq.

B. The LW Antigens

The original experiments identifying the Rh groups on the red blood cells of the rhesus monkey were wrong; these cells do not bear the Rh antigens. The antigen that was detected is the LW antigen [106], which is related to the Rh antigens by the following facts: (a) The expression of the LW antigens (LW^a and LW^b) is greater on D (Rh$_0$) positive cells than on D negative cells [61]; this accounts for the error in equating the antigen on the rhesus cells with the D antigen, (b) Red cells of the Rh$_{null}$ phenotype always lack the LW antigens [107]. On the other hand, cells may lack the LW antigen but not the Rh antigens.

The LW antigens are found on a protein of ~40 kDa [108,109]. Like the Rh protein, the protein bearing LW antigens is attached to the cytoskeleton and may contribute to the integrity of the cell [105]. Unlike the Rh protein, this protein is N-glycosylated and removal of some of the sugars diminishes its reactivity with antibody [105].

The exact relationship of the LW protein to the Rh protein is not clear. There appear to be approximately one-tenth the number of copies of the LW protein as of the Rh protein [105].

V. OTHER PROTEIN ANTIGENS

A number of antigens appear to be present on other membrane proteins that probably do not bear as close a relationship to membrane lipid as the Rh antigens. Many of the antigens have been grouped into "families" that can be shown by population studies to be related or linked; in most cases, the linkage is probably due to the fact that all the antigens of a group are present on the same protein. In several cases, there appear to be a series of sites with allelic expression that may correspond to alternative amino acid residues in a polypeptide.

A. The Kell Antigens

The Kell antigens are a group of related antigens that frequently cause alloimmunization. These antigens occur on a protein of 93 kDa [110–113]. They consist of a series of allelic antigens whose arrangement on the Kell protein is not known (Table 2). These allelic antigens are presumably the result of differences in a single or a few amino acids, possibly also resulting in differences in glycosylation.

Table 2 Known Allelic Antigens of the Kell System

Infrequent antigen	Frequent antigen
Kell: K, K1	Cellano: k, K2
Penney: Kp^a, K3	Rautemberg: Kp^b, K4
Levay: Kp^c, K21	
Sutter: Js^a, K6	Matthews: Js^b, K7
Weeks: Wk^a, K17	Cote: Cote, K11

Table 3 The Allelic Antigens of
the Lutheran System

Common	Rare
Lu^b (Lu2)	Lu^a (Lu1)
Lu6	Lu9
Lu8	Lu14
Lu5	Lu10

The protein on which the Kell antigens are located appears to be glycosylated [109,114] and to require the integrity of internal disulfide bonds, because treatment of the red blood cell with agents able to disrupt these bonds causes loss of antigenic reactivity [115]. The red blood cells of rare individuals lack all Kell antigens, probably because of deletion of the gene encoding the Kell protein [116]; this phenotype is called K_O. These cells are not noticeably abnormal in structure or function.

Normal expression of Kell antigens also requires the function of a gene on the X chromosome, which is presumed to encode an antigen called K_X [117]; when the K_X antigen is missing, the Kell antigens are weakly expressed. However, K_O cells express the K_X antigen. Absence of the protein bearing the K_X antigen results in the McLeod phenotype [118], in which the red blood cells are acanthocytic (actually echinocytic) in form and have an alteration in the exchange of phosphatidylcholine across the membrane [119,120]. They leak cations and have a shortened red blood cell survival. This suggests that the protein containing K_X but not the protein bearing the Kell antigens is important to the integrity of the cell.

The McLeod syndrome is associated in some cases with chronic granulomatous disease [121]. It is assumed that this is due to a deficiency of the K_X protein on neutrophils, which normally express it. On the other hand, the association may be at the genetic level, because the genes for CGD and Duchenne's dystrophy are located near the gene for the erythrocyte K_X protein.

B. The Lutheran Antigens

The Lutheran antigens are similar in many ways to the Kell antigens. They also comprise a large series of antigens, of which three pairs have been shown to be allelic. In each of these pairs, one of the alleles is the "wild type" and is common, and the other is rare [120] (Table 3). A number of other common ("high frequency") and rare ("low frequency") antigens have been shown to be part of or related to the Lutheran system, although their allelic antigens have not been

identified. The protein bearing the Lutheran antigens has not been fully characterized. Immunoblotting with a murine monoclonal anti-Lu[b] has detected two
proteins of 85 kDa and 78 kDa [122].

Like the Kell groups, the gene for Lutheran antigens may also be deleted.
When this occurs, no apparent abnormality of the red blood cells occurs; this is
the recessively inherited Lutheran negative phenotype (Lu$_0$) [123].

A distinct gene, called *In(Lu)*, which segregates separately from the gene encoding the Lutheran antigens, influences the expression of the Lutheran genes
[124,125]. The presence of a single copy of the gene results in a markedly
weakened expression of the Lutheran antigens, giving rise to the so-called dominant or *In(Lu)* Lu(a-b-) phenotype. The presence of the *In(Lu)* gene is invariably associated with markedly decreased expression of a membrane glycoprotein
of 80 kDa [126-128]; this protein is present in normal amounts on the cells of
the Lu$_0$ phenotype [129]. Cells from people bearing the *In(Lu)* gene also have
diminished expression of the P$_1$, i, and Auberger antigens [124] as well as of
numerous other high incidence antigens [130,131]. Because the P$_1$ and i antigens are of known oligosaccharide structures, this effect has led some researchers
to hypothesize that the *In(Lu)* locus encodes a transferase that affects carbohydrate antigens and that Lutheran antigen determinants are, at least in part,
dependent on such carbohydrates [132]. The protein(s) bearing Lu[b] and the
In(Lu)-related 80-kDa protein have both been shown to require internal disulfide bonds for antigenic activity [121,126]; in addition, the *In(Lu)*-related protein, as well as the protein identified by monoclonal anti-Lu$_b$ contain N-
linked oligosaccharide [121,126]. The relationship of the *In(Lu)*-related 80-
kDa protein to the Lutheran protein or to the other antigens affected by the
In(Lu) gene is not clear.

C. The Duffy Antigens

The Duffy antigens reside on a molecule with a mobility of 35–43 kDa that
has one and perhaps more *N*-linked polysaccharide complexes on it [84,133,
134]. There are three commonly expressed antigens: Fy[a] and Fy[b] (which are
allelic) and Fy[3], which is present when either Fy[a] or Fy[b] is present; Fy[3] appears to be a nonpolymorphic antigen on the portion of the molecule nearer
the membrane surface, because Fy[a] or Fy[b] are removed by chymotrypsin treatment [131] but Fy[3] is not. The red blood cells of most nonblack individuals
possess Fy[a], Fy[b], or both, whereas many black individuals lack all three Duffy
antigens and apparently lack the protein [135]. This has led to the theory that
the Fy(a-b-) phenotype protects against malaria, an effect that has been demonstrated for several species of *Plasmodium*. Experiments with the primate parasite, *P. knowlesi*, showed that this was the case [136]. This parasite is able to in-
vade red blood cells bearing the Fy[a], Fy[b], or Fy[3] antigen but not those lacking
the protein bearing these three antigens. The merozoites are able to attach to

the red blood cells lacking the Duffy protein but are not able to form the critical junction with the cell that permits invasion. *P. vivax* appears to resemble *P. knowlesi* in this respect because individuals with red blood cells of the phenotype Fy(a-b-) do not readily become infected with this parasite [137].

Nothing is known of the function of the protein bearing the Duffy antigens. There is no major change in the function of the red blood cells when it is missing. Clearly, its function is less important than the danger of its presence when vivax malaria is a threat.

VI. CONCLUSIONS

There are many more antigens of the red blood cell surface. Some have been identified to reside on proteins whose functions are understood, whereas many remain of unknown biochemical structure. However, the analysis of the antigens in relationship to one another and in relationship to the structures of which they are part will give further insight into genetic, functional, and biochemical relationships of surface structures of the red blood cell.

ACKNOWLEDGMENT

Supported by Grants DK31379 (W.F.R.) and HL 33572 (M.J.T.) from the National Institutes of Health and CH-371 (MJT) from the American Cancer Society.

REFERENCES

1. Marcus, D. M. (1969). The ABO and Lewis blood-group system. Immunochemistry, genetics and relation to human disease. *New Engl. J. Med.* 280: 994.
2. Watkins, W. M. (1966). Blood group substances in the ABO system, the genes control the arrangement of sugar residues that determine blood group specificity. *Science* 152:172.
3. Hakomori, S.-I. (1977). Blood group ABH and Ii antigens of human erythrocytes: Chemistry, polymorphism, and their developmental change. *Semin. Hematol.* 18:39.
4. Watkins, W. M. (1977). The glycosyltransferase products of the *A, B, H,* and *Le* genes and their relationship to the structure of the blood group antigens. In *Human Blood Groups,* J. F. Mohn, R. W. Plunkett, R. K. Cunningham, R. M. Lambert (Eds.). S. Karger, Basel, p. 135.
5. Rosenblum, B. B., Judd, W. J., and Carey, T. E. (1986). Biochemical characterization of a monoclonal antibody to the H type-2 antigen: comparison to other ABH antibodies. *Hybridoma* 5:117.
6. Abe, K., Levery, S. B., and Hakomori, S. (1984). The antibody specific to type 1 chain blood group A determinant. *J. Immunol.* 132:1951.

7. Koscielak, J., Piasek, A., Gorniak, H., Gardas, A., and Gregor, A. (1973). Structures of fucose-containing glycolipids with H and B blood group activity and of sialic acid and glucosamine-containing glycolipid of human erythrocyte membrane. *Eur. J. Biochem.* 37:214.

8. Stellner, K., Watanabe, K., and Hakomori, S. (1973). Isolation and characterization of glycosphingolipids with human blood group H specificity from membranes of human erythrocytes. *Biochemistry* 12:656.

9. Marcus, D. M., and Cass, L. E. (1969). Glycosphingolipids with Lewis blood group activity: uptake by human erythrocytes. *Science* 164:553.

10. Tilley, C. A., Crookston, M. C., Brown, B. L., and Wherrett, J. R. (1975). A and B and $A_1 Le^b$ substances in glycosphingolipid fractions of human serum. *Vox Sang.* 28:25.

11. Garretta, M., Muller, A., Courouce-Pauty, A. M., and Moullec, J. (1974). *In vitro* transformation of group O red blood cells by A and B serum substances. *Biomedicine* 21:114.

12. Hakomori, S., Stellner, K., and Watanabe, K. (1972). Four antigenic variants of blood group A glycolipids: examples of highly complex, branched chain glycolipids of animal cell membrane. *Biochem. Biophys. Res. Commun.* 49:1061.

13. Kannagi, R., Levery, S. B., and Hakomori, S. (1985). Le^a-active heptaglycosylceramide, a hybrid of type 1 and type 2 chain, and the pattern of glycolipids with Le^a, Le^b, X (Le^x), Y (Le^y) determinants in human blood cell membranes (ghosts). Evidence that type 2 chain can elongate repetitively but type 1 chain cannot. *J. Biol. Chem.* 260:6410.

14. Watanabe, K., Hakomori, S. I., Childs, R. A., and Feizi, T. (1979). Characterization of a blood group I active ganglioside. Structural requirements for the I and i specificities. *J. Biol. Chem.* 254:3221.

15. Bhende, Y. M., Deshpande, E. K., Bhatia, H. L. M., Sanger, R., Race, R. R., Morgan, W. T. J., and Watkins, W. M. (1952). A "new" blood-group character related to the ABO system. *Lancet* 1:903.

16. Schenkel-Brunner, H., Chester, M. A., and Watkins, W. M. (1972). (Alpha)-L-Fucosyltransferases in human serum from donors of different ABO, secretor, and Lewis blood group phenotypes. *Eur. J. Biochem.* 30:269.

17. Bhatia, H. M. (1977). Serologic reactions of ABO and O$_h$ (Bombay) phenotypes due to variations in H antigens. In *Human Blood Groups*, J. F. Mohn, R. W. Plunkett, R. K. Cunningham, R. M. Lambert (Eds.). S. Karger, Basel p. 296.

18. Oriol, R., Danliovs, J., and Hawkins, B. R. (1981). A new genetic model proposing that the *Se* gene is a structural gene closely linked to the *H* gene. *Am. J. Hum. Genet.* 33:421.

19. Kobata, A., Grollman, E. F., and Ginsburg, V. (1968). An enzymatic basis for blood type A in humans. *Arch. Biochem. Biophys.* 124:609.

20. Kobata, A., Grollman, E. F., and Ginsburg, V. (1968). An enzymatic basis for blood type B in humans. *Biochem. Biophys. Res. Commun.* 32:272.

21. Race, C., Ziderman, D., and Watkins, W. M. (1968). An α-D-galactosyltransferase associated with the blood group B character. *Biochem. J.* 107:733.

22. Frederick, J., Hunter, J., Greenwell, P., Winger, K., and Gottschall, J. L. (1985). The $A_1 B$ genotype expressed as $A_2 B$ on the red cells of individuals with strong B gene-specific transferases. Results from two paternity cases. *Transfusion* 25:30.

23. Salmon, C., Cartron, J. P., Lopez, M., Rahuel, C., Badet, J., and Janot, C. (1984). Level of the A, B, and H blood group glycosyltransferases in red cell membranes from patients with malignant hemopathies. *Rev. Fr. Transfus. Immunohematol.* 27:625.

24. Yoshida, A., Yamato, K., Dave, V., Yamaguchi, H., and Okubo, Y. (1982). A case of weak blood group B expression (B_m) associated with abnormal blood group galactosyltransferase. *Blood* 59:323.

25. Goldstein, J., Siviglia, G., Hurst, R., Lenny, L., and Reich, L. (1982). Group B erythrocytes enzymatically converted to group O survive normally in A, B and O individuals. *Science* 215:168.

26. Dybus, S., and Aminoff, D. (1983). Action of alpha-galactosidase from Clostridium sporogenes and coffee beans on blood group B antigen of erythrocytes. The effect on the viability of erythrocytes in circulation. *Transfusion* 23:244.

27. Mohn, J. F., Cunningham, R. K., and Bates, J. F. (1977). Qualitative distinctions between subgroups A_1 and A_2. In *Human Blood Groups*, J. F. Mohn, R. W. Plunkett, R. K. Cunningham, R. M. Lambert (Eds.). S. Karger, Basel, p. 316.

28. Schachter, H., Michaels, M. A., Tilley, C. A., Crookston, M. C., and Crookston, J. H. (1973). Qualitative differences in the *N*-acetyl-D-galactosaminyl-transferases produced by human A^1 and A^2 genes. *Proc. Natl. Acad. Sci. USA* 70:220.

29. Clausen, H., Watanabe, K., Kannage, R., Levery, S. B., Nudelman, E., Arao-Tomono, Y., and Hakomori, S. (1984). Blood group A glycolipid (Ax) with globo-series structure which is specific for blood group A_1 erythrocytes: one of the chemical bases for A_1 and A_2 distinction. *Biochem. Biophys. Res. Comm.* 124:523.

30. Clausen, H., Levery, S. B., Kannagi, R., and Hakomori, S. (1986). Novel blood group H glycolipid antigens exclusively expressed in blood group A and AB erythrocytes (Type 3 Chain H). I. Isolation and chemical characterization. *J. Biol. Chem.* 261:1380.

31. Clausen, H., Holmes, E., and Hakomori, S. (1986). Novel blood group H glycolipid antigens exclusively expressed in blood group A and AB erythrocytes (Type 3 Chain H). II. Differential conversion of different H substrates by A_1 and A_2 enzymes, a type 3 chain H expression in relation to secretor status. *J. Biol. Chem.* 261:1388.

32. Chester, M. A., and Watkins, W. M. (1969). α-L-Fucosyltransferases in human submaxillary gland and stomach tissues associated with the H, Le[a], and Le[b] blood-group characters and ABH secretor status. *Biochem. Biophys. Res. Commun.* 34:835.

33. Grollman, E. F., Kobata, A., and Ginsburg, V. (1969). An enzymatic basis for Lewis blood types in man. *J. Clin. Invest.* 48:1489.

34. Naiki, M., and Marcus, D. M. (1974). Human erythrocyte P and P^k blood group antigens: identification as glycophingolipids. *Biochem. Biophys. Res. Commun.* 60:1105.
35. Naiki, M., Fong, J., Ledeen, R., and Marcus, D. M. (1975). Structure of the human erythrocyte blood group P_1 glycosphingolipid. *Biochemistry* 14: 4831.
36. Marcus, D. M., Kundu, S. K., and Suzuki, A. (1981). The P blood group system: recent progress in immunochemistry and genetics. *Semin. Hematol.* 18:63.
37. Kijimoto-Ochiai, S., Naiki, M., and Makita, A. (1977). Defects of glycosyltransferase activities in human fibroblasts of P^k and p blood group phenotypes. *Proc. Natl. Acad. Sci.* 74:5407.
38. Kannagi, R., Nudelman, E., Levery, S. B., and Hakomori, S. (1982). A series of human erythrocyte glycosphingolipids reacting to the monoclonal antibody directed to a developmentally regulated antigen SSEA-1. *J. Biol. Chem.* 257:14865.
39. Hakomori, S. I., Nudelman, E., Levery, S., Solter, D., and Knowles, B. B. (1981). The hapten structure of a developmentally regulated glycolipid antigen (SSEA-1) isolated from human erythrocytes and adenocarcinoma; A preliminary note. *Biochem. Biophys. Res. Commun.* 100:1578.
40. Fredman, P., Richert, N. D., Magnani, J. L., Willingham, M. C., Pastan, I., and Ginsburg, V. (1983). A monoclonal antibody that precipitates the glycoprotein receptor for epidermal growth factor is directed against the human blood group H type 1 antigen. *J. Biol. Chem.* 258:11206.
41. Gooi, H. C., Picard, J. K., Hounsell, E. F., Gregoriou, M., Rees, A. R., and Feizi, T. (1985). Monoclonal antibody (EGR/C49) reactive with the epidermal growth factor receptor of A431 cells recognized the blood group ALe^b and ALe^y structures. *Mol. Immunol.* 22:689.
42. Marcus, D. M. (1984). A review of the immunogenic and immuno-modulatory properties of glycosphingolipids. *Mol. Immunol.* 21:1083.
43. Hakomori, S. (1984). Blood group glycolipid antigens and their modifications as human cancer antigens. *Am. J. Clin. Pathol.* 82:635.
44. Hakomori, S. (1984). Tumor-associated carbohydrate antigens. *Ann. Rev. Immunol.* 2:103.
45. Anstee, M. N. (1981). The blood group MNSs-active sialoglycoproteins. *Semin. Hematol.* 18:1.
46. Tomita, M., and Marchesi, V. T. (1975). Amino acid sequence and oligosaccharide attachment sites of human erythrocyte glycophorin. *Proc. Natl. Acad. Sci. USA* 72:2964.
47. Chasis, J. A., Mohandas, N., and Shohet, S. B. (1985). Erythrocyte membrane rigidity induced by glycophorin A-ligand interaction: evidence for a ligand-induced association between glycophorin A and skeletal proteins. *J. Clin. Invest.* 75:1919.
48. Furthmayr, H. (1978). Structural comparison of glycophorins and immunochemical analysis of genetic variants. *Nature* 271:519.

49. Blumenfeld, OO, and Adamy, A. M. (1978). Structural polymorphism within the amino-terminal region of MM, MN, NN glycoproteins (glycophorins) of the human erythrocyte membrane. *Proc. Natl. Acad. Sci. USA* 75:2727.

50. Springer, G. F., and Ansell, N. J. (1958). Inactivation of human erythrocyte agglutinogens M and N by influenza viruses and receptor-destroying enzyme. *Proc. Natl. Acad. Sci. USA* 44:182.

51. Dahr, W., Beyreuther, K., Gallasch, E., Kruger, J., and Morel, P. (1981). Amino acid sequence of the blood group Mg-specific major human erythrocyte membrane sialoglycoprotein. *Hoppe Seylers Z. Physiol. Chem.* 362:81.

52. Furthmayr, H., Metaxas, M. N., and Metaxas-Buhler, M. (1981). Mg and Mc: mutations within the amino-terminal region of glycophorin A. *Proc. Natl. Acad. Sci. USA* 78:631.

53. Dahr, W., Kordowicz, M., Beyreuther, K., and Kruger, J. (1981). The amino acid sequence of the Mc-specific major red cell membrane sialoglycoprotein —an intermediate of the blood group M- and N-active molecules. *Hoppe Seylers Z. Physiol. Chem.* 362:363.

54. Bell, C. A., and Zwicjer, H. (1978). Further studies on the relationship of anti-En(a) and anti-Wr(b) in warm auto-immune haemolytic anemia. *Transfusion* 18:572.

55. Darnborough, J., Dunsford, I., and Wallace, J. (1969). The En[a] antigen and antibody. A genetical modification of human red cells affecting their blood grouping reactions. *Vox Sang.* 17:241.

56. Furuhjelm, U., Myllyla, G., Nevanlinna, H. R., Nordling, S., Pirkola, A., Gavin, J., Gooch, A., Sanger, R., and Tippett, P. (1969). The red cell phenotype En(a–) and anti-En[a]: serological and physicochemical aspects. *Vox Sang.* 17:256.

57. Furuhjelm, U., Nevanlinna, H. R., and Pirkola, A. (1973). A second Finnish En(a–) propositus with anti-En[a]. *Vox Sang.* 24:545.

58. Pasvol, G., Wainscoat, S. J., and Weatherall, D. J. (1982). Erythrocytes deficient in glycophorin resist invasion by the malarial parasite, *Plasmodium falciparum*. *Nature* 297:64.

59. Dahr, W., Beyreuther, K., Steinbach, H., Gielen, W., and Kruger, J. (1980). Structure of the Ss blood group antigens, II: A methionine/threonine polymorphism within the N-terminal sequence of the Ss glycoprotein. *Hoppe Seylers Z. Physiol. Chem.* 361:895.

60. Dahr, W., Kordowicz, M., Judd, W. J., Moulds, J., Beyreuther, K., and Kruger, J. (1984). Structural analysis of the Ss sialoglycoprotein specific for Henshaw blood group from human erythrocyte membranes. *Eur. J. Biochem.* 141:51.

61. Dahr, W., Uhlenbruck, G., Issitt, P. D., et al. (1975). SDS-polyacrylamide gel electrophoretic analysis of the membrane glycoproteins from S-s-U-erythrocytes. *J. Immunogenet.* 2:249.

62. Dahr, W., Issitt, P. D., and Uhlenbruck, G. (1977). New concepts in the MNSs blood group system. In: *Human Blood Groups, Fifth International Convocation on Immunology*. Basel, Karger, pp. 197–205.

63. Tanner, M. J. A., Anstee, D. J., and Judson, P. A. (1977). A carbohydrate deficient membrane glycoprotein in human erythrocytes of phenotype S-s-. *Biochem. J.* 165:157.
64. Tanner, M. J. A., Anstee, D. J., and Mawby, W. J. (1980). A new human erythrocyte variant (Ph) containing an abnormal membrane sialoglycoprotein. *Biochem. J.* 187:493.
65. Mawby, W. J., Anstee, D. J., and Tanner, M. J. A. (1981). Immunochemical evidence for hybrid sialoglycoproteins of human erythrocytes. *Nature* 291:161.
66. Anstee, D. J., Mawby, W. J., Parsons, S. F., Tanner, M. J. A., and Giles, C. M. (1982). A novel hybrid sialoglycoprotein in St(a) positive human erythrocytes. *J. Immunogenet.* 9:51.
67. Siebert, P. D., and Fukuda, M. (1986). Isolation and characterization of human glycophorin A cDNA clones by a synthetic oligonucleotide approach: Nucleotide sequence and mRNA structure. *Proc. Natl. Acad. Sci. USA* 83:1665.
68. Siebert, P. D., and Fukuda, M. (1986). Human glycophorin A and B are encoded by separate, single copy genes coordinately regulated by a tumor-promoting phorbol ester. *J. Biol. Chem.* 261:12433.
69. Anstee, D. J., Parsons, S. F., Ridgwell, K., et al. (1984). Two individuals with elliptocytic red cells apparently lack three minor erythrocyte membrane sialoglycoproteins. *Biochem. J.* 218:615.
70. Booth, P. B., and McLoughlin, L. (1972). The Gerbich blood group system, especially in Melanesians. *Vox Sang.* 22:73.
71. Daniels, G. L., Shaw, M.-A., Judson, P. A., et al. (1986). A family demonstrating inheritance of the Leach phenotype: A Gerbich negative phenotype associated with elliptocytosis. *Vox Sang.* 50/117.
72. Mueller, T., and Morrison, M. (1981). Glycoconnectin (PAS 2); a membrane attachment site for the human erythrocyte cytoskeleton. In: Erythrocyte membranes II: *Recent Clinical and Experimental Advances.* New York, Alan R. Liss, pp. 95–112.
73. Pasvol, G., Anstee, D. J., and Tanner, M. J. A. (1984). Glycophorin C and the invasion of red cells by *Plasmodium falciparum* malaria [Letter]. *Lancet* 1:907.
74. Anstee, D. J., Ridgwell, K., Tanner, M. J. A., et al. (1984). Individuals lacking the Gerbich blood group antigen have alterations in the human erythrocyte membrane sialoglycoproteins beta and gamma. *Biochem. J.* 221:97.
75. Springer, G. F., Desai, P. R. (1974). Common precursors of human blood group MN specificities. *Biochem. Biophys. Res. Commun.* 61:470.
76. Dahr, W., Uhlenbruck, G., Gunson, H. H., et al. (1975). Molecular basis of Tn-polyagglutinability. *Vox Sang.* 29:36.
77. Cartron, J.-P., Andreu, G., Cartron, J., et al. (1978). Demonstration of T-transferase deficiency in Tn-polyagglutinable blood samples. *Eur. J. Biochem.* 92:111.
78. Bird, G. W. G., Wingham, J., Pippard, M. J., et al. (1976). Erythrocyte membrane modifications in malignant diseases of myeloid and lymphoretic-

ular tissue. I. Tn-polyagglutination in acute myelocytic leukemia. *Br. J. Haematol.* 33:289.

79. Cartron, J.-P., Blanchard, D. (1982). Association of human erythrocyte membrane glycoproteins with blood group Cad specificity. *Biochem. J.* 207:497.

80. Blanchard, D., Cartron, J.-P., Fournet, B., Montreuil, J., Van Halbeek, H., and Vlieganthart, J. F. G. (1983). Primary structure of the oligosaccharide determinant of blood group Cad specificity. *J. Biol. Chem.* 258:7691.

81. Cartron, J.-P., Prou, O., Leilier, M., and Soulier, J. P. (1983). Susceptibility to invasion of some human erythrocytes carrying rare blood group antigens. *Br. J. Haematol.* 55:639.

82. Green, F. A. (1965). Studies on the Rh(D) antigen. *Vox Sang.* 10:32.

83. Plapp, F. V., Kowalski, M. M., Evans, J. P., Tilzer, L. L., and Chiga, M. (1980). The role of membrane phospholipids in expression of erythrocyte $Rh_O(D)$ antigen activity. *Proc. Soc. Exp. Biol. Med.* 164:561.

84. Moore, S., Woodrow, C. F., and McClelland, D. B. L. (1982). Isolation of membrane components associated with human red cell antigens Rh(D), (C), (E) and Fy(a). *Nature* 285:529.

85. Gahmberg, C. G. (1982). Molecular identification of the human $Rh_O(D)$ antigen. *FEBS Lett.* 140:93.

86. Ridgwell, K., Roberts, S. J., Tanner, M. J. A., and Anstee, D. J. (1983). Absence of two membrane proteins containing extracellular thiol groups in Rh_{null} human erythrocytes. *Biochem. J.* 213:267.

87. Gahmberg, C. G., Karhi, K. K. (1984). Association of $Rh_O(D)$ with the membrane skeleton in $Rh_O(D)$-positive human red cells. *J. Immunol.* 133: 334.

88. Paradis, G., Bazin, R., and Lemieux, R. (1986). Protective effect of the membrane skeleton on the immunologic reactivity of the human red cell Rho(D) antigen. *J. Immunol.* 137:240.

89. Sinor, L. T., Brown, P. J., Evans, J. P., and Plapp, F. V.(1984). The Rh antigen specificity of erythrocyte proteolipid. *Transfusion* 24:179.

90. Agre, P. Personal communication.

91. Rosenfield, R. E., Vogel, P., Gibbel, N., Sanger, R., and Race, R. R. (1953). A "new" Rh antibody, anti-f. *Br. Med. J.* 1:975.

92. Rosenfield, R. E., and Haber, G. V. (1958). An Rh blood factor, rh^i (Ce), and its relationship to hr (ce). *Am. J. Hum. Genet.* 10:474.

93. Dunsford, I. (1961). A new Rh antibody—anti-ce. *Proceedings of 8th Congress of European Society of Haematology*, Paper no. 491.

94. Gold, E. R., Gillespie, E. M., and Tovey, G. H. (1961). A serum containing 8 antibodies. *Vox Sang.* 6:157.

95. Tate, H., Cunningham, C., McDade, M. G., Tippett, P., and Sanger, R. (1960). An Rh gene complex Dc-. *Vox Sang.* 5:398.

96. Race, R. R., Sanger, R., and Selwyn, J. G. (1950). A probable deletion in a human Rh chromosome. *Nature* 166:520.

97. Unger, L. J., Wiener, A. S., and Wiener, L. (1959). New antibody (anti-Rh^B) resulting from blood transfusion in an Rh-positive patient. *J. Am. Med. Assoc.* 170:1380.

98. Unger, L. J., and Wiener, A. S. (1959). A "new" antibody, anti-Rh^C, resulting from isosensitization by pregnancy, with special reference to the heredity of a new Rh-Hr agglutinogen Rh$_2$ ^{C}. *J. Lab. Clin. Med.* 54:835.

99. Ceppellini, R., Dunn, L. C., and Turri, M. (1955). An interaction between alleles at the Rh locus in man which weakens the reactivity of the Rh$_0$ factor (D^u). *Proc. Natl. Acad. Sci. USA* 41:283.

100. Race, R. R., and Sanger, R. (1975). *Blood Groups in Man*, Ed. 6. Blackwell, Oxford.

101. Callender, S. T., and Race, R. R. (1946). A serological and genetical study of multiple antibodies formed in response to blood transfusion in a patient with lupus erythematosus diffusus. *Ann. Eugen.* 13:102.

102. Issitt, P. D., and Pavone, B. G. (1978). Critical re-examination of the specificity of auto-anti-Rh antibodies in patients with a positive direct antiglobulin test. *Br. J. Haematol.* 38:63.

103. Ishimori, T., Hasekura, H. (1967). A Japanese with no detectable Rh blood group antigens due to silent Rh alleles or deleted chromosomes. *Transfusion* 7:84.

104. Levine, P., Cellano, M. J., Falkowski, F., Chamber, J. W., Hunter, O. B., Jr., and English, C. T. (1965). A second example of −/− or Rh$_{null}$ blood. *Transfusion* 5:492.

105. Sturgeon, P. (1970). Hematologic observations on the anaemia associated with blood type Rh$_{null}$. *Blood* 36:310.

106. Levine, P., Cellano, M., Fenichel, R., and Singher, A. (1970). A "D"-like antigen in Rhesus red blood cells and in Rh-positive and Rh-negative red cells. *Science* 133:332.

107. Levine, P., Cellano, M., Vos, G. H., and Morrison, J. (1962). The first human blood −/−, which lacks the "D"-like antigen. *Nature* 194:304.

108. Moore, S. (1983). Identification of red cell membrane components associated with rhesus blood group antigen expression. In *Red Cell Glycoconjugates and Related Genetic Markers*, Cartron, J. P., Rouger, P., Salmon, Ch. (Eds.). Librarie Arnette, Paris, pp. 97–106.

109. Mallinson, G., Martin, P. G., Anstee, D. J., Tanner, M. J. A., Merry, A. H., Tills, D., and Sonneborn, H. H. (1986). Identification and partial characterization of the human erythrocyte membrane components that express the antigens of the LW blood group system. *Biochem. J.* 234:649.

110. Redman, C. M., Marsh, W. L., Mueller, K. A., Avellino, G. P., and Johnson, C. L. (1984). Isolation of Kell-active protein from the red cell membrane. *Transfusion* 24:176.

111. Wallas, C., Simon, R., Sharpe, M. A., and Byler, C. (1986). Isolation of a Kell-reactive protein from red cell membranes. *Transfusion* 26:173.

112. Redman, C. M., Avellino, G., Pfeffer, S. R., Mukherjee, T. K., Nichols, M., Rubinstein, P., and Marsh, W. L. (1986). Kell blood group antigens are part of a 93,000-dalton red cell membrane protein. *J. Biol. Chem.* 261:9521.

113. Marsh, W. L., Redman, C. L., Kessler, L. A., DiNapoli, J., Scarborough, A. L., Philipps, A. G., and Mody, K. L. (1987). K23: a low-incidence anti-

gen in the Kell blood group system identified by biochemical characterization. *Transfusion* 27:36.

114. Marsh, W. L., Nichols, M. E., Oyen, R., Thayer, R. S., Deere, W. L., Freed, P. J., and Schmelter, S. E. (1978). Naturally occurring anti-Kell stimulated by *E. coli* enterocolitis in a 20-day-old child. *Transfusion* 18:149.

115. Advani, H., Zamor, J., Judd, W. J., Johnson, C. L., and Marsh, W. L. (1982). Inactivation of Kell blood group antigens by 2-aminoethylisothiouronium bromide. *Br. J. Haematol.* 51:107.

116. Chown, B., Lewis, M., and Kaita, H. (1957). A "new" Kell blood group phenotype. *Nature* 180:711.

117. Marsh, W. L., Oyen, R., Nichols, M. E., Allen, F. H., Jr. (1975). Chronic granulomatous disease and the Kell blood groups. *Br. J. Haematol.* 29:247.

118. Allen, F. H., Krabbe, Sissel, M. R., and Corcoran, P. A. (1961). A new phenotype (McLeod) in the Kell blood-group system. *Vox Sang.* 6:555.

119. Wimer, B. M., Marsh, W. L., Taswell, H. F., and Galey, W. R. (1977). Haematologic changes associated with the McLeod phenotype of the Kell blood group system. *Br. J. Haematol.* 36:219.

120. Kuypers, F., Linde-Sibenius-Trip, M., Roelofsen, B., et al. (1985). The phospholipid organisation in the membrane of McLeod and Leach phenotype erythrocytes. *FEBS Lett.* 184:20.

121. Issitt, P. D. (1985). Ed. 3. *Applied Blood Group Serology*, Montgomery Scientific, Miami.

122. Parsons, S. F., Mallinson, G., Judson, P. A., Anstee, D. J., Tanner, M. J. A., and Daniels, G. L. (1987). Evidence that the Lu[b] blood group antigen is located on red cell membrane glycoproteins of 85 and 78 kd. *Transfusion* 27:61.

123. Brown, F., Simpson, S., Cornwall, S., Moore, B. P. L., Oyen, R., Marsh, W. L. (1974). The recessive Lu(a–b–) phenotype. *Vox Sang.* 26:259.

124. Crawford, M. N., Greenwalt, T. J., Sasaki, T., Tippett, P., Sanger, R., and Race, R. R. (1961). The phenotype Lu(a–b–) together with unconventional Kidd groups in one family. *Transfusion* 1:228.

125. Taliano, V., Guevin, R.-M., and Tippett, P. (1973). The genetics of a dominant inhibitor of the Lutheran antigens. *Vox Sang.* 24:42.

126. Telen, M. J., Eisenbarth, G. S., and Haynes, B. F. (1983). Regulation of expression of a novel erythrocyte surface antigen by the inhibitor Lutheran In(Lu) gene. *J. Clin. Invest.* 71:1878.

127. Telen, M. J., Palker, T. J., and Haynes, B. F. (1984). The In(Lu) gene regulates expression of an antigen on an 80-kilodalton protein of human erythrocytes. *Blood* 64:599.

128. Telen, M. J., Rogers, I., and Letarte, M. (1987). Further characterization of the In(Lu)-related p80 and the defect of In(Lu) Lu(a–b–) erythrocytes. *Blood* 70:1475.

129. Telen, M. J. Unpublished observations.

130. Daniels, G. L., Shaw, M. A., Lomas, C. G., Leak, M. R., and Tippett, P. (1986). The effect of *In(Lu)* on some high incidence antigens. *Transfusion* 26:171.

131. Marsh, W. L., Brown, P. J., DiNapoli, J., Beck, M. L., Wood, M., Wojcicki, R., and de la Camera, D. (1983). Anti-Wj: an autoantibody that defines a high-incidence antigen modified by the *In(Lu)* gene. *Transfusion* 23: 128.
132. Marcus, D. M., Kundu, S. K., and Suzuki, A. (1981). The P blood group system: recent progress in immunochemistry and genetics. *Semin. Hematol.* 18:63.
133. Hadley, T. J., David, P. H., McGinness, M. H., and Miller, L. H. (1984). Identification of an erythrocyte component carrying the Duffy group Fy[a] antigen. *Science* 223:597.
134. Anstee, D. J. (1986). Blood group active components of the human red cell membrane. In *Red Cell Antigens and Antibodies*, G. Garratty (Ed.). American Association of Blood Banks, Arlington, Virginia, pp. 1-15.
135. Sanger, R., Race, R. R., Jack, J. (1955). The Duffy blood groups of New York Negroes: the phenotype Fy(a-b-). *Br. J. Haematol.* 1:370.
136. Miller, L. H., Aikaw, M., Johnson, J. G., Shiroishi, T. (1979). Interaction between cytochalasin B-treated malarial parasites and erythrocytes: attachment and junction formation. *J. Exp. Med.* 149:172.
137. Hadley, T. J., McGinniss, M. H., Klotz, F. W., Miller, L. H. (1986). Blood group antigens and invasion of erythrocytes by malaria parasites. In *Red Cell Antigens and Antibodies*, G. Garratty (Ed.). American Association of Blood Banks, Arlington, Virginia, pp. 17-33.

12

The Transferrin Receptor

ROSE M. JOHNSTONE *McGill University, Montreal, Quebec, Canada*

I. INTRODUCTION

Although studies on the transferrin receptor have appeared in the scientific literature since the pioneering observations made by Jandl and Katz [1], the last
seven or eight years have seen a marked increase in the intensity of interest in
this receptor. Several reviews on the receptor have appeared recently [2-7]. To
an appreciable extent, heightened intensity has developed from the availability
of antibodies, both monoclonal and polyclonal, directed against the transferrin
receptor that can be used to probe the receptor in different species [8-19]. Two
research groups independently and almost simultaneously [11,18] recognized
that an antibody directed against a prominent plasma membrane marker on
rapidly growing or transformed cells was directed against the transferrin receptor and not against a specific marker of malignant transformation. A fallout of
these observations has been the recognition that the transferrin receptor is (a)
very antigenic, and (b) quantitatively a major constituent of the glycoprotein
makeup of the surface of growing cells.

A curious feature of the majority of the antibodies against the transferrin
receptor is that they do not block transferrin binding or endocytosis [3,19].
It is quite likely that the earlier studies failed to recognize antireceptor antibodies because the assay involved measuring inhibition of transferrin binding
[20]. It needs to be stressed, however than antibodies do exist that interfere
with transferrin binding [3,19] and these are generally toxic.

The last five to seven years have also been witness to a tremendous increase
in the understanding of endocytic mechanisms and membrane recycling (for reviews see [21-23]). As in most aspects of receptor structure and function, the
transferrin receptor has stood out as having unusual and atypical features. If

the exception proves the rule, the behavior of the transferrin receptor qualifies for the designation of the exception.

Receptors can be subdivided into two classes, signal delivery and nutrient delivery. The majority of receptors, such as the insulin receptor, the EGF receptor, or the PDGF receptor deliver a signal to the cell to alter one or a cascade of activities. There is no compelling evidence that the ligands themselves have intracellular functions.

The transferrin receptor and the LDL receptor are the two major examples of receptors that are involved in nutrient delivery, iron [1] and cholesterol [24], respectively. Although most low-molecular-weight nutrients such as ions, amino acids, vitamins, and sugars are delivered to the cell by carriers [25], the two insoluble nutrients have been adapted to receptor-medicated endocytosis by virtue of combining to specific binding proteins (transferrin and LDL), in the circulating fluid. Whether these receptors or their ligands have additional roles to play in signal transmission is still moot, although such a role for transferrin has been proposed [26-30].

Although most ligands and some receptors are degraded after internalization, both transferrin [1,31-34] and its receptor [33-36] undergo numerous cycles of internalization and externalization. The half-life of the receptor [37-38] is on the order of hours and that for transferrin [31] is on the order of days, whereas recycling time is reckoned in minutes [33-36,39-41]. The early recognition that transferrin undergoes many cycles of iron delivery without degradation may have given rise to the belief that iron transfer occurred at the surface of the cell rather than intracellularly [1]. This view may still be held by some investigators [40,41,81] because iron release is reported to be a very early event after transferrin binding. As a mechanism for iron delivery the transferrin receptor–transferrin complex is probably one of the most efficient delivery systems known. Each receptor (a dimer) binds two molecules of transferrin. Each molecule of transferrin in turn binds two atoms of Fe^{3+}. Thus, four atoms of Fe^{3+} may be delivered for each term of the cycle. The number of atoms of iron delivered per reticulocyte have been reported at $\sim 8 \times 10^3$/min [33,40]. This value makes this one of the most efficient delivery systems for receptor-mediated endocytosis approaching the rate of transport of low-molecular-weight solutes such as Gly, at nanomolar concentrations in a rapidly growing malignant cell, the Ehrlich cell (R. M. Johnstone, unpublished data, 1984).

II. STRUCTURE AND ORGANIZATION OF THE TRANSFERRIN RECEPTOR

The transferrin receptor of many mammalian cells, including mammalian reticulocytes, appears to be a dimer of identical monomers of 90-95 kDa, the monomers being cross-linked by thiol bridges [6,37,42-45]. Although the diagram

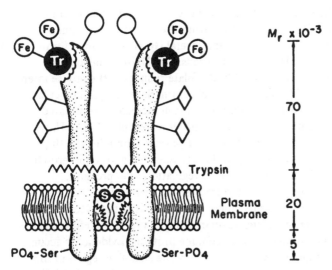

Figure 1 *Schematic representation of the cell surface transferrin receptor.* The fatty acid, phosphoserine and disulphide bonds have now been assigned to amino acid residues 62, 24 and 88-98. Reprinted with permission from Trowbridge & Newman, Biochem. Pharmacol. (1984) 33, 925.

(Figure 1) shows only a single S-S bridge, recent work by Jing and Trowbridge [81a] shows that two intermolecular sulfhydryl bridges occur at residues 89 and 98, respectively. This may explain why a dimer is frequently found after gel electrophoresis even under reducing conditions. One of these bridges is extracellular and one (at 89) is at the junction between the extracellular region and transmembrane spanning region. In fowl cells, the receptor appears to be smaller [46], and exists partly as a dimer and partly as a monomer [48]. Mice show a relatively larger receptor [47]. The small size of the chicken receptor may pertain to embryonic cells because adult chicken cells have a receptor of typical size, 95 kDa monomer [48].

That each monomer spans the membrane has now been shown by several lines of evidence: (a) Proteolytic cleavage of the cell surface shows that a peptide of ~70 kDa is released, capable of binding transferrin but lacking the phosphate group and fatty acids incorporated into the receptor [37,43,45,49], (b) In sealed reticulocytes ghosts, the transferrin receptor can be labeled by iodination of either the cytoplasmic surface or the external surface (Pan et al. [16]), (c) With the cloning of the receptor, hydropathy maps of the deduced amino acid sequences show the presence of hydrophobic stretch of 26 amino acid residues (63-88 inclusive) consistent with a single membrane spanning region per monomer [50,51].

In a recent report, the structure of the cloned gene has been examined by Miskimins et al. [52]. A 365 base-pair region from the 5' end of the cloned transferrin receptor contains a TATA box and several GC-rich regions. It promotes the expression of bacterial CAT in 3T3 cells. The DNA sequence has homology with the promoter region of dihydrofolate reductase, the mouse interleukin 3 gene as well as the SV 40 origin of replication. Several high-molecular-weight proteins that interact with the promoter region increase in stimulated 3T3 cells and their increase precedes a rise in transferrin receptor RNA. These events all precede entry of the cell into S phase. The gene's homology with promotor regions of other genes, whose expression is associated with rapid growth, may reflect structures of DNA subject to acute regulation of expression.

Although a transmembrane protein, the transferrin receptor is among a growing number of transmembrane proteins whose N terminal is at the cytoplasmic surface [50,51]. The mechanism of insertion into the bilayer is therefore different from that of many well-studied transmembrane proteins. Evidence is accumulating that the hydrophobic segment of the membrane spanning region in the protein may act as a recognition site for insertion into the membrane [53]. The general significance, if any, to cellular function of transmembrane proteins with their N-terminal residues at the cytoplasmic surface is unknown.

As mentioned above, much of the peptide chain (~70%) of the transferrin receptor is external to the cell surface. Because cell surface proteolysis releases monomers, the disulphide bonds linking the monomers appear to be close to the membrane and distal to the trypsin cleavage site [43–45] in line with the current new information on the sites of the S-S bridges [81a]. Only ~6 kDa of the peptide monomer appears on the cytoplasmic surface. The transferrin receptor, like transferrin, is a glycosylated protein containing three N-linked glycan units per monomer, both high mannose and complex carbohydrates, all of which face the external milieux. The carbohydrate residues contribute ~15 kDa to the molecular size of the monomer [37,43,45].

Several investigators have now shown that the transferrin receptor is a phosphoprotein [43,45,54–60] with the phosphate groups facing the cytosolic milieux. Recently Davis et al. [60] reported that only a single Ser residue (24) is phosphorylated per receptor monomer on the human receptor. Earlier workers [54,56] had proposed that more than a single site was phosphorylated based on the appearance of several limiting phosphopeptides or several peptides with different isoelectric points after proteolytic digestion of the receptor. It is not yet clear whether the difference in conclusion is artifactual (nesting peptides rather than limiting peptides), or due to differences in receptors from different cells.

In addition to phosphorylation and glycosylation, the transferrin receptor is also acylated [37,61]. The acylation site has now been shown to be at Cys residue (62) just outside the membrane barrier at the cytoplasmic surface [81a]. Treatment with hydroxylamine at neutral pH removes most of the fatty acid liberating a fatty alcohol. This lability is consistent with the acylation of a Cys residue of the human transferrin receptor [50,51,81a].

A. Transferrin Processing and Iron Use

The intensity of iron use by animal cells is generally reflected in the number of transferrin receptors per cell and the distribution of the receptors in the cell. Not surprisingly, the developing red blood cell has been reported to have the highest number of receptors—80 X 10^3/cell—in the mouse normoblast [62,63]. The peripheral reticulocyte from a number of species including rabbit, rat, mouse, humans, and sheep, has been reported to have ~100-150 X 10^3 sites/cell [1,62-66]. The placental trophoblast also expresses large numbers of receptors per cell (~400 X 10^3/cell) [9,10]. The K562 cell line expresses a larger number of receptors than most other cultured lines (~500 X 10^3/cell [67,68]. Other cultured lines including fibroblasts, lymphocytes, kidney cells, and malignant cell lines express ~50 X 10^3 to 100 X 10^3 receptors/cell during exponential growth [11,12, 18,35,36,69,70].

Current views on the pathway for iron use in mammals are summarized by the diagram in Figure 2. A recent study with chicken reticulocytes presents a very similar, if not identical pathway [46]. At neutral pH, ferrotransferrin has a high affinity for the transferrin receptor at the cell surface, with a kDa of ~10^{-9} reported for many systems [34,35,42,67,68,71-78]. Binding to the cell surface is an energy- and temperature-independent process and is usually complete by ~15 min at 0-4°C [79]. This observation has been made by countless investigators and is now used to distinguish between surface-associated and internalized transferrin. Transferrin bound to its receptor is concentrated in clathrin-coated pits on the cell surface. The receptor–ligand complex is internalized into the cell, the clathrin-coated pits bud into the cell, forming clathrin-coated vesicles. After clathrin removal, the vesicles fuse to become endosomes (for reviews see [22, 80].

Although the common view is that the coated pit becomes a coated vesicle after its detachment from the plasma membrane [22,80], some investigators have concluded that coated vesicles remain attached to the plasma membrane and only the unfrocked endosomes are in the cytoplasm [81].

There does not appear to be consensus amongst investigators whether clustering of transferrin receptors into coated pits or internalization of the transferrin receptor is triggered by the presence of ligand. Several investigators have demonstrated the existence of coated pits containing the transferrin receptor on the cell surface in absence of transferrin and at 0°C [82-86] in both cultured cell lines and reticulocytes. Other studies have shown clustering of receptor only after adding transferrin and raising the temperature [87,88].

Similarly, differences of opinion exist on the role of transferrin in initiating receptor internalization. Using antigen binding (FAB)-fragments of an antibody

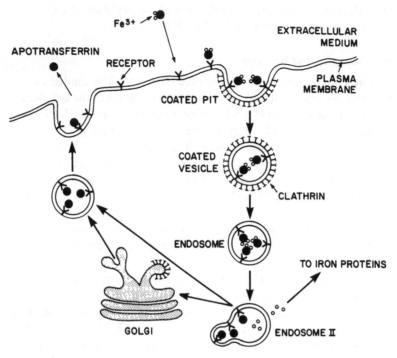

Figure 2 *Transferrin cycle.* Ferrotransferrin is bound to surface receptor in coated pits on the cell surface. The coated pit becomes a coated vesicle. After removal of the coat, the endosome is formed. The endosome contains a H^+-pumping ATP-ase which acidifies the internal milieux. Vesicle population with different internal pH's exist. Iron is removed from a compartment of relatively high pH ($\sim$6.0).

The apotransferrin, still attached to receptor, is returned to the plasma membrane. At the surface, at neutral pH, the apotransferrin dissociates from the receptor. In some cases, the receptor containing vesicle may route through the Golgi before returning to the plasma membrane.

to the transferrin receptor, Enns et al. [88] showed a stimulation of internalization of the FAB fragment only after addition of transferrin, suggesting that internalization depends on transferrin. Klausner et al. [67,75] showed a decrease in surface receptor on addition of transferrin, also suggesting that transferrin triggers internalization. Watts [84], Hopkins and Trowbridge [86,89], Sullivan et al. [82], and more recently Stein and Sussman [90] have shown evidence for receptor internalization in the absence of transferrin. Although some of these differences may be associated with the different types of cells used (reticulocytes and a variety of cultured cells), conflicts exist even in studies on the K562 cell lines [67,84,90]. The reasons for these differences may be due to experimental conditions that may affect the rate of recycling or the stability of the receptor [91].

It has been estimated that surface pits may occupy 1-2% of the entire cell surface [22,80]. More than one ligand can be found within a single coated pit or an internalized endosome [77,83,92,93]. Thus, the site of internalization is not specific for a particular ligand receptor complex. Once internalized, the process of sorting, segregating, and selective recycling must occur in the endosomal network.

One of the characteristics of the endosomal network is that the membranes appear to possess an H^+-adenosine triphosphatase (ATPase), which maintains an internal acidic environment [94-96]. It now appears that there is a gradation in pH in the endosomal network ranging from ~6.5 down to <5 in the lysosomes. It is well known that Fe^{3+} dissociates readily from transferrin as the pH is lowered [97]. In the acidified endosomal compartment, Fe^{3+} dissociates from transferrin, but apotransferrin remains attached to the receptor at mildly acid pH [98-100]. It has been shown repeatedly that mild acidification (>pH 5) does not cause dissociation of apotransferrin from its receptor, in contrast with other ligands [99,100]. Iron dissociated from transferrin does not appear to leave the endosome in absence of an iron acceptor [100a]. The immediate fate of the iron released from transferrin remains unknown. How iron leaves the endosome, and what are the intermediate iron acceptors in the cytosol are issues that have escaped resolution. Eventually iron will appear in heme, iron proteins, or ferritin. It is likely that Fe^{3+} is removed in an endosomal compartment of relatively high pH (~6.0), because Fe^{3+} uptake is very sensitive to low concentrations of agents that raise the pH, such as monensin or chloroquine [73,101] under conditions that do not affect transferrin uptake or recycling. That iron removal is not required for recycling is shown by the fact that concentrations of chloroquine that abolish iron removal do not prevent recycling of transferrin [101], and that ferrotransferrin is returned to the cell surface in reticulocytes. It has also been proposed that iron returns independently of transferrin [81].

The signals that permit ligand receptor sorting in the endosomal network are not known. During this passage, some ligands and receptors will be targeted to the lysosomes for degradation, as for example epidermal growth factor (EGF) and its receptor in KB cells [83]. The asialoglycoprotein receptor will be separated from its ligand, and the receptor cycled, but the ligand may be targeted to lysosomes [102,103]. The transferrin cycle shows atypical but not unique properties, with both receptor and ligand recycling to the plasma membrane, normally after iron depletion. The endocytosis and exocytosis of ligand and receptor may be analogous to early reports of diacytosis [104].

On refusion of the apotransferrin-receptor complex with the plasma membrane, and exposure to the neutral pH of the medium, the apotransferrin dissociates from its receptor and a new cycle is ready to commence with ferrotransferrin. It has been shown that the binding constant for apotransferrin at neutral pH is over an order of magnitude less than for ferrotransferrin [99,100].

Several investigators have suggested that the Golgi network is part of the transferrin receptor recycling route [49,83,105-107]. Whether the Golgi route is essential for all transferrin receptor recycling is not clear, although it appears unlikely. In cultured cell lines and reticulocytes, a number of investigators have failed to find evidence for Golgi involvement in transferrin receptor recycling [81,96,109-112]. Recent reports [113,114] have shown that the transferrin receptor can be resialylated as well as undergoing repair of the complex and high mannose polysaccharide chains. Passage through the Golgi may be slow relative to the route for iron delivery and the possibility exists that passage of the receptor through the Golgi is required for the repair of a "damaged" receptor [114] but is not part of the recycling pathway for iron delivery.

In reticulocytes, the half-time for recycling of transferrin [39,40,101,109, 110] is generally shorter than reported for cultured cell lines [35,49,55,75,90, 93,111] with reticulocytes showing recycling times of 4-5 min, and other cell lines 8-10 min. In several cell lines, evidence has also been obtained for short (2-10 min) and long cycles (~30 min) [89,90,107,110,111] for both the transferrin receptor and asialoglycoprotein receptor [115].

How do the endosomes migrate through the cell? Do they move like railroad cargo on a track with a number of sidings at which different populations of vesicles get shunted? Or is the movement more random?

Willingham and colleagues [83] have observed that endosomes move by saltatory motion along microtubule tracks to the area of transreticular Golgi where sorting of vesicles and contents takes place. Evidence for the involvement of microtubules has received support from other groups [54,116] in cultured cells and reticulocytes.

Some investigators have questioned the role of microtubules in directing vesicular traffic because agents that cause depolymerization of the tubular network do not appear to influence endocytosis and membrane processing [117]. The concentration of vinca alkaloids required to inhibit transferrin uptake was 10 times greater than required to disrupt microtubule assembly in a lymphoid cell line [117].

Reticulocytes do not show evidence for the presence of microtubules. Although the possibility exists that hemoglobin obscures microtubule assemblies, it is equally possible that microtubules are not essential for recycling of the transferrin receptor-transferrin complex in reticulocytes, even though the structures may play such a role in other cells.

The fact that a steady-state level of surface receptors can be achieved with most cells during Fe^{3+} delivery suggests that there may be some signal to externalize and internalize the same number of receptors. If recycling is not continuous and signals do exist to export a vesicle with receptor in response to an incoming receptor, a box-car replacement on a limited length of track has a certain appeal.

In summary, many aspects of the transferrin cycle have become clearer over the last five years, although controversy remains in a number of specific aspects of the cycle. Most investigators appear to agree that transferrin is internalized along with its receptor via coated pits (Fig. 2). After defrocking, and endosome formation, iron is removed in a mildly acid compartment, which is not lysosomal. Aportransferrin remains bound to the receptor and the complex is segregated from the pool of other internalized ligands and receptors whose destination is other than a return to the plasma membrane. The remarkable difference in the association constant for ferrotransferrin and apotransferrin as a function of pH assures that at the cell surface apotransferrin will dissociate, allowing ferrotransferrin to commence a new cycle.

Although the cellular journey of transferrin has become clearer in the last five or so years, the same cannot be said for iron. Once dissociated from transferrin, the route traveled by iron is unknown. How it leaves the vesicle, what carries it, and where it goes remain shrouded in uncertainty. Because ferric iron is virtually insoluble at neutral pH, the iron must be reduced or chelated in some manner to expedite its transfer to the sites of heme and/or ferritin synthesis. The elucidation of the journey taken by iron remains a challenge.

III. RECEPTOR POOLS

The fact that receptors are endocytosed and not degraded leads to the obvious conclusion that there is at least one pool of intracellular transferrin receptors. With receptors like EGF, the majority of the cellular receptors exist on the cell surface, because much of the internalized receptor is degraded [118-121]. However, even with EGF, the experimental evidence suggests that in liver there are pools of EGF receptors, some of which may be cryptic in the sense that they come into play only under specific conditions [122]. Intracellular pools have been described for other receptors [123-129], and acute up- and downregulation of surface receptors may be associated with the transfer of receptor from the intracellular pool to the surface.

Most investigators agree that the cellular pool of transferrin receptors is at least as large as that on the surface and in many cases, much larger [49,55,66, 77,101,108]. This appears to be true with reticulocytes and cultured cells. Several investigators have reported that surface binding of transferrin at steady-state at $37°C$ represents only ~20% of transferrin binding with 80-90% of the receptors being intracellular [66,77,101,108]. This may be the normal in vivo distribution of the transferrin receptor because the transferrin concentration in the circulation (~2 mg/ml plasma in humans, ~20 μM) is sufficient to saturate the receptor binding sites (nanomolar concentrations) [59,66,109] of most cells.

It also appears self-evident that the intracellular pool of receptors must be heterogeneous. The incoming pools of receptors bearing ferrotransferrin must be differentiated from outgoing, iron-depleted, apotransferrin-receptor pools. In some cells, transferrin receptors can be found in at least two regions of the Golgi, which may represent further differentiation of the intracellular pools [83,105].

Recent reports have provided additional evidence for heterogenous pools of intracellular transferrin receptors. In K562 cells, treatment with an antibody against the receptor (OKT9) seems to remove a fraction of the receptor from the recycling pool [91,130]. Stein and Sussman [90] have shown that there are two types of intracellular recycling pools, one sensitive to monensin and the other relatively insensitive, although both are thought to provide cellular iron. In maturing reticulocytes, there is evidence for a pool of transferrin receptors that is no longer undergoing recycling but in which the receptors are capable of transferrin and antibody binding [66,85,110,131]. In sheep reticulocytes [66] it was impossible to label the total number of cellular receptors by incubating reticulocytes with saturating amounts of transferrin at $37°C$. At steady-state, at $37°C$, less than half of the total estimated cellular receptors bound exogenous transferrin, suggesting a cryptic pool of receptors outside the recycling pool.

Although the kinetic studies and localization studies of the transferrin receptor have provided evidence for the presence of different pools of receptor, the studies have generally been consistent in concluding that the affinity of the receptor for transferrin is constant, irrespective of cellular location. These observations contrast with studies on EGF receptors, which show binding sites of different affinities [122].

IV. DE NOVO SYNTHESIS OF THE RECEPTOR: POSTTRANSLATIONAL MODIFICATION AND FACTORS CONTROLLING RECEPTOR SYNTHESIS

The de novo synthesis of the receptor has now been examined in a number of cultured cell lines [37,43,45,56,132] as well as in reticulocytes [59a,133,133a]. These studies have shown that the time required to complete the synthesis of the mature receptor can be as long as 2 hr. At early times (~10 min) peptides of ~78 kDa are detectable and in the course of time, these are chased into the higher-molecular-weight forms [43,45,132].

The primary translation product can be formed in a cell-free system [132] and core glycosylation of the receptor can be obtained in the cell-free systems, if dog pancreatic microsomes are added [132]. The mature form of the receptor is believed to contain three oligosaccharide chains per monomer, of which two are high mannose [37,45]. That the receptor is made as a transmembrane protein is suggested by the insensitivity of the receptor to tryptic digestion during its de novo synthesis in the presence of microsomes [45]. Whether tunicamycin prevents expression of the receptor at the cell surface appears to be characteristic of the cell type. Thus, it has been shown that Molt 4 and CCRF-CEM cells expressed an unglycosylated receptor, whereas Nalm and A431 cells did not [37,45,214]. Snider and Rogers [113,114] have also shown that inhibition of glycosylation by deoxymannojirimycin does not prevent the expression of a functional receptor at the cell surface in erythroleukemia cells. Moreover, the receptor can be "repaired" if desialylyated or when inhibitors of carbohydrate processing are removed [114].

The role of the oligosaccharide side chains on the receptor has not been established. Indeed, to date no role has been found for any of the posttranslational modifications.

Because the phosphate moiety appears to be incorporated after glycosylation is complete [45], and glycosylation is apparently not essential for plasma membrane localization in all cases [37,45], a functional role for receptor phosphorylation during iron delivery seems unlikely.

Acylation is also associated with the mature form of the receptor [37]. In this instance it has been shown that, in tunicamycin-treated cells, despite the absence of completed oligosaccharide chains, acylation of the receptor still occurs, although it is much reduced. The significant feature is that acylation can occur on an incomplete receptor, suggesting that normally the glycosylated receptor is not an essential substrate for acylation.

All three modifications of the receptor—glycosylation, phosphorylation, and acylation—may occur long after protein synthesis is complete [37,45,114]. Omary and Trowbridge [37] have shown that in cultured cells acylation can occur as long as 48 hr after protein synthesis has terminated. Adam et al. [61] have shown acylation of the sheep reticulocyte receptor in isolated plasma membranes.

The incorporation of ^{32}P into the receptor can also occur in absence of de novo protein synthesis [45-49,133b]. The enhanced, rapid phosphorylation of the transferrin receptor by β phorbol esters also testifies to the fact that phosphorylation occurs in absence of de novo receptor synthesis [54-59,133b].

Although ^{32}P incorporation into the receptor can be measured in isolated cells and in isolated plasma membranes, in sheep reticulocytes the stoichiometry ^{32}P incorporation suggests that only a small percentage of the total number of receptors is undergoing phosphorylation [57]. Because it has been impossible to dephosphorylate the receptor before ^{32}P incorporation, it is unclear whether the low level of phosphorylation in vitro signifies the absence of sites for phosphorylation or a conformational state of receptor (e.g., binding to another protein), which prevents receptor phosphorylation.

V.   REGULATION OF THE SYNTHESIS OF TRANSFERRIN RECEPTORS

Increased numbers of transferrin receptors correlate with increased cellular growth [11,12,64,134-139]. The cellular signal that appears to increase receptor synthesis is a diminishing cellular iron level. Conversely an increase in the cellular iron level appears to be involved in decreasing the rate of transferrin receptor synthesis in cultured cells [64,76,136,140-146].

Two points of view are expressed in the literature about cellular iron as a mediator of receptor synthesis, namely that elevated free iron [136,142,144-147], or elevated heme [76,133,140,143] acts as the signal to reduce the synthesis of the receptor. That heme itself can be a contributor to the free iron pool has also been shown [145]. Because the solubility product of ferric hydroxide is $\sim10^{-18}$ at neutral pH, the free iron concentration is vanishingly small under physiological condition. Any ferrous iron, unless complexed, is likely to be oxidized to the ferric state at normal cellular pH. If free iron is the signal for regulation of receptor synthesis this raises the question of signal detection at these low iron concentrations. What is the cytosolic chelator that maintains nonheme iron at a significant level, or is the recognition mechanism so sensitive as to detect iron in the range of its solubility in water? As yet, this interesting and complex problem, which is at the root of the role of transferrin in biology, remains obscure.

Terminal, peripheral reticulocytes have recently been shown to synthesize the transferrin receptor [59a,133,133a]. Unlike other cultured, growing cell lines, reticulocytes increase transferrin receptor synthesis in presence of heme. Thus, Cox et al. [133] showed that inhibition of heme synthesis diminished transferrin receptor synthesis and that the inhibition could be overcome by exogenous heme. Ahn and Johnstone [133a] found that 10 μM heme increased receptor synthesis in sheep reticulocytes. The stimulation by heme may be a general stimulating effect of heme on protein synthesis in reticulocytes [59a, 133], but a more specific effect on receptor formation has not been ruled out. It is curious that reticulocytes respond atypically to the effect of heme on transferrin receptor synthesis. In growing cells, heme regulates receptor synthesis by altering mRNA levels [144,147]. In reticulocytes, devoid of nuclei and de novo mRNA synthesis, only translational controls are possible. The question arises whether the difference in behavior in growing cells and reticulocytes with respect to heme is due to the masking of the translational controls by transcriptional controls in growing cells or whether the erythroid cell, with its commitment to hemoglobin synthesis, upregulates transferrin receptor synthesis in an anomalous way.

Both Cox et al. [133] and Ahn and Johnstone [133a] find that the ^{35}S-methionine incorporated into the transferrin receptor in reticulocytes is membrane bound. However, Ahn and Johnstone have found that little of the ^{35}S-receptor is translocated to the plasma membranes [133a]. The ^{35}S incorporated is largely insensitive to proteolysis at either 4°C or 37°C, whereas the majority of the preexisting receptor is destroyed by incubating cells with trypsin at 37°C. The data imply that the receptor is membrane-bound intracellularly but does not reach the plasma membrane or the recycling pools. This conclusion differs from that of Cox et al. [133]. Furthermore, the newly synthesized receptor in sheep reticulocytes is 2-3 kDa smaller than the mature receptor. In rabbit reticulocytes, no difference in size of the newly synthesized and preexisting receptor was noted [133]. Evidence shows that the newly synthesized receptor binds transferrin in both rabbit [133] and sheep reticulocytes [J. Ahn and R. M. Johnstone, 1988 (submitted)].

Rabbit reticulocytes generally appear younger than sheep peripheral reticulocytes as judged by size (relative to the mature red blood cell) and intensity of methylene blue staining. Based on the differences in the capacity to synthesize a mature transferrin receptor, it is tempting to suggest that the membranous organelles required to complete the synthesis of mature receptor are lost before the loss of ribosomes and the translational machinery. Thus, both sheep and rabbit peripheral reticulocytes retain the capacity to synthesize the peptide, but only the rabbit cells retain the machinery to convert the protein into the mature form.

In this regard, it is interesting to note that the newly synthesized receptor of the sheep reticulocyte is not externalized, along with the preexisting receptor, into vesicles during in vitro maturation of sheep reticulocytes (Ahn and Johnstone [133a]).

VI. ACUTE UP- AND DOWNREGULATION OF THE TRANSFERRIN RECEPTOR

As previously noted, the cellular level of the transferrin receptor appears to be linked to the intracellular iron supply, although the specific form of the iron is still under investigation. The rate of new receptor synthesis is slow by comparison to the rate of recycling. Other forms of regulation exist that can bring about a redistribution of the receptors to modify the rate of iron use. Many disparate metabolic conditions have been shown to trigger up- and downregulation of the transferrin receptor.

Transferrin itself has been proposed by several investigators to decrease the steady-state number of surface receptors in a number of cultured cell lines [55, 67,87,148], and thereby regulate the rate of iron uptake. Other investigators have found no evidence of a change in receptor distribution by transferrin [35, 82,84-86,112]. Because transferrin is a major protein in the plasma, whose concentration is (a) not known to vary under normal conditions, and (b) nearly an order of magnitude higher than that normally required to saturate the receptor, it would appear somewhat unlikely that transferrin levels would have a role in regulating the cellular distribution of the receptor. Unlike other growth-regulating peptides such as insulin, EGF, platelet derived growth factor (PDGF), etc., transferrin secretion, synthesis, and degradation are not subject to acute regulation. It is possible, however, that in the organism, transferrin receptors are chronically "downregulated" and that under conditions in which transferrin fails to be synthesized, cells have the capacity to upregulate their transferrin receptors and thereby maintain normal rates of iron transport for a significant period, as demanded by the prevailing circumstances.

Recent reports have shown that both EGF and insulin may alter the number of surface transferrin receptors in a variety of cultured cell lines [149-152]. Heightened interest in mechanisms for acute up- and downregulation of transferrin receptors has appeared since β phorbol esters were shown to alter receptor distribution and phosphorylation [54,55,58,59,74,116,148,153]. In most systems reported, β phorbol esters appeared to decrease the number of surface receptors by ~40%, concomitant with a three to fourfold increase in receptor phosphorylation. In sheep reticulocytes the effect of phorbols on receptor internalization is small (~20%) and can only be measured by antibody, but not transferrin binding, to the cell surface (Adam and Johnstone, unpublished data). Hebbert and Morgan, however, did observe an increase in transferrin endocytosis in rat reticulocytes of ~15% in the presence [153a] of phorbol esters.

In contrast with reticulocytes and cultured cells, Kaplan and colleagues [149, 150] have shown that in macrophages, surface transferrin binding increased after treatment with β phorbol esters. The physiological significance of the difference in response to β phorbols in macrophages and other cultured cells is not known, nor, is it clear whether the change in receptor distribution and/or phosphorylation brings about significant changes in receptor recycling and the rate of iron delivery. In one report, β phorbols were shown not to increase Fe^{3+} uptake [55]. It is also curious that trifluoperazine, which normally inhibits kinase activity, stimulates transferrin receptor phosphorylation and internalization in K562 cells [154].

The absence of a role for phosphorylation in transferrin receptor processing for iron delivery has recently been reported. Rothenberg et al. [154a], using a cloned, mutated T$_f$R, showed that the absence of phosphorylation sites on the receptor had little effect on endocytosis or Fe^{3+} delivery. The role of increased receptor phosphorylation in response to agents such as heme or phorbol esters becomes even less clear.

Not unexpectedly, the iron supply has been implicated as a major factor in regulating the rate of iron and transferrin uptake [155,156] and the distribution of surface receptors [133,146,157]. Both the free iron level or heme iron have been implicated in these observations. Much inconsistency is evident in the literature on the effect of heme on transferrin uptake, especially in reticulocytes. Iacopetta and Morgan [157] showed a decreased rate of transferrin uptake with little effect on the steady-state distribution of the receptor. Cox et al. [133], also using rabbit reticulocytes, showed a marked increase in the surface receptor level by heme, but a decreased rate of internalization, the rate and extent of transferrin binding to the surface being increased by heme. Ponka and Schulman, however, concluded that iron uptake from transferrin is inhibited by heme [155,156], with less important effects on transferrin binding.

Decreases in surface transferrin receptors have been shown to occur during mitotic activity [139,158]. Intracellular vesicular traffic in general is believed to decrease during mitosis. In HeLa cells, it has been shown that the t½ for release of transferrin may be increased from 5–6 min to >50 min [159] during mitotic activity.

The relative distribution of transferrin receptors on the cell surface and intracellularly in a virus infected erythroid cell line also varies with the developmental stage of the cell, with more internal receptors as the development proceeds [160]. Because the rate of iron uptake appears to reach a limiting rate in these cells, independent of the number of surface receptors, the observations suggest that under some conditions, the rate of endocytosis may limit the rate of iron uptake [140].

It is clear that many factors appear to regulate the acute distribution of transferrin receptors and the rate of recycling. Although the role of iron in this process is relatively simple to understand, the physiological significance of most of the other potential regulators is not self-evident and requires further study.

VII. TRANSFERRIN RECEPTORS AND THE DEVELOPMENT OF ERYTHROID CELLS

The first recognition of the transferrin receptor as an entity established the fact that mature red blood cells lose the ability to bind transferrin and the competence to take up iron [1,161].

It has also been clear that as reticulocytes develop in vivo and in vitro, the maturation is associated with a decrease in the number of functional transferrin receptors [1,16,33,63,64,162], whereas differentiation into the erythroid cell line is associated with an increase in receptor number [63,160,163,164].

The maturation of reticulocytes into red blood cells has fascinated investigators for years. The fact that these cells undergo the physical loss of organelles during maturation was shown by the classic work of Bessis [165] on the loss of the nucleus. Since these early observations, a number of investigators have shown evidence for the extrusion of organelles such as plasma membrane, lysosomes, mitochondria, etc. during the maturation process [166-170]. The idea that some surface membrane components, destined for removal at maturation, aggregate at the cell surface and are removed by reactions involving the spleen has been proposed by several investigators [168-171]. Because reticulocyte

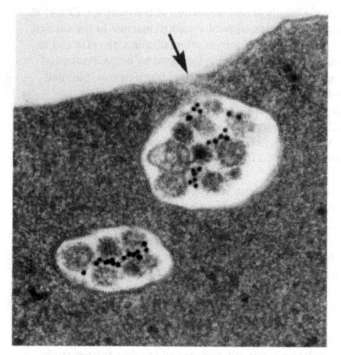

Figure 3a *MVB's forming after 1 hr incubation of reticulocytes with gold labelled second antibody to the transferrin receptor.* Note beginning of membrane fusion site (arrow). (R. M. Johnstone, unpublished data, 1984.)

maturation can occur in vitro [167,172,173] it is clear that the spleen cannot be essential. Although many changes in structure and function of red blood cells have been associated with maturation for half a century, no data have been available on the fate of any specific membrane component during the course of maturation.

The observation that a membrane protein, characteristic of reticulocytes, is physically released from the cell in a specialized manner during maturation, has added a new dimension to the understanding of reticulocyte development.

VIII. THE TRANSFERRIN RECEPTOR AND RETICULOCYTE MATURATION

Since the classical demonstration that the reticulocytes but not erythrocytes bind serum transferrin [1], the fate of the transferrin receptor during reticulocyte maturation remained elusive. Absence of function need not mean absence of the specific protein. Early investigators speculated on the possible loss of transferrin binding by inactivation of the receptor [174]. With the advent of

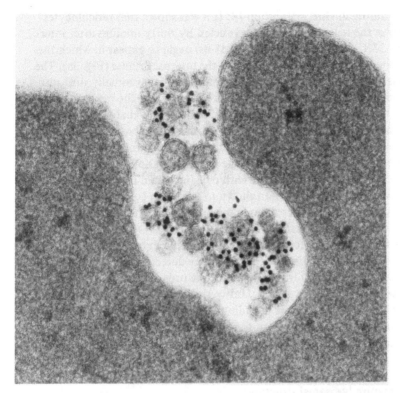

Figure 3b *Fusion of MVBs with the plasma membrane and release of 50nm vesicles.* The figure shows exocytosis into the medium of small dense bodies labeled with gold-conjugated antibody after 18 hr incubation at 37°C. The gold label is only present on the 50-nm bodies which are inside sacs of ~800 nm in diameter. The limiting membrane of the MVB is devoid of label. Reprinted with permission from Pan et al. J. Cell Biol. 1986 101:942.

antibodies against the transferrin receptor, the opportunity arose to assess whether the transferrin receptor remained in the cell or was lost. Such studies led Pan et al. [16,175] to conclude that the receptor is physically removed from the cell and extruded as a small vesicle (the exosome) (Pan and Johnstone, [175]) by a process akin to exocytosis. It was shown that ^{125}I-surface-labeled transferrin receptor was recovered in the exosomes as sheep reticulocytes matured in vitro. As the amount of transferrin receptor protein in the cells decreased with maturation in vitro, the amount of receptor recovered in vesicles increased [162]. However, in the best estimates to date, only ~40% of the lost cellular receptor has been recovered in vesicles [176]. Using a second antibody labeled with colloidal gold to follow the kinetics of the cellular disposition of the receptor during in vitro maturation [85], it was shown that reticulocytes first internalize the receptor into simple vesicles. By thirty minutes after internalization at 37°C, multivesicular bodies (MVBs) begin to appear in which the receptor is present at the external surface of the internal vesicles (Fig. 3a). The multivesicular bodies appear to grow larger by fusion and eventually fuse with the plasma membrane, thereby releasing their contents into the medium (Fig. 3b). Similar figures were detected in pre-fixed preparations that were probed with antireceptor antibody after fixation of the cells. The control studies underscore the conclusion that the MVBs are neither artifacts of fixation nor an abnormal route of processing the receptor due to its unphysiologic association with a large complex, consisting of the antibody and a gold-labeled second antibody [85]. Such problems have been shown to arise with gold-labeled antitransferrin antibody, or antireceptor antibody in which transferrin was routed to lysosomes, a pathway abnormal to this ligand [89,177].

The time frame for MVB formation in sheep reticulocytes is much longer than for receptor recycling. Few labeled MVBs are detected prior to 30 min at 37°C, whereas the half-time for recycling is of the order of 5 min in reticulocytes from several species (see above). Multivesicular bodies in maturing reticulocytes have been noted by others including Harding et al. [110,131] and Iacopetta and Morgan [178]. Although at first considered an alternate recycling pathway, Harding and Stahl [131] have also concluded the MVB formation is a route for shedding of transferrin receptors during red blood cell maturation and have shown that the MVBs are free from acid phosphatase, and therefore unlikely to be mature lysosomal structures.

The protein composition of the vesicle as visualized by Coomassie Blue staining is striking [175]. By visual inspection of the stained gels (Figure 4), the externalization is highly specific. Two major proteins are found (M_R ~94 kDa and ~70 kDa), neither of which is a major membrane component in the plasma membranes. A dimer of the 94-kDa peptide is frequently detected, despite the use of reducing agents. Immunoprecipitation of vesicle proteins with antibody against the transferrin receptor has shown that the 94-kDa protein is exclusively the transferrin receptor. This control is of some importance because the anion exchange protein is a major protein in red blood cell membranes with molecular size in the same range of sodium dodecylsulfate (SDS) gels [179,180]. After immunoprecipitation, the supernatant is devoid of a 94-kDa species. The iodotyrosyl peptide map as well as the molecular size of the externalized receptor are indistinguishable from the native receptor by available techniques [175]. Moreover, the vesicles have a right-side-out orientation as shown by the ability to isolate the intact vesicles by an immunoaffinity column [175]. Surface labeling of the vesicles with immobilized iodinating reagents labels the transferrin receptor, but not the 70-kDa protein (R. M. Johnstone, 1985, unpublished observations).

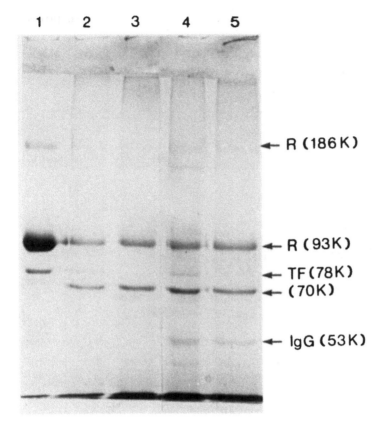

Figure 4 *Release of the Transferrin Receptor in Presence and Absence of Antibody and Transferrin.* Sheep reticulocytes were isolated and treated to remove bound transferrin and a portion was incubated with reticulocyte-specific rabbit antiserum at 0°C. Control cells were treated with nonimmune serum. After 90 min at 0°C, the cells were washed, resuspended in fresh incubation medium (serum free), with or without 50 μg/ml transferrin, and incubated at 37°C under O_2/CO_2. After 21 hr, the cells were removed by centrifugation and the cell-free culture medium was centrifuged at 100,000 xg for 90 min. The pellets thus obtained were washed once in phosphate-buffered saline (pH 7.4), recentrifuged, and lyophilized, and the material was subjected to SDS gel electrophoresis. (Lane 1) Transferrin receptor isolated by a transferrin affinity column. 186 kDa and 94 kDa are the transferrin receptor. 78 kDa is transferrin. (Lane 2) Peptides from the culture medium of cells preincubated with nonimmune serum. The incubation medium contained transferrin. (Lane 3) is the same as lane 2, but without transferrin. (Lane 4) The cells were preincubated with immune serum, and the incubation medium contained transferrin. (Lane 5) Is the same as lane 4, but without transferrin. Note that incubation with transferrin leads to isolation of a peptide with a molecular weight characteristic of transferrin (78 kDa). Incubation with antibody leads to the isolation of a peptide with a molecular weight characteristic of the heavy chain of immunoglobulin (53 kDa). In all cases peptides of 94 and 186 kDa appear. R: receptor. TF: transferrin. IgG: heavy chain of IgG. Reprinted with permission from Pan & Johnstone, Cell, (1983) 33; 967.

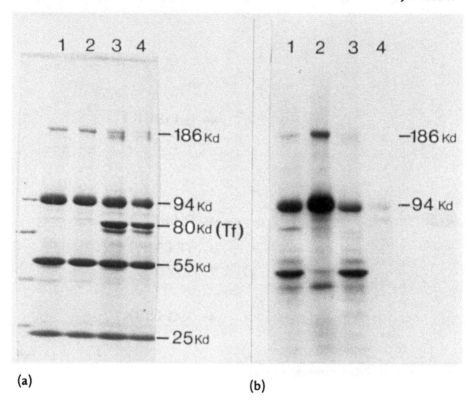

Figure 5 *Phosphorylation of immunoprecipitated receptors from fresh membranes and exocytosed vesicles by using purified protein kinase C and cyclic AMP-dependent protein kinase.* Immunoprecipitates of the transferrin receptor from plasma membranes and released vesicles were phosphorylated with exogenous kinases for 30 min at 22°C. The immunoprecipitates were subjected to SDS gel electrophoresis and radioautography. Note that in lanes 3 and 4 (panel a) that transferrin (Tf; 80 kDa) coprecipitates with immunoprecipitates of the receptor obtained from vesicles. (a) Coomassie Blue stain of immunoprecipitates from plasma membranes (lanes 1 and 2) and from exocytosed vesicles (lanes 3 and 4). Lane 1, immunoprecipitates from plasma membranes phosphorylated with cyclic AMP-dependent protein kinase; lane 2, immunoprecipitates from plasma membranes phosphorylated with protein kinase C; lane 3, immunoprecipitates from vesicles phosphorylated with cyclic AMP-dependent protein kinase; lane 4, immunoprecipitates from vesicles phosphorylated with protein kinase C. (b) Radioautograph corresponding to (a). Reprinted with permission from Adam and Johnstone, Biochem. J. (1987) 242, 191.

The externalized receptor retains transferrin binding activity [162] and the fatty acyl groups that were incorporated during incubation in vitro [61]. Although the transferrin receptor of sheep reticulocytes can be labeled posttranslationally with $^{32}P_i$ in intact cells (or with γ-^{32}P-ATP in isolated plasma membranes), cells labeled with ^{32}P and cultured for up to 40 hr at 37°C do not show a labeled receptor in the externalized vesicles [133a]. Moreover, although exogenous protein kinase C will phosphorylate the immunoprecipitated transferrin receptor obtained from plasma membranes, almost no phosphorylation is observed with protein kinase C if immunoprecipitates of the receptor from the externalized vesicles are used as substrate [133a] (Figure 5).

Phosphorylation of the immunoprecipitated receptor is not refractive to phosphorylation by all kinases because cAMP activated kinase does phosphorylate the externalized receptor of the vesicles after immunoprecipitation, albeit to a lesser extent than with immunoprecipitates from the plasma membrane as substrate. The data are consistent with but do not prove that the segregated receptor, prepared for externalization, has lost the capacity to undergo phosphorylation by the endogenous kinase and exogenous protein kinase C [133b]. The physiological significance of the change is unclear except to suggest that a subtle change in the receptor has occurred and may be involved in signaling the segregation of the receptor between the recycling pool and the pool designated for externalization. It is self-evident that during the final hours of reticulocyte maturation, iron uptake continues at a decreasing rate as the transferrin receptor and the iron-utilizing machinery are lost from the cell.

The 70–72 kDa protein always accompanies the receptor in the externalized vesicle and the relative amounts of 94 kDa and ~70–72 kDa are approximately equal. With better resolution, two species of the 70–72 kDa are observed: 70–71 kDa and 72 kDa, respectively. Fractionation of vesicles by an immunoaffinity column has shown that the 70–72 kDa protein is in the same population of vesicles as the receptor [175]. There does not appear to be a covalent linkage between the receptor and the 70–72-kDa peptides, and the reason for their close association has not been established. Both 70 and 72 kDa peptides have almost identical iodotyrosyl peptide maps [181]. The identity of the 70–72 kDa protein has now been established as the clathrin uncoating ATPase [181]. We have speculated that the association of the 70–72-kDa protein with vesicles containing the receptor or the receptor itself may target the vesicles and receptor for externalization (Figure 6). This question requires further probing.

Although the major peptides in the exosomes are the uncoating ATPase and the transferrin receptor, the population of vesicles contains many other minor proteins detectable by silver staining of SDS gels. No function has been specifically assigned to any of the other peptides, but binding assays and measurements of enzyme activity have shown that a number of known plasma membrane activities, which diminish in the cell during maturation, are found in the vesicles.

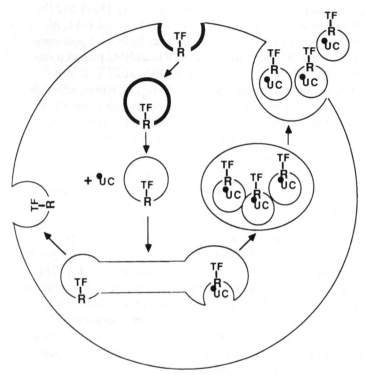

Figure 6 *Possible scheme for the externalization of the transferrin receptor in vesicles.* The erythrocyte protein related to the clathrin-uncoating protein (UC) first uncoats endocytosed transferrin receptor (R) and transferrin (TF) containing coated vesicles (clathrin coat represented by a bold line) which then fuse to form larger vesicles. UC then tags aged or senescent receptors, targeting them for a reendovesiculation event forming multivesicular elements. The multivesicular structures then fuse with the plasma membrane with subsequent secretion of the small vesicles containing externally oriented receptor and internal uncoating protein. Reprinted with permission from J. Q. Davis et al. J. Biol. Chem. (1986) 261; 15371.

These include protein kinase activity, acetylcholinesterase, cytochalasin B binding (glucose carrier) amino acid transport activity, and the nucleoside carrier (Johnstone et al. [176]).

It is noteworthy that all activities detected in the exosomes have earlier been shown to decrease in red blood cells of various species during maturation [173, 182–185] including sheep red blood cells. The insulin receptor has also been detected in the exosomes (C. C. Yip and R. M. Johnstone, unpublished observations, 1986) and is known to diminish during red blood cell maturation [186]. Although the β adrenergic and the fibronectin receptors are also known to decrease during reticulocyte maturation [187,188], these receptors have not yet been shown in the exosomes. It is probably significant that a major membrane protein (the anion exchange protein), which has not been shown to diminish during red blood cell maturation [189], does not appear in the externalized vesicles (R. M. Johnstone, unpublished observations, 1985).

The full extent of the activities in the exosomes is not known. Whether exosomes also contain components of other subcellular structures is not yet definitive. Evidence has been obtained for lysosomal but not mitochondrial enzymes in the sheep exosomes [190]. An entirely different mechanism for the degradation and removal of mitochondria in maturing reticulocytes has been proposed by Rapoport (for complete review see [170]). No evidence for the presence of cytosolic enzymes (such as lactic dehydrogenase glucose 6-phosphate dehydrogenase) in the exosome fraction has been obtained, although hemoglobin is present [176].

Not all activities externalized appear to follow the same time course of loss from the cell. Thus, transferrin receptor loss follows the pattern of overall maturation (RNA loss) [162], whereas lysosomal enzymes from sheep reticulocytes seem to be lost with a $t_{1/2}$ of ~24 hr [190]. If all vesicles are externalized by the same mechanism, the composition of the vesicles would be expected to show variations with time, with early and late vesicles showing different proportions of the different activities. No experimental evidence is as yet available in answer to this question. Nor is it known whether each exosome carries a mixture of all the activities, a limited number of activities, or whether each one carries a single activity.

The phospholipid composition of the exosome from sheep reticulocytes is remarkably similar to that of the sheep reticulocyte plasma membrane with its high sphingomyelin content, which differs from other cellular elements of the blood [176]. The presence of the characteristic red blood cell membrane lipids in the exosome attests to the fact that the exosomes are of red blood cell origin.

Because a substantial number of membrane-bound activities are extruded with maturation, the physiological role of the exocytic process may be to shed proteins superfluous to the mature cell. For example, the retention of the transferrin receptor and an endocytic pathway might be toxic in cells no longer capable of synthesizing hemoglobin. Although lysosomal enzymes are detectable in reticulocytes [191], these activities are also diminishing rapidly and the exocytic pathway may be the ultimate route by which remaining intracellular membranes are removed [192]. Thus, the mature red blood cell retains the minimal activity required to sustain its existence until further changes designate it for removal from the circulation. It would be interesting to determine at what stage of the maturation cycle exosome formation starts (early in the establishment of the erythroid cell line, after the final stage of cell division, or after the nucleus is expelled) and whether exocytosis is a major route for remodeling the membrane during maturation. In this regard, it is interesting to note that Wraith and Chesterton [193] concluded that the major change in red blood cell membrane protein composition occurs during transformation of the reticulocyte into the erythrocyte. Early erythroid cells and reticulocytes show great similarity in membrane protein composition.

Investigators have long recognized that the peripheral reticulocyte undergoes substantial changes as it becomes the erythrocyte. Several investigators have shown membrane shedding and release of vesicles, which they have associated with the maturation process [166–169]. In contrast to earlier work, the exosome route described for the externalization of the transferrin receptor in sheep reticulocytes involves intracellular processing and sorting to form specialized vesicles to prepare for receptor externalization. This proposal is different from the blebbing and surface protein aggregation, which do not depend on intracellular processing and the intermediate formation of MVBs.

Quantitatively, however, it has not yet been possible to assess whether exosome formation is the major or a relatively minor route for removing plasma membrane components. Does cellular degradation of plasma membrane proteins contribute to the overall loss of the transferrin receptor during maturation? Studies on Na/K ATPase loss have suggested a mechanism of intracellular proteolysis by ATP-dependent proteases [185]. However, no evidence was actually shown for protein degradation [185]. Exosomes, however, have been reported to bind an antibody against the Na/K ATPase (R. Blostein, personal communication, 1987, and ouabain binding has been detected in them [176]).

The systems known to be involved in red blood cell protein degradation [194, 195] have not been shown to degrade membrane proteins. Rapoport and colleagues [170] have reported that lipid oxidizing enzymes, induced in the late stages of development, prime the mitochondria for destruction and thus make protein synthesis essential for mitochondrial destruction. With the transferrin receptor, there is no evidence that inhibition of protein synthesis affects exosome formation (R. M. Johnstone, unpublished observations, 1984).

Whether intracellular, MVBs bearing the transferrin receptor exist in cells other than reticulocytes is not known. It may be that these bodies are unique to terminally differentiated cells such as red blood cells (and perhaps lens cells), which lose intracellular organelles as they mature. These terminal structures may represent a type of prelysosomal structure from which cellular membrane components are exocytosed. It is noteworthy in this regard that the transferrin receptor can be detected in a 100,000 g pellet obtained from the plasma of phlebotomized sheep, but not unbled animals [176]. These data are consistent with the observation that exocytosis is a normal mechanism for shedding of the transferrin receptor.

IX. THE TRANSFERRIN RECEPTOR IN
DIAGNOSIS AND MEDICINE

Because the transferrin receptor is a highly antigenic molecule, it was interesting to learn [196] that an autoimmune anemia has been detected in humans due to the presence of antibodies against the transferrin receptor. Because iron use is essential for life, the antibody cannot completely suppress iron uptake. However, the existence of alternate (and normally minor pathways) for iron uptake might also contribute to cell survival in the absence of functional transferrin receptors. An example of a mutant cultured cell line capable of survival in absence a normal pathway for endocytosis of transferrin has been reported [197].

Attempts have been made to use the fact that most cells have transferrin receptors to deliver toxins into the cell [198-203,203a] by conjugating either transferrin or antireceptor antibodies to well-known toxins such as ricin A or the diptheria toxin. Because growing cells generally have increased numbers of transferrin receptors, toxin delivery would affect the rapidly growing population first. The use of mouse antibodies in humans (whether conjugated or not) would present a toxicity problem in its own right. The presence of a natural human antibody against the receptor might present a better source of material for making toxic conjugates. If cancerous cells were less discriminating than normal cells vis-à-vis binding of heterologous transferrin, conjugates of the latter might present some opportunity for selective uptake of the toxin. A concise review of attempts to manufacture toxic conjugates using antireceptor antibodies or transferrin has appeared [203]. The high concentration of transferrin receptors on erythroid precursor cells may always preclude the safe use of such ligands because of their potential toxicity to the blood-forming elements. Shannon et al. have also shown that antibodies against transferrin or the receptor selectively inhibit burst formation [204].

The transferrin receptor has been implicated as a recognition site for killer cells [205,206]. There has not been universal acceptance of this idea, with several investigators failing to see a significant correlation between killer cell targeting and the specific expression of the transferrin receptor [207,208].

X. DIAGNOSIS OF MALIGNANCY
AND ERYTHROPOIESIS

The presence of transferrin receptor in the circulation has also been proposed as a diagnostic tool [203,203a,203b,209]. During some malignant states in humans an abnormally high level of transferrin receptors has been detected in the circulation, unassociated with cells in the circulation. In additional studies, circulating receptors in human sera have been correlated with elevated erythropoiesis. Not only were increased receptors found during the expression of the anemia, but values fell to normal range when treatment was successful. It was proposed that circulating transferrin receptors in serum in humans could provide an index of bone marrow erythropoiesis [203a]. An increased level of transferrin receptor-positive lymphocytes has been associated with an increased rate of rejection in heart-transplant patients [209]. Conditions that increase circulating reticulocytes may thus lead to an increase in circulating cell-free transferrin receptors as well as cell-associated receptors and may provide a diagnostic tool for reticulocytosis (see above).

XI. THE TRANSFERRIN RECEPTOR AND CELL DIFFERENTIATION

The recycling of the transferrin receptor appears to be essential for differentiation of erythroid cells in culture. Inhibition of receptor recycling by an antibody to the transferrin receptor leads to arrest of cell development followed by cellular disintegration [210].

The expression of the transferrin receptor in normal T-lymphocytes appears to be required for lymphocyte activation by the interleukin-2 (IL-2) receptor. If transferrin receptor expression was blocked by antibodies against the receptor, DNA synthesis was blocked although IL-2 expression was normal. Furthermore, DNA synthesis was reduced only when anti–IL-2 antibodies were added before the expression of transferrin receptors [211]. Thus, the normal immune response may be tied to the expression of the transferrin receptor. Such an observation puts at great risk the idea of using toxins conjugated to transferrin for selective therapeutic effects unless highly specific means are available to direct the toxins to specific types of cells and to avoid the lymphocytes and blood-forming elements.

Not only higher animal cells, but parasites that develop in higher organisms, appear to depend on transferrin receptors. Malarial parasites grow in a sea of hemoglobin, but apparently do not use heme iron. The malarial parasite synthesizes its own transferrin receptor and reinserts its transferrin receptor into the red blood cell membrane, thus restoring to the cell a function it lost in its adolescence [212].

It would be of some interest to determine whether (and how) the newly synthesized receptor differs from the normal receptor and whether its synthesis is regulated in an analogous manner to that of the developing red blood cell.

XII. CONCLUSION

In less than 10 years of intense study, the description of the transferrin receptor has evolved from a descriptive science to a science at the molecular level. The gene structure is beginning to be understood, the location of the gene on chromosome 3 has been shown [213], the regulation of receptor expression is being unfolded and the cellular route of the receptor and transferrin are clearer. Little of the new and exciting information has been particularly informative on the mechanisms for getting iron out of the endosome and into iron proteins. Very recent work has shown that iron is removed from transferrin at a very early stage and that when iron use is prevented, Fe^{3+} and transferrin recycle independently, an observation that is contrary to most of current thinking on receptor/transferrin processing [81]. Moreover, iron does not appear to leave the endosome in absence of an iron acceptor in the cell [100a].

Until there is satisfactory understanding of iron processing, there remains a major gap in our understanding of how the transferrin receptor and transferrin provide cells with iron. Thus, those of us who have an interest in iron metabolism can look forward to exciting new developments.

REFERENCES

1. Jandl, J. H., and Katz, J. H. (1963). The cell to plasma transferrin cycle. *J. Clin. Invest.* 42:314–326.
2. May, W. S., Jr., and Cuatrecasas, P. (1985). Transferrin receptor: its biological significance. *J. Membr. Biol.* 88:205–215.
3. Trowbridge, I. S., Newman, R. A., Domingo, D. L., and Sauvage, C. (1984). Transferrin receptors: structure and function. *Biochem. Pharmacol.* 33:925–932.
4. Hanover, J. A., and Dickson, R. B. (1985). *The Transferrin Receptor in Endocytosis*, I. Pastan and M. C. Willingham (Eds.). Plenum Press, New York, pp. 131–162.
5. *Proteins of Iron Storage and Transport* (1985). G. Spik, J. Montreuil, R. R. Chrichton, and J. Mazurier (Eds.). Elsevier Science Publishers, Amsterdam.
6. Newman, R., Schneider, C., Sutherland, R., Vodinelich, L., and Greaves, M. F. (1982). The transferrin receptor. *Trends Biochem. Sci.* 7:397–400.
7. Dautry-Varsat, A. (1986). Receptor-mediated endocytosis: the intracellular journey of transferrin and its receptor. *Biochimie* 68:375–381.
8. Robinson, J., Sieff, C., Delia, D., Edwards, P. A., and Greaves, M. (1981). Expression of cell-surface HLA-DR, HLA-ABC and glycophorin during erythroid differentiation. *Nature* 289:68–71.
9. Hamilton, T. A., Wada, H. G., and Sussman, H. H. (1979). Identification of transferrin receptors in the surface of human cultured cells. *Proc. Natl. Acad. Sci. USA* 76:6406–6410.
10. Enns, C. A., Shindelman, J. E., Tonik, S. E., and Sussman, H. H. (1981). Radioimmunochemical measurement of the transferrin receptor in human trophoblast and reticulocyte membranes with a specific anti-receptor antibody. *Proc. Natl. Acad. Sci. USA* 78:4222–4225.
11. Trowbridge, I. S., and Omary, M. B. (1981). Human cell surface glycoprotein related to cell proliferation is the receptor for transferrin. *Proc. Natl. Acad. Sci. USA* 78:3039–3043.
12. Larrick, J. W., and Cresswell, P. (1979). Modulation of cell surface iron transferrin receptors by cellular density and state of activation. *J. Supramol. Struc.* 11:579–586.

13. Seligman, P. A., Schleicher, R. B., and Allen, R. H. (1979). Isolation and characterization of the transferrin receptor from human placenta. *J. Biol. Chem.* 254:9943–9946.

14. Judd, W., Poodry, C. A., and Strominger, J. L. (1980). Novel surface antigen expressed on dividing cells but absent from nondiving cells. *J. Exp. Med.* 152:1430–1435.

15. Reinherz, E. L., Kung, P. C., Goldstein, G., Levey, R. H., and Schlossman, S. F. (1980). Discrete stages of human intrathymic differentiation: analysis of normal thymocytes and leukemic lymphoblasts of T-cell lineage. *Proc. Natl. Acad. Sci. USA* 77:1588–1592.

16. Pan, B. T., Blostein, R., and Johnstone, R. M. (1983). Loss of the transferrin receptor during the maturation of sheep reticulocytes in vitro. *Biochem. J.* 210:37–47.

17. Lebman, D., Trucco, M., Bottero, L., Lange, B., Pessano, S., and Rovera, G. (1982). A monoclonal antibody that detects expression of transferrin receptors in human erythroid precursor cells. *Blood* 59:671–678.

18. Sutherland, R., Delia, D., Schneider, C., Newman, R., Kemshead, J., and Greaves, M. (1981). Ubiquitous cell-surface glycoprotein on tumor cells is proliferation-associated receptor for transferrin. *Proc. Natl. Acad. Sci. USA* 78:4515–4519.

19. Trowbridge, I. B., and Newman, R. A. (1984). Monoclonal antibodies to transferrin receptors. In *Antibodies to Receptors*, Chapman and Hall, London, Series B, Vol. 17, pp. 237–261.

20. Schulman, H. M., and Nelson, R. A. (1969). Antibody to rabbit reticulocytes. *Nature* 223:623–624.

21. Evered D., and Collins, G. M. (Eds.). (1982). Membrane recycling. In: *CIBA Sympodium* Vol. 92, Pitman Press, London, pp. 1–254.

22. Pastan, I., and Willingham, M. C. (1985). Pathway of Endocytosis. In: *Endocytosis*, I. Pastan and M. C. Willingham (Eds.). Plenum Press, New York, pp. 1–44.

23. Mellman, I., Fuchs, S., and Helenius, A. (1986). Acidification of the endocytic and exocytic pathways. *Annu. Rev. Biochem.* 55:663–700.

24. Anderson, R. G., Goldstein, J. L., and Brown, M. S. (1976). Localization of low density lipoprotein receptors on plasma membrane of normal human fibroblasts and their absence in cells from a familial hypercholesterolemia homozygote. *Proc. Natl. Acad. Sci. USA* 73:2434–2438.

25. Martonosi, A. N. (1982). *Membranes and Transport*, Vol. 1, Plenum Press, New York.

26. Tormey, D. C., Imrie, R. C., and Mueller, G. C. (1972). Identification of transferrin as a lymphocyte growth promoter in human serum. *Exp. Cell. Res.* 74:163–169.

27. Barnes, D., and Sato, G. (1980). Serum-free cell culture: a unifying approach. *Cell* 22:649–655.

28. Dillmer-Centerlind, M-L., Hammarstrom, S., and Perlman, P. (1980). Transferrin can replace serum for in vitro growth of mitogen-stimulated T lymphocytes. *Eur. J. Immunol.* 9:942–948.

29. Popiela, H., Taylor, D., Ellis, S., Beach, R., and Festoff, B. (1984). Regulation of mitotic activity and the cell cycle in primary chick muscle cells by neurotransferrin. *J. Cell Physiol.* 119:234–240.
30. Ekblom, P., Thesleff, I., Saxen, L., Miettinen, A., and Timpl, R. (1983). Transferrin as a fetal growth factor; acquisition of responsiveness related to embryonic induction. *Proc. Natl. Acad. Sci. USA* 80:2651–2655.
31. Awai, M., and Brown, E. B. (1963). Studies of the metabolism of I-131-labeled human transferrin. *J. Lab. Chem. Med.* 61:363–396.
32. Paoletti, C., Durand, M., Gosse, C. H., and Boiron, M. (1958). Absence de consommation de la siderophiline au cours de la synthèse de hemoglobin in vitro. *Rev. Fr. Etude Clin. Biol.* 3:259–261.
33. van Bockxmeer, F. M., and Morgan, E. H. (1979). Transferrin receptors during rabbit reticulocyte maturation. *Biochim. Biophys. Acta* 584:76–83.
34. Paterson, S., and Morgan, E. H. (1980). Effect of changes in the ionic environment of reticulocytes on the uptake of transferrin-bound iron. *J. Cell Physiol.* 105:489–502.
35. Karin, M., and Mintz, B. (1981). Receptor-mediated endocytosis of transferrin in developmentally totipotent mouse teratocarcinoma stem cells. *J. Biol. Chem.* 256:3245–3252.
36. Ciechanover, A., Schwartz, A. L., Dautry-Varsat, A., and Lodish, H. F. (1983). Kinetics of internalization and recycling of transferrin and the transferrin receptor in a human hepatoma cell line. Effect of lysosomotropic agents. *J. Biol. Chem.* 258:9681–9689.
37. Omary, M. B., and Trowbridge, I. S. (1981). Biosynthesis of the human transferrin receptor in cultured cells. *J. Biol. Chem.* 256:12888–12892.
38. Klausner, R. D., Harford, J. B., Rao, K., Mattia, E., Weissman, A. M., Rouault, T., Ashwell, G., and van Renswoude, J. (1985). Molecular aspects of the regulation of cellular iron metabolism. In *Protein of Iron Storage and Transport*, G. Spik, J. Montreuil, R. R. Chrichton, and J. Mazurier (Eds.). Elsevier Press, Amsterdam, pp. 111–122.
39. Morgan, E. H., and Laurell, B. (1963). Studies on the exchange of iron between transferrin and reticulocytes. *Br. J. Haematol.* 9:471–483.
40. Nunez, M. T., and Glass, J. (1983). The transferrin cycle and iron uptake in rabbit reticulocytes. Pulse studies using ^{59}Fe, ^{125}I-labeled transferrin. *J. Biol. Chem.* 258:9676–9680.
41. Glass, J., and Nunez, M. T. (1986). Amines as inhibitors of iron transport in rabbit reticulocytes. *J. Biol. Chem.* 261:8298–8302.
42. Wada, H. G., Hass, P. E., and Sussman, H. H. (1979). Transferrin receptor in human placental brush border membranes. Studies on the binding of transferrin to placental membrane vesicles and the identification of a placental brush border glycoprotein with high affinity for transferrin. *J. Biol. Chem.* 254:12629–12635.
43. Omary, M. B., and Trowbridge, I. S. (1981). Covalent binding of fatty acid to the transferrin receptor in cultured human cells. *J. Biol. Chem.* 256:4715–4718.

44. Hu, H. Y., and Aisen, P. (1978). Molecular characteristics of the transferrin-receptor complex of the rabbit reticulocyte. *J. Supramol. Struc.* 8: 349–360.
45. Schneider, C., Sutherland, D. R., Newman, R. A., and Greaves, M. F. (1982). Structural features of the cell surface receptor for transferrin that is recognized by the monoclonal antibody OKT9. *J. Biol. Chem.* 257: 8516–8522.
46. Markelonis, G. J., Oh, T. H., Park, L. P., Cha, C. Y., Sofia, C. A., Kim, J. W., and Azari, P. (1985). Synthesis of the transferrin receptor by cultures of embryonic chicken spinal neurons. *J. Cell Biol.* 100:8–17.
47. Van Agthoven, A., Goridis, C., Naquet, P., Pierres, A., and Pierres, M. (1984). Structural characteristics of the mouse transferrin receptor. *Eur. J. Biochem.* 140:433–440.
48. Schmidt, J. A., Marshall, J., and Hayman, M. J. (1985).. Identification and characterization of the chicken transferrin receptor. *Biochem. J.* 232:735–741.
49. Bleil, J. D., and Bretcher, M. S. (1982). Transferrin receptor and its recycling in HeLa cells. *EMBO J.* 1:351–355.
50. McClelland, A., Kunn, L. C., and Ruddle, F. H. (1984). The human transferrin receptor gene: genomic organization, and the complete primary structure of the receptor deduced from a cDNA sequence. *Cell* 39:267–274.
51. Schneider, C., Owen, M. J., Banville, D., and Williams, J. G. (1984). Primary structure of human transferrin receptor deduced from the mRNA sequence. *Nature* 311:675–678.
52. Miskimins, W. K., McClelland, A., Roberts, M. P., and Ruddle, F. H. (1986). Cell proliferation and expression of the transferrin receptor gene. *J. Cell Biol.* 103:1781–1788.
53. Zerial, M., Melancon, P., Schneider, C., and Garoff, H. (1986). The transmembrane segment of the human transferrin receptor functions as a signal peptide. *EMBO J.* 5:1543–1550.
54. May, W. S., Sahyoun, N., Jacobs, S., Wolf, M., and Cuatrecasas, P. (1985). Mechanism of phorbol diester-induced regulation of surface transferrin receptor involves the action of activated protein kinase C and an intact cytoskeleton. *J. Biol. Chem.* 260:9419–9426.
55. Klausner, R. D., Harford, J., and van Renswoude, J. (1984). Rapid internalization of the transferrin receptor in K562 cells is triggered by ligand binding or treatment with a phorbol ester. *Proc. Natl. Acad. Sci. USA* 81:3005–3009.
56. Hunt, R. C., Ruffin, R., and Yang, Y. S. (1984). Alterations in the transferrin receptor of human erythroleukemic cells after induction of hemoglobin synthesis. *J. Biol. Chem.* 259:9944–9952.
57. Johnstone, R. M., Adam, M., Turbide, C., and Larrick, J. (1984). Phosphorylation of the transferrin receptor in isolated sheep plasma membranes. *Can. J. Biochem. Cell Biol.* 62:927–934.
58. May, W. S., Lapetina, E. G., and Cuatrecasas, P. (1986). Intracellular activation of protein kinase C and regulation of the surface transferrin receptor

by diacylglycerol is a spontaneously reversible process that is associated with rapid formation of phosphatidic acid. *Proc. Natl. Acad. Sci. USA* 83: 1281–1284.

59. May, W. S., Jacobs, S., and Cuatrecasas, P. (1984). Association of phorbol induced hyperphosphorylation and reversible regulation of transferrin membrane receptors in HL-60 cells. *Proc. Natl. Acad. Sci. USA* 81:2016–2020.

59a. Cox, T. M., O'Donnell, M. W., Aisen, P., and London, I. M. (1985). Hemin inhibits internalization of transferrin by reticulocytes and promotes phosphorylation of the membrane transferrin receptor. *Proc. Natl. Acad. Sci. USA* 82:5170–5174.

60. Davis, R. J., Johnson, G. L., Kelleher, D. G., Anderson, J. K., Mole, J. E., and Czech, M. P. (1986). Identification of serine 24 as the unique site on the transferrin receptor phosphorylated by protein kinase C. *J. Biol. Chem.* 261:9034–9041.

61. Adam, M., Rodriquez, A., Turbide, C., Larrick, J., Meighen, E., and Johnstone, R. M. (1984). In vitro acylation of the transferrin receptor. *J. Biol. Chem.* 259:15460–15463.

62. Nunez, M. T., Glass, J., Fisher, S., Larides, L. M., Lenk, E. M., and Robinson, S. H. Transferrin receptor in developing murine erythroid cells. *Br. J. Haematol.* 36:519–526.

63. Kailis, S. G., and Morgan, E. H. (1974). Transferrin and iron uptake by rabbit bone marrow cells in vitro. *Br. J. Haematol.* 28:37–52.

64. Frazier, J. L., Caskey, J. H., Yoffe, M., and Seligman, P. A. (1982). Studies of the transferrin receptor on both human reticulocytes and nucleated human cells in culture: comparison of factors regulating receptor density. *J. Clin. Invest.* 69:853–865.

65. Iacopetta, B. J., Morgan, E. H., and Yeoh, G. C. T. (1982). Transferrin receptors and iron uptake during erythroid cell development. *Biochim. Biophys. Acta* 687:204–210.

66. Adam, M., Wu, C., Turbide, C., Larrick, J., and Johnstone, R. M. (1986). Evidence for a pool of non recycling transferrin receptors in peripheral sheep reticulocytes. *J. Cell Physiol.* 127:8–16.

67. Klausner, R. D., van Renswoude, J., Ashwell, G., Kempf, C., Schechter, A. N., Dean, A., and Bridges, K. R. (1983). Receptor-mediated endocytosis of transferrin in K562 cells. *J. Biol. Chem.* 258:4715–4274.

68. Schulman, H. M., Wilczynska, A., and Ponka, P. (1981). Transferrin and iron uptake by human lymphoblastoid and K-562 cells. *Biochem. Biophys. Res. Commun.* 100:1523–1530.

69. Faulk, W. P., Hsi, B-L, and Stevens, P. J. (1980). Transferrin and transferrin receptors in carcinoma of the breast. *Lancet* 2:390–392.

70. Goding, J. W., and Burns, G. F. (1981). Monoclonal Antibody OKT-9 recognizes the receptor for transferrin on human acute lymphocytic leukemia cells. *J. Immunol.* 127:1256–1258.

71. Fernandez-Pol, J. A., and Klos, D. J. (1980). Isolation and characterization of normal rat kidney cell membrane proteins with affinity for transferrin. *Biochemistry* 19:3904–3912.

72. Ciechanover, A., Schwartz, A. L., and Lodish, H. F. (1983). The asialogly-coprotein receptor internalizes and recycles independently of the transferrin and insulin receptors. *Cell* 32:267–275.

73. Harding, C., and Stahl, P. (1983). Transferrin recycling in reticulocytes: pH and iron are important determinants of ligand binding and processing. *Biochem. Biophys. Res. Commun.* 113:650–658.

74. Pellicci, P. G., Testa, U., Thomopoulos, P., Tabilio, A., Vainchenker, W., Titeux, M., Gourdin, M. F., and Rochant, H. (1984). Inhibition of transferrin binding and iron uptake of hematopoietic cell lines by phorbol esters. *Leuk. Res.* 8:597–609.

75. van Renswoude, J., Bridges, K. R., Harford, J. B., and Klausner, R. D. (1982). Receptor-mediated endocytosis of transferrin and the uptake of Fe in K562 cells: identification of a nonlysosomal acidic compartment. *Proc. Natl. Acad. Sci. USA* 79:6186–6190.

76. Ward, J. H., Kushner, J. P., and Kaplan, J. (1982). Transferrin receptors of human fibroblasts. Analysis of receptor properties and regulation. *Biochem. J.* 208:19–26.

77. Lim, B. C., and Morgan, E. H. (1984). Transferrin endocytosis and the mechanism of iron uptake by reticulocytes in the toad (Bufo marinus). *Comp. Biochem. Physiol. A.* 79:317–323.

78. Taetle, R., Rhyner, K., Castagnola, J., To, D., and Mendelsohn, J. (1985). Role of transferrin, Fe, and transferrin receptors in myeloid leukemia cell growth. Studies with an antitransferrin receptor monoclonal antibody. *J. Clin. Invest.* 75:1061–1067.

79. Morgan, E. H. (1964). The interaction between rabbit, human and rat transferrin and reticulocytes. *Br. J. Haematol.* 10:442–452.

80. Anderson, R. G., Brown, M. S., and Goldstein, J. L. (1977). Role of the coated endocytic vesicle in the uptake of receptor-bound low density lipoprotein in human fibroblasts. *Cell* 10:351–364.

81. Nunez, M. T., and Glass, J. (1985). Iron uptake in reticulocytes. Inhibition mediated by the ionophores monensin and nigericin. *J. Biol. Chem.* 260: 14707–14711.

81a. Jing, S., and Trowbridge, I. S. (1987). Identification of the intermolecular disulfide bonds of the human transferrin receptor and its lipid attachment site. *EMBO J.* 6:327–331.

82. Sullivan, A. L., Grasso, J. A., and Weintraub, L. R. (1976). Micropinocytosis of transferrin by developing red cells: an electron-microscopic study utilizing ferritin-conjugated transferrin and ferritin-conjugated antibodies to transferrin. *Blood* 47:133–143.

83. Willingham, M. C., Hanover, J. A., Dickson, R. B., and Pastan, I. (1984). Morphologic characterization of the pathway of transferrin endocytosis and recycling in human KB cells. *Proc. Natl. Acad. Sci. USA* 81:175–179.

84. Watts, C. (1985). Rapid endocytosis of the transferrin receptor in the absence of bound transferrin. *J. Cell Biol.* 100:633–637.

85. Pan, B. T., Teng, K., Wu, C., Adam, M., and Johnstone, R. M. (1985). Electron microscopic evidence for externalization of the transferrin receptor in vesicular form in sheep reticulocytes. *J. Cell. Biol.* 101:942–948.

86. Hopkins, C. R. (1985). The appearance and internalization of transferrin receptors at the margins of spreading tumor cells. *Cell* 40:199–208.

87. Larrick, J. W., Enns, C., Raubitschek, A., and Weintraub, H. (1985). Receptor-mediated endocytosis of human transferrin and its cell surface receptor. *J. Cell Physiol.* 124:287–287.

88. Enns, C. A., Larrick, J. W., Suomalainen, H., Schroder, J., and Sussman, H. H. (1983). Co-migration and internalization of transferrin and its receptor on K562 cells. *J. Cell Biol.* 97:579–585.

89. Hopkins, C. R., and Trowbridge, I. S. (1983). Internalization and processing of transferrin and the transferrin receptor in human carcinoma A431 cells. *J. Cell Biol.* 97:508–521.

90. Stein, B. S., and Sussman, H. H. (1986). Demonstration of two distinct transferrin receptor recycling pathways and transferrin-independent receptor internalization in K562 cells. *J. Biol. Chem.* 261:10319–10331.

91. Wiessman, A. M., Klausner, R. D., Rao, K., and Harford, J. B. (1986). Exposure of K562 cells to anti-receptor monoclonal antibody OKT9 results in rapid redistribution and enhanced degradation of the transferrin receptor. *J. Cell Biol.* 102:951–958.

92. Geuze, H. J., Slot, J. W., Strous, G. J., Lodish, H. F., and Schwartz, A. L. (1983). Intracellular site of asialoglycoprotein receptor-ligand uncoupling: double-label immunoelectron microscopy during receptor-mediated endocytosis. *Cell* 32:277–287.

93. Hanover, J. A., Willingham, M. C., and Pastan, I. (1984). Kinetics of transit of transferrin and epidermal growth factor through clathrin-coated membranes. *Cell* 39:283–293.

94. Tycko, B., and Maxfield, F. R. (1982). Rapid acidification of endocytic vesicles containing alpha 2-macroglobulin. *Cell* 28:643–651.

95. Helenius, A., Kartenbeck, J., Simons, K., and Fries, E. (1980). On the entry of Semliki forest virus into BHK-21 cells. *J. Cell Biol.* 84:404–420.

96. Yamashiro, D. J., Fluss, S. R., and Maxfield, F. R. (1983). Acidification of endocytic vesicles by an ATP-dependent proton pump. *J. Cell Biol.* 97: 929–934.

97. Morgan, E. H. (1979). Studies on the mechanism of iron release from transferrin. *Biochim. Biophys. Acta* 580:312–326.

98. Ecarot-Charrier, B., Grey, V. L., Wilczynska, A., and Schulman, H. M. (1980). *Can. J. Biochem.* 58:418–426.

99. Dautry-Varsat, A., Ciechanover, A., and Lodish, H. F. (1983). pH and the recycling of transferrin during receptor-mediated endocytosis. *Proc. Natl. Acad. Sci. USA* 80:2258–2262.

100. Klausner, R. D., Ashwell, G., van Renswoude, J., Harford, J. B., and Bridges, K. R. (1983). Binding of apotransferrin to K562 cells: explanation of the transferrin cycle. *Proc. Natl. Acad. Sci. USA* 80:2263–2266.

100a. Bakkeren, D. L., deJeu-Jaspars, C. M. H., Kroos, M. J., and van Eijk, H. G. (1987). *Int. J. Biochem.* 19:179–186.

101. Paterson, S., Armstrong, N. J., Iacopetta, B. J., McArdle, H. J., and Morgan, E. H. (1984). Intravesicular pH and iron uptake by immature erythroid cells. *J. Cell Physiol.* 120:225–232.

102. Baenziger, J. U., and Fiete, D. (1982). Recycling of the hepatocyte asialo-glycoprotein receptor does not require delivery of ligand to lysosomes. *J. Biol. Chem.* 257:6007–6009.

103. Bridges, K., Harford, J., Ashwell, G., and Klausner, R. D. (1982). Fate of receptor and ligand during endocytosis of asialoglycoproteins by isolated hepatocytes. *Proc. Natl. Acad. Sci. USA* 79:350–354.

104. Regoeczi, E., Chindemi, P. A., Debanne, M. T., and Hatton, M. W. (1982). Dual nature of the hepatic lectin pathway for human asialotransferrin type 3 in the rat. *J. Biol. Chem.* 257:5431–5436.

105. Woods, J. W., Doriaux, M., and Farquhar, M. G. (1986). Transferrin receptors recycle to cis and middle as well as trans Golgi cisternae in Ig-secreting myeloma cells. *J. Cell Biol.* 103:277–286.

106. Iacopetta, B. J., Morgan, E. H., and Yeoh, G. C. (1983). Receptor-mediated endocytosis of transferrin by developing erythroid cells from the fetal rat liver. *J. Histochem. Cytochem.* 31:336–344.

107. Stein, B. S., Bensch, K. G., and Sussman, H. H. (1984). Complete inhibition of transferrin recycling by monensin in K562 cells. *J. Biol. Chem.* 259:14762–14772.

108. Lamb, J. E., Ray, F., Ward, J. H., Kushner, J. P., and Kaplan, J. (1983). Internalization and subcellular localization of transferrin and transferrin receptors in HeLa cells. *J. Biol. Chem.* 258:8751–8758.

109. Iacopetta, B. J., and Morgan, E. H. (1983). The kinetics of transferrin endocytosis and iron uptake from transferrin by rabbit reticulocytes. *J. Biol. Chem.* 258:9108–9115.

110. Harding, C., Heuser, J., and Stahl, P. (1983). Receptor-mediated endocytosis of transferrin and recycling of the transferrin receptor in rat reticulocytes. *J. Cell Biol.* 97:329–339.

111. Hopkins, C. R. (1983). Intracellular routing of transferrin and transferrin receptors in epidermoid carcinoma A431 cells. *Cell* 35:321–330.

112. Ajioka, R. S., and Kaplan, J. (1986). Intracellular pools of transferrin receptors result from constitutive internalization of unoccupied receptors. *Proc. Natl. Acad. Sci. USA* 83:6445–6449.

113. Snider, M. D., and Rogers, O. C. (1985). Intracellular movement of cell surface receptors after endocytosis: resialylation of asialo-transferrin receptor in human erythroleukemia cells. *J. Cell Biol.* 100:826–834.

114. Snider, M. D., and Rogers, O. C. (1986). Membrane traffic in animal cells: cellular glycoproteins return to the site of Golgi mannosidase. I. *J. Cell Biol.* 103:265–275.

115. Weigel, P. H., Clarke, B. L., and Oka, J. A. (1986). The hepatic galactosyl receptor system: two different ligand dissociation pathways are mediated by distinct receptor populations. *Biochem. Biophys. Res. Commun.* 140:43–50.

116. Hebbert, D., and Morgan, E. H. (1985). Calmodulin antagonists inhibit and phorbol esters enhance transferrin endocytosis and iron uptake by immature erythroid cells. *Blood* 65:758–763.

117. Hedley, D. W., and Musgrove, E. A. (1986). Transferrin receptor cycling by human lymphoid cells: lack of effect from inhibition of microtubule

assembly. *Biochem. Biophys. Res. Commun.* 138:1216–1222.

118. Beguinot, L., Lyall, R. M., Willingham, M. C., and Pasten, I. (1984). Down-regulation of the epidermal growth factor receptor in KB cells is due to receptor internalization and subsequent degradation in lysosomes. *Proc. Natl. Acad. Sci. USA* 81:2384–2388.

119. Carpenter, G., and Cohen, S. (1976). ^{125}I-labeled human epidermal growth factor. Binding, internalization, and degradation in human fibroblasts. *J. Cell Biol.* 71:159–171.

120. Vlodavsky, I., Brown, K. D., and Gospodarowicz, D. (1978). A comparison of the binding of epidermal growth factor to cultured granulosa and luteal cells. *J. Biol. Chem.* 253:3744–3750.

121. Fox, C. F., Wrann, M., Linsley, P., and Vale, R. (1979). Hormone-induced modification of EGF receptor proteolysis in the induction of EGF action. *J. Supramol. Struct.* 12:517–531.

122. Dunn, W. A., Connolly, T. P., and Hubbard, A. L. (1986). Receptor-mediated endocytosis of epidermal growth factor by rat hepatocytes: receptor pathway. *J. Cell Biol.* 102:24–36.

123. Corvera, S., and Czech, M. P. (1985). Mechanism of insulin action on membrane protein recycling: a selective decrease in the phosphorylation state of insulin-like growth factor II receptors in the cell surface membrane. *Proc. Natl. Acad. Sci. USA* 82:7314–7318.

124. Cushman, S. W., and Wardzala, L. J. (1980). Potential mechanism of insulin action on glucose transport in the isolated rat adipose cell. Apparent translocation of intracellular transport systems to the plasma membrane. *J. Biol. Chem.* 255:4758–4762.

125. Olefsky, J. M., Marshall, S., Berhanu, P., Saekow, M., Heidenreich, K., and Green, A. (1982). Internalization and intracellular processing of insulin and insulin receptors in adipocytes. *Metabolism* 31:670–690.

126. Krupp, M. N., and Lane, D. M. (1982). Evidence for different pathways for the degradation of insulin and insulin receptor in the chick liver cell. *J. Biol. Chem.* 257:1372–1377.

127. Fehlmann, M., Carpentier, J. L., van Obberghen, E., Freychet, P., Thamm, P., Saunders, D., Brandenburg, D., and Orci, L. (1982). Internalized insulin receptors are recycled to the cell surface in rat hepatocytes. *Proc. Natl. Acad. Sci. USA* 79:5921–5925.

128. Tietze, C., Schlesinger, P., and Stahl, P. (1982). Mannose-specific endocytosis receptor of alveolar macrophages: demonstration of two functionally distinct intracellular pools of receptor and their roles in receptor recycling. *J. Cell Biol.* 92:417–424.

129. Stahl, P., Schlesinger, Ph. H., Sigardson, E., Rodman, J. S., and Lee, Y. C. (1980). Receptor-mediated pinocytosis of mannose glycoconjugates by macrophages: characterization and evidence for receptor recycling. *Cell* 19:207–215.

130. Cheng, T. P. (1986). Redistribution of cell surface transferrin receptors prior to their concentration in coated pits as revealed by immunoferritin labels. *Cell Tissue Res.* 244:613–619.

131. Harding, C., and Stahl, P. (1984). Endocytosis and intracellular processing of transferrin and colloidal gold-transferrin in rat reticulocytes: demonstration of a pathway for receptor shedding. *Eur. J. Cell Biol.* 35:256-263.

132. Schneider, C., Asser, U., Sutherland, D. R., and Greaves, M. F. (1983). In vitro biosynthesis of the human cell surface receptor for transferrin. *FEBS Lett.* 158:259-264.

133. Cox, T. M., O'Donnell, M. W., Aisen, P., and London, I. M. (1985). Biosynthesis of the transferrin receptor in rabbit reticulocytes. *J. Clin. Invest.* 76:2144-2150.

133a. Ahn, J., and Johnstone, R. M. (1987). Incomplete synthesis and plasma membrane translocation of the transferrin receptor in peripheral sheep reticulocytes. In *Eighth International Conference on Proteins of Iron Transport and Storage.* Montebello, Quebec, Canada, p. 112.

133b. Adam, M., and Johnstone, R. M. (1987). Protein kinase C does not phosphorylate the externalized form of the transferrin receptor. *Biochem. J.* 242:151-161.

134. Hamilton, T. A. (1982). Regulation of transferrin receptor expression in concanavalin A stimulated and Gross virus transformed rat lymphoblasts. *J. Cell Physiol.* 113:40-46.

135. Breitman, T. R., Collins, S. J., and Keen, B. R. (1980). Replacement of serum by insulin and transferrin supports growth and differentiation of the human promyelocytic cell line, HL-60. *Exp. Cell Res.* 126:494-498.

136. Rudolph, N. S., Ohlsson-Wilhelm, B. M., Leary, J. F., and Rowley, P. T. (1985). Regulation of K562 cell transferrin receptors by exogenous iron. *J. Cell Physiol.* 122:441-450.

137. Tei, I., Makino, Y., Kodofuku, T., Kanamaru, I., and Konno, K. (1984). Increase of transferrin receptors in regenerating rat liver cells after partial hepatectomy. *Biochem. Biophys. Res. Commun.* 121:717-721.

138. Galbraith, G. M., Goust, J. M., Mercurio, S. M., and Galbraith, R. M. (1980). Transferrin binding by mitogen-activated human peripheral blood lymphocytes. *Clin. Immunol. Immunopathol.* 16:387-395.

139. Musgrove, E., Rugg, C., Taylor, I., and Hedley, D. (1984). Transferrin receptor expression during exponential and plateau phase growth of human tumour cells in culture. *J. Cell Physiol.* 118:6-12.

140. Pelicci, P. G., Tabilio, A., Thomopoulos, P., Titeux, M., Vainchenker, W., Rochant, H., and Testa, U. (1982). Hemlin regulates the expression of transferrin receptors in human hematopoietic cell lines. *FEBS Lett.* 145:350-354.

141. Louache, F., Testa, U., Pelicci, P., Thomopoulos, P., Titeux, M., and Rochant, H. (1984). Regulation of transferrin receptors in human hematopoietic cell lines. *J. Biol. Chem.* 259:11576-11582.

142. Louache, F., Pelosi, E., Titeux, M., Peschle, C., and Testa, U. (1985). Molecular mechanisms regulating the synthesis of transferrin receptors and ferritin in human erythroleukemic cell lines. *FEBS Lett.* 183:223-227.

143. Ward, J. H., Jordan, I., Kushner, J. P., and Kaplan, J. (1984). Heme regulation of HeLa cell transferrin receptor number. *J. Biol. Chem.* 259:13235-13240.

144. Mattia, E., Rao, K., Shapiro, D. S., Sussman, H. H., and Klausner, R. D. (1984). Biosynthetic regulation of the human transferrin receptor by desferrioxamine in K562 cells. *J. Biol. Chem.* 259:2689–2692.
145. Rouault, T., Rao, K., Harford, J., Mattia, E., and Klausner, R. D. (1985). Hemin, chelatable iron, and the regulation of transferrin receptor biosynthesis. *J. Biol. Chem.* 260:14862–14866.
146. Bridges, K. R., and Cudkowicz, A. (1984). Effect of iron chelators on the transferrin receptor in K562 cells. *J. Biol. Chem.* 259:12970–12977.
147. Rao, K. K., Shapiro, D., Mattia, E., Bridges, K., and Klausner, R. (1985). Effects of alterations in cellular iron on biosynthesis of the transferrin receptor in K562 cells. *Mol. Cell. Biol.* 5:595–600.
148. Fallon, R. J., and Schwartz, A. L. (1986). Regulation by phorbol esters of asialoglycoprotein and transferrin receptor distribution and ligand affinity in a hepatoma cell line. *J. Biol. Chem.* 261:15081–15089.
149. Wiley, H. S., and Kaplan, J. (1984). Epidermal growth factor rapidly induces a redistribution of transferrin receptor pools in human fibroblasts. *Proc. Natl. Acad. Sci. USA* 81:7456–7460.
150. Ward, D. M., McVey, D. M., and Kaplan, J. (1986). Mitogenic agents induce redistribution of transferrin receptors from internal pools to the cell surface. *Biochem. J.* 238:721–728.
151. Davis, R. J., and Czech, M. P. (1986). Regulation of transferrin receptor expression at the cell surface by insulin-like growth factors, epidermal growth factor and platelet-derived growth factor. *EMBO J.* 5:653–658.
152. Davis, R. J., Corvera, S., and Czech, M. P. (1986). Insulin stimulates cellular iron uptake and causes the redistribution of intracellular transferrin receptors to the plasma membrane. *J. Biol. Chem.* 5:8708–8711.
153. Buys, S. S., Keogh, E. A., and Kaplan, J. (1984). Fusion of intracellular membrane pools with cell surfaces of macrophages stimulated by phorbol esters and calcium ionophores. *Cell* 38:569–576.
153a. Hebbert, D., and Morgan, E. H. (1985). Calmodulin antagonists inhibit and phorbol esters enhance transferrin endocytosis and iron uptake by immature erythroid cells. *Blood* 65:758–763.
154. Hunt, R. C., and Marshall-Carlson, L. (1986). Internalization and recycling of transferrin and its receptor. *J. Biol. Chem.* 261:3681–3686.
154a. Rothenberger, S., Iacopetta, B., and Kuhn, L. (1987). Endocytosis of the transferrin receptor requires the cytoplasmic domain but not its phosphorylation sites. *Cell* 49:423–431.
155. Ponka, P., and Schulman, H. M. (1985). Acquisition of iron from transferrin regulates reticulocyte heme synthesis. *J. Biol. Chem.* 25:14717–14721.
156. Ponka, P., and Schulman, H. M. (1985). Regulation of heme synthesis in erythroid cells: hemin inhibits transferrin iron utilization but not protoporphyrin synthesis. *Blood* 65:850–857.
157. Iacopetta, B., and Morgan, E. (1984). Heme inhibits transferrin endocytosis in immature erythroid cells. *Biochim. Biophys. Acta* 805:211–216.

158. Warren, G., Davoust, J., and Cockcroft, A. (1984). Recycling of transferrin receptors in A431 cells is inhibited during mitosis. *EMBO J.* 3:2217–2225.

159. Sager, P. R., Brown, P. A., and Berlin, R. D. (1984). Analysis of transferrin recycling in mitotic and interphase HeLa cells by quantitative fluorescence microscopy. *Cell* 39:275–282.

160. Sawyer, S. T., and Krantz, S. B. (1986). Transferrin receptor number, synthesis, and endocytosis during erythropoietin-induced maturation of Friend virus-infected erythroid cells. *J. Biol. Chem.* 261:9187–9195.

161. Jandl, J. H., Inman, J. K., Simmons, R. L., and Allen, D. W. (1959). Transfer of iron from serum iron binding proteins to human reticulocytes. *J. Clin. Invest.* 38:161–185.

162. Johnstone, R. M., Adam, M., and Pan, B. T. (1984). The fate of the transferrin receptor during maturation of sheep reticulocytes in vitro. *Can. J. Biochem. Cell Biol.* 62:1246–1254.

163. Wilczynska, A., and Schulman, H. M. (1980). Friend erythroleukemia cell membrane transferrin receptors. *Can. J. Biochem.* 58:935–940.

164. Wilczynska, A., Ponka, P., and Schulman, H. M. (1984). Transferrin receptors and iron utilization in DMSO-inducible and uninducible Friend erythroleukemia cells. *Exp. Cell Res.* 154:561–566.

165. Bessis, M. (1973). *Living Blood Cells and Their Ultrastructure*. Springer-Verlag, New York.

166. Gasko, O., and Danon, O. (1974). Endocytosis and exocytosis in membrane remodelling during reticulocyte maturation. *Br. J. Haematol.* 28:463–470.

167. Gronowicz, G., Swift, H., and Steck, T. L. (1984). Maturation of the reticulocyte in vitro. *J. Cell Sci.* 71:177–197.

168. Lux, S. E., and John, K. M. (1977). Isolation and partial characterization of a high molecular weight red cell membrane protein complex normally removed by the spleen. *Blood* 50:625–641.

169. Zweig, S. E., Tokuyasu, K. T., and Singer, S. J. (1981). Membrane associated changes during erythropoiesis: on the mechanism of maturation of reticulocytes to erythrocytes. *J. Supramol. Struc.* 17:163–181.

170. Rapoport, S. M. (1986). *The Reticulocyte*. CRC Press, Boca Raton, Florida.

171. Kent, G., Minnick, O. T., Volini, F. I., Orfei, E., and Madera-Orsini, F. (1965). Autophagic vacuoles (lysosomes) in human erythrocytes: their role in red cell maturation and the effect of the spleen on their disposal. *J. Cell Biol.* 27:51A–52A.

172. Baldini, M., and Pannacciuli, I. (1960). The maturation rate of reticulocytes. *Blood* 15:614–629.

173. Benderoff, S., Blostein, R., and Johnstone, R. M. (1978). Changes in amino acid transport during red cell maturation. *Membr. Biochem.* 1:89–106.

174. Leibman, A., and Aisen, P. (1977). Transferrin receptor of the rabbit reticulocyte. *Biochemistry* 16:1268–1272.

175. Pan, B. T., and Johnstone, R. M. (1983). Fate of the transferrin receptor during maturation of sheep reticulocytes in vitro: selective externalization of the receptor. *Cell* 33:967-978.
176. Johnstone, R. M., Adam, M., Orr, L., Hammond, J., and Turbide, C. (1987). Vesicle formation during reticulocyte maturation: association of plasma membrane activities with released vesicles (exosomes). *J. Biol. Chem.* 262:9412-9420.
177. Neutra, M. R., Ciechanover, A., Owen, L. S., and Lodish, H. F. (1985). Intracellular transport of transferrin- and asialoorosomucoid-colloidal gold conjugates to lysosomes after receptor-mediated endocytosis. *J. Histochem. Cytochem.* 33:1134-1144.
178. Iacopetta, B. J., and Morgan, E. H. (1983). An electron-microscope auto-radiographic study of transferrin endocytosis by immature erythroid cells. *Eur. J. Cell Biol.* 32:17-23.
179. Steck, T. L. (1974). The organization of proteins in the human red blood cell membrane. A review. *J. Cell Biol.* 62:1-19.
180. Cabantchik, Z. I., Knauf, P. A., and Rothstein, A. (1978). The anion transport system of the red blood cell. The role of membrane protein evaluated by the use of "probes." *Biochim. Biophys. Acta* 29:239-302.
181. Davis, J. Q., Dansereau, D., Johnstone, R. M., and Bennett, V. (1986). Selective externalization of an ATP binding protein structurally related to the clathrin-uncoating ATPase/Heat Shock protein in vesicles containing terminal transferrin receptors during reticulocyte maturation. *J. Biol. Chem.* 261:15368-15371.
182. Jarvis, S. M., and Young, J. D. (1982). Nucleoside translocation in sheep reticulocytes and fetal erythrocytes: a proposed model for the nucleoside transporter. *J. Physiol. (Lond.)* 324:47-66.
183. Zeidler, R. B., and Kim, H. D. (1982). Pig reticulocytes, IV. In vitro maturation of naturally occurring reticulocytes with permeability loss to glucose. *J. Cell Physiol.* 112:360-366.
184. Blostein, R., Drapeau, P., Benderoff, S., and Weigensberg, A. M. (1983). Changes in Na+-ATPase and Na, K-pump during maturation of sheep reticulocytes. *Can. J. Biochem. Cell Biol.* 61:23-28.
185. Inaba, M., and Maede, Y. (1986). Na, K-ATPase in dog red cells. Immunological identification and maturation-associated degradation by the proteolytic system. *J. Biol. Chem.* 261:16099-16105.
186. Thomopoulos, P., Berthellier, M., and Laudat, M-H. (1978). Loss of insulin receptors on maturation of reticulocytes. *Biochem. Biophys. Res. Comm.* 85:1460-1465.
187. Montandon, J. B., and Porzig, H. (1984). Changes in beta-adrenoceptor binding properties and receptor-cyclase coupling during in vitro maturation of rat reticulocytes. *J. Recept. Res.* 4:91-102.
188. Patel, V. P., Ciechanover, A., Platt, O., and Lodish, H. F. (1985). Mammalian reticulocytes lose adhesion to fibronectin during maturation to erythrocytes. *Proc. Natl. Acad. Sci. USA* 82:440-444.

189. Foxwell, B. M., and Tanner, M. J. (1981). Synthesis of the erythrocyte anion-transport protein. Immunochemical study of its incorporation into the plasma membrane of erythroid cells. *Biochem. J.* 195:129–137.

190. Orr, L., and Johnstone, R. M. (1987). Externalization of membrane-bound activities during reticulocyte maturation is temperature and ATP dependent. *Biochem. Cell Biol.* 65:1080–1090.

191. Kornfeld, S., and Gregory, W. (1969). The identification and partial characterization of lysosomes in human reticulocytes. *Biochim. Biophys. Acta* 177:615–624.

192. Holtzman, E. (1976). The lysosomes: a survey: In *Cell Biology Monographs*, No. 3 Springer-Verlag, New York, pp. 79–84.

193. Wraith, D. C., and Chesterton, C. J. (1982). Cell-surface remodelling during mammalian erythropoiesis. *Biochem. J.* 208:239–242.

194. Hershko, A., and Ciechanover, A. (1982). Mechanisms of intracellular protein breakdown. *Annu. Rev. Biochem.* 51:335–364.

195. Ciechanover, A., Finley, D., and Varshavsky, A. (1984). Ubiquitin dependence of selective protein degradation demonstrated in the mammalian cell cycle mutant ts85. *Cell* 37:57–66.

196. Larrick, J. W., and Hyman, E. S. (1984). Acquired iron-deficiency anemia caused by an antibody against the transferrin receptor. *N. Engl. J. Med.* 311:214–218.

197. Klausner, R. D., van Renswoude, J., Kempf, C., Rao, K., Bateman, J. L., and Robbins, A. R. (1984). Failure to release iron from transferrin in a Chinese hamster ovary cell mutant pleiotropically defective in endocytosis. *J. Cell Biol.* 98:1098–1101.

198. Trowbridge, I. S., and Domingo, D. L. (1981). Anti-transferrin receptor monoclonal antibody and toxin-antibody conjugates affect growth of human tumour cells. *Nature* 294:171–173.

199. FitzGerald, D. J., Trowbridge, I. S., Pastan, I., and Willingham, M. C. (1983). Enhancement of toxicity of antitransferrin receptor antibody-Pseudomonas exotoxin conjugates by adenovirus. *Proc. Natl. Acad. Sci. USA* 80:4134–4138.

200. Lesley, J., Domingo, D. L., Schulte, R., and Trowbridge, I. S. (1984). Effect of an anti-murine transferrin receptor-ricin A conjugate on bone marrow stem and progenitor cells treated in vitro. *Exp. Cell Res.* 150:400–407.

201. O'Keefe, D. O., and Draper, R. K. (1985). Characterization of a transferrin-diphtheria toxin conjugate. *J. Biol. Chem.* 25:932–937.

202. Raso, V., and Basala, M. (1984). A highly cycotoxic human transferrin-ricin A chain conjugate used to select receptor-modified cells. *J. Biol. Chem.* 259:1143–1149.

203. Kohgo, Y., Niitsu, Y., Nishisato, T., Urushizaki, Y., Kondo, H., Fukushima, M., Tsushima, N., and Urushizaki, I. (1985). Transferrin receptors of tumor cells–potential tools for diagnosis and treatment in proteins of iron storage and transport. In *Proteins of Iron Storage and Transport*, G. Spik, J. Montreuil, R. R. Crichton, and J. Magurier (Eds.). Elsevier, Amsterdam, pp. 155–169.

203a. Niitsu, Y., Kohgo, Y., Katoh, J., Sasaki, K., and Urushizaki, I. (1987). Receptor mediated delivery of anticancerous drug by transferrin neocazinostatin conjugate. In *Eighth International Conference on Proteins of Iron Transport and Storage*. Montebello, Quebec, Canada, p. 70.

203b. Kohgo, Y., Niitsu, Y., Nishisato, T., Kondo, H., and Urushizaki, I. (1986). Detection of the circulating transferrin receptor with a sensitive radioimmunoassay system. *Br. J. Haematol.* 64:277–281.

204. Shannon, K. M., Larrick, J. W., Fulcher, S. A., Burck, K. B., Pacely, J., Davis, J. C., and Ring, D. B. (1986). Selective inhibition of the growth of human erythroid bursts by monoclonal antibodies against transferrin or the transferrin receptor. *Blood* 67:1631–1638.

205. Vodinelich, L., Sutherland, R., Schneider, C., Newman, R., and Greaves, M. (1983). Receptor for transferrin may be a "target" structure for natural killer cells. *Proc. Natl. Acad. Sci. USA* 80:835–839.

206. Lazarus, A. H., and Baines, M. G. (1985). Studies on the mechanism of specificity of human natural killer cells for tumor cells: correlation between target cell transferrin receptor expression and competitive activity. *Cell Immunol.* 96:255–266.

207. Dokhelar, M. C., Garson, D., Testa, U., and Tursz, T. (1984). Target structure for natural killer cells: evidence against a unique role for transferrin receptor. *Eur. J. Immunol.* 14:340–344.

208. Bridges, K. R., and Smith, B. R. (1985). Discordance between transferrin receptor expression and susceptibility to lysis by natural killer cells. *J. Clin. Invest.* 76:913–918.

209. Hoshinaga, K., Wood, N. L., Wolfgang, T., Szentpetery, S., Lee, H. M., Lower, R. R., and Mohanakuma, T. M. (1986). Clinical usefulness of monitoring for transferrin receptor positive circulating lymphocytes in cardiac transplant recipients. *Transplant. Proc.* 18:743–744.

210. Schmidt, J. A., Marshall, J., Hayman, M. J., Ponka, P., and Beug, H. (1986). Control of erythroid differentiation: possible role of the transferrin cycle. *Cell* 46:41–51.

211. Neckers, L.M., and Cossman, J. (1983). Transferrin receptor induction in mitogen-stimulated human T lymphocytes is required for DNA synthesis and cell division and is regulated by interleukin 2. *Proc. Natl. Acad. Sci. USA* 80:3494–3498.

212. Rodriguez, M. H., and Jungery, M. (1986). A protein on Plasmodium falciparum-infected erythrocytes functions as a transferrin receptor. *Nature* 324:388–391.

213. Enns, C. A., Suomalainen, H. A., Gebhardt, J. E., Shroder, J., and Sussmann, H. H. (1982). Human transferrin receptor: expression of the receptor is assigned to chromosome 3. *Proc. Natl. Acad. USA* 79:3241–3245.

214. Reckhow, C. L., and Enns, C. A. (1988). Characterization of the transferrin receptor in Tunicamycin-treated A431 cells. *J. Biol. Chem.* 263:7297–7301.

215. Casey, J. L., Di Jeso, B., Rao, K., Klausner, R. D., and Harford, J. B. (1988). *Proc. Natl. Acad. USA* 85:1787–1791.

13

Red Blood Cell Membrane Protein and Lipid Diffusion

DAVID E. GOLAN   *Harvard Medical School and Brigham and Women's Hospital, Boston, Massachusetts*

I. INTRODUCTION

Cell membranes exist in a dynamic state. Measurements of membrane protein and lipid diffusion provide information concerning the molecular details of protein and lipid interactions that govern membrane structure and that mediate cell–cell interactions. The human red blood cell (RBC) membrane is a unique model for studying the dynamics of membrane protein and lipid interactions, for several reasons. First, this membrane contains a limited number of well-characterized proteins and lipids (reviewed in [1,2]). Second, analogues to RBC membrane proteins have been identified in many different nonerythroid cell types (reviewed in [1]). Third, protein and lipid interactions within the RBC membrane have been found to be altered specifically in RBCs from patients with hereditary hemolytic anemias (reviewed in [3,4]). Finally, altered protein mobility and distribution may underlie the recognition of sickle and senescent RBCs by autologous antibodies and by cells of the reticuloendothelial system (RES), and thereby provide a model for cell–cell interactions. (The term RES is herein used for convenience to define cells that line the human vasculature, including vascular endothelial cells and bone marrow–derived mononuclear phagocytes).

The human RBC membrane is composed of an asymmetric lipid bilayer, in which integral membrane proteins are embedded, and a multicomponent protein skeleton that laminates the inner bilayer surface (reviewed in [1]) (Figure 1). The two major integral membrane glycoproteins, band 3 and glycophorin A, span the bilayer. Band 3, present in $\sim 1 \times 10^6$ copies per cell [5], mediates anion exchange across the membrane [6]. Glycophorin A, present in $\sim 5 \times 10^5$ copies

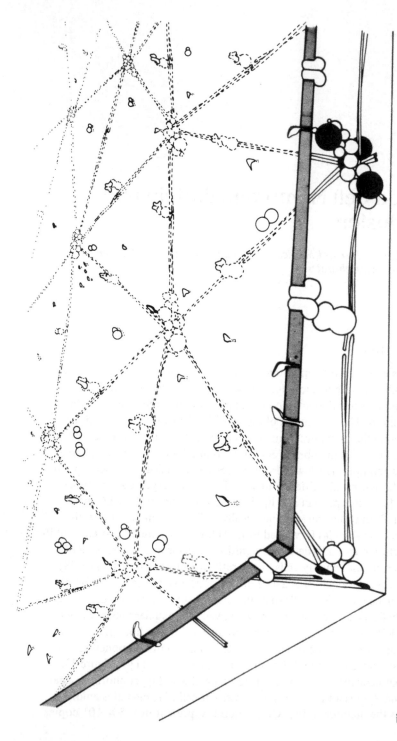

Figure 1 Schematic diagram of human RBC membrane. The lipid bilayer is represented by the hatched band. Transmembrane proteins include dimers of band 3, glycophorin A, and glycophorin B. Some transmembrane proteins are involved in high-affinity binding interactions with membrane skeletal proteins (viz. band 3 with ankyrin and glycophorin A with band 4.1). Other transmembrane proteins, in contrast, are free to diffuse laterally. Underlying the lipid bilayer is the membrane skeletal meshwork, comprised of spectrin heterotetramers cross-linked by junctional complexes in which actin oligomers, band 4.1, tropomyosin, and other proteins are found.

per cell [7], is the major sialoglycoprotein. The membrane skeleton determines RBC stability, deformability, and shape [8,9]. Its major protein, spectrin, is attached to the overlying bilayer through interactions involving: (a) ankyrin, which links spectrin to a portion (10–15%) of band 3; (b) band 4.1, which links spectrin to glycophorin A, glycophorin C, and band 3; and, possibly, (c) negatively charged inner-leaflet phospholipids (reviewed in [1]).

II. MODELS OF MEMBRANE PROTEIN AND LIPID DIFFUSION

A. Fluid-Mosaic Model

Models for macromolecular interactions in cell membranes have been generated by experimental determinations of protein and lipid dynamics in artificial and biological membranes. The fluid-mosaic model [10] postulates that randomly distributed transmembrane proteins and lipids diffuse in the membrane bilayer at rates determined by the viscosity of a homogeneous lipid milieu. Based on this model, membrane lipids should diffuse at approximately the same rate in biological and model membranes, and membrane proteins should diffuse at rates

---

Figure 1 (continued)   KEY TO PROTEINS:

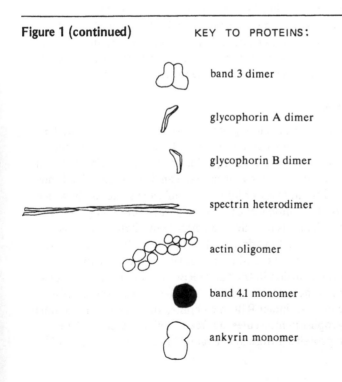

band 3 dimer

glycophorin A dimer

glycophorin B dimer

spectrin heterodimer

actin oligomer

band 4.1 monomer

ankyrin monomer

Table 1 Lateral Mobility of Fl-PE in Human Cell Membranes[a]

Cell type	D[b]	f[c]
Red blood cells	4.1 ± 1.1	99 ± 6
EBV-transformed B cells (JY)	4.9 ± 2.3	93 ± 6
Umbilical vein endothelial cells	8.2 ± 2.0	96 ± 5

Fl-PE, fluorescein phosphatidylethanolamine; EBV, Epstein-Barr virus.
[a]Fl-PE was incubated with RBCs and JY cells for 30 min at room temperature, and with endothelial cells for 45 min at 35°C. Lateral mobility
at 37°C was measured by fluorescence photobleaching recovery.
[b]D, diffusion coefficient, $\times 10^9$ cm^2 sec^{-1}.
[c]f, fractional mobility, %.

only slightly lower than those of membrane lipids [11,12]. The lateral diffusion
rates of fluorescent lipid analogues in many biological membranes are, in fact,
similar to one another. We have found, for example, that the phospholipid analogue fluorescein phosphatidylethanolamine (Fl-PE) diffuses at approximately
the same rate in human RBCs, B lymphoblastoid cells, and endothelial cells, and
that the probe is 90–100% mobile in all of these cell types (Table 1). Furthermore, the diffusion of lipid analogues is comparable in biological and model
membranes with similar lipid compositions, although lipid mobility in biological
membranes is governed to some extent by the fraction of bilayer surface area occupied by the relatively less mobile transmembrane proteins [13,14].

B. Mobility Restraint Mechanisms

Purified transmembrane proteins, including both glycophorin [15] and band 3
[16], diffuse laterally at rates comparable to those of lipid analogues on reconstitution into model membranes (reviewed in [15]). The same proteins diffuse at
much slower rates, however, in their native membrane environments. This fundamental observation has led to models of biological membrane structure in which
mechanisms exist to restrict the mobility of transmembrane proteins [17–19].
In general terms, these mechanisms include interactions with heterogeneously
distributed membrane lipids (e.g., lipid domains) and proteins (e.g., membrane
skeletal, cytoskeletal, transmembrane, and extracellular matrix proteins).

Band 3 and glycophorin diffusion in the native RBC membrane, like the diffusion of many transmembrane proteins in biological membranes, is markedly restricted [20–23]. In the normal, intact RBC membrane, 40–60% of band 3 molecules and 60–80% of glycophorin molecules are free to diffuse laterally. The
mobile fractions of both proteins diffuse at the rate of 1–3 $\times 10^{-11}$ cm^2 sec^{-1}.

Table 2 Lateral Mobility of Glycophorin, Band 3, and Fl-PE in Human
Red Blood Cell (RBC) Membranes[a]

Cell type	Glycophorin		Band 3		Fl-PE	
	D^b	f^c	D	f	D	f
Intact RBCs	1.9–3.1	66–90	1.2–2.4	45–78	300–530	91–99
RBC ghosts	1.1–3.1	59–82	1.1–2.6	14–47	300–510	95–101

Fl-PE, fluorescein phosphatidylethanolamine.

[a]Glycophorin was labeled with fluorescein-5-thiosemicarbazide, band 3 was labeled with
eosin-5-maleimide, and RBCs were labeled with Fl-PE. Red blood cell ghosts were prepared
by hypotonic lysis. Lateral mobility at $37°C$ was measured by fluorescence photobleaching
recovery.

[b]D, diffusion coefficient, $\times 10^{-11}$ cm^2 sec^{-1}.

[c]f, fractional mobility, %.

In contrast, phospholipid and cholesterol analogues have fractional mobilities of
90-100% and diffusion coefficients of $2-4 \times 10^{-9}$ cm^2 sec^{-1} in the RBC membrane [13,24] (Table 2).

Mechanisms restricting transmembrane protein diffusion in the human RBC
membrane can be divided into four general classes (Table 3). We list here the
general classes and whether or not they are thought to be operative in the normal, unperturbed RBC membrane. In sections IV and V of this review we discuss both naturally occurring and artificially induced membrane perturbations
that alter protein mobility by these mechanisms.

The membrane skeleton has been shown to restrict the diffusion of band 3
and glycophorin in the native RBC membrane [21,23,25-27]. Mobility restraint
by this mechanism is mediated by specific binding interactions between transmembrane and membrane skeletal proteins, although nonspecific steric hindrance interactions may also be involved (Figure 1). Extracellular constraints do
not appear to retard the mobility of proteins or lipids in the native RBC membrane, because the normal RBC is a nonadherent cell that does not elaborate
either an extracellular matrix or a glycocalyx. Similarly, heterogeneous domains
of membrane lipid have not been observed in the native RBC membrane [13,21,
28-32]. Finally, lateral associations of band 3 [33-39] and of glycophorin [40]
into dimers, tetramers, and possibly higher-order oligomers have been demonstrated to occur in the native RBC membrane. It is not clear whether heterocom-
plexes between band 3 and glycophorin molecules exist in this membrane.

Table 3 Mobility Restraint Mechanisms in Human Red Blood Cell Membranes

Mechanism	Native membrane	Increase mobility	Decrease mobility
Membrane skeleton			
Specific binding	Yes	Ankyrin fragment	Band 3-ankyrin
		Trypsin treatment	?Glycophorin-band 4.1
		Abnormal membrane skeleton (HS, HE, HPP)	
Steric hindrance	?	Low ionic strength	Diamide
		High ionic strength	Hemin + peroxide
		Polyanions (2,3-DPG, ATP, PIP$_2$)	Phenylhydrazine
			Cell dehydration
			?Polyamines
Extracellular constraints	No	–	Polylysine
			Wheat germ agglutinin
Domain formation	No	–	Lysophosphatidylcholine
Lateral association			
Homocomplex	Yes	–	Band 3 oligomers
			Glycophorin oligomers
Heterocomplex	?	–	?Band 3/glycophorin complexes

HS, hereditary spherocytosis; HE, hereditary elliptocytosis; HPP, hereditary pyropoikilocytosis; DPG, diphosphoglycerate; ATP, adenosine triphosphate; PIP$_2$, phosphatidylinositol-4,5-biphosphate.

III. MEASUREMENTS OF MEMBRANE PROTEIN AND LIPID DIFFUSION

A. Lateral Diffusion

Fluorescence photobleaching recovery (FPR) is used to measure lateral diffusion of fluorescently labeled molecules on cell surfaces [41] (Figure 2). In this technique, a single cell membrane containing fluorescently labeled protein or lipid is

PreBleach

PostBleach

Recovery Phase

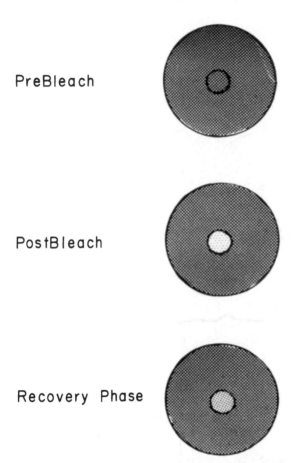

Figure 2 Schematic diagram of fluorescence photobleaching recovery experiment. In the pre-bleach phase of the experiment, proteins or lipids in a cell membrane are uniformly fluorescently labeled (dark stipples). Fluorescence intensity is quantified in a small patch of membrane (area within the inner dark circle) by using a laser beam to excite fluorescence and a photomultiplier tube to detect fluorescence emission. The bleaching pulse photochemically destroys ∼50–70% of the fluorophores within the bleached area. The fluorescence intensity immediately after the bleaching pulse is therefore ∼30–50% of its initial value (light stipples). During the recovery phase of the experiment, fluorescently labeled molecules diffuse laterally from the surrounding membrane into the previously bleached area. Fluorescence therefore recovers in the membrane patch sampled by the laser beam (medium stipples). The rate and extent of fluorescence recovery can be quantified, yielding the diffusion coefficient, D, and fractional mobility, f, of the fluorescently labeled protein or lipid.

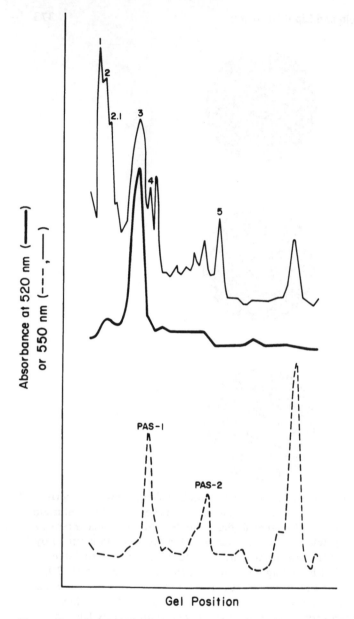

Figure 3 Band 3 labeling by eosin-5-maleimide. SDS-polyacrylamide gel electrophorogram of ghost membranes from RBCs labeled with eosin-5-maleimide [24]. Eosin absorbance was quantified by scanning the unstained gel with a densitometer (dark solid line). Membrane proteins were identified by Coomassie blue staining (light solid line). Membrane glycoproteins were identified by periodic acid-Schiff base (PAS) staining (dashed line). Major protein and glycoprotein bands are identified for convenience in the Coomassie blue and PAS-stained gels, respectively. Greater than 80% of the eosin absorbance co-migrated with band 3.

observed in a fluorescence microscope using a focused laser beam as the excitation source. One region of the membrane is exposed to a brief intense pulse of laser light, causing irreversible bleaching of the fluorophore in that region. The rate and total amount of recovery of the fluorescence in the bleached region, which results from lateral diffusion of unbleached fluorophore into the bleached area, are monitored with a photomultiplier tube. Analysis of fluorescence recovery curves determines both the fraction of mobile fluorophore (f), and the lateral diffusion coefficient of the mobile fraction (D). The experimental apparatus and analytical methods currently used in our laboratory are described in detail elsewhere [24].

RBC band 3 is specifically labeled with the covalent fluorescent probe eosin-5-maleimide [24] (Figure 3). RBC glycophorin is specifically labeled through limited periodate oxidation of its extracellular sialic acid moieties and conjugation of the covalent fluorescent probe fluorescein-5-thiosemicarbazide to the resultant aldehydes [24] (Figure 4). The fluorescent phospholipid analogues Fl-PE [24] and N-(7-nitro-2,1,3-benzoxadiazol-4-yl) phosphatidylethanolamine (NBD-PE) [13] are incorporated directly into RBC membranes. The fluorescent cholesterol analogue N^1-cholesterylcarbamoyl-N^8-(7-nitro-2,1,3-benzoxadiazol-4-yl)-3,6-dioxaoctane-1,8-diamine (NBD-Chol) is exchanged into RBC membranes from multilamellar liposomes prepared from extracted RBC lipids [13].

Examples of experimental FPR curves illustrating the lateral mobilities of glycophorin and Fl-PE in control RBC membranes are shown in Figure 5. The 200-fold difference between the diffusion coefficients of these macromolecules is evident by inspection of the time scales of the FPR curves. Furthermore, the difference between the fractional mobilities of these membrane components is shown by the extent to which the post-bleach fluorescence recovers in the two experiments (Figure 5a and b).

B. Rotational Diffusion

Flash photolysis is used to measure rotational diffusion of labeled proteins in an aqueous suspension of biological membranes [42]. In this technique a membrane protein is labeled with a dye that has a significant quantum yield for triplet formation. A brief pulse of polarized light excites to the triplet state chromophores having absorption transition dipoles parallel to the plane of polarization. The pulse thus depletes preferentially ground state (singlet) molecules with parallel absorption dipoles, and populates preferentially excited state (triplet) molecules with parallel absorption dipoles. Anisotropic depletion of the ground state can be measured by the dichroism of singlet–singlet absorption signals. Alternatively, anisotropic population of the excited state can be quantified by the dichroism of triplet–triplet absorption signals or of triplet-singlet phosphorescence signals. Decay of absorption of phosphorescence anisotropy is determined by the triplet lifetime of the chromophore and the rotational relaxation time of

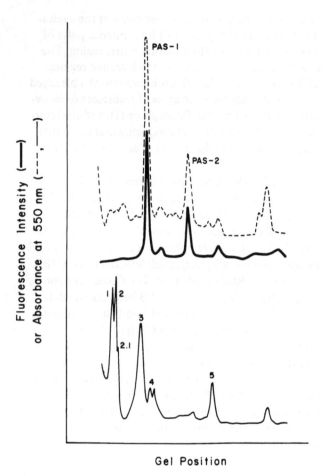

Figure 4 Glycophorin labeling by fluorescein-5-thiosemicarbazide. SDS-polyacrylamide gel electrophorogram of ghost membranes from RBCs labeled with
fluorescein-5-thiosemicarbazide [24]. Fluorescein fluorescence was quantified
by scanning the unstained gel with a densitometer equipped for fluorescence detection (dark solid line). Membrane proteins were identified by Coomassie blue
staining (light solid line). Membrane glycoproteins were identified by PAS staining (dashed line). Major protein and glycoprotein bands are identified for convenience in the Coomassie blue and PAS-stained gels, respectively. Greater than
80% of the fluorescein fluorescence comigrated with the glycophorin bands
PAS-1,2,3,4; of this, greater than 75% co-migrated with the glycophorin A bands
PAS-1,2.

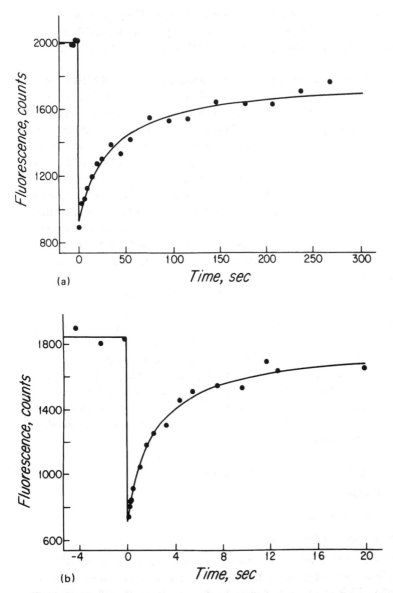

Figure 5 Representative FPR curves of glycophorin (a) and F1-PE (b) lateral diffusion in normal RBC membranes. Glycophorin was labeled with fluorescein-5-thiosemicarbazide. Fl-PE was incorporated directly into RBC membranes. Crosses represent fluorescence counts at various times before and after the photobleaching pulse, which occurred at time = 0. Recovery curves were fit by nonlinear least squares analysis to theoretical curves for diffusion with a circularly symmetric Gaussian laser beam. (a) Diffusion coefficient $D = 1.9 \times 10^{-11}$ cm^2 sec^{-1}; fractional mobility f = 79%. (b) $D = 3.6 \times 10^{-9}$ cm^2 sec^{-1}; f = 95%.

the protein to which the chromophore is rigidly attached. These parameters can be determined from decay of anisotropy curves. (See [42–44] for a full discussion of the theory, data analysis, and experimental methods used in the flash photolysis technique.)

Polarized fluorescence depletion is used to measure rotational diffusion of fluorescently labeled proteins in single cell membranes [45,46]. As in the flash photolysis technique, a membrane protein is specifically labeled with a suitable triplet probe and a brief pulse of polarized light is used for photoselection. The fraction of molecules that either remain in or return to the ground state is monitored over time by the fluorescence excited by a probe beam, which is alternatively polarized parallel and perpendicular to the polarization of the pulse beam. Recovery of fluorescence after the ground state depletion pulse is dependent on the triplet lifetime of the fluorophore and the rotational relaxation time of the protein to which the fluorophore is attached. Both of these parameters can be determined from fluorescence recovery curves. (See [45,46] for a full discussion of the theory, data analysis, and experimental methods used in the polarized fluorescence depletion technique.)

Technical constraints limit the applicability of flash photolysis and polarized fluorescence depletion to measuring the rotational diffusion of band 3 in RBC membranes. Eosin-5-maleimide labels band 3 specifically at a single site [47,48]. Furthermore, the triplet lifetime of this probe is comparable to the rotational relaxation time of band 3 [43–45,48]. Finally, the bound fluorophore is rotationally immobile with respect to band 3 [47,48]. In contrast, glycophorin has not been successfully labeled with a triplet probe that is rotationally immobile with respect to the protein [49]. The rotational relaxation times of fluorescently labeled phospholipid and cholesterol are too short to be measured by single-cell methods currently available.

IV. BAND 3 AND GLYCOPHORIN DIFFUSION

A. Lateral Diffusion

Membrane Skeletal Interactions

  Structurally Important Linkages

Effect of low ionic strength.   Abundant evidence has implicated the RBC membrane skeleton in the control of transmembrane protein lateral mobility. In our initial studies, FPR was used to measure band 3 mobility in RBC ghost membranes [21]. We reasoned that the same experimental parameters causing complete dissociation of the RBC cytoskeleton from the membrane, i.e., low ionic strength and high temperature, might also cause a rearrangement of membrane skeletal proteins, intermediate between tight binding to the membrane and complete dissociation from it, that would allow unrestricted mobility of band 3. It was known from previous biochemical studies that many interactions in the RBC membrane skeleton, including spectrin–ankyrin [50], spectrin–actin [51], spectrin–band 4.1 [50], spectrin dimer–dimer [52,53], and possibly ankyrin–band 3 [54; but see also 55] interactions, are weakened by low-ionic-strength treatment. We found that low temperature (21°C) and moderate ionic strength (46 mM NaPO$_4$) favored immobilization of band 3 (f = 10% mobile) and slow diffusion of the mobile fraction (D = 4 $\times$ 10^{-11} cm^2 sec^{-1}). Increasing temperature (to 37°C) and decreasing ionic strength (to 13 mM NaPO$_4$) led to increases in both the fraction of mobile band 3 (to f = 90%) and the diffusion rate of the mobile fraction (to D = 190 $\times$ 10^{-11} cm^2 sec^{-1}). Increases in D and f were dissociated from one another in that the f increased at higher ionic strength and lower temperature than the ionic strength and temperature at which the D increased. This dissociation was manifested kinetically on prolonged ghost incubation at constant ionic strength and temperature: the f increased immediately to maximal values, whereas the D increased slowly at first and more rapidly after the initial lag period.

These experiments were repeated, using an improved and recalibrated FPR apparatus, in the presence of the protease inhibitors ethylenediamine tetraacetic acid (EDTA), phenylmethylsulfonyl fluoride, and pepstatin A [25,26]. The same relative dependencies of diffusion coefficient and fractional mobility on ionic strength and temperature were found, although all diffusion coefficients were two- to fourfold lower than previously observed (i.e., extreme values of 1 $\times$ 10^{-11} and 50 $\times$ 10^{-11} cm^2 sec^{-1}). These effects were shown to be due neither to complete dissociation of spectrin from the membrane nor to proteolysis of any major RBC membrane protein.

Effects of hypertonic ionic strength, exogenously added ankyrin fragments, and trypsin treatment. Further experiments have suggested a central role for ankyrin in mediating the restriction of band 3 mobility [25,26]. Unlike low-ionic-strength treatment, hypertonic ionic strength selectively destabilizes ankyrin–band 3 and ankyrin–spectrin linkages, leaving the spectrin–actin–band 4.1 "shell" intact [50,54–58]. In the presence of protease inhibitors, addition of hypertonic KCl (320 mM) to 40 mM NaPO$_4$ led to maximal increases in both D (50 $\times$ 10^{-11} cm^2 sec^{-1}) and f (90% mobile). As seen in the low-ionic-strength experiments, increases in D and f were dissociated from one another. On prolonged incubation of ghosts at constant ionic strength and temperature, there was an immediate and maximal increase in f followed by a more gradual increase in D.

Band 3 mobility has been examined in RBC ghosts incubated with intact ankyrin and in ghosts incubated with a 72,000-dalton ankyrin fragment containing the high-affinity binding site for spectrin but not for band 3. Incubation with intact ankyrin led to a modest decrease in the fractional mobility of band 3 [27]. In contrast, incubation with the ankyrin fragment induced threefold increases in both the D and the f of band 3 [25,26,59]. The apparent discrepancy between these findings can be resolved by assuming that exogenously added in-

tact ankyrin binds to free sites on both spectrin and band 3, whereas exogenously added ankyrin fragment displaces spectrin–ankyrin linkages in the membrane skeleton. Intact ankyrin would therefore increase the fraction of band 3
molecules bound to the membrane skeleton, whereas the ankyrin fragment
would free some band 3 molecules from their ankyrin–spectrin skeletal attachment sites.

Finally, band 3 mobility has been examined in RBC ghosts treated with trypsin at concentrations that selectively release from the membrane the 43-kDa
ankyrin-binding domain of band 3 and, possibly, fragments of ankyrin as well
[27]. Under these conditions ~50% of band 3 molecules manifested Ds of
40×10^{-11} cm^2 sec^{-1} [27], identical to the maximal value observed on membrane skeletal perturbation by low and hypertonic ionic strength treatment (see
above). Taken together, the data presented thus far suggest strongly that high-
affinity spectrin–ankyrin and ankyrin–band 3 linkages mediate band 3 mobility
restriction in the normal RBC membrane.

Effect of membrane skeletal protein deficiency. Spectrin-deficient RBCs
from patients with hereditary spherocytosis (HS) and pyropoikilocytosis (HPP)
manifest rapid diffusion of both band 3 and glycophorin. In these preliminary
experiments, we measured the lateral mobility of band 3, glycophorin, and Fl-
PE in intact RBCs from two patients with severe, recessively inherited HS, two
patients with HPP, and six controls with normal RBC morphology. The four patients with HS or HPP, as well as one control without hematologic disease, had
been splenectomized. Two of the controls were HPP carriers and two were HS
carriers: the results in RBCs from these carriers did not differ from those in
RBCs from the two unrelated normal individuals. Similarly, the results in RBCs
from the splenectomized control were identical to those in RBCs from the unsplenectomized controls. Spectrin-to-band-3 ratios were calculated from densitometer scans of sodium dodecyl sulfate (SDS)-polyacrylamide gels of normal
and abnormal RBC membranes. These ratios were ~0.7 for RBCs from the pa-
teints with HS or HPP and 1.0 for RBCs from the controls and HS or HPP carriers.

Control values for band 3 and glycophorin Ds were $(1-3) \times 10^{-11}$ cm^2
sec^{-1}. The Ds observed for the HS and HPP RBCs, in contrast, were $\sim 10 \times$
10^{-11} cm^2 sec^{-1}. Especially interesting was the observation that HPP RBCs
had the same Ds for both band 3 and glycophorin as recessive HS RBCs with
the same degree of spectrin deficiency. Fluorescein phosphatidylethanolamine
Ds were $(300-700) \times 10^{-11}$ cm^2 sec^{-1}, and did not differ consistently among
any of the RBC subsets. These data suggest that: (a) both band 3 and glycophorin mobility are regulated either by the spectrin content of the membrane
skeleton or by something that changes in parallel with the spectrin content in
HS and HPP RBCs, such as the ankyrin, actin, or band 4.1 content; (b) the spectrin dimer–dimer interaction is unimportant in regulating band 3 and glyco-

phorin mobility, because HPP (and HPP carrier) RBCs that are defective in this interaction manifest mobility identical to that of HS (and HS carrier) RBCs with the same degree of spectrin deficiency but without the defect in spectrin oligomerization; and (c) at least in certain RBCs, band 3 and glycophorin cannot be linked in a tight complex in the membrane because they do not diffuse at the same rate. Further, these data confirm in human RBCs the earlier studies of Sheetz et al. [23] on cytoskeleton-deficient RBCs from mice with severe, recessively inherited spherocytosis. In these studies, transmembrane proteins (predominantly band 3 and glycophorin) were labeled with the covalent fluorescent probe dichlorotriazinylaminofluorescein (DTAF). As measured by FPR, transmembrane protein diffusion coefficients were 50-fold greater in spherocytic RBCs than in control RBCs [23].

In preliminary experiments, RBCs deficient in band 4.1 appear to exhibit rapid diffusion of glycophorin but not band 3. The lateral mobility of band 3, glycophorin, and Fl-PE was measured in intact RBCs from 1 patient with hereditary ovalocytosis. RBC membranes from this pateint had ~50% of the amount of normal band 4.1, and 100% of the amount of all other normal membrane components, on SDS-polyacrylamide gel electrophoresis (P. Agre, personal communication). Band 3 and F1-PE Ds were identical to the control values in these membranes. The glycophorin Ds, in contrast, was four- to fivefold greater than control values. These data are consistent with the hypothesis that band 4.1 serves as a specific glycophorin-binding protein in the intact RBC membrane [60,61].

Effects of membrane skeletal perturbation by polyamines and diamide. Polyamines such as spermine have been shown to stabilize spectrin tetramers, both in solution and on RBC ghost membranes [27]. Treatment of RBC ghosts with 0.2–0.6 mM spermine resulted in immobilization [62] or 10-fold slowing of diffusion [27] of fluorescently labeled band 3. Both Schindler et al. [62] and Tsuji and Ohnishi [27] have interpreted their data in terms of band 3 mobility restriction mediated by the increased oligomerization or aggregation state of spectrin. It is also possible, however, that polyamines affect band 3 mobility by inducing transmembrane protein and/or membrane skeletal rearrangements secondary to cross-linking of sialic acid–bearing glycophorin molecules at the extracellular membrane surface. This effect would be analogous to that of polylysine on transmembrane protein mobility, which appears to be mediated by extracellular rather than membrane skeletal cross-linking reactions (see Extracellular Interactions, below). The latter interpretation is further supported by data concerning the effect of the oxidant diamide on band 3 and glycophorin mobility. This agent has been shown to cross-link spectrin on the membranes of intact RBCs by means of intermolecular disulfide bonds [63]. Diamide treatment slowed diffusion of both band 3 and glycophorin by 30–40%, without affecting the fractional mobility of these transmembrane proteins [24,63]. Because the effect of spermine on transmembrane protein mobility is similar to that of polylysine but much greater than that of diamide, the primary effect of spermine may be the extracellular cross-linking of glycophorin rather than the intracellular cross-linking of spectrin.

Models of membrane skeletal restraint.   As is evident from the above review, the intermolecular interactions responsible for membrane skeletal restriction of RBC transmembrane protein mobility have begun to be elucidated. High-affinity binding of a fraction of band 3 to ankyrin [64] and, possibly, band 4.1 [65], and of a fraction of glycophorin to band 4.1 [60,61], probably accounts for the 20–60% immobile fractions of these proteins under physiological conditions. Other mechanisms must be invoked, however, for the marked restriction of lateral diffusion of the mobile species. Competing models for this regulation include: (a) *specific, low-affinity(K_d ~ 1-10 μM) binding* of the cytoplasmic portions of band 3 and/or glycophorin to membrane skeletal components, (b) *nonspecific entanglement* of the cytoplasmic portion of band 3 in a dynamic membrane skeletal matrix [66], and (c) lateral association (i.e., *complex formation*) between band 3 and glycophorin. As noted above, we have begun to examine the lateral mobility of band 3 and glycophorin in the membranes of RBCs with membrane skeletons deficient in the absolute amount and/or binding function of particular protein components. Specific skeletal defects are defined in RBCs from patients with HS (partial spectrin deficiency; defective spectrin–band 4.1 interaction), and hereditary elliptocytosis (HE) (defective spectrin oligomerization; band 4.1 deficiency) [3,4,67–78]. Correlations between transmembrane protein mobility and specific biochemical defects have provided preliminary evidence concerning the three models outlined above. RBCs from HPP and HS patients with the same degree of spectrin deficiency manifest identical diffusion coefficients for both band 3 and glycophorin (see Effect of Membrane Skeletal Protein Deficiency, above). These data suggest that the spectrin dimer–dimer interaction, which is deficient in HPP but not HS RBCs [4], is unimportant in regulating transmembrane protein mobility. This argues against the steric hindrance model [(b) above], which postulates that transmembrane protein lateral diffusion rates are governed by the rate at which spectrin dimer–dimer bonds break and reform [9,27,66]. RBCs from one patient with hereditary ovalocytosis and partial band 4.1 deficiency exhibit abnormally rapid glycophorin mobility but normal band 3 mobility (see Effects of Membrane Skeletal Protein Deficiency, above), suggesting that specific interactions between band 4.1 and glycophorin are at least partially responsible for the retardation of glycophorin diffusion. Finally, some abnormal RBCs manifest significant differences between the diffusion coefficients of glycophorin and band 3 suggesting that these two proteins are not linked in a tight complex in the membrane.

Regulatory Agents and Pathophysiologic Effects

Effect of polyphosphates.   The diffusion coefficient of DTAF-labeled transmembrane proteins was increased two to fourfold by treatment of RBC ghosts

with adenosine triphosphate (ATP) or 2,3-diphosphoglycerate [62] at levels
that are sufficient to destabilize isolated RBC membrane skeletons [79], and
twofold by treatment of ghosts with phosphatidylinositol-4,5-bisphosphate at
levels that are incorporated significantly into ghost membrane lipid [80].
Sheetz et al. [62,79,80] have postulated that polyphosphates increase transmembrane protein mobility by disrupting linkages in the RBC membrane skeleton. Polyphosphate levels could serve to regulate membrane skeletal structure,
and thereby such characteristics as cell shape and deformability, in the intact
RBC as well [80].

Effects of elevated intracellular calcium and altered red blood cell volume.
RBC dehydration, ATP depletion, and calcium (Ca) accumulation may all contribute to the removal of abnormal RBCs from the circulation. We have examined the effects of these treatments on the lateral mobility of glycophorin
in normal RBC membranes, reasoning that alterations in the distribution and
dynamics of negative charges at the RBC surface might affect RBC interactions
with cells of the RES. In an initial study [81], the ionophore A23187 was used
for rapid Ca loading of intact RBCs. Control RBCs incubated in phosphate buffered saline (PBS) without A23187 had a diffusion coefficient for glycophorin
of 2×10^{-11} cm^2 sec^{-1} and a fractional mobility of 61–66%. Calcium treatment
at 10–100 μM induced a 50% decrease in f without a change in diffusion coefficient. The immobilization was not due to ionophore alone, because RBCs incubated with ionophore and EDTA yielded results identical to those incubated
without ionophore.

Calcium loading under the conditions of this experiment [81], however, also
changed cell ATP, water, and Mg content. Further experiments dissected the
role of these factors in Ca-induced glycophorin immobilization. Selective depletion of ATP was achieved by incubation of RBCs for 12 hr at 37°C in a glucose-
and phosphate-free buffer containing 1 mM ethylene glycol-bis(aminoethyl
ether) tetraacetic acid and 1 mM MgCl$_2$. Glycophorin fractional mobility was
60%, which was not different from the control value in RBCs incubated at 0°C.
Selective Ca loading without ATP deletion, cell dehydration, or Mg depletion
was obtained by incubation of RBCs in two different EDTA buffer systems containing 80 mM KCl, 10 mM NaPO$_4$, 10 μM A23187, 0–100 μM free CaCl$_2$, and
0.15 mM free MgCl$_2$. The first buffer contained 50–100 μM sodium orthovana-
date, which blocks Ca-activated ATPase, whereas the second buffer contained 10
mM each of glucose and inosine to maintain intracellular ATP levels in the face
of ATPase stimulation. In both buffers, glycophorin f in Ca-treated RBCs was
not significantly different from that in control RBCs. Selective perturbation of
RBC volume without change in cell ATP, Ca, or Mg content was achieved by incubation or RBCs in phosphate buffers containing 127 mM KCl, 13 mM NaCl,
40 μg/ml nystatin, and 20 or 150 mM sucrose. Nystatin loading in the presence
of 150 mM sucrose was accompanied by a 50% reduction in mean RBC volume,

as well as a 50% reduction in glycophorin f. This was the same degree of reduction in f observed in the initial studies, suggesting that the Ca-induced immobilization of glycophorin was mediated by RBC dehydration.

Detailed examination of the effect of RBC volume on protein mobility revealed a striking discordance between the behavior of glycophorin and that of band 3. Red blood cell volume was regulated by incubation of RBCs for 17 hr at 4°C in phosphate buffers containing 130 mM KCl, 10 mM NaCl, 10 mM glucose, and 20–250 mM sucrose. Glycophorin fractional mobility decreased from 65% to 23% as the sucrose concentration in the buffer was increased from 20 mM to 250 mM, whereas band 3 f was 61% and 58% in buffers containing 20 and 250 mM sucrose, respectively. Further, the decrease in glycophorin mobility was fully reversible on restoration of RBC volume to normal by incubation of sucrose buffer-treated RBCs for 17 hr at 4°C in PBS with 10 mM glucose (PBS/glc). The f of glycophorin was 63% and 60% in RBCs that had been incubated in buffers containing 20 and 250 mM sucrose, respectively, and then removed into PBS/glc.

Effects of hemichrome and hemin binding.   Clustering of band 3 by hemichromes has been postulated to occur as hemoglobin denatures in senescent RBCs [82,83]. This clustering may provide recognition sites for autologous IgG antibodies, which bind to senescent RBCs and facilitate their removal by cells of the RES. In preliminary experiments, we have measured the lateral mobility of band 3, glycophorin, and Fl-PE in RBCs incubated with 15 mM phenylhydrazine, a reagent that catalyzes the denaturation of hemoglobin and induces Heinz body formation in intact RBCs. The f of both band 3 and glycophorin was decreased by 50–80% after treatment with phenylhydrazine, without significant change in D. Neither the f nor the D of Fl-PE was affected by such treatment. These data suggest that hemichrome formation induces immobilization of transmembrane proteins in normal RBC membranes. In addition, clusters of band 3 observed over sites of Heinz body binding to the membrane may be immobile in the plane of the membrane.

Increased hemin binding to the membrane has also been found in senescent RBCs [84]. In preliminary experiments, we have measured the lateral mobility of glycophorin and Fl-PE in RBC ghosts incubated with 5–20 μM hemin, before and after hydrogen peroxide treatment. In the absence of peroxide, low concentrations of hemin did not significantly affect the mobility of either glycophorin or Fl-PE. After peroxide treatment, however, the fractional mobility of glycophorin was decreased by 50–70%, without significant change in D. The lateral mobility of Fl-PE was not altered by treatment with hemin and peroxide. These data suggest that membrane-bound hemin catalyzes an oxidative reaction that induces glycophorin immobilization in normal RBC membranes.

Sickle RBCs are heterogeneous with respect to size, shape, and membrane functional properties. We have measured the lateral mobility of glycophorin in

RBCs from six patients with sickle cell anemia and that of band 3 in RBCs from two patients with disease. Compared with values in control RBCs examined at the same time, the f of glycophorin in randomly chosen sickle RBCs was decreased by 16%, and that of band 3 was decreased by 19%. Ds were not significantly different in any of the comparisons. RBCs from one patient were separated visually into those of normal size (diameter, 7–8 μm) and those of small size (diameter, 5 μm). In this preliminary study, glycophorin and band 3 each had a f identical to control values in sickle RBCs of normal size and a f 25% less than control values in small sickle RBCs. Again, Ds were not significantly different in any of the comparisons. These data suggest that the mechanisms controlling glycophorin and band 3 lateral diffusion are altered in sickle RBC membranes. Reductions in protein mobility may correlate with sickle RBC size, which is a function of the age and density of these cells.

Physiologic and pathophysiologic correlates. The molecular mechanisms responsible for adherence of sickle RBCs to cells of the RES and for removal of sickle and senescent RBCs from the circulation remain to be elucidated. Clues are provided by treatments of normal RBCs that induce abnormal interactions between these cells and either autologous antibody or RES cells. Calcium loading causes normal RBCs to adhere to endothelial cells (ECs) [85], although these experiments were performed in high-sodium, low-potassium buffers in which RBCs would be expected to become dehydrated as well as Ca-loaded. Membrane oxidation by peroxide or malonyldialdehyde causes normal RBCs to adhere to and become phagocytosed by macrophages [86]. Phenylhydrazine-induced hemicrhome formation induces increased binding of autologous IgG to the membrane [82]. Calcium accumulation, cell dehydration, membrane oxidation, and hemoglobin binding to the membrane are present in sickle and senescent RBCs. It is difficult to imagine how any of these defects could be the ultimate cause of abnormal interactions with antibody and RES cells, however, because these presumably involve a change in the RBC surface itself. We hypothesize that sickle and senescent RBC interactions with autologous antibody and RES cells result from abnormalities in the mobility and distribution of proteins and lipids at the surface of these RBCs. RBC-associated factors such as cell dehydration, membrane oxidation, and membrane hemoglobin binding may perturb interactions between membrane skeletal proteins and transmembrane proteins or lipids, resulting in altered mobility and distribution of these membrane components. Band 3 aggregation at the RBC surface, induced by hemichrome binding to band 3 and/or oxidant damage to the membrane, may promote autologous IgG binding and thereby facilitate RBC adherence to and phagocytosis by macrophages [82,87]. Autologous IgG is, in fact, found on the membranes of sickle, senescent, and other abnormal RBCs [86,88-92], and these RBCs are preferentially phagocytosed by macrophages [88,90,92,93]. Surface charge redistribution accompanying clustering of glycophorin may promote abnormal RBC adherence to ECs [85]. Plasma factors such as calcium and fibronectin may also be involved in this process [94].

Preliminary data have been presented concerning the lateral diffusion of band 3, glycophorin, and Fl-PE in specifically perturbed normal RBC and in sickle RBC membranes. Red blood cell dehydration, which is found in both senescent and sickle RBCs, induces reversible immobilization of glycophorin but not band 3 in intact RBCs. Denatured hemoglobin products, also seen in senescent and sickle RBCs, cause immobilization of both glycophorin and band 3. In at least a fraction of sickle RBCs, glycophorin and band 3 manifest reduced fractional mobilities. These observations may be relevant to the pathophysiology of senescent and sickle RBC removal from the circulation and of sickle RBC adherence to vascular endothelial cells, by mechanisms that are described above. Clearly much work remains to be done. The results of such studies are expected to elucidate further molecular defects in sickle and senescent RBC membranes that are directly responsible for abnormal interactions between these RBCs and autologous antibody or cells of the RES. Such interactions are likely to be significant pathophysiologically. For example, the degree of sickle RBC adherence to endothelium correlates with the frequency and clinical severity of vaso-occlusive crisis [95], and the degree of sickle RBC adherence to macrophages correlates with the clinical severity of hemolysis [86].

Extracellular Interactions

Both wheat germ agglutinin [96] and polylysine are polyvalent ligands for glycophorin at the extracellular surface of the RBC membrane. Glass slides coated with either of these agents induced 80-95% immobilization of both band 3 and glycophorin in RBCs adherent to these slides [24]. Transmembrane protein immobilization could be mediated directly, by wheat germ agglutinin- or polylysine-induced cross-linking of glycophorin, or indirectly, by membrane skeletal rearrangements induced by ligand binding to the extracellular portion of glycophorin [97]. The native RBC membrane does not possess an extracellular matrix or glycocalyx capable of restricting transmembrane protein mobility. Extracellular ligands in the form of RES cell membranes or antibodies could, however, perturb RBC transmembrane protein mobility and thereby dictate interactions between RBCs and RES cells or antibodies (see Physiologic and Pathophysiologic Correlates, above).

Lipid Domain Formation

Effect of red blood cell adherence to schistosomula of Schistosoma mansoni. Human RBCs adhere to and are lysed by schistosomula of *S. mansoni* [98]. We have investigated the mechanism of RBC lysis by comparing the dynamic properties of transmembrane protein and lipid probes in adherent ghost membranes with those in control RBCs and in RBCs treated with various membrane perturbants [24]. Fluorescence photobleaching recovery was used to measure the lateral mobility of band 3, glycophorin, and two lipid analogs, Fl-PE and carbocyanine dyes, in RBCs and ghosts adherent to schistosomula. Adherent ghosts manifested 95–100% immobilization of both transmembrane proteins and 45–55% immobilization of both lipid probes. A unique pattern of protein and lipid immobilization, identical to that found in ghosts adherent to schistosomula, was observed in RBCs lysed with egg lysophosphatidylcholine (lysoPC) at a concentration of 8.4 μg/ml. Furthermore, we found that schistosomula incubated with labeled palmitate in vitro release lysoPC into the culture medium at a rate of 1.5 fmol/hr/10^3 organisms. This rate of release is sufficient to account for the in vitro effects of lysoPC on RBC lysis and membrane component immobilization. These data suggest that lysoPC is transferred from schistosomula to adherent RBCs, causing their lysis. Lysophosphatidylcholine may also be involved in other parasite-mediated phenomena, such as lysis of human eosinophils, fusion with human neutrophils, prevention of degranulation by adherent rat mast cells, release of parasite membrane components, and acquisition of host membrane components.

Effect of monopalmitoyl phosphatidylcholine. The effects of lysoPC on human RBC ghost morphology, transmembrane protein and lipid lateral mobility, and membrane lipid composition were further studied to elucidate mechanisms by which lysoPC immobilizes ghost membrane components [99]. Under standardized conditions 1.0–1.5 μg/ml egg lysoPC lysed 50% of RBCs and induced, in some ghosts, the formation of large patches of wrinkled membrane. Patches exhibited complete immobilization of glycophorin and band 3 and partial immobilization of the phospholipid analogue Fl-PE, whereas adjacent smooth membrane areas manifested only partial immobilization of proteins and no immobilization of Fl-PE. Supralytic concentrations of lysoPC induced both progressive, homogeneous wrinkling of RBC ghost membranes and concentration-dependent decreases in the lateral mobilities of glycophorin, band 3, and Fl-PE. Complete immobilization of glycophorin and band 3 occurred at 8.4 μg/ml and of Fl-PE at 16.8 μg/ml lysoPC. Monopalmitoyl phosphatidylcholine (MPPC), the major component of egg lysoPC, induced a concentration-dependent decrease in Fl-PE mobility, leading to complete immobilization in wrinkled ghosts at 10 μg/ml. Other synthetic lysoPCs did not completely immobilize Fl-PE, although some caused membrane wrinkling. MPPC was incorporated into ghost membranes with a linear dependence ($r = 0.97$) on MPPC concentration. Relative to total membrane lipid, the lysoPC mole fraction increased from 0.2 ± 0.1% at 0 μg/ml MPPC to 25 ± 2% at 16 μg/ml MPPC. The molar ratio of cholesterol to phospholipid (exclusive of lysoPC) in MPPC-treated ghosts was inversely dependent on MPPC concentration, decreasing from 1.0 ± 0.1 at 0 μg/ml MPPC to 0.7 ± 0.2 at 8 μg/ml MPPC. This ratio did not decrease further at 12 and 16 μg/ml MPPC. Treatment with MPPC did not affect the relative amounts of the major RBC phospholipid classes. These results suggest that MPPC causes concentration-dependent changes in the composition and organization of RBC ghost membranes. Lysophosphatidylcholine insertion and cholesterol depletion may induce lipid domain formation, which causes membrane protein and lipid immobilization (see also [100]). Because cholesterol depletion alone does not induce lipid probe immobilization [101], lysoPC insertion is likely to be the dominant force leading to membrane reorganization and membrane component immobilization.

Protein Complex Formation

Band 3 and glycophorin manifest identical Ds in normal RBC membranes under physiologic conditions [24]. Further, the rotational diffusion of band 3 is reduced after antibody-mediated cross-linking of glycophorin A [102], and the lateral diffusion of band 3 is reduced after polylysine- or wheat germ agglutinin-induced cross-linking of glycophorin [24]. These and other observations [103-106] have suggested that band 3 and glycophorin may be linked in a tight complex on the membrane. We have found, however, that there are several sets of conditions under which the lateral mobilities of these two proteins are differentially regulated. These conditions include RBCs from unsplenectomized patients with dominant HS and HPP (unpublished observations), RBCs from one patient with hereditary ovalocytosis and band 4.1 deficiency (see Effects of Membrane Skeletal Protein Deficiency, above), and normal RBCs incubated in hypertonic buffer (see Effects of Elevated Intracellular Calcium and Altered Red Blood Cell Volume, above). There are two possible interpretations of these data. First, band 3 and glycophorin may not be linked in a tight complex in either normal or abnormal RBC membranes. Restriction of band 3 diffusion by anti-glycophorin antibodies [102] or other cross-linking agents [24] may result from nonspecific trapping of band 3 in an extracellular meshwork consisting of glycophorin and the cross-linking antibody, protein, or peptide. Alternatively, band 3 and glycophorin may be tightly linked in normal membranes but uncoupled from one another under a variety of abnormal conditions. Further comparisons between the mobilities of band 3 and glycophorin in normal and abnormal RBC membranes will be necessary to distinguish between these two possiblities.

B. Rotational Diffusion

Band 3 rotation has been studied in macroscopic samples of normal RBC ghosts [34,43,44,48,102,107-110]. In the initial studies of Cherry et al. [44], the average rotational correlation time of band 3 was found to be ~1 msec at room temperature, and this value was unaffected by depleting ghosts of spectrin at very low ionic strengths [44] or by cross-linking the cytoplasmic domains of band 3 dimers using copper-phenanthroline [34]. More recent studies have suggested that rapidly rotating, slowly rotating, and, possibly, immobilized forms of band 3 coexist in the membrane. The equilibrium among these forms is dependent on temperature, as is the rotational relaxation time of the slowly but not the rapidly rotating form [48]. At 37°C, it is estimated that ~25% of band 3 rotates with a relaxation time of 150 μsec, ~50% rotates with a relaxation time of 3.4 msec, and ~25% is rotationally immobile on the time scale of the experiment [43,48]. Proteolytic cleavage of the cytoplasmic domain of band 3 shifts this equilibrium in favor of the rapidly rotating form, as does membrane depletion of ankyrin and band 4.1 by low-ionic-strength/high-ionic-strength extraction [108]. These data have suggested that the rotational mobility of band 3 is restricted by linkages to ankyrin and by self-association into tetramers and higher-order oligomers [48,108]. Many of these studies were performed, however, on RBC ghosts and membrane fragments in 5 mM phosphate buffer at 37°C. Using lateral mobility measurements, we have found that the dynamic properties of band 3 differ between RBC ghosts and intact cells (Table 2), and that band 3 mobility is increased in ghosts under low-ionic-strength, high-temperature conditions [21]. The physiological relevance of the rotational mobility studies is therefore uncertain.

Band 3 rotation has also been studied in microscopic samples of normal RBC ghosts [45]. Average rotational relaxation times on the order of 1 msec were observed [45], similar to the results of the studies on macroscopic ghost samples. Low signal-to-noise ratios prevented both a more precise determination of average rotational relaxation times and the deconvolution of rotational relaxation times for more than one component from these data [45]. The rotational diffusion of band 3 has not been studied in intact RBCs.

Measurements of band 3 lateral and rotational mobilities on the same individual RBCs would yield important information concerning membrane skeletal mechanisms controlling band 3 mobility. For example, if band 3 is restricted in its mobility by specific binding to a particular skeletal protein (e.g., ankyrin), both lateral and rotational diffusion should be affected. If, on the other hand, its mobility is retarded by nonspecific (i.e., steric) interactions with a dynamic membrane skeletal matrix, lateral but not rotational diffusion should be affected. Similarly, if band 3 is induced to aggregate in pathologic RBC membranes without developing new membrane skeletal attachments, then its rotational diffusion would be greatly altered but its lateral diffusion would be only minimally changed. This may be the case in sickle RBCs, for example, in which band 3 may be clustered and the ankyrin–band 3 attachment may be weakened.

V. PHOSPHOLIPID AND CHOLESTEROL DIFFUSION

A. Measurements of Lipid Diffusion

The lateral mobilities of the lipid probe diI, the phospholipid analogues fluorescein-PE, NBD-PE, NBD-PC, and NBD-PS, and the cholesterol analogue NBD-cholesterol have been measured in normal RBCs and ghosts [13,24,101,111-114]. At 35-40°C the diffusion rate of diI in ghost membranes is variable, $D = (2-15) \times 10^{-9}$ cm^2 sec^{-1} [24,101,111,112], whereas that of phospholipid and cholesterol analogues is more consistent, $D = (2-5) \times 10^{-9}$ cm^2 sec^{-1} [13,24, 113,114]. This difference is reproduced in intact RBCs, in which the diI diffusion rate, $D = 2 \times 10^{-8}$ cm^2 sec^{-1} [24,111], is considerably faster than the Fl-PE rate, $D = (3-5) \times 10^{-9}$ cm^2 sec^{-1} [24, unpublished observations]. DiI probes may disrupt membrane architecture, leading to bilayer defects that permit faster diffusion of the probe [24,111,112].

The lateral diffusion rates of NBD-PE and NBD-cholesterol are identical in RBC ghost membranes over the range of temperatures from 15-37°C. Furthermore, the diffusion rates of NBD-PE and NBD-cholesterol exhibit a weak temperature dependence, increasing only twofold over the range 15-37°C [13]. Curves describing the relationship between lipid probe diffusion rate and temperature do not manifest a "break point" that would signify a membrane lipid phase transition over the range 15-37°C [13,101,111,112]. These data suggest that phospholipid and cholesterol probes sample the same membrane environment, and that phospholipid- or cholesterol-rich domains do not exist in the native RBC membrane [13]. The latter conclusion is consistent with data obtained in other studies (see Section IIB).

Inner and outer leaflet RBC lipids may diffuse at different rates. Morrot et al. [113] found that phospholipid analogue diffusion in intact RBC membranes was modeled best using biexponential curves described by a fast component with $D = (8-9) \times 10^{-9}$ cm^2 sec^{-1} and a slow component with $D = (1-2) \times 10^{-9}$ cm^2 sec^{-1}. Depending on the analogue used, the fast component comprised 40-80% of the total. Intracellular ATP appeared to be required for the fast component to become manifested, because this component was present in ghosts resealed in the presence of ATP but absent in ghosts resealed in buffer alone [113]. Although these data have been interpreted to indicate that the viscosity of the inner, aminophospholipid-containing membrane leaflet is lower than that of the outer leaflet [113], this interpretation is controversial [115].

B. Comparison with Measurements of Protein Diffusion

Under several sets of conditions characterized by destabilization of the RBC membrane skeleton, the diffusion coefficients of band 3 and glycophorin increase to a "rapid diffusion limit" of $D \sim 5 \times 10^{-10}$ cm^2 sec^{-1} [21,25-27; see also Section IVA]. This limiting diffusion rate can be compared with the diffusion, in the same membrane, of fluorescent phospholipid and cholesterol analogues. At 37°C the diffusion coefficient for NBD-PE and NBD-cholesterol was 2×10^{-9} cm^2 sec^{-1}, or ~fourfold greater than the limiting diffusion rate of band 3 in similarly prepared ghost membranes. This ratio of Ds agrees with that expected for two molecules, of relative sizes corresponding to band 3 and phospholipid, which experience the same effective membrane viscosity [13]. That is, in the "rapid diffusion limit," the lateral mobility of band 3 appears to be restricted only by the intrinsic viscosity of the lipid bilayer.

C. Effect of Protein Removal

Both NBD-PE and NBD-cholesterol diffused fourfold faster in liposomes prepared from chloroform/methanol extracts of ghost membranes ($D = 8 \times 10^{-9}$ cm^2 sec^{-1} at 37°C) than in the ghost membranes themselves. This difference was not due to changes in the membrane phospholipid/cholesterol ratio. Rather, these data suggest that transmembrane and/or membrane skeletal proteins directly restrict both phospholipid and cholesterol mobilities in RBC membranes [13]. The RBC membrane can be modeled as a two-dimensional fluid in which solid-phase proteins, occupying an area fraction of ~17% [13], are inserted in a random distribution [116,117]. Using this model, a three to fourfold difference is predicted between lipid diffusion coefficients in the native, protein-containing membrane and the equivalent fluid-phase membrane without protein [116], in excellent agreement with the experimental result. Protein-mediated restriction of lipid mobility has also been demonstrated in human fibroblast plasma membranes, in which the lipid probe diI diffused fourfold slower than in multibilayers reconstituted from plasma membrane lipid [14].

VI. CONCLUSION AND FUTURE PROSPECTS

The RBC membrane is a dynamic structure that both induces and responds to changes in its intracellular and extracellular environments. Protein and lipid interactions regulate the structure of the RBC membrane and may mediate interactions between RBCs and autologous antibodies or cells of the reticuloendothelial system. Although it has been recognized that studying such interactions in the native membrane environment would be important, techniques for approaching such study have only recently become available. Fluorescence photobleaching recovery, flash photolysis, and polarized fluorescence depletion have begun to be used to differentiate among protein and lipid interactions that control the lateral and rotational diffusion of band 3, glycophorin, phospholipid, and cholesterol, including specific binding interactions, steric hindrance interactions, complex formation, and domain formation. In such studies, we and others have found that the lateral diffusion of band 3 and glycophorin is ~200-fold slower than that of phospholipid and cholesterol in normal RBC membranes. This protein diffusion rate is ~50-fold less than that predicted on the basis of molecular size. Furthermore, band 3 and glycophorin exhibit significant immobile fractions, ~50%, whereas phospholipid and cholesterol are 90-100% mobile. Perturbations of spectrin-ankyrin and ankyrin-band 3 linkages within the normal membrane skeleton induce 50-fold increases in the diffusion rates and twofold increases in the fractional mobilities of band 3 and glycophorin. The diffusion rates of band 3 and glycophorin are also increased in abnormal RBCs from patients with hereditary anemias characterized by membrane skeletal deficiency of spectrin or band 4.1. These and other data from a large number of studies on normal and abnormal RBCs suggest that: (a) specific protein–protein linkages involving band 3, glycophorin, spectrin, ankyrin, and band 4.1 constrain transmembrane protein diffusion, (b) contrary to one published model [66], the spectrin dimer–dimer interaction site appears not to be involved in the control of band 3 and glycophorin diffusion, (c) band 4.1 serves as a specific glycophorin-binding protein in the RBC membrane, and (d) at least in certain RBCs, band 3 and glycophorin cannot be linked in a tight complex in the membrane because they do not diffuse at the same rate.

The study of red blood cell membrane protein and lipid diffusion may also elucidate molecular mechanisms governing cell–cell adhesive and cytotoxic interactions, such as those between abnormal RBCs and vascular endothelial cells and RBCs or granulocytes and the parasite *S. mansoni*. In comparison to the studies of specific protein-protein linkages that regulate transmembrane protein mobility, studies of regulatory and pathophysiologic mechanisms are in their infancy. These mechanisms may, however, be central to the physiology of senescent red blood cell removal from the circulation, and they may dictate critical pathophysiologic events such as vaso-occlusive crisis and RBC hemolysis in sickle cell anemia and host defense evasion by *S. mansoni*. They are therefore worthy of intensive further investigation.

ACKNOWLEDGMENTS

The author gratefully acknowledges the contributions of his mentor, the late William R. Veatch, and of the members of his laboratory, Adam D. Abroms, Carl S. Brown, Yung J. Han, Peter A. Singer, Alan H. Stolpen, and Patrick W. Yacono, and of his collaborators, Peter C. Agre, M. Robert Alecio, John P. Caulfield, Catherine M. L. Cianci, Stephen T. Furlong, Jiri Palek, Robert R. Rando, and Nurith Shaklai. This work was supported in part by National Institutes of Health grants AI-15311 and HL-32854. The author was a Research Fellow of The Medical Foundation, Inc. (Boston, Massachusetts) during the course of this work.

REFERENCES

1. Bennett, V. (1985). The membrane skeleton of human erythrocytes and its implications for more complex cells. *Annu. Rev. Biochem.* 54:273–304.
2. Cohen, C. M. (1983). The molecular organization of the red cell membrane skeleton. *Semin. Hematol.* 20:141–158.

3. Becker, P. S., and Lux, S. E. (1985). Hereditary spherocytosis and related disorders. *Clin. Haematol.* 14:15–43.
4. Palek, J. (1985). Hereditary elliptocytosis and related disorders. *Clin. Haematol.* 14:45–87.
5. Steck, T. L. (1974). The organization of proteins in the human red blood cell membrane. *J. Cell Biol.* 62:1–19.
6. Cabantchik, Z. I., Knauf, P. A., and Rothstein, A. (1978). The anion transport system of the red blood cell: the role of membrane protein evaluated by the use of "probes." *Biochim. Biophys. Acta* 515:239–302.
7. Marchesi, V. T., Furthmayr, H., and Tomita, M. (1976). The red cell membrane. *Annu. Rev. Biochem.* 45:667–698.
8. Chasis, J. A., and Shohet, S. B. (1987). Red cell biochemical anatomy and membrane properties. *Annu. Rev. Physiol.* 49:237–248.
9. Sheetz, M. P. (1983). Membrane skeletal dynamics: role in modulation of red cell deformability, mobility of transmembrane proteins, and shape. *Semin. Hematol.* 20:175–188.
10. Singer, S. J., and Nicolson, G. L. (1972). The fluid mosaic model of the structure of cell membranes. *Science* 175:720–731.
11. Hughes, B. D., Pailthorpe, B. A., White, L. R., and Sawyer, W. H. (1982). Extraction of membrane microviscosity from translational and rotational diffusion coefficients. *Biophys. J.* 37:673–676.
12. Saffman, P. G., and Delbruck, M. (1975). Brownian motion in biological membranes. *Proc. Natl. Acad. Sci. USA* 72:3111–3113.
13. Golan, D. E., Alecio, M. R., Veatch, W. R., and Rando, R. R. (1984). Lateral mobility of phospholipid and cholesterol in the human erythrocyte membrane: effect of protein-lipid interactions. *Biochemistry* 23:332–339.
14. Jacobson, K., Hou, Y., Derzko, Z., Wojcieszyn, J., and Organisciak, D. (1981). Lipid lateral diffusion in the surface membrane of cells and in multibilayers formed from plasma membrane lipids. *Biochemistry* 20:5268–5275.
15. Vaz, W. L. C., Derzko, Z. I., and Jacobson, K. A. (1982). Photobleaching measurements of the lateral diffusion of lipids and proteins in artificial phospholipid bilayer membranes. *Cell Surf. Rev.* 8:83–136.
16. Chang, C. H., Takeuchi, H., Ito, T., Machida, K., and Ohnishi, S. (1981). Lateral mobility of erythrocyte membrane proteins studied by the fluorescence photobleaching recovery technique. *J. Biochem. (Tokyo)* 90:997–1004.
17. Edelman, G. M. (1976). Surface modulation in cell recognition and cell growth. *Science* 192:218–226.
18. Jacobson, K., Ishihara, A., and Inman, R. (1987). Lateral diffusion of proteins in membranes. *Annu. Rev. Physiol.* 49:163–175.
19. Nicolson, G. (1976). Transmembrane control of the receptors on normal and tumor cells. *Biochim. Biophys. Acta* 457:57–108.
20. Fowler, V., and Branton, D. (1977). Lateral mobility of human erythrocyte integral membrane proteins. *Nature* 268:23–26.

21. Golan, D. E., and Veatch, W. (1980). Lateral mobility of band 3 in the human erythrocyte membrane studied by fluorescence photobleaching recovery: evidence for control by cytoskeletal interactions. *Proc. Natl. Acad. Sci. USA* 77:2537–2541.
22. Peters, R., Peters, J., Tews, K. H., and Bahr, W. (1974). A microfluorimetric study of translational diffusion in erythrocyte membranes. *Biochim. Biophys. Acta* 367:282–294.
23. Sheetz, M. P., Schindler, M., and Koppel, D. E. (1980). The lateral mobility of integral membrane proteins is increased in spherocytic erythrocytes. *Nature* 285:510–512.
24. Golan, D. E., Brown, C. S., Cianci, C. M. L., Furlong, S. T., and Caulfield, J. P. (1986). Schistosomula of *Schistosoma mansoni* use lysophosphatidylcholine to lyse adherent human red blood cells and immobilize red cell membrane components. *J. Cell Biol.* 103:819–828.
25. Golan, D. E. (1982). *Dynamic Interactions in the Human Erythrocyte Membrane: Studies on the Lateral Mobility of Band 3, Phospholipid, and Cholesterol*, Ph.D. Thesis, Yale University, New Haven, Connecticut, pp. 1–282.
26. Golan, D. E., and Veatch, W. R. (1982). Lateral mobility of band 3 in the human erythrocyte membrane: control by ankyrin-mediated interactions. *Biophys. J.* 37:177a.
27. Tsuji, A., and Ohnishi, S. (1986). Restriction of the lateral motion of band 3 in the erythrocyte membrane by the cytoskeletal network: dependence on spectrin association state. *Biochemistry* 25:6133–6139.
28. Gottlieb, M. H., and Eanes, E. D. (1974). On phase transitions in erythrocyte membranes and extracted membrane lipids. *Biochim. Biophys. Acta* 373:519–522.
29. Karnovsky, M. J., Kleinfeld, A. M., Hoover, R. L., and Klausner, R. D. (1982). The concept of lipid domains in membranes. *J. Cell Biol.* 94:1–6.
30. Kinosita, K., Kataoka, R., Kimura, Y., Gotoh, O., and Ikegami, A. (1981). Dynamic structure of biological membranes as probed by 1,6-diphenyl-1,3,5-hexatriene: a nanosecond fluorescence depolarization study. *Biochemistry* 20:4270–4277.
31. Klausner, R. D., Kleinfeld, A. M., Hoover, R. L., and Karnovsky, M. J. (1980). Lipid domains in membranes. Evidence derived from structural perturbations induced by free fatty acids and lifetime heterogeneity analysis. *J. Biol. Chem.* 255:1286–1295.
32. Ladbrooke, B. D., and Chapman, D. (1969). Thermal analysis of lipids, proteins, and biological membranes. *Chem. Phys. Lipids* 3:304–356.
33. Kiehm, D. J., and Ji, T. H. (1977). Photochemical cross-linking of cell membranes. *J. Biol. Chem.* 252:8524–8531.
34. Nigg, E. A., and Cherry, R. J. (1979). Dimeric association of band 3 in the erythrocyte membrane demonstrated by protein diffusion measurements. *Nature* 277:493–494.
35. Peters, K., and Richards, F. M. (1977). Chemical cross-linking: reagents and problems in studies of membrane structure. *Annu. Rev. Biochem.* 46:523–551.

36. Steck, T. L. (1972). Cross-linking the major proteins of the isolated erythrocyte membrane. *J. Mol. Biol.* 66:295–305.
37. Weinstein, R. S., Khodadad, J. K., and Steck, T. L. (1978). Fine structure of the band 3 protein in human red cell membranes: freeze-fracture studies. *J. Supramol. Struct.* 8:325–335.
38. Yu, J., and Branton, D. (1976). Reconstitution of intramembrane particles in recombinants of erythrocyte protein band 3 and lipid: effects of spectrin-actin association. *Proc. Natl. Acad. Sci. USA* 73:3891–3895.
39. Yu, J., Fischman, D. A., and Steck, T. L. (1973). Selective solubilization of proteins and phospholipids from red blood cell membranes by nonionic detergents. *J. Supramol. Struct.* 1:233–248.
40. Furthmayr, H., and Marchesi, V. T. (1976). Subunit structure of human erythrocyte glycophorin A. *Biochemistry* 15:1137–1144.
41. Axelrod, D., Koppel, D. E., Schlessinger, J., Elson, E., and Webb, W. W. (1976). Mobility measurements by analysis of fluorescence photobleaching recovery kinetics. *Biophys. J.* 16:1055–1069.
42. Cherry, R. J. (1979). Rotational and lateral diffusion of membrane proteins. *Biochim. Biophys. Acta* 559:289–327.
43. Austin, R. H., Chan, S. S., and Jovin, T. M. (1979). Rotational diffusion of cell surface components by time-resolved phosphorescence anisotropy. *Proc. Natl. Acad. Sci. USA* 76:5650–5654.
44. Cherry, R. J., Burkli, A., Busslinger, M., Schneider, G., and Parish, G. R. (1976). Rotational diffusion of band 3 proteins in the human erythrocyte membrane. *Nature* 263:389–393.
45. Johnson, P., and Garland, P. B. (1981). Depolarization of fluorescence depletion. A microscopic method for measuring rotational diffusion of membrane proteins on the surface of a single-cell. *FEBS Lett.* 132:252–256.
46. Yoshida, T. M., and Barisas, B. G. (1986). Protein rotational motion in solution measured by polarized fluorescence depletion. *Biophys. J.* 50:41–53.
47. Macara, I. G., Kuo, S., and Cantley, L. C. (1983). Evidence that inhibitors of anion exchange induce a transmembrane conformational change in band 3. *J. Biol. Chem.* 258:1785–1792.
48. Nigg, E. A., and Cherry, R. J. (1979). Influence of temperature and cholesterol on the rotational diffusion of band 3 in the human erythrocyte membrane. *Biochemistry* 18:3457–3465.
49. Cherry, R. J., Nigg, E. A., and Beddard, G. S. (1980). Oligosaccharide motion in erythrocyte membranes investigated by picosecond fluorescence polarization and microsecond dichroism of an optical probe. *Proc. Natl. Acad. Sci. USA* 77:5899–5903.
50. Tyler, J. M., Hargreaves, W. R., and Branton, D. (1979). Purification of two spectrin-binding proteins. *Proc. Natl. Acad. Sci. USA* 76:5192–5196.
51. Fairbanks, G., Steck, T. L., and Wallach, D. F. H. (1971). Electrophoresis analysis of the major polypeptides of the human erythrocyte membrane. *Biochemistry* 10:2606–2617.
52. Liu, S.-C., and Palek, J. (1980). Spectrin tetramer-dimer equilibrium and the stability of erythrocyte membrane skeletons. *Nature* 285:586–588.

53. Ungewickell, E., and Gratzer, W. B. (1978). Self-association of human spectrin. *Eur. J. Biochem.* 88:379–385.
54. Bennett, V., and Stenbuck, P. J. (1980). Association between ankyrin and the cytoplasmic domain of band 3 isolated from the human erythrocyte membrane. *J. Biol. Chem.* 255:6424–6432.
55. Hargreaves, W. R., Giedd, K. N., Verkleij, A., and Branton, D. (1980). Reassociation of ankyrin with band 3 in erythrocyte membranes and in lipid vesicles. *J. Biol. Chem.* 255:11965–11972.
56. Bennett, V., and Stenbuck, P. J. (1979). The membrane attachment site for spectrin is associated with band 3 in human erythrocyte membranes. *Nature* 280:468–473.
57. Bennett, V., and Stenbuck, P. J. (1980). Human erythrocyte ankyrin. *J. Biol. Chem.* 255:2540–2548.
58. Sheetz, M. P. (1979). Integral membrane protein interaction with triton cytoskeletons of erythrocytes. *Biochim. Biophys. Acta* 557:122–134.
59. Fowler, V., and Bennett, V. (1978). Association of spectrin with its membrane attachment site restricts lateral mobility of human erythrocyte integral membrane proteins. *J. Supramol. Struct.* 8:215–221.
60. Anderson, R. A., and Lovrien, R. E. (1984). Glycophorin is linked by band 4.1 protein to the human erythrocyte membrane skeleton. *Nature* 307:655–658.
61. Anderson, R. A., and Marchesi, V. T. (1985). Regulation of the association of membrane skeletal protein 4.1 with glycophorin by a polyphosphoinositide. *Nature* 318:295–298.
62. Schindler, M., Koppel, D. E., and Sheetz, M. P. (1980). Modulation of membrane protein lateral mobility by polyphosphates and polyamines. *Proc. Natl. Acad. Sci. USA* 77:1457–1461.
63. Smith, D. K., and Palek, J. (1982). Modulation of lateral mobility of band 3 in the red cell membrane by oxidative cross-linking of spectrin. *Nature* 297:424–425.
64. Bennett, V. (1982). The molecular basis for membrane-cytoskeleton association in human erythrocytes. *J. Cell. Biochem.* 18:49–65.
65. Pasternack, G. R., Anderson, R. A., Leto, T. L., and Marchesi, V. T. (1985). Interactions between protein 4.1 and band 3. An alternative binding site for an element of the membrane skeleton. *J. Biol. Chem.* 260:3676–3683.
66. Koppel, D. E., Sheetz, M. P., and Schindler, M. (1981). Matrix control of protein diffusion in biological membranes. *Proc. Natl. Acad. Sci. USA* 78:3576–3580.
67. Agre, P., Casella, J. F., Zinkham, W. H., McMillan, C., and Bennett, V. (1985). Partial deficiency of erythrocyte spectrin in hereditary spherocytosis. *Nature* 314:380–383.
68. Alloisio, N., Morle, L., Bachir, D., Guetarni, D., Colonna, P., and Delaunay, J. (1985). Red cell membrane sialoglycoprotein beta in homozygous and heterozygous 4.1(-) hereditary elliptocytosis. *Biochim. Biophys. Acta* 816:57–62.
69. Coetzer, T. L., and Palek, J. (1986). Partial spectrin deficiency in hereditary pyropoikilocytosis. *Blood* 67:919–924.

70. Garbarz, M., Lecomte, M. C., Dhermy, D., Feo, C., Chaveroche, I., Gautero, H., Bournier, O., Picat, C., Goepp, A., and Boivin, P. (1986). Double inheritance of an alpha I/65 spectrin variant in a child with homozygous elliptocytosis. *Blood* 67:1661-1667.
71. Lawler, J., Coetzer, T. L., Palek, J., Jacob, H. S., and Luban, N. (1985). Sp alpha I/65: a new variant of the alpha subunit of spectrin in hereditary elliptocytosis. *Blood* 66:706-709.
72. Lawler, J., Liu, S.-C., Palek, J., and Prchal, J. (1984). A molecular defect of spectrin in a subset of patients with hereditary elliptocytosis. Alterations in the alpha-subunit domain involved in spectrin self-association. *J. Clin. Invest.* 73:1688-1695.
73. Lawler, J., Palek, J., Liu, S.-C., Prchal, J., and Butler, W. M. (1983). Molecular heterogeneity of hereditary pyropoikilocytosis: identification of a second variant of the spectrin alpha-subunit. *Blood* 62:1182-1189.
74. Marchesi, S. L., Knowles, W. J., Morrow, J. S., Bologna, M., and Marchesi, V. T. (1986). Abnormal spectrin in hereditary elliptocytosis. *Blood* 67:141-151.
75. Morle, L., Garbarz, M., Alloisio, N., Girot, R., Chaveroche, I., Boivin, P., and Delaunay, J. (1985). The characterization of protein 4.1 Presles, a shortened variant of RBC membrane protein 4.1. *Blood* 65:1511-1517.
76. Ohanian, V., Evans, J. P., and Gratzer, W. B. (1985). A case of elliptocytosis associated with a truncated spectrin chain. *Br. J. Haematol.* 61:31-39.
77. Palek, J., and Lux, S. E. (1983). Red cell membrane skeletal defects in hereditary and acquired hemolytic anemias. *Semin. Hematol.* 20:189-224.
78. Conboy, J., Mohandas, N., Tchernia, G., and Kan, Y. W. (1986). Molecular basis of hereditary elliptocytosis due to protein 4.1 deficiency. *N. Engl. J. Med.* 315:680-685.
79. Sheetz, M. P., and Casaly, J. (1980). 2,3-Diphosphoglycerate and ATP dissociate erythrocyte membrane skeletons. *J. Biol. Chem.* 255:9955-9960.
80. Sheetz, M. P., Febbroriello, P., and Koppel, D. E. (1982). Triphosphoinositide increases glycoprotein lateral mobility in erythrocyte membranes. *Nature* 296:91-93.
81. Golan, D. E., Singer, P. A., and Veatch, W. R. (1984). Calcium-induced immobilization of glycophorin in the human erythrocyte membrane. *Biophys. J.* 45:202a.
82. Low, P. S., Waugh, S. M., Zinke, K., and Drenckhahn, D. (1985). The role of hemoglobin denaturation and band 3 clustering in red blood cell aging. *Science* 227:531-533.
83. Waugh, S. M., Willardson, B. M., Kannan, R., Labotka, R.J., and Low, P. S. (1986). Heinz bodies induce clustering of band 3, glycophorin, and ankyrin in sickle cell erythrocytes. *J. Clin. Invest.* 78:1155-1160.
84. Shaklai, N., Shviro, Y., Rabizadeh, E., and Kirschner-Zilber, I. (1985). Accumulation and drainage of hemin in the red cell membrane. *Biochim. Biophys. Acta* 821:355-366.
85. Hebbel, R. P., Yamada, O., Moldow, C. F., Jacob, H. S., White, J. G., and Eaton, J. W. (1980). Abnormal adherence of sickle erythrocytes to cultured

vascular endothelium: possible mechanism for microvascular occlusion in sickle cell disease. *J. Clin. Invest.* 65:154–160.

86. Hebbel, R. P., and Miller, W. J. (1984). Phagocytosis of sickle erythrocytes: immunologic and oxidative determinants of hemolytic anemia. *Blood* 64:733–741.

87. Hebbel, R. P., Schwartz, R. S., and Mohandas, N. (1985). The adhesive sickle erythrocyte: cause and consequence of abnormal interactions with endothelium, monocytes/macrophages and model membranes. *Clin. Haematol.* 14:141–161.

88. Galili, U., Clark, M. R., and Shohet, S. B. (1986). Excessive binding of natural anti-alpha-galactosyl immunoglobulin G to sickle erythrocytes may contribute to extravascular cell destruction. *J. Clin. Invest.* 77:27–33.

89. Galili, U., Korkesh, A., Kahane, I., and Rachmilewitz, E. A. (1983). Demonstration of a natural antigalactosyl IgG antibody on thalassemic red blood cells. *Blood* 61:1258–1264.

90. Kay, M. M. B., Bosman, G. J. C. G. M., Shapiro, S. S., Bendich, A., and Bassel, P. S. (1986). Oxidation as a possible mechanism of cellular aging: vitamin E deficiency causes premature aging and IgG binding to erythrocytes. *Proc. Natl. Acad. Sci. USA* 83:2463–2467.

91. Petz, L. D.. Yam, P., Wilkinson, L., Garratty, G., Lubin, B., and Mentzer, W. (1984). Increased IgG molecules bound to the surface of red blood cells of patients with sickle cell anemia. *Blood* 64:301–304.

92. Singer, J. A., Jennings, L. K., Jackson, C. W., Dockter, M. E., Morrison, M., and Walker, W. S. (1986). Erythrocyte homeostasis: antibody-mediated recognition of the senescent state by macrophages. *Proc. Natl. Acad. Sci. USA* 83:5498–5501.

93. Khansari, N., and Fudenberg, H. H. (1986). Phagocytosis of senescent erythrocytes by autologous monocytes: requirement of membrane-specific autologous IgG for immune elimination of aging red blood cells. *Cell. Immunol.* 78:114–121.

94. Mohandas, N., and Evans, E. (1985). Sickle erythrocyte adherence to vascular endothelium. Morphologic correlates and the requirement for divalent cations and collagen-binding plasma proteins. *J. Clin. Invest.* 76:1605–1612.

95. Hebbel, R. P., Boogaerts, M. A. B., Eaton, J. W., and Steinberg, M. H. (1980). Erythrocyte adherence to endothelium in sickle-cell anemia. A possible determinant of disease severity. *N. Engl. J. Med.* 302:992–995.

96. Adair, W. L., and Kornfeld, S. (1974). Isolation of receptors for wheat germ agglutinin and the *Ricinus communis* lectins from human erythrocytes using affinity chromatography. *J. Biol. Chem.* 249:4676–4704.

97. Chasis, J. A., Mohandas, N., and Shohet, S. B. (1985). Erythrocyte membrane rigidity induced by glycophorin A-ligand interaction. Evidence for a ligand-induced association between glycophorin A and skeletal proteins. *J. Clin. Invest.* 75:1919–1926.

98. Caulfield, J. P., and Cianci, C. M. L. (1985). Human erythrocytes adhering to schistosomula of *Schistosoma mansoni* lyse and fail to transfer membrane components to the parasite. *J. Cell Biol.* 101:158–166.

99. Golan, D. E., Furlong, S. T., Brown, C. S., and Caulfield, J. P. (1988). Monopalmitoyl phosphatidylcholine incorporation into human erythrocyte ghost membranes causes protein and lipid immobilization and cholesterol depletion. *Biochemistry* 27:2661-2667.

100. Yechiel, E., and Edidin, M. (1987). Micrometer-scale domains in fibroblast plasma membranes. *J. Cell Biol.* 105:755-760.

101. Thompson, N. L., and Axelrod, D. (1980). Reduced lateral mobility of a fluorescent lipid probe in cholesterol-depleted erythrocyte membrane. *Biochim. Biophys. Acta* 597:155-165.

102. Nigg, E. A., Bron, C., Girardet, M., and Cherry, R. J.(1980). Band 3-glycophorin A association in erythrocyte membranes demonstrated by combining protein diffusion measurements with antibody-induced cross linking. *Biochemistry* 19:1887-1893.

103. Cabantchik, Z. I., and Rothstein, A. (1974). Membrane proteins related to anion permeability of human red blood cells. II. Effects of proteolytic enzymes on disulfonic stilbene sites of surface proteins. *J. Membr. Biol.* 15: 227-248.

104. Pinto da Silva, P., Douglas, S. D., and Branton, D. (1971). Location of A antigen sites on human erythrocyte ghosts. *Nature* 232:194-195.

105. Pinto da Silva, P., and Nicolson, G. L. (1974). Freeze-etch localization of concanavalin A receptors to the membrane intercalated particles of human erythrocyte ghost membranes. *Biochim. Biophys. Acta* 363:311-319.

106. Tillack, T. W., Scott, R. E., and Marchesi, V. T. (1972). The structure of erythrocyte membranes studied by freeze-etching. *J. Exp. Med.* 135:1209-1227.

107. Muhlebach, T., and Cherry, R. J. (1982). Influence of cholesterol on the rotation and self-association of band 3 in the human erythrocyte membrane. *Biochemistry* 21:4225-4228.

108. Nigg, E. A., and Cherry, R. J. (1980). Anchorage of a band 3 population at the erythrocyte cytoplasmic membrane surface: protein rotational diffusion measurements. *Proc. Natl. Acad. Sci. USA* 77:4702-4706.

109. Nigg, E. A., Gahmberg, C. G., and Cherry, R. J. (1980). Rotational diffusion of band 3 proteins in membranes from En(a-) and neuraminidase-treated normal human erythrocytes. *Biochim. Biophys. Acta* 600:636-642.

110. Nigg, E. A., Kessler, M., and Cherry, R. J. (1979). Labeling of human erythrocyte membranes with eosin probes used for protein diffusion measurements. *Biochim. Biophys. Acta* 550:328-340.

111. Bloom, J. A., and Webb, W. W. (1983). Lipid diffusibility in the intact erythrocyte membrane. *Biophys. J.* 42:295-305.

112. Kapitza, H. G., and Sackmann, E. (1980). Local measurement of lateral motion in erythrocyte membranes by photobleaching technique. *Biochim. Biophys. Acta.* 595:56-64.

113. Morrot, G., Cribier, S., Devaux, P. F., Geldwerth, D., Davoust, J., Bureau, J. F., Fellmann, P., Herve, P., and Frilley, B. (1986). Asymmetric lateral mobility of phospholipids in the human erythrocyte membrane. *Proc. Natl. Acad. Sci. USA* 83:6863-6867.

114. Rimon, G., Meyerstein, N., and Henis, Y. I. (1984). Lateral mobility of phospholipids in the external and internal leaflets of normal and hereditary spherocytic human erythrocytes. *Biochim. Biophys. Acta* 775:283–290.
115. Cogan, U., and Schachter, D. (1981). Asymmetry of lipid dynamics in human erythrocyte membranes studied with impermeant fluorophores. *Biochemistry* 20:6396–6403.
116. Saxton, M. J. (1982). Lateral diffusion in an archipelago. Effects of impermeable patches on diffusion in a cell membrane. *Biophys. J.* 39:165–173.
117. Eisinger, J., Flores, J., and Petersen, W. P. (1986). A milling crowd model for local and long-range obstructed lateral diffusion. Mobility of excimeric probes in the membrane of intact erythrocytes. *Biophys. J.* 49:987–1001.

14

Red Blood Cell Shape

BRIAN S. BULL   *School of Medicine, Loma Linda University, Loma Linda, California*

DOUGLAS BRAILSFORD   *School of Allied Health Professions, Loma Linda University, Loma Linda, California*

I. INTRODUCTION

The mammalian erythrocyte is a most persistently enigmatic entity. Its main function, that of oxygen transport, is obvious enough. Anatomically it is simple compared with most other mammalian cells. In its mature state it has no nucleus and no cytoskeleton. Consequently, the factors that determine its shape ought to be simple to decipher. Not so! Even though a great deal has been learned in the last two decades concerning the molecular structure of the membrane, the way in which the constituent molecules interact to control the shape of the red blood cell is still imperfectly understood. There is no consensus yet on how red blood cell shape is maintained.

The first difficulty lies in the ability of the red blood cell to assume several stable, interconvertible, but very dissimilar shapes. Before considering these alternative shapes, however, the characteristic resting shape, a smooth, biconcave disk, is a good starting point and, for the moment, it will provide challenge enough.

Inherent in a discussion of the shape of any object are two questions. What are the forces that operate to produce the shape and on what elements of the object do they act? It is known that any system of forces can be resolved into just two kinds: forces that tend to extend/compress the object and forces that tend to produce curvature deformation. For reasons that are not immediately obvious but that are nicely explained in Chapter 15 of this book and in further detail elsewhere [1,2] the two kinds of deformation are referred to as shear and bending, respectively. In the case of the red blood cell the only solid constituent on which these forces can act is the plasma membrane. Also, because the various shapes described are virtually static equilibrium states, viscous membrane behavior and plastic flow effects need not be considered.

II. STRUCTURE OF THE RED BLOOD
CELL MEMBRANE: I

A notion of the structure of the plasma membrane of the red blood cell can be gleaned from Figure 1. The top half of the membrane, from the phospholipid bilayer outwards, has been redrawn from a model by Grant [3]. From the phospholipid bilayer inwards the interrelationships of the molecules are similar to those indicated by Elgsaeter et al. [4] but with a 60-nm-thick spectrin layer as shown by Bull et al. [5].

The extraordinary complexity of the red blood cell membrane is obvious from Figure 1. It will be necessary to explore some of the mechanical consequences of this complexity later but for the moment it will suffice to consider the red blood cell membrane as a phospholipid bilayer sandwiched between outer and inner elements as shown schematically in Figure 2. The outer layer

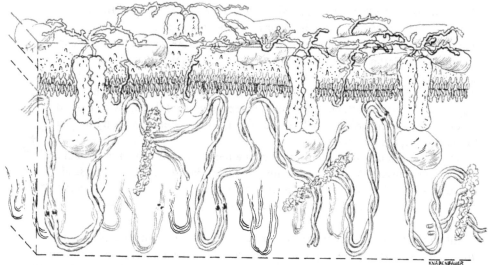

Figure 1 The complex structure of the red cell membrane can, for mechanical purposes, be considered as a phospholipid bilayer sandwiched between an outer layer of adsorbed albumin (and occasional globulin molecules) and an inner layer of spectrin. The upper half of the illustration has been redrawn from Grant [3] the thickness of the spectrin layer from Bull et al. [5] and the organization of the spectrin layer as depicted by Elgsaeter et al. [4].

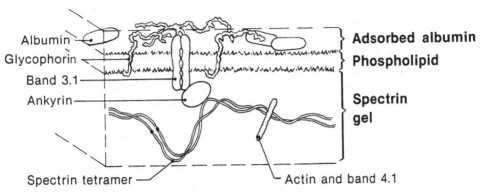

Albumin
Glycophorin
Band 3.1
Ankyrin

Adsorbed albumin
Phospholipid

Spectrin
gel

Spectrin tetramer — Actin and band 4.1

Figure 2 The four layers detailed in Fig. (1) are schematically shown to underscore that the red cell membrane may function mechanically as a *duplex bilayer couple* with band 3.1 molecules ensuring that the whole structure functions as a single mechanical unit.

is largely composed of adsorbed albumin interspersed between glycophorin and sialic acid residues. The inner element is largely composed of spectrin, probably in the form of tetramers interconnected in a netlike array and almost certainly functioning as an ionic gel [4,6-15].

III. BENDING OR SHEAR?

Bending and shear forces operate on a sheet of material such as the membrane of the red blood cell in characteristic ways. Bending forces will cause it to curve into either a cylindrical or spheroidal form. Shear forces, by definition, transform a square portion of membrane surface into a rhombus or parallelogram. Neither bending nor shear bring about any appreciable change in surface area. The red blood cell membrane, unlike a rubber balloon or a soap bubble, is formed around a bimolecular leaflet and thus tends to conserve its thickness. As a consequence, its surface area remains constant except under the influence of large dilational forces, such as occur during hypotonic lysis [16].

Having now examined in a general way the anatomy of the membrane and considered briefly the two kinds of forces to which it is subject, what conclusions can be drawn regarding the biconcave discoid resting shape of the healthy red blood cell? There are two possibilities. One well-supported viewpoint in red blood cell membrane research is that the biconcave disk is a reference shape [1] into which the membrane is cast much as a latex rubber glove is cast in the shape of a human hand. If this is the case then at any time that the red blood cell is distorted into some other shape, both shear and bending forces will be involved and elastic energy will be stored as the membrane is deformed. Left to itself, the red blood cell will resume its reference shape, the energy for this return coming from the elastic energy stored as the deformation took place. This possibility is easy to visualize because it is the normal behavior of a homogeneous material such as rubber, but it presupposes that elastic shear energy is the dominant factor in maintaining the discoid shape.

The alternative view is that the biconcave disk is not just a reference but a dynamic equilibrium form controlled by the minimization of bending energy and that elastic shear energy is relatively small.

The seminal work on the relationship between membrane bending and the resting shape of the red blood cell is an article by Canham [17]. In this article Canham documented that the partial collapse of a spherical, thin-walled shell would result in a biconcave discoidal shape provided that the shell was made of appropriate, although unusual, material. The material, he posited, had little resistance to shear distortion but significantly greater resistance to bending forces. Thus, on partial collapse the red blood cell membrane would eschew wrinkles, creases, and sharp changes of curvature because it is in these deformations that stored bending energy is highest. A biconcave disk shaped like the red blood cell would result because it is the shape that absorbs the least amount of bending energy.

The mathematical predictions of Canham were confirmed by the production of macroscopic, mechanical models of the red blood cell membrane by Brailsford and Bull [2,18,19]. The models were constructed in such a fashion that the material representing the membrane was highly anisotropic. That is, its physical characteristics were not the same in all three dimensions.

Why is anisotropy important? If the membrane is a shell of normal isotropic material then bending resistance cannot dominate its behavior and the reference hypothesis is, by default, the only viable explanation for the discoid resting shape. In isotropic material the bending resistance is proportional to the third power of the thickness of the membrane whereas resistance to shear is proportional to only the first power. It follows that when thin sheets of isotropic materials are deformed, bending resistance is vanishingly low. In a suitably constructed anisotropic model, however, these relationships can be manipulated at will.

Computations based on the concept of minimum bending energy have been carried out by Zarda [20] using realistic elastic moduli for membrane bending and for membrane shear. The computations show that the biconcave discocyte can, in fact, be produced by partial deflation of a membrane whose resting shape is that of a sphere and that only a small negative pressure is required to maintain the shape. Similar calculations have been performed by other investigators [21–23], who have reached essentially similar conclusions.

We thus have two apparently conflicting explanations for the biconcave discord resting shape. It may be a reference shape in which the membrane is stress-free so that the cell will return to that shape whenever circumstances permit. Alternatively, the membrane may possess the unusual characteristic of storing more energy in bending than in shear whenever the red blood cell shape is other than a sphere. Because the normal biconcave disk possesses insufficient volume to assume a spherical shape, it is forced to dispose of some 40% excess membrane in accordance with minimum energy requirements. The biconcave discoid shape is the outcome.

Either explanation for the biconcave discoid shape of the red blood cell appears to be convincing when considered from a limited point of view. The reference shape concept that is implied in Chapter 15 has been systematically developed by an able group of experimenters and theoreticians into a sophisticated system of mechanics and thermodynamics, a system basic to the study of the viscoelastic properties and rheology of red blood cells. On the other hand, the several shapes which the red blood cell so readily assumes, the *morphological* versus the *rheological* evidence, leads to a different set of conclusions. These shapes can be so successfully mimicked by anisotropic shells and by calculations based on the minimum bending energy hypothesis that this viewpoint is difficult to ignore. Because both viewpoints are the outcome of careful experimentation, they must be reconcilable. This present chapter on the morphology of the red blood cell is an attempt to accomplish that task—to show that for at least some red blood cell shapes the two hypotheses are indistinguishable whereas for other shapes the process of measurement itself may be the cause for some of the discrepant data.

One piece of evidence that has been brought against the minimum bending energy hypothesis is that during its development the red blood cell does not undergo a deflation from a sphere to a biconcave disk. Rather, the erythroblast undergoes unequal division with one daughter cell – the nucleus and a small amount of cytoplasm—being phagocytosed while the majority of the membrane and cytoplasm remains to enter the circulation as a reticulocyte [24]. The reticulocyte certainly not spherical nor is it a biconcave disk. It is usually a puckered, oblate spheroid.

Does the absence of clear-cut deflation from sphere to deflated sphere and deflated sphere to biconcave disk in the maturation of the red blood cell invalidate the minimum bending energy hypothesis? We think not. The concept is valid no matter what the developmental history of the red blood cell, provided that the membrane, once assembled, possesses the appropriate mechanical characteristics of low shear and high bending resistance. Indeed, this assembly is now known to be almost geometrically regular. Thus, the production of a uniform membrane shell with the required preponderance of bending resistance would seem to be readily achievable without the developing red blood cell ever actually going through a spherical, maturation stage.

There is also strong experimental evidence that the biconcave shape is not a reference shape but is dependent on the intrinsic membrane properties. If the biconcave shape were truly a reference shape it would follow that the biconcavities would be fixed to certain portions of the red blood cell membrane shell just as the tip of the thumb in a latex rubber glove is fixed to a particular domain of the rubber. That the biconcavities are thus restricted is not borne out by experiment. There is substantial evidence from the flow cell experiments [19] that the biconcavities and rim of the biconcave disk can be placed anywhere on the red blood cell surface. The same conclusions can also be reasonably inferred from the fact that it is possible to force the membrane of a biconcave discoid red blood cell to "tank tread" smoothly in shear flow [25]. Thus, the membrane must constitute a shell that, in its mechanical properties (at least to a first approximation), is essentially uniform over its entire surface. Careful experiments on the morphological stability of the red blood cell under osmotic sphering also confirm that shear elasticity alone is insufficient to account for the behavior of the cell [22]. We, therefore, prefer the minimum bending energy hypothesis as an explanation for the discoid red blood cell shape.

It should, however, be emphasized that giving bending energy priority in the control of the red blood cell resting shape does not exclude the possibility that the dimples and the rim may have a preferred location on the red blood cell surface. It would take only a very minor nonuniformity in the pattern of the spectrin skeleton to provide the red blood cell membrane with a memory that would return the dimples and the rim to some preferred location. Just such a secondary memory seems to be indicated by the experiments of Fischer et al. [26].

It would be unfair to proceed from this point without underscoring the importance to the study of erythrocyte morphology of the concept that the resting shape of the red blood cell results from a minimization of elastic bending energy. The decision is a fundamental one for at least one of the explanations later to be offered for red blood cell crenation and cupping. Even the calculations involved in determining the elastic moduli of membrane bending and membrane shear are affected by this choice. If this choice is in error then the entire superstructure concerning red blood cell morphology will be faulty. With this caveat let us proceed on the assumption that the minimization of elastic energy of bending is the major factor causing the resting discoid red blood cell to assume its familiar biconcave shape.

IV. THE SOURCE OF MEMBRANE BENDING RESISTANCE

It has long been realized that a liquid crystal array of phospholipid molecules will have bending elasticity without shear resistance [17,27]. Furthermore, it is generally accepted that the phospholipids of the membrane are arranged in the form of a liquid crystal bilayer.

The difficult task of measuring the bending modulus of elasticity of the whole red blood cell membrane has been accomplished [28] and is explained in Chapter 15. A modulus in this sense is simply a constant that scales the elastic energy to the dimensions of the membrane and so provides a quantitative measure of the elastic properties. A liquid crystal bilayer can resist bending because curvature deformation expands the surface which becomes convex and compresses the surface which becomes concave. The modulus of area expansion/compression is more easily measured than the membrane bending modulus because fewer assumptions are involved [16]. From this value the modulus of bending can be calculated, making certain assumptions, and compared to the measured value. If the resistance to area expansion is considered to occur only in the two parallel planes formed by the polar head groups of the lipid bilayer, the bending elasticity will be proportional to the *square* of the membrane thickness [23]. This gives much too high a modulus relative to the value measured directly by Evans et al. already referred to. If, on the other hand, the two monolayers are considered separately and the attraction between the molecules in each monolayer is considered to act uniformly along the whole length of the molecules, the bending elasticity will be proportional to the *cube* of the effective monolayer thickness [29]. This gives much too small a modulus of elasticity by the same process of reasoning. By taking a position somewhere between these two extremes it is conceivable that all the bending elasticity required could be accounted for in the phospholipid bilayer.

Although this conclusion is mechanically reasonable it is probably not correct in that it would relieve the spectrin layer from making any contribution to the bending resistance of the intact membrane. Biconcave shapes have not been observed to occur in deflated phospholipid vesicles nor in pathological red blood cells in which the spectrin submembrane skeleton is defective [8]. Thus, it seems likely that the bilayer does contribute to the bending energy necessary for the formation of the biconcave shape but is not alone in that role. The spectrin net is also an essential contributor and possibly also, albumin adsorbed to the outer surface. Before dealing with these complications, however, it may be best to consider some other forms the cell can take. Without a listing of the bewildering variety of forms that the red blood cell can assume, it may not be apparent what use all this membrane complexity may subserve.

V. ALBUMIN, pH, AND RED BLOOD CELL SHAPE

Any treatment of red blood cell shape must, if it is to be complete, address the other morphological forms that the membrane so readily assumes. Additionally, a choice must be made as to which of the inciting agents that produce these forms are going to be considered, because several hundred shape-changing agents have been identified. It is not, however, unreasonable to place albumin and pH at the top of the list of such agents. Albumin is strongly adsorbed to the outer surface of the red blood cell membrane [3]. So much is adsorbed, in fact, that it can be said to be a normal constituent of the red blood cell in vivo [30]. The effects of albumin on red blood cell shape have been studied exhaustively, beginning with the epochal work of Furchgott and Ponder [31].

Crenated red blood cells (echinocytes) and cupped red blood cells (stomatocytes) must be given priority in any list of the morphological forms that a red blood cell may assume. The crenated cell is composed of a spherical central portion studded by 15–30 fingerlike rounded projections. Bessis and Brecher [24] in a most useful attempt to confer uniformity on discussions of red blood cell shape, have designated this type of crenated cell an echinocyte II if the underlying shape is still identifiably a disk or an echinocyte III if it is approximately spherical. Before proceeding to the cup-shaped erythrocyte it is worth noting that these same authors also identified another, earlier form of crenated cell that they designated the echinocyte I. This cell is still basically a biconcave disk but the toroid-shape rim of the disk displays several large, low protuberances that give it a scalloped appearance. This is a stable red blood cell shape, not merely a transient form on the way to an echinocyte II. Proceeding from echinocyte back to discocyte and beyond the next shape encountered is a discocyte with unusually steep-sided dimples. This cell is difficult to distinguish from the discocyte unless it is viewed in optical cross section such as when it is suspended on edge from a coverslip. This cell is sometimes designated as bow-tie because of this characteristic cross-sectional shape. Finally, there is the cup-shape cell or stomatocyte. This form, as the name implies, has a convexity on one side and a deeper-than-usual concavity on the opposite side. Although there are rare and unusual circumstances in which a cup cell may show crenations, this form is generally taken to be at the far left of the morphological spectrum with the crenated cell at the far right and the biconcave disk occupying a centrist position. Such an arrangement would place the echinocyte I just to the right of center and the bow-tie to the left Figure 3.

In the absence of albumin, red blood cells suspended in buffered isotonic saline usually take the shape either of a disk or of an echinocyte I. Albumin is a most effective antiechinocytogenic agent and when it is present in the suspending medium the red blood cells move rapidly to the left of the morphological spectrum. If sufficient albumin is added, ~10 g/L, some 20% of the cells become cup-shape, all echinocytes disappear, and the majority of the cells are either bow-tie shape or are smooth, biconcave disks [32].

At physiological pH in isotonic saline as already noted the average red blood cell is slightly to the right of center. As the pH is raised the red blood cell becomes first an echinocyte I and then an echinocyte II. Lowering the pH produces the opposite effect, returning the cells to the discocyte stage and eventually forming cups. A particularly engaging experiment of this type was performed many years ago [33]. A red blood cell in an albumin-free medium was

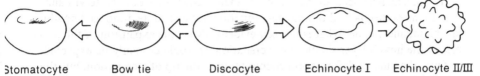

Stomatocyte   Bow tie   Discocyte   Echinocyte I   Echinocyte II/III

Figure 3 The several stable morphological forms which the red cell can assume. The bow tie cell is so called because a cross section through the center of the upper and lower dimples reveals them to be much steeper-sided than the dimples of the normal, biconcave disk.

caught on a micropipette. Under video camera observation a micromanipulator was used to bring a glass rod into the immediate environment of the trapped red blood cell. As the glass rod approached (and the immediate environment of the red blood cell became more alkaline) the cell crenated. As it was withdrawn the cell reverted to a biconcave disk. The experiment could be repeated until the experimenter tired yet the red blood cell was apparently unharmed! Thus, both albumin and pH changes are highly effective in modifying red blood cell shape and as a first approximation an elevation in albumin has the opposite effect from an elevation in pH.

The crenated red blood cell has always been most difficult to accommodate within a comprehensive explanation of red blood cell shape. It is mathematically intractable because it is not axisymmetrical. Thus, investigators who attempt to model it mathematically are usually content to approximate the shape of a single crenation arising from a small portion of the red blood cell membrane shell. A cupped red blood cell, because it is axisymmetric, is mathematically much more tractable.

VI. CRENATION AND CUPPING—THE BILAYER COUPLE HYPOTHESIS

One of the most productive proposals in regard to the mechanisms responsible for crenation/cupping is that generally known as the "bilayer couple hypothesis" [34]. According to this hypothesis the various shape-changing agents preferentially perturb either the inner or the outer half of the phospholipid bilayer. The perturbed membrane then deforms so as to place the expanded half-layer on the convex side of the deformation. A very similar notion invoking chemically induced, localized bending moments has also been advanced by Evans [35]. Drug-induced crenation, according to this hypothesis, is due to an expansion of the outer half-layer of the red blood cell membrane while cupping results from expansion of the inner half-layer [35]. This hypothesis provides a plausible explanation for the effects of a wide variety of pharmacological agents on red blood cell shape [36–38].

The mechanical behavior of a coupled membrane is easy enough to visualize because most mechanical thermostats function because of heat-induced bending in a bimetallic couple. What is not self-evident is why the expansion of the outer half of the lipid bilayer should favor crenation. It would certainly be expected to favor the kind of deformation encountered at the tip of a crenation, but the skirtlike base is oppositely curved and of larger area. The two curvatures must therefore compete, especially when the crenation is developing.

The hypothesis would be significantly strengthened if it could be demonstrated mathematically that a crenated cell does indeed show an overall increase in the area of the outer half-layer as the usual formulation of the hypothesis requires. Unfortunately, as already mentioned, the crenated cell is a mathematically intractable shape and the approximations chosen for both the curvature of the starting membrane surface as well as for the shape of the crenation significantly influence the outcome of the calculations [2,39]. On balance, however, the weight of evidence in favor of the bilayer couple hypothesis is considerable and its explanatory powers are impressive.

VII. CRENATION AND CUPPING: THE PROTEIN GEL—LIPID BILAYER HYPOTHESIS

A provocative new twist to the bilayer couple hypothesis was provided by Stokke et al. [15,40]. These workers, drawing on the observations of Tanaka et al. [14] relative to phase transitions in ionic gels, suggested that the spectrin net could also be considered an ionic gel. In effect, they added a third component to the mechanical makeup of the red blood cell membrane. If, indeed, spectrin behaved like an ionic gel then it would be expected to substantially affect the mechanical properties of the resultant complex. The power of the concept was underscored by its ability to predict both the elastic shear modulus and the maximum elastic extension ratio of the red blood cell membrane on theoretical grounds. Furthermore, because ionic gels exhibit critical phenomena and phase transitions it seemed possible that its contraction could account for crenation in harmony with the conclusions drawn from the bilayer couple hypothesis. Such behavior is, in fact, observed in isolated red blood cell ghosts when contraction of the spectrin gel is brought about by lowering the pH of the suspending medium [4]. What is puzzling about the latter observation is that intact red blood cells, in contrast, crenate in an alkaline environment where the spectrin gel would be expected to expand. Although it is possible that the effects of pH in an intact cell may differ from the effects in red cell ghosts (Branton: personal communication) there is an alternative, logically consistent and mathematically tenable possibility. Before dealing with this possible explanation a further consideration of the complexity of membrane structure is in order.

VIII. STRUCTURE OF THE RED BLOOD
CELL MEMBRANE: II

Spectrin occurs in two chains, α and β, which combine to form a heterodimer. This molecule is highly flexible and forms higher-order polymers. It is likely that much of the spectrin in the submembrane skeleton is in the form of tetramers because hemoglobin, with which it is closely associated, stabilizes the spectrin tetramer [12,41,42]. There are binding sites for spectrin on protein 4.1 and it is known that the spectrin polymers attach, via an intermediate protein ankyrin, to band 3.1, an amphipathic transmembrane protein. Direct visualization of the resulting spectrin net has been achieved recently [6,43].

How is the spectrin gel attached to the lipid bilayer? The answer to this question is more significant than may at first appear. From Figure 2 it is clear that the spectrin gel is attached to band 3.1, which is free to move in the plane of the bilayer but cannot be detached from it. As a consequence the outer ends of the band 3.1 molecules, together with the extramembranous portion of glycophorin and adherent proteins such as albumin and IgG, will tend to diffuse randomly to cover the surface uniformly. But the band 3.1 molecules are attached to the spectrin gel below and are thus restrained. Curvature will therefore alter the local surface concentration of the proteins on the outer aspect of the membrane. That is, the proteins will spread out on a convex surface and condense on a concave surface. That there is actual movement of the oligosaccharides attached to the external proteins may be inferred from the observation that crenated cells treated with wheat germ lectin (which binds to the external molecules) cannot any longer return to the discoid state [44]. Similarly, a discoid cell treated with wheat germ lectin cannot subsequently be crenated. Curvature will at the same time affect the local density of the spectrin gel on the inner aspect of the membrane because in this case it is a continuous convoluted network. Because the diffusion of molecules is a thermodynamic process, both the spectrin gel below and the albumin moderated layer above can act as two-dimensional entropic springs; that is, they can both manifest surface elasticity. This is a very important concept because the upper and lower layers are kept a constant and relatively large distance apart by the phospholipid bilayer and are strongly coupled by band 3.1 proteins and their attached sugar residues. It is much easier, therefore, than in the case of the phospholipid bilayer to see how this structure can be sensitive to a variety of chemical agents and produce strong intrinsic curvature effecting red blood cell shape.

The picture that emerges is that of a four-layer (duplex bilayer) structure as shown diagrammatically in Figure 2. The phospholipid bilayer controls the surface area of the cell membrane, which as far as curvature deformation is concerned is constant. Thus, the phospholipid bilayer forms one of the two complete systems, having its own curvature elasticity. Substances that can insert themselves into either of the phospholipid monolayers can cause intrinsic precurvature, which will affect the whole membrane. Outside of the phospholipid bilayer lies the layer of adsorbed albumin, and inside is the spectrin net. It is from the latter layer that the shear elasticity of the membrane arises. In addition, because of its netlike arrangement the resistance to shear deformation of the spectrin net rises rapidly as the membrane approaches its elastic deformation limits.

Intrinsic precurvature can be defined as the curvature a portion of the membrane would adopt if it were able to exist on its own, separate from the intact cell membrane. If the whole cell were sphered and every portion of the cell membrane were trying to curl up into an even smaller radius of curvature it would be said to have positive intrinsic precurvature. If it were trying to evert and curl outwards it would be said to have negative intrinsic precurvature.

The outer red blood cell membrane layer consists of sugar moieties and adsorbed albumin molecules. It is, in effect, coupled to the spectrin layer via the band 3.1 protein molecules. These two layers thus form the second system having curvature elasticity. Intrinsic precurvature can be produced in this system by change of surface tension in the outer layer and changes in the gel pressure/tension in the inner layer. Both layers are sensitive to pH, and the outer layer is sensitive to the presence or absence of albumin.

It is this recognition that in the red blood cell membrane there are two linked systems, each sensitive to a different set of shape-changing agents, that makes the task of unraveling the complicated morphological behavior of the red blood cell much simpler.

IX. CRENATION AND CUPPING: A SPECULATIVE HYPOTHESIS BASED ON THE DUPLEX BILAYER MODEL

One way out of the apparent impasse occasioned by pH effects on the spectrinionic gel layer is that proposed by Brailsford et al. [45,46]. Mathematical modeling of the beginnings of a crenation led to the counter-intuitive conclusion that *expansion* of the inner half-leaflet would provide the kind of bumpy surface on a disk-shape red blood cell that is characteristic of an echinocyte I. The same modeling process uniformly failed to produce any surface irregularities at all as a result of outer half-leaflet expansion. What was not clear was the origin of the forces capable of transforming these low-level bumps into fully formed crenations. The insights of Elgsaeter and colleagues relative to the spectrin–ionic gel concept and the pictures of Byers and Branton [43] depicting the regular, netlike organization of the spectrin molecules have provided a plausible although highly speculative source for the missing crenating force.

We start with the proposition that the resting biconcave red blood cell represents a stressed uniform shell (minimum bending energy hypothesis). We then assume that the membrane material is positively precurved with the same radius of curvature as the osmotically sphered cell. If such a cell is sphered it will initially have zero internal pressure with respect to the fluid outside. To get the cell into the biconcave shape it must now be osmotically deflated. This deforms the membrane by both changing its curvature and putting it into shear. Elastic energy of both kinds is therefore required. The work done to overcome the elastic resistance is equal to the product of the volume and the mean pressure change inside the cell. To produce this kind of change, the pressure inside the cell must finish up lower than the outside pressure. Moreover, because the required elastic moduli are known, the pressure difference between the inside and outside of the cell can be calculated. Such a pressure difference puts the whole membrane shell into tangential compression.

Although the pressure difference itself is very small it acts on the cross-sectional area of the whole cell, which is large. The large resulting force is supported by only the thin ring of membrane at the circumference. The compressive pressure in the membrane is thus many times greater than the pressure difference across the membrane. The magnitude of the resultant compressive force in the membrane can be calculated. Using one assumed value [21] for the bending modulus, the resultant compressive force turns out to be such that if it acted in one direction it would shorten the membrane by ~30%. Because it acts almost equally in all directions the forces balance out, but such a compressive force has the potential to buckle the membrane. Consider now a round hole appearing in the membrane and that the compressive forces in the membrane are maintained even though, in reality, the hole would relieve the pressure difference. The radial inwardly directed forces in the membrane around the hole will now be unbalanced and will move the membrane inwards to try to close the hole. To do so, however, the material around the circumference of the hole has to shrink, that is, be put into shear. Such a stress is called a hoop stress and is similar to the state of affairs when a portion of the membrane is sucked into a micropipette. If instead of a hole a perturbation such as a small surface bump is substituted, the curvature of the skirtlike base of the bump deflects the radial compressive forces outwards. This also produces a hoop stress similar to that produced by a hole. Because there is no actual hole in this case the pressure differential is maintained and the protuberance continues to grow just as though it is being pushed into an invisible micropipette. Growth stops only when the shear deformation in the successive hoop elements sets up a reaction force equal to the deforming force or reaches the maximum that the spectrin net-like array can tolerate. Because of this limiting mechanism, all protuberances will be of approximately the same height.

The normal biconcave red blood cell is obviously a stable form so the membrane must normally have enough bending rigidity to resist the crenation-forming process outlined above. That is, the pressure difference across the membrane is insufficient, even when the surface is perturbed, to propagate crenation. However, the pressure differential is a function of the intrinsic precurvature. The case

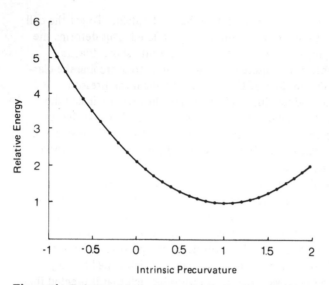

Figure 4 Relationship between the intrinsic pre-curvature and the relative amount of bending energy stored in the membrane of a biconcave disc. Zero on the x axis denotes a membrane that is flat in its resting state. +1 denotes a membrane that is intrinsically, positively precurved to the same extent as the surface of a sphered cell; –1 denotes intrinsic negative precurvature of similar degree. In order to emphasize the shape of the relationship between the energy stored in the deflated shell and the starting precurvature of the membrane the y axis values have simply been plotted in energy units relative to the minimum value for this particular biconcave shape which is E. A. Evans' average shape obtained by interference holography [50].

chosen above, where the precurvature was positive and the same as that of an osmotically swollen cell, is almost the least energetic. This state of affairs is indicated as 1 on the x axis of Figure 4 where 0 represents a flat membrane and –1 a membrane with negative curvature equivalent to that of the sphere but in the opposite direction. Thus, strong negative intrinsic precurvature both increases the pressure differential and causes low-level perturbations to appear such as are required to initiate crenation. Buckling of the red blood cell membrane is the result.

Proceeding from a state of strong negative precurvature through a flat membrane to positive precurvature has the opposite effect on the energy stored in the shell and consequently on the likelihood that the membrane shell will buckle. As intrinsic negative precurvature lessens, the energy required to deflate the membrane shell decreases. The pressure difference across the membrane drops, the buckling forces dissipate and membrane shear plays an ever more important role in determining red blood cell shape.

As the ratio of bending to shear energy drops, the cross-sectional profile of a discoid red blood cell becomes more bow-tie shape [23]. This is a very low (bending) energy form and thus is virtually indistinguishable from a reference shape. Eventually, the bending energy is insufficient to hold the red blood cell in a discoid configuration, shear forces take over and the cup-shape cell is the result. The interconvertability of the cup and the bow-tie is most clearly evident in albumin-rich suspending media [32].

So far no mention has been made of the effect of high pH on a cell that has had its adherent albumin depleted or removed. Albumin has been identified as the antisphering agent, so named by Furchgott and Ponder [31], because without it red blood cells not only crenate but shrink to smooth spheres. This is the ultimate stage before hemolysis when crenated cells have developed into the spheroechinocyte, which is a sphere but covered with many sharp, almost hair-like crenations. The sphereoechinocyte stage is very difficult to explain because it must be produced by a *loss* of membrane surface area. We have supposed that the surface area of the red blood cell is maintained constant by the phospholipid bilayer with its stable acyl chain region. To make the area smaller, phospholipid material must be squeezed out. Moreover, the sphereoechinocyte is reversible so that the excess lipid material must be temporarily sequestered rather than permanently lost. The explanation for crenation given above has provided for a negative internal pressure that subjects the membrane to considerable compressive force. But once the cell has assumed a spherical shape the elastic energy of deformation is no longer present, so there must be some other source of compressive energy to take over. The source of this energy is suggested by the arrangement of the layers in Figure 2, where the spectrin layer is many times thicker than the phospholipid bilayer or the layer of adsorbed albumin. Evidence in support of a significant thickness to the spectrin layer has recently been advanced by Bull et al. [5] and has considerable significance in this connection. If the whole of the spectrin layer is considered as a gel then the pressure developed in it to produce a given amount of bending moment, with respect to the neutral plane within the acyl chain region of constant area, can be very small compared with the tension developed in the albumin-modulated layer on the outer surface of the outer phospholipid monolayer. Thus, although it has been suggested that at high pH the expansion of the spectrin gel and the compression of the outer phospholipid monolayer, when albumin is deficient, both produce negative intrinsic precurvature, the magnitude of the tension due to condensation of the outer phospholipid monolayer can be much greater than the expansive force of the gel. There is thus the possibility of a strong net compressive force in the membrane under the conditions of high pH and low albumin concentration. This may well be strong enough to extrude lipid material from the bilayer and thus decrease the surface area of the cell without affecting its volume. This sequence of events would explain the otherwise puzzling observations of Furchgott and

Ponder [31] regarding the spheroechinocyte. The hypothesis that crenation is the result of negative intrinsic curvature of the membrane thus holds up over the entire range of shape changes through which the red blood cell passes from stomatocyte to spheroechinocyte.

To summarize the steps involved in this hypothesis of red blood cell crenation/cupping: (a) The red blood cell membrane consists of a phospholipid bilayer sandwiched between an outer layer of albumin and an inner spectrin net. (b) The outermost and the innermost layers of the four layered complex are linked through the phospholipid bilayer via band 3.1 molecules. (c) *Negative* intrinsic precurvature induced by raising the pH and removing albumin from the outer surface of the membrane produces the low-level protrusions on the rim of a discoid red blood cell (echinocyte I). More intense negative precurvature raises the pressure differential across the red blood cell membrane. This, translated into compressive forces, buckles the membrane at the sites of the rounded protrusions. The protrusions become full crenations and are stopped by the mechanical constraints of a rapidly rising shear resistance in the spectrin net. (d) Raising the pH still more and adsorbing away the attached albumin introduces large surface tension forces into the interface between the outer phospholipid molecules and the suspending fluid. This breaks down the phospholipid bilayer and produces reversible, isovolumetric sphering. (e) Lessening the intrinsic precurvature by lowering the pH or counteracting it by adding albumin to the outer layer restores the discoid form once more. (f) Finally, as the intrinsic negative precurvature approaches the zero to slightly positive region, shear forces in the red blood cell membrane become dominant and the cup-shape cell is now the minimum energy shape.

X. THE SPECTRIN NET

The ionic gel characteristics of the spectrin layer have been emphasized in the preceding discussion. The remarkable regularity of the molecular arrangement of this layer invites some speculation as to the functions that it subserves. One possible function is that of shear elasticity. It seems likely that membrane shear elasticity resides almost exclusively in the spectrin gel [4]. Because the molecules of this region are arranged in the form of a highly convoluted and thermodynamically agitated net, shear deformation must tend to stretch the net out and put the molecules into a higher state of order. There is clearly a well-defined limit to which such a net can be stretched and the force required to do so must increase rapidly as the molecules approach the fully extended condition. The theory of elasticity of such gels has been worked out by Flory [47] and Treloar [48] and this suggests that the shear modulus of the red blood cell membrane should increase rapidly with extension.

Measurements made by micropipette aspiration give no hint of a rapidly rising shear modulus with membrane extension, but measurements made on the

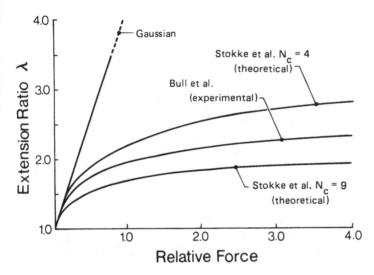

Figure 5 The relationship between membrane tension and consequent extension ratio for three theoretical cases and the experiments performed by Bull et al. using direct traction. The purpose of the graph is to show the marked discrepancy between the micropipet aspiration experiments (approximately equal to the Gaussian relationship) and the direct traction experiments (a non-Gaussian relationship). The theoretical curves are for a junction functionality of between 4 and 6. Nc is the effective chain length factor.

deformation of a red blood cell draped over a spider's web do [49]. The graph of extension versus tension from such experiments is shown in Figure 5 together with two theoretical curves calculated for a junction functionality of between four and six and values for the chain length factor (Nc) of 4 and 9, respectively. The latter two curves had been rescaled and redrawn from Stokke et al. [15]. The junction functionality represents the number of chainlike molecules joined together at a junction and the factor Nc represents the number of rodlike molecules that make up a chain joining nearest neighbor junctions. The experimental curve corresponds very closely to theory if the factor Nc is equal to 6. This gives considerable support to the view that the rubber elasticity of the spectrin net is non-Gaussian and that the linear relationships between extension and force shown by micropipette experiments is somehow in error.

Unfortunately, the portion of the graph of extension versus tension below an extension ratio of 1.5 is inaccessible to both the pipette and direct traction methods, yet the extension ratio involved in producing the biconcave discoid shape is only ~1.15. A low modulus of bending in this region could correspond to an even lower shear modulus of the membrane, as has been noted by other investigators [1,22] and thus account for the relatively high ratio of bending to shear energy required to produce a stable, biconcave shape.

XI. CONCLUSION

The most noticeable differences between micropipette experiments and direct traction experiments are the large force required in the latter before the onset of plastic flow and the absence of the large extension ratio measured in pipette experiments. The elegance and precision of micropipette experiments may be misleading because the crenating red blood cell is quite capable of spontaneously forming many protuberances that bear a striking resemblance to the portion of membrane aspirated into a micropipette. It seems, therefore, that the future is open to the performance of definitive experiments that will resolve these intriguing anomalies. Thus, in spite of the scrutiny to which it has been subjected and the substantial number of secrets it has relinquished, the red blood cell remains enigmatic.

REFERENCES

1. Evans, E. A., and Skalak, R. (1980). *Mechanics and Thermodynamics of Biomembranes*. CRC Press, Inc., Boca Raton.
2. Brailsford, J. D. (1983). Mechanoelastic properties of biological membranes. In *Membrane Fluidity in Biology*, Vol. 1, R. Aloia (Ed.). Academic Press, New York, p. 291.
3. Grant, C. W. M. (1983). Lateral phase separations and the cell membrane. In *Membrane Fluidity in Biology*, Vol. 1, R. Aloia (Ed.). Academic Press, New York, p. 131.
4. Elgsaeter, A., Stokke, B. T., Mikkelsen, A., and Branton, D. (1986). The molecular basis of erythrocyte shape. *Science* 234:1217.
5. Bull, B. S., Weinstein, R. S., and Korpman, R. A. (1986). On the thickness of the red cell membrane skeleton: quantitative electron microscopy of maximally narrowed isthmus regions of intact cells. *Blood Cells* 12:25.
6. Tsukita, S., Tsukita, S., and Ishikawa, H. (1980). Cytoskeletal network underlying the human erythrocyte membrane: thin-section electron microscopy. *J. Cell Biol.* 85:567.
7. Sheetz, M. P. (1983). Membrane skeleton dynamics: role in modulation of red cell deformability, mobility of transmembrane proteins, and shape. *Semin. Hematol.* 20:175.
8. Shohet, S. B. (1979). Reconstitution of spectrin-deficient, spherocytic mouse erythrocyte membranes. *J. Clin. Invest.* 64:483.
9. Stokke, B. T., Mikkelsen, A., and Elgsaeter, A. (1985). Human erythrocyte spectrin dimer intrinsic viscosity: Temperature dependence and implications for the molecular basis of the erythrocyte membrane free energy. *Biochim. Biophys. Acta* 816:102.

10. Pinder, J. C., Clark, S. E., Baines, A. J., Morris, E., and Gratzer, W. B. (1981). The construction of the red cell cytoskeleton. In *The Red Cell: Fifth Ann Arbor Conference.* Alan R. Liss, New York, p. 343.

11. Liu, S. C., and Palek, J. (1980). Spectrin tetramer-dimer equilibrium and the stability of erythrocyte membrane skeletons. *Nature* 285:586.

12. Liu, S. C., and Palek, J. (1982). Hemoglobin stabilizes spectrin in a tetrameric form in erythrocyte membranes. *J. Cell Biol.* 95:257a.

13. Litman, D., Hsu, C. J., and Marchesi, V. T. (1980). Evidence that spectrin binds to macromolecular complexes on the inner surface of the red cell membrane. *J. Cell Sci.* 42:1.

14. Tanaka, T., Fillmore, D., Sun, S. T., Nishio, I., Swislow, G., and Shah, A. Phase transitions in ionic gels. *Phys. Rev. Lett.* 45:1636.

15. Stokke, B. T., Mikkelsen, A., and Elgsaeter, A. (1986). The human erythrocyte membrane skeleton may be an ionic gel. I. Membrane mechanochemical properties. *Eur. Biophys. J.* 13:203.

16. Evans, E. A., Waugh, R., and Melnik, L. (1976). Elastic area compressibility modulus of red cell membrane. *Biophys. J.* 16:585.

17. Canham, P. B. (1970). The minimum energy of bending as a possible explanation of the biconcave shape of the human red blood cell. *J. Theor. Biol.* 26:61.

18. Brailsford, J. D., and Bull, B. S. (1973). The red cell—a macromodel simulating the hypotonic-sphere isotonic-disc transformation. *J. Theor. Biol.* 39: 325.

19. Bull, B. (1972). Red cell biconcavity and deformability a macromodel based on flow chamber observations. *Nouv. Rev. Fr. Hematol.* 12:835.

20. Zarda, P. R. (1975). *Large Deformations of an Elastic Shell in a Viscous Fluid.* Ph.D. thesis, Columbia University, New York.

21. McMillan, D. E., Mitchell, T. P., and Utterback, N. G. (1986). Deformational strain energy and erythrocyte shape. *J. Biomech.* 19:275.

22. Fischer, T. M., Haesr, C. W. M., Stohr-Liesen, M., Schmid-Schonbein, H., and Skalak, R. (1981). The stress-free shape of the red blood cell membrane. *Biophys. J.* 34:409.

23. Brailsford, J. D., Korpman, R. A., and Bull, B. S. (1976). The red cell shape from discocyte to hypotonic spherocyte—a mathematical delineation based on a uniform shell hypothesis. *J. Theor. Biol.* 60:131.

24. Bessis, M., (1973). *Living Blood Cells and Their Ultrastructure.* Springer-Verlag, New York.

25. Schmid-Schonbein, H., and Wells, R. (1969). Fluid drop-like transitions of erythrocytes under shear. *Science* 165:288.

26. Fischer, T. M., Stohr-Liesen, M., Secomb, T. W., and Schmid-Schonbein, H. (1981). Does the dimple of the human red cell have a stable equilibrium position on the membrane? *Biorheology* 18:46.

27. Dintenfass, L. (1969). The internal viscosity of the red cell and the structure of the red cell membrane. Considerations of the liquid crystalline structure of the red cell interior and membrane from rheological data. In *Molecular Crystals and Liquid Crystals* Gordon and Breach Science Publishers, England, p. 101.

28. Evans, E. A. (1983). Bending elastic modulus of red blood cell membrane derived from buckling instability in micropipet aspiration tests. *Biophys. J.* 43:27.

29. Fung, Y. C. (1966). Theoretical considerations of the elasticity of red cell and small blood vessels. *Fed. Proc.* 25:1761–1772.

30. Ketis, N. V., and Grant, C. W. M. (1982). Co-operative binding of concanavalin A to a glycoprotein in lipid bilayers. *Biochim. Biophys. Acta* 689:194.

31. Furchgott, R. F., and Ponder, E. (1940). Disk-sphere transformation in mammlaian red cells. II. the nature of the anti-sphering factor. *J. Exp. Biol.* 17:117.

32. Jay, A. W. L. (1975). Geometry of the human erythrocyte. I. Effect of albumin on cell geometry. *Biophys. J.* 15:205.

33. Bessis, M., and Prenant, M. (1972). Topographie de l'apparition des spicules dans les erythrocytes creneles (echinocytes). *Nouv. Rev. Fr. Hematol.* 12:351.

34. Sheetz, M. P., and Singer, S. J. (1974). Biological membranes as bilayer couples. A molecular mechanism of drug-erythrocyte interactions. *Proc. Natl. Acad. Sci. USA* 71:4457.

35. Evans, E. A. (1974). Bending resistance and chemically induced moments in membrane bilayers. *Biophys. J.* 14:923.

36. Sheetz, M. P., and Singer, S. J. (1976). Equilibrium and kinetic effects of drugs on the shapes of human erythrocytes. *J. Cell Biol.* 70:247.

37. Deuticke, B. (1968). Transformation and restoration of biconcave shape of human erythrocytes induced by amphiphilic agents and changes of ionic environment. *Biochim. Biophys. Acta* 163:494.

38. Matayoshi, E. D. (1980). Distribution of shape-changing compounds across the red cell membranes. *Biochemistry* 19:3414.

39. Beck, J. S. (1978). Relations between membrane monolayers in some red cell shape transformations. *J. Theor. Biol.* 75:487.

40. Stokke, B. T., Mikkelsen, A., and Elgsaeter, A. (1986). The human erythrocyte membrane skeleton may be an ionic gel. II. numerical analyses of cell shapes and shape transformations. *Eur. Biophys. J.* 13:219.

41. Sayare, M., and Fikiet, M. (1981). Cross-linking of hemoglobin to the cytoplasmic surface of human erythrocyte membranes: identification of band 3 as a site for hemoglobin binding in Cu^{2+}-o-phenanthroline catalyzed cross-linking. *J. Biol. Chem.* 256:13152.

42. Cassoly, R. (1982). Interaction of hemoglobin with the red blood cell membrane a saturation transfer electron paramagnetic resonance study. *Biochim. Biophys. Acta* 689:203.

43. Byers, T. J., and Branton, D. (1985). Visualization of the protein associations in the erythrocyte membrane skeleton. *Proc. Natl. Acad. Sci. USA* 82:6153.

44. Lovrien, R. E., and Anderson, R. A. (1980). Stoichiometry of wheat germ agglutinin as a morphology controlling agent and as a morphology protective agent for the human erythrocyte. *J. Cell Biol.* 85:534.

45. Brailsford, J. D., Korpman, R. A., and Bull, B. S. (1980). Crenation and cupping of the red cell: a new theoretical approach. Part I. Crenation. *J. Theor. Biol.* 86:513.
46. Brailsford, J. D., Korpman, R. A., and Bull, B. S. (1980). Crenation and cupping of the red cell: a new theoretical approach. Part II. Cupping. *J. Theor. Biol.* 86:531.
47. Flory, P. J. (1953). *Principles of Polymer Chemistry*. Cornell University Press, Ithaca, New York.
48. Treloar, L. R. G. (1975). *The Physics of Rubber Elasticity*. Clarendon Press, Oxford, England.
49. Bull, B. S., and Brailsford, J. D. (1976). Red cell membrane deformability: new data. *Blood* 48:663.
50. Evans, E. A., and Fung, Y. C. (1972). Improved measurements of the erythrocyte geometry. *Microvasc. Res.* 4:335.

15

Viscoelastic Properties and Rheology

DAVID A. BERK* and ROBERT M. HOCHMUTH *Duke University, Durham, North Carolina*

RICHARD E. WAUGH *University of Rochester School of Medicine and Dentistry, Rochester, New York*

I. INTRODUCTION

The viability of the red blood cell within the circulation depends on the mechanical properties of its membrane. The membrane must be strong enough to prevent fragmentation of the cell and flexible enough to permit transit through capillaries. Obviously, the structure of the cell membrane must be compatible with the mechanical demands placed on it.

With proper theoretical interpretation, measurement of membrane deformation and flow provides a nondestructive method of probing membrane structure. The field of study concerned with the systematic, quantitative analysis of deformation and flow is known as rheology. Beyond a mere descriptive characterization like "flexible," rheological studies of the red blood cell membrane have measured some remarkable material properties that are a direct consequence of its own unique structure.

The field of membrane rheology can be subdivided into three separate activities: development of experimental techniques for directly measuring the state of deformation and force on the membrane, use of continuum-mechanical analyses to identify and calculate membrane material properties based on experimental results, and interpretation of membrane properties with respect to an overall structural model of the membrane.

In one sense, the red blood cell is well suited for studies of membrane rheology because the membrane is the only structural component of the cell. No

---

Current affiliation: The University of British Columbia, Vancouver, British Columbia, Canada

additional load-bearing elements complicate the experimental design or the analysis. The small size of the red blood cell does pose problems for the experimenter. The application of a small but precisely measured force on the cell is not a trivial task, and measurement of the corresponding deformation is especially problematic due to the limits of optical resolution. For example, when red blood cells are swollen osmotically into spheres, a further increase in cell diameter with increasing internal pressure is barely detectable.

The most important technological advancement for overcoming these difficulties and making precise measurements of applied forces and the resulting membrane deformation has been the development and implementation of micropipets as tools for deforming individual cells. Single-cell experiments have significant advantages over other "bulk" approaches such as filtration or viscosity measurements. The forces on the cells and the resulting deformation are known with far greater precision, making it possible to determine intrinsic material properties and to compare the properties of one cell to another. Although bulk measurements may reflect changes in membrane properties, they are also affected by extraneous factors such as cell surface area and volume. Rand and Burton [1] introduced the use of a micropipet to apply a precise suction pressure to the cell, and Evans et al. [2] applied the technique to measure very small surface deformations. An alternative experimental technique for measuring the properties of single cells uses fluid shear stress to stretch a cell that has attached to a surface [3], but the micropipet method reigns as the most versatile and rational experimental approach to deforming single cells. Some variation of the method can be used to measure virtually any membrane property, and it is often the most precise method available. For these reasons, most of the experimental techniques mentioned in this chapter involve the use of a micropipet.

The experimental determination of applied force and resulting cell deformation is not in itself an adequate description of the membrane properties. Proper analysis of the experiment requires consideration of the force distribution within the membrane and establishment of a relation between the intrinsic stress resultant at a point in the membrane and the local deformation (and rate of deformation) at that point.

Classical mechanics provides the mathematical formalisms for describing the force distribution within the membrane and the geometry of the membrane as it undergoes deformation. It is the task of the membrane rheologist to formulate a model that specifies the relationship between deformation and the stress within an element of membrane. The mathematical statement of that model is called the constitutive equation. By describing membrane behavior, the constitutive equation identifies an intrinsic membrane material property. If the model states that the *extent* of deformation is proportional to stress, the constant of proportionality is an elastic modulus. If the model states that the *rate* of deformation is proportional to stress, then the constant of proportionality is a coefficient of viscosity.

Essential for any analysis is the principle of mechanical equilibrium, which requires a balance of internal and external forces. The equations of mechanical equilibrium assume a special form when the body is a thin-walled shell. Fung [4] recognized the red blood cell membrane as a type of thin-walled shell and used the associated equilibrium equations to show that the cell in its normal biconcave disk shape cannot support a pressure gradient across the membrane. The initial application of shell theory by Fung and Tong [5] accounted for the large deformations experienced by the membrane. In retrospect, their constitutive equation was inappropriate because it failed to account for the structural peculiarities of the membrane. Subsequent analyses by Skalak et al. [6] and Evans [7,8] distinguished between deformations involving area dilation and those involving extension without any increase in membrane surface area. In the first case, the membrane is highly resistant to deformation; in the latter case, the membrane is quite easily deformed. Each mode of deformation is characterized by a separate elastic modulus. This analytical development immediately established the basis for analyzing some previously paradoxical experimental observations. Subsequent theoretical developments have included the identification of bending as a third basic mode of deformation and the formulation of constitutive equations that describe the membrane resistance to rate of deformation.

In the remainder of this chapter, we discuss the membrane constitutive equations, the experimental measurement of red blood cell membrane properties, and the structural interpretation of these properties.

II. MEMBRANE DEFORMATION: THEORY AND EXPERIMENT

A. Two-Dimensional Continuum

Clearly, the red blood cell membrane shown in Figure 1 has a complex and heterogeneous structure. A plane of the membrane contains mixtures of phospholipids, cholesterol, and proteins, whereas in cross section the membrane consists of two layers of phospholipids and cholesterol associated with a complex spectrin network or membrane skeleton on its inner surface. To describe this complex structure as a continuum, we must select a large enough element of area such that the individual molecular components are unobservable. Thus, imagine a thin membrane with a surface element that is at least 200 nm on its side. Because an individual molecule in such an element would occupy <0.5% of the total area and because the total thickness of the membrane is <10 nm, we can neglect the molecular nature and thickness of a membrane element of this size and can consider it as a two-dimensional continuum.

Figure 2 demonstrates the treatment of the force distribution within the membrane surface. In general, a square element of material is subject to "force resultants" acting on its edges. These resultants have units of force per unit

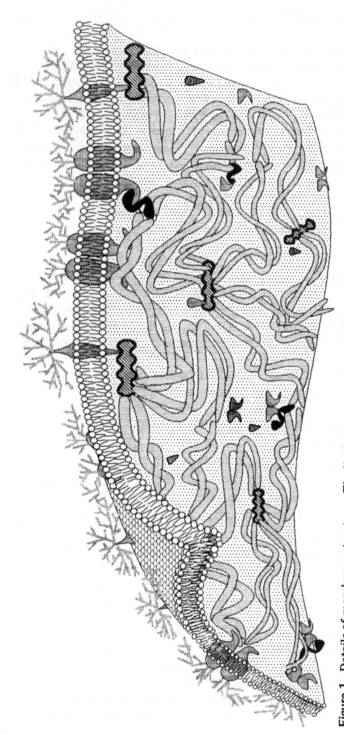

Figure 1 Details of membrane structure. The lipid bilayer contains integral proteins and is associated with a layer of "skeletal" proteins at the cytoplasmic face of the bilayer. Molecule sizes and densities are not "to scale" but are intended to illustrate major features such as spectrin folding and attachment to transmembrane proteins via ankyrin and band 4.1. (See previous chapters for details.)

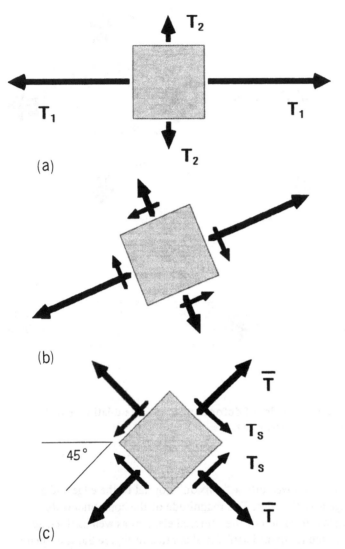

Figure 2 Stress distribution in a two-dimensional continuum. (a) In the principal axis system, there is no shear stress. The normal force resultants T_1 and T_2 are the principal tensions. (b) As the imaginary element is rotated away from the principal axes, the shear resultant increases and the difference between the two tensions decreases. (c) At $45°$ from the principal axes, the same distribution is described by the isotropic tension $\bar{T}$ and the maximum shear resultant T_S.

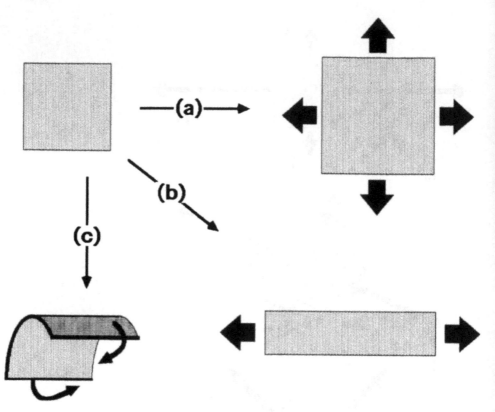

Figure 3 Three independent modes of deformation: (a) area dilation, (b) shear (extension at constant area), (c) bending.

length. A resultant can be resolved into a "tension" normal to the edge and a "shear resultant" tangent to the edge. The magnitude of the components depends on the coordinate orientation of the material element as well as the magnitude and direction of the resultant itself. As illustrated in Figure 2a, the square element can be oriented so that the shear resultant is zero. This defines the "principal axes," and the tensions T_1 and T_2 are the *principal stress resultants*. If the imaginary element were rotated an angle of 45° from the principal axis orientation, the normal stress resultants would become equal; all sides experience the same *isotropic tension* $\overline{T}$ and the *maximum shear stress resultant* T_S (Fig. 2c). Either pair of resultants T_1, T_2 or $\overline{T}$, T_S can specify the same state of membrane stress. They are related by

$$\overline{T} = \frac{T_1 + T_2}{2} \tag{1}$$

$$T_S = \frac{T_1 - T_2}{2} \tag{2}$$

In some cases, the third dimension of the membrane intrudes into the mechanical analysis. When the stress varies across the thickness of the membrane, a torque is created. This feature can be incorporated into the two-dimensional model by assigning to the element edge a moment resultant M with units of force · length per unit length.

As shown in Figure 3, a membrane element can be deformed in three ways [9] : (a) it can be dilated isotropically (which, for a volumetrically incompressible material results in a concomitant decrease in thickness), (b) it can be sheared, and (c) it can be curved. These deformations—area dilation, shear, and bending—are mathematically separable and in principle each can be performed independently of the other two. Thus, each of these deformations, when done at a sufficiently slow rate such that it is thermodynamically reversible (i.e., there is no viscous dissipation or "hysteresis"), is characterized by an elastic modulus: (a) an area expansion modulus K with units of N/m, (b) a shear modulus μ with units of N/m, and (c) a bending modulus B with units of N · m. In general, a large modulus indicates a greater resistance to that particular form of deformation.

B. Area Expansivity

As illustrated in Figure 3a, when a membrane element is subjected to an isotropic stress resultant, $\overline{T}$, it undergoes a relative expansion in area, α. This phenomenon is represented by a simple, linear constitutive equation [2,9] :

$$\overline{T} = K\alpha \tag{3}$$

where K is the area expansion modulus with a value of $\sim$450 mM/m [10] for a red blood cell membrane at room temperature. Phospholipid bilayers have similar values for K [11] and the magnitude of these values depends strongly on the amount of cholesterol in the membrane [12].

The relatively large value for the area expansion modulus in comparison with the shear modulus discussed in the next section is caused by the strong cohesive forces that exist among the oriented phospholipids in the bilayer. The cohesive forces arise from the "hydrophobic effect" [13]. Expanding the membrane even slightly causes water molecules to be displaced into hydrophobic regions of the lipid. However, only a small amount of expansion in surface area can occur before the membrane "fails." Failure (lysis) occurs when "holes" are opened that are of sufficient size to permit the rapid flux of water molecules through the membrane. A typical value for the isotropic tenstion, $\overline{T}$, at failure is 20 mN/m [14]. Thus, for K = 450 mN/m, the calculated value from Eq. 3 for the relative increase in area, α, at failure would be $\sim$4%.

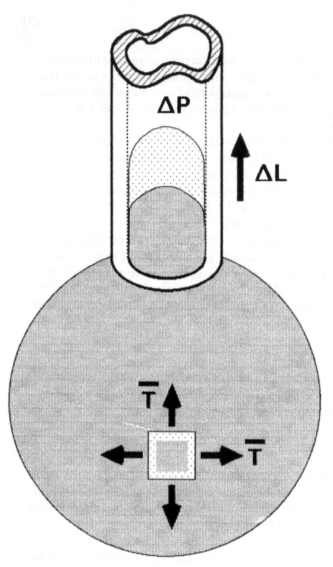

Figure 4 Membrane area dilation using micropipet suction. Isotropic tension is proportional to suction pressure, and area dilation is evidenced by an increase in the length of the aspirated bump.

To measure a value for K, red blood cells are swollen in a hypotonic medium to a nearly spherical state and then aspirated into a micropipet as shown in Figure 4. This creates an isotropic tension in the membrane that is directly proportional to the aspiration pressure, P [14]:

$$\bar{T} = \frac{PR_p}{2(1 - R_p/R_c)} \tag{4}$$

where R_p is the radius of the pipet and R_c is the radius of the spherical part of the cell outside of the pipet.

As the isotropic tension in the membrane is increased by increasing the aspiration pressure (Eq. 4), the area of the membrane is increased according to Eq. 3. As discovered by Evans et al. [2], this increase in area manifests itself as an increase in the length, ΔL, of that portion of the membrane that has been aspirated into the pipet (Figure 4). If we assume that the volume of water inside the cell remains constant during aspiration (this is precisely true only at high osmolarities), then a simple linear approximation can be written relating α and ΔL [15]:

$$\alpha = \frac{\Delta A}{A_0} = (1 - \frac{R_p}{R_c}) \frac{2\pi R_p \cdot \Delta L}{A_0} \tag{5}$$

where A_0 is the original area of the cell and R_c is the radius of the cell. For a relative increase in area of 4%, the relative increase in the radius of a spherical cell will be ~2%, which is negligible. However, for an increase in area of 4% and for typical values for R_p, R_c, and A_0 of 1 μm, 3.2 μm, and 140 μm^2, the change in length of the membrane within the pipet as calculated by Eq. 5 is 1.3 μm. This is readily measurable with an optical, visible-light microscope, whereas a change in R_c of 0.06 μm (2%) is not. Thus, the pipet acts as an "area amplifier" and permits very small changes in surface area or "area per molecule" to be measured with precision in experiments involving changes in isotropic tension, temperature, and osmolarity.

C. Shear Elasticity

When an element of membrane is subjected to a nonisotropic stress resultant (force per unit width of membrane), the element undergoes a significant elongation (Figure 3b). From Eq. 2 and Fig. 2, it is evident that a large maximum shear stress is equivalent to a large difference between principal stress resultants. Because the membrane has a large resistance to changes in surface area, the elongation produced by the nonisotropic component, or shear stress component, occurs at essentially constant area. That is, $\lambda_1 \lambda_2 = 1$ where λ_1 and λ_2 are the extension ratios (deformed length/original length) in the two principal directions.

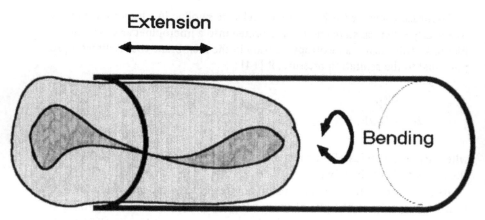

Figure 5 Preference of the cell for constant area of deformation. As the red blood cell enters a narrow passage, it first folds, then stretches to fit. Finally a small degree of area dilation may occur.

A simple first-order constitutive equation between the shear stress resultant and the deformation is given by [8]:

$$T_S = \frac{\mu}{2} \left(\lambda_1{}^2 - \lambda_2{}^2 \right) \tag{6}$$

where $\lambda_2 = \lambda_1{}^{-1}$ and μ is the elastic shear modulus for the membrane surface. A number of investigators have measured a value for μ. Typical values for human red blood cells at room temperature are $6\text{-}9 \times 10^{-3}$ mN/m [16].

The small value for μ in comparison to the value for K indicates the relatively "soft" resistance to elongation due to the deformation of the membrane skeleton in comparison to the very strong resistance to expansion due to cohesive interaction among polar lipids in a bilayer formation. Given a choice, the cell will readily elongate rather than dilate its area. This allows the normal disklike red blood cell, with its excess of membrane surface area for the enclosed volume, to move through capillary vessels with diameters that are significantly less than the cell's major diameter. The cell does this by folding and stretching along the axis of the capillary (Figure 5). However, if the capillary diameter is so small that the membrane surface area is inadequate to enclose the cell volume within the capillary, isotropic stresses are produced in the membrane and the surface area of the membrane must increase for further deformation to occur. Hemolysis follows after area dilations of only a few percent. Such area dilations will occur only in the presence of extremely large surface stress resultants (on the order of 20 mN/m).

To measure a value for μ, a portion of the dimple region of the membrane from a flaccid red blood cell is aspirated into a pipet (Figure 6). An analysis of

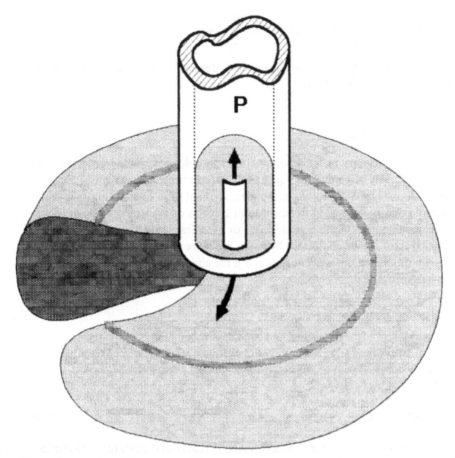

Figure 6 Membrane elongation (shear) using micropipet suction. Material is elongated in the radial direction on the surface of a flaccid cell. Material in the pipet can experience extension ratios of over 200%.

cell membrane deformation during the aspiration process [8] using the equations of membrane equilibrium and the constitutive equation (Eq. 6) gives a simple relation between the aspiration pressure, P, and the length, L, of that portion of the membrane that has been aspirated into the pipet [10]:

$$P \frac{R_p}{\mu} = \frac{2L}{R_p} - 1 + \ln \frac{2L}{R_p} \qquad L \geqslant R_p \qquad (7)$$

where R_p is the pipet radius. Eq. 7 is easily linearized [17]:

$$P \frac{R_p}{\mu} = 2.45 \left(\frac{L}{R_p}\right) - 0.63 \quad 1 \leqslant \frac{L}{R_p} \leqslant 4 \qquad (8)$$

It is readily seen in Eq. 8 that a value for μ is determined from a measurement of the slope of the P versus L line:

$$\mu = \frac{R_p^2}{2.45} \frac{\Delta P}{\Delta L} \qquad (9)$$

The major uncertainty in the determination of a value for μ is the uncertainty in the measurement of a value for R_p. A 20% error in the measurement of the radius will produce a 40% error in the measurement of μ. In studying cells under different conditions (e.g., at different temperatures or at different pH) or in comparing diseased cells (e.g., sickle cells or hereditary spherocytes) to normal cells, this error can be eliminated by using the same pipet in a given experiment and by normalizing the measurements of μ for abnormal or altered cells with that for normal, control cells.

D. Bending Elasticity

As illustrated in Figure 3c, when an element of membrane is subjected to a bending moment resultant, M (the moment or torque per unit width of membrane), curvatures are produced according to the following constitutive equation [9, 18,19]:

$$M = B(C_1 + C_2 - C_0) \qquad (10)$$

where C_1 and C_2 are the curvatures (inverse of radius of curvature) in two principal directions, C_0 is the curvature in the stress-free state and B is the bending modulus. For red blood cell membrane a value for B is $\sim 1.8 \times 10^{-19}$ N $\cdot$ m [20].

Although very small, the resistance of the membrane to bending stabilizes the biconcave shape of the cell during the early phases of the swelling process [21, 22] and during micropipet aspiration [23]. Bending resistance occurs in the bilayer membrane because of differential expansion and compression of adjacent layers within the membrane. If we assume the two halves of the bilayer to be coupled to each other, then the resistance to bending arises from the expansion or compression of one monolayer relative to the other. For this case, the relation between the bending modulus B, the area expansion modulus of the bilayer, K, and the bilayer separation distance, h, is [9]

$$B = \frac{Kh^2}{4} \qquad (11)$$

(Here the monolayers have zero thickness and are separated by a distance h).

In general, the individual layers of the bilayer are not coupled to each other, and some lateral redistribution of molecules in one layer relative to the other probably lessens the resistance to curvature change. The expansion and compression of individual layers due to local curvature changes can be distributed over the entire membrane surface. However, because bilayer membranes exist as closed capsules, the amount of redistribution that can occur is limited, and the individual layers can behave as if they are coupled.

For uncoupled layers in situations of very high local curvature, the resistance of individual monolayers to curvature change will be important. To account for the monolayer deformation when an uncoupled bilayer is deformed into a long cylinder or "tether," in which the radius of curvature of the tether is comparable to the membrane thickness, Waugh and Hochmuth [24] have modeled the membrane as two adjacent, thick, anisotropic liquid shells and derived the following coefficient for bending resistance:

$$B = \frac{Kh'^2}{12} \tag{12}$$

In this case, h' is the thickness of a monolayer and the bilayer separation distance is taken as zero. In the previous case (Eq. 11), the bilayer separation distance is h and the thickness is taken as zero. If we assume that $K = 450$ mN/m and h' is 2.0 nm, then the value for B calculated from Eq. 12 is

$$B = 1.5 \times 10^{-19} \text{ N} \cdot \text{m}$$

Eqs. 11 and 12 will give identical results if $h = h'/\sqrt{3}$.

The close agreement between the theoretically predicted value for B based on the differential expansion in area of a monolayer (or coupled bilayer) suggests that the resistance to bending of the red blood cell membrane resides in the lipid component of the membrane and not in the membrane skeleton (spectrin). This theory can be tested by producing changes in the area expansion modulus, perhaps by adding cholesterol, and seeing if a corresponding change is produced in B in accordance with Eqs. 11 or 12.

In an experiment designed to measure a value for B in Eq. 10 [20], a red blood cell is aspirated into a pipet until the circumferential compressive load on the membrane causes it to buckle (Figure 7). A perturbation analysis of the onset of buckling shows that

$$\frac{B}{PR_p^3} = f\left(\frac{R_p}{R_0}\right) \tag{13}$$

where f is a given function of the ratio of the pipet radius, R_p, to the radius of the rim of the cell, R_0. As can be seen from Eq. 13, for a given value for f, the

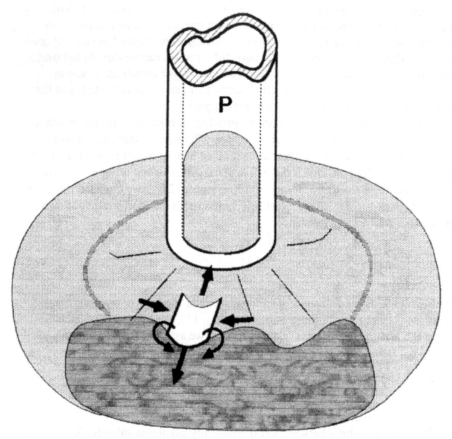

Figure 7 Membrane buckling provides a measure of bending stiffness. As material is extended, the circumferential force resultant becomes compressive, tending to cause the thin membrane to buckle. Bending rigidity stabilizes the planar shape until a critical extension is reached.

aspiration pressure, P, at which buckling occurs varies inversely as R_p^3. Thus, small changes in R_p can produce significant changes in P. Also, to obtain an accurate measurement for B, the radius of the pipet must be measured with extreme accuracy. Evans [20] accomplished this by inserting a calibrated, tapered microneedle into the pipet and measuring the depth or length of insertion. From the calibration of length versus width of the microneedle, the radius of the pipet is determined accurately.

E. Membrane Viscosity

The elastic moduli for area dilation, shear, and bending discussed in the previous section represent measurements made at equilibrium. All of the work of deformation is stored as an elastic (Helmholtz) free energy. However, if deformation occurs rapidly, some of the work of deformation will go into overcoming viscous resistance. By analogy to Eqs. 3 and 10, the viscous components for area dilation and bending can be characterized by the following phenomenological equations:

$$\overline{T}_V = \kappa \, \frac{\partial \alpha}{\partial t} \tag{14}$$

$$M_V = \beta \, \frac{\partial (C_1 + C_2)}{\partial t} \tag{15}$$

where the subscript v refers to the viscous components of the isotropic stress resultant and bending moment resultant and κ and β represent the viscous resistances to area dilation and bending. Evans and Hochmuth [9] estimated a very small value for κ on the order of 10^{-10} N · s/m. If the relation between the bending modulus and area expansion modulus (Eq. 12) can be applied in an analogous way to dissipative processes, then

$$\beta = \frac{\kappa h'^2}{12} \tag{16}$$

and β will be on the order of 10^{-29} N · m · s when κ is 10^{-10} N · s/m and h' is 2 nm. For dissipative processes to be significant, the characteristic time of deformation, τ, must be on the order of

$$\tau = \frac{\kappa}{K} = \frac{\beta}{B} = 10^{-9} \, s$$

Only extraordinarily rapid deformation processes could occur within characteristic times as small as a nanosecond and, therefore, dissipative processes can be safely neglected when the membrane surface is either dilated or curved. However, shear deformations occurring over times on the order of 0.1 s will produce significant viscous dissipation in the membrane.

Evans and Hochmuth [25] accounted for dissipation in a viscoelastic solid membrane undergoing shear deformation by adding to the elastic term in Eq. 6 a linear (in the sense of finite deformation) viscous term:

$$T_s = \frac{\mu}{2} \, (\lambda_1{}^2 - \lambda_1{}^{-2}) + 2\eta \, \frac{\partial ln\lambda_1}{\partial t} \tag{17}$$

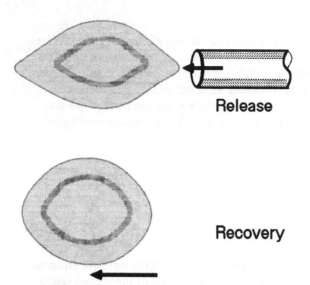

Release

Recovery

Figure 8 Viscoelastic recovery of the whole cell. The cell is stretched, then re-leased. Membrane elastic forces are resisted by viscous forces.

where t is time and η is the coefficient of surface viscosity. A characteristic time constant t_c is obtained in Eq. 17 simply by dividing η by μ:

$$t_c = \frac{\eta}{\mu} \tag{18}$$

Hochmuth et al. [26] and Chien et al. [17] obtained a value for t_c of ~0.1 s at room temperature. Thus, the value for the shear viscosity η is $\sim 6 - 9 \times 10^{-4}$ mN · s/m.

To measure a value for t_c, Hochmuth et al. [26] devised a simple experiment (Figure 8) in which a red blood cell adheres to a surface at a point and is elonga-ted by using a pipet to pull on a diametrically opposite point on the red blood cell rim. When the rim of the cell is released from the pipet, the cell recovers its original, undeformed shape. This recovery process is described by the integra-tion of Eq. 17 with $T_s = 0$ [26]:

$$\frac{L/W - (L/W)_\infty}{L/W + (L/W)_\infty} \frac{(L/W)_m + (L/W)_\infty}{(L/W)_m - (L/W)_\infty} = \exp\frac{-t}{t_c} \tag{19}$$

where L/W, the length-width ratio, is proportional to the extension ratio squared (λ_1^2). The subscript m stands for the initial (maximum) value at t = 0, and the subscript ∞ stands for the final equilibrium value. The best fit of the experimental data to Eq. 18 gives a value for t_C. Standard deviations in the measurements for t_C are approximately ± 20%.

F. Membrane Failure and Permanent Deformation

Under a wide range of conditions, the mechanical behavior of the flaccid red blood cell is well described by the viscoelastic model. However, when the force resultants within the membrane become excessive, the elastic limits of the material are exceeded. We consider four separate situations in which membrane behaves nonelastically. Three of those cases can be labeled "failure" because the material abruptly loses its elastic behavior, indicative of a breakage of some structural element. Another type of nonelastic behavior is the permanent deformation that occurs when a moderate shear resultant is applied for a long time.

In view of the very different membrane shear and area elasticities, it should come as no surprise that the maximum supportable force resultants are different in the two cases of shear and isotropic loading of the membrane. Failure due to excessive area dilation results in lysis, an obvious breach of the membrane integrity. The experimental measurement of the tension and area dilation at lysis is simply a continuation of the micropipet experiment previously presented (Figure 4). Suction pressure is applied to the aspirated bump until the membrane lyses and is sucked out of sight into the micropipet. As before, the isotropic tension in the membrane is related to the pipet suction pressure, and the increase in surface area is related to the height of the bump within the pipet. The isotropic tension required to produce lysis is on the order of 20 mN/m [14], and the corresponding increase in area is ~3–4%.

Two completely different modes of membrane failure occur in shear deformation. One involves the yield and fragmentation of the membrane skeleton; the other involves separation of the membrane bilayer from the skeleton. In each case, the membrane retains its physical integrity as a barrier between the cytoplasm and extracellular fluid, but it loses its apparent shear rigidity. Figure 9 illustrates an experiment in which this transformation to liquidlike behavior was first observed [27]. A red blood cell is allowed to adhere to the bottom of a parallel plate flow channel. As the shear stress of the external fluid causes the cell to stretch, the membrane experiences elastic deformation while maintaining constant surface area. Local deformations are especially large near the point of attachment. When the force on the cell is great enough, material flows from the cell body onto an apparently hollow filament. As a consequence, the cell body moves downstream with the fluid flow, still connected to the original point of attachment by the growing cylinder of membrane material, dubbed a "tether."

Evans and Hochmuth [28] originally analyzed tether formation by modeling the membrane as a two-dimensional "Bingham plastic." The model identifies two intrinsic material properties that characterize the flow process: the yield

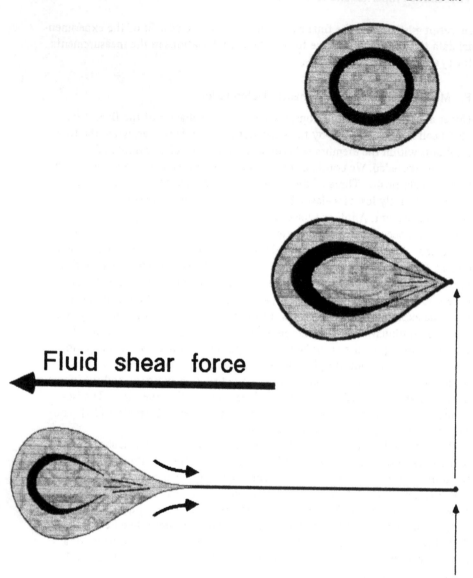

Fluid shear force

Attachment

Figure 9 Membrane extensional failure. Under the action of an external fluid force, a point-attached red cell first stretches elastically then undergoes an apparent yield process near the point of attachment. The cell body moves downstream, still attached by a thin strand of membrane material.

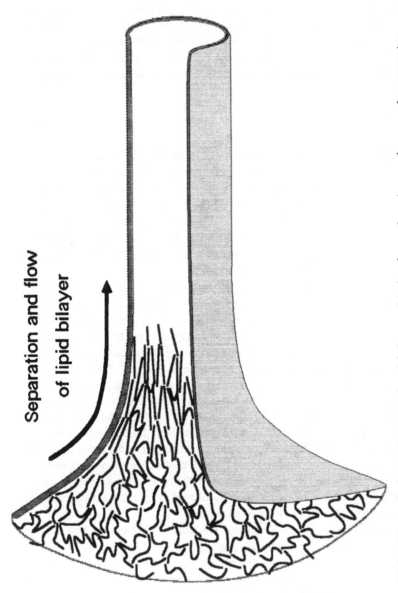

Figure 10 One scenario for extensional failure and tether formation. As membrane undergoes extensions in excess of 300%, maximally extended spectrin molecules collect in the yield region, and the lipid bilayer flows over the resulting "log jam."

shear resultant T_y and the viscosity η. The first parameter is the local tension at which the membrane exceeds its ability to deform elastically and begins to flow like a liquid. The second term is the resistance of the membrane to flow. Specifically, the model states that no flow occurs if the shear resultant is less than the yield value, and that the rate of flow is proportional to the excess tension:

$$T_s - T_y = 2\eta \; \partial ln\lambda/dt \quad (T_s > T_y)$$

Calculations by Evans and Hochmuth [28] place T_y in the range of 1.6–4.0×10^{-2} mN/m. From more recent measurements, Waugh [29] has calculated $T_y = 2 \times 10^{-2}$ mN/m. This value is three orders of magnitude less than the tension required for lysis. Estimates of the extension ratio at which yield occurs give a value of $\lambda \approx 3$. Thus, shear failure occurs at extensions of 300% compared with lysis at an area increase of ~3%. Again, the entirely different natures of membrane extension and dilation are demonstrated. The two modes of failure clearly involve different mechanisms.

Recently Berk and Hochmuth [30] have observed rapid rates of lateral diffusion of integral proteins in tethers. These observations indicate that membrane tethers lack an intact skeleton because proteins in a membrane with an associated skeleton diffuse 100 times more slowly than proteins in tethers. Preliminary evidence based on electron microscopic examination of tethers by one of us (R. E. W.) also indicates that tethers lack skeleton. Thus, the original interpretation of tether formation as representing a yield and plastic flow of the membrane skeleton may be incorrect. Instead, it now appears that tether formation involves the dissociation of the membrane bilayer from the membrane skeleton. A detailed mechanical analysis of bilayer–skeletal dissociation has not been performed yet. One possible interpretation of this mode of failure is illustrated by the "log-jam" model shown in Figure 10. Here the membrane skeleton is unable to flow with the bilayer component of the membrane as it forms the tether. As the tether is increased in length, more skeletal material accumulates at the junction of the cell body with the tether.

Even though the analysis of Evans and Hochmuth [28] may not be appropriate for tether formation, the rheological studies of yield and plastic flow of membrane still represent an important aspect of membrane behavior. Yield and fragmentation of red blood cells has been observed both by micropipet and by ektacytometry (Chapter 7). Unfortunately, these approaches are semiquantitative and cannot be interpreted in terms of intrinsic membrane coefficients.

A final example of deviation of membrane behavior from the simple viscoelastic model is the observation of permanent deformation. When a single state of deformation is imposed on a red blood cell for a prolonged period, the cell begins to assume that shape as its new "reference" shape. If a portion of a cell is aspirated into a micropipet and the deformation is maintained for several minutes, a "bump" remains after the cell is expelled from the pipet [31]. Markle et al. [32] determined that the magnitude of permanent deformation is proportional to the applied force and the duration of the imposed extension. To account for the process of creep and force relaxation occurring on this slow time scale, the membrane can be modeled as an elastic component in series with a viscous component [33]. Application of this model to the data of Evans and LaCelle [31] and Markle et al. [32] reveals a "creep viscosity" several orders of magnitude larger than the viscosities associated with elastic deformation or the yield and plastic flow phenomenon. As noted by Markle et al. [32], the mechanism for viscous dissipation in this case is distinct from the frictional dissipation involved in the other relatively rapid deformations. Instead, creep viscosity reflects the gradual molecular reorganization of the solid component of the membrane.

III. STRUCTURAL IMPLICATIONS

The rheological studies described in the previous section follow a continuum approach to the analysis of the membrane. The membrane is treated as a homogeneous two-dimensional material devoid of all structural detail, yet the results of these rheological studies are intimately related to the study of membrane structure. By determining the physical properties of the featureless membrane substance, structural models can be devised and tested. Most importantly, rheological measurements are nondestructive and offer a means of monitoring dynamic, often subtle, aspects of membrane structure.

Biochemical techniques have been quite successful in identifying the discrete structural components of the red blood cell membrane and in establishing many of the interactions between the components, described in previous chapters. Figure 1 illustrates some of the features of the current model for red blood cell membrane structure [34,35]. The "fluid mosaic" part of the membrane consists of a lipid bilayer containing integral proteins free to diffuse about within the plane of the two-dimensional solution. Associated with the cytoplasmic face of the bilayer is a mesh of skeletal protein. Filamentous spectrin tetramers and higher-order oligomers are linked together by a complex of actin, band 4.1, and other proteins to form an extensive network. The network is linked to the fluid mosaic by specific interactions: ankyrin binds spectrin to the band 3 integral protein and band 4.1 binds the actin complex to a glycophorin integral protein. Details of this skeletal ultrastructure have been viewed with electron microscopy [36,37] (Chap. 10).

Continuum methods complement these structure-identifying approaches by establishing the functional role of the structural elements. During the time that spectrin and other skeletal proteins were being identified [38] and models for their organization were proposed [39], mechanical studies were establishing the

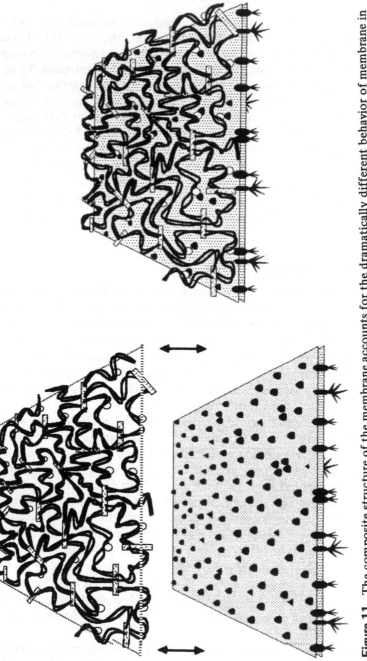

Figure 11 The composite structure of the membrane accounts for the dramatically different behavior of membrane in area dilation and shear.

presence of a solid structure capable of conferring shear rigidity to the membrane. Another useful continuum approach is the study of lateral diffusion within the membrane, presented in depth in chapter 14 of this volume. Studies of integral protein diffusion have also provided evidence of the solid membrane component [40] and have revealed important aspects of the interaction between the membrane skeleton and the integral proteins.

The mechanical properties of the membrane are the consequence of the composite liquid–solid structure, illustrated in Figure 11, in which a lipid bilayer is coupled to the dense skeletal mesh previously described. Such a model accounts for the disparity between membrane deformation in extension and dilation. Membrane behavior related to area expansion is largely the expression of the properties of the lipid bilayer. The membrane skeleton is very extensible and can easily accommodate the small expansions associated with an isotropic state of tension. Indeed, removal of the lipid bilayer allows the membrane skeleton to be expanded to several times its original area [36]. The real expenditure of energy is in increasing the area per molecule within the bilayer. Similarly, the tension at lysis is determined by the bilayer. Because the membrane bending modulus is related to compressibility of different layers of the composite structure, the highly incompressible bilayer will determine that value as well. In all these cases, the contribution, if any, of the spectrin layer is more likely to be chemical than mechanical. The small dimensional changes required of the spectrin filaments are unlikely to involve significant resistance, but the large concentration of spectrin at the cytoplasmic face of the bilayer creates a different chemical environment at that surface. Thus, alterations in structure or composition of the membrane skeleton could conceivably change the energy needed to expand the bilayer area. Studies of spectrin-deficient cells, discussed below, provide some evidence of a skeletal role in area modulus and bending stiffness.

The shear elasticity of the red blood cell membrane clearly indicates the existence of a solid component, the membrane skeleton. A lipid bilayer, in its liquid state, has absolutely no shear elasticity. In fact, the viscous character of membrane behavior in shear is attributable to the skeleton, because the viscosity calculated from the elastic recovery experiment [25] is at least 100 times greater than the viscosity calculated from experiments involving the flow of lipid bilayer [41]. The spectrin tetramers appear to function as springs capable of storing energy during deformation and capable of frictionally dissipating energy as they slide against each other as well as other membrane components and, possibly, hemoglobin molecules. Electron micrographs of the spectrin tetramer show a flexible filament ~200 nm long [42]. Within the network, the average end-to-end distance of a tetramer is calculated to be much less than the 200 nm maximal length. During deformation the molecules are capable of extending in one direction while oligomers aligned in the perpendicular direction undergo a commensurate reduction in their end-to-end lengths (Figure 12). From estimates of

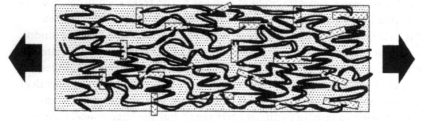

Figure 12 Large elastic deformations are attributable to extension of spectrin tetramers. Extensional failure results when the maximal length is exceeded.

protein concentrations at the membrane surface, Waugh [29] has calculated an average tetramer unstressed length and hence the amount of extension that is permissible. The estimated ratio of full length to unstressed length is comparable to the membrane extension ratio at which shear elasticity is lost and the material yields ($\lambda \approx 3$).

The temperature dependence of the red blood cell shear elastic modulus has interesting thermodynamic implications regarding the nature of the energy storage mechanism of the spectrin network. A polymeric mesh could be expected to store elastic energy in the form of decreased entropy (an imposition of more order as elements are aligned). The observed decrease in shear elastic modulus with temperature indicates that the entropy of the membrane structure is lowest in the undeformed state and actually increases with extension [10]. The deformed spectrin network is in a state of greater disorder and higher energy than the undeformed network.

Hemoglobin, the major protein of the red blood cell interior, has been shown to affect the rheological behavior of the membrane. Calculations of the viscous dissipation rate in the cytoplasm show that the bulk viscosity of hemoglobin is unlikely to have a significant role in the extensional recovery experiment used to determine recovery time and hence the membrane viscosity [25]. However, under some conditions, hemoglobin has a significant effect on the viscoelastic properties of red blood cell membrane. The value for the shear modulus in normal cells is independent of hemoglobin concentration, but the viscosity increases with interior hemoglobin concentration [43,44]. Dense (high hemoglobin concentration) red blood cells have a membrane viscosity that is 50% greater than the least dense cells from the same individual. The effect of hemoglobin is mediated through the membrane skeleton and probably involves binding of hemoglobin to the membrane, specifically the band 3 protein [45,46]. The formation of a hemoglobin gel layer within the interstices of the spectrin mesh could account for the increased membrane viscosity.

In addition to spectrin and the other skeletal proteins, the transmembrane proteins also have a role in determining membrane mechanical properties. The lectin wheat germ agglutinin binds to the integral protein glycophorin and produces dramatic alterations in membrane elasticity and viscosity. Small concentrations (~0.1 μg/ml) can increase shear viscosity by a factor of three [47], and slightly greater concentrations (up to 2 μg/ml) can increase the shear elastic modulus by an order of magnitude [48]. The mechanism for this effect has not been established, but it appears to involve the interaction of glycophorin with the skeletal proteins [49,50]. The coupling of the membrane skeleton with the lipid bilayer is a crucial, but not well-understood, determinant of the rheological properties of the membrane.

The molecular events associated with membrane extensional failure remain to be elucidated. Once the easily extended spectrin elements reach their maximal length, another element must then accommodate the continued extension. The model of membrane structure (Figure 1) provides a number of structural connections that could be the "weak link" that fails and allows material yield and liquidlike flow. The band 4.1 protein has recently been identified as a critical structural element preventing cell fragmentation [51]. As the "node" in the spectrin network, the actin–band 4.1 complex is likely to have a major role during the failure process.

The molecular components associated with permanent membrane deformation may be the same ones involved in the more abrupt yield and flow phenomenon. The viscosity of permanent deformation identified by Markle et al. [32] reflects a biochemical process in which the spectrin network assumes a new "resting" or unstressed state. That process must involve the dissolution of at least one interaction, either between spectrin dimers or between spectrin and the actin complex. Based on the dynamic equilibrium among different oligomers of spectrin, Morrow and Marchesi [52] postulated that the skeleton undergoes continuous but gradual reorganization, a view that is certainly consistent with mechanical observation.

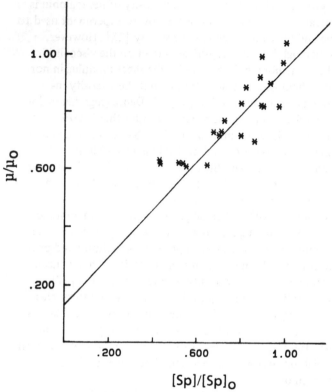

Figure 13 Shear modulus, expressed as fraction of control vs. fractional reduction in spectrin density. Solid line is linear regression to data with spectrin density greater than 60 percent of control (slope = 0.83, intercept = 0.13). Data with density below 0.6 was omitted because geometric abnormalities prevented unbiased sampling of the population during mechanical measurements. Data obtained from Waugh and Agre [54].

IV. STRUCTURAL DEFECTS

Conclusions about the contributions of the different structural elements of the membrane to its proper mechanical function are based on a growing body of evidence about how structural properties change with changing molecular composition. One of the most valuable sources of information about the importance of specific molecular components for proper mechanical function has been the study of red blood cell membrane properties in individuals with inherited abnormalities in the membrane skeleton. For example, a series of experiments by Takakuwa et al. [51] on membranes deficient in protein 4.1 has led to an understanding of the important role that protein 4.1 plays in maintaining the stability of the membrane when it is subjected to large stresses. (This work is described in greater detail in Chapter 7.)

Of more immediate relevance toward understanding the molecular basis of the intrinsic membrane properties described in the present chapter is a recent series of experiments by Waugh [53] and Waugh and Agre [54] investigating the effects of spectrin deficiency and other molecular abnormalities on the viscoelastic properties of the red blood cell membrane. The spectrin deficiency resulted from a variety of inherited defects, some of which are described in the present volume and others that remain as yet unidentified. The degree of deficiency was assessed by measuring the ratio of spectrin to band 3 on polyacrylamide gels. Assuming that the membrane area is proportional to its band 3 content, this ratio provides a measure of the reduction in spectrin density for the abnormal membranes. The membrane shear modulus, bending stiffness, and viscosity were measured for 20 patients with varying degrees of spectrin deficiency. A strong correlation was observed between the surface density of spectrin and the membrane shear modulus (Figure 13). These results represent the first quantitative evidence that spectrin is responsible for the shear rigidity of the membrane, and suggest that the membrane shear modulus is directly proportional to the density of spectrin on the surface. Thermodynamically, these observations are consistent with the idea that the elastic work done (energy required) to deform the membrane is stored by the spectrin molecules. Mechanical deformation of the membrane imposes changes in the arrangement of the spectrin molecules, disturbing them from their "natural" low energy state and increasing the average energy per molecule in the surface. When external forces are removed the molecules return to their natural arrangement, returning the membrane to its natural geometry.

A useful analogy for understanding the molecular basis of the decreased shear modulus in the abnormal membranes is to think of the spectrin molecules as molecular springs. The data indicate that the spring constant of the molecules is not significantly affected by these disorders, but because there are fewer springs per unit area on the abnormal membrane it takes less energy to deform the abnormal surface (Figure 14).

The strong correlation between shear modulus and spectrin density is in sharp contrast to the lack of correlation between membrane viscoelastic properties and other molecular abnormalities of the membrane skeleton. For example, protein 4.1 function appears to have no direct effect on viscoelastic properties. A patient completely deficient in protein 4.1 [55] had similar reductions in membrane shear modulus to a patient with full complement of 4.1, but a 40% deficiency in spectrin–4.1 binding [56]. In spite of the binding abnormality, the reduction in shear modulus of membranes of the latter patient correlated closely with the fractional reduction in spectrin density. Finally, two patients with a high-molecular-weight variant of protein 4.1 and no deficiency in spectrin [57, 58] exhibited normal membrane shear elasticity.

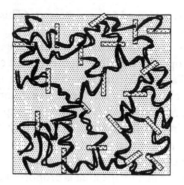

Figure 14 Reduced elastic modulus of spectrin-deficient membrane is related to the reduced number of "springs," allowing a greater extension for a given force compared to normal membrane. (Compare Fig. 12.)

In addition to the correlation between spectrin density and shear modulus, there was also a strong correlation between the reduction in spectrin density and the reduction in membrane bending stiffness, as indicated by the pressure at which creases formed in the cell surface during micropipet aspiration [20]. Because bending stiffness arises from the resistance of the membrane to area expansion [33], these observations indicate that the area expansivity modulus of spectrin-deficient membrane is probably reduced. Preliminary measurements in the laboratory of one of us (R. M. H.) indicate that the area modulus is, in fact, reduced in spectrin deficient membranes. These results are somewhat surprising, because (as discussed in the preceding sections) the membrane skeleton is not expected to contribute significantly to the area modulus or the bending stiffness. One hypothesis is that spectrin deficiency affects the area modulus indirectly by altering the composition or organization of the membrane bilayer. Indeed, there is a preliminary report of abnormal lipid organization in spectrin-deficient membranes [59]. Existing experimental evidence is insufficient to draw firm conclusions on this question, and further experiments will be needed to determine whether spectrin makes a direct contribution to the area modulus and bending stiffness or if the changes in these properties are the result of alterations in bilayer structure caused by the spectrin deficiency.

Membrane viscosity is also reduced in spectrin-deficient membranes, but the correlation between viscosity and spectrin deficiency is not as strong as the correlations for shear modulus or bending stiffness. Generally, the fractional reduction in viscosity was greater than the fractional reduction in spectrin density, as might be expected based on molecular theories of viscosity [60]. According to these theories, viscosity should depend on the density of molecules and the frequency of their entanglements. Because the frequency of entanglements should increase with increasing density, the viscosity is expected to depend on the density raised to a power greater than one. Furthermore, it should be recognized that membrane viscosity is highly sensitive to environmental factors such as salt concentration and temperature [61], and (as discussed in the Section III) it depends very sensitively on the intracellular hemoglobin concentration [62].

V. CONCLUSION

A great deal remains to be learned about the function of many of the membrane skeletal proteins and how they contribute to the mechanical behavior of the membrane. Measurement of changes in specific membrane rheological properties in single cells having known defects in molecular structure has proven to be an invaluable approach for identifying structure–function relationships within the membrane skeleton. It is an active area of research and one that is likely to provide many new insights in the years to come.

ACKNOWLEDGMENT

This work was supported by National Institutes of Health grants HL 23728, HL 31524, and HL 18208.

REFERENCES

1. Rand, R. P., and Burton, A. C. (1964). Mechanical properties of the red cell membrane: I. Membrane stiffness and intracellular pressure. *Biophys. J.* 4: 115.
2. Evans, E. A., Waugh, R., and Melnik, L. (1976). Elastic area compressibility modulus of red cell membrane. *Biophys. J.* 16:585.
3. Hochmuth, R. M., and Mohandas, N. (1972). Uniaxial loading of the red-cell membrane. *J. Biochem.* 5:501.
4. Fung, Y. C. (1966). Theoretical considerations of the elasticity of red cells and small blood vessels. *Fed. Proc.* 25:1761.
5. Fung, Y. C., and Tong, P. (1968). Theory of the sphering of red blood cells. *Biophys. J.* 8:175.
6. Skalak, R., Tözeren, A , Zarda, P., and Chien, S. (1973). Strain energy function of red blood cell membranes. *Biophys. J.* 13:245.
7. Evans, J. E. (1973). A new material concept for the red cell membrane. *Biophys. J.* 13:926.
8. Evans, E. A. (1973). New membrane concept applied to the analysis of fluid shear- and micropipette-deformed red blood cells. *Biophys. J.* 13:941.

9. Evans, E. A., and Hochmuth, R. M. (1978). Mechanochemical properties of membranes. In *Current Topics in Membranes and Transport*, F. Bronner and A. Kleinzeller (Eds.). Academic Press, New York, pp. 1–64.

10. Waugh, R., and Evans, E. A. (1979). Thermoelasticity of red blood cell membrane. *Biophys. J.* 26:115.

11. Kwok, R., and Evans, E. (1981). Thermoelasticity of large lecithin bilayer vesicles. *Biophys. J.* 35:637.

12. Evans, E., and Needham, D. (1986). Giant vesicle bilayers composed of mixtures of lipids, cholesterol and polypeptides: thermochemical and (mutual) adherence properties. *Faraday Discuss. Chem. Soc.* 81:267.

13. Tanford, C. (1973). *The Hydrophobic Effect*. John Wiley & Sons, New York.

14. Rand, R. P. (1964). Mechanical properties of the red cell membrane: II. Viscoelastic breakdown of the membrane. *Biophys. J.* 4:303.

15. Evans, E. A., and Waugh, R. (1977). Osmotic correction to elastic area compressibility measurements on red cell membrane. *Biophys. J.* 20:307.

16. Hochmuth, R. M. (1987). Properties of red blood cells. *Handbook of Bioengineering*, R. Skalak and S. Chien (Eds.). McGraw Hill, New York, pp. 12.1–12.17.

17. Chien, S., Sung, K-L. P., Skalak, R., Usami, S., and Tözeren, A. (1978). Theoretical and experimental studies on viscoelastic properties of erythrocyte membrane. *Biophys. J.* 24:463.

18. Helfrich, W. (1973). Elastic properties of lipid bilayers: theory and possible experiments. *Z. Naturforsch.* 28c:693.

19. Evans, E. A. (1974). Bending resistance and chemically induced moments in membrane bilayers. *Biophys. J.* 14:923.

20. Evans, E. A. (1983). Bending elastic modulus of red blood cell membrane derived from buckling instability in micropipet aspiration tests. *Biophys. J.* 43:27.

21. Zarda, P. R., Chien, S., and Skalak, R. (1977). Elastic deformations of red blood cells. *J. Biomech.* 10:211.

22. Fischer, T. M., Haest, C. W. M., Stohr-Liesen, M., Schmid-Schönbein, H., and Skalak, R. (1981). The stress-free shape of the red blood cell membrane. *Biophys. J.* 34:409.

23. Evans, E. A. (1980). Minimum energy analysis of membrane deformation applied to pipet aspiration and surface adhesion of red blood cells. *Biophys. J.* 30:265.

24. Waugh, R. E., and Hochmuth, R. M. (1987). The mechanical equilibrium of thick, hollow, liquid membrane cylinders. *Biophys. J.* 52:391.

25. Evans, E. A., and Hochmuth, R. M. (1976). Membrane viscoelasticity. *Biophys. J.* 16:1.

26. Hochmuth, R. M., Worthy, P. R., and Evans, E. A. (1979). Red cell extensional recovery and the determination of membrane viscosity. *Biophys. J.* 26:101.

27. Hochmuth, R. M., Mohandas, N., and Blackshear, P. L. (1973). Measurement of the elastic modulus for red cell membrane using a fluid mechanical technique. *Biophys. J.* 13:747.

28. Evans, E. A., and Hochmuth, R. M. (1976). Membrane viscoplastic flow. *Biophys. J.* 16:13.
29. Waugh, R. E. (1982). Temperature dependence of the yield shear resultant and the plastic viscosity coefficient of erythrocyte membrane: implications about molecular events during membrane failure. *Biophys. J.* 39:273.
30. Berk, D. A., and Hochmuth, R. M. (1986). Diffusion in erythrocyte membrane tethers. *Biophys. J.* 49:312a.
31. Evans, E. A., and LaCelle, P. L. (1975). Intrinsic material properties of the erythrocyte membrane indicated by mechanical analysis of deformation. *Blood* 45:29.
32. Markle, D. R., Evans, E. A., and Hochmuth, R. M. (1973). Force relaxation and permanent deformation of erythrocyte membrane. *Biophys. J.* 42:91.
33. Evans, E. A., and Skalak, R. (1980). *Mechanics and Thermodynamics of Biomembranes.* CRC Press, Boca Raton, Florida.
34. Branton, D., Cohen, C. M., and Tyler, J. (1981). Interactions of cytoskeletal proteins on the human erythrocyte membrane. *Cell* 24:24.
35. Bennett, V. (1985). The membrane skeleton of human erythrocytes and its implications for more complex cells. *Annu. Rev. Biochem.* 54:273.
36. Byers, T. J., and Branton, D. (1985). Visualization of the protein associations in the erythrocyte membrane skeleton. *Proc. Natl. Acad. Sci. USA* 82:6153.
37. Shen, B. W., Josephs, R., and Steck, T. L. (1986). Ultrastructure of the intact skeleton of the human erythrocyte membrane. *J. Cell Biol.* 102:997.
38. Marchesi, V. T., Steers, E., Tillack, T. W., and Marchesi, S. L. (1969). Properties of spectrin: fibrous protein isolated from red cell membranes. In *Red Cell Membrane*, G. A. Jamieson and T. J. Greenwalt (Eds.). Lippincott, Philadelphia, p. 117.
39. Steck, T. L. (1974). The organization of proteins in the human red blood cell membrane. *J. Cell Biol.* 62:1.
40. Peters, R. J., Peters, J., Tews, K. H., and Bähr, W. (1974). A microfluorimetric study of translational diffusion in erythrocyte membranes. *Biochim. Biophys. Acta* 367:282.
41. Waugh, R. E. (1982). Surface viscosity measurements from large bilayer vesicle tether formation: II. Experiments. *Biophys. J.* 38:39.
42. Shotton, M. D., Burke, B. E., and Branton, D. (1979). The molecular structure of human erythrocyte spectrin. Biophysical and electron microscopic studies. *J. Mol. Biol.* 131:303.
43. Linderkamp, O., and Meiselman, H. J. (1982). Geometric, osmotic, and membrane mechanical properties of density-separated human red cells. *Blood* 59:1121.
44. Nash, G. B., and Meiselman, H. G. (1983). Red cell and ghost viscoelasticity. *Biophys. J.* 43:63.
45. Chetrite, G., and Cassoly, R. (1985). Affinity of hemoglobin for the cytoplasmic fragment of human erythrocyte membrane band 3. *J. Mol. Biol.* 185:639.

46. Eisinger, J., Flores, J., and Salhany, J. M. (1982). Association of cytosol hemoglobin with the membrane in intact erythrocytes. *Proc. Natl. Acad. Sci. USA* 79:408.

47. Smith, L., and Hochmuth, R. M. (1982). Effect of wheat germ agglutinin on the viscoelastic properties of erythrocyte membrane. *J. Cell Biol.* 94:7.

48. Evans, E., and Leung, A. (1984). Adhesivity and rigidity of erythrocyte membrane in relation to wheat germ agglutinin binding. *J. Cell Biol.* 98: 1201.

49. Chasis, J. A., Mohandas, N., and Shohet, S. B. (1985). Erythrocyte membrane rigidity induced by glycophorin A-ligand interaction. *J. Clin. Invest.* 75:1919.

50. Lovrien, R. E., and Anderson, R. A. (1980). Stoichiometry of wheat germ agglutinin as a morphology controlling agent and as a morphology protective agent for the human erythrocyte. *J. Cell Biol.* 85:534.

51. Takakuwa, Y., Tchernia, G., Rossi, M., Benabadji, M., and Mohandas, N. (1986). Restoration of normal membrane stability to unstable protein 4.1-deficient erythrocyte membranes by incorporation of purified protein 4.1. *J. Clin. Invest.* 78:85.

52. Morrow, J. S., and Marchesi, V. T. (1981). Self-assembly of spectrin oligomers in vitro: a basis for a dynamic cytoskeleton. *J. Cell Biol.* 88:463.

53. Waugh, R. E. (1987). Effects of inherited membrane abnormalities on the viscoelastic properties of erythrocyte membrane. *Biophys. J.* 51:363.

54. Waugh, R. E., and Agre, P. (1988). Reductions in erythrocyte membrane viscoelastic coefficients reflect spectrin deficiencies in hereditary spherocytosis. *J. Clin. Invest.* 81:133.

55. Mueller, T. J., Williams, J., Wang, W. C., and Morrison, M. (1981). Cytoskeletal alterations in hereditary elliptocytosis. (Abstr.) *Blood* 58:47a.

56. Wolfe, L. C., John, K. M., Falcone, J. C., Bynre, A. M., and Lux, S. E. (1982). A genetic defect in the binding of protein 4.1 to spectrin in a kindred with hereditary spherocytosis. *New Engl. J. Med.* 307:1367.

57. Conboy, J., Kan, Y. S., Agre, P., and Mohandas, N. (1982). Molecular characterization of hereditary elliptocytosis due to an elongated protein 4.1. (Abstr). *Blood* 68:34a.

58. Letsinger, J. T., Agre, P., and Marchesi, S. L. (1986). High molecular weight protein 4.1 in the cytoskeletons of hereditary elliptocytes (Abstr.). *Blood* 68:38a.

59. Lubin, B., Chiu, D., Schwartz, R. S., Cooper, B., John, K., Wolf, L., and Lux, S. E. (1983). Abnormal membrane phospholipid organization in spectrin-deficient human red cells (Abstr.). *Blood* 62:34a.

60. Meares, P. (1965). *Polymers, Structure and Bulk Properties.* Van Nostrand Reinhold, London, pp. 347–365.

61. Hochmuth, R. M., Buxbaum, K. L., and Evans, E. A. (1980). Temperature dependence of the viscoelastic recovery of red cell membrane. *Biophys. J.* 29:177.

62. Evans, E., Mohandas, N., and Leung, A. (1984). Static and dynamic rigidities of normal and sickle erythrocytes. Major influence of cell hemoglobin concentration. *J. Clin. Invest.* 73:477.

16

Active Transport of Sodium and Potassium

JACK H. KAPLAN   *University of Pennsylvania, Philadelphia, Pennsylvania*

I. INTRODUCTION

The central importance of the role of the sodium pump in almost all eukaryotic cells is by now well appreciated. In cells of the central nervous system the sodium pump is responsible for maintaining the low intracellular [Na] following the inward Na currents of action potentials. In cells of the kidney and gastro-intestinal system and in some (nucleated) red blood cells the sodium pump is required to maintain an inwardly directed Na gradient, the energy of which is used to accumulate solutes, and in many cells the sodium pump in intimately involved in the regulation and control of cell volume. The structure, function, and mechanism of the sodium pump has been the subject of several recent reviews [1–6]. In the present chapter the focus will be on the sodium pump of human red blood cells. The substantial and important literature on other systems that has contributed to our understanding of the sodium pump will only be drawn on to understand the red blood cell phenomena.

   Much of what we know about the sodium pump is derived from studies using human red blood cells. Why should this be, because, as mentioned above, the sodium pump is present in the plasma memberane of most eukaryotic cells? The value of the red blood cell (and in particular the human red blood cell) lies in several experimental advantages. These include the following: (a) it is readily available in convenient quantity; (b) it is a cell with no internal compartments, so that compositional and transport studies are more straightforward; (c) the proximate energy source, adenosine triphosphate (ATP), is derived only from glycolysis so that metabolic manipulations are simpler than in cells in which both aerobic and anaerobic metabolism occurs; and (d) both intracellular and extracellular compartments are readily and independently accessible for the experimental alteration of cation or substrate concentrations. This last point has been pivotal: It has been possible either by using reversibly acting sulfhydryl reagents [7,8] or ionophores [9] in intact cells, by preparing resealed red blood cell ghosts by reversible hemolysis [10–13], or by using a preparation of inside-out vesicles [14,15] to characterize many of the substrate and cation interactions at both the cytoplasmic and extracellular faces of the sodium pump. However, because of the low number of copies (74) of the sodium pump protein per human red blood cell (approximately 250, compared with, for example, 1×10^6 copies of band 3, the anion exchanger), this has not been a very useful system for detailed studies of the properties of the enzyme and phosphoenzyme intermediates nor of the effects of chemical modification of the protein. It is only in a fair small number of studies that biochemical properties of the pump have been characterized in this system. One of these was the estimation of the molecular weight of the phosphorylated catalytic subunit and its pronase insensitivity (80). Current models of the reaction mechanism of the sodium pump that relate the biochemical reactions it catalyzes to the associated transport processes result from a fusion of experimental results from diverse systems. It is a rare situation in which biochemical and transport studies have been performed on the same system under comparable conditions. Many of the apparently plausible and even convincing mechanistic interpretations of transport data rely heavily on *inferences* drawn from other systems and *assumptions* about the generalities (sometimes in detail) of mechanism of all sodium pumps. These caveats are especially important when the conclusions drawn from enzyme studies relate to the sidedness of action of activating cations. The enzyme preparations contain no barrier separating internal and external aspects of the pump protein, so that the sum of ionic effects is always seen on both surfaces. On the other hand, the effects of activating cations on small reconstituted proteoliposomes or on membrane vesicles with a high pump density, although having a barrier separating extracellular from intracellular compartments, have the added complication that relatively few pump turnovers may generate relatively large membrane potentials. The generated membrane potential may feed back with its own effects on subsequent turnovers and transport rates.

The sodium pump and particularly the human red blood cell sodium pump has played an especially important role in the elucidation of active transport mechanisms. The description of the cycle of intermediates and their involvement in transport phenomena was first proposed for the sodium pump. It is clear from a variety of transport and mechanistic studies that great similarities exist between the reaction mechanisms of the sodium pump, the calcium pump (from sarcoplasmic reticulum, sarcolemma, and probably endoplasmic reticulum) the proton pump from gastric mucosa, the proton pump from *Neurospora*, and a K pump from *Escherichia coli*. This functional similarity is reflected in the recent developments from molecular cloning approaches in which the structural basis (at least at the primary structure, i.e., sequence level) for these similarities has been shown by the high degree of amino acid sequence homologies of the Na,K-ATPase (from a variety of species and tissues), sarcoplasmic reticulum Ca^- ATPase, and gastric H,K-ATPase [16-18].

II. TRANSPORT MODES OF THE SODIUM PUMP

The major physiological role of the sodium pump is in the active transport of Na^+ ions out of the cell and of K_3 ions into the cell against steep electrochemical gradients. In round figures the cellular cation composition of most mammalian and avian red blood cells is approximately 130 mM K^+ and 8 mM Na^+, whereas the plasma composition is approximately 140 mM Na^+ and 4 mM K^+. This cellular composition is essentially unaltered throughout the time that the cells spend in circulation. Considerations of Donnan equilibria and estimates of the red blood cell membrane potential (for a discussion see [19]) reveal that although Cl^-, the major anion, is passively distributed across the red cell membrane, Na^+ and K^+ ions are far from equilibrium. It was established in 1941 that either cold storage or glucose depletion of human red blood cells resulted in a loss of cell K and a gain in cell Na, and that warming or glucose addition dramatically reversed these changes [20,21]. This metabolically dependent uptake or extrusion of cations against their electrochemical potential gradients and the obligatory coupling of Na and K movements form the basis of many subsequent studies on the mechanism of the sodium pump. I will refer to the major physioloigcal mode of transport mediated by the pump as $Na_i:K_o$ exchange, indicating that intracellular Na (Na_i) is exchanged for extracellular K (K_o).

Under physiological conditions, to a limited extent, the pump also mediates other cation transport processes. These other transport modes can be made the major or only transport reaction taking place by experimentally manipulating the cation compositions of the intracellular and extracellular compartments and also by altering the substrate composition of the intracellular medium. The existence and characterization of these other transport modes has been performed extensively in human red blood cells and no comparably complete data exist in any other system. The occurrence and basic characteristics of the major pump transport modes were first described in a series of articles published some 20 years ago [22-26]. The characterization of these other transport modes and their relationship to the variety of biochemical partial reactions catalyzed by the sodium pump protein have contributed significantly to our understanding of the relationship between the coupling of the scalar biochemical reactions catalyzed by the pump and the vectorial transport processes that the pump mediates.

A. $Na_i : K_0$ Exchange

General Properties

Early studies on net Na and K movements in human red blood cells established
that three Na ions were expelled for every two K ions that were taken up [27],
and that unidirectional, metabolically dependent Na efflux required extracellular
K. The proximate energy source for the active transport was established as ATP
[28] from experiments using resealed red blood cell ghosts prepared from
depleted cells, and the rate of lactate production (a measure of ATP synthesis
and thus use via the glycolytic cycle) correlated with the rate of active transport.
Other parallel and important developments were the discovery and characterization of a sodium plus potassium activated ATPase (Na,K-ATPase) in microsomal
preparations from crab leg nerve [29], an activity later described in red blood
cell membranes [30,31]. The crucial earlier report [32] of the specific inhibition of energy-dependent Na and K transport by cardiotonic steroids such as
ouabain or the aglycones such as strophanthidin provided an essential tool for
the subsequent elucidation of the sodium pump mechanism. The almost
uniquely selective action of ouabain allowed the phrase "ouabain-sensitive" to
be treated synonymously as "sodium-pump–mediated," The portion of cation
fluxes or ATPase activity that were ouabain-sensitive were taken as those parts
of the total flux or total activity attributed to the sodium pump. It was later
shown that ouabain or strophanthidin acted only at the outer surface of red
blood cell ghosts [33]. The factors affecting ouabain binding have more recently
been the subject of several studies (see, for example, [34,35]). From these basic
observations (condensing the work of more than two decades) a simple equation
for the overall reaction may be written as follows:

$$3Na_i{}^+ + 2K_0{}^+ + ATP_i \rightarrow 3Na_0{}^+ + 2K_i{}^+ + ADP_i + P_i$$

This description immediately leads to several important consequences. The
equation describes the movement of three positive charges out of the cell and
two positive charges into the cell per cycle. If no other compensatory ionic
movements of another cation, perhaps a proton or the outward movement of an
anion, perhaps Cl^-), then the pump should operate electrogenically. In other
words, as the pump turns over, its operation leads to the generation of a transmembrane potential. The size of this potential depends in any particular cell on
the pump density (number of pumps per unit membrane surface area), turnover rate in the specified experimental conditions, and the electrical resistance
of the cell membrane. The electrogenic nature of the sodium pump was first
established using microelectrodes in large molluscan neurons [36]. The consequences of pump electrogenicity and a discussion of its implications for pump
models is the subject of a review [37]. The electrogenic operation of the sodium
pump in red blood cells has also been established [38], and the contribution of

the sodium pump in human red blood cells to the (low) resting membrane potential is small (approximately 1 mV). However, its characterization and study in several of the other transport modes have questioned the constancy of the electrogenic nature of the red blood cell sodium pump. We will return to this point in discussion of the so-called uncoupled Na efflux. An assumption that appears in this simple equation is that cytoplasmic ATP is the fuel for the active transport of monovalent cations. Although in general this seems to be true, there is a considerable body of evidence that suggests that compartmentalization of ATP may take place and that a pool of ATP separate from the total cytoplasmic ATP may be preferentially used by the Na pump in human red blood cells [39]. The sodium pump protein may be a member of a multienzyme complex whose detailed organization is not yet clear. However, experimental evidence has accumulated over the past 20 years that suggests that in human red blood cells the complex probably consists of glyceraldehyde-2-phosphate dehydrogenase and phosphoglycerate kinase [40-42] in association with the sodium pump and perhaps also the calcium pump [39,40,42]. An elegant and convincing demonstration of the existence of such a pool of compartmentalized ATP that was specifically accessible to the sodium pump was also made using inside-out vesicles [43].

The most generally accepted picture of the sodium pump reaction mechanism is derived largely from the studies of Albers and co-workers [44] on brain enzyme and of Post and co-workers [45], who used kidney enzyme. The model has been recently extended to include the explicit cation loading and transport steps by Glynn and co-workers [46]. One form of this model is shown in Fig. 1. The cycle describes the sequence of events taking place on the enzyme during one cycle of $Na_i:K_o$ exchange. It is also possible to account for many of the features of other transport reactions using this scheme. The cycle is composed of events associated with sequential phosphorylation and dephosphorylation of the pump protein. Phosphorylation of the protein is catalyzed by Na_i binding to the protein with high affinity and an adenosine diphosphate (ADP)-sensitive, K-insensitive phosphoenzyme $E_1 P (Na_3)$ is formed that has three Na ions tightly bound (or occluded). Occlusion is a term used to signify lack of accessibility of the bound cation to either cytoplasmic or extracellular aqueous compartments. It is distinguished from merely tight binding by the fact that both access to and dissociation from occluded sites by the cation is a process in which barriers must be overcome in each direction. We know from studies of the purified enzyme that in $E_1 P$ a particular aspartyl residue is phosphorylated to yield an acyl phosphate intermediate. A conformational transition spontaneously takes place and $E_1 P$ is converted to $E_2 P$, an ADP-insensitive and K-sensitive form of phosphoenzyme. The same aspartyl group is involved in both $E_1 P$ and $E_2 P$—the phosphoenzyme has undergone a change in conformation (shape) modifying the solvent (water) accessibility of the acylphosphate. In $E_1 P$ the phosphoenzyme reacts more readily with ADP than with water, but in $E_2 P$ this situation is re-

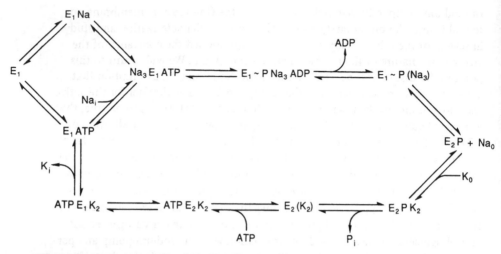

Figure 1 Reaction cycle of Na,K-ATPase. This scheme illustrates the sequence of events occurring on the enzyme as the cycle of phosphorylation and dephosphorylation accompanies the efflux of three Na ions and the associated uptake of two K ions. Na ions and ATP may bind randomly, but the inner cycle probably occurs under physiological conditions.

versed. After the phosphoenzyme conformational transition, K_0 binds to $E_2 P$, which then dephosphorylates, and the K ions become occluded, forming $E_2 (K_2)$. The K ions are subsequently released to the inside of the cell. At physiological (millimolar) ATP levels the K release occurs rapidly, after ATP binds to $E_2 (K)$ with relatively low affinity (approximately 100 μM). At low (micromolar) levels of ATP, the K release is extremely slow and rate-limiting in the overall cycle. The release is so slow that if micromolar levels of ATP are present, the addition of K ions to the enzyme suspended in Na-containing medium results in an inhibition of ATPase activity [47,48]. This apparently anomalous property of an (Na + K)-ATPase enzyme and its resolution by Post and co-workers [49] led to the realization of the central importance of occlusion in the pump mechanism and its role in $Na_i:K_0$ exchange. Under physiological conditions of temperature, pH, substrate saturation, etc., it is likely that two steps of approximately equal rate contribute to the overall rate-limitation of $Na_i:K_0$ exchange (and Na,K-ATPase activity), those are, the conformational transitions $E_1 P \rightarrow E_2 P$ and $E_2 (K_2) \rightarrow E_1$. A detailed discussion of the kinetics of the Albers-Post model and a compilation of the values of various rate constants from the literature appear in the monograph by Stein [50].

The depiction of the sequence of enzymatic steps in Fig. 1 and similar models contains the implicit assumption that the cations activating the enzymatic conversions are the cations that are transported. This leads to the idea that the transport process follows a ping-pong type of mechanism. In other words, Na binds, is transported, and leaves, before K binding and being transported. The question of whether or not it is possible to establish that this transport mechanism is correct, rather than one in which both Na and K ions are bound simultaneously at opposite faces and then are transported, received considerable attention in red blood cell studies during the past several years. The problem has been considered in greatest depth in the work of Sachs, who concluded that if allowance is made for the uncoupled Na efflux mode, the available data are consistent with a ping-pong model [51,52]. The existence of separate partial "transport" modes involving only Na or K limbs for exchange processes and the sequential nature of the biochemical model strongly suggest a ping-pong model; however, in the normal $Na_i:K_0$ exhcange mode measured under steady-state conditions the complex set of interacting ligands and sites makes an unambiguous demonstration difficult. In recent studies using reconstituted kidney enzyme in proteoliposomes at $0°C$ (to slow rates) or in kidney microsomal vesicles, using caged ATP to synchronize pump turnover, single turnover transport studies yield data that are also consistent with a ping-pong mechanism for $Na_i:K_0$ exchange [53,54].

Substrate Specificity and Affinities

The nucleoside triphosphate requirement for $Na_i:K_0$ exchange or Na,K-ATPase activity in red blood cells is almost exclusively satisfied by ATP. Some studies have examined other nucleotides in resealed ghosts in a variety of contexts [55, 56]. However because in red blood cells there are extremely high levels of nucleoside diphosphokinase activity, only micromolar levels of ADP would be rapidly converted to ATP by this activity in the presence of other nucleoside triphosphate compounds. Thus, until these types of studies are performed in the presence of inhibitors of nucleoside diphosphokinase (such as trypan red, trypan blue, etc.), the conclusions are somewhat ambiguous [57]. The binding affinity for ATP at a high affinity site on kidney enzyme is approximately 0.1 μM and the $K_{0.5}$ for maximal phosphorylation is also in this range. Following the description above of the catalytic effect of ATP at a low-affinity site in speeding the rate-limiting release of K from the occluded $E_2(K_2)$ form, it is not surprising that two affinities, one in the micromolar range, the other at approximately 100 μM, have been reported for the Na,K-ATPase in resealed human red blood cell ghosts [58]. The higher affinity corresponds to the catalytic phosphorylation site, the lower to the nonhydrolytic K-release site. The question of whether or not these "sites" represent a single ATP binding domain on the protein with differeing affinities in the E_1 Na and E_2 P or $E_2(K_2)$ forms or two independent and separate domains on the sodium pump protein is still unresolved.

The intracellular monovalent cation binding sites that stimulate phosphorylation of the pump by ATP show a high selectivity for Na ions. Until recently, it was thought that Na alone could activate the pump at its inner surface. However, recent studies have demonstrated that if cell Na is extremely low or absent it is possible to measure $Li_i:K_0$ exchange in human red blood cells, where the inner Na sites are activated by Li ions with a low affinity [59]. There is also growing evidence that in human red blood cells H^+ ions can activate an ATPase activity, which is ouabain-sensitive, and in this situation protons are substituting for Na ions at the cytoplasmic pump sites [60]. These studies were performed using inside-out vesicles and more recently in this system it has been demonstrated that a proton flux occurs through the pump associated with this ATPase activity [61]. In the absence of intracellular K ions, the intracellular Na sites are saturated when Na is approximately 2 mM level. However, under physiological conditions in which K_i is approximately 120 mM and Na_i approximately 8 mM, the pump is far from saturated and increases in Na_i (through increases in Na permeability) result in a pump stimulation. Indeed, this effect is probably the basis of many reported "modulations" of pump activity in the literature where, for example, agent X or hormone Y causes an increase in ouabain-sensitive ^{86}Rb uptake. Because of this lack of saturation of the pump by Na_i, it is always important to be sure that Na_i is not increased by X or Y and that the pump stimulation is not merely a result of an increase in cell Na. The extracellular sites where dephosphorylation of E_2P is activated can be considered as "not-Na" sites rather than K sites, because a variety of monovalent cations can activate the pump at these sites and produce a $Na_i:X_0$ exchange. In human red blood cells, the sequence of selectivity at these sites has been shown as Rb > K > Cs > Li [62,63]. Interestingly, Li at the outer surface appeared to be a K-congener while at the inner surface it acted as a Na-congener. Recent studies on the K:K exchange process (see below) have established that Li can behave at the cytoplasmic surface as both a K or Na substitute (92); presumably similar effects occur at the outer surface. The $K_{0.5}$ for K at the outer surface of the pump is approximately 2 mM, when physiological Na_0 (140 mM) is present. In the presence of choline the $K_{0.5}$ falls to below 0.2 mM, reflecting the competition between Na_0 and K_0 at external pump sites [23]. Under physiological conditions, typical rates of ouabain-sensitive Na efflux from human red blood cells are approximately 3 mmol $1.cells^{-1} hr^{-1}$.

In Fig. 2, the underlying biochemical steps associated with Na exit and K entry during $Na_i:K_0$ exchange are shown. Sodium exit steps in the ATP phosphorylation limb are shown in an oval, and K limb reactions are shown in a rectangle. This convention will be adopted in depicting the various partial transport modes in later sections of this article, where the shape of the symbol surrounding the steps define which route is being followed by the particular cation.

B. $K_i:Na_0$ Exchange

This transport mode of the sodium pump is characterized by a Na_0-activated K efflux from red blood cells that is inhibited by ouabain. This mode was first

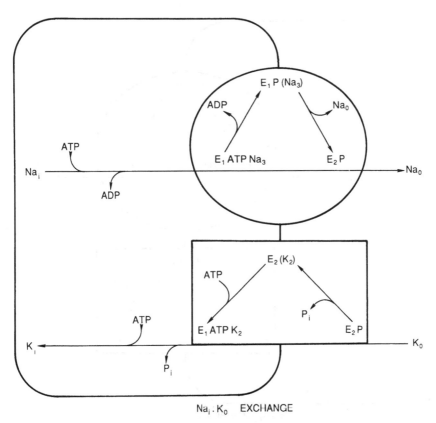

$Na_i \cdot K_o$ EXCHANGE

Figure 2 $Na_i:K_o$ exchange. This figure illustrates the association between the movements catalyzed by the pump apparatus and the events taking place on the protein. The large open symbol (left) that represents the cell containing intracellular cations and substrates. The oval symbol on the cell boundary contains the events associated with (outward) Na movements through the Na limb of the enzyme pathway. The rectangle contains the events associated with the (inward) movement of K ions through the K ion pathway in the enzyme.

demonstrated in red blood cells and fulfills the thermodynamic expectation of reversibility of the entire pump cycle [26]. If the cation gradients are sufficiently steep, then the pump apparatus is driven backwards and a net synthesis of ATP from ADP and P_i is observed. The association of the discrete biochemical steps with the ion transport pathway are depicted in Fig. 3.

C. $Na_i:Na_o$ Exchange

When red blood cells are suspended in a K-free isotonic NaCl solution, an ouabain-sensitive isotopic Na efflux can be measured that is dependent on the pres-

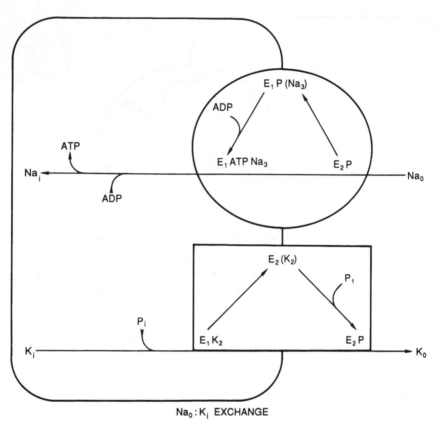

Figure 3 Na_0:K_i exchange or pump reversal. The convention is as described for Figure 2 but the direction of the sequence of events is reversed in both the Na and K limbs. Thus, ATP is synthesized from ADP plus P_i in this way rather than being used as in Figure 2.

ence of Na_0. It was shown that the transport reaction consists of a 1:1 exchange of intracellular and extracellular Na [22], which is dependent on the presence of ATP [64], is stimulated by ADP and ATP [65], and takes place in the absence of net ATP hydrolysis [25]. This transport mode was also observed in muscle [66], where it is enhanced by metabolic poisoning (presumably as ADP rises). The substrate requirements and lack of ATP hydrolysis led to the suggestion that the underlying biochemical reaction associated with this transport mode was the ATP:ADP exchange reaction [65] described earlier by Albers and co-workers in studies on the brain enzyme [67]. In this reaction, phosphoenzyme generated by the Na-dependent phosphorylation of E_1 from ATP (i.e., E_1P) subsequently rephosphorylates ADP to regenerate ATP. The Na_i-dependent steps

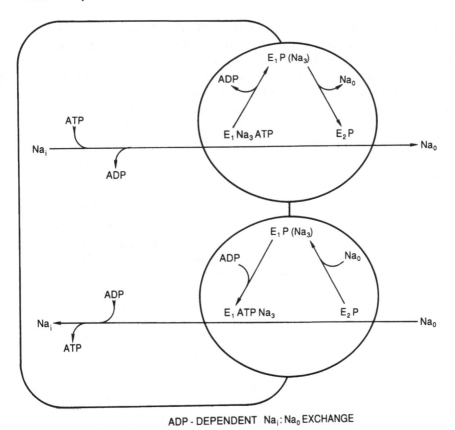

ADP - DEPENDENT $Na_i : Na_0$ EXCHANGE

Figure 4 ADP-dependent $Na_i : Na_0$ exchange. In this transport mode, Na influx and Na efflux occur through Na ion pathways (oval symbols). Na influx produces ATP, and Na efflux uses ATP.

leading to $E_2 P$ are associated with Na exit and these steps are reversed as Na enters and the sequence of events $Na_3 E_2 P \rightarrow E_1$ taks place (Fig. 4). The ATP: ADP exchange reaction is usually measured as the rate of production of radio-labeled ATP from labeled ADP and unlabeled ATP [57]. This reaction was of central importance in the development of the Albers-Post model, because N-ethyl maleimide and oligomycin are agents that completely inhibit Na,K-ATPase but stimulate ATP:ADP exchange. These observations led to the suggestion of a sequence of phosphoenzymes involving ADP-sensitive and insensitive forms in series in which the agents acted by blocking their interconvention, producing only ADP-sensitive $E_1 P$ (stimulating ATP:ADP exchange) and preventing the formation of ADP-insensitive, K-sensitive $E_2 P$ (inhibiting Na,K-ATPase activity). To establish that the ATP:ADP exchange reation does underly Na:Na exchange it was necessary to measure and characterize ATP:ADP exchange in resealed red blood cell ghosts and compare the sidedness and affinities of cation activation of the bhichemical and transport reactions. The measurement of ATP:ADP exchange was performed using the photorelease of ATP from caged ATP to initiate the reaction in resealed human red blood cell ghosts [13,57]. The overall characteristics of the cation activation of the biochemical reaction agreed well with the cation activation of the transport [13]. Na_i:Na_o exchange, measured via isotopic Na fluxes, had been previously shown to be activated with a high affinity by Na_i and have a biphasic activation by Na_o [22]. Na_o in the range 0-5 mM inhibited Na efflux, whereas Na_o at higher levels stimulated the rate with a relatively low affinity and the activation remained approximately linear in the physiological range. These characteristics were mirrored in the Na activation of ATP:ADP exchange [57]. K_o, which had been shown to inhibit Na influx (Na_i:Na_o exchange) with the same affinity with which it activated Na efflux (Na_i:K_o exchange), had a half-maximal effect at approximately 1 mM K. The ATP:ADP exchange reaction was similarly inhibited by K_o, whereas K_i was without effect in this concentration range [13]. Thus the measurement of ATP:ADP exchange in resealed ghosts confirmed that the cation activation (and inhbition) pattern was the same as for Na_i:Na_o transport, and that the complex multiphasic Na-activation curve for ATP:ADP exchange in porous membranes could be dissected into the separate activating and inhibiting effects of intracellular and extracellular Na ions. An additional significant finding from these studies was that when Na_o was 0, and Na_i:Na_o could not take place, there was a significant rate of ATP:ADP exchange [13,57]. This implies that although ATP:ADP exchange and Na_i:Na_o exchange are linked, they are probably not tightly coupled and the biochemical phosphorylation dephosphorylation cycle can take place in the absence of transport. It would be of interest to know whether when Na_o was 0, a Na efflux occurred in association with the forward limb of the ATP:ADP exchange reaction. The lack of tight coupling between the transport and biochemical reactions had previously been suggested by the observation that oligomycin, which stimulates ATP:ADP exchange, inhibits Na_i:Na_o exchange [68]. An extremely interesting result has recently been reported involving an unexpected effect of Na and K interactions at the outer surface of the sodium pump. In a situation in which ghosts contain high ADP concentrations and in which extracellular K is clearly saturating ($>$10 mM K_o), the removal of normal extracellular Na resulted in a stimulation of Na efflux [101]. It is surprising to see any effect of Na_o where K_o is saturating; the basis of this result is unclear. Because it occurs only with high ADP, it is tempting to believe that the $E_1 P \rightleftharpoons E_2 P$ interconversion may be involved in this Na_o effect.

The sidedness of the effects of cations on both Na transport and ATP:ADP exchange agreed, but what is the rationalization of the effects of the cations on ATP:ADP exchange? Here most of the information is derived from enzyme from other sources. Na_0 inhibits ATP:ADP exchange between 0 and 5 mM by slowing the rate of dephosphorylation of E_2P so that more of the enzyme is held up in the E_2P form and is out of commission for ATP:ADP exchange. Such effects have been reported on dephosphorylation of purified enzyme at low [Na] although the biphasic nature of the dephosphorylation rates reported in this work clouds their interpretation [69]. The increase in ATP:ADP exchange as Na_0 is elevated beyond 5 mM corresponds to an increase in the ratio E_1P: E_2P as Na_0 increases the rate of $E_2P \rightarrow E_1P$. Such effects have also been observed in kidney enzyme [69,70]. A unique feature of red blood cell enzyme is, however, in the temperature sensitivity of the phosphenzyme conformational equilibrium. When red blood cell membranes are cooled to $0°C$ the conformational transition $E_1P \rightarrow E_2P$ is blocked and all of the phosphoenzyme is ADP-sensitive and K-insensitive [68,71,72]. The Na-activation curve of ATP:ADP exchange at $0°C$ is a simple hyperbola with a $K_{0.5}$ for Na of >1 mM [72]. The shift from a complex multiphasic curve to a simple hyperbola is seen in kidney enzyme after N-ethyl maleimide treatment, which also blocks $E_1P \rightarrow E_2P$ [69]. However, the observation in red blood cells enables us to make an important conclusion. In the sodium pump model shown in Fig. 1, it is assumed that it is E_2P (rather than E_1P) that has external sites for Na. The disappearance of inhibition of ATP:ADP exchange by Na_0 in the range of 0-5 mM a $0°C$, when all the phosphoenzyme is E_1P, the first direct demonstration that E_2P is the form with external sites for Na ions [72].

The occlusion of Na ions on the sodium pump protein follows phosphorylation and the release of ADP. The return pathways for Na entry presumably involve deocclusion of Na, after the binding of ADP to $(Na_3)E_1P$ and dephosphorylation. It is not clear at this point whether the rate-limiting step for ATP: ADP exchange in the ATP synthesizing direction resides mainly at the Na_3 $E_2P \rightarrow (Na_3)E_1P$ transition or at the release of ATP from Na_3E_1 ATP. If the latter alternative is correct, then the rate limitation for Na entry is separated from the associated biochemical pathway. In contrast to the occlusion of K ions on $E_2(K_2)$, the measurement of the direct occlusion of Na ions (in E_1P) has been difficult and so far has been achieved only when $E_1P \rightleftharpoons E_2P$ is blocked by modification of the kidney enzyme with either chymotrypsin or N-ethylmaleimide [73].

ADP-independent $Na_i:Na_0$ exchange has also been described. This transport model is associated with ATP hydrolysis and is not stimulated by ADP. Sodium efflux is observed in K-free media containing Na and appears to be the pump operating in its physiological $Na_i:K_0$ mode in which Na_0 acts as a poor substitute for K_0 [75] (Fig. 5). The maximal rate of transport in this mode appears to be only approximately 10% of the rate of $Na_i:K_0$ exchange. Presumably this transport is taking place under circumstances in which Na_0 was shown to stimulate Na,ATPase in resealed ghosts [58]. In the $Na_i:Na_0$ exchange mode, where

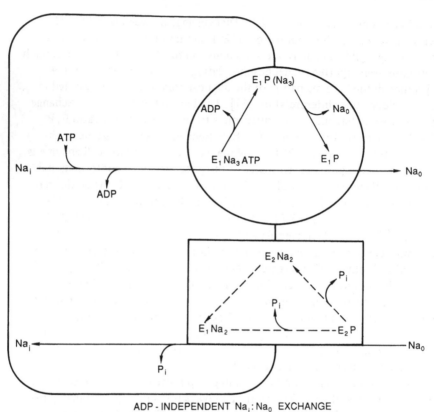

ADP - INDEPENDENT $Na_i : Na_0$ EXCHANGE

Figure 5 ADP-independent $Na_i:Na_0$ exchange. In this mode Na_0 acts as a poor K_0. Thus, Na entry occurs through the normal K entry pathway (rectangle). It isnot known whether Na enters via formation of an $E_2 Na_2$ form (just like K, but not being occluded) or if an alternate route to $E_1 Na_2$ exists.

ATP hydrolysis occurs, the transport (and ATPase) rate is limited by the rate of dephosphorylation of $E_2 P$, which is stimulated by Na_0 with lower affinity and to a much lesser extent than K_0 in the physiological range. In a sense, this transport mode is merely a reflection of the lack of selectivity of the external monovalent cation sites on the pump. It is considered here as a separate mode to emphasize that the measurement of a Na_0-dependent Na efflux with an associated Na influx in red blood cells is not automatically to be taken as a transport corrollary of ATP:ADP exchange. Another difference between these two Na_i: Na_0 exchanges is in the expected electrogenicity. Adenosine diphosphate-dependent $Na_i:Na_0$ exchange is a 1:1 exchange and is electroneutral. $Na_i:Na_0$ exchange associated with ATP hydrolysis would result in a net efflux (resulting from the 3:2 stoichiometry) and would be electrogenic.

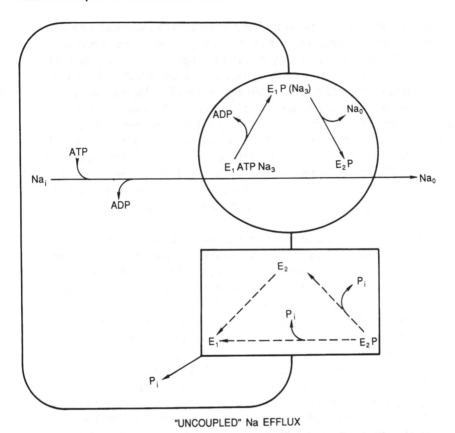

"UNCOUPLED" Na EFFLUX

Figure 6 Uncoupled Na efflux. In this transport mode, there is no extracellular counter cation to occupy the K entry pathway (rectangle). The possible complexities of associated anion movements have been omitted from the figure (see text).

D. Na_i Efflux

The so-called uncoupled Na efflux is seen when red blood cells (or resealed ghosts) are suspended in media containing no Na, K, or other monovalent cation congeners [22,23]. The studies of this mode have often been performed in choline Cl or $MgCl_2$. The ouabain-sensitive Na efflux that can be measured is associated with net ATP hydrolysis and the stoichiometry of this mode is approximately three Na ions transported per ATP molecule hydrolyzed (Fig. 6). The fact that this value is 3:1 [58] when the transport is dissipative (energetically downhill), just as in the case of Na_i:K_o exchange, which is steeply uphill, supports our notion that this stoichiometry (3:1) is built into the machinery of active Na transport and is not a consequence of the thermodynamics of actively pumping against Na and K gradients. The ATPase activity associated with the Na efflux has a high affinity for ATP (micromular) and in studies in resealed ghosts was shown to be inhibited by Na_0 in the range 0–5 mM and inhibited as Na_0 increased [58]. It now seems likely (see above) that at higher Na_0 values (5–150 mM) an associated Na_i:Na_0 exchange occurs [75]. Uncoupled Na efflux when supported by low (micromolar) levels of ATP is also inhibited by K_0, as a consequence of formation of $E_2 (K_2)$ and the slow release of occluded K to the intracellular compartments [55]. The inhibition by low Na_0 is probably related to the inhibition of dephosphorylation referred to above in the section dealing with Na_i:Na_0 exchange. In this transport mode if Na_0 and K_0 are 0, it is not known whether the return half of the cycle (with empty pump sites) occurs via the normal dephosphorylation route $E_2 P \rightarrow E_2 \rightarrow E_1$ or whetner $E_2 P \rightarrow E_1$ occurs.

The Na_i:K_0 transport mode in human red blood cells is electrogenic and the change in membrane potential, $\Delta\psi$, as the pump is activated by K_0 or inhibited by ouabain can be monitored using voltage-sensitive dyes [38]. When the electrogenicity of the pump operating in the absence of Na_0 or K_0 was examined, no indication of electrogenicity was observed [76]. Because at 37°C the rate of uncoupled Na efflux is 10% of the rate of Na efflux via Na_i:K_0 exchange, the size of the voltage change in the uncoupled mode should be approximately 30% of the voltage change in the Na_i:K_0 mode, if no accompanying anions or compensating cations move through the pump. These observations have led to a complex set of phenomena of pump operations, which suggests that the sodium pump in some special circumstances can mediate anion movements [76,77]. There is little precedent for such fluxes from enzymatic studies and so far a systematic study of such phenomena has only been performed in (human) red blood cells [76,77]. The complex set of phenomena include the observations that, if all cell and media chloride ions are replaced by sulfate and the band 3 anion exchanger blocked with DIDS, the pump mediates a cotransport of Na and sulphate ions. Further investigation of these phenomena has added to the complexity. While ouabain completely blocks the Na efflux, a small proportion (up to 20%) of the efflux remains when Na_0 is 5 mM. It appears that the residual Na_0-insensitive Na efflux is coupled to a different type of anion flux. The major portion of the Na efflux is coupled with SO_4^{2-} exit so that the transport is electroneutral and two Na ions move fore each SO_4^{2-} ions. This SO_4^{2-} flux disappears if there is a low level (5 mM) Cl^- in the medium. The minor portion of Na efflux (the Na_0-insensitive part) seems to be associated with the transmembrane transport, through the pump, of phosphate released from the terminal phosphate of pump-hydrolyzed ATP. This represents the first experimental claim that P_i can be transported through the Na pump after its release from ATP. It should be emphasized that these anion movements do not occur in the physiological Na_i:K_0 mode. The mechanisms involved in

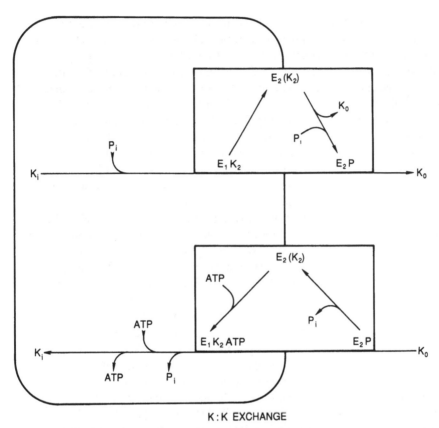

Figure 7 K_iK_o exchange. In this transport mode, K exit and K entry occur via K pathways (rectangles). The roles of ATP and P_i are shown. In this mode ADP can substitute for ADP (see text), and arsenate for P_i. The very slow K:K exchange transport when ATP (or ADP) and P_i (or arsenate) are absent follows presumably identical pathways but the rate constants for the transitions are much slower.

such profound changes in reaction pathway and how they are regulated are of considerable interest. The resolution of these problems may come more rapidly if these pump modes are characterized in a cell with a greater pump density.

E. $K_i:K_o$ Exchange

An ouabain-sensitive isotopic K efflux can be measured in suspensions of intact cells or resealed red blood cell ghosts, which is dependent on the presence of K_o [78] (in contrast to the reversed mode, $K_i:Na_o$ exchange) and is greatly stimulated by the presence of P_i (inorganic phosphate) and ATP [79]. This transport consists of a 1:1 exchange of intracellular for extracellular K ions. Although ATP is required for optimal rates, no significant hydrolysis occurs [81] and the ATP requirement can be satisfied by nonhydrolyzable ATP analogs [82], or ADP with a slightly lower affinity [83]. Under optional conditions the rate of K:K exchange may reach approximately 20% of the rate of $Na_i:K_0$ exchange. The P_i requirement for K:K exchange can be met by inorganic arsenate with approximately the same affinity [84]. The ATP requirement shows a $K_{0.5}$ of approximately 100 μM and corresponds to the low-affinity ATP site in $Na_i:K_0$ exchange [81]. It is evidently involved in accelerating the release of K ions from the $E_2(K_2)$ occluded form in the K entry steps (Fig. 7) [85]. The P_i requirement has been assumed to be involved in a reversal of the usual K entry steps of $Na_i:K_0$ exchange in which E_2P is hydrolyzed with P_i release [78]. Thus the P_i requirement for K:K exchange was assumed to be involved in phosphory-lating the enzyme; however, until recently it was not clear that phosphorylation was associated with K:K exchange or whether P_i binding alone was adequate. Recent studies have resolved this issue by measuring the ^{31}P nuclear magnetic resonance spectra of suspensions of resealed red blood cell ghosts performing K:K exchange [86]. The spectra were used to measure the rate of exchange of O^{16} atoms of $[O^{16}]$-labeled inorganic phosphate with O^{18} atoms of solvent water molecules. Such an exchange process requires the involvement of phos-phorylation events. It was shown that phosphorylation of the pump protein by P_i does indeed occur during K:K exchange and that the characteristics (asymmetry of K activation, inhibition by Na_i and oligomycin, etc.) of the P_i: H_2O exchange reaction paralleled those of $K_i:K_0$ exchange [86]. However, in a manner reminiscent of the situation for ATP:ADP exchange and $Na_i:Na_0$ ex-change, although the phosphorylation:dephosophorylation reactions were linked to K:K exchange, the reactions were not tightly coupled and significant rates of $P_i:H_2O$ exchange occurred in the absence of transport.

The potassium activation of $K_i:K_0$ exchange shows a striking asymmetry, high-affinity stimulation ($K_{0.5}$ approximately 0.25 mM) at the outer surface of the pump (in the absence of Na; [78]) and lower-affinity stimulation by K_i at the cytoplasmic surface ($K_{0.5}$ approximately 10 mM with 0.7 mM cell Na; [81]). On the basis of predictions originally made by Stein [89] and subse-quently investigated in red blood cells and resealed ghosts [90-92] and in proteoliposomes containing reconstituted renal enzyme [87], it became clear that a form of the enzyme $E_2(K_2)$ is able to bind ATP and P_i simultaneously. The important transport intermediate the occluded enzyme K complex, $E_2(K_2)$, was involved in both the inward and outward transport limbs of K:K exchange [94]. The existence and characterization of small ouabain or vanadate-sensitive K fluxes in proteoliposomes [95] and resealed red blood cell ghosts [96] in the absence of either P_i or ATP enabled a detailed study to be performed on the K limb of the sodium pump cycle. In resealed red blood cell ghosts, these K or Rb fluxes occur at $< 1\%$ of the rate of normal pump turnover [92]. Several important points emerged from these studies. In the absence of P_i and ATP [92,96], Rb_0 activated a $Rb_i:Rb_0$ exchange with a high affinity (approximately 25 μM). This result demonstrated that the existence of high-affinity cation sites for K at the outer aspect of the sodium pump was not a consequence of phosphorylation, because in the absence of phosphorylation such sites already existed. In addition, the separate and additive effects of ADP (or ATP) and P_i on $K_i:K_0$ exchange and congener fluxes were examined to establish at which transitions in the hetero-exchange processes ($Rb_i:Cs_0$, $Li_i:Rb_0$, etc.) were the rate-linking steps [92].

An important assumption in the models of K transport and especially for $K_i:K_0$ exchange (see, for example, Figs. 1 and 7) was that the occluded cation form $E_2(K_2)$ was a central intermediate and that ATP (or ADP) only speeds release to the inside of the cell and P_i only to the outside. From studies on the effects of ligands on the rates of deocclusion carried out with renal enzyme [97–99] it was not possible to establish that the release was limited to only one direction or whether the ligand merely destabilized the $E_2(K_2)$ form and stimulated K release. Studies in resealed ghosts were able to establish that indeed the accelerated release of occluded cations occurred only in the predicted direction [100]. These studies used the observation that Li could act as a K congener in the $K_i:K_0$ exchange mode [92] and that Li was not significantly occluded or its release from the occluded state was rapid [49] in the absence of ATP (or as it turned out P_i). When ^{86}Rb uptake was measured into Li-containing ghosts, we would expect that ATP would stimulate the flux and P_i would be without effect. This is because the inward release of Rb should be speeded by ATP, but the outward release of Li, which is not tightly occluded, should be unaffected by P_i. This expectation was fulfilled [100]. Clearly, if ATP or P_i merely destabilized the occluded $E_2(K_2)$ form, then either agent should have stimulated the rate of ^{86}Rb uptake. Similar experiments using Cs and also measuring the effects of [ATP] on Li^0 or Rb^0-activated Na efflux confirmed that ATP speeds the release of cations to the inside only (at a low-affinity site) and that the ligand is without effect if the cation is not tightly bound [100]. An observation made in these and earlier studies awaits rationalization. This is the apparent lower V_{max} for Na efflux when Li is the external cation compared with K. $Na_i:Li_0$ exchange is approximately 65–70% of the rate of $Na_i:K_0$ exchange. This would not be expected from the simple models we have described.

III. REGULATION OF THE SODIUM PUMP

There is little evidence of physiological regulation of the sodium pump in human red blood cells. However, recent studies involving the cytoplasmic inhibition of the sodium pump by Ca ions suggest the intermediacy of regulatory protein, calnaktin, which dramatically affects the affinity with which Ca inhibits active

Na pumping [105]. The protein has been detected in other cells so that calnaktin may be a more widely occurring regulator than previously suspected.

IV. CONCLUDING REMARKS

It is apparent that there is a great deal of information available on the various properties and actions of the sodium pump in human red blood cells. The material covered in this chapter is only one subjective digest of this wide area. There are several topics that have not been covered but that provide further information on sodium pump characterization. These include the description of the inhibitory actions of vanadate anion(s) (for example, [102]), the involvement of Mg ions in the red blood cell pump cycle [88,103], and the well-characterized dimorphism in HK and LK red cells of sheep and goats (see, for example [104]).

The basic transport properties of the sodium pump were first described in human red blood cells 20 years ago and since that time the red blood cell sodium pump has increasingly displayed its versatility. To rationalize many of the basic phenomena, we sometimes choose to ignore the inconvenient complexities. Ultimately, these complexities will, we hope, enhance rather than threaten our models.

ACKNOWLEDGMENTS

I thank Dr. Linda J. Kenney for helpful advice during the preparation of this article. I am also grateful to Dr. John Parker for the invitation to write this chapter, but more grateful for his patience in waiting for its completion. Studies described here that were performed in the author's laboratory were supported by National Institutes of Health grant HL 30315.

REFERENCES

1. Glynn, I. M., and Karlish, S. J. D. (1975). The sodium pump. *Annu. Rev. Physiol.* 37:13–55.
2. Jorgensen, P. L. (1980). Sodium potassium ion pump in kidney tubules. *Physiol. Rev.* 60:864–917.
3. Kaplan, J. H. (1983). Sodium ions and the sodium pump: transport and enzymatic activity. *Am. J. Physiol.* 245:G327–333.
4. Cavieres, J. D. (1977). The sodium pump in human red cells. In *Membrane Transport in Red Cells*, J. C. Ellory (Ed.). Academic Press, New York, pp. 1–38.
5. Jorgensen, P. L., and Andersen, J. P. (1988). Structural basis for E1-E2 conformational transitions in Na,K-pump and Ca-pump protein. *J. Membr. Biol.* 103:95–120.
6. Glynn, I. M. (1985). The Na,K-transporting adenosine triphosphatase. In *The Enzymes of Biological Membranes*, Vol. 3, 2nd ed., A. Martonosi (Ed.). Plenum Press, New York, pp. 35–114.

7. Garrahan, P. J., and Rega, A. F. (1967). Cation loading of red cells. *J. Physiol.* 193:459–466.
8. Sachs, J. R. (1977). Kinetics of the inhibition of the Na, K pump by external sodium. *J. Physiol. (Lond.)* 264:449–470.
9. Cass, A., and Dalmark, M. (1973). Equilibrium dialysis of ions in nystatin treated cells. *Nature New Biol.* 244:47–49.
10. Hoffman, J. F. (1962). The active transport of sodium by ghosts of human red blood cells. *J. Gen. Physiol.* 45:837–859.
11. Schwoch, G., and Passow, H. (1973). Preparations and properties of human erythrocyte ghosts. *Mol. Cell. Biochem.* 2:197–218.
12. Wood, P. G., and Passow, H. (1981). Techniques for the modification of the intracellular composition of red blood cells. In *Techniques in Cellular Physiology*, P. F. Baker (Ed.). Elsevier/North-Holland. London, pp. 1–43.
13. Kaplan, J. H. (1982). Sodium pump mediated ATP:ADP exchange: the sided effects of sodium and potassium ions. *J. Gen. Physiol.* 80:915–937.
14. Blostein, R., and Chu, L. (1977). Sidedness of (sodium, potassium)-adenosine triphosphatease of inside-out red cell membrane vesicles. *J. Biol. Chem.* 205:3035–3043.
15. Blostein, R., Pershadsingh, H. A., Drapeau, P., and Chu, L. (1979). Side-specificity of alkali cation interactions with Na,K-ATPase. In *Na,K-ATPase*, J. C. Skou, and J. G. Norby (Eds.). Academic Press, New York, pp. 233–245.
16. Shull, G. E., Schwartz, A., and Lingrel, J. B. (1985). Aminoacid sequence of the catalytic subunit of Na,K-ATPase. *Nature* 316:691–695.
17. Maclennan, D. H., Brandle, C. J., Korzak, B., and Green, N. M. (1985). Aminoacid sequence of a Ca^{2+} & Mg^{2+}-dependent ATPase from rabbit muscle sarcoplasmic reticulum, deduced from its complementary DNA sequence. *Nature* 316:695–700.
18. Shull, G. E., and Lingrel, J. B. (1988). Molecular cloning of the rat stomach $(H^+ + K^+)$-ATPase. *J. Biol. Chem.* 261:16788–16791.
19. Hoffman, J. F. (1986). Active transport of Na^+ and K^+ by red blood cells. In *Physiology of Membrane Disorders*, 2nd Ed., T. E. Andreoli, J. F. Hoffman, D. F. Fanestil, and S. G. Schultz (Eds.). Plenum Press, New York, pp. 221–231.
20. Harris, J. E. (1941). The influence of the metabolism of human erythrocytes on their potassium content. *J. Biol. Chem.* 141:479–595.
21. Danowski, T. S. (1941). The transfer of potassium across the human blood cell membrane. *J. Biol. Chem.* 139:693–705.
22. Garrahan, P. J., and Glynn, I. M. (1967). The behaviour of the sodium pump in red cells in the absence of external potassium. *J. Physiol. (Lond.)* 192:159–174.
23. Garrahan, P. J., and Glynn, I. M. (1967). The sensitivity of the sodium pump to external sodium. *J. Physiol. (Lond.)* 192:175–188.
24. Garrahan, P. J., and Glynn, I. M. (1967). Factors affecting the relative magnitudes of the sodium:potassium and sodium:sodium exchanges catalysed by the sodium pump. *J. Physiol. (Lond.)* 192:189–216.

25. Garrahan, P. J., and Glynn, I. M. (1967). The stoichiometry of the sodium pump. *J. Physiol. (Lond.)* 192:217-235.
26. Garrahan, P. J., and Glynn, I. M. (1967). The incorporation of inorganic phosphate into adenosine triphosphate by reversal of the sodium pump. *J. Physiol. (Lond.)* 192:237-256.
27. Post, R. L., and Jolly, P. C. (1957). The linkage of sodium, potassium and ammonium active transport across the human erythrocyte membrane. *Biochem. Biophys. Acta.* 25:118-128.
28. Sen, A. K., and Post, R. L. (1964). Stoichiometry and localization of adenosine triphosphate-dependent sodium and potassium transport in the erythrocyte. *J. Biol. Chem.* 239:345-352.
29. Skou, J. C. (1957). The influence of some cations on an adenosine triphosphatase from peripheral nerve. *Biochim. Biophys. Acta* 23:394-401.
30. Post, R. L., Merritt, L. C., Kinsolving, C. R., and Albright, C. D. (1960). Membrane adenosine triphosphatase as a participant in the active transport of sodium and potassium in the human erythrocyte. *J. Biol. Chem.* 235: 2796-1802.
31. Dunham, E. T., and Glynn, I. M. (1961). Adenosine triphosphatase activity and the active movements of alkali metal ions. *J. Physiol. (Lond.)* 156:274-293.
32. Schatzmann, H. J. (1953). Herzglykoside als Hemmstoff fur den aktiven Kalium und Natrium-transport durch die Erythrocyten-membrane. *Helv. Physiol. Pharmacol. Acta* 11:346-354.
33. Hoffman, J. F. (1966). The red cell membrane and the transport of sodium and potassium. *Am. J. Med.* 41:666-668.
34. Bodemann, H. H., and Hoffman, J. F. (1976). Side-dependent effects of internal versus external Na and K on ouabain binding to reconstituted human red blood cell ghosts. *J. Gen. Physiol.* 67:497-525.
35. Sachs, J. R. (1974). Interaction of external K, Na and cardioactive steroids with the Na, K pump of the human red blood cell. *J. Gen. Physiol.* 63:123-243.
36. Thomas, R. C. (1972). Electrogenic sodium pump in nerve and muscle cells. *Physiol. Rev.* 52:563-594.
37. DeWeer, P., Gadsby, D. C., and Rakowski, R. F. (1968). Voltage dependence of the Na-K pump. *Ann. Rev. Physiol.* 50:235-241.
38. Hoffman, J. F., Kaplan, J. H., and Callahan, T. J. (1979). The Na:K pump in red cells is electrogenic. *Fed. Proc.* 38:2440-2441.
39. Proverbio, F., Shoemaker, D. L., and Hoffman, J. F. (1988). Functional consequences of the membrane pool of ATP associated with the human red blood cell Na/K pump. In *The Na+, K+-Pump, Part A: Molecular Aspects*, AR Liss, New York, pp. 561-567.
40. Schrier, S. L. (1967). ATP synthesis in human erythrocyte membranes. *Biochim. Biophys. Acta* 135:591-598.
41. Proverbio, F., and Hoffman, J. F. (1977). Membrane compartmentalized ATP and its preferential use by the Na,K-ATPase of human red cell ghosts. *J. Gen. Physiol.* 69:605-632.

42. Parker, J. C., and Hoffman, J. F. (1967). The role of membrane phospho-glycerate kinase in the control of glycolytic rate by active transport in human red blood cells. *J. Gen. Physiol.* 50:893–916.
43. Mercer, R. W., and Dunham, P. B. (1981). Membrane-bound ATP fuels the Na/K pump. *J. Gen. Physiol.* 78:547–568.
44. Alberts, R. W., Koval, G. J., and Siegel, G. J. (1968). Studies on the interaction of ouabain and other cardioactive steroids with Na,K-activated ATPase. *Mol. Pharmacol.* 4:324–336.
45. Post, R. L., Kume, S., Tobin, T., Orcutt, B., and Sen, A. K. (1969). Flexibility of an active center in sodium-plus-potassium adenosine triphosphatase. *J. Gen. Physiol.* 54:306s–326s.
46. Karlish, S. J. D., Yates, D. W., and Glynn, I. M. (1978). Conformational transitions between Na-bound and K-bound forms of (Na + K)-ATPase studied with formycin nucleotides. *Biochim. Biophys. Acta* 525:252–264.
47. Neufeld, A. H., and Levy, H. M. (1969). A second ouabain-sensitive sodium-dependent adenosine triphosphatase in brain microsomes. *J. Biol. Chem.* 244:6493–6497.
48. Czerwinski, A., Gitelman, H. J., and Welt, L. G. (1967). A new member of the ATPase family. *Am. J. Physiol.* 213:786–792.
49. Post, R. L., Hegyvary, C., and Kume, S. (1972). Activation by adenosine triphosphate in the phosphorylation kinetics of sodium and potassium ion transport adenosine triphosphatase. *J. Biol. Chem.* 247:6530–6540.
50. Stein, W. D. (1986). Primary active transport: chemiosmosis. In *Transport and Diffusion Across Cell Membranes*. Academic Press, Orlando, Florida, pp. 475–612.
51. Sachs, J. R. (1980). The order of release of sodium and addition of potassium in the sodium-potassium pump reaction mechanism. *J. Physiol. (Lond.)* 302:219–240.
52. Sachs, J. R. (1986). The order of addition of sodium and release of potassium at the inside of the sodium pump of the human red cell. *J. Physiol. (Lond.)* 381:149–168.
53. Karlish, S. J. D., and Kaplan, J. H. (1985). Pre-steady-state kinetics of Na^+ transport through the Na,K pump. In *The Sodium Pump* I. M. Glynn and C. Ellory (Eds.). The Company of Biologists, Cambridge, England, pp. 501–506.
54. Forbush, B., III (1984). Na^+ movements in a single turnover of the Na pump. *Proc. Natl. Acad. Sci. USA* 81:5310–5314.
55. Karlish, S. J. D., and Glynn, I. M. (1979). An uncoupled efflux of sodium ions from human red cells, probably associated with Na-dependent ATPase activity. *Ann. NY Acad. Sci.* 242:461–470.
56. Bodemann, H. H., and Hoffman, J. F. (1976). Comparison of the side-dependent effects of Na and K on orthophosphate, UTP and ATP-promoted ouabain binding to reconstituted human red blood cell ghosts. *J. Gen. Physiol.* 67:527–545.
57. Kaplan, J. H., and Holis, R. J. (1980). External Na dependence of ouabain-sensitive ATP:ADP exchange initiated by photolysis of intracellular caged-ATP in human red cell ghosts. *Nature* 288:587–589.

58. Glynn, I. M., and Karlish, S. J. D. (1976). ATP hydrolysis associated with an uncoupled Na efflux through the sodium pump: evidence for allosteric effects of intracellular ATP and extracellular Na. *J. Physiol. (Lond.)* 256: 465–496.

59. Rodland, K. D., and Dunham, P. B. (1980). Kinetics of Li efflux through the(Na,K) pump of human erythrocytes. *Biochim. Biophys. Acta* 602: 376–388.

60. Blostein, R. (1985). Proton-activated rubidium transport catalyzed by the sodium pump. *J. Biol. Chem.* 260:839–843.

61. Polvani, C., and Blostein, R. (1988b). Protons as congeners of sodium and potassium in human red cell Na,K-ATPase. *Biophys. J.* xxx:xxx–xxx.

62. McConaghey, P. D., and Maizels, M. (1962). Cation exchanges of lactose-treated human red cells. *J. Physiol. (Lond.)* 162:485–509.

63. Sachs, J. R., and Welt, L. G. (1967). The concentration dependence of active potassium transport in the human red blood cell. *J. Clin. Invest.* 46: 1433–1441.

64. Cavieres, J. D., and Glynn, I. M. (1979). Sodium-sodium exchange through the sodium pump, the roles of ATP and ADP. *J. Physiol. (Lond.)* 297:637–645.

65. Glynn, I. M., and Hoffman, J. F. (1971). Nucleotide requirements for sodium: sodium exchange catalyzed by the sodium pump in human red cells. *J. Physiol. (Lond.)* 218:239–256.

66. Kennedy, B. G., and DeWeer, P. (1976). Strophanthidin-sensitive sodium fluxes in metabolically poisoned frog skeletal muscle. *J. Gen. Physiol.* 68: 405–420.

67. Fahn, S., Koval, G. J., and Albers, R. W. (1966). Sodium-potassium-activated adenosine triphosphatase of electrophorus electric organ. I. An associated sodium-activated transphosphorylation. *J. Biol. Chem.* 241:1882–1889.

68. Blostein, R. (1968). Relationships between erythrocyte membrane phosphorylation and adenosine triphosphate hydrolysis. *J. Biol. Chem.* 243: 1957–1965.

69. Beauge, L. A., and Glynn, I. M. (1979). Sodium ions, acting at high-affinity extracellular sites, inhibit sodium-ATPase activity of the sodium pump by slowing dephosphorylation. *J. Physiol. (Lond.)* 289:17–31.

70. Hara, Y., and Nakao, M. (1981). Sodium ion discharge from pig kidney Na,K-ATPase Na dependency of the E1P E2P equilibrium in absence of KCl. *J. Biochem. (Tokyo)* 90:923–931.

71. White, B., and Blostein, R. (1982). Comparison of red cell and kidney (Na + K)-ATPase at 0°. *Biochim. Biophys. Acta* 688:685–690.

72. Kaplan, J. H., and Kenney, L. J. (1985). Temperature effects on sodium pump phosphoenzyme distribution in human red blood cells. *J. Gen. Physiol.* 85:123–136.

73. Glynn, I. M., Hara, Y., and Richards, D. E. (1984). The occlusion of sodium ions within the mammalian sodium-potassium pump: its role in sodium transport. *J. Physiol. (Lond.)* 351:531–547.

74. Hoffman, J. F. (1969). The interaction between tritiated ouabain and the Na-K pump in red blood cells. *J. Gen. Physiol.* 54:343s–350s.

75. Blostein, R. (1983). Sodium pump catalyzed Na:Na exchange associated with ATP hydrolysis. *J. Biol. Chem.* 248:7948-7953.
76. Dissing, S., and Hoffman, J. F. (1983). Anion-coupled Na efflux mediated by the Na/K pump in human red blood cells. In *Current Topics in Membranes and Transport*, Vol. 19. Academic Press, New York, pp. 693-695.
77. Martin, R., and Hoffman, J. F. (1988). Two different types of ATP-dependent anion coupled Na transport are mediated by the human red blood cell Na/K-pump. In *The Na$^+$, K$^+$ pump, Part A: Molecular Aspects*, J. C. Skou, J. G. Norby, A. B. Maunsbach, and M. Esmann (Eds.). A. R. Liss, New York, pp. 539-544.
78. Glynn, I. M., Lew, V. L., and Luthi, U. (1970). Reversal of the potassium entry mechanism in red cells, with and without reversal of the entire pump cycle. *J. Physiol. (Lond.)*:207:371-391.
79. Simons, T. J. B. (1974). Potassium-potassium exchange catalyzed by the sodium pump in human red cells. *J. Physiol. (Lond.)* 237:123-155.
80. Krauf, P. A., Proverbio, F., and Hoffman, J. F. (1974). Chemical characterization and pronase susceptibility of the Na:K pump-associated phosphoprotein of human red blood cells. *J. Gen. Physiol.* 63:305-323.
81. Simons, T. J. B. (1974). Potassium:potassium exchange catalyzed by the sodium pump in human red cells. *J. Physiol. (Lond.)* 237:123-155.
82. Simons, T. J. B. (1975). The interaction of ATP-analogues possessing a blocked phosphate group with the sodium pump in human red cells. *J. Physiol. (Lond.)* 244:731-739.
84. Kaplan, J. H., and Kenney, L. J. (1982). ADP supports ouabain-sensitive K:K exchange in human red blood cells. *Ann. NY Acad. Sci.* 402:292-295.
84. Kenney, L. J., and Kaplan, J. H. (1988). Arsenate substitutes for phosphate in the human red cell sodium pump and anion exchanger. *J. Biol. Chem.* 263:7954-7960.
85. Beauge, L. A., and Glynn, I. M. (1980). The equilibrium between different conformations of the unphosphorylated sodium pump: effects of ATP and of potassium ions, and their relevance to potassium transport. *J. Physiol. (Lond.)* 299:367-383.
86. Kaplan, J. H., Kenney, L. J., and Webb, M. R. (1985). In *The Sodium Pump*, J. C. Ellory and I. M. Glynn (Eds.). The Company of Biologists, Cambridge, England, pp. 415-421.
87. Karlish, S. J. D., Lieb, W. R., and Stein, W. D. (1982). Combined effects of ATP and phosphate on rubidium exchange mediated by Na-K-ATPase reconstituted into phospholipid visicles. *J. Physiol. (Lond.)* 328:333-350.
88. Sachs, J. R. (1988). Interaction of magnesium with the sodium pump of the human red cell. *J. Physiol. (Lond.)*:400:574-591.
89. Stein, W. D. (1979). Half-of-the-sites reactivity and the Na,K-ATPase. In *Na,K-ATPase, Structure and Kinetics*, J. C. Skou and J. G. Norby (Eds.). Academic Press, London, pp. 475-486.
90. Sachs, J. R. (1986). Potassium-potassium exchange as part of the over-all reaction mechanism of the sodium pump of the human red blood cell. *J. Physiol. (Lond.)* 374:221-244.

91. Eisner, D. A., and Richards, D. E. (1983). Stimulation and inhibition by ATP and orthophosphate of the potassium-potassium exchange in resealed red cell ghosts. *J. Physiol. (Lond.)* 335:495–506.

92. Kenney, L. J. (1987). Characterization of potassium transport by the sodium pump in human red blood cells. Ph.D. Thesis, University of Pennsylvania.

93. Karlish, S. J. D., and Stein, W. D. (1982). Effects of ATP or phosphate on passive rubidium fluxes mediated by the Na,K-ATPase reconstituted into phospholipid vesicles. *J. Physiol. (Lond.)* 328:317–331.

94. Glynn, I. M., and Karlish, S. J. D. (1982). Conformational changes associated with K^+ transport by the Na^+/K^+-ATPase. In *Membranes and Transport*. A. N. Martonosi (Ed.). Plenum Press, New York, pp. 529–536.

95. Karlish, S. J. D., and Stein, W. D. (1982). Passive rubidium fluxes mediated by Na-K-ATPase reconstituted into phospholipid vesicles when ATP- and phosphate-free. *J. Physiol. (Lond.)* 328:295–316.

96. Kenney, L. J., and Kaplan, J. H. (1985). Arsenate replaces phosphate in ADP-dependent and ADP-independent Rb-Rb exchange mediated by the red cell sodium pump. In *The Sodium Pump*, I. M. Glynn and C. Ellory (Eds.). The Company of Biologists, Cambridge, England, pp. 535–539.

97. Forbush, B., III (1987). Rapid release of [42]K and ^{86}Rb from an occluded state of the Na,K-pump in the presence of ATP or ADP. *J. Biol. Chem.* 262:11104–11115.

98. Forbush, B., III (1988). Rapid ^{86}Rb release from an occluded state of the Na,K-pump reflects the rate of dephosphorylation or dearsenylation. *J. Biol. Chem.* 263:7961–7969.

99. Glynn, I.M., and Richards, D. E. (1982). Occlusion of rubidium ions by the sodium-potassium pump: its implications for the mechanism of potassium transport. *J. Physiol. (Lond.)* 330:17–43.

100. Kenney, L. J., and Kaplan, J. H. (1988). The vectorial effect of ligands on the occluded intermediate in red cell sodium pump transport. In *The Na^+, K^+-Pump, Part A: Molecular Aspects*, J. C. Skou, J. G. Norby, A. B. Maunsbach, and M. Esmann (Eds.). A. R. Liss, New York, pp. 525–530.

101. Kennedy, B. G., Lunn, G., and Hoffman, J. F. (1986). Effects of altering the ATP/ADP ratio on pump-mediated Na/K and Na/Na exchanges in resealed human red blood cell ghosts. *J. Gen. Physiol.* 87:47–72.

102. Sachs, J. R. (1987). Inhibition of the Na, K pump by vanadate in high-Na solutions. *J. Gen. Physiol.* 90:291–320.

103. Flatman, P., and Lew, V. L. (1981). The magnesium dependence of sodium pump-mediated sodium-potassium and sodium-sodium exchange in intact human red cells. *J. Physiol. (Lond.)* 315:421–446.

104. Dunham, P. B., and Anderson, C. (1987). On the mechanism of stimulation of the Na/K pump of LK sheep erythrocytes by anti-L antibody. *J. Gen. Physiol.* 90:3–25.

105. Yingst, D. R. (1988). Modulation of the Na,K-ATPase by Ca and intracellular proteins. *Annu. Rev. Physiol.* 50:291–303.

17

The Plasma Membrane Calcium Pump
The Red Blood Cell as a Model

FRANK F. VINCENZI   *University of Washington, Seattle, Washington*

I. INTRODUCTION

A. Historical Perspective

It has been 20 years since the initial reports concerning the plasma membrane (PM) Ca pump. I was fortunate enough to be in on the "ground floor" of the idea and to have witnessed the tremendous progress in this area over these 20 years. What follows will be a personal and hopefully factual account of my view of this field. For any shortcomings of fact, improper citations, or less than accurate recollections, I must take responsibility. Credit belongs to a wide range of collaborators and competitors over many years. Unless otherwise noted, reference will be to the human red blood cell (RBC), which has been such an extremely useful model cell in the development of the ideas considered here.

In the mid 1960s I had the privilege of being a postdoctoral fellow in the laboratory of H. J. Schatzmann at the Pharmacology Institute of the University of Berne, Switzerland. Schatzmann had already distinguished himself by demonstrating the selective inhibition of Na-K transport by digitalis-related compounds [1]. In the mid 1960s interest in Ca transport was growing and he and I discussed the possibility that digitalis might exert its positive inotropic effect by inhibiting Ca extrusion from cardiac cells. At that time it was unclear whether inhibition of Na-K transport was the basis of the positive inotropic effect of digitalis [2]. The issue appears to be more clearly resolved today and decreased extrusion of Ca from cardiac cells appears to be involved in digitalis inotropism, but the mechanism appears to involve altered Na/Ca exchange rather than direct inhibition of the Ca pump [3].

Schatzmann and I were convinced that the models of Ca transport available in the 1960s were rather primitive and not well defined. Thus, we sought to find a simple model of Ca transport and in particular to find a model of PM Ca transport. We planned to test the hypothesis that digitalis would inhibit such Ca transport. Because of his experience with them, Schatzmann suggested that we should try human RBCs. We were unaware then that Passow [4,5] had already reported that human RBCs do not transport Ca. At the time of our discussions about the possibility of a Ca pump in RBCs I was involved in a project examining the relationship between cardiac muscle excitation and the rate of onset of digitalis action [6]. Schatzmann, working with RBCs, carried out the initial Ca transport experiments [7]. To my knowledge, this was the first report of the active transport of Ca across a PM.

B. Early Findings

Encouraged by his apparent initial success, we began to pursue criteria to unequivocally demonstrate the hypothetical PM Ca pump. Among our expectations, by analogy with the then rather well developed area of Na-K transport, was the idea that an active pump ought to be associated with the activity of a membrane-bound adenosine triphosphatase (ATPase), and that the ATPase ought to be activated by the transported ion on the side of the membrane from which it would be transported. Because it was beginning to be appreciated even in the 1960s that intracellular Ca is maintained at very low levels, we set as a criterion of any supposed Ca pump ATPase that it ought to be activated by low levels of Ca. It was known from the work of Dunham and Glynn [8] and others that RBC membranes contain an ATPase activity that is activated by Ca. But the sensitivity to Ca had not been determined. We found that the threshold for activation of the ATPase was $\sim 10^{-7}$ M Ca [9]. Thus, an important initial criterion was met.

A major task to prove the existence of a Ca pump in the PM was to demonstrate an association between Ca transport and ATP use (similar to what had been done so brilliantly by a number of workers for the Na-K pump). This was accomplished in large part during the balance of my postdoctoral fellowship, during which Schatzmann became Professor of Veterinary Pharmacology and we moved from the Pharmacology Institute of the Medical School to the Veterinary-Pharmacology Institute of the Veterinary School of the University of Berne. Our major work demonstrated that the extrusion of Ca from resealed RBC ghosts is accelerated by glucose addition or by inclusion of ATP when ghosts are loaded with Ca, that the transport occurs "uphill," is highly dependent on temperature, and does not occur in cells deficient in ATP. In addition, it was demonstrated that ATP is split in association with Ca transport and that the activation of the ATPase is restricted to the inner surface of the membrane. It it was also shown, but not conclusively, that the transport as well as the ATPase activity is dependent on Mg in the cells. Even with this rather convincing demonstration of a Ca pump in the PM, Schatzmann was reticent to boldly proclaim a "pump" in so many words and thus our major work had a rather bland title: "Calcium movements across the membrane of human red blood cells" [10].

After the early papers on the PM Ca pump ATPase from the laboratory of Schatzmann [7,10], two other similar articles appeared with essentially the same findings [11,12]. All of these articles agreed: There is an ATPase-based Ca pump in the PM of human RBCs. It now appears that the PM Ca pump is more or less ubiquitous, as is the maintenance of low intracellular Ca; at least in normal living cells. Since these initial reports there have been many papers confirming the existence in a wide variety of PMs of an ATP-dependent Ca pump. Thus, just as they had been so useful as a model for Na-K transport, so data from the RBC have been extremely useful in the field of Ca transport.

C. A Personal Reflection

It may be noted in retrospect that by a delay in publishing and by my own naiveté, Schatzmann and I were nearly "scooped" in this first definitive description of the PM Ca pump. Although much of the work had been accomplished before I left Switzerland in the summer of 1967, there was still a substantial amount of work to be done. Furthermore, the manuscript went through many revisions across the 6,000 or so miles between Berne and Seattle, where I had accepted a position as an Assistant Professor of Pharmacology. And, because of a conjunction of geographic location and parallel scientific interest, I had the opportunity to tell professor Kwang Soo Lee about the experiments I had performed with Schatzmann when (in 1967 or 1968) Lee passed through Seattle on one of his many sojourns to Korea. The net result is that a paper by Lee and Shin [11] appeared very shortly after the paper by Schatzmann and Vincenzi [10].

When Lee and his co-workers presented results of related work at a Fall Meeting of the American Society for Pharmacology and Experimental Therapeutics [13] I recall having expressed some confusion because the authors reported ATP-dependent Ca *uptake* in isolated RBC "membrane fragments." This was a natural for a laboratory that had been working with sarcoplasmic reticulum (SR) (and which routinely measured Ca uptake), but it seemed antithetical to the extrusion that Schatzmann and I had described. Of course, the explanation is that Cha et al. had (unwittingly) made inside-out vesicles (IOVs) of human RBC membranes, a technique that later proved to be extremely valuable in the study of this pump [14].

II. SOME HARD FACTS CONCERNING THE
PLASMA MEMBRANE CALCIUM PUMP

A. No Specific Inhibitor

Not incidentally, it must be pointed out that one initial reason for performing experiments on the PM Ca pump resulted in a "negative" finding: that is, digitalis-related compounds exert no effect on the Ca pump or ATPase. Although this fact helped establish the separate nature of the Ca pump from the Na-K pump, an interrelated fact has plagued this research for more than 20 years, at least from my perspective. There is no "ouabain" for the Ca pump. That is, there is no potent, selective inhibitor of the Ca pump. As stated in the last sentence of the discussion in our 1969 article: "It is a serious handicap that no specific inhibitor for Ca transport is available that could play the part that ouabain, as a specific inhibitor, has played in the elucidation of Na-K transport" [10].

B. Inhibitors

Early work on potential inhibitors of the Ca pump was consistent with the idea that a sulfhydryl group was involved in the activity of the enzyme [10,15]. Rega and Garrahan [16] did a fine job of reviewing the subject of inhibitors. Most inhibitors studies have been devoted to the anti-CaM drugs, which I will consider separately below. Recently, an elegant piece of work on inhibition of ATPase and its dependence on unsaturated lipid has appeared [17].

C. Stoichiometry

The stoichiometry of the Ca pump ATPase has been considered by several groups, and the issue is well summarized by Rega and Garrahan [16]. We found that the apparent stoichiometry of the pump is one Ca pumped per ATP split [18]. Others have concluded that the stoichiometry may be two Ca per ATP [19], or even a variable stoichiometry [20]. In the early days of this research it was important to ascertain the stoichiometry of the PM Ca pump to help establish its sameness or uniqueness relative to the SR pump, the stoichiometry of which is generally accepted to be two Ca per ATP.

   As work on the two pumps progressed it became clear in the mind of this author that they differ in stoichiometry as well as molecular weight. Rega and Garrahan, based on their review of approximabely 15 papers on the subject, consider the matter of stoichiometry an open question. Thus, although these pumps perform similar functions, they are not identical in location of molecular structure. Penniston and co-workers have created an antibody to the PM Ca pump ATPase. It is noteworthy that the antibody does not cross-react with the ATPase of the SR [21]. This serves to reinforce the notion that these two pumps are distinct molecules, and may even have evolved from different progenitors.

III. ACTIVATION OF THE CALCIUM PUMP
BY CALMODULIN

A. Protein Factor

Bond and Clough [22] published a report that described activation of the Ca
pump ATPase of human RBC membranes by something in RBC hemolysates. At
the time we considered the report interesting, but dismissed hemolysates as too
complex to consider a specific source of any "activator" of the RBC Ca pump.
Somewhat later, Martha Farrance, then a graduate student in my laboratory,
somewhat accidentally found evidence for an apparently proteinaceous material
that (under some conditions) binds to RBC membranes during their isolation
and that increased the activity of the Ca pump ATPase. She also found that the
material could be removed from the membranes by washing in ethyleneglycol
tetraacetic acid [23,24].

In 1977 I attended a conference in Keystone, Colorado. At the conference I
presented a poster paper on Martha Farrance's data, hoping that someone at
the conference could offer advice on how we might best proceed to identify
the unknown "activator protein" [25]. Dr. Judith Ryan Gunsalus, then a graduate student in the laboratory of Dr. Daniel Storm, who was at that time at University of Illinois, took one look at our poster paper and said that our activator
sounded like "phosphodiesterase (PDE) activator," a protein that was receiving
attention in some (then small) circles as a Ca-dependent activator of cyclic
nucleotide PDE. I was really excited abou the idea that a cytoplasmic protein
might be involved in the regulation of the membrane-bound Ca pump ATPase.
I recall now what animated discussions I had with the man who happened to be
my roommate at the Keystone conference, Dr. John Penniston. He was skepti-
cal, but as I was to learn later, he must also have been intrigued by the idea.

B. Calmodulin as the Ca Pump Activator

When I returned to Seattle, I discussed the suggestion with my colleagues, including Dr. Rama Gopinath, a postdoctoral fellow in my laboratory. We therefore asked Dr. Edmond Fischer of the Department of Biochemistry for a small
sample of this PDE activator (named calmodulin and reviewed shortly thereafter
by Cheung [26]) protein which he kindly supplied along with several related
proteins. It only took a few experiments to convince ourselves that this activator
protein was a potent and selective activator of the Ca pump ATPase. About the
time we were ready to submit this observation to Biochemical and Biophysical
Research Communications (BBRC), I got a call from John Penniston. He and
his student, Harry Jarrett, were also ready to submit findings on the binding of
CaM to the Ca pump ATPase. We agreed that we would request of the editors
of BBRC that if both papers were acceptable, then we would prefer to have

them published back-to-back. The result was a pair of articles that first called attention in the RBC field to a potential role of CaM in RBC physiology [27,28].

From that time Penniston's lab tended to pursue biochemical features of CaM and its interaction with the Ca pump. Jarrett later published further elegant work on the isolation of CaM from human RBCs [29]. By contrast, my laboratory tended to pursue what I consider somewhat physiological features of the interaction. Thus, we shortly thereafter reported that CaM increases the rate of membrane Ca transport [30]. But it is quite clear that the Ca pump can operate in the absence of CaM [14], and the significance of CaM in regulation of the pump in vivo is not clear to this author. We presented "ballpark" calculations in support of the suggestion that CaM could be considered a subunit of the pump in vivo [31]. On the other hand, Scharff and Foder [32] suggested that CaM is not bound to the Ca pump ATPase under normal physiological conditions in the human RBC. Their data were interpreted to mean that a significant increase in intracellular Ca results in relatively slow binding of the CaM $(Ca^{2+})_n$ complex to the enzyme. This results in hysteretic activation and deactivation of the pump, according to Scharff and Foder [32]. It seems fair to suggest that in cells undergoing extreme Ca loading, CaM becomes involved in binding of Ca and activation of the pump, but just what the role of CaM is under normal physiological conditions remains to be clearly elucidated.

C. Calmodulinomimetic Agents

Negative charge and hydrophobicity seem to be the minimal structural requirements for an agent to mimic CaM. I coined the term "calmodulinomimetic" to describe this activity [33], but for obvious reasons, the term did not catch on. Tokumura et al. [34] studied the structural requirements of lipids to activate the Ca pump ATPase. Results were compatible with the simple interpretation that hydrophobicity and a negative charge are necessary and sufficient features of a calmodulinomimetic. The authors also found that the nature of the hydrocarbon chain as well as the lyso structure were critical to the effect [34]. It could be shown that free fatty acids such as oleic acid stimulate not only the enzyme, but within limits of its tendency to break IOVs, also increase the rate of transport [35].

Activation of the ATPase by acidic phospholipids and related materials was known for many years [eg., 36,37], but it was not immediately appreciated that such activation could mimic (and even mask) the activation by CaM. This fact obscured the significance of the early work of Wolf, who used acidic phospholipids to help purify the ATPase. The enzyme retained high activity, but displayed no CaM activation [38]. Thus, it was not quite clear that the real pump enzyme had been isolated.

IV. ANTICALMODULIN AGENTS AS INHIBITORS OF THE CALCIUM PUMP

A. Many Types of Anticalmodulin Drugs

Raess and Vincenzi [39] showed that trifluoperazine (TFP) and certain other antipsychotic agents were effective antagonists of CaM activation of the Ca pump ATPase, but the notion that there was a relationship to antipsychotic activity was not supported. In 1982 we reported that a wide range of pharmacologic agents all acted as somewhat selective anti-CaM agents, although the antagonism appeared to be more apparently noncompetitive than competitive, as might have been expected [40]. Among the agents that we found to be anti-CaM was dibucaine, the local anesthetic agent that Volpi et al. [41] had reported to be anti-CaM, and on which basis they suggested that local anesthetic activity might be related to antagonism of CaM. This suggestion, which was similar to the earlier suggestion of Levin and Weiss [42] that antipsychotic activity of TFP and related compounds was related to their anti-CaM properties, we found to be too nonspecific considering the extremely broad range (pharmacologically speaking) of agents expressing anti-CaM activity [40].

A wide variety of drugs bind to CaM, especially in the presence of Ca. Beginning with the classical works of Wolff and Brostrom [43] and Levin and Weiss [42], it became apparent that phenothiazines, such as TFP, selectively antagonized certain CaM-dependent enzymes. These preliminary insights now have been expanded to show that a wide range of amphipathic cations, including Ca entry blockers [44,45] and tetracyclines [45] can bind to and antagonize the effects of CaM in many systems, including the RBC Ca pump ATPase [46,47]. Also, while recognizing explicitly the severe limitations of an agent with so many nonspecific effects, one can nevertheless obtain evidence using TFP that is consistent with inhibition of the Ca pump in intact RBCs loaded with Ca^{2+} [48].

Minocherhomjee and Roufogalis [49] provided another reminder of the non-specificity of drugs. They found PDE inhibition by nifedipine and related compounds, but inhibitory potencies did not correlate with potencies to inhibit Ca channels. It was concluded that pharmacological actions of nifedipine-related Ca entry blockers are unlikely to be due to intracellular inhibition of CaM at concentrations normally used to block Ca entry. This is not surprising because it has been known for some time that anti-CaM agents include a wide variety of pharmacological agents, but that there is little or no correlation between clinical efficacy and anti-CaM activity [40,50]. In any event, these and other data support the notion that, in addition to binding CaM and/or the Ca pump ATPase, such drugs bind to other sites.

B. The Mechanisms of Anticalmodulin Drugs

Most anti-CaM agents are thought to prevent the binding of the $CaM(Ca^{2+})_n$ complex to its various effectors. Roufogalis [51] noted that anti-CaM drugs do not antagonize Ca binding to CaM. They could bind to the $CaM(Ca^{2+})_n$ complex and/or to the enzyme in question. It was suggested that evidence favors the interpretation that most such agents interact with CaM and not the enzyme. For a somewhat different view see Vincenzi et al. [33].

The findings of Roufogalis et al. [52] and others fit with the notion that phenothiazines and antipsychotics bind to CaM, probably at two hydrophobic sites that are Ca dependent [53]. These are apparently not stereospecific sites, and TFP is not stereospecific. There is also evidence that TFP, in addition to binding to CaM, binds to the activated states of CaM sensitive enzymes, in particular the Ca pump ATPase. Rather than thinking of drug sites of action being located on either CaM or its binding sites one may have to consider the more holistic enzyme/environment as the drug "receptor."

C. Compound 48/80 as an Inhibitor

Compound 48/80 is a mixture of polycationic polymers with average molecular weight of ~1300, which Gietzen et al. [54] reported to be selective and potent CaM antagonist. That is, compound 48/80 appears to not inhibit the basal (CaM independent) Ca pump ATPase activity. Presumably, compound 48/80 is a more selective inhibitor of CaM activation of the RBC Ca pump ATPase of isolated RBC membranes that drugs such as TFP, because it enters the lipid environment of the membrane to a lesser extent. The same property probably also limits the usefulness of compound 48/80 as an inhibitor of CaM in intact cells. Although the effects reported by Gietzen et al. [54] are quite convincing, it may be noted that inhibition of the Ca pump ATPase by compound 48/80 may not be entirely simple and straightforward. Thus, we recently found that a fraction of compound 48/80 separated by CaM affinity chromatography was a potent inhibitor of both the CaM activated and basal Ca pump ATPase activities of RBC membranes [55]. The fact that this fraction was equipotent as an inhibitor of the basal and CaM-activated ATPases, respectively, shows that its inhibition cannot be accounted for on the basis of CaM antagonism alone.

D. Calmodulin as a Regulator of Red Blood Cell Shape?

As mentioned above, it is clear that the $CaM(Ca^{2+})_n$ complex binds to the Ca pump Ca pump ATPases with high affinity, much higher than it binds to other membrane proteins, for example, spectrin [56,57]. The idea that CaM binds to RBC membrane proteins other than the Ca pump ATPase was extended by Nelson et al. [58]. These authors advanced the idea that CaM is a potential regulator of RBC shape. The idea is based on the observation that, almost without exception, anti-CaM drugs (which are amphipathic cations) are all "cup formers," whereas "crenators" of RBCs are neutral or anionic compounds (and not CaM antagonists). In our opinion the evidence for this point of view is not entirely straightforward. It was argued that CaM antagonists favor cup morphology by preserving ATP (by inhibiting an ATP using enzyme?), and it was argued that drugs bind to $CaM(Ca^{2+})_n$ and prevent Ca release, keeping Ca low. The fact that the affinity of CaM for spectrin is some 1000-fold less than for the Ca pump ATPase was not emphasized.

V. THE CALCIUM PUMP IN MANY CELLS WITH OTHER PUMPS

A. Other Cells

I will not attempt to cite all the different cell types in which PM Ca transport has been sought and found. Reference to one of several reviews or monographs [16,59–62] will guide the reader to such information. Now the PM Ca pump ATPase is so well accepted that it is stated as fact without references [e.g., 63]. Suffice to say that the RBC appears to have served as a useful model of cellular biology in this respect. The ubiquity of the PM Ca pump can be appreciated by noting that a CaM-activated Ca Pump ATPase exists in plant [64] as well as animal cells.

B. Other Ion Pumps

It seems likely to this author that when the structure of the Ca pump ATPase is determined it will bear close relationship to the Na-K ATPase and possibly to the (SR) Ca pump ATPase [65]. On the other hand, the evidence presently available does not support the idea of a molecular relationship between the PM and SR pumps. Antibodies against the PM Ca pump ATPase do not cross-react with the SR enzyme [21]. It may be, for example, that the SR Ca pump ($\sim$100 kDa) evolved as a smaller molecule lacking (or having lose?) the CaM binding region (present on the $\sim$140 kDa PM Ca pump). Phospholamban, a 22 [66] or 11 [67] kDa protein in heart SR membranes, increases the activity of the SR Ca pump [66]. The activation by phospholamban depends on its phosphorylation by a CaM-dependent or cyclic adenosine monophosphate-dependent kinase [66, 67]. It is as if CaM activation of the SR Ca pump became separated from the main pumping part of the molecule. Possibly, because the concentration of Ca pump ATPase in SR membranes is so high, the extra activity that could be imparted by CaM was not necessary (except, perhaps, in the heart, which must move substantial quantities of Ca with each beat [68].

It seems reasonable to assume until shown otherwise (an event likely to occur in the near future considering the cloning and sequencing efforts of several laboratories) that the Ca pump ATPase and the Na-K pump ATPase are related. I assume that the Ca pump ATPase evolved early in evolution, partly for reasons so well articulated by Kretsinger [69]. Paraphrased, these postulates inclde that all resting eukaryotic cells maintain free Ca between 10^{-8} and 10^{-7} M;

that the function of Ca in the cytosol is to transmit information, that the information is received by proteins containing the so-called "EF-hand," structure and that cells initially extruded Ca so they could use phosphate as their basic energy currency [i.e., $Ca_3(PO_4)_2$ is insoluble]. The fact that some RBCs (such as the dog RBC) exist without Na-K transport, but, to our knowledge, none exist without maintaining extremely low intracellular Ca, tends to support this point of view. Another set of observations also are in agreement with this point of view: That is, it appears that accumulation of intracellular Ca may be a final common pathway in most if not all [70] cell injury and death [71,72].

VI. ISOLATION OF THE CALCIUM PUMP ATPase AND ITS SEQUELAE

A. Isolation: A Dead Heat?

After it was clearly demonstrated that CaM interacts with the Ca pump ATPase, Penniston began an extremely fruitful collaboration with Carafoli and co-workers in Zurich, Switzerland. From that collaboration came the application of CaM affinity chromatography for the isolation of the Ca pump ATPase protein in pure form [37,73]. Similar work was carried out independently and at almost exactly the same time (personal accounts by the authors differ somewhat) by Gietzen et al. in Ulm, West Germany. Importantly, the work of both groups depended not only on CaM affinity chromatography but also on early work of Wolf, who had pioneered the use of more traditional biochemical techniques in the purification of the ATPase from RBC membranes [38]. Solubilization of the intrinsic membrane protein was achieved with Triton X-100 [75] and with deoxycholate when stabilized with Tween 20 [74].

Carafoli and his co-workers isolated [37,73] and reconstituted the enzyme [37] and showed that it is activated by CaM when the enzyme is suspended in neutral phospholipids, but that CaM does not activate the ATPase when it is suspended in an acidic phospholipid such as phosphatidylserine [37,76]. The latter fact was also a critical insight appreciated by Gietzen et al. [74], who independently isolated the pump by CaM affinity chromatography. Reconstitution of the Ca pump was also performed by Haaker and Racker [77] and Gietzen et al. [78].

B. Proteolysis

In an excellent series of papers, Carafoli and co-workers have, by the use of the isolated enzyme, elucidated a number of important features of the Ca pump ATPase. For example, they have used proteolysis to examine its molecular structure [79]. For detailed description of the pattern of proteolysis, the reader is referred to the original work. The important message emerging from these papers is that partial proteolysis (by trypsin) of the Ca pump ATPase results in a decrease in molecular weight from ~138,000 to ~90,000. Associated with this partial proteolysis, the enzyme activity increases, and at the same time loses its CaM dependence. The suggestion is that the CaM binding region of the enzyme normally exhibits some inhibitory influence on the enzyme, unless CaM is bound to it. On the other hand, if the CaM binding sequence is removed by partial proteolysis, then the enzyme becomes "disinhibited," similar to its CaM activated state. Similar changes have been observed in other CaM dependent enzymes. Evidence that an active CaM binding region is removed by partial trypsinization has been put forth by Emelyanenko et al; [80].

Enzymes other than trypsin have been used to examine the pump ATPase. It is known that proteolysis with chymotrypsin results in rapid loss of activity and that treatment of the isolated enzyme with calpain results in loss of a 24,000-kDa fragment without loss of CaM dependence [79].

VII. THE STATUS OF THE PLASMA MEMBRANE CALCIUM PUMP IN THE RED BLOOD CELL

A. Calcium Pump Subunit Structure

Although detailed analysis of the isolated Ca pump ATPase is important, so too is consideration of the state of the pump as it exists in the RBC membrane. At least two groups have, by quite different approaches, come to the conclusion that the Ca pump exists as a dimer in the membrane in vivo. Hinds and Andreasen [56] noted the 1:1 stoichiometry of the interaction of the ~150 kDa ATPase with CaM. They also noted the positive cooperativity of the activation of the ATPase by CaM. A simple interpretation is that the binding of one CaM to one of two identical subunits of a dimeric functional pumping unit facilitates the binding of CaM to the second subunit. Of course, other explanations of positive cooperativity are possible, but further supportive evidence was obtained when the azido[125]I-CaM-labeled ATPase was solubilized in Triton X-100. The apparent molecular weight of the solubilized material using nondenaturing gels was ~470,000. Subtracting 100,000 for the weight of the micelle alone left ~370,000 for the weight of the nondenatured protein complex. If it is assumed that the molecular weight of the nomomeric CaM-ATPase complex is 170,000, then this is consistent with the inclusion of two ATPase subunits, but not three or more [81]. These data are thus consistent with, but do not prove, that the Ca pump ATPase functional unit is a homodimer.

By a completely different approach Minocherhomjee et al. [82] concluded that the Ca pump ATPase exists as a dimer. These authors employed radiation inactivation, a technique based on the average apparent target size, for inactivation of biological activity by ionizing radiation.

B. The Number of Calcium Pumps per Cell

We described a photoactivated marker CaM, azido-^{125}I-CaM, which reacts co-valently and almost exclusively with the Ca pump ATPase in isolated RBC membranes [56,83] with a 1:1 stoichiometry. That is, in the presence of ultraviolet light one photoaffinity probe molecule (i.e., azido-^{125}I-CaM) covalently cross-links with one ATPase of ~150 kDa. I will refer to the ATPase as having a sub-unit molecular weight of 150 because that is the value we have obtained in our laboratory. Other investigators report that the molecular weight as somewhat less. For example, Carafoli and co-workers [79] report that the ATPase is 138 kDa. Because the crosslinking is irreversible, this results in irreversible activation of the ATPase. Although the crosslinking reaction is less than 100% efficient, the reaction does not destroy the activation of any nonreacted sites. Thus, addition of unlabeled CaM produces "full activation," even in membranes treated with the photoactivated probe. This happy constellation of events allowed Hinds and Andreasen [56] to estimate the number of Ca pump ATPase units per RBC.

Based on the assumption that azido-^{125}I-CaM is as effective as CaM when bound to the enzyme, the number of 150 kDa ATPase units per RBC was esti-mated to be ~1600–2000. This is the first good estimate of the number of sites per RBC, in this author's opinion. Earlier attempts at quantification of the num-ber of sites was based on phosphorylation. Drickamer [84], for example, con-cluded that there were 400 Ca pump ATPases per RBC. These results, although discordant, are at least within the same order of magnitude. It must be noted that the problem is much more difficult than quantification of the number of Na-K pump sites per RBC, which has for years been based on the selective and high-affinity binding of labeled ouabain, and is generally agreed as being equal to about 200 sites per RBC. As noted above, there is a pressing need for a potent and specific inhibitor of the PM Ca pump that could serve quantification and other roles played by ouabain in elucidation of the Na-K active transport system.

Taken together, these results may be interpreted to suggest that ~1000 homodimeric Ca pump sites are present in the normal RBC. By comparison it may be noted that it is generally accepted that there are ~200 Na-K pump ATPase sites per RBC. Considering the average specific activity of the fully activated Ca pump ATPase (~50–70 nmol · min^{-1} mg^{-2} in our laboratory) in comparison to the Na-K pump ATPase (5–7 nmol · min^{-1} mg^{-1}) it is reasonable to conclude that the turnover numbers of the two ATPases are comparable.

C. Regulation of the Plasma Membrane Calcium Pump in Situ

What really regulates the Ca pump in situ? It is not CaM (which is in great ex-cess). It is not ATP (also in great excess). It is almost certainly Ca at the inner surface of the PM that regulates the rate of the Ca pump. How does the Ca get to the inner surface? Mainly by influx. In most PM systems it is fashionable to consider Ca channels as the major source of Ca influx. Although binding sites for certain surface active agents may be found on RBCs by binding assays, there is little evidence that they contain active, receptor-coupled Ca channels. Based on the low membrane potential of the RBC ($\sim$9 mV negative inside [85]), it is also unlikely that the RBC contains (functional) voltage-operated Ca channels. Varecka and Carafoli [86] reported that verapamil inhibits Ca influx in vanadate-treated RBCs. Because half maximal inhibition occurs only at $\sim$100 μM, it seems unlikely that the effect proves the existence of Ca channels in RBCs.

Does Ca even get into the RBC? The maximal activity of the Ca pump ATPase has been considered in earlier reviews [31,87]. We have been impressed that the capacity of the pump is enormous compared with the passive transport of Ca. Based on figures from a number of laboratories it appears that the passive flux of Ca across the human RBC membrane in vitro is $\sim$17–170 nmol $\cdot$ min^{-1} L^{-1} RBC. The capacity of the active Ca pump is on the order of 83,000–600,000 nmol $\cdot$ min^{-1} L^{-1} RBC [88]. From numbers summarized by Rega and Garrahan [16] it can be estimated that the Ca pump of human RBCs runs at <1% of its capacity, at least in vitro.

D. Mechanically Induced Calcium Permeability

Based on these considerations, a speculative model of physiological regulation of the RBC Ca pump has been presented [89]. It is suggested that the most significant inward leak of Ca into RBCs is stimulated by mechanical deformation of RBCs as they pass through the microcirculation. Evidence has been put forth to support the idea that both Ca [90] and Na [90,91] permeabilities are increased on mechanical deformation of RBCs in vitro. Thus, it is suggested that each pass through the microcirculation results in a transient but significant increase in passive Ca influx in RBCs. Whether this constitutes a "physiological Ca signal" (as is a transient increase in intracellular Ca in other cells) or is merely a side effect of the circulatory status of the RBC remains to be determined. In any event, the high capacity of the RBC Ca pump appears well suited to carry out rapid clearing of Ca gained on microcirculatory (or other) stimulated Ca influx. Accumulation of Ca in cells can initiate cell injury and/or death [71,72] and the high capacity of the Ca pump therefore probably serves to limit the damage associated with the microcirculatory-induced Ca fluxes.

E. An Exception to Test the Rule

In general, it appears that PM Ca pump ATPases are capable of being activated by CaM. One apparent exception is the Ca pump ATPase of dog RBCs. We have suspected the Ca-activated protease, calpain [92,93], as the culprit in this phenomenon. On the other hand, this suggestion does not match the qualitative effects of calpain on the isolated Ca pump ATPase [79]. Carafoli et al. found that calpain caused significant reduction of molecular weight of the ATPase, apparently by a single proteolytic "clip," but no loss of CaM activation. Further work is needed to resolve this question.

Another apparent exception to the CaM dependence of PM Ca pumps has been pursued in our laboratory. Isolated membranes from dog RBCs express low Ca-activated ATPase activity [94]. In fact, the activity is so low that Parker suggested that the Ca pump ATPase of dog RBCs might be "utterly dependent" on CaM [95]. This was such an intriguing suggestion that we pursued it and (at least at first) were very surprised by the results. Thus, we confirmed the low specific activity of the Ca pump ATPase in isolated dog RBC membranes, but also found that the activity is not increased by addition of CaM [96]. We have suggested that proteolysis during membrane isolation also causes loss of CaM dependence of the pump enzyme in dog RBC membranes, because the pump exhibits CaM dependence in intact cells [48].

F.   Assay of Calcium Pump ATPase in Intact Cells

According to the idea considered above, regulation of the Ca pump is via "leak" (i.e., influx of Ca [87,90]). Recent experiments in our laboratory tend to support this view and consideration of these may provide additional insight into the workings of the Ca pump. These experiments are based on assaying the activity of the Ca pump ATPase in intact RBCs [48]. Development of this assay was prompted by apparently anomalous behavior of dog RBCs. When isolated membranes are prepared from dog RBCs, they express little or no Ca pump ATPase activity [94,97] and the activity is not activated by CaM [96,98]. Because this behavior is different from other species, we attempted to assay the Ca pump ATPase without isolating the membrane. Our approach was based on the use of the ionophore, A23187. A23187 increases the passive flux of Ca across membranes. If the RBCs contain an active Ca pump ATPase, then such a pump would be expected to respond to increased Ca influx by hydrolyzing ATP and extruding Ca (which then leaks back in in the presence of the ionophore). Essentially, A23187 should "uncouple" the Ca pump ATPase.

It had been known that ATP breaks down fairly rapidly in RBCs exposed to A23187 [99–101] but the rate of breakdown had not been quantified. We quantified ATP breakdown in human RBCs in pilot experiments and found that if iodoacetic acid (IAA) is included to prevent ATP resynthesis, then after the addition of A23187 the amount of ATP decreases rapidly, initially according to first-order kinetics. In addition to 1 mM IAA, the incubation normally contains N-2-hydroxyethylpiperazine-N'-2-ethanesulfonic acid 20 mM, pH 7.4; KCl, 120 mM, MgCl$_2$, 2 mM, CaCl$_2$, 0.1 mM. First-order kinetics are followed very closely for up to 12 min, or during the loss of up to ~90% of the original ATP content of the RBCs. In a typical human RBC sample the rate constant of the first-order process is ~ -0.2 min^{-1}. This is equivalent to a half-life of ATP of ~ 3.5 min. In intact RBCs exposed to A23187 in the absence of Ca, the amount of ATP is quite stable [48]. Adenosine triphosphate is also stable in the absence of added ionophore. Thus, in intact cells the major ATP-consuming reaction under the conditions of this assay is a Ca- and flux-dependent process, that is, the Ca pump ATPase.

We applied this same approach to the study of the Ca pump ATPase in intact dog RBCs. In dog RBCs under such assay conditions, the typical rate constant is -0.094 min^{-1}, equivalent to a half-life of ATP of ~ 7.4 min. Thus, in the intact RBC assay, dog RBCs exhibit a Ca pump ATPase capacity that is approximately half that of human RBCs.

It was shown that the Ca pump ATPase activity of intact dog RBCs is sensitive to inhibition by TFP [48]. These data are compatible with, but do not prove, that the dog ATPase (like that of human RBCs) is activated by CaM. The apparently anomalous behavior of dog RBCs is thus not inherent in the Ca pump ATPase, but emerges as an artifact of membrane isolation. The mechanism of this artifactual change in the activity of the Ca pump ATPase during membrane preparation is currently under investigation in our laboratory.

Several features of the assay of the Ca pump ATPase of intact RBCs are noteworthy. First, when exposed to A23187 in buffered NaCl containing Ca, human RBCs consume ATP, but they also undergo considerable shrinkage (Vincenzi et al., unpublished data). This shrinkage is presumably due to accumulation of intracellular Ca, activation of the Gardos channel (Ca-activated K conductance), and the movement of K down its electrochemical gradient. This results in the loss of K and water from the cells. To prevent shrinkage of human RBCs exposed to A23187, further experiments (such as those reported here) were carried out in media containing KCl as the major osmotic constituent.

Another effect was noted early in the development of this assay. That is, there was an initial burst of activity after addition of ionophore, and then the ATPase activity slowed considerably. This was prevented by the inclusion in the medium of 2 mM MgCl$_2$. Apparently, when exposed to A23187, RBCs not only gain Ca but they also lose Mg [102]. This is because A23187 is not a Ca-specific ionophore. Thus, although it can promote the entry of Ca, it can also promote the loss of Mg. Loss of Mg can be prevented by inclusion of Mg in the medium. Because the Ca pump ATPase is actually a (Ca^{2+} Mg^{2+})-ATPase, it is important to maintain intracellular levels of Mg for maximal expression of its activity. Although we have not optimized the levels of various constituents in the incubation medium, the assay is sufficiently well developed that it can be used to estimate the ATPase activity of intact cells on a comparative basis. This has been done recently in a hemolytic anemia of beagle dogs. The ATPase activity of the anemic beagle is approximately half that of normal beagles [103].

G. Change in Calcium Pump ATPase Activity Associated With Red Blood Cell Aging

Another useful feature of the assay of the Ca pump ATPase in intact RBCs is that the assay can be performed on a very small quantity of cells (usually 10–20 μL of packed RBCs/ml incubation medium). As little as 5 μL in a 0.5 mL incubation has been used with success. Because the assay of the Ca pump ATPase in intact RBCs uses such a small quantity of cells we were able to measure the activities in RBCs fractionated by density in a microhematocrit tube. Approximately 60 μL of packed RBCs were centrifuged for 30 min in a microhematocrit capillary tube. Based on results of cell fractionation with phthalate oils of different densities [104], it is known that RBCs distribute according to their densities in a microhematocrit tube. The top and bottom 5 μL of packed cells were removed with a positive displacement pipet. The least (top) and most (bottom) dense RBCs exhibited respective k values of -0.195 and -0.169 min^{-1} (Vincenzi et al., unpublished data). The data demonstrate the short half-life of ATP in human RBCs exposed to A23187 and Ca and reasonably good fit to a first-order process of the loss of ATP. In addition, it is shown that the half-life of ATP under such conditions is ~15% shorter in less dense (and presumably younger) RBCs than in more dense (and presumably older) RBCs. An inference that may be drawn from these data is that there is a modest but significant loss of maximal Ca pump ATPase associated with aging of the human RBC. Whether such loss of activity is causally related to the senescence of the most dense cells remains to be determined. It has been recognized for some years that the accumulation of Ca in cells reduces their deformability [105] and it is generally thought that decreased deformability is a signal for termination of RBC life, but whether Ca accumulates to a significant extent in RBCs with a modestly impaired Ca pump is unknown. Considering the immense capacity of the Ca pump [87], this simple inference, attractive as it may be, seems unlikely. It should be noted that in contrast to the present results with intact RBCs, membrane isolated from density-separated RBCs showed no change in Ca pump ATPase actively associated with cell aging [106].

H. Some Teleology

The RBC does not require "Ca signals" as a trigger for the basis for muscular contraction, release of neurotransmitters, etc. In fact, it is not really clear whether there is a "physiological" response to a transient increase in intracellular Ca in RBCs as there is in essentially all other cells. There are a number of clearly detrimental responses to a sustained increase in intracellular Ca in the RBC. These include K and water loss, increased passive permeability to Na as well as K, reported decreases in RBC deformability [105], and even membrane protein crosslinking [107].

One of the "reasons" we initially offered for the existence of the Ca pump probably does not apply universally. Thus, it was suggested that intracellular Ca must be maintained at low levels so as not to inhibit the Na-K pump ATPase [108]. It had been known that Ca inhibits the Na-K pump ATPase and we found the K_i to be $\sim 10^{-4}$ M at a site presumed to be on the inner surface of the membrane. Now that we understand that Na transport is not necessary for RBC viability (e.g., dog RBCs have no Na-K pump [95;109], this suggestion might be applied only to suggest that low intracellular Ca is permissive of active Na-K transport in those cells that contain the Na-K pump.

So, like all cells, RBCs maintain low intracellular Ca. It is clear that prolonged elevations in intracellular Ca are detrimental to RBCs and to other cells. Thus, the RBC Ca pump, and the condition it brings about and insures, appears to be a useful model in cell biology in general. Irrespective of the molecular approaches that nature employs the overall goal seems clear. That is, all living cells maintain low intracellular Ca, at least in their normal state [72]. When they do not, possibly irreversible cell damage is initiated. As in other cells, there are a number of detrimental changes in RBCs when they accumulate intracellular Ca above "normal" levels, including possible proteolysis [92,93] as well as protein crosslinking [107]. In this sense, the RBC has served as an excellent model of other cells. Among the most sensitive changes is an increase in passive K conductance, the so-called "Gardos effect" [110,111]. Under physiological conditions, this effect results in loss of K and water and decreased RBC deformability. The latter effect may be a signal for premature destruction of RBCs in certain hemolytic anemias. One of the most characteristic features of a normal RBC is its high deformability. Even at relatively low concentrations, accumulation of intracellular Ca results in decreased deformability, a "pathological" response of the cell. This appears to be a general feature of intracellular Ca. Low concentrations and/or transient increases elicit "physiological" responses in cells, but prolonged elevations of intracellular Ca result in cellular injury and/or death [72]. The mechanisms by which cell injury occur are not yet elucidated, although some intriguing results are available [70]. Whether transient elevations of intracellular Ca result in "physiological" responses in RBCs of some or all species is a question this author has considered for a long time, with no satisfactory answer.

ACKNOWLEDGMENTS

Research in my laboratory has been supported in part by USPHS Grant AM-16436 and by Grants from the Cystic Fibrosis Foundation, the American Diabetes Association, and the National Dairy Board administered in cooperation with the National Dairy Council. I thank many colleagues and competitors for their stimulating discussions over the years. Particular thanks must be extended to Dr. Thomas R. Hinds for his spirited insights which, among other advances in our laboratory, led to the development of the assay of the Ca pump ATPase in intact cells, as presented herein. Professor Ernesto Carafoli and his colleague, Dr. Joachim Krebs, kindly supplied and discussed a preprint of their work on the effects of calpain on the isolated Ca pump ATPase.

REFERENCES

1. Schatzmann, H. J. (1953). Herzglykoside als Hemmstoffe fuer den Aktiven Kalium-und Natriumtransport durch die Erythrozyteenmembran. *Helv. Physiol. Pharmacol. Acta* 11:346–354.
2. Lee, K. S., and Klaus, W. (1971). The subcellular basis for the mechanism of inotropic action of cardiac glycosides. *Pharmacol. Rev.* 23:193–261.
3. Akera, T. (1986). The function of Na^+, K^+-ATPase and its importance for drug action. In *Cardiac Glycosides 1785-1985*, E. Erdman, K. Greeff, and J. C. Skou (Eds.). Steinkopff Verlag, Darmstadt, F. R. G., pp. 19–25.
4. Passow, H. (1961). Zusammenwirken von Membranstruktur und Zellstoffwechsel bein der Regulierung der Ionenpermeabilitaet roter Blutkoerperchen. *Colloq. Ges. Physiol. Chem.* 12:54–55.
5. Passow, H. (1963). Metabolic control of passive cation permeability in human red blood cells. In *Cell Interphase Reactions*, H. D. Brown (Ed.). Scholars Library, New York, pp. 57–107.
6. Vincenzi, F. F. (1967). Influence of myocardial activity on the rate of onset of ouabain action. *J. Pharmacol. Exper. Ther.* 155:279–287.
7. Schatzmann, H. J. (1966). ATP-dependent Ca^{++} extrusion from human red cells. *Experientia* 22:364–365.
8. Dunham, E. T., and Glynn, I. M. (1961). Adenosinetriphosphatase activity and active movements of alkali metal ions. *J. Physiol. (Lond.)* 156:274–293.
9. Vincenzi, F. F., and Schatzmann, H. J. (1967). Some properties of Ca-activated ATPase in human red cells. *Helv. Physiol. Pharmacol. Acta* 25: CR233–CR234.
10. Schatzmann, H. J., and Vincenzi, F. F. (1969). Calcium movements across the membrane of human red cells. *J. Physiol. (Lond.)* 201:369–395.
11. Lee, K. S., and Shin, B. C. (1969). Studies on the active transport of calcium in human red cells. *J. Gen. Physiol.* 54:713–729.
12. Olson, E. J., and Cazort, R. J. (1969). Active calcium and strontium transport in human erythrocyte ghosts. *J. Gen. Physiol.* 63:590–600.
13. Cha, Y. N., Shin, B. C., and Lee, K. S. (1970). Ca^{++} uptake and Ca^{++} stimulated ATPase of red blood cell membrane fragments (abstract). *Pharmacologist* 12:241.
14. Hinds, T. R., and Vincenzi, F. F. (1983). The red blood cell as a model for calmodulin-dependent Ca^{2+} transport. In *Methods in Enzymology, Volume 102, Hormone Action, Part G, Calmodulin and Calcium-Binding Proteins*. A. R. Means and B. W. O'Malley (Eds.). Academic Press, New York, pp. 47–62.
15. Vincenzi, F. F. (1968). The calcium pump of erythrocyte membrane and its inhibition by ethacrynic acid. *Proc. West. Pharmacol. Soc.* 11:58–60.

16. Rega, A. F., and Garrahan, P. J. (1986). *The Ca^{2+} Pump of Plasma Membranes*. CRC Press, Boca Raton, Florida.

17. Hebbel, R. P. (1986). Erythrocyte antioxidants and membrane vulnerability. *J. Lab. Clin. Med.* 107:401–404.

18. Larsen, F. L., Hinds, T. R., and Vincenzi, F. F. (1978). On the red blood cell Ca^{2+} pump: an estimate of stoichiometry. *J. Membr. Biol.* 41:361–376.

19. Quist, E. E., and Roufogalis, B. D. (1975). Determination of the stoichiometry of the calcium pump in human erythrocytes using lanthanum as a selective inhibitor. *FEBS Lett.* 50:135–139.

20. Sarkadi, B. (1980). Active calcium transport in human red cells. *Biochim. Biophys. Acta* 604:159–190.

21. Verma, A. K., Penniston, J. T., Muallem, S., and Lew, V. L. (1984). Effects of affinity-purified antibodies on the Ca^{2+} pumping ATPase of erythrocyte membranes. *J. Bioenerg. Biomembr.* 16:365–378.

22. Bond, G. H., and Clough, D. L. (1973). A soluble protein activator of $(Mg^{2+} + Ca^{2+})$-dependent ATPase of human red blood cells. *Biochim. Biophys. Acta* 323:592–599.

23. Farrance, M. L., and Vincenzi, F. F. (1977). Enhancement of $(Ca^{2+} + Mg^{2+})$-ATPase activity of human erythrocyte membranes by hemolysis in isosmotic imidazole buffer. I. General properties of variously prepared membranes and the mechanism of the isosmotic imidazole effect. *Biochim. Biophys. Acta* 471:49–58.

24. Farrance, M. L., and Vincenzi, F. F. (1977). Enhancement of $(Ca^{2+} + Mg^{2+})$-ATPase activity of human erythrocyte membranes by hemolysis in isosmotic imidazole buffer. II. Dependence on calcium and a cytoplasmic activator. *Biochim. Biophys. Acta* 471:59–66.

25. Vincenzi, F. F., and Farrance, M. L. (1977). Interaction between cytoplasmic $(Ca^{2+} + Mg^{2+})$-ATPase activator and the erythrocyte membrane. *J. Supramolec. Struct.* 7:301–306.

26. Cheung, W. Y. (1979). Calmodulin plays a pivotal role in cellular regulation. *Science* 207:19–27.

27. Gopinath, R. M., and Vincenzi, F. F. (1977). Phosphodiesterase protein activator mimics red blood cell cytoplasmic activator of $(Ca^{2+} + Mg^{2+})$-ATPase. *Biochem. Biophys. Res. Comm.* 77:1203–1209.

28. Jarrett, H. W., and Penniston, J. T. (1977). Partial purification of the $Ca^{2+} - Mg^{2+}$ ATPase activator from human erythrocytes: its similarity to the activator of 3':5'-cyclic nucleotide phosphodiesterase. *Biochem. Biophys. Res. Commun.* 77:1210–1216.

29. Jarrett, H. W., and Kyte, J. (1979). Human erythrocyte calmodulin. *J. Biol. Chem.* 254:8237–8244.

30. Larsen, F. L., and Vincenzi, F. F. (1979). Ca^{2+} transport across the plasma membrane: stimulation by calmodulin. *Science* 204:306–309.

31. Vinzenzi, F. F., and Hinds, T. R. (1980). Calmodulin and Plasma Membrane Calcium Transport. In *Calcium and Cell Function*, Vol 1, W. Y. Cheung (Ed.). Academic Press, New York, pp. 127–165.

32. Scharff, O., and Foder, B. (1982). Rate constants for calmodulin binding to Ca^{2+}-ATPases in erythrocyte membranes. *Biochim. Biophys. Acta* 691: 133–143.

33. Vincenzi, F. F., Adunyah, E. S., Niggli, V., and Carafoli, E. (1982). Purified red blood cell Ca^{2+}-pump ATPase: evidence for direct inhibition of by presumed anti-calmodulin drugs in the absence of calmodulin. *Cell Calcium* 3:545–559.

34. Tokumura, A., Mostafa, M. H., Nelson, D. R., and Hanahan, D. J. (1985). Stimulation of $(Ca^{2+} + Mg^{2+})$-ATPase activity in human erythrocyte membranes by synthetic lysophosphatidic acids and lysophosphatidylcholines. Effects of chain length and degree of unsaturation of the fatty acid groups. *Biochim. Biophys. Acta* 812:568–574.

35. Pine, R. W., Vincenzi, F. F., and Carrico, C. J. (1982). Apparent inhibition of the plasma membrane Ca^{2+} pump by oleic acid. *J. Trauma* 23:366–371.

36. Ronner, P., Gazzotti, P., and Carafoli, E. (1977). A lipid requirement for the $(Ca^{2+} + Mg^{2+})$-activated ATPase of erythrocyte membranes. *Arch. Biochem. Biophys.* 179:578–583.

37. Niggli, V., Adunyah, E. S., Penniston, J. T., and Carafoli, E. (1981). Acidic phospholipids, unsaturated fatty acids, and limited proteolysis mimic the effect of calmodulin on the purified erythrocyte Ca^{2+}-ATPase. *J. Biol. Chem.* 256:8858–8592.

38. Wolf, H. U., Diekvoss, G., and Lichtner, R. (1977). Purification and properties of high affinity Ca^{2+} ATPase of human erythrocyte membranes. *Acta Biol. Med. Ger.* 36:847–858.

39. Raess, B. U., and Vincenzi, F. F. (1980). Calmodulin activation of red blood cell $(Ca^{2+} + Mg^{2+})$-ATPase and its antagonism by phenothazines. *Mol. Pharmacol.* 18:253–258.

40. Vincenzi, F. F. (1982). The pharmacology of calmodulin antagonism: a reappraisal. In *Calmodulin and Intracellular Ca^{++} Receptors*, S. Kakiuchi, H. Hidaka, and A. R. Means (Eds.). Plenum Press, New York, pp. 1–17.

41. Volpi, M., Sha'afi, R. I., Epstein, P. M., Andrenyak, D. M., and Feinstein, M. B. (1981). Local anesthetics, mepacrine, and propranolol are antagonists of calmodulin. *Proc. Natl. Acad. Sci. USA* 78:795–799.

42. Levin, R. M., and Weiss, B. (1977). Binding of trifluoperazine to the calcium-dependent activator of cyclic nucleotide phosphodiesterase. *Mol. Pharmacol.* 13:690–697.

43. Wolff, D. J., and Brostrom, C. O. (1976). Calcium-dependent cyclic nucleotide phosphodiesterase from brain: identification of phospholipids as calcium-independent activators. *Arch. Biochem. Biophys.* 173:720–731.

44. Bostrom, S.-L., Ljung, B., Mardh, S., Forsen, S., and Thulin, E. (1981). Interaction of the antihypertensive drug felodipine with calmodulin. *Nature* 292:777–778.

45. Schlondorff, D., and Satriano, J. (1985). Interactions with calmodulin: potential mechanism for some inhibitory actions of tetracyclines and calcium channel blockers. *Biochem. Pharmacol.* 34:3391–3393.

46. Vincenzi, F. F. (1981). Calmodulin pharmacology. *Cell Calcium* 2:387–409.

47. Weiss, B., Sellinger-Barnette, M., Winkler, J. D., Schechter, L. E., and Pro-
    zialeck, W. C. (1985). Calmodulin antagonists: structure-activity relation-
    ships. In: *Calmodulin Antagonists and Cellular Physiology*, H. Hidaka, and
    D. J. Hartshorne (Eds.). Academic Press, New York, pp. 45-62.
48. Hinds, T. R., and Vincenzi, F. F. (1986). Evidence for a calmodulin acti-
    vated Ca^{2+} pump ATPase in dog erythrocytes. *Proc. Soc. Exp. Biol. Med.*
    181:542-549.
49. Minocherhomjee, A. M., and Roufogalis, B. D. (1983). Antagonism of cal-
    modulin and phosphodiesterase by nifedipine and related calcium blockers.
    Cell Calcium 5:57-63.
50. Norman, J. A., Drummond, A. H., and Moser, P. (1979). Inhibition of cal-
    cium-dependent regulator-stimulated phosphodiesterase activity by neuro-
    leptic drugs is unrelated to their clinical efficacy. *Mol. Pharmacol.* 16:1089-
    1094.
51. Roufogalis, B. D. (1982). Specificity of trifluoperazine and related pheno-
    thiazines for calcium binding proteins. In *Calcium and Cell Function*, W. Y.
    Cheung (Ed.). Academic Press, New York, pp. 129-159.
52. Roufogalis, B. D., Minocherhomjee, A.-E.-V. M., and Al-Jobore, A. (1983).
    Pharmacological antagonism of calmodulin. *Can. J. Biochem. Cell Biol.* 61:
    927-933.
53. LaPorte, D. D., Wierman, B. W., and Storm, D. R. (1980). Calcium-induced
    exposure of a hydrophobic surface on calmodulin. *Biochemistry* 19:3814-
    3819.
54. Gietzen, K., Adamczyk-Engelmann, A., Wuthrich, A., Konstantinova, A.,
    and Bader, H. (1983). Compound 48/80 is a selective and powerful inhibitor
    of calmodulin-regulated functions. *Biochim. Biophys. Acta* 736:109-118.
55. Hinds, T. R., Di Julio, D., and Vincenzi, F. F. (1987). Red blood cell Ca^{2+}
    pump ATPase: inhibition by compound 48/80. *Proc. West. Pharmacol. Soc.*
    30:93-95.
56. Hinds, T. R., and Andreasen, T. J. (1981). Photochemical cross-linking of
    azidocalmodulin to the $(Ca^{2+} + Mg^{2+})$-ATPase of the erythrocyte mem-
    brane. *J. Biol. Chem.* 256:7877-7882.
57. Sobue, K., Muramoto, Y., Fujita, M., and Kakiuchi, S. (1981). Calmodulin-
    binding protein in erythrocyte cytoskeleton. *Biochim. Biophys. Res. Com-
    mun.* 100:1063-1070.
58. Nelson, G. A., Andrews, M. L., and Karnovsky, M. J. (1983). Control of
    erythrocyte shape by calmodulin. *J. Cell Biol.* 96:730-735.
59. Penniston, J. T. (1983). Plasma membrane Ca^{2+}-ATPases as active Ca^{2+}
    pump. In *Calcium and Cell Function*, vol. 4, W. Y. Cheung (Ed.). Academic
    Press, New York, pp. 99-149.
60. Carafoli, E. (1984). Molecular, mechanistic, and functional aspects of the
    plasma membrane calcium pump. In *Epithelial calcium and phosphate trans-
    port: Molecular and cellular aspects*, vol. 168, F. Bronner and M. Peterlik
    (Eds.). Alan R. Liss, New York, pp. 13-17.
61. Carafoli, E. (1984). Calmodulin-sensitive calcium-pumping ATPase of plas-
    ma membranes: isolation, reconstitution, and regulation. *Fed. Proc.* 43:
    3005-3010.

62. Bader, H., Gietzen, K., Rosenthal, J., Ruedel, R., and Wolf, H. U. (Eds.). (1986). *Intracellular Calcium Regulation*. Manchester University Press, Manchester, England.
63. Campbell, A. K. (1978). Intracellular calcium: friend or foe? *Clin. Sci.* 72: 1–10.
64. Marme, D., and Dieter, P. (1983). Role of Ca^{2+} and calmodulin in plants. In *Calcium and Cell Function*, vol. 4, W. Y. Cheung (Ed.). Academic Press, New York, pp. 263–311.
65. Shull, G. E., Schwartz, A., and Lingrel, J. B. (1985). Amino-acid sequence of the catalytic subunit of the $(Na^+ + K^+)$-ATPase deduced from a complementary DNA. *Nature* 316:691–695.
66. Tada, M., and Katz, A. M. (1982). Phosphorylation of the sarcoplasmic reticulum and sarcolemma. *Ann. Rev. Physiol.* 44:401–423.
67. Wegener, A. D., and Jones, L. R. (1984). Phosphorylation-induced mobility shift in phospholamban in sodium dodecyl sulfate-polyacrylamide gels. *J. Biol. Chem.* 259:1834–1841.
68. Carafoli, E. (1985). The homeostasis of calcium in heart cells. *J. Mol. Cell Cardiol.* 17:203–212.
69. Kretsinger, R. H. (1977). Evolution of the informational role of calcium in eukaryotes. In *Calcium-Binding Proteins and Calcium Function*, R. H. Wasserman, R. A. Corradino, E. Carafoli, R. H. Kretsinger, D. H. MacLennan, and R. L. Siegel (Eds.). North-Holland, New York, pp. 63–72.
70. Starke, P. E., Hoek, J. B., and Farber, J. L. (1986). Calcium-dependent and calcium-independent mechanisms of irreversible cell injury in cultured hepatocytes. *J. Biol. Chem.* 261:3006–3012.
71. Schanne, F. A. X., Kane, A. B., Young, E. E., and Farber, J. L. (1979). Calcium dependence of toxic cell death: a final common pathway. *Science* 206:700–702.
72. Farber, J. L. (1981). The role of calcium in cell death. *Life Sci* 29:1289–1295.
73. Niggli, V., Penniston, J. T., and Carafoli, E. (1979). Purification of the $(Ca^{2+} + Mg^{2+})$-ATPase from human erythrocyte membranes using a calmodulin affinity column. *J. Biol. Chem.* 254:9955–9958.
74. Gietzen, K., Tejcka, M., and Wolf, H. U. (1980). Calmodulin affinity chromatography yields a functional purified erythrocyte $(Ca^{2+} + Mg^{2+})$-dependent adenosine triphosphatase. *Biochem. J.* 189:81–88.
75. Wolf, H. U., and Gietzen, K. (1974). The solubilization of high-affinity Ca^{2+}-ATPase of human erythrocyte membranes (abstract). *Hoppe Seylers Z. Physiol. Chem.* 355:1272.
76. Niggli, V., Adunyah, E. S., Penniston, J. T., and Carafoli, E. (1981). Purified $(Ca^{2+} -Mg^{2+})$-ATPase of erythrocyte membrane. *J. Biol. Chem.* 26:395–401.
77. Haaker, H., and Racker, E. (1979). Purification and reconstitution of the Ca^{2+}-ATPase from plasma membranes of pig erythrocytes. *J. Biol. Chem.* 254:6598–6602.

78. Gietzen, K., Seiler, S., Fleischer, S., and Wolf, H. U. (1980). Reconstitution of the Ca^{2+}-transport system of human erythrocyte. *Biochem. J.* 88:47–54.

79. Carafoli, E., Fischer, R., James, P., Krebs, J., Maeda, M., Enyedi, A., Morelli, A., and de Flora, A. (1987). The calcium pump of the plasma membrane. Recent studies on the purified enzyme and on its proteolytic fragments, with particular attention to the calmodulin binding domain. In *Calcium-Binding Proteins in Health and Disease*, A. W. Norman, T. C. Vanaman and A. R. Means (Eds.). Academic Press, New York, pp. 78–91.

80. Emelyanenko, E. I., Shakhparonovi, M. I., and Modyanov, N. N. (1985). Limited proteolysis of human erythrocyte Ca^{2+}-ATPase in membrane-bound form. Identification of calmodulin-binding fragments. *Biochem. Biophys. Res. Commun.* 126:214–219.

81. Hinds, T. R., Shattuck, R. L., and Vincenzi, F. F. (1985). Elucidation of a possible multimeric structure of the human RBC Ca^{2+}-pump ATPase. *Acta Physiol. Lat. Am.* 32:97–98.

82. Minocherhomjee, A. M., Beauregard, B., Potier, M., and Roufogalis, B. D. (1983). The molecular weight of the calcium-transport, ATPase of the human red blood cell determined by radiation inactivation. *Biochem. Biophys. Res. Commun.* 116:895–900.

83. Vincenzi, F. F., Andreasen, T. J., and Hinds, T. R. (1981). Azido-[125] I-calmodulin, a photoaffinity probe for the study of calmodulin as a regulator of plasma membrane Ca^{2+} transport. In *Calcium and Phosphate Transport Across Biomembranes*, F. Bronner and M. Peterlik (Eds.). Academic Press, New York, pp. 45–50.

84. Drickamer, L. K. (1975). The red cell membrane contains three different adenosine triphosphatases. *J. Biol. Chem.* 250:1952–1954.

85. Jay, A. W. L., and Burton, A. C. (1969). Direct measurement of potential difference across the human red blood cell membrane. *Biophys. J.* 9:115–121.

86. Varecka, L., and Carafoli, E. (1982). Vanadate-induced movements of Ca^{2+} and K^+ in human red blood cells. *J. Biol. Chem.* 257:7414–7421.

87. Vincenzi, F. F., and Hids, T. R. (1988). Drug effects on the plasma membrane calcium pump. In *Calcium in Drug Actions*, P. F. Baher (Ed.). Springer-Verlag, New York, pp. 147–162.

88. Lew, V. L., and Ferreira, H. G. (1978). Calcium transport and the properties of a calcium-activated potassium channel in red cell membranes. *Curr. Top. Membr. Transp.* 10:217–277.

89. Vincenzi, F. F. (1988). Regulation of the plasma membrane Ca^{2+} pump. In *The Red Cell Membrane: A Model for Solute Transport*, B. U. Raess and G. Tunicliff (Eds.). Humana Press, Clifton, New Jersey, in press.

90. Vincenzi, F. F., and Cambareri, J. J. (1985). Apparent ionophoric effects of red blood cell deformation. In *Cellular and Molecular Aspects of the Red Cell as a Model*, Vol. 195, J. W. Eaton, D. K. Konzen, and J. G. White (Eds.). Alan R. Liss, New York, pp. 213–222.

91. Lubowitz, H., Harris, F., Mehrjardi, M. H., and Sutera, S. P. (1974). Shear-induced changes in permeability of human RBC to sodium. *Trans. Am. Soc. Artif. Intern. Organs* 20:470–473.

92. Murachi, T., Tanaka, K., Hatakana, M., and Murakami, T. (1981). Intra-cellular Ca^{2+}-dependent protease (calpain) and its high-molecular-weight endogenous inhibitor. In *Advances in Enzyme Regulation*, Vol. 19, G. Weber (Ed.). Pergamon Press, New York, pp. 407–424.

93. Pontremoli, S., and Melloni, E. (1986). Extralysosomal protein degrada-tion. *Ann. Rev. Biochem.* 55:455–481.

94. Rega, A. F., Richards, D. E., and Garrahan, P. J. (1974). The effects of Ca^{2+} on ATPase and phosphatase activities on erythrocyte membranes. *Ann. NY Acad. Sci.* 242:317–323.

95. Parker, J. C. (1979). Active and passive Ca movements in dog red blood cells and resealed ghosts. *Am. J. Physiol.* 237:C10–C16.

96. Schmidt, J. W., Hinds, T. R., and Vincenzi, F. F. (1985). On the failure of calmodulin to activate the Ca^{2+} pump ATPase of dog red blood cells. *Comp. Biochem. Physiol.* 92[A]:601–607.

97. Vincenzi, F. F. (1981). Red blood cell calmodulin and Ca^{2+} pump ATPase: preliminary results of a species comparison. In *The Red Cell: Fifth Ann Arbor Conference*, G. Brewer (Ed.). Alan R. Liss, New York, pp. 363–378.

98. Schmidt, J. W., Vincenzi, F. F., and Hinds, T. R. (1982). Dog red blood cell Ca^{2+}-pump ATPase: lack of activation by calmodulin. *Acta Physiol. Lat. Am.* 32:250–252.

99. Edmondson, J. W., and Li, T. K. (1976). The effects of ionophore A23187 on erythrocytes: relationship of ATP and 2,3-diphosphoglycerate to cal-cium-binding capacity. *Biochim. Biophys. Acta* 443:106–113.

100. Taylor, D., Baker, R., and Hochstein, P. (1977). The effects of calcium ionophore A23187 on the ATP level of human erythrocytes. *Biochem. Biophys. Res. Commun.* 76:205–211.

101. Eaton, J. W., Berger, E., White, J. G., and Nelson, D. (1978). Interspecies variation in erythrocyte calcium responses. In *Erythrocyte Membranes. Recent Clinical and Experimental Advances*, Vol. 20, W. C. Kruckberg, J. W. Eaton, and G. J. Brewer (Eds.). Alan R. Liss, New York, pp. 37–46.

102. Hinds, T. R., and Vincenzi, F. F. (1985). The effect of ETH 1001 on ion fluxes across red blood cell membranes. *Cell Calcium* 6:265–279.

103. Hinds, T. R., Hammond, W. P., Price, L. M., Dodson, R. A., and Vincenzi, F. F. (1989). The activity of the red blood cell Ca pump is decreased in hemolytic anemia of the beagle dog. *Blood Cells*, (in press).

104. Danon, D., and Marikovsky, Y. (1964). Determination of density distribu-tion of red cell population. *J. Lab. Clin. Med.* 64:668–674.

105. Weed, R. I., and Chailley, B. (1973). Calcium-pH interactions in the pro-duction of shape changes in erythrocytes. In *Red Cell Shape*, M. Bessis, R. I. Weed, and P. F. LeBlond (Eds.). Springer-Verlag, Berlin, pp. 55–68.

106. Clark, M. R. (1985). Selected ionic and metabolic characteristics of human red cell populations. In *Cellular and Molecular Aspects of the Red Cell as a Model*, Vol. 195, J. W. Eaton, D. K. Konzen, and J. G. White (Eds.). Alan R. Liss, New York, pp. 381–386.

107. Lorand, L., and Conrad, S. M. (1984). Transglutaminases. *Mol. Cell Bio-chem.* 58:9–35.

108. Davis, P. W., and Vincenzi, F. F. (1971). Ca-ATPase activation and NaK-ATPase inhibition as a function of calcium concentration in human red cell membranes. *Life Sci.* 10:401–406.
109. Parker, J. C. (1977). Solute and water transport in dog and cat red blood cells. In *Membrane Transport in Red Cells,* J. C. Ellory, and V. L. Lew (Eds.). MIC Press, New York, pp. 427–465.
110. Gardos, G. (1958). The role of clacium in the potassium permeability of human erythrocytes. *Acta Physiol. Acad. Sci. Hung.* 15:121–125.
111. Simons, T. J. B. (1976). Calcium-dependent potassium exchange in human red cell ghosts. *J. Physiol. (Lond.)* 256:227–244.

18

Passive Cation Transport

JOHN C. PARKER *University of North Carolina at Chapel Hill School of Medicine, Chapel Hill, North Carolina*

PHILIP B. DUNHAM *Syracuse University, Syracuse, New York*

I. INTRODUCTION

The high K and low Na concentrations in eukaryotic cells is a consequence of the operation of the Na/K pump working against "downhill" fluxes of these ions (Chapter 16). Most mammalian erythrocytes also maintain these cation gradients, although there are notable exceptions. The voltage-gated passive Na and K channels in electrically excitable cells have for more than half a century been understood to be selective. Nevertheless, fluxes of Na and K independent of the Na and K pump in nonexcitable cells were regarded until about 20 years ago to be simple, electrodiffusional fluxes through nonspecific pathways. The pump-leak hypothesis was first proposed in passing by R. B. Dean in 1940 in a discussion after a presentation by Davson [1]. It was later presented briefly by August Krogh in 1946 [2], and then in detailed form by Tosteson and Hoffman [3]. We now understand that most of the traffic of inorganic ions through the membrane "leaks" is specific, complex, and regulated. The component of transport of cations that is electrodiffusional, nonspecific, and a linear function of concentration, the original conception of leaks, is a small fraction of total transmembrane transport of all ions.

Not only is much of the transport of cations that is independent of adenosine triphosphate (ATP)-driven pumps specific, it is also complex in the sense that flows of one ion are often directly dependent on flows of another ion. Furthermore, this type of dependence is often in the form of obligate, direct coupling of flows of ions either in opposite directions or the same direction.

The term "passive transport" in the title of this chapter is misleading in the following sense. The complementary term "active transport" means membrane transport processes capable of work in the thermodynamic sense (transport of a substance from region of low to high electrochemical potential); at one time all active transport processes in eukaryotic cells were regarded as being driven directly by the hydrolysis of ATP. However, it is now clear that a number of transport processes that are clearly "uphill" thermodynamically are not fueled by ATP, but rather are driven by direct coupling to the flow of another solute down its electrochemical potential gradient. The term "primary active transport" applies to processes fueled directly by metabolism and "secondary active transport" to active transport processes driven by coupling to the flow of another solute. Although our title contains "passive transport," several of the phenomena we present are secondary active transport, coupled flows both in the same direction (cotransport) and in opposite directions (countertransport).

Mammalian erythrocytes present an astonishing variety of patterns of cation transport. This is true not only of the Na/K pump, but also of the other pathways. For example, Na/K/Cl cotransport has a rate more than an order of magnitude greater in ferret red blood cells than in human or rat cells, and is absent in sheep cells. The prominent Na/Ca exchange system of carnivore red blood cells is negligible in cells from most other mammals. The Ca-activated K channel is also variable in its distribution, and absent in sheep cells.

The variability in erythrocyte cation transport among mammalian species is much more striking than the variability in rates or types of anion transport in red blood cells. The probable maximum range in anion transport rates among mammals is approximately two-fold [4]. In addition to this range of quantitative differences, there are no known major qualitative differences in anion transport mechanisms among mammals that correspond to the qualitative differences in cation transport. The difference between anion and cation transport may be a consequence of the central and essential role of anion transport in erythrocyte function in comparison to that of cation transport. Anion transport is critical to the contribution of red blood cells to the respiratory function of the organism (Chapter 19). In consequence, selection pressure has apparently been intolerant of tendencies toward variation in anion transport in red blood cells during mammalian evolution. In contrast, cation transport in red blood cells plays no major role in the function of the organism; rather, it performs "local" functions, such as maintenance of cell volume and ion concentrations, with perhaps roles in maturation and senescence.

Also contributing to the variability in erythrocyte cation transport among mammals is the obvious lack of requirement for high cellular K. A mutation that led to low K and high Na concentrations in nerve or muscle cells, and perhaps all nucleated cells, would presumably be lethal. This has obviously not been the case with red blood cells. Bovids (sheep, cows, goats, etc.) are dimorphic for this character and some individuals have cell K concentrations only a few fold greater than that in plasma. The Na and K concentrations in red blood cells from carnivores are nearly the same as in plasma; these cells lack Na/K pumps altogether.

Thus, in a sense mammalian red blood cells are *museums* for cation transport pathways. They display an assortment of systems, not necessarily of importance to them, but essential to their ancestral cells.

In this chapter we treat the following cation transport systems: Cl-dependent cotransport, Na/K/Cl, and K/Cl; Na/Ca countertransport; Ca-activated K transport, Na/H and Na/Na exchanges, and cation fluxes through the anion exchanger. Adenosine triphosphate-driven, primary active transport systems for Na, K, and Ca are dealt with in other chapters in this volume. For a detailed discussion of the residual nonspecific fluxes of cations through red blood cell membranes, sometimes referred to as "ground permeability," the interested reader should consult Lew and Beaugé [7].

II. COTRANSPORT

Cotransport is a transport pathway in which the flows of more than one solute are tightly coupled in the same direction. The coupling is tight or "chemical" in the sense that it depends on the direct interaction of the transported solutes with the same elements of the membrane. In contrast, in "electrical" coupling, the flow of one ionic species drives another in consequence of an electrical potential difference generated by the flow of the first.

In mammalian red blood cells, several cotransport systems have been studied. The first reported was Na/K cotransport, in human red blood cells. Subsequently, this cotransport system was shown to require Cl, and was concluded by analogy with other systems, and with a modicum of pharmacological and other indirect evidence, to be Na/K/Cl cotransport. Subsequently, sheep red blood cells were shown to lack Na/K cotransport, but to have a Cl-dependent K transport. A similar pathway was reported to coexist with Cl-dependent Na/K cotransport in human red blood cells under some circumstances. This was also concluded to be a cotransport system, K/Cl cotransport, by analogy with other systems, and with indirect evidence. These are the two cotransport systems dealt with here. The interesting and potentially important Na-amino acid cotransport systems in red blood cells are not treated here (see [5] and [6] for reviews).

A. Na/K/Cl Cotransport

The earliest evidence for Na/K cotransport in human red blood cells was a ouabain-insensitive Na influx dependent on extracellular K [8,9], which was inhibited by furosemide and by metabolic depletion [10]. A K influx stimulated by Na [10] was inhibited by another diuretic, ethacrynic acid [11,12]. The first clear proposal for Na/K cotransport was made by Wiley and Cooper [13], in which they showed that, in the presence of ouabain, the influxes of the two ionic species were interdependent. In addition, both were inhibited by furosemide. These various results in human red blood cells were probably the first clear evidence in any system for the direct coupling of the transport of Na and K in the same direction. It was proposed at about the same time that with the appropriate orientation of the gradients of Na and K, the flow of one ionic species could promote active transport of the other [14], now called secondary active transport.

A further characteristic of this cotransport system in human red blood cells was published from two laboratories in 1980: the dependence of Na/K cotransport on Cl [15,16]. Earlier, Na/Cl cotransport had been proposed for epithelial cells [17]. After Geck et al. [18] demonstrated with thermodynamic rigor Na/K/2Cl cotransport in Ehrlich ascites cells, it seemed possible that the dependence of Na/K cotransport on Cl in human red blood cells was evidence for coupling of Na and K to Cl transport. Subsequently, many of the Na/Cl cotransport systems in epithelia were found to require K, and to be, as in Ehrlich ascites cells, electroneutral Na/K/2Cl cotransport ([19–24]; see [25] and [26] for reviews). However, there is Na/Cl cotransport, independent of K, in rabbit gallbladder [27] and in other epithelia under some circumstances [28]. Electroneutral Na/K/Cl cotransport has been demonstrated convincingly in some other cell types, including squid axon [29,30] and duck red blood cells [31,32]. In addition to humans, among mammals Na/K/Cl cotransport has been shown in red blood cells of ferrets [33,34] rats [35], and guinea pigs [36]. The magnitudes of the fluxes are similar in human and guinea pig cells, are several-fold higher in rat cells, and an order of magnitude or so higher in ferret cells [37]. Furosemide-sensitive K flux has also been measured in rabbit reticulocytes [38], but this has not been characterized to the extent that it can be designated as Na/K/Cl cotransport. Na/K/Cl cotransport is absent in mature sheep erythrocytes [39].

The specificity of the anion dependence of the Na-dependent K influx has been studied in human red blood cells. From measurements of unidirectional K influx in the presence of Na, ± furosemide, there was little or no K influx greater than the furosemide-independent flux in the presence of iodide, propionate, methylsulfate, nitrate, or acetate, all permeant monovalent anions [15]. Only bromide served as a reasonable substitute, supporting ~75% of the K influx observed in Cl medium.

Wiley and Cooper [13] estimated some of the kinetic constants of Na/K cotransport. With 50 mM [K]$_0$, furosemide-inhibitable Na influx was a saturable function of [Na]$_0$ with an apparent K_m of 24 mM. In complementary experiments the apparent K_m for furosemide-inhibitable (ouabain-insensitive) K influx was 7 mM [K]$_0$. Dunham et al. [15] reported a lower apparent K_m for Na influx (8 mM), but the method entailed measuring Cl dependence of Na influx rather than furosemide sensitivity, and NO_3 has a secondary effect on Na influx (see below). The apparent K_m for Cl-dependent K influx, 5 mM [14], was similar to the earlier estimate, 7 mM, for furosemide-inhibitable K influx [13].

It was clear in the work of Wiley and Cooper [13] that a significant fraction of furosemide-inhibitable K influx did not require Na. Kaji and Kahn [40] have now made separate estimates of the K_m of Na-dependent and Na-independent components of K influx that are Cl-dependent, and they are ~4 mM and 17 mM, respectively. Thus, the apparent affinity for K is higher in the presence of Na. The two possible interpretations of this are, first, that there are two separate pathways for K influx, and second, that there is modulation by Na of the mode of operation of a single pathway. It was once held that Na-independent, Cl-dependent K influx in normal human erythrocytes is K/Cl cotransport [41]. It now appears more likely to be a mode of operation of the Na/K/Cl cotransporter [42], although K/Cl cotransport can be observed in human red blood cells under some circumstances (see below).

Lacking are good estimates of the kinetics of mutual stimulation between Na and K. In a simple mechanism of a cotransport system, Na should activate K influx with the same apparent affinity that it activates its own influx; the same should hold for activation by K. If there were ordered addition of substrates, or some other complexity, this simple scheme need not apply (see below).

Cl-activation of influxes of Na and K has been looked at, using both NO_3 and methyl sulphate as Cl substitutes [15]. Cl seemed to activate Na influx with simple saturation kinetics and a $K_{1/2}$ of ~40 mM. Activation of K influx by Cl was complex, maybe sigmoid with a $K_{1/2}$ of ~80 mM. Owing to the high permeability to Cl relative to permeability to Na and K, even with the anion exchanger inhibited more than 95%, it has not yet been possible to measure Na, K-dependent Cl fluxes and their kinetics in human red blood cells.

The anion dependence of K-stimulated Na influx has been more difficult to determine because of an enhancement of Na influx by substitution for Cl with one of several monovalent permeant substitute anions [43]. This phenomenon, best described with NO_3, is specific for Na, in that in the absence of Na, K influx is stimulated little or not at all by NO_3. Two lines of evidence indicate that this NO_3-induced (or Cl-free-induced) Na influx is not mediated by the cotransport pathway. One is the specificity for Na over K. The second has to do with the relative dependencies of Na influx on metabolism in the presence of Cl and NO_3, respectively. These two lines of evidence are illustrated in Figure 1. The upper panel shows unidirectional Na influx in human red blood cells with either Cl or NO_3 the principal anion with external K concentration, $[K]_o$, varied. At low $[K]_o$ Na influx is higher in NO_3. Raising $[K]_o$ activates Na influx in Cl with a $K_{1/2}$ of ~20 mM, consistent with Na/K cotransport. Raising $[K]_o$ in NO_3 has no effect on Na influx, consistent with a separate pathway for Na influx in NO_3 specific for Na.

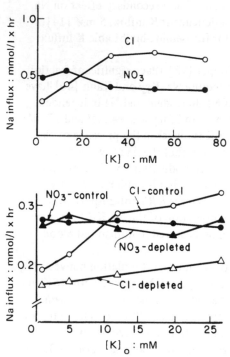

Figure 1 Unidirectional Na influxes in human red cells as a function of external K concentration, $[K]_0$. Cells were equilibrated in isotonic media with Cl as the principal anion, or with all Cl replaced by NO_3. Media were buffered with Tris-HCl or Tris-HNO$_3$ (10 mM) at pH 7.4, and contained ouabain (0.1 mM). [Na] was 20 mM in all media; [choline]$_0$ and $[K]_0$ were varied reciprocally so that the total cation concentration was 150 mM. Influxes were measured using ^{22}Na as a tracer. After exposure for 1 hour, cells were washed free of ^{22}Na in the medium by three steps of centrifugation and resuspension. Radioactivities were determined of lysates of pellets of the washed cells. Volumes of cells in the pellets were calculated from hemoglobin concentrations in the lysates. Results shown are means of triplicate determinations expressed as mmol per liter of packed cells per hour. *Upper panel*: Na influx with varying $[K]_0$ in Cl and NO_3 cells. *Lower panel*: The experiment in the upper panel repeated (control) with, in addition, half of the cells metabolically depleted prior to replacement of Cl by incubation for 6 hours at 37°C in a glucose-free medium. (Dunham, unpublished results.)

The lower panel of Figure 1 compares another property of Na influxes in Cl and NO_3 media: dependence on metabolism. The circles show the experiment in the upper panel repeated: Na influx activated by K in Cl (open circles) and NO_3 (closed circles). The triangles show Na influxes measured at the same time, but in cells that had been metabolically depleted by incubation in the absence of glucose. In Cl, Na influx is inhibited in depleted cells, and has greatly reduced dependence on K. This dependence of Na/K cotransport on metabolism will be discussed further below. In contrast, Na influx in NO_3 was completely unaffected by depletion, consistent with separate pathways for Na influx in Cl and NO_3 media.

In the experiments in Figure 1 and in an earlier publication [15], Na influxes in high K, NO_3-media were lower than in the corresponding Cl media. However, it is more common that Na influx in NO_3 exceeds that in Cl, even with high $[K]_0$. These quantitative differences may be due to differences in properties of cells among donors, but this has not been verified. The specific Na influx pathway induced in NO_3 (Cl-free) media has not been characterized further in human red blood cells. We are unaware of reports of it in other cells, or of any possible physiological relevance, apart from confounding experiments on Na/K/Cl cotransport.

Interestingly, NO_3 may under some circumstances be able to interact with the Na/K/Cl cotransporter. Indeed, NO_3 may be able to serve as an alternate substrate for Cl in human red blood cells when the cotransporter mediates effluxes [44,45]. In shark rectal gland, NO_3 may substitute for Cl in Na/K/Cl cotransport in the influx mode as well [46]. One attractive, but untested, possibility is that the two Cl sites on each transporter differ in their affinity for Cl, and NO_3 may be able to substitute for Cl at one of them. Furthermore, the influx and efflux modes of operation of the cotransporter may differ in their anion specificities [44,45].

The stoichiometry of the Na/K cotransport has been studied mainly by measuring tracer fluxes. The estimates have varied from 5K:1Na to 1K:3Na [15,41, 47,48]. The Na and K fluxes mediated by the cotransporter have been measured as either the furosemide-inhibitable, the Cl-dependent, or the mutually (Na and K) dependent fluxes. No one criterion gives clearer results than the others. In a recent study in which furosemide-inhibitable tracer influxes were measured, it was concluded that the stoichiometry is variable, depending on the Na and K concentrations, and is 3K:2Na under physiological conditions [47]. In another recent study, Na/K stoichiometry was estimated from net fluxes. Mutual dependence of fluxes (Na-dependent K influx and K-dependent Na influx) was used as a criterion for cotransport, rather than furosemide-sensitivity, and stoichiometry was very close to 1Na:1K [45]. In a few experiments, both furosemide-sensitivity and Cl-dependence were employed as criteria for cotransport, and again, based on net influxes, the stoichiometry was 1Na:1K. However, all of these experiments were carried out with both $[K]_0$ and $[Na]_0$ at 50 mM and with low intracellular Na and K concentrations, distinctly unphysiological conditions, so if stoichiometry varies as a function of concentration, and is 3K:2Na under physiological conditions, these results would not have been obtained. Furosemide-inhibitable tracer fluxes may not be a good way to measure stoichiometry of cotransport if there are exchange fluxes superimposed on net cotransport. However, Brugnara et al. [47] argue that exchange fluxes do not invalidate their approach.

All Na/K/Cl cotransport systems which have been tested rigorously have been shown to be electroneutral. Most of these have a stoichiometry of 1Na:1K:2Cl (for epithelial cells, see [19-24,26]; for Ehrlich ascites cells, [18]). Na/K/Cl cotransport in squid axons seem to have a different electroneutral stoichiometry, 2Na:K:3Cl [29]. The same stoichiometry has been claimed for ferret red blood cells [33], but this estimate is uncertain because it is based on tracer influxes, so 1:1 exchanges of Na or K could confound the conclusion on stoichiometry. In ferret red blood cells, with cotransport fluxes much higher than in other mammalian red blood cells, it has been possible to measure Cl fluxes mediated by the cotransporter [33], even though their stoichiometric relation to Na and K fluxes is uncertain.

In human red blood cells the large Cl permeability has so far precluded measuring stoichiometry of Na/K/Cl cotransport directly; indeed, rigorous demonstration of coupling of Cl to Na/K cotransport has not been made. Several attempts have been made to measure stoichiometry in human red blood cells indirectly by determining whether cotransport is sensitive to changes in membrane potential. Electroneutrality would argue for a stoichiometry of 1Na:1K:2Cl if indeed 1Na:1K is the stoichiometry of the cation fluxes [45]. Chipperfield and Shennan [50] have measured furosemide inhibitable Na and K effluxes. The usefulness of the measurements in this study in characterizing cotransport is based on the assumption that cotransport in human red blood cells is symmetrical to the extent that coupled Na/K/Cl fluxes are readily mediated in the efflux direction, and is reflected by the furosemide-inhibitable effluxes. Evidence will be presented below that this assumption is subject to question.

More recently a systematic study of furosemide-inhibitable unidirectional Na influxes was made in human red blood cells over a voltage range of 140 mV [49]. The membrane potential was varied using valinomycin, to make it a K equilibrium potential, and by varying the K gradients. In all experiments Na concentrations were 10 mM both inside and outside the cells. Furosemide-inhibitable Na influxes varied little over the 140 mV range, with a slight increase in the influx with increased intracellular positive potentials. If the Na influx were carrying positive current, over the 140 mV range the increased inside positive voltage would have reduced Na influx by 37-fold. The small increase (0.4 μmol/L $\times$ hr $\times$ mV) is consistent with electroneutrality of the cotransporter and with

1Na:1K:2Cl stoichiometry if the conclusion of Kracke et al. [45] about the
1Na:1K stoichiometry is correct. The interpretation of the results on voltage-
dependence of furosemide-inhibitable Na influxes [49] is subject to question if
there had been significant furosemide-inhibitable Na/Na exchange. There was no
transstimulation by intracellular Na of furosemide inhibitable influxes, and
therefore no Na/Na exchange through the cotransporter under the conditions
employed in these experiments [49]. Under other conditions, furosemide-inhibitable Na/Na exchange has been observed [44].

It has now been proposed that furosemide-inhibitable K/K exchange in human red blood cells, observed by some [44], but not others [45], is a partial
reaction of the Na/K/Cl cotransporter, and may account for observations of variable stoichiometries and stoichiometries different from 1:1, when fluxes are
measured using tracers [42]. When K/K exchange was observed, it appeared to
require intracellular Na (see McManus in ref. [37] for references and discussion).
Kracke et al. [45] observed stimulation of ^{86}Rb efflux by intracellular Na, but
the flux was as great in K-free (choline) medium as in medium with 50 mM K.
Therefore, there appeared to be no Na-dependent K/K exchange under the conditions of these experiments.

A more complete analysis has been made of the mechanism of Na/K/Cl cotransport in duck red blood cells. Bumetanide-sensitive exchanges of both Na for
Na and K for K have been observed [37,51]. Furthermore, each exchange pathway has specific, and different, requirements for the presence of ions on the two
sides of the membrane: K/K exchange requires (in addition to K on both sides)
Cl outside and Na and Cl inside. Na/Na exchange requires (in addition to Na) K
and Cl outside [37]. These results are consistent with a model proposed by Mc-
Manus [37] in which there is ordered binding of the four substrate ions on one
side of the membrane (Na, K, 2Cl) and ordered release on the other side of the
membrane. Furthermore, the results are entirely consistent with the order of
binding and the order of release being the same, i.e., Na first on and first off, Cl
second and fourth on and off, and K third on and off [37,42]. This "first on,
first off" ordering has been called "glide symmetry" by Stein [52]. Figure 2
shows a minimal kinetic model as proposed by McManus for the glide symmetry
of Na/K/Cl cotransport in duck red blood cells. Furosemide-sensitive K/K exchange would be the forward and backward running of the mechanism between
the addition (and release) of K on the left and the release (and addition) of K
on the right, and would require Cl$_0$, Na$_i$, and Cl$_i$, as McManus and his group have
found [37,51]. This elegant proposal, which fits all the data available for duck
red blood cells, is an important advance in that it provides a framework for at-
tempting to explain existing data for other systems, and should promote further
tests for the model in these other systems.

The model in Figure 2 predicts that the cotransporter should mediate coupled
Na/K/Cl cotransport both inward and outward as it does in duck cells. Wiley and

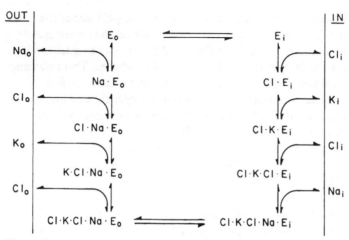

Figure 2 Minimal kinetic model for Na/K/2Cl cotransport in duck red cells as proposed by McManus [37,51]. Model shows "first on, first off" binding and release, or glide symmetry (redrawn from ref. 42).

Cooper [13] observed that furosemide inhibited Na and K effluxes in human red blood cells, from which it could be concluded that the cotransporter is reversible. Garay et al. [53] confirmed these observations and extended them by showing that furosemide-inhibitable Na efflux was activated by K_i and furosemide-inhibitable K efflux was activated by Na_i. Reversible Na/K/Cl cotransport was also concluded from results on cultured kidney cells (MDCK cells), again based on inhibition by furosemide of effluxes, as well as influxes, of Na and K [54]. However, furosemide sensitivity of effluxes may not be a sufficient criterion for coupled outward cotransport, at least in human red blood cells. There are indeed furosemide-inhibitable components of fluxes of both Na and K in both directions [44,45,53]. The cotransporter mediates a Na-stimulated K efflux, but K-stimulated Na efflux was not observed [44,45]. If this observation is correct, then the cotransporter in human red blood cells does not readily mediate outward coupled transport, and its function is difficult to fit to the model in Figure 2.

The evidence on Na/Na and K/K exchanges is not consistent among several studies [44,45,49], as discussed above. If the cotransporter does not readily mediate the two exchange fluxes, then this is an additional difficulty in fitting the model to human red blood cells.

Canessa et al. [44] have proposed a model for furosemide-inhibitable Na and K fluxes in human red blood cells. The phenomenological complexity of this transport system in human red blood cells is reflected by the inability of the model, with 12 reversible reactions, to account for all of the six modes of furosemide-inhibitable fluxes Canessa et al. [44] have observed. These authors propose that several of the modes of furosemide-inhibitable flux take place through a membrane pathway distinct from the one responsible for Na/K/Cl cotransport. It does not seem likely that this view will prevail. Nevertheless, a simple model is difficult to construct that fits either the results and views of Canessa et al. [44], or the somewhat different results and views of Kracke et al. [45].

Na/K/Cl cotransport in squid axon also has some properties that make difficult fitting it to the glide symmetry model for duck red blood cells [30]. For example, Cl/Cl exchange appeared not to be mediated by the cotransport in squid axon, and this is a transport mode predicted by the glide symmetry model (Figure 2).

As McManus suggests [37], if the fully loaded and unloaded forms of the transporter crossed the membrane with equal rates, the obligate exchanges representing partial reactions would probably not be observed, and the transport pathway might appear to have one primary mode: net salt transport.

There were early reports of metabolic dependence of cation fluxes in human red blood cells, modes of transport that, in retrospect, were undoubtedly mediated by the Na/K/Cl cotransporter [10,11,55,56]. Confirmation has been reported recently in studies of inhibition by metabolic depletion of various modes of operation of Na/K/Cl cotransporter [57]; see also Figure 1. The critical metabolite was presumed to be ATP in both the early and recent studies. Although our understanding of the complexity of the transporter has increased since the early studies, our understanding of its metabolic dependence has not. One theme of most of the studies is that, although ATP may be necessary, it is not used stoichiometrically to fuel the transport of the ions. The clearest demonstration of this is in Ehrlich ascites cells [18].

Garay [58] tested the effect of cyclic AMP on furosemide-inhibitable Na effluxes in human red blood cells, and observed substantial, although variable, inhibition. This effect is in striking contrast to the stimulatory effect of cyclic adenosine monophosphate (cAMP) on the cotransporter in avian red blood cells [59]. Although human red blood cells lack the adenylate cyclase and β-adrenergic receptors found in avian red blood cells, they do contain cAMP-dependent kinases (see [58] for references). Although there is no explanation for these opposite effects of cAMP, it is important to note that in a variety of other tissues, cases of both inhibition and stimulation of Na/K/Cl cotransport by cAMP have been reported (reviewed in [26]). (For a detailed treatment of regulation of cotransport in red cells, see Chapter 22.)

There are now a few reports of isolation of proteins purported to be the Na/K/Cl (or Na/Cl) cotransporter, or part of it [60-62]. Most recently a protein of 150 kDa has been isolated from duck red blood cells using a tritiated photoactivated analog of bumetanide as a probe [62]. Proof by reconstitution of the protein in a functional form, e.g., in artificial vesicles, has not been reported.

There was a preliminary report of reconstitution of Na/K/Cl cotransport activity from renal medulla into artificial membranes [63], but the protein was not purified or otherwise characterized. With the availability of increasingly sophisticated biochemical techniques, and of the techniques of modern gene cloning, it should not be long before complete primary protein structures are available for Na/K/Cl cotransporters. Expression systems, mutagenesis, and tests of kinetic models of mechanism should not be far behind.

B. K/Cl Cotransport

At about the same time that the Cl dependence of Na/K cotransport was found in human red blood cells, Cl dependence of K transport was reported in red blood cells from sheep of the LK phenotype [39,64]. This was an important observation because it demonstrated a new type of cotransport pathway. Sheep cells lack Na/K/Cl cotransport [39]. The Cl-dependent, Na-independent K transport in LK sheep cells is apparently K/Cl cotransport. As with Na/K/Cl cotransport in human red blood cells, there has been no direct demonstration of coupling of Cl and K fluxes in sheep red blood cells, and cotransport has been inferred from the specificity of dependence on Cl (and Br), and from the inhibition by furosemide. Recently Brugnara et al. have obtained results indicating that Cl-dependent K flux is electroneutral in young (low density) human red blood cells [65]. This was observed in experiments in which the membrane potential was "clamped" away from the Cl equilibrium potential by partial replacement of Cl with NO_3 or thiocyanate, more permeant anions than Cl. An outwardly directed Cl gradient promoted a net K efflux against a K gradient. This preliminary observation strengthens the case for electroneutral K/Cl cotransport in red blood cells.

K/Cl cotransport has been demonstrated with thermodynamic rigor in the basolateral membrane of *Necturus* gallbladder [66,67]. It has also been proposed for the basolateral membrane of the thick ascending limb in rabbit nephrons [68]. Ehrlich ascites cells normally perform regulatory volume decrease after hypotonic challenge by loss of K and Cl through separate conductive channels [69]. By contrast, Thornhill and Laris [70] demonstrated K/Cl cotransport accompanying regulatory volume decrease in Ehrlich ascites cells. Kramhøft et al. [71] resolved the apparent discrepancy, showing that K/Cl cotransport could be activated in Ehrlich cells by Ca deprivation and reduced pH. In the history of the attempts to distinguish K and Cl effluxes through conductive and electroneutral (cotransport) pathways, it is important to have the demonstration of both classes of transport in a single cell type.

It was shown early that, when duck red blood cells are swollen in a hypotonic medium, they lose KCl, and restore their initial volume, a regulatory volume decrease [72]. This KCl efflux from duck cells was later shown to be K/Cl cotransport by Haas and McManus [73], who entertained the possibility that swelling-induced K/Cl cotransport and shrinkage-induced Na/K/Cl cotransport [31] in duck cells are alternate modes of the same transport process. However, their results could not rule out the possibility that there are two separate, distinct pathways. More recently, preliminary results have shown qualitative pharmacological differences between the Na-dependent and Na-independent pathways of Cl-dependent K transport in duck red blood cells, strengthening the case for two separate pathways [74].

A striking feature of K/Cl cotransport in LK sheep red blood cells is its sensitivity of osmotically induced changes in cell volume. Swelling of 10% caused an increase of up to sevenfold in K influx; the increase was entirely dependent on the presence of Cl [39]. Another interesting feature of K/Cl cotransport in LK sheep red blood cells is its inhibition by an alloimmune antibody to the L_1 antigen, a blood group antigen specific for cells of the LK phenotype and associated with the K/Cl cotransporter [75,76]. The anti-L_1-inhibitable K influx is identical with the Cl-dependent, volume-sensitive K transport pathway in that the effects of Cl removal and antibody are seen in swollen cells but not in shrunken ones, as shown in Figure 3. The inhibition of the cotransport by anti-L_1 antibody could be interpreted to mean that the antigen is an endogenous promoter of cotransport, and binding of the antibody to the antigen inhibits by relieving this promotion. However, there is no direct evidence for this suggestion. This antigen–antibody interaction should provide a useful system for probing the mechanism of cotransport, but so far it has not been particularly fruitful. For a discussion of the relation of the L_1 antigen to other aspects of the regulation of K/Cl cotransport in sheep cells, see [77].

Lauf and Theg [78] made the interesting and important observation that N-ethylmaleimide (NEM) stimulates K transport in LK sheep red blood cells. The effect was specific for K in that Na transport was unaffected. In subsequent studies it was shown that the NEM-enhanced K transport is identical to the Cl-dependent pathway [79], and to the antibody-inhibited pathway [80]. The stimulation of K/Cl cotransport in LK sheep cells by NEM has been described in intricate detail by Lauf in an extensive series of papers, culminating in a study in which it is shown that the oxidizing agent diamide has a similar effect to that of the alkylating agent NEM [81] (see reviews by Lauf in [82] and [83]). In spite of the detailed description of the stimulation of K/Cl cotransport by sulfhydryl-active agents, it has brought us only a little closer to an understanding of the mechanism K/Cl cotransport and its regulation.

Lauf and Bauer [84] presented evidence that sheep reticulocytes from both LK and HK animals reduce their volumes after hypotonic challenge by a process that requires Cl, apparently K/Cl cotransport. This may be responsible for the reduction in cell volume that occurs during maturation of red blood cells. Mature red blood cells from the sheep of the HK phenotype lack K/Cl cotransport. Thus, inactivation of the K/Cl cotransport is a feature of maturation. The co-

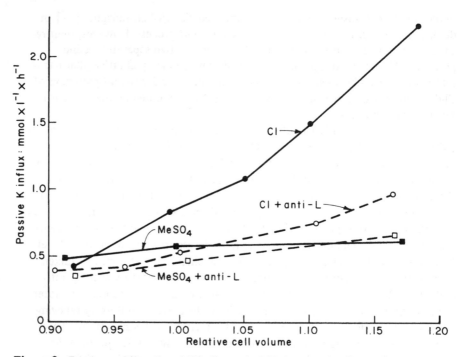

Figure 3 Passive unidirectional K influxes in LK sheep red cells equilibrated in media of osmolarities varied by dilution of control media (290 mOsm/kg), or by addition of sucrose. [K] in all media was 10 mM. Relative cell volumes were calculated from hematocrits with volume in control medium set at 1.00. Prior to altering cell volumes aliquots of cells were treated with alloimmune anti-L antiserum. Then aliquots of untreated and anti-L-treated cells were made Cl-free by incubation and washing in media with all Cl replaced by methyl sulphate (MeSO$_4$). Shown are means of quadriplicate determinations of K influxes in mmol per liter of packed cells per hour obtained using ^{86}Rb as a tracer. Incubation was for 30 min. The remainder of the experiment was carried out as described for the experiment in Fig. 1. (Dunham and Ellory, unpublished results; similar results are in refs. 39, 64).

transporter remains in a latent form in HK cells, but can be reactivated by a variety of treatments, including NEM and osmotic swelling [85].

The possibility has been entertained that K/Cl cotransport coexists in human erythrocytes with Na/K/Cl cotransport. The first evidence was the observation of Wiley and Cooper [13] that in human red blood cells, there is a fraction K influx inhibitable by furosemide in Na-free media. Dunham et al. [15] reported that Cl-dependent K influx exceeded Cl-dependent Na influx. N-ethylmaleimide promoted K influx, but not Na influx [41]. These observations led to the view that Na/K/Cl and K/Cl cotransport coexist as separate transport pathways [41]. Lauf et al. [86] confirmed and extended the observation that NEM selectively activates Cl-dependent K influx in human red blood cells. However, these authors favored the view that NEM was activating a latent K/Cl cotransport pathway, rather than enhancing an already existing pathway. Subsequent reports from several laboratories supported this view of Lauf et al. [86]. In mature, normal human erythrocytes, there is probably no K/Cl cotransport. The Na-independent, furosemide-inhibitable, Cl-dependent fluxes in these cells have been held to reflect a mode of operation of the Na/K/Cl cotransporter [42], as discussed above, perhaps K/K exchange.

Nevertheless, K/Cl cotransport can be demonstrated in human red blood cells under a number of conditions; it can be "unmasked" by NEM, as just noted [86]. It is also present in human erythrocytes swollen by 10% or more, as shown recently by Kaji [87]. These two observations raise the possibility that swelling activates K/Cl cotransport by a mechanism that depends on sulfhydryl groups, but this possibility has not been clarified.

K/Cl cotransport has been reported for young human red blood cells (separated by gradient centrifugation) [88,89]. It was suggested that this transporter plays a role in the reduction in cell volume and increase in cell density that occur early in maturation of human red blood cells [90]. Indeed, when young cells (of low density) were caused to swell, there ensued a reduction of cell volume with a concomitant loss of KCl [90]. Volume regulation required Cl, and was partially inhibited by furosemide, but not by bumetanide. The same role for K/Cl cotransport was suggested for the maturation of HK sheep reticulocytes, as discussed above [84]. In human red blood cells, as in sheep cells, a feature of maturation is inactivation of the K/Cl cotransporter involving changes in sulfhydryl groups and subject to reactivation by NEM, or other treatments [90].

K/Cl cotransport has also been demonstrated in red blood cells from human subjects with abnormal hemoglobins, hemoglobin S (the hemoglobin of sickle cell anemia), and hemoglobin C [89,91,92]. K/Cl cotransport in cells with hemoglobin C or S is similar in many respects to K/Cl transport in reticulocytes from normal human subjects (with hemoglobin A) and in LK sheep erythrocytes. The high K/Cl cotransport in hemoglobin C cells is not due to a preponderance of young cells in circulation because C cells have a life span like those of normal (hemoglobin A) cells. Interestingly, transport rates in C and S cells are about the same even though reticulocytosis is high in patients with hemoglobin S. The causal relationship between abnormal hemoglobins and high K/Cl cotransport has not been explained. Binding of the positively charged S and C hemoglobins to the membranes has been invoked [91,92], but this cannot serve as a satisfactory explanation without further results, such as effects of S and C hemoglobins on membrane sulfhydryl groups associated with the cotransporter.

Finally K/Cl cotransport has also been demonstrated in resealed ghosts made
from human red blood cells [93]. Cl-dependent K influx in ghosts was much
greater than in intact cells, and was not enhanced further by NEM treatment. It
may be provoked by oxidation occurring during preparation of the ghosts because inclusion of dithiothreitol during lysis, resealing, and the measurement of
the flux prevented the activation of the large K flux normally seen in ghosts.
Obviously the appearance of Cl-dependent K transport in resealed ghosts, caused
by oxidation of the same sulfhydryl groups affected by NEM in intact cells, or
by some other mechanism, is another example of activation of K/Cl cotransport,
latent in normal erythrocytes.

These observations are further support for the view that the putative K/Cl
cotransporter, latent in normal erythrocytes, is all the same pathway, whether
induced by NEM, cell swelling, formation of ghosts, or found in young cells or
cells with abnormal hemoglobin. In a result consistent with this view, K/Cl
cotransport in resealed ghosts is enhanced by osmotic swelling of the ghosts
[93], as shown in Figure 4. Some of these results on K/Cl cotransport in resealed ghosts have recently been confirmed and extended [94].

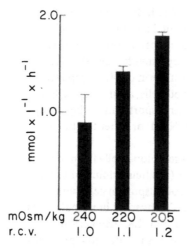

Figure 4   Volume sensitivity of furosemide-inhibitable unidirectional K influxes
in resealed human red cell ghosts. The osmolarity of the medium in which the
ghosts were resealed was 240 mOsm/kg. This was reduced to 220 or 205 by dilution, except that [K] in all three media was 20 mM. Relative volumes of the
ghosts (r.c.v.) were determined from hematocrits of ghosts with the r.c.v. of
ghosts in 240 mOsm/kg medium set at 1.0. Shown are means of differences between total influxes (measured in triplicate), and influxes in 0.25 mM furosemide. Errors are standard deviations of differences. (Taken from Table 4, ref. 93.)

Na/K/Cl and K/Cl cotransporters can be distinguished on the basis of a pharmacological difference as well as their difference in dependence on Na. The first clue was the observation from several laboratories that K/Cl cotransport in LK sheep red blood cells was less sensitive to inhibition by furosemide than was Na/K/Cl cotransport in human red blood cells, but this could easily have been a species difference. The pharmacological difference was clarified after the demonstration of the activation of K/Cl cotransport in human cells by NEM. It had been reported in 1982 that the $K_{1/2}$ for the inhibition of Na/K/Cl cotransport in human red blood cells was ~0.2 μM for bumetanide and 9 μM for furosemide [48]. The NEM-induced K influx was not only inhibited by much higher concentrations of bumetanide and furosemide, but the $K_{1/2}$ for inhibition were the same, 2 mM [48]. It was subsequently shown that NEM was activating a latent K/Cl pathway, as discussed above [42,86]; these vastly different drug sensitivities could then be understood in terms of two separate transport pathways. Recent studies on swelling-induced K efflux from young human red blood cells have confirmed and extended these earlier results [90]. No effect on the K efflux was observed at 0.1 mM bumetanide. At higher drug concentrations, bumetanide was a somewhat less effective inhibitor of K efflux than furosemide. Similar results were also obtained recently by Kaji for swelling-induced K influx in human red blood cells [87]. The corresponding opposite order of effectiveness for furosemide and bumetanide in inhibiting K/Cl and Na/K/Cl cotransport in duck red blood cells was recently presented in a preliminary report, as mentioned above [75]. The higher sensitivity of Na/K/Cl cotransport to bumetanide and furosemide than K/Cl (and Na/Cl) cotransport in other cell systems was recently reviewed [26]. The pharmacological difference may be sufficiently widespread and consistent to be employed as a diagnostic test for determining which type of cotransporter a Cl-dependent K flux is. This could be particularly useful for a K efflux when it is difficult to manipulate intracellular Na.

Dog red blood cells normally have Na and K concentrations similar to those in plasma [95]. The driving forces for Na/K/Cl and K/Cl cotransport are not present physiologically, and these pathways are not evident. However, if dog red blood cells are modified experimentally to have a high K concentration, there is a K efflux analogous to those in human and LK sheep red blood cells in that it requires Cl and is inhibited by furosemide [96]. The apparent presence of a pathway for K/Cl cotransport in cells essentially lacking a K gradient, cell to plasma, underscores our view of mammalian red cells collectively as museums for membrane transport pathways. (See [97] for a brief comparative consideration of volume-sensitive K/Cl cotransport in mammalian red blood cells.)

The metabolic dependence of K/Cl cotransport in mammalian red blood cells has been studied in both human and LK sheep red blood cells. Lauf [98] reported no effect on the Cl-dependent K *efflux* from LK cells by depletion of intracellular ATP. In contrast Logue et al. [80] reported an increase in passive K *influx* in LK cells after brief metabolic depletion. Lauf [77] later confirmed that Cl-dependent K influx was increased by depletion, but efflux was not. It is difficult to interpret these results; however, they do provide a warning about possible asymmetry of the K/Cl transport system in LK cells.

More striking results were obtained on metabolic dependence of the NEM-stimulated component of K/Cl cotransport in LK sheep cells. Lauf [77] showed that it is abolished by metabolic depletion and can be restored by metabolic repletion. Influx and efflux were affected in the same way. Logue et al. [80] obtained a similar result: in LK cells depleted briefly (2 hr), there was a reduction in the stimulation of K influx by NEM. Thus, there seems to be a modification in metabolic dependence concomitant with the enhancement of K/Cl cotransport by NEM, although the specific nature of either the dependence or its modification is unknown (see [82] and [83] for reviews).

Lauf et al. [99] have also investigated the metabolic dependence of NEM-activated K/Cl cotransport in human red blood cells. In cells incubated in 2-deoxyglucose for 6 hr, ATP was reduced by about an order of magnitude to less than 0.1 mmol/L cells, and the NEM-activated component of K influx was reduced eightfold. As with sheep cells, there is no direct evidence, beyond a reasonable supposition, that ATP is the critical metabolite. The metabolic dependence of K/Cl cotransport has also been looked at in resealed ghosts of human red blood cells [94]. It was shown that cotransport depended on MgATP (not free ATP), and the apparent affinity was surprisingly high, with a $K_{1/2}$ for MgATP < 15 μM. Preliminary evidence was presented for the failure of nonhydrolyzable analogues of ATP serving to support cotransport, which led to the suggestion that the transporter, or an associated regulatory element, is phosphorylated [94]. If this observation is correct, it is the second step, long after the first step, toward an understanding of the dependence of cotransport on metabolism.

As with all volume-sensitive transport pathways, the mechanisms by which a cell senses its volume change, and then transmits a signal to the K/Cl cotransporter, are completely unknown. Likewise the mechanism by which the transporter responds to the signal is completely obscure. Two obvious possible mechanisms for the volume sensor are: (a) a change in the concentration of an intracellular solute in consequence of cell shrinkage or swelling, or (b) a change in contacts or interposition between the membrane and another structure, such as the cytoskeleton, due to mechanical changes in the membrane [100]. Because relatively small volume changes (~10%) can cause large changes in transport (~ sevenfold; [39]), a change in concentration of a cytoplasmic solute, a second messenger, seems unlikely. The concentration of a solute serving as a volume sensor seems particularly unlikely in view of the volume sensitivity of cotransport in resealed ghosts, in which intracellular solutes are essentially replaced by the constituents of the resealing solution. A high-gain transduction mechanism seems necessary, because of the large change in transport rate for a small volume change. A high gain might be provided by a change in membrane/ cytoskeleton contacts involving an area of the membrane far exceeding the area occupied by the cotransporters [101]. Nevertheless a change in a solute concentration, even a major intracellular solute such as Cl, has not entirely been ruled out as volume sensor [100]. Another possible mechanism for a volume sensor is a combination of the above two mechanisms. A change in membrane/cytoskeleton contacts could cause a local redistribution of a critical solute. Candidate solutes include ATP, known to reside in a membrane-associated compartment [102], and divalent cations. Lauf [103] has shown that Ca and Mg inhibit cotransport in LK cells and removal of divalent cations from the cells using A23187 activates cotransport. A change in the spatial relation between charged groups on the membrane and the cytoskeleton could cause redistribution of the anion ATP or cations Ca or Mg. Charged solutes associated with the membrane and/or cytoskeleton would be largely retained in resealed ghosts. There are various possible ways in which high gain could be achieved in such a scheme. Nevertheless, all these suggestions are largely (or entirely) speculative.

III. CALCIUM-DEPENDENT K CHANNEL (THE GARDOS PATHWAY)

Most animal cell types respond to an elevation in cytosolic free Ca by activating a K conductance pathway in the plasma membrane. The phenomenon was discovered in red blood cells and is referred to as the "Gardos effect" because Gardos was the first to recognize the link between Ca and K permeability [104]. Although calcium is the most physiologically interesting activator of the mechanism, strontium [105] and lead [106-109] can trigger the channel as well [110]. Over the years, red blood cell experimentalists have invested much thought and imagination in characterizing this transport pathway, as is documented in recent reviews [104,110,111]. The emphasis in the present account will be on work reported since 1983.

The effect has been studied in several red blood cell preparations. Intact cells manifest the selective increase in K flux under various experimental circumstances designed to increase the cytosolic free calcium level. Typical manipulations include: (a) raising external Ca to very high levels [112], (b) puncturing the cell with a microelectrode [112], (c) depleting the cell of ATP so that the Ca extrusion pump cannot compensate for the inward leak of Ca [104], (d) using a Ca pump inhibitor such as vanadate to permit Ca to accumulate [105, 113], (e) treating the cells with external Ca plus an agent, such as A23187 or propranolol, that increases Ca influx [114,115], or (f) making use of an endogenous Ca-Na exchanger to increase passive Ca entry [96,116]. Resealed, right-side-out ghosts manifest the Gardos effect. This preparation has been useful in clarifying the sidedness of the action of Ca [117] and in permitting the

Table 1 Electrophysiology of the Gardos Pathway

	Resting, normal human red blood cell in plasmalike medium	Same cell with Ca activation of Gardos K pathway
E_K	-94 mV	-94 mV
E_{Cl}	-8 mV	-8 mV
P_K	10^{-10} cm/sec	10^{-7} cm/sec
P_{Cl}	10^{-8} cm/sec	10^{-8} cm/sec
V_m	-10 mV	-70 mV

Calculations are for a human red blood cell containing K and Cl of 140 and 85 mmol/kg cell water, respectively, bathed in a medium that is 4 mM K and 115 mM Cl. E_K and E_{Cl} are the equilibrium potentials for K and Cl. P_K and P_{Cl} are the permeabilities for K and Cl. V_m is the membrane potential.

incorporation of dyes that report free Ca levels [118]. The process of lysing and resealing cells may by itself activate the Gardos pathway [119]. Inside-out vesicles can be prepared that show the effect. In this preparation the behavior of the K pathway in response to graded concentrations of free Ca can be observed, and an estimate can be made of the number of K transport routes per cell [120]. Patches of red blood cell membrane have recently been found to contain K channels that are opened by Ca at the cytoplasmic face [121,122].

The cation movements that are activated by internal Ca occur through conductive channels that have a K/Na selectivity of 17:1 to 100:1 [122]. The only other monovalent cation that permeates the Gardos channel as well as K is Rb. Normally the large K gradient between cytoplasm and plasma contributes little to the red blood cell's membrane potential. This is because the cell is 100 times less permeable to K than to small anions. When the Gardos channel is activated, K permeability rises to such a degree that the membrane potential is dominated by the K gradient, and if the cell is in a physiologic situation with high internal and low external K, the membrane hyperpolarizes (Table 1) [123]. Net outward movements of K under these circumstances are restrained by the hyperpolarization. Inhibitors of chloride permeability, such as 4,4'-diisothiocyanostilbene-2,2'-disulfonic acid (DIDS) accentuate the membrane hyperpolarization that occurs with the activation of Gardos channels and thus retard net K efflux [124]. Pre-equilibration of cells with anions that are more conductive than Cl, such as nitrate or thiocyanate, reduces the change in membrane potential caused by the activation of the K conductance pathway and speeds the movements of K in response to concentration gradients [96,116].

Some insights into the system have come from a preparation of one-step inside-out vesicles conceived by Lew [120]. These structures contain 0-1 Na-K pumps and 0-1 Gardos channel per vesicle, some having both transport systems.

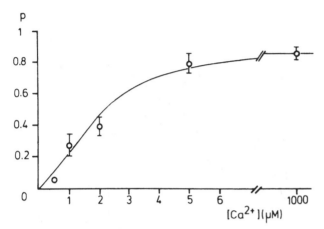

Figure 5 Probability, p, that a K channel will be open, as a function of free calcium, $[Ca^{2+}]$. Data points are from membrane patches. Solid line represents normalized rate constants of K loss from red cell ghosts. From Grygorczyk et al. [122], with permission.

The density of Ca-activated K channels on the cell membrane is of the same order of magnitude as the number of Na-K pumps, being ~100-200 per cell. The K channels were shown to exhibit heterogeneity of response to a given level of calcium (see below).

A powerful way to study the Ca-activated K channel is the single channel recording or patch clamp technique, first reported in red blood cells by Hamill [121,125] and pursued by Grygorczyk et al. [122] and Furhmann et al. [126]. Correlative data from patch clamp studies and tracer Rb fluxes in intact cells have been interpreted to show that the Ca-activated K flux occurs through aqueous channels that are intermittently open or closed. A single channel when open has a zero current conductance of 22 pS. As the cytosolic Ca increases the probability of a channel being open increases. In human blood 75% of red blood cells have one to five Gardos channels each, and 25% have 11-55 channels each. There is evidence that movements of K through the Gardos pathway are by single-file electrodiffusion, with an average of 2.7 K ions in the channel at any one time [127].

The level of free Ca required to activate a K channel has been studied in intact cells, resealed ghosts, inside-out vesicles, and membrane patches. Half-maximum K fluxes are seen at anywhere from 0.4 [128] to 2 μM Ca^{2+}, and maximum activation of the channel is seen above 6 μM Ca^{2+} [118,129] (Figure 5). Under normal conditions human red blood cells have a total Ca concentration of 10-50 $\mu mol/L$ cells, but most of this is bound. The free Ca level in normal, ATP-replete human red blood cells has recently been estimated at 10-70 nM

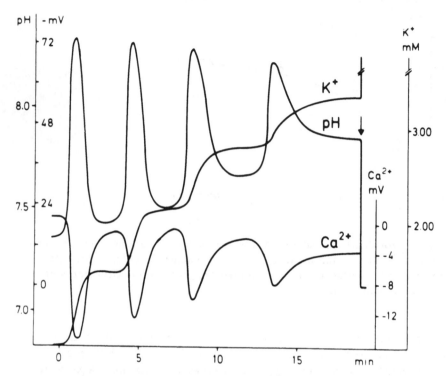

Figure 6 Extracellular pH, K activity, and $[Ca^{2+}]$, all as functions of time. At zero time the calcium ionophore was added to a suspension of intact red cells. The pH reflects the membrane potential. Note oscillations, as calcium moves in and out of the cells, causing hyperpolarization and a stepwise accumulation of K in the medium as this cation is lost from cells during the pulses of activation of the Gardos pathway. From Vestergaard-Bogind and Bennekou [132], with permission.

[130,131]. The Gardos pathway is reversible, i.e., it can be turned off. This was especially well demonstrated by Vestergaard-Bogind and Bennekou [132], who showed that alternate activation and deactivation of the Gardos pathway in human red blood cells leads to oscillations of K conductance and membrane potential (Figure 6). Because K flux through the channel may cease despite sustained elevations of cytosolic Ca, Vestergaard-Bogind [133] suspects that activation of the channel may depend more on dCa^{2+}/dt than on any particular free Ca level.

Many studies attest to the heterogeneous responses of a red blood cell population to a submaximal induction of the Gardos effect. Early techniques could not distinguish between differences in levels of cytosolic free Ca among cells, and

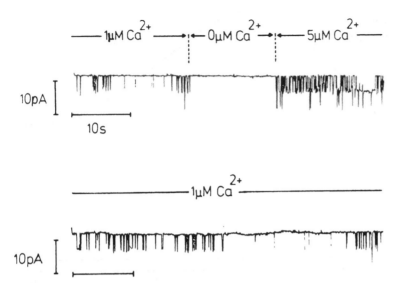

Figure 7 Effect of different Ca^{2+} concentrations on the single-channel current in cell-free membrane patches. Note increasing open time (downward deflection) with increasing $[Ca^{2+}]$. From Grygorczyk et al. [122], with permission.

differences among cells in response to a given calcium level [118]. Some interpretations of the data have lead to the suggestion that K channels respond to a rise in the cytosolic free Ca concentration either maximally or not at all [134, 135]; this "all-or-none" idea has been supported by studies with inside-out vesicles [120,134]. In contrast to this notion, the data from patch clamping studies indicate that K permeability in red blood cells is capable of a graded response: although the conductance of each channel is constant, and although each channel is either open or closed, the probability of the open state varies with the level of cytosolic free Ca [122] (Figures 5 and 7). How the patch clamp observations can be reconciled with the apparent all-or-none behavior of channels observed in inside-out vesicles is a matter of debate. Derivative preparations, e.g., vesicles, ghosts, and patches, may show responses that do not reflect the physiology of intact cells.

Adorante and Macey [136] report an intriguing maneuver by which they think they can transiently externalize the Ca triggering site for the Gardos channel. This is done simply by bathing cells in a low-ionic-strength medium. Addition of Ca to such a suspension causes a 2–6 min pulse of conductive K efflux that is dependent on small concentrations of external K and inhibited by oligomycin. The authors make a convincing case that the channel was activated by external rather than internal Ca during the transient.

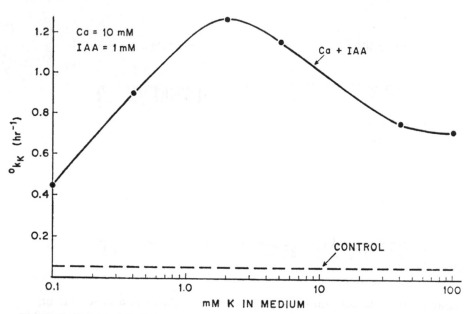

Figure 8 Outward K flux rate constant as a function of medium K concentration. Gardos effect stimulated by incorporation in the medium of calcium and iodoacetate. Note the stimulation of K efflux by external K, and the optimum at 2 mM. From Blum and Hoffman [117], with permission.

Sodium and K ions have complex mediating effects on the Gardos pathway. In the presence of optimum levels of cytoplasmic free Ca the channel will not open unless there is K on the outer face of the membrane. This effect of external K reaches an optimum at 2-3 mM [111,118] (Figure 8). Internal Na exerts an inhibitory effect on the channel at low external K [118]. Magnesium ions raise the threshold for Ca activation of the channel in inside-out vesicles [134]. External Ca has no modulating effect on the channel [118]. A high proton concentration is inhibitory to Ca-activated K movements, as demonstrated by Stampe [137] (Figure 9). Below pH 5.8 there is virtually no K flux. As internal pH is increased, K conductance rises in a sigmoid fashion and appears to saturate at pH 6.8. Extracellular pH changes had no effect on the channel.

The K flux in cells exposed to an increase in internal Ca is modulated by a variety of treatments that influence the redox potential of the cell [126,138–140]. The system behaves as if it had a membrane component with a standard redox potential at pH 7.5 of 47 mV (Figure 10). When the postulated mediator is converted from its oxidized to its reduced state, the K channel undergoes a 15-fold increase in affinity for cytoplasmic Ca. Although the data suggest involvement of a membrane flavoprotein, such as reduced nicotinamide adenine

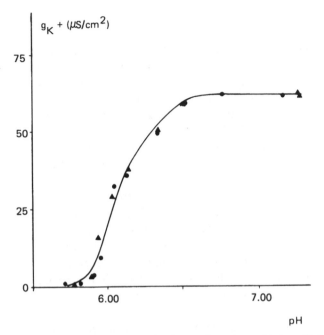

Figure 9 K conductance (g_K) as a function of intracellular pH (pH_c) in intact red cells treated so as to activate the Gardos pathway. From Stampe and Vestergaard-Bogind [137], with permission.

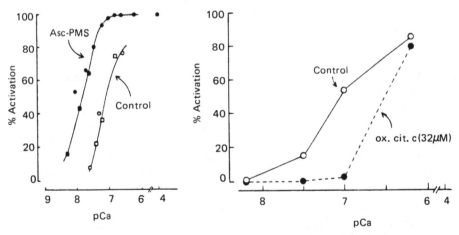

Figure 10 Per cent activation of the Ca-dependent K channel in inside-out red cell vesicles, as a function of ionized calcium (negative log scale). The effect of severe reducing conditions (ascorbate + phenazine methosulfate) is shown in the left panel. The effect of severe oxidizing conditions (oxidized cytochrome C) is shown in the right panel. Note that the affinity of the pathway for Ca increases with reducing conditions and decreases with oxidizing conditions. From Alvarez et al. [138], with permission.

dinucleotide (NADH), ferricyanide reductase or NADH-cytochrome c reductase, a species comparison of Gardos channel activity showed no correlation with the content of these enzymes. It seems unlikely that the data can be explained on the basis of oxidation and reduction of cysteine groups [140].

There are several lines of evidence in favor of the possibility that a cytoplasmic protein mediates the activation of the Gardos channel. Studies with calmodulin inhibitors have shown divergent results. Either the calmodulin inhibitors reduce the Ca-activated K flux, suggesting that calmodulin modulates the opening of the channels [118,141], or the calmodulin inhibitors accentuate the Ca-activated K flux, suggesting that the drug effects are due to an increase in cytosolic free Ca caused by an interference with calmodulin's activation of the Ca pump [142]. Pape et al. [143] found a threefold stimulation of the channel when calmodulin was applied to inside-out red blood cell ghosts. Plishker [144] recently reported that a 23,000-D cytoplasmic protein, distinct from calmodulin, becomes associated with the cell membrane when internal free Ca rises. Antibodies to this protein inhibit Ca-activated K transport in resealed ghosts.

A number of compounds inhibit the Ca-activated K channel in red blood cells, as tabulated by Sarkadi and Gardos [104]. Quinine and quinidine appear to do so by displacing K from its external binding site [145]. Carbocyanine dyes, useful for the measurement of membrane potential in red blood cells, inhibit the Gardos pathway [146]. Cetiedil, an agent that inhibits hemoglobin S red blood cells from sickling in vitro, is a Gardos pathway inhibitor [147,148], as is doxorubicin, an anticancer drug [149]. Barium ions block K movements through the channel [136]. Vanadium compounds react in a complex way with the Gardos channel, depending on the oxidation state of the metal [126,150]. Iodoacetate, an agent often used to deplete red blood cells of ATP to facilitate Ca loading, has an inhibitory action on the Gardos mechanism that Plishker [151] attribute to carboxymethylation of a 23,000-Da cytoplasmic protein that may be necessary for activation of the channel (vide supra). A component of the venom of the Israeli scorpion (*Leiurus quniquestriatus*) has been found to inhibit the Gardos channel [152]. Apamin, a bee venom toxin that inhibits some classes of Ca-activated K channels in excitable tissues, has no effect on the red blood cell channel [152]. Oligomycin is a well-known Gardos channel blocker [124].

Calcium-activated potassium movements do not occur in the red blood cells of all species [140,153–156]. Humans, dogs, rats, guinea pigs, *Amphiuma*, frogs, fish, and chickens all have the channel in one form or another, whereas sheep, cow, and goat red blood cells lack it. Fetal sheep have it, but the pathway is lost in adult sheep, even in reticulocytes.

The membrane molecule that is responsible for Ca-activated K currents in red blood cells has not been characterized. Early suggestions that the channel represents a mode of function of the Na-K [124] or Ca pump have been disputed [154,157,158].

The role of Gardos channels in pathologic red blood cells is discussed in Chapter 23. The existence and physiological function of the pathway in excitable and secretory [159] cells is well reviewed by Sarkadi and Gardos [104]. It is paradoxical that a widespread and important transport system like Ca-induced K conductance should have been discovered and perhaps best characterized in red blood cells, where its contribution to normal or abnormal physiology is still debated.

IV. PASSIVE CALCIUM MOVEMENTS AND Ca/Na EXCHANGE

Estimates of passive Ca movements under physiological conditions are difficult. The ATP- and Mg-dependent Ca extrusion pump is so dominant that strategies have to be devised to inhibit it, so that residual fluxes can be measured. There is always the concern that the means used to block the Ca pump may have altered the passive pathways of interest. At present there is no inhibitor for the Ca pump as specific—and as free from influence on other cell functions—as is ouabain for the Na-K pump.

Inhibition of the Ca pump in studies of passive Ca movements has been accomplished by the use of lanthanum [160], vanadate [105,113,161,162], ATP-depletion [163,164], and magnesium depletion [161,164]. Recently, Ca chelating agents have become available that can be introduced into intact red blood cells in the form of permeant esters. Once in the cell the chelators are de-esterified and rendered impermeant. Such compounds are widely used to measure cytosolic free Ca: their fluorescence, or their nuclear magnetic resonance spectrum, reflects the degree to which they are liganded with Ca [131,165]. Lew et al. [130] conceived the strategy of using compounds of this sort to lower cytosolic free Ca levels to such a degree that the Ca pump is virtually inactive. The initial rate of Ca accumulation in cells loaded with such agents then gives an estimate of passive Ca influx. Although the chelating esters generate formaldehyde, the metabolic lesion induced by this substance [166,167] can be bypassed if pyruvate is included in the cell suspension [168]. As promising as the use of these esters appears, there may still be artifacts, as suggested by recent reports on squid axon [169,170]. Using this approach Lew et al. [130] and McNamara and Wiley [164] have found that under physiological, steady-state conditions Ca flux is 20-40 μmol/L cells $\times$ hr. A curve for Ca influx rate versus external Ca appears to have both a linear and a saturable component, the latter with a $K_{1/2}$ of 0.95 mM and a V_{max} of 42 μmol/L cells $\times$ hr [164] (Figure 11).

Many questions and suggestions with regard to passive Ca movements in human red blood cells remain to be investigated. What is the specificity of the

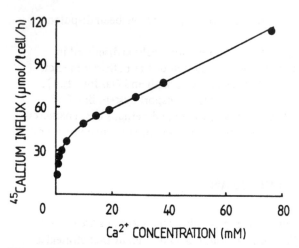

Figure 11 Dependence of Ca influx on external Ca in human red cells loaded with quin-2, showing both saturable and non-saturable components. From McNamara and Wiley [164], with permission.

transporter? Are any of the components of this flux conductive? Can the transporter mediate obligate Ca–Ca exchange? Are there Ca channels in red blood cells like those in excitable cells, and can a response to agents such as verapamil [113] be confirmed? What are the influences of monovalent cations [113], divalent cations [105], metabolic state of the cell [130], sulfhydryl reagents [105], and cell volume [171]?

The red blood cells of some carnivores transact Ca–Na exchange, a pathway first described in heart muscle [172] and squid axon [173]. Although unrecognized at the time, the influence of this transporter in red blood cells can be discerned in the experiments of Davson [116], who in 1942 reported that a sample of defibrinated dog blood placed in 10–15 times its volume of isotonic KCl would rapidly hemolyze. Apparently, this phenomenon did not occur with the blood of other species. We now know that when Davson placed dog blood into KCl, the low external Na concentration stimulated Ca influx into the high-Na cells via Ca–Na exchange. The rise in cytosolic free Ca triggered the Gardos pathway (vide supra), and the cells underwent colloid osmotic hemolysis due to entry of K from the medium [96]. Omachi et al. [174] in 1961 showed that incubation of dog red blood cells in an all-K medium caused a large influx of Ca. Niedergerke [175] had recently published experiments on frog hearts, showing that the inotropic effect of replacing Na in the medium is due to a stimulation of Ca entry. Omachi postulated that Ca accumulation in dog red blood cells in zero-Na media might be due to a transport system similar to the one in frog heart.

It has been known for many years that mature red blood cells from most dogs have a high internal Na/K concentration ratio [95] and lack a Na-K pump [176]. Nucleated precursors of dog red blood cells have Na,K-ATPase, which is proteolytically degraded as the cell matures [177]. The elegant pump-leak formulation for red blood cell volume control conceived by Tosteson and Hoffman [3] could not apply to a cell that lacks Na-K pumps. Experiments were therefore undertaken that demonstrated that mature dog red blood cells have a mechanism, distinct from the Na-K pump, that functions both in vivo [178] and in vitro [179] to extrude Na and regulate cell volume. Active Na transport in dog red blood cells occurs only if Ca is included in the medium [179,180].

The early studies of Omachi, cited above, and the observations on Ca-dependent Na efflux stimulated a search in dog red blood cells for the presence of a Ca–Na exchanger [161,181-185]. Passive Ca movements are stimulated by trans Na and inhibited by cis Na. Passive Na efflux is stimulated by external Ca. Ca pumped into inside-out vesicles of dog red blood cells is released in the presence of extravesicular Na [185]. Although these fluxes are consistent with the idea that dog red blood cells have a Ca–Na exchanger, the stoichiometry and electrogenicity [186] of the system remain to be clarified. Calcium influx via the Ca–Na exchanger is activated by cell swelling and alkalinization [181,182], but Ca efflux through the pathway seems not to respond to changes in cell volume or pH [161]. Amrinone, a cardiotonic agent, stimulates the Na–Ca exchanger [183], as does the replacement of Cl with nitrate or thiocyanate. The anion substitution effects might be interpreted as indicating electrogenicity [184]. Inhibitors include quinidine [181,187] and doxorubicin [149]. The cells have an ATP-dependent Ca pump [161,182,185].

Recently it has been possible to obtain a 10-fold stimulation of the Ca–Na exchanger in dog red blood cells by pretreating them with small amounts of diamide, a thiol oxidizing agent. The conditions for stimulation are quite specific and require that at the time of exposure to diamide the cells be in a Na-free medium containing >1 μM Ca. The effect is increased if the cells are swollen at the time of pretreatment. Dithiothreitol reverses the effect of diamide treatment [188,189]. These results are in some respects similar to those of Reeves et al. [190], who showed stimulation of Ca–Na exchange in sarcolemmal vesicles with a combination of redox agents, including ferric ions and reduced glutathione. The ionic conditions favoring oxidative activation of Ca-Na exchange in sarcolemmal vesicles are different from the ones with dog red blood cells (Na is required, Ca inhibits). The data suggest that the interconversion of -SH and -SS-groups is critical, either for the activity of the Na–Ca exchanger itself or for a "switch" that turns the transporter on.

Calcium-sodium exchange has been demonstrated in the red blood cells of another carnivore, the ferret, in which measurement of the Ca and Na fluxes shows that outward movements of Na through the system are sufficient to keep the cells in a steady state. Calcium that enters the cell in exchange for Na is promptly extruded by the ATP-dependent Ca pump [162,191]. Murine erythroleukemia cells apparently have a Ca–Na exchanger that has been claimed to mediate stimuli toward terminal differentiation [192], but subsequent work on those cells has resulted in alternative views [193].

V. BAND 3-MEDIATED CATION MOVEMENTS

The notion that cation movements can occur through anion-selective channels was conceived by Wieth [194], who observed that Na and Li fluxes in human red blood cells were selectively increased in the presence of bicarbonate ions, particularly in alkaline media. Wieth postulated that the bicarbonate effect is due to the formation of negatively charged ion pairs, according to the following scheme.

In any bicarbonate solution, some carbonate is present; the dissociation of bicarbonate is favored by a high pH.

$$HCO_3^- \longrightarrow H^+ + CO_3^{2-}$$

Highly hydrated cations, such as Li^+ and Na^+, but not K^+ or Rb^+, combine with carbonate to form negatively charged ion pairs:

$$Li^+ + CO_3^{2-} \longrightarrow LICO_3^-$$

$$Na^+ + CO_3^{2-} \longrightarrow NaCO_3^-$$

The dissociation constants for the two ion pairs are such that in 150 mM solutions of the respective bicarbonates, $LiCO_3^-$ would be 2 mM and $NaCO_3^-$ would be 0.5 mM [195]. Because anions can exchange across the red blood cell membrane approximately a million times faster than most passive cation fluxes, even these relatively small concentrations of ion pairs can account for large fluxes of Na and Li. Thus, in high bicarbonate media, Na and Li move faster via the pathway available to the ion pairs than via cation permeability routes.

The hypothesis about the movement of Na and Li as ion pairs antedated the discovery of specific inhibitors for anion exchange and the identification of band 3 as the anion transport protein. Subsequent work has, however, amply confirmed Wieth's original formulation. DIDS (4,4'-diisothiocyanostilbene-2,2'-disulfonic acid), SITS (4-acetoamido-4'-isothiocyanostilbenne-2,2'-disulfonic acid) dipyridamole, phloretin, and other anion exchange blockers strongly inhibit bicarbonate-stimulated Na and Li fluxes [195–198].

One does not need to incubate cells in 150 mM bicarbonate to see the effects of ion pair formation on steady-state red blood cell cation contents. The effect can be discerned under pathophysiological circumstances in vivo. Patients with metabolic alkalosis have increased steady-state Na levels within their red blood cells, reversible with correction of the serum bicarbonate [199]. The ion-pair phenomenon can be used to load human red blood cells with solute and water in preparation for in vivo physiological studies [200].

Other nonphysiological anions such as sulfite [201], oxalate, phosphite, and phthalate, can form ion pairs with Na and Li that are transported by anion exchange [196,197]. Lead ions can cross the red blood cell membrane by interacting with band 3. More than 90% of passive lead flux is inhibited by anion transport blockers, and lead fluxes are proportional to the bicarbonate concentration [108]. The ion species that lead forms when it is transported via band 3 is not clear, however, inasmuch as there is evidence that the process involves net charge transfer.

Quite a different mode of cation movement through the band-3 anion transporter is suggested by the reinvestigation of a long-known [202,203] phenomenon, namely, the massive increase in K flux that occurs when human red blood cells are suspended in solutions of low ionic-strength. Jones and Knauf [204] found that the cation movements seen under these conditions are sensitive to DIDS and other anion exchange inhibitors, and that there is no selectivity of Na over K. The results were interpreted in terms of possible perturbations of the band 3 anion transport protein that might alter its selectivity for anions over cations. The critical influences of nonionic media on the red blood cell membrane are thought to include: (a) low external Cl, causing a recruitment of anion binding sites to the cytoplasmic face of the band 3 protein, (b) alteration of the charge density at the cell surface [205], and (c) reversal of the membrane potential. No data are available regarding the mechanism of the cation movements through band 3 under these circumstances—whether conductive or coupled or saturable—nor it is clear whether these observations are relevant to physiological or pathological cation movements. A report that anion transport inhibitors block the rise in Na and K permeability seen when red blood cells containing hemoglobin S are caused to sickle [206] requires verification.

Other more circumstantial evidence for mediation of cation movements by band 3 under unusual circumstances is discussed by Grinstein and Rothstein [207] and Solomon et al. [208].

VI. Na(Li)–Na(Li) COUNTERTRANSPORT

Tosteson and Hoffman [3] reported that sheep red blood cells shuttle Na for Na in a 1:1 ratio, via a mechanism distinct from the Na/K pump. Bovine [209] and human [11,12,55] red blood cells were found to have a similar transport process. Because no net gain or loss of Na could occur through such a transporter, the physiological importance of Na–Na exchange was a matter of conjecture—a phenomenon to be aware of when interpreting the results of isotope flux experiments. Many features of the system were described in bovine red blood cells by Motais and Sola [210,211].

Table 2 Comparison of Na and K Contents and Li-Na Exchange Capacity
Among Various Species of Red Blood Cell

Species	Cell cation contents (mmol/L cells)		Li–Na exchange (mmol/L cells X hr)
	Na	K	
Man	9	96	0.12
HK sheep	10	84	0.76
LK sheep	89	12	1.16
Rabbit	12	100	2.60
LK cow	62	21	3.50

Adapted from ref. 216, used with permission.

In the early 1970s, as lithium therapy for bipolar manic-depression became
popular, interest in the Na-Na shuttle was greatly stimulated when it was reported that this transport system could also carry Li and could exchange Li
for Na in human [212,213], bovine [214], rabbit [215], and sheep [216] red
blood cells.

Table 2 shows that human red blood cells have a relatively weak Na(Li)–
Na(Li) exchanger, compared to the red blood cells of some other mammals.
The mechanism follows strict Michaelis-Menten kinetics [210,217] with apparent K_m as shown in Table 3 for human cells. Kinetic values in other species
are comparable [210,214,216]. The mechanism is inhibited by low pH [210,
214–216], but protons are not transported by it [215], indeed, Na and Li are
the only ions that interact with the transporter. By varying systematically both
cis and trans Na and Li, Hannaert and Garay [218] showed that the system in
human red blood cells behaves according to a consecutive or "ping-pong"

Table 3 Affinities of the Na(Li)–Na(Li)
Exchanger in Human Red Blood Cells

	Apparent K_m (mM)	
	Na	Li
Outside	25	1.5
Inside	9	0.5

Adapted from ref. 217, used with permission.

model. Arrhenius plots for the transporter show a break at 20–30°C. Below the break the activation energy is 28 kcal/mol; above it the value is 8 kcal/mol [219].

Sulfhydryl reagents inhibit the Na(Li)–Na(Li) exchanger. N-ethylmaleimide, a permeant alkylating agent, inhibits 60–80% of the exchange without changing the affinities for Na or Li. The rate of inhibition is strongly influenced by the ionic environment of the cells during N-ethylmaleimide exposure, being increased in media that contain transportable substrates, i.e., Na or Li. P-chlormercuribenzene sulfonate, a sparingly permeant agent, depresses Na(Li)–Na(Li) exchange in human red blood cells only after long incubation times; bovine red blood cells, by contrast, are quite susceptible to this agent [211,214,220–222]. Phloretin and its analogues are potent inhibitors of the system [214,216,222, 223]. Furosemide was found to be an inhibitor by Sarkadi et al. [217], but this diuretic is not thought to be effective in other laboratories [216]. The discrepancy may be explained by the action of the Na + K + 2Cl cotransport system in some of the experiments of Sarkadi et al. [217].

Suggestions that the Na(Li)–Na(Li) exchanger represents a mode of function of some other, more physiologically plausible system, have been refuted. Thus, the mechanism does not participate in Na–H exchange [215], or volume regulation [224]. Although the Ca–Na exchanger in swollen dog red blood cells is capable of Na–Na exchange [181], bovine cells show no reactivity of Ca with their high capacity Na(Li)–Na(Li) exchanger [214]. The exchange is not likely mediated by negatively charged ion pairs of Li and Na (vide supra), because it is not inhibited by band 3 blockers [216]. Although Na–Na exchange can be transacted by the Na–K pump, such fluxes are ouabain-sensitive [225,226]. Most studies of the Na(Li)–Na(Li) exchanger are done in the presence of ouabain, so as to remove any influence of the Na–K pump.

In human (and bovine) red blood cells substantial variations in the capacity of the Na(Li)–Na(Li) exchanger are found from individual to individual. It is because of these differences in countertransport capacity that, among patients receiving long-term Li therapy, the steady-state red blood cell Li content varies [223,227,228]. People with a high countertransport capacity have a cell/plasma Li concentration ratio that approaches the ratio for Na, with R values ($Li_{cell}/Li_{plasma} = R \times Na_{cell}/Na_{plasma}$) of 1.7 to 2.1, whereas individuals with a low exchange capacity have values for R of up to 5.4 [216,223,224]. This heterogeneity among human donors has lead many investigators to seek correlations between Na(Li)–Na(Li) exchange and familial disorders such as bipolar manic-depressive disease and essential hypertension. The relationships are complex and controversial, as recently reviewed [230]. Activity of the transporter seems to be increased in pregnancy [231] and transiently suppressed in some patients after hemodialysis [232,233].

Table 4   Na-H Exchange in Red Blood Cells of Various Species

Species	Activating agent
Amphiuma	Cell shrinkage [237,242]
Frog	Catecholamines[a], cell shrinkage [244]
Trout	Catecholamines[a] [245,248], hypoxia [262]
Dog	Cell acidification, cell shrinkage, internal Li [254]
Rabbit	Cell acidification, cell shrinkage [256]
Human	Cell acidification, internal Ca [257,258]

[a]Effects of catecholamines duplicated by cyclic adenosine monophosphate.

VII. Na-H EXCHANGE

Except for an unconfirmed report [234], the first account of an inhibitory effect of amiloride on Na movements in red blood cells was that of Siebens and Kregenow [235]. Amiloride is known to inhibit both conductive Na movements and electroneutral Na-H exchange [236]. Cala [237] showed in *Amphiuma* red blood cells that the amiloride-sensitive fluxes are electroneutral; thus emerged the concept that red blood cells are capable of Na-H exchange. To date Na-H exchange or amiloride-sensitive Na-H fluxes have been found in the red blood cells of *Amphiuma* [100,235,237-243], frog [244], trout [245-250], dog [188,189,251-255], rabbit [215,256], and human [257,258] (Table 4).

The Na-H exchanger in red blood cells transports Na, H, and Li [242,254]. The stoichiometry of countertransport has been hard to measure accurately, because in all species studied pH gradients established by the operation of the exchanger tend to be dissipated rapidly via Cl-HCO_3 exchange, even in the presence of inhibitors such as dipyridamole [255]. Nevertheless, in *Amphiuma* cells a 1:1 stoichiometry was approximated, and a curve of amiloride-sensitive Na influx versus external Na looks as if it would have a Hill coefficient of unity, making a stoichiometry of, e.g., 2:2 or 3:3 unlikely [243].

In support of 1:1 coupling of the Na-H counterflow are Cala's observations, based on electrophysiological data in *Amphiuma* cells, that activation of the Na-H pathway affects neither the membrane potential nor the change in membrane potential in response to valinomycin. Thus, when the Na-H exchanger is operating the partial conductances of the membrane for Na or H are not altered: the transporter is electrically silent [237]. A similar conclusion was reached by less direct means in dog cells [255]. Furthermore, Cala [238,239] has formulated a thermodynamic force-flow analysis for cation movements that, when applied to the Na-H exchanger, is compatible with an electroneutral countertransport model.

Protons, in addition to being preferred substrates for the Na-H exchanger, have modulatory effects on the transport mechanism that were first described in epithelial tissues, as reviewed by Aronson [259]. Thus, if the cytoplasm is made acidic, Na-H exchange is stimulated, while acidification of the medium inhibits the pathway, as demonstrated in dog [254] and rabbit [256] red blood cells. In dog cells Li ions act like protons in causing stimulation of Na-H exchange when applied to the cytoplasmic surface and inhibition from the exterior. The actions of lithium and protons may not be on the same site, however: dog red cell ghosts made acidic conduct Na-H exchange, but this pathway is not stimulated by internal Li in the ghost preparation (Parker, unpublished observations).

Na-H exchange in red blood cells is a "regulated" transport system. Unlike the Na/K pump, the Ca pump, and the anion exchanger, all of which are poised to transport whatever substrate is presented to them, the Na-H exchanger in red blood cells must be "switched on." One way to activate the mechanism is to acidify the cell interior selectively, but this is not easy to do in red blood cells because pH gradients are dissipated rapidly via Cl-bicarbonate exchange. Osmotic shrinkage is a powerful way to activate the exchanger in some red blood cell types. It has been postulated in lymphocytes that activation of Na-H exchange by cell shrinkage is mediated via the pH trigger, which changes its affinity for protons as a function of volume [260,261]. A similar relationship has been discussed in rabbit red blood cells [256]. The stimulation of Na-H exchange by hypoxia in trout red blood cells has been shown to be related to the change in hemoglobin conformation rather than to the lack of oxygen. It is postulated that connections between hemoglobin and the membrane are responsible for this effect [262]. Table 4 shows a list of animal species whose red blood cells have Na-H exchange (or amiloride-sensitive Na fluxes). Also shown are the means for triggering the countertransporter in the red blood cells of each species. In nonerythroid cells Na-H exchange is activated by growth factors, insulin, vasopressin, bradykinin, angiotensin, phorbol esters, diacylglycerol, fertilization, lectins, and miscellaneous agents [260], none of which, to our knowledge, affect red blood cells.

Inhibitors of the exchanger include amiloride and its analogues [236], quinidine [252,263,264], external Li and protons [254,256], and substitution of nitrate or thiocyanate for Cl [239,252,256]. The latter effect may be an indirect one, relating to the activation of the exchanger by volume stimuli, as discussed below. Phloretin, an inhibitor of Na(Li)-Na(Li) exchange, does not affect Na-H countertransport [257,258], and the two pathways have been carefully dissected in rabbit red blood cells [215]. In trout red blood cells, but not in dog cells, alkyl tin derivatives inhibit Na-H exchange [248,265].

Because activation of Na-H exchange seems to occur in a wide variety of cells when they are stimulated to grow, divide, or differentiate, attempts have been made to put this regulated transporter into the perspective of "second messengers" [260,266]. An important caveat against equating "amiloride sensitive" and "Na-H exchange mediated" has been issued [267]. Data in lymphocytes, fibroblasts, and other cell types have linked Na-H exchange with various protein kinases, inositol phospholipid metabolism, and Ca transients encountered in the process of cell stimulation. Red blood cell investigations have not contributed much to this endeavor. Although some red blood cells appear to have protein kinase C [268], we are not aware of any species of red blood cell that responds like lymphocytes, by turning on Na-H exchange in response to phorbol esters [269].

There are, nevertheless, interesting links between stimulus and response in red blood cells that require explanation. In *Amphiuma* red blood cells, for example, a delay period of several minutes is seen between the application of a stimulus (shrinkage) and the maximum response of the Na-H exchanger. If shrinkage is imposed under conditions (presence of amiloride, absence of external Na) that prevent the Na-H exchanger from responding, and if then those restraints are suddenly removed, there is a massive and rapid "overshoot" of Na-H countertransport [242]. These observations suggest that *Amphiuma* cells have an intermediate signalling step that takes time to develop. By manipulating the experimental conditions the cells can be primed, so that the influence of the accumulated mediator are exerted rapidly and effectively, as measured by Na-H exchange.

Another line of evidence that suggests an intermediate step between stimulus and response of the Na-H exchanger comes from manipulations that irreversibly activate the countertransporter. Glutaraldehyde and the non-crosslinking agent, *N*-phenylmaleimide, if applied to dog red blood cells when they are shrunken, render the Na-H exchanger effectively locked in the "on" position, so that reswelling of the cells now fails to turn off Na-H exchange. Na-H exchange activity in these "locked on" cells is not inhibited by replacing Cl with nitrate or thiocyanate [253]. This same anion replacement will strongly inhibit the activation of Na-H exchange by osmotic shrinkage in untreated red blood cells from dogs, *Amphiuma*, and rabbits [239,252,256,270] and by catecholamines in trout red blood cells [248]. Thus, nitrate and thiocyanate must be inhibitors, not of Na-H exchange per se, but of the process by which cell shrinkage or catecholamines activate Na-H exchange.

ATP depletion inhibits the activation by shrinkage of Na-H countertransport in dog cells [271], and there are indications that the system in human cells requires ATP [257,258]. These observations suggest that the linkage between stimulus and response involves some sort of kinase, as has been shown in lymphocytes [269,272].

VIII. MONOVALENT CATION CONDUCTANCES INDUCED BY EXTERNAL ATP

External ATP opens up a cation-specific conductance pathway in dog red blood cells through which Na, K, and Li (but not choline or divalent species) can pass [273-275]. Cells in which this channel is activated have a Na permeability of about 10^{-7} cm/sec, whereas normal Na permeability is about 10^{-10} cm/sec [276]. This action of ATP was first described in ascites tumor cells [277] and subsequently in a variety of other cell types, including mast cells and macrophages. So far as we are aware, the red blood cells of no other species—not even other carnivores, such as the ferret and cat—show this phenomenon. The ATP effect is maximal at about 0.1 mM and is reversible. If equimolar amounts of divalent cations (calcium or magnesium) are included in the medium, the ATP-induced conductance is prevented or abruptly terminated. The permeabilization does not depend on the catabolism of ATP or the transfer of a phosphate.

The importance of this instantaneous and dramatic effect is obscure. It would seem that no cell is ever bathed with a medium free of divalent cations. Adenine nucleotides have, however, been detected in extracellular fluid, where they have been postulated to function as neurotransmitters [278]. It may be that dog red blood cells have on their surface formes frustes of purinergic receptors.

IX. CONCLUDING REMARKS

One of the themes of this survey of "passive" cation transport pathways in mammalian red blood cells is diversity. We have emphasized striking examples among various species of the presence or absence of Na/K/Cl and K/Cl cotransport, Ca-activated K transport, Na/Na exchange, and Na/H and Na/Ca countertransport. This diversity has helped efforts to define the features of each pathway, but the use of red blood cells has its limitations. Mammalian red blood cells are unlikely to serve as an important source for the isolation of cation transport proteins. Although each red blood cell has 1,000,000 copies of band 3, the major anion exchange protein, cation transporters are much less abundant. There are 100-200 Gardos channels/cell [120], and estimates based on maximum fluxes and probable turnover numbers suggest that other passive cation transporters are present in approximately the same low abundance. (Ferrets may, according to an unverified estimate have 12,000 Na/K/Cl cotransporters per red blood cell [279].

The usefulness of future cation transport studies in red blood cells would appear to lie in further characterization of transport mechanisms and particularly in the elucidation of regulatory phenomena. There is scattered evidence, some of which we have presented, and some of which is reviewed in Chapter 22, that red blood cell transporters can be turned on and off by physiologically important stimuli, such as changes in volume, pH, PO_2, and the presence or absence of hormones. Studies in red blood cells may therefore afford insights of general interest relating to cytoplasmic pH control [280], epithelial function [281], cell growth and differentiation [282-284], and signal transduction [104,159].

REFERENCES

1. Davson, H. (1940). The permeability of the erythrocyte to cations. *Cold Spring Harbor Symp. Quant. Biol.* 8:255-268.
2. Krogh, A. (1946). Croonian lecture. The active and passive exchanges of inorganic ions through the surfaces of living cells and through living membranes generally. *Proc. R. Soc. Lond.* [Biol.] 133:140-200.
3. Tosteson, D. C., and Hoffman, J. F. (1960). Regulation of cell volume by active cation transport in high and low potassium sheep red cells. *J. Gen. Physiol.* 44:169-194.
4. Wieth, J. O., Funder, J., Gunn, R. B., and Brahm, J. (1974). Passive transport pathways for chloride and urea through the red cell membrane. In *Comparative Biochemistry and Physiology of Transport*, L. Bolis, K. Bloch, S. E. Luria, and F. Lynen (Eds.). North Holland, Amsterdam/London, pp. 317-337.
5. Young, J. D. (1983). Erythrocyte amino acid and nucleoside transport. In *Red Blood Cells of Domestic Animals*, N. S. Agar and P. G. Board (Eds.). Elsevier, Amsterdam, pp. 271-290.
6. Young, J. D., and Ellory, J. C. (1977). Red cell amino acid transport. In *Membrane Transport in Red Cells*, J. C. Ellory and V. L. Lew (Eds.). Academic Press, London, pp. 301-325.
7. Lew, V. L., and Beaugé, L. (1979). Passive cation fluxes in red cell membranes. In *Membrane Transport in Biology*, G. Giebisch, D. C. Tosteson, and H. H. Ussing (Eds.). Springer-Verlag, Berlin, pp. 81-115.
8. Glynn, I. M. (1957). The action of cardiac glycosides on sodium and potassium movements in human red cells. *J. Physiol. (Lond.)* 136:148-173.
9. Garrahan, P. J., and Glynn, I. M. (1967). Factors affecting the relative magnitudes of the sodium:potassium and sodium:sodium exchanges by the sodium pump. *J. Physiol. (Lond.)* 192:189-216.
10. Sachs, J. R. (1971). Ouabain-insensitive sodium movements in the human red blood cell. *J. Gen. Physiol.* 57:259-282.
11. Hoffman, J. F., and Kregenow, F. M. (1966). The characterization of new, energy-dependent cation transport processes in red blood cells. *Ann. NY Acad. Sci.* 137:566-576.
12. Lubowitz, H., and Whittam, R. (1969). Ion movements in human red cells independent of the Na pump. *J. Physiol. (Lond.)* 202:111-131.
13. Wiley, J. S., and Cooper, R. A. (1974). A furosemide-sensitive cotransport of sodium plus potassium in the human red cell. *J. Clin. Invest.* 53:745-755.
14. Sachs, J. R., Knauf, P. A., and Dunham, P. B. (1975). Transport through red cell membranes. In *The Red Blood Cell*, Vol. II, 2nd ed., D. M. Surgenor (Ed.). Academic Press, New York, pp. 613-703.

15. Dunham, P. B., Stewart, G. W., and Ellory, J. C. (1980). Chloride-activated passive potassium transport in human erythrocytes. *Proc. Natl. Acad. Sci. U.S.A.* 77:1711–1715.

16. Chipperfield, A. R. (1980). An effect of chloride on (Na + K) cotransport in human red blood cells. *Nature* 286:281–282.

17. Frizzell, R. A., Field, M., and Schultz, S. G. (1979). Sodium-coupled chloride transport by epithelial tissues. *Am. J. Physiol.* 236:F1–F8.

18. Geck, P., Pietrzyk, C., Burkhardt, B.-C., Pfeiffer, B., and Heinz, E. (1980). Electrically silent cotransport of Na, K, and Cl in Ehrlich cells. *Biochem. Biophys. Acta* 600:432–447.

19. Greger, R., and Schlatter, E. (1980). Presence of luminal K^+, a prerequisite for active NaCl transport in the cortical thick ascending limb of Henle's loop of rabbit kidney. *Pflugers Arch.* 392:92–94.

20. Musch, M. W., Orellana, S. A., Kimber, L. S., Field, M., Halm, D. R., Krasny, E. J., and Frizzell, R. A. (1982). Na^+-K^+-Cl^- cotransport in the intestine of a marine teleost. *Nature* 300:351–353.

21. Hannafin, J., Kinne-Saffran, E., Friedman, D., and Kinne, R. (1983). Presence of a sodium-potassium-chloride cotransport system in the rectal gland of *Squalus acanthias*. *J. Membr. Biol.* 75:73–83.

22. Oberleithner, H., Guggino, W., and Giebisch, G. (1983). The effect of furosemide on luminal sodium, chloride, and potassium transport in the early distal tubule of *Amphiuma* kidney. *Pflugers Arch.* 396:27–33.

23. McRoberts, J. A., Erlinger, S., Rindler, M. J., and Saier, M. H. (1982). Furosemide-sensitive salt transport in the Madin-Darby canine kidney cell line. Evidence for cotransport of Na^+, K^+, and Cl^-. *J. Biol. Chem.* 257:2260–2266.

24. Brown, C. D. A., and Murer, H. (1985). Characterization of a Na:K:2Cl cotransport system in the apical membrane of a renal epithelial cell line (LLC-PK$_1$). *J. Membr. Biol.* 87:131–139.

25. Chipperfield, A. R. (1986). The (Na^+-K^+-Cl^-) cotransport system. *Clin. Sci.* 71:465–476.

26. O'Grady, S. M., Palfrey, H. C., and Field, M. (1987). Characteristics and functions of Na/K/Cl cotransport in epithelial cells. *Am. J. Physiol.* 253: C177–C192.

27. Cremaschi, D., Meyer, G., Botta, G., and Rossetti, C. (1987). The nature of the neutral Na^+-Cl^- coupled entry at the apical membrane of rabbit gallbladder epithelium. II. Na^+-Cl^- symport is independent of K^+. *J. Membr. Biol.* 95:219–228.

28. Eveloff, J. L., and Warnock, D. G. (1987). Activation of ion transport systems during cell volume regulation. *Am. J. Physiol.* 252:F1–F10.

29. Russell, J. M. (1983). Cation-soupled chloride influx in squid axon. Role of potassium and stoichiometry of the transport process. *J. Gen. Physiol.* 81:909–925.

30. Altimirano, A. A., and Russell, J. M. (1987). Coupled Na/K/Cl efflux. "Reverse" unidirectional fluxes in squid giant axons. *J. Gen. Physiol.* 89:669–686.

31. Haas, M., Schmidt, W. F., III, and McManus, T. J. (1982). Catecholamines-stimulated ion transport in duck red cells. Gradient effects in electrically neutral [Na+K+Cl] cotransport. *J. Gen. Physiol.* 80:125–147.
32. Kregenow, F. M. (1981). Osmoregulatory salt transporting mechanisms: control of cell volume in anisotonic media. *Annu. Rev. Physiol.* 43:493–505.
33. Hall, A. C., and Ellory, J. C. (1985). Measurement and stoichiometry of bumetanide sensitive (2Na:1K:3Cl) cotransport in ferret red cells. *J. Membr. Biol.* 85:205–213.
34. Flatman, P. W. (1983). Sodium and potassium transport in ferret red cells. *J. Physiol. (Lond.)* 341:545–557.
35. Duhm, J., and Göbel, B. O. (1984). Na^+-K^+ transport and volume of rat erythrocytes under dietary K^+ deficiency. *Am. J. Physiol.* 246:C20–C29.
36. Hall, A. C., and Willis, J. S. (1984). Differential effects of temperature on three components of passive permeability to potassium in rodent red cells. *J. Physiol. (Lond.)* 348:629–643.
37. Lauf, P. K., McManus, T. J., Haas, M., Forbush, B., III, Duhm, J., Flatman, P. W., Sailer, M. H., Jr., and Russell, J. M. (1987). Physiology and biophysics of chloride and cation cotransport across cell membranes. *Fed. Proc.* 46:2377–2394.
38. Panet, R., and Atlan, H. (1980). Characterization of a potassium carrier in rabbit reticulocyte cell membranes. *J. Membr. Biol.* 52:273–280.
39. Dunham, P. B., and Ellory, J. C. (1981). Passive potassium transport in low potassium sheep red cells: dependence upon cell volume and chloride. *J. Physiol. (Lond.)* 318:511–530.
40. Kaji, D., and Kahn, T. (1985). Kinetics of Cl-dependent K influx in human erythrocytes with and without external Na: effect of NEM. *Am. J. Physiol.* 249:C490–C496.
41. Wiater, L. A., and Dunham, P. B. (1983). Passive transport of K^+ and Na^+ in human red blood cells: sulfhydryl binding agents and furosemide. *Am. J. Physiol.* 245:C348–C356.
42. Duhm, J. (1987). Furosemide-sensitive K^+ (Rb^+) transport in human erythrocytes: modes of operation, dependence on extracellular and intracellular Na^+, kinetics, pH dependency and the effect of cell volume and N-ethylmaleimide. *J. Membr. Biol.* 98:15–32.
43. Funder, J., and Wieth, J. O. (1967). Effects of some monovalent anions on fluxes of Na and K, and on glucose metabolism of ouabain treated human red cells. *Acta Physiol. Scand.* 71:168–185.
44. Canessa, M., Brugnara, C., Cusi, D., and Tosteson, D. C. (1986). Modes of operation and variable stoichiometry of the furosemide-sensitive Na and K fluxes in human red cells. *J. Gen. Physiol.* 87:113–142.
45. Kracke, G. R., Anatra, M. A., and Dunham, P. B. (1988). Asymmetry of Na-K-Cl cotransport in human erythrocytes. *Am. J. Physiol.* 254:C243–C250.
46. Silva, P., Myers, M., Landsberg, A., Silva, P., Jr., Silva, P. J., Silva, M., Brown, R., and Epstein, F. H. (1983). Stoichiometry of sodium chloride transport by the shark rectal gland. *Bull. Mt. Desert Island Biol. Lab.* 23:47–50.

47. Brugnara, C., Canessa, M., Cusi, D., and Tosteson, D. C. (1986). Furosemide-sensitive Na and K fluxes in human red cells. Net uphill Na extrusion and equilibrium properties. *J. Gen. Physiol.* 87:91–112.
48. Ellory, J. C., Dunham, P. B., Logue, P. J., and Stewart, G. W. (1982). Anion-dependent cation transport in erythrocytes. *Phil. Trans. R. Soc. Lond.* [Biol.] 299:483–495.
49. Kracke, G. R., and Dunham, P. B. (1987). Effect of membrane potential on furosemide-inhibitable sodium influxes in human red blood cells. *J. Membr. Biol.* 98:117–124.
50. Chipperfield, A. R., and Shennan, D. B. (1986). The influence of pH and membrane potential on passive Na^+ and K^+ fluxes in human red blood cells. *Biochim. Biophys. Acta* 886:373–382.
51. Lytle, C., Haas, M., and McManus, T. J. (1986). Chloride-dependent obligate cation exchange: A partial reaction of [Na + K + 2 Cl] co-transport [abstract]. *Fed. Proc.* 45:548.
52. Stein, W. D. (1986). Intrinsic, apparent, and effective affinities of co- and countertransport systems. *Am. J. Physiol.* 250:C523–C533.
53. Garay, R., Adragna, N., Canessa, M., and Tosteson, D. C. (1981). Outward sodium and potassium cotransport in human red cells. *J. Membr. Biol.* 62: 169–174.
54. Saier, M. H., and Boyden, D. A. (1984). Mechanism, regulation and physiological significance of the loop diuretic-sensitive NaCl/KCl symport system in animal cells. *Mol. Cell. Biochem.* 59:11–32.
55. Dunn, M. J. (1973). Ouabain-uninhibited sodium transport in human erythrocytes. Evidence against a second pump. *J. Clin. Invest.* 52:658–670.
56. Beaugé, L. A., and Adragna, N. (1971). The kinetics of ouabain inhibition and the partition of rubidium influx in human red blood cells. *J. Gen. Physiol.* 57:576–592.
57. Dagher, G., Brugnara, C., and Canessa, M. (1985). Effect of metabolic depletion on the furosemide-sensitive Na and K fluxes in human red cells. *J. Membr. Biol.* 86:145–155.
58. Garay, R. P. (1982). Inhibition of the Na^+/K^+ cotransport system by cyclic AMP and intracellular Ca^{2+} in human red cells. *Biochim. Biophys. Acta* 688:786–792.
59. Riddick, D. H., Kregenow, F. M., and Orloff, J. (1971). The effect of norepinephrine and dibutyryl cyclic adenosine monophosphate on cation transport in duck erythrocytes. *J. Gen. Physiol.* 57:752–766.
60. Haas, M., and Forbush, B. III. (1987). Photolabelling of a 150 kDa (Na + K + Cl) cotransport protein from dog kidney with a bumetanide analogue. *Am. J. Physiol.* 253:C243–C250.
61. Feit, P. W., Hoffman, E. K., Schiødt, M., Kristensen, P., Jessen, F., and Dunham, P. B. (1988). Purification of proteins of the Na/Cl cotransporter from membranes of Ehrlich ascites cells using a bumetanide-Sepharose affinity column. *J. Membr. Biol.* 103:135–147.
62. Haas, M., and Forbush, B. III. (1988). Photoaffinity labelling of a 150 kDalton (Na + K + Cl) cotransport protein from duck red cells with an analog of bumetanide. *Biochim. Biophys. Acta* 939:131–144.

63. Burnham, C., Karlish, S. J. D., and Jørgensen, P. L. (1985). Identification and reconstitution of a Na/K/Cl cotransporter and K channel from luminal membranes of renal red outer medulla. *Biochim. Biophys. Acta* 821:461–469.

64. Ellory, J. C., and Dunham, P. B. (1980). Volume dependent passive potassium transport in LK sheep red cells. In *Membrane Transport in Erythrocytes*, Alfred Benzon Symposium 14, U. V. Lassen, H. H. Ussing, and J. O. Wieth (Eds.). Munksgaard, Copenhagen, pp. 409–427.

65. Brugnara, C., Van Ha, T., and Tosteson, D. C. (1988). Role of Cl in K transport through a volume-dependent KCl cotransport system in human red cells [abstract]. *J. Gen. Physiol.* 92:42a.

66. Reuss, L. (1983). Basolateral KCl cotransport in a NaCl-absorbing epithelium. *Nature* 305:723–726.

67. Corcia, C., and Armstrong, W. McD. (1983). KCl cotransport: a mechanism for basolateral chloride exit in *Necturus* gallbladder. *J. Membr. Biol.* 76: 173–182.

68. Greger, R., and Schlatter, E. (1983). Properties of the basolateral membrane of the cortical thick ascending limb of Henle's loop of rabbit kidney. A model for secondary active chloride transport. *Pflugers Arch.* 396:325–334.

69. Hoffman, E. K., Lambert, I. H., and Simonsen, L. O. (1986). Separate Ca^{2+}-activated K^+ and Cl-transport pathways in Ehrlich ascites tumor cells. *J. Membr. Biol.* 91:227–224.

70. Thornhill, W. G., and Laris, P. C. (1984). KCl loss and cell shrinkage in the Ehrlich ascites tumor cell induced by hypotonic media, 2-deoxyglucose, and propranolol. *Biochim. Biophys. Acta* 773:207–218.

71. Kramhøft, B., Lambert, I. H., Hoffmann, E. K., and Jørgensen, F. (1986). Activation of Cl-dependent K transport in Ehrlich ascites tumor cells. *Am. J. Physiol.* 251:C369–379.

72. Kregenow, F. M. (1971). The response of duck erythrocytes to nonhemolytic hypotonic media. Evidence for a volume-controlling mechanism. *J. Gen. Physiol.* 58:372–395.

73. Haas, M., and McManus, T. J. (1985). Effect of norepinephrine on swelling induced potassium transport in duck red cells. Evidence against a volume-regulatory decrease under physiological conditions. *J. Gen. Physiol.* 85: 649–669.

74. Lytle, C., and McManus, T. J. Effect of loop diuretics and stilbene derivatives on swelling-induced K-Cl cotransport [abstract]. *J. Gen. Physiol.* 90: 28a.

75. Dunham, P. B. (1979). Passive potassium transport in LK sheep red cells: effect of anti-L antibody and intracellular potassium. *J. Gen. Physiol.* 68: 567–581.

76. Dunham, P. B. (1976). Anti-L serum. Two populations of antibodies affecting cation transport in LK erythrocytes of sheep and goats. *Biochim. Biophys. Acta* 443:219–226.

77. Lauf, P. K. (1984). Thiol-dependent passive K^+ Cl^- transport in sheep red blood cells. VI. Functional heterogeneity and immunologic identity with volume-stimulated K^+ (Rb^+) fluxes. *J. Membr. Biol.* 82:167–175.

78. Lauf, P. K., and Theg, B. E. (1980). A chloride-dependent K^+ flux induced by N-ethylmaleimide in genetically low K^+ sheep and goat erythrocytes. *Biochem. Biophys. Res. Commun.* 92:1422–1428.

79. Lauf, P. K. (1983). Thiol-dependent passive K/Cl transport in sheep red cells. I. Dependence on chloride and external $K^+[Rb^+]$ ions. *J. Membr. Biol.* 73:237–246.

80. Logue, P., Anderson, C., Kanik, C., Farquharson, B., and Dunham, P. (1983). Passive potassium transport in LK sheep and red cells. Modification by *N*-ethyl maleimide. *J. Gen. Physiol.* 81:861–885.

81. Lauf, P. K. (1988). Thiol-dependent K:Cl transport in sheep red cells: VIII. Activation through metabolically and chemically reversible oxidation by diamide. *J. Membr. Biol.* 101:179–188.

82. Lauf, P. K. (1986). Chloride-dependent cation cotransport and cellular differentiation: a comparative approach. *Curr. Top. Membr. Trans.* 27:89–125.

83. Lauf, P. K. (1985). $K^+:Cl^-$ cotransport: sulfhydryls, divalent cations, and the mechanism of volume activation in a red cell. *J. Membr. Biol.* 88:1–13.

84. Lauf, P. K., and Bauer, J. (1987). Direct evidence for chloride-dependent volume reduction in monocytic sheep reticulocytes. *Biochem. Biophys. Res. Commun.* 144:849–855.

85. Fujise, H., and Lauf, P. K. (1987). Swelling, NEM, and A23187 activate Cl-dependent K^+ transport in high-K^+ sheep red cells. *Am. J. Physiol.* 252: C197–C204.

86. Lauf, P. K., Adragna, N. C., and Garay, R. P. (1984). Activation by *N*-ethymaleimide of a latent K^+ Cl^- flux in human red blood cells. *Am. J. Physiol.* 246:C385–C390.

87. Kaji, D. (1986). Volume-sensitive K transport in human erythrocyte. *J. Gen. Physiol.* 88:719–738.

88. Hall, A. C., and Ellory, J. C. (1986). Evidence for the presence of volume-sensitive KCl transport in young human red cells. *Biochim. Biophys. Acta* 858:317–320.

89. Canessa, M., Fabry, M. E., Blumenfeld, N., and Nagel, R. L. (1987). Volume-stimulated Cl-dependent K^+ efflux is highly expressed in young human red cells containing normal hemoglobin or HbS. *J. Membr. Biol.* 97:97–105.

90. Brugnara, C., and Tosteson, D. C. (1987). Cell volume, K transport and cell density in human erythrocytes. *Am. J. Physiol.* 252:C269–C276.

91. Brugnara, C., Bunn, H. F., and Tosteson, D. C. (1986). Regulation of erythrocyte cation and water content in sickle cell anemia. *Science* 232:388–390.

92. Brugnara, C., Kopin, A. K., Bunn, H. F., and Tosteson, D. C. (1985). Regulation of cation content and cell volume in erythrocytes from patients with homozygous hemoglobin C disease. *J. Clin. Invest.* 75;1608–1617.

93. Dunham, P. B., and Logue, P. J. (1986). Potassium chloride cotransport in resealed human red cell ghosts. *Am. J. Physiol.* 250:C578–C583.

94. Sachs, J. R. (1988). Volume-sensitive K influx in human red cell ghosts. *J. Gen. Physiol.* (in press).

95. Bernstein, R. E. (1954). K and Na balance in mammalian red cells. *Science* 120:459–460.

96. Parker, J. C. (1983). Hemolytic action of potassium salts on dog red blood cells. *Am. J. Physiol.* 244:C313–C317.

97. Ellory, J. C., Hall, A. C., and Stewart, G. W. (1985). Volume-sensitive cation fluxes in mammalian red cells. *Mol. Physiol.* 8:235–246.

98. Lauf, P. K. (1983). Thiol-dependent passive K^+–Cl^- transport in sheep red blood cells. V. Dependence on metabolism. *Am. J. Physiol.* 245:C445–C448.

99. Lauf, P. K., Perkins, C. M., and Adragna, N. C. (1985). Cell volume and metabolic dependence of the N-ethylmaleimide activated K^+Cl^- flux in human red cells. *Am. J. Physiol.* 249:C124–C128.

100. Siebens, A. W. (1985). Cellular volume control. In *The Kidney: Physiology and Pathophysiology*, D. W. Seldin and G. Giebisch (Eds.). Raven Press, New York, pp. 91–115.

101. Sachs, F. (1987). Baroreceptor mechanisms at the cellular level. *Fed. Proc.* 46:12–16.

102. Mercer, R. W., and Dunham, P. B. (1981). Membrane-bound ATP fuels the Na/K pump. Studies on membrane-bound glycolytic enzymes on inside-out vesicles from human red cell membranes. *J. Gen. Physiol.* 78:547–568.

103. Lauf, P. K. (1985). Passive K^+–Cl^- fluxes in low-K^+ sheep erythrocytes: modulation by A23187 and bivalent cations. *Am. J. Physiol.* 249:C271–C278.

104. Sarkadi, B., and Gardos, G. (1985). Calcium-induced potassium transport in cell membranes. In *The Enzymes of Biological Membranes* 2nd Ed., vol. 3, A. N. Martonosi (Ed.). Plenum Press, New York, pp. 193–234.

105. Varecka, L. E., Peterajova, and J. Pogady, Inhibition by divalent cations and sulfhydryl reagents of the passive Ca transport in human red blood cells observed in the presence of vanadate. *Biochim. Biophys. Acta* 856:585–594.

106. Simons, T. J. B. (1985). Influence of lead ions on cation permeability in human red cell ghosts. *J. Membr. Biol.* 84:61–71.

107. Simons, T. J. B.(1986). Passive transport and binding of lead by human red blood cells. *J. Physiol. (Lond.)* 378:267–286.

108. Simons, T. J. B. (1986). The role of anion transport in the passive movement of lead across the human red cell membrane. *J. Physiol. (Lond.)* 378:287–312.

109. Shields, M., Gregorczyk, R., Fuhrmann, G. F., Schwarz, W., and Passow, H. (1985). Lead-induced activation and inhibition of potassium-selective channels in the human red blood cell. *Biochim. Biophys. Acta* 815:223–232.

110. Lew, V. L., and Ferriera, H. G. (1978). Calcium transport and the properties of a calcium-activated potassium channel in red cell membranes. *Curr. Top. Membr. Transp.* 10:217–277.

111. Schwarz, W., and Passow, H. (1983). Ca-activated K channels in erythro-
     cytes and excitable cells. *Annu. Rev. Physiol.* 45:359–374.
112. Lassen, U. V., Pape, L., and Vestergaard-Bogind, B. (1976). Effect of cal-
     cium on the membrane potential of *Amphiuma* red cells. *J. Membr. Biol.*
     26:51–70.
113. Varecka, L., and Carafoli, E. (1982). Vanadate-induced movements of Ca
     and K in human red blood cells. *J. Biol. Chem.* 257:7414–7421.
114. Reed, P. W. (1976). Effects of the divalent cation ionophore, A23187, on
     potassium permeability of rat erythrocytes. *J. Biol. Chem.* 251:3489–
     3494.
115. Manninen, V. (1970). Movements of sodium and potassium ions and their
     tracers in propranolol-treated red cells and diaphragm muscle. *Acta
     Physiol. Scand.* 355(suppl):1–37.
116. Davson, H. (1942). The haemolytic action of potassium salts. *J. Physiol.
     (Lond.)* 101:265–283.
117. Blum, R. M., and Hoffman, J. F. (1971). The membrane locus of Ca-stim-
     ulated K transport in energy-depleted human red blood cells. *J. Membr.
     Biol.* 6:315–328.
118. Yingst, D. R., and Hoffman, J. F. (1984). Ca-induced K transport in hu-
     man red blood cell ghosts containing arsenazo III. Trans-membrane inter-
     actions of Na, K, and Ca and the relationship to the functioning Na-K
     pump. *J. Gen. Physiol.* 83:19–46.
119. Wood, P. G. (1984). The spontaneous activation of a potassium channel
     during the preparation of resealed human erythrocyte ghosts. *Biochim.
     Biophys. Acta* 774:103–109.
120. Lew, V. L., Muallem, S., and Seymour, C. A. (1982). Properties of the Ca-
     activated K channel in one-step inside-out vesicles from human red cell
     membranes. *Nature* 296:742–744.
121. Hammill, O. P. (1983). Potassium and chloride channels in red blood cells.
     In *Single Channel Recording*, B. Sackmann and E. Neher (Eds.). Plenum,
     New York, pp. 451–471.
122. Grygorczyk, R., Schwarz, W., and Passow, H. (1984). Ca-activated K chan-
     nels in human red cells. *Biophys. J.* 45:693–698.
123. Glynn, I. M., and Warner, A. E. (1972). Nature of the calcium-dependent
     potassium leak induced by (+)-propranolol and its possible relevance to
     the drug's anti-arrhythmic effect. *Br. J. Pharmacol.* 44:271–278.
124. Hoffman, J. F., and Knauf, P. A. (1973). The mechanism of the increased
     K transport induced by Ca in human red blood cells. In *Erythrocytes,
     Thrombocytes, Leukocytes*, E. Gerlach, K. Moser, E. Deutsch, and W.
     Willmanns (Eds.). Georg Thieme, Stuttgart, pp. 66–70.
125. Hammill, O. P. (1981). Potassium channel currents in human red blood
     cells. *J. Physiol. (Lond.)* 314:125P.
126. Fuhrmann, G. F., Schwarz, W., Kersten, R., and Sdun, H. (1985). Effects
     of vanadate, menadione, and menadione analogues on the Ca-activated K
     channels in human red cells. Possible relations to membrane-bound oxido-
     reductase activity. *Biochim. Biophys. Acta* 820:223–234.

127. Vestergaard-Bogind, B., Stampe, P., and Christophersen, P. (1985). Single-file diffusion through the Ca-activated K channel of human red blood cells. *J. Membr. Biol.* 88:67–76.

128. Simons, T. J. B. (1976). Calcium-dependent potassium exchange in human red cell ghosts. *J. Physiol. (Lond.)* 256:227–244.

129. Porzig, H. (1977). Studies on the cation permeability of human red cell ghosts. *J. Membr. Biol.* 31:317–349.

130. Lew, V. L., Tsien, R. Y., Miner, C., and Bookchin, R. M. (1982). Physiological [Ca]$_i$ level and pump-leak turnover in intact red cells measured using an incorporated Ca chelator. *Nature* 298:478–481.

131. Murphy, E., Levy, L., Berkowitz, L. R., Orringer, E. P., Gabel, S. A., and London, R. E. (1986). Nuclear magnetic resonance measurement of cytosolic free calcium levels in human red blood cells. *Am. J. Physiol.* 251: C496–C504.

132. Vestergaard-Bogind, B., and Bennekou, P. (1982). Calcium-induced oscillation in K conductance and membrane potential of human erythrocytes mediated by the ionophore A23187. *Biochim. Biophys. Acta* 688:37–44.

133. Vestergaard-Bogind, B. (1983). Spontaneous inactivation of the Ca-sensitive K channels of human red cells at high intracellular Ca activity. *Biochim. Biophys. Acta* 730:285–294.

134. Garcia-Sancho, J., Sanchez, A., and Herreros, B. (1982). All or none response of the Ca-dependent K channel in inside-out vesicles. *Nature* 296: 744–746.

135. Lew, V. L., Muallem, S., and Seymour, C. A. (1983). The Ca-activated K channel of human red cells: all or none behaviour of the Ca-gating mechanism. *Cell Calcium* 4:511–517.

136. Adorante, J. S., and Macey, R. I. (1986). Calcium-induced transient potassium efflux in human red blood cells. *Am. J. Physiol.* 250:C55–C64.

137. Stampe, P., and Vestergaard-Bogind, B. (1985). The Ca-sensitive K conductance of the human red cell membrane is strongly dependent on cellular pH. *Biochim. Biophys. Acta* 815:313–321.

138. Alvarez, J., Garcia-Sancho, J., and Herreros, B. (1984). Effects of electron donors on Ca-dependent K transport in one-step inside-out vesicles from the human erythrocyte membrane. *Biochim. Biophys. Acta* 771:23–27.

139. Alvarez, J., Camaleno, J., Garcia-Sancho, J., and Herreros, B. (1986). Modulation of Ca-dependent K transport by modifications of the NDA/NADPH ratio in intact human red cells. *Biochim. Biophys. Acta* 856: 408–411.

140. Miner, C., Lopez-Burillo, S., Garcia-Sancho, J., and Herreros, B. (1983). Plasma membrane NADH dehydrogenase and Ca-dependent potassium transport in erythrocytes of several animal species. *Biochim. Biophys. Acta* 727:266–272.

141. Lackington, I., and Orrega, F. (1981). Inhibition of calcium-activated potassium conductance of human erythrocytes by calmodulin inhibiting drugs. *FEBS Lett.* 133:103–106.

142. Plishker, G. A. (1984). Phenothiazine inhibition of calmodulin stimulates calcium-dependent potassium efflux in human red blood cells. *Cell Calcium* 5:177–185.
143. Pape, L., and Kristensen, B. I. (1984). A calmodulin-activated Ca-dependent K channel in human erythrocyte membrane inside-out vesicles. *Biochim. Biophys. Acta* 770:1–6.
144. Plishker, G. A., White, P. H., and Cadman, E. D. (1986). Involvement of a cytosolic protein in calcium-dependent potassium efflux in red blood cells. *Am. J. Physiol.* 251:C535–C540.
145. Reichstein, E., and Rothstein, A. (1981). Effects of quinine on Ca-induced K efflux from human red blood cells. *J. Membr. Biol.* 59:57–63.
146. Simons, T. J. B. (1979). Actions of a carbocyanine dye on calcium-dependent potassium transport in human red cell ghosts. *J. Physiol. (Lond.)* 288:481–507.
147. Berkowitz, L. R., and Orringer, E. P. (1981). Effect of cetiedil, an in vitro sickling agent, on erythrocyte membrane cation permeability. *J. Clin. Invest.* 68:1215–1220.
148. Berkowitz, L. R., and Orringer, E. P. (1984). An analysis of the mechanism by which cetiedil inhibits the Gardos phenomenon. *Am. J. Hematol.* 17:217–223.
149. Harper, J. R., and Parker, J. C. (1979). Adriamycin inhibits Ca permeability and Ca-dependent K movements in red blood cells. *Res. Commun. Chem. Pathol. Pharmacol.* 26:277–284.
150. Fuhrmann, G. F., Hutterman, J., and Knauf, P. A. (1984). The mechanism of vanadium action on selective K permeability in human erythrocytes. *Biochim. Biophys. Acta* 769:130–140.
151. Plishker, G. A. (1985). Iodoacetate inhibition of calcium-dependent potassium efflux in red blood cells. *Am. J. Physiol.* 248:C419–C424.
152. Abia, A., Lobaton, C. D., Moreno, A., and Garcia-Sancho, J. (1986). Leiurus quinquestriatus venom inhibits different kinds of Ca-dependent K channels. *Biochim. Biophys. Acta* 856:403–407.
153. Jenkins, D. M. G., and Lew, V. L. (1973). Ca uptake by ATP depleted red cells from different species with and without associated increase in K permeability. *J. Physiol. (Lond.)* 234:41P–42P.
154. Richardt, H-W., Fuhrmann, F., and Knauf, P. A. (1979). Dog red blood cells exhibit a Ca-stimulated increase in K permeability in the absence of Na,K-ATPase activity. *Nature* 279:248–250.
155. Marino, D., Sarkadi, B., Gardos, G., and Bolis, L. (1981). Calcium-induced alkali cation transport in the nucleated red cells. *Mol. Physiol.* 1:295–300.
156. Brown, A. M., Ellory, J. C., Young, J. D., and Lew, V. L. (1978). A calcium activated potassium channel present in foetal red cells of sheep but absent from reticulocytes and mature red cells. *Biochim. Biophys. Acta* 511:163–175.
157. Karlish, S. J. D., Ellory, J. C., and Lew, V. L. (1981). Evidence against Na-pump mediation of Ca-activated K transport and diuretic-sensitive Na/K cotransport. *Biochim. Biophys. Acta* 646:353–355.

158. Verma, A. K., and Penniston, J. T. (1985). Evidence against involvement of the human erythrocyte plasma membrane Ca-ATPase in the Ca-dependent K transport. *Biochim. Biophys. Acta* 815:135–138.

159. Peterson, O. H., and Maruyama, Y. (1984). Calcium-activated potassium channels and their role in secretion. *Nature* 307:693–696.

160. Szasz, I., Sarkadi, B., Shubert, A., and Gardos, G. (1978). Effects of lanthanum on calcium-dependent phenomena in human red cells. *Biochim. Biophys. Acta* 512:331–340.

161. Altamirano, A. A., and Beauge, L. (1985). Calcium transport mechanisms in dog red blood cells from measurements of initial flux rates. *Cell Calcium* 6:503–525.

162. Milanick, M., and Hoffman, J. F. (1986). Na/Ca exchange and Ca pump fluxes in ferret red blood cells. *J. Gen. Physiol.* 88:39a.

163. Ferriera, H. G., and Lew, V. L. (1977). Passive Ca transport and cytoplasmic Ca buffering in intact red cells. In *Membrane Transport in Red Cells*, J. C. Ellory and V. L. Lew (Eds.). Academic, London, pp. 53–91.

164. McNamara, M. K., and Wiley, J. S. (1986). Passive permeability of human red blood cells to calcium. *Am. J. Physiol.* 250:C26–C31.

165. Tsien, R. Y. (1976). A non-disruptive technique for loading calcium buffers and indicators into cells. *Nature* 290:527–528.

166. Orringer, E. P., and Mattern, W. D. (1976). Formaldehyde-induced hemolysis during chronic hemodialysis. *New Engl. J. Med.* 294:1416–1420.

167. Tiffert, T., Garcia-Sancho, J., and Lew, V. L. (1984). Irreversible ATP depletion caused by low concentrations of formaldehyde and calcium-chelator esters in intact human red cells. *Biochim. Biophys. Acta* 773:143–156.

168. Garcia-Sancho, J. (1985). Pyruvate protects the ATP depletion caused by formaldehyde or calcium chelator esters in the human red cell. *Biochim. Biophys. Acta* 814:148–150.

169. Allen, T. J. A., and Baker, P. F. (1985). Intracellular Ca indicator Quin-2 inhibits Ca inflow via Na_i/Ca_o exchange in squid axon. *Nature* 315:755–756.

170. DiPolo, R., and Beaugé, L. (1986). Reverse Ca-Na exchange requires internal Ca and/or ATP in squid axons. *Biochim. Biophys. Acta* 854:298–306.

171. Plishker, G., and Gitelman, H. J. (1976). Calcium transport in human erythrocytes. *J. Gen. Physiol.* 68:29–42.

172. Reuter, H., and Seitz, N. (1968). Dependence of calcium efflux from cardiac muscle on the temperature and external ion composition. *J. Physiol. (Lond.)* 195:451–470.

173. Baker, P. F., Blaustein, M. P., Hodgkin, A. L., and Steinhardt, R. A. (1969). The influence of calcium on sodium efflux in squid axons. *J. Physiol. (Lond.)* 200:431–458.

174. Omachi, A., Markel, R. P., and Hegarty, H. (1961). 45-Ca uptake by dog erythrocytes suspended in sodium and potassium chloride solutions. *J. Cell. Compar. Physiol.* 57:95–100.

175. Niedergerke, R. (1959). Calcium and the activation of contraction. *Experientia* 15:128-130.
176. Chan, P. C., Calabrese, V., and Thiel, L. S. (1964). Species difference in the effect of sodium and potassium ions on the ATPase of erythrocyte membranes. *Biochim. Biophys. Acta* 79:424-430.
177. Inaba, M., and Maede, Y. (1986). Na,K-ATPase in dog red cells. Immunological identification and maturation-associated degradation by the proteolytic system. *J. Biol. Chem.* 261:16099-16105.
178. Parker, J. C. (1973). Dog red blood cells. Adjustment of density in vivo. *J. Gen. Physiol.* 61:146-157.
179. Parker, J. C. (1973). Dog red blood cells. Adjustment of salt and water content in vitro. *J. Gen. Physiol.* 62:147-156.
180. Parker, J. C., Gitelman, H. J., Blosson, P. S., and Leonard, D. L. (1975). Role of calcium in volume regulation by dog red blood cells. *J. Gen. Physiol.* 65:84-96.
181. Parker, J. C. (1978). Sodium and calcium movements in dog red blood cells. *J. Gen. Physiol.* 71:1-17.
182. Parker, J. C. (1979). Active and passive calcium movements in dog red blood cells and resealed ghosts. *Am. J. Physiol.* 237:C10-C16.
183. Parker, J. C., and Harper, J. R. (1980). Effects of amrinone, a cardiotonic drug, on calcium movements in dog red blood cells. *J. Clin. Invest.* 66:254-259.
184. Parker, J. C. (1983). Passive calcium movements in dog red blood cells: Anion effects. *Am. J. Physiol.* 244:C318-C323.
185. Ortiz, O. E., and Sjodin, R. A. (1984). Sodium and adenosine triphosphate-dependent calcium movements in membrane vesicles prepared from dog erythrocytes. *J. Physiol. (Lond.)* 354:287-301.
186. Eisner, D. A., and Lederer, W. J. (1985). Na-Ca exchange: stoichiometry and electrogenicity. *Am. J. Physiol.* 248:C189-C202.
187. Requina, J., Whittembury, J., Tiffert, T., Eisner, D. A., and Mullins, L. J. (1985). The influence of chemical agents on the level of ionized Ca in squid axons. *J. Gen. Physiol.* 85:789-804.
188. Parker, J. C. (1987). Diamide stimulates calcium-sodium exchange in dog red blood cells. *Am. J. Physiol.* 253:C580-C587.
189. Parker, J. C. (1986). Calcium-sodium (Ca-Na) and sodium-proton (Na-H) exchange in dog red blood cells (RBC): Fixation of the activating mechanisms with sulfhydryl crosslinkers. *J. Gen. Physiol.* 88:45A.
190. Reeves, J. P., Bailey, C. A., and Hale, C. C. (1986). Redox modification of sodium-calcium exchange activity in cardiac sarcolemmal vesicles. *J. Biol. Chem.* 261:4948-4955.
191. Milanick, M. A., and Hoffman, J. F. (1986). Ion transport and volume regulation in red blood cells. *Ann. N.Y Acad. Sci.* 488:174-186.
192. Smith, R. L., Macara, I. G., Levenson, R., Housman, D., and Cantley, L. (1982). Evidence that a Na/Ca antiport system regulates murine erythroleukemia cell differentiation. *J. Biol. Chem.* 257:773-801.

193. Lannigan, D. A., and Knauf, P. A. (1985). Decreased intracellular Na concentrationis an early event in murine erythroleukemic cell differentiation. *J. Biol. Chem.* 260:7322–7324.

194. Wieth, J. O. (1970). Effects of monovalent cations on sodium permeability of human red blood cells. *Acta Physiol. Scand.* 79:76–87.

195. Funder, J., Tosteson, D. C., and Wieth, J. O. (1978). Effects of bicarbonate on lithium transport in human red cells. *J. Gen. Physiol.* 71:721–746.

196. Duhm, J., and Becker, B. F. (1978). Studies on Na-dependent Li countertransport and bicarbonate-stimulated Li transport in human erythrocytes. In *Cell Membrane Receptors for Drugs and Hormones. A multidisciplinary Approach*, R. W. Straub and L. Bolis (Eds.), Raven Press, New York, pp. 281–299.

197. Becker, B. F., and Duhm, J. (1978). Evidence for anionic cation transport of lithium, sodium, and potassium across the human erythrocyte membrane induced by divalent anions. *J. Physiol. (Lond.)* 282:149–168.

198. Callahan, T. J., and Goldstein, D. A. (1978). Anion-inhibitor sensitive unidirectional sodium movement in the human erythrocyte. *J. Gen. Physiol.* 72:87–100.

199. Funder, J., and Wieth, J. O. (1974). Human red cell sodium and potassium in metabolic alkalosis. *Scand. J. Clin. Lab. Invest.* 34:49–59.

200. Orringer, E. P., Roer, M. S., and Parker, J. C. (1980). Cell density profile as a measure of erythrocyte hydration: therapeutic alteration of salt and water content in normal and SS red blood cells. *Blood Cells* 6:345–353.

201. Parker, J. C. (1969). Influence of 2,3-diphosphoglycerate metabolism on sodium-potassium permeability in human red blood cells. Studies with bisulfite and other redox agents. *J. Clin. Invest.* 48:117–125.

202. Mond, R. (1927). Umkehr der Anionpermeabilitat der roten Blutkorperchen in eine elektive Durchlassigkeit fur Kationen. Ein Beitrag zur Analyse der Zellmembranen. *Pflugers Arch. Ges. Physiol.* 217:618–630.

203. Donlon, J. A., and Rothstein, A. (1969). The cation permeability of erythrocytes in low ionic strength media of various tonicities. *J. Membr. Biol.* 1:37–52.

204. Jones, G. S., and Knauf, P. A. (1985). Mechanisms of the increase incation permeability of human erythrocytes in low-chloride media. Involvement of the anion transport protein capnorphorin. *J. Gen. Physiol.* 86:721–738.

205. Bernhatdt, I., Donath, E., and Glaser, R. (1984). Influence of surface charge and transmembrane potential on rubidium-86 efflux of human red blood cells. *J Membr. Biol.* 78:249–255.

206. Joiner, C. H., Platt, O. S., and Lux, S. E. (1986). Cation depletion by the sodium pump in red cells with pathologic cation leaks. Sickle cells and xerocytes. *J. Clin. Invest.* 78:1487–1496.

207. Grinstein, S., and Rothstein, A. (1978). Chemically-induced cation permeability in red cell membrane vesicles. The sidedness of the response and the proteins involved. *Biochim. Biophys. Acta* 508:236–245.

208. Solomon, A. K., Chasan, B., Dix, J. A., Lucakovic, M. F., Toon, M. R., and Verkman, A. S. (1983). The aqueous pore in the red cell membrane. Band 3 as a channel for anions, cations, nonelectrolytes, and water. *Ann. N.Y. Acad. Sci.* 414:97–124.

209. Sorensen, A. L., Kirschner, L. B., and Barker, J. (1962). Sodium fluxes in the erythrocytes of swine, ox, and dog. *J. Gen. Physiol.* 45:1031–1047.

210. Motais, R. (1973). Sodium movements in high sodium beef red cells: properties of a ouabain-insensitive exchange diffusion. *J. Physiol. (Lond.)* 233:395–422.

211. Motais, R., and Sola, F. (1973). Characteristics of a sulphydryl group essential for sodium exchange in beef erythrocytes. *J. Physiol. (Lond.)* 233:423–438.

212. Haas, M., Schooler, J., and Tosteson, D. C. (1975). Coupling of lithium to sodium transport in human red cells. *Nature* 258:424–427.

213. Duhm, J., Eisenried, F., Becker, B. F., and Greil, W. (1976). Studies on the lithium transport across the red cell membrane. I. Li uphill transport by the Na-dependent Li counter-transport system of human erythrocytes. *Pflugers Arch. Ges. Physiol.* 364:147–155.

214. Funder, J., and Wieth, J. O. (1978). Coupled lithium-sodium exchange in bovine red cells. In *Cell Membrane Receptors for Drugs and Hormones. A multidisciplinary Approach*, R. W. Straub and L. Bolis (Ed.). Raven Press, New York, pp. 271–279.

215. Jennings, M. L., Adams-Lackey, M., and Cook, K. W. (1985). Absence of specific sodium-hydrogen exchange by rabbit erythrocyte sodium-lithium countertransporter. *Am. J. Physiol.* 249:C63–C68.

216. Duhm, J., and Becker, B. F. (1979). Studies on the lithium transport across the red cell membrane. V. On the nature of the Na-dependent Li countertransport system of mammalian erythrocytes. *J. Membr. Biol.* 51:263–286.

217. Sarkadi, B., Alifimoff, J. K., Gunn, R. B., and Tosteson, D. C. (1978). Kinetics and stoichiometry of Na-dependent Li transport in human red blood cells. *J. Gen. Physiol.* 72:249–264.

218. Hannaert, P. A., and Garay, R. P. (1986). A kinetic analysis of Na-Li countertransport in human red blood cells. *J. Gen. Physiol.* 87:353–368.

219. Levy, R., and Livne, A. (1984). The erythrocyte membrane in essential hypertension. Characterization of the temperature dependence of Li efflux. *Biochim. Biophys. Acta* 769:41–60.

220. Becker, B. F., and Duhm, J. (1979). Studies inthe lithium transport across the red cell membrane. VI. Properties of a sulfhydryl group involved in ouabain-resistant Na-Li (and Na-Na) exchange in human and bovine erythrocytes. *J. Membr. Biol.* 51:287–310.

221. Levy, R., and Livne, A. (1984). Erythrocyte Li-Na countertransport system. Inhibition by *N*-ethylmaleimide probes for a conformational change of the transport system. *Biochim. Biophys. Acta* 777:157–166.

222. Canessa, M., Bize, I., Adragna, N., and Tosteson, D. C. (1982). Cotransport of lithium and potassium in human red cells. *J. Gen. Physiol.* 80:149–168.

223. Duhm, J., and Becker, B. F. (1977). Studies in the lithium transport across the red cell membrane. IV. Interindividual variaions in the Na-dependent Li countertransport system of human erythrocytes. *Pflugers Arch. Ges. Physiol.* 370:211–219.

224. Adragna, N. C., and Tosteson, D. C. (1984). Effect of volume changes on ouabain-insensitive net outward cation movements in human red cells. *J. Membr. Biol.* 78:43–50.

225. Garrahan, P. J., and Glynn, I. M. (1967). The sensitivity of the sodium pump to external sodium. *J. Physiol. (Lond.)* 192:175–188.

226. Kaplan, J. H., Dunham, P. B., Logue, P. J., and Kenney, L. J. (1984). Na/Na exchange through the Na/K pump of HK sheep erythrocytes. *J. Gen. Physiol.* 84:839–844.

227. Duhm, J., and Becker, B. F. (1977). Studies in the lithium transport across the red cell membrane. II. Characterization of ouabain-sensitive and ouabain-insensitive Li transport. Effects of bicarbonate and dipyridamole. *Pflugers Arch. Ges. Physiol.* 367:211–219.

228. Duhm, J., and Becker, B. F. (1977). Studies in the lithium transport across the red cell membrane. III. Factors contributing to the interindividual variability of the in vivo Li distribution across the human red cell membrane. *Pflugers Arch. Ges. Physiol.* 203–208.

229. Mendels, J., and Frazer, A. (1974). Alterations in cell membrane activity in depression. *Am. J. Psych.* 131:1240–1246.

230. Parker, J. C., and Berkowitz, L. R. (1986). Genetic variants affecting the structure and function of the human red cell membrane. In *Physiology of Membrane Disorders*, T. E. Andreoli, J. F. Hoffman, and D. D. Fanestil (Eds.). Plenum Press, New York, pp. 785–813.

231. Worley, R. J., Hentschel, W. M., Cormier, C., Nutting, S., Pead, G., Zlenkos, K., Smith, J. B., Ash, K. D., and Williams, R. R. (1982). Increased sodium-lithium countertransport in erythrocytes of pregnant women. *New Engl. J. Med.* 307:412–416.

232. Woods, J. W., Parker, J. C., and Watson, B. S. (1983). Perturbation of sodium-lithium countertransport in red cells. *New Engl. J. Med.* 308:1258–1261.

233. Trevisan, M., DeSanto, N., Laurenzi, M., DiMuro, M., De Chiara, F., Latte, M., Franzese, A., Iacone, R., Capodicassa, G., and Giordano, G. (1986). Intracellular ion metabolism in erythrocytes and uremia. The effect of different dialysis treatments. *Clin. Sci.* 71:545–552.

234. Aceves, J., and Cereijido, M. (1973). The effect of amiloride on sodium and potassium fluxes in red cells. *J. Physiol. (Lond.)* 229:707–718.

235. Siebens, A., and Kregenow, F. M. (1978). Volume regulatory responses of salamander red cells incubated in anisosmotic media: effect of amiloride. *Physiologist* 21:110.

236. Benos, D. J. (1982). Amiloride: a molecular probe of sodium transport in tissues and cells. *Am. J. Physiol.* 242:C131–C145.

237. Cala, P. (1980). Volume regulation by Amphiuma red blood cells. The membrane potential and its implications regarding the nature of the ion-flux pathways. *J. Gen. Physiol.* 76:683–708.

238. Cala, P. M. (1985). Volume regulation by Amphiuma red blood cells. Strategies for identifying alkali metal-H exchange. *Fed. Proc.* 44:2500–2507.

239. Cala, P. (1983). Volume regulation by red blood cells: mechanisms of ion transport. *Mol. Physiol.* 4:33–52.
240. Cala, P. M. (1983). Cell volume regulation by Amphiuma red blood cells. The role of Ca as a modulator of alkali metal/H exchange. *J. Gen. Physiol.* 82:761–784.
241. Cala, P. M., Mandel, L. J., and Murphy, E. (1986). Volume regulation by Amphiuma red blood cells. Cytosolic free Ca and alkalai metal/H exchange. *Am. J. Physiol.* 250:C423–C429.
242. Siebens, A. W., and Kregenow, F. M. (1985). Volume-regulatory responses of Amphiuma red cells in anisotonic media. The effect of amiloride. *J. Gen. Physiol.* 86:527–564.
243. Kregenow, F. M., Caryk, T., and Siebens, A. W. (1985). Further studies of the volume-regulatory response of Amphiuma red cells in hypertonic media. Evidence for amiloride-sensitive Na/H exchange. *J. Gen. Physiol.* 86:565–584.
244. Palfrey, H. C., Stapleton, A., and Greengard, P. (1980). Activation of amiloride-sensitive Na permeability in frog erythrocytes by cyclic AMP and other stimuli. *J. Gen. Physiol.* 76:25a.
245. Nikinmaa, M. (1982). Effects of adrenaline on red cell volume and concentration gradient of protons across the red cell membrane in the rainbow trout, Salmo gairdneri. *Mol. Physiol.* 2:287–297.
246. Baroin, A., Garcia-Romeu, F., Lamarre, T., and Motais, R. (1984). Hormone-induced co-transport with specific pharmacological properties in erythrocytes of rainbow trout, Salmo gairdneri. *J. Physiol. (Lond.)* 350: 137–157.
247. Baroin, A., Garcia-Romeu, F., Lamarre, T., and Motais, R. (1984). A transient sodium-hydrogen exchange system induced by catecholamines in erythrocytes of rainbow trout, Salmo gairdneri. *J. Physiol. (Lond.)* 356: 21–31.
248. Borgese, F., Garcia-Romeu, F., and Motais, R. (1986). Catecholamine-induced transport systems in trout erythrocyte. Na/H countertransport or NaCl cotransport? *J. Gen. Physiol.* 87:551–566.
249. Bourne, P. H., and Cossins, A. R. (1984). Sodium and potassium transport in trout (Salmo gairdneri) erythrocytes. *J. Physiol. (Lond.)* 347:361–375.
250. Cossins, A. R., and Richardson, P. A. (1985). Adrenaline-induced Na/H exchange in trout erythrocytes and its effects upon oxygen carrying capacity. *J. Exp. Biol.* 118:229–246.
251. Parker, J. C. (1986). Na-proton exchange in dog red blood cells. *Curr. Top. Membr. Transp.* 26:101–114.
252. Parker, J. C. (1983). Volume-responsive sodium movements in dog red blood cells. *Am. J. Physiol.* 244:C324–C330.
253. Parker, J. C. (1984). Glutaraldehyde fixation of sodium transport in dog red blood cells. *J. Gen. Physiol.* 84:789–804.
254. Parker, J. C. (1986). Interactions of lithium and protons with the sodium-poroton exchanger of dog red blood cells. *J. Gen. Physiol.* 87:189–200.

255. Parker, J. C., and Castranova, V. (1984). Volume-responsive sodium and proton movements in dog red blood cells. *J. Gen. Physiol.* 84:379–402.

256. Jennings, M. L., Douglas, S. M., and McAndres, P. E. (1986). Amiloride-sensitive sodium-hydrogen exchange in osmotically shrunken rabbit red blood cells. *Am. J. Physiol.* 251:C32–C40.

257. Escobales, N., and Canessa, M. (1985). Ca-activated Na fluxes in human red cells. *J. Biol. Chem.* 260:11914–11923.

258. Escobales, N., and Canessa, M. (1986). Amiloride-sensitive Na transport in human red cells. Evidence for a Na/H exchange system. *J. Membr. Biol.* 90:21–28.

259. Aronson, P. S. (1985). Kinetic properties of the plasma membrane Na-H exchanger. *Annu. Rev. Physiol.* 47:545–560.

260. Grinstein, S., and Rothstein, A. (1986). Mechanisms of regulation of the Na/H exchanger. *J. Membr. Biol.* 90:1–12.

261. Grinstein, S., Rothstein, A., and Cohen, S. (1985). Mechanism of activation of Na/H exchange in rat thymic lymphocytes. *J. Gen. Physiol.* 85: 765–788.

262. Motaís, R., Garcia-Romeu, F., and Borgese, F. (1987). The control of Na^+/H^+ exchange by molecular oxygen in trout erythrocytes. A possible role of hemoglobin as a transducer. *J. Gen. Physiol.* 90:197–207.

263. Parker, J. C. (1981). Heterogeneity among dog red blood cells. *J. Gen. Physiol.* 78:141–150.

264. Mahnensmith, R. L., and Aronson, P. S. (1985). Interrelationships among quinidine, amiloride, and lithium as inhibitors of the renal Na-H exchanger. *J. Biol. Chem.* 260:12586–12592.

265. Funder, J., Parker, J. C., and Wieth, J. O. (1987). Further evidence for coupling of sodium and proton movements in dog red blood cells. *Biochim. Biophys. Acta* 899:311–312.

266. Macara, I. G. (1985). Oncogenes, ions, and phospholipids. *Am. J. Physiol.* 248:C3–C11.

267. Besterman, J. M., May, W. S., Levine, H., Cragoe, E. J., and Cuatrecasas, P. (1985). Amiloride inhibits phorbol ester-stimulated Na/H exchange and protein kinase C. An amiloride analogue selectively inhibits Na/H exchange. *J. Biol. Chem.* 1155–1159.

268. Cohen, C. M., and Foley, S. F. (1986). Phorbol ester- and Ca-dependent phosphorylation of human red cell membrane skeletal proteins. *J. Biol. Chem.* 261:7701–7709.

269. Grinstein, S., Cohen, S., Goetz, J. D., and Rothstein, A. (1985). Osmotic and phorbol ester-induced activation of Na/H exchange. Possible role of protein phosphorylation in lymphocyte volume regulation. *J. Cell Biol.* 101:269–276.

270. Adorante, J. S., and Cala, P. M. (1987). Activation of electroneutral K flux in *Amphiuma* red cells by N-ethylmaleimide. Distinction between K/H exchange and KCl transport. *J. Gen. Physiol.* 40:209–227.

271. Parker, J. C., and Hoffman, J. F. (1965). Interdependence of cation permeability, cell volume, and metabolism in dog red cells. *Fed. Proc.* 24:589.

272. Grinstein, S., Goetz-Smith, J. D., Stewart, D., Beresford, B. J., and Mellors, A. (1986). Protein phosphorylation during activation of Na/H exchange by phorbol esters and by osmotic shrinkage. *J. Biol. Chem.* 261: 8009–8016.

273. Parker, J. C., and Snow, R. L. (1972). Influence of external ATP on permeability and metabolism of dog red blood cells. *Am. J. Physiol.* 223: 888–893.

274. Brown, A. M., and Obaid, A. L. (1976). ATP-induced swelling of dog erythrocytes: The roles of Na and cellular metabolism [abstract]. *The Physiologist* 19:140.

275. Romualdez, A., Volpi, M., and Sha'afi, R. I. (1976). Effect of exogenous ATP on sodium transport in mammalian red cells. *J. Cell. Physiol.* 87: 297–306.

276. Parker, J. C., Castranova, V., and Goldinger, J. M. (1977). Dog red blood cells: Na and K diffusion potentials with extracellular ATP. *J. Gen. Physiol.* 69:417–430.

277. Hempling, H. G., Stewart, C. C., and Gasic, G. (1964). The effect of exogenous ATP on the electrolyte content of TA3 ascites tumor cells. *J. Cell. Physiol.* 73:133–140.

278. Burnstock, G. (1977). The purinergic nerve hypothesis. In *Symposium on Purine and Pyrimidine Metabolism*, Ciba Foundation Symposium 48 (new series). Elsevier/Excerpta Medica/North Holland, Amsterdam, pp. 295–314.

279. Mercer, R. W., and Hoffman, J. F. (1985). Bumetanide-sensitive Na/K cotransport in ferret red blood cells [abstract]. *Biophys. J.* 47:157.

280. Moolenar, W. H. (1986). Effects of growth factors on intracellular pH regulation. *Annu. Rev. Physiol.* 48:363–376.

281. Greger, R. (1988). Chloride transport in thick ascending limb, distal convolution and collecting duct. *Annu. Rev. Physiol.* 50:111–112.

282. Paris, S., and Pouysségur, J. (1986). Growth factors activate the bumetanide-sensitive $Na^+/K^+/Cl^-$ cotransport in hamster fibroblasts. *J. Biol. Chem.* 261:6177–6183.

283. O'Brien, T. G., and Prettyman, R. (1987). Phorbol esters and mitogenesis: Comparison of the proliferative response of parent and Na^+ K^+ Cl^--cotransport defective BALB/c 3T3 to 12-O-tetradecanoylphorbol-13-acetate. *J. Cell. Physiol.* 130:377–381.

284. Amsler, K., Donahue, J. J., Slayman, C. W., and Adelberg, E. A. (1986). Stimulation of bumetanide-sensitive K^+ transport in Swiss 3T3 fibroblasts by serum and mitogenic hormones. *J. Cell Physiol.* 123:257–263.

19

Anion Transport

ROBERT B. GUNN, OTTO FRÖHLICH, PATRICIA A. KING,* and
DAVID G. SHOEMAKER *Emory University School of Medicine, Atlanta,
Georgia*

I. FUNCTIONS OF ANION TRANSPORT

A. Anion Exchange on Band 3

There are $\sim 10^6$ copies of the band 3 protein per normal human red blood cell
[1]. The principal function of this transport protein is to mediate the rapid exchange of extracellular chloride (Cl^-) and intracellular bicarbonate (HCO_3^-) ions
during the red blood cell's passage through the capillaries of the tissue, and the
reverse exchange during their transit through the alveolar capillaries. This rapid
exchange is of central importance in the physiology of CO_2 transport. Hamburger has been credited with first proposing, in 1892, that Cl and HCO_3 are
exchanged across the red blood cell membrane [2]. This "Hamburger shift" is
the classical mechanism that explains the observation that venous plasma Cl is
1-2 mM lower than arterial plasma Cl concentration. The overall mechanism
(Figure 1) was first clearly described by Jacobs and Stewart [3]. As the erythrocyte enters a tissue capillary, it encounters a higher pCO_2 (46 mm Hg) than in
the arterial circulation (40 mm Hg). The pCO_2 of the plasma and intraerythrocyte compartments rises rapidly to 46 mm Hg, and takes CO_2 and carbonic
acid (H_2CO_3) out of equilibrium. As part of reestablishing the equilibrium, some
CO_2 is hydrated to carbonic acid, which instantaneously dissociates to bicarbonate and protons. The hydration reaction occurs much more rapidly within the
erythrocyte than in the plasma because of intracellular carbonic anhydrases
(EC 4.2.1.1). The new protons are mostly buffered by hemoglobin and this
protonation reaction facilitates oxygen unloading because protonated hemo-

---

Current affiliation: The University of Vermont College of Medicine, Burlington, Vermont

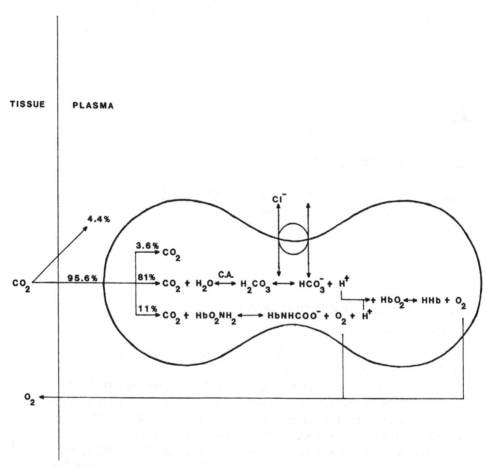

Figure 1   Erythrocyte uptake of CO_2 in peripheral capillaries. Tissue CO_2 is transported as dissolved CO_2 (8%) in both cytoplasm and plasma, as carbamino hemoglobin ($HbNHCOO^-$, 11%) in the cytoplasm and as bicarbonate (81%) in cytoplasm (24%) and plasma (57%). This latter bicarbonate is the result of intracellular hydration catalyzed by carbonic anhydrase (C. A.) followed by dissociation into a proton and bicarbonate and its exchange with extracellular chloride on band 3. A proton is produced by each CO_2 not carried as dissolved CO_2 and these protons are largely buffered by oxyhemoglobin (HbO_2) and promote O_2 dissociation and delivery to the tissue.

globin has a decreased affinity for O_2 (Bohr shift). The new bicarbonate ions accumulate within the cell and would remain there were not band 3 present to export them in exchange for plasma chloride. The removal of this bicarbonate from the cells allows the chain of reactions including the hydration of CO_2 and the subsequent dissociation of carbonic acid to continue until the plasma bicarbonate reaches equilibrium as well. By carrying HCO_3 in the plasma as well as the cytoplasm, the blood greatly increases its total CO_2 carrying capacity (see below).

Carbon dioxide is transported in three principal forms: dissolved CO_2, combined with amino groups on proteins, and as HCO_3. Dissolved CO_2 only accounts for 8% of the total CO_2 transported from the tissues to the lungs. The venous-arterial difference in pCO_2 is only 6 mm Hg (46–40 mm Hg) but would have to be 70 mm Hg if all of the CO_2 were to be transported in dissolved form. The carbamino groups formed by the reaction of CO_2 with the N-terminal valines of the α and β chains of hemoglobin account for 11% of the total CO_2 transported to the lungs [4]. Many textbooks continue to quote a figure of 30%, which was calculated from experiments using outdated bank blood with diminished concentrations of 2,3-diphosphoglycerate. In fresh blood the higher concentrations of 2,3-diphosphoglycerate, which binds to hemoglobin, greatly reduced the formation of carbamino groups and thus the ability of hemoglobin to transport CO_2 in this form. The third form of CO_2 transport as bicarbonate ions constitutes 81% of the total. Of this bicarbonate, 83% is carried in the plasma after being made in the red blood cell, which means that 67% ($0.83 \times 0.81 = 0.67$) of the total net CO_2 transport requires the exchange of chloride and bicarbonate. Given that 15–20 mol of CO_2 are exhaled each day (resting man), band 3 protein mediates 20–27 mol/day of net exchange because the exchange must be performed twice: once in the tissues as discussed here and once in the lungs, where the entire process is reversed as the pCO_2 of the plasma and erythrocyte cytoplasm is lowered to 40 mm Hg.

Although the activity of Cl–HCO_3 exchange is very high, its rate is low compared with the physiologically ineffective exchange of intraerythrocyte Cl for plasma Cl (or HCO_3 for HCO_3) that band 3 mediates at a rate of 150 mol $\cdot$ min^{-1} kg^{-1} of dry cell solids (216,000 mol/day). A 70-kg man has $\sim 3 \times 10^{13}$ red blood cells, which contain 1 kg of dry cell solids, including 900 g of hemoglobin. These exchange modes, although physiologically accomplishing nothing, are the expression of the system idling so that at the instant of a pCO_2 shift the system is ready to transfer HCO_3 in exchange for Cl and thus use the plasma space for CO_2 transport to the lungs. Consider how overdesigned the system appears to be. The rapid increase of blood pCO_2 on entering the tissue capillary results in a 2 mM increase in the intracellular HCO_3 concentration. During the ensuing reequilibration of intracellular and extracellular HCO_3, it is estimated that 66% of the cycles mediate Cl–Cl exchange and 4% of the cycles mediate HCO_3–HCO_3 exchange. The Cl_i–HCO_{3-o} exchange (14.4%) and HCO_{3-i}–Cl_o

exchange (15.9%) have different rates, which means that 1.5% of the exchange cycles are physiologically effective. This design seems to be required to have rapid net exchange in *both* directions so that blood uptake of CO_2 in the tissue capillaries and delivery of CO_2 to the alveoli will take place within the transit time of the red blood cells through the respective capillary beds (in the range of seconds).

B. Anion Conductance

Two greatly differing values of the red blood cell chloride permeability can be calculated from the chloride flux and the concentration gradient, depending on the type of the underlying transport experiment. A *unidirectional flux* is measured in tracer flux experiments, in which the opposite side of the membrane is tracer-free but may contain exchangeable anions. A *net flux* is measured by following the total chemical cellular content of a given compound as a function of time, or as the difference between the oppositely directed unidirectional fluxes. In the human erythrocyte, the rate of tracer chloride efflux (in the presence of extracellular exchangeable, nonradioactive Cl) is much more rapid than the rate of net chloride loss: at 37°C, the rate constant of tracer chloride equilibration is ~50 ms [5], corresponding to a tracer (or unidirectional) permeability of 3 X 10^{-4} cm s^{-1} and a flux of 150 mol (kg dry cell solids · min)$^{-1}$. For comparison, the net (or conductive) permeability is only 2 X 10^{-8} cm s^{-1} [6]. This very large difference is indicative of the efficiency of coupling between the influx of extracellular anions and the efflux of intracellular anions.

Nevertheless, this relatively low net permeability of chloride constitutes the major conductance pathway in the erythrocyte membrane: the chloride conductance is at least two orders of magnitude larger than the potassium or sodium conductance of the intact red blood cell membrane (Figure 2). As a consequence, in unmodified cells the membrane potential is essentially identical to the chloride equilibrium potential. Measurements of the membrane potential have been made with microelectrodes in the case of large amphibian red blood cells [7] and with voltage-sensitive fluorescent dyes in the case of smaller human red blood cells [8,9]. These experiments have shown that the membrane potential of erythrocytes at physiological pH lies in the range of -6 to -10 mV, consistent with the measurements that the intracellular Cl concentration is 0.7 to 0.8 times the extracellular Cl concentration: $V_m = E_{Cl} = RT\ F^{-1}\ ln(Cl_{in}/Cl_{out})$ = 26 mV ln (0.7–0.8). In other words, the intracellular and extracellular Cl concentrations are at electrochemical equilibrium, due to the high Cl conductance of the erythrocyte membrane.

A comment is in order on this equilibrium distribution of anions across the erythrocyte membrane. At times this equilibrium situation has been referred to as a Donnan equilibrium. Such a description implies that the lower intracellu-

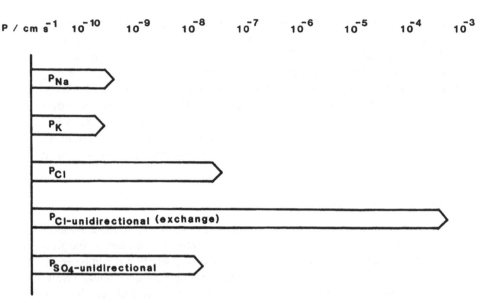

Figure 2 Permeabilities of the red cell membrane. The permeabilities, P, in cm/s of Na^+ and K^+ are 100-fold less than the permeability of chloride. The obligatory exchange of chloride has an apparent permeability at least 10^4-fold larger than the true conductive permeability for chloride. The cation permeabilities were estimated from the ouabain-insensitive fluxes at $37°C$. The anion permeabilities were estimated from the non-exchange-restricted net flux and the tracer fluxes at $37°C$.

lar anion concentration is a consequence of the presence of impermeant anions in the cytoplasm (mainly hemoglobin, but also organic phosphates). It also implies that the sum of intracellular and extracellular cation concentrations (mainly Na and K) are equal and that these cations are impermeant. The lower intracellular anion concentration would result from the need to maintain overall electroneutrality. This Donnan equilibrium description needs to be clarified because it cannot be correct in a strict sense. It is important to note that the Donnan equilibrium consideration does not take into account osmotic phenomena. A cell can therefore only be at Donnan equilibrium if the osmotic contribution of the impermeant anion is the same as that of the permeant anion it replaces. This is clearly not the case with hemoglobin, whose net charge can change with changes of the intracellular pH but whose osmotic contribution remains essentially constant over the same pH range [9]. A cell at simple Donnan equilibrium would therefore not be at osmotic equilibrium; it would swell and lyse. To protect itself from lysis, the cell must possess volume regulatory mecha-

nisms that control the intracellular cation concentrations and make up for osmotic imbalances that originate from the impermeant anionic species. These
mechanisms include the Na-K-ATPase that uses metabolic energy to maintain
concentration gradients, and, depending on the cell type, volume-sensitive cation–anion cotransporters or cation–proton exchangers coupled to a Cl^--HCO_3^-
exchanger. It is therefore reasonable to neglect the Donnan equilibrium notion
and to consider the intracellular anion concentration of the erythrocyte as a
consequence of the rates of the different cation transport and cation–anion cotransport systems that depend on cell volume, while anions also follow to maintain charge neutrality.

C. Phosphate Transport

Phosphate is important in intermediary metabolism in the red blood cell as well
as in other cells. Inorganic phosphate enters the glycolytic pathway at the glyc-
eraldehyde-3-phosphate dehydrogenase (GAPDH) step where it is combined at
the 1-position of glyceraldehyde-3-phosphate, forming 1,3-biphosphoglycerate
(1,3-DPG). This 1-position phosphate of 1,3-DPG is then immediately distributed between two important organophosphorus compounds of the red blood
cell, 2,3-DPG and adenosine triphosphate (ATP). The increasing concentration
of 2,3-DPG in response to respiratory alkalosis (such as occurs at high altitude)
can decrease the affinity of hemoglobin for oxygen both directly and indirectly
through enhancing the Bohr effect of protons. 2,3-DPG levels are also known to
increase in anemia, and cardiac and pulmonary disease. Adenosine triphosphate
is used as the major energy source of the cell to fuel metabolic reactions as well
as to maintain the membrane cytoskeleton and the ionic gradients across the
membrane by providing energy for the Na/K pump and Ca pump.

Until recently phosphate transport was believed to be mediated exclusively
by the anion exchange protein and by the divalent species of phosphate
(HPO_4^{2-}). Both of these beliefs appear to be false.

Phosphate is present in both monovalent and divalent forms at physiological
pH because its pK is 6.8 at physiological conditions of temperature and ionic
strength. The impression that divalent phosphate was the transported anion was
based on two observations. First, phosphate transport is slow (40,000 times
slower than chloride [10]) like sulfate and thus should be handled like sulfate.
Second, as the pH of red blood cell suspensions decreases from 8.5, the phosphate–phosphate exchange flux increases, peaks at pH 6.2, and then declines
again, just like sulfate–sulfate exchange and unlike chloride–chloride exchange
and that of other monovalent anions whose fluxes decline as pH is lowered from
pH 8.5 to 5.3 [11]. One experimental problem is that the pH dependence of Cl
transport must be due to titration of the transporter, band 3, but the pH dependence of phosphate flux must combine the reciprocal changes in the concen-

trations of $H_2PO_4^-$ and $HPO_4^=$ as well as changes in the titration of band 3. An additional problem is that changes in suspension pH also change internal pH; thus pH changes on both sides of the membrane could be contributing to the pH-dependent characteristics of the flux. It is now known that if phosphate influx is measured in exchange for internal Cl, the rapidly transported Cl is not rate-limiting and the flux characteristics are solely those of the phosphate influx half-cycle of the exchange. Furthermore, when under these conditions the phosphate anion concentrations are varied in two ways—by varying total phosphate concentration at constant external pH, or by varying external pH at constant total phosphate—the influx of phosphate is a single simple Michaelis-Menten saturating curve if graphed as a function of $H_2PO_4^-$ concentration [12]. However, if graphed as a function of $HPO_4^=$ or total phosphate concentration, a different curve is obtained at each external pH value. The external phosphate transport site seems to behave as if monovalent phosphate is the transported species. Also supporting the notion that monovalent phosphate is transported is the finding that at very low external pH (<5.3) phosphate-chloride exchange is *inhibited*, similarly to chloride–chloride exchange [12], and unlike sulfate-chloride exchange, which is only activated at very low external pH [13].

Whether this phosphate transport occurs as $H_2PO_4^-$ or as cotransport of H^+ and $HPO_4^=$ is not easy to determine, but if it is cotransport, it is different from H^+-$SO_4^=$ cotransport in that $H_2PO_4^-$ transport can be further inhibited by protons at very low pH, which has not been observed for H^+-$SO_4^=$ cotransport under the appropriate conditions. Furthermore, in potassium salt solutions monovalent phosphate is the form that is the competitive inhibitor of chloride-chloride exchange at $0°C$ [14], again suggesting that it is the monovalent form that binds to band 3. If there is any inhibition by divalent phosphate, it is at least 10 times less potent than the monovalent form. Therefore, there is good evidence that phosphate transport by band 3 is entirely as the monovalent form, $H_2PO_4^-$.

There now appears to be at least *three* phosphate transport mechanisms in the red blood cell membrane under physiological conditions ($[Na]_o = 140$ mM, $[PO_4]_o = 1.0$ mM, pH 7.4, 37°C) [15]. The main mechanism is mediated by band 3 and contributes 75% of the total phosphate transport; it is inhibited by stilbenedisulfonates [4,4'-diisothiocyanostilbene-2-2'-disulfonate and 4,4'-dinitrostilbene-2,2'-disulfonate (DNDS)] and is insensitive to the presence of sodium (Figure 3). The next mechanism is a sodium-phosphate cotransport pathway that is insensitive to stilbenedisulfonates; it makes up 20% of the total phosphate transport under physiological conditions. In addition there is a third phosphate transport pathway whose rate is linear with phosphate concentrations up to 2.0 mM and which contributes the residual 5% of the phosphate uptake. These transport systems together provide the cytosol with extracellular phosphate at a rate of 0.75 mmol (1 cells)$^{-1}$ hr^{-1} under physiological conditions.

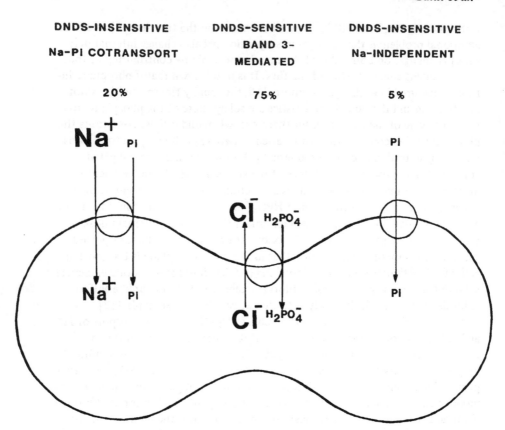

Figure 3 Phosphate transport. Phosphate influx is divided among three pathways at 37°C, pH 7.4, 1 mM phosphate and 140 mM Na$^+$. Band 3 mediates 75%, only transports monovalent phosphate, and is fully inhibited by 4,4'-dinitro-2,2'-stilbenedisulfonate (DNDS). Newly discovered Na-activated pathways mediate 20%. In the absence of extracellular Na and in the presence of high concentrations of DNDS (200 μM) 5% of the flux remains.

The sodium-phosphate (Na-P$_i$) cotransport pathway has a stoichiometry of 1.7 Na to 1.0 P$_i$ over a pH range of 6.90–7.75 but has a Hill coefficient for extracellular Na activation not significantly different from 1.0. These results can be reconciled by a single transport system that 30% of the time carries monovalent P$_i$ phosphate together with Na and 70% of the time carries the Na plus NaHPO$_4$$^-$ ion pair. The results could also be explained by the existence of multiple Na-P$_i$ cotransport systems in the red blood cell. The current data indicate that this red blood cell Na-P$_i$ cotransport system may be different from the one described for the renal proximal tubule cells where Hill coefficients of 2.0 are obtained for extracellular Na activation. However, recent reports indicate there may be some heterogeneity along the proximal tubule with regard to the mechanism of P_i transport [16].

The sodium-phosphate cotransport system seems to have a special relationship with the phosphorylation of adenosine diphosphate (ADP) to ATP by the membrane associated GAPDH in red blood cells. Substitution of extracellular Na with either potassium or N-methyl-D-glucamine inhibits 90% of the incorporation of extracellular ^{32}P incorporated into membrane-bound nucleotides ADP and ATP, whereas the presence of DNDS exhibits no such effect. This result is interesting in light of reports indicating that the ATP synthesized by the membrane-associated GAPDH reaction is preferentially delivered to the membrane-bound pool of ATP that is available to drive and phosphorylate the Na/K pump [17-19]. The spatial and functional relationships between this special Na-P_i cotransporting system, the glycolytic enzymes, and the Na pump remain to be worked out, but the possibility of an assembly of these elements in and adjacent to the cytoplasmic aspect of the erythrocyte membrane needs further experimentation to determine its physiological relevance.

D. Monocarboxylate Transport: Lactate and Pyruvate

Erythrocytes derive their metabolic energy from glycolysis. One might therefore expect that they would possess transport systems for the intake of their energy source (glucose) and for the export of the metabolic end product (lactate). For a brief period it had been surmised that the band 3 protein in its original or a modified form might also transport glucose. It is now established from the analysis of the peptide sequence of the cloned anion and glucose transport proteins that they are completely different molecules [20,21]. Lactate, on the other hand, is transported by the anion transporter and also by an additional transport system for lactate, pyruvate, and other related compounds (Figure 4) [22]. Owing to its specificity, the latter transporter has been referred to as the monocarboxylate transport system; this pathway is inhibited by p-chloromercuribenzenesulfonate. In rabbit erythrocytes, where it exists at a much higher density, the protein responsible for this transport pathway has been characterized as a peptide of 43,000 Da [23]. In an analysis of the relative transport rates of lactate by band 3 or its specific transporter, band-3 mediated lactate transport exhibited both a high K_m and V_{max}, whereas the lactate transporter exhibited both a low K_m and V_{max} [24]. Under physiological conditions of low lactate concentrations, the monocarboxylate transport pathway is by far the most prevalent one and mediates 90% of the total lactate movement, whereas band 3 mediates only 6% [25]. Another 4% are presumed to permeate the membrane as the undissociated acid.

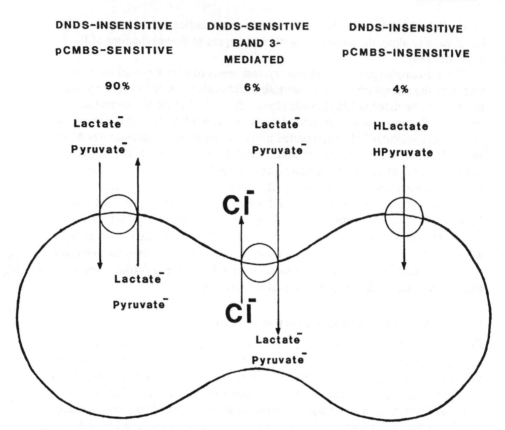

Figure 4 Lactate and pyruvate transport. Monocarboxylic acids are transported by three mechanisms. About 90% of the transport is sensitive to sulfhydryl reagents like p-chloromercuribenzenesulfonate (pCMBS) and insensitive to 4,4'-dinitro-2,2'-stilbenedisulfonate (DNDS) and is probably transported on a 43,000 Da protein. Only 6% is DNDS-sensitive and is transported in exchange for chloride on band 3. The remaining 4% is probably diffusion of the neutral undissociated acid across the red cell lipid bilayer membrane.

E. Amino Acid Transport

Glutathione (GSH) is the major nonprotein thiol in the red blood cell and is responsible for maintaing the oxidation-reduction state of the red blood cell proteins. Because GSH has a half-life of three to four days, there is continued synthesis of this tripeptide using enzymes produced during the nuclear stage and retained in the mature cell. One consequence of this is the need for continued amino acid transport into the cell, in particular the constituent amino acids Glu, Cys, and Gly. Gly appears to be transported by a number of pathways (Figure 5)

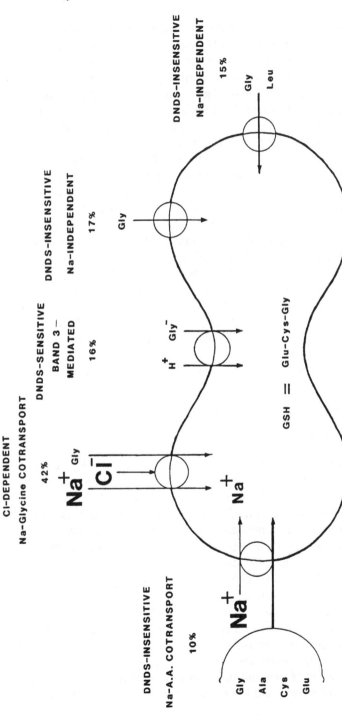

Figure 5 Amino acid transport. Glycine (gly) uptake is required for the continual synthesis of glutathione (GSH) by the mature erythrocyte. There are five pathways for glycine. Two pathways are shared with other amino acids: one is Na$^+$ cotransport (10%) and one is Na$^+$-independent (15%). The major pathway (42%) is both Na$^+$- and Cl$^-$-dependent. Band 3 cotransports (16%) the glycine anion and a proton. Finally there is a Na-independent pathway that is not shared with leucine (leu) and is not inhibited by 4,4-dinitro-2,2′-stilbenedisulfonate (DNDS). It accounts for 17% of the ^{14}C-glycine influx under physiological conditions: 0.2 mM glycine, 140 mM NaCl, pH 7.4, 37°C.

[26] with the major route being a Na- and Cl-dependent transport pathway. Recent studies of this Na-dependent influx [27] have shown that two Na ions are cotransported with the Gly zwitterion, the form that makes up 99% of the Gly at physiological pH (7.4) and carries both a positively charged amino group (pK 9.4) and a negatively charged carboxyl group (pK < 3.0). Transport is stimulated by the presence of Cl (compared with NO_3), which binds to the transporter and appears to increase the affinity of one of the Na sites. These experiments suggest that the substrates bind in an ordered fashion with Gly first, then Na, then Cl, and finally the second Na to form the fully loaded complex that crosses the membrane to transfer Gly into the cell. In addition, a significant influx of Gly occurs by a Na-independent, stilbenedisulfonate-sensitive pathway, presumably on band 3, which transports the Gly anion. Although the anionic form of Gly accounts for only 1% of the Gly in the physiological þH range, this pathway accounts for 16% of the total Gly influx at plasma Gly concentrations. King and Gunn [28] have shown that this pathway is stimulated by protons. Thus it appears that the glycine anion and protons are cotransported by band 3.

Much less is known about Cys and Glu transport. Cysteine is transported by a Na-dependent pathway not stimulated by Cl (the ASC system [29]) that shows a high affinity for this amino acid [39]. Human red blood cells are essentially impermeable to Glu. As a result, cellular Glu is likely formed from the deamidation of Gln or transamination of Ala, which enter the cell by the Na-dependent pathway described for Cys [30,31].

II. CHARACTERISTICS OF BAND 3 TRANSPORT

A. Binding/Transport Site

Although the maximum rates of exchange transport of different anions varies from Cl and bicarbonate at 150 mol (kg dry cell solid · min)$^{-1}$ to sulfate at 3 mmol (kg dry cell solid · min)$^{-1}$, the affinity of band 3 for the different anions is nearly the same. When the apparent affinities of chloride, bromide, iodide, and sulfate are measured under comparable conditions, that is, as Cl_i-X_O exchange with the anion transporter mostly recruited into the outward-facing state, the concentration of external anion (Cl^-, Br^-, I^-, $SO_4^=$) required for half-activation of the flux is between 1 and 5 mM [13,32] (bicarbonate half-activation is < 10 mM [33]). From recent nuclear magnetic resonance measurments of the rates of Cl binding and dissociation from band 3, it can be concluded that the transport steps, not the binding or dissociation steps, are rate-limiting [34] and thus the apparent affinity as measured above closely approximates the affinity of these anions (note that in the cases of Cl and bicarbonate the anion transport site is primarily inward facing and the value of $K\frac{1}{2}$ probably results from a combination of the true affinity and the translocation asymmetry of the ex-

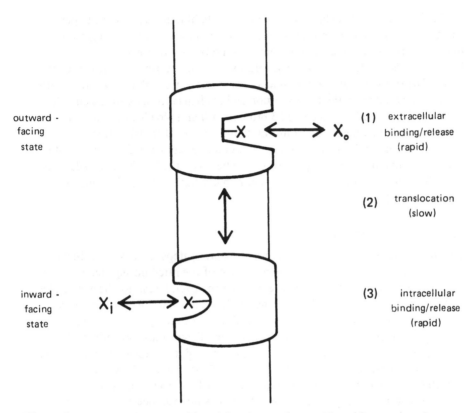

Figure 6 Ping-pong kinetics of band 3 anion exchange. The obligatory exchange of inorganic anions on band 3 involves three classes of reactions, shown here. 1) The rapid binding or release of anion between the extracellular medium and a "site" within the protein that is embedded in the phospholipid membrane. 2) The slow and usually rate-limiting translocation reaction which reorients the bound anion so that it can dissociate into the opposite medium. This translocation may have several steps which, for the present, can be lumped together. It does not reorient the entire protein but only a very small subregion. Only when the "site" is complexed with a transportable anion can the translocation reaction occur. 3) The rapid binding or release of anion between the intracellular medium and the inward-facing "site."

change mechanism [35]). Consequently one important characteristic that an ultimate detailed description of the binding site will have to explain is the binding nonselectivity among a large range of inorganic anions. It appears that the band 3 protein is a general anion-binding protein but the anion translocation rate is very dependent on the chemical nature of the specific anion bound to the site.

The band 3 protein behaves as if it has a single binding/transport site per 95,000 Da monomer (Figure 6). The exchange is almost entirely obligatory in that an anion from the cis-solution binds to the transport site and is transferred and dissociated into the trans-solution on the opposite side of the membrane. The transport site cannot bind a second anion from the cis-solution until it binds an anion from the trans-solution and exchanges it back to the cis-solution. Rarely, if at all, can the band 3 protein with an unloaded or empty binding/transport site undergo the conformational change of translocation. Thus, most of the time only the complex between band 3 and a small anion can alternate the access of the binding/transport site between the two bathing solutions. This obligatory, alternate access mechanism has what is technically called ping-pong kinetics [36].

B. Temperature Dependence

The temperature dependence of a series of reaction steps such as occur in anion exchange reflects the temperature dependence of the rate-limiting step because only by accelerating the rate-limiting step can the overall reaction be accelerated. As stated above, the transporting (translocation) step and not the binding of anions has been shown to limit the maximal rate of exchange. Consequently the very high temperature dependence of band 3-mediated transport reflects the activation energy of the transporting step. In general, the exchange of anions has an activation energy of $110-130$ kJ mol^{-1} or $25-35$ kcal mol^{-1}, which is energetically *equivalent* to the making and breaking of five to seven hydrogen bonds during influx and during efflux, although there is no evidence to date to implicate an involvement of hydrogen bonds in the conformational change. There are differences in the measured activation energies among anions but the reasons for these are unclear. In part these differences may reflect differences in the titration of band 3 by protons because pKs are temperature-dependent. In general, for the rapidly transported anions a nonlinear behavior is found in the Arrhenius diagram, with a lower activation energy above $17-22°C$. In recent studies this nonlinearity could be modeled from three constant activation energies for binding to the anion binding site, for binding to the self-inhibition site, and for the transport step [36]. Thus it is likely, but unproven, that the complex temperature dependence of exchange is solely the result of interactions between the temperature dependence of these three components of the process and not a reflection of "phase transitions" in the membrane or a shift between rate-limiting steps as has been previously proposed [5,37].

C. Ping-Pong Kinetics

The characterization of band 3-mediated exchange by a ping-pong kinetic mechanism was preceded by two important discoveries. First, it was found that self-exchange of chloride was a saturating Michaelis-Menten function of the form

$$\text{Flux} = V_{max}/(1 + K_m/\text{Cl})$$

where V_{max} is the maximum flux at saturating concentrations of Cl and K_m is the apparent Michaelis constant (~3 mM if Cl concentration is changed on the outside and 20-60 mM if Cl is changed on the inside). This kinetic observation contributed to the realization that anion transport was not by diffusion through a porous fixed charge network as was believed between 1927 and 1972. Second, it was shown that anion transport was an obligatory exchange of one anion for another even when cation transport was not rate-limiting [38]. Thus, the mechanism seems able only to transfer anions back and forth but mediates little net transport by returning empty or by the tunneling mechanism (see below).

Two general classes of obligatory exchange transport mechanisms can be envisioned. The first mechanism, described by ping-pong kinetics, involves only a single binding/transport site that has alternating access to the ions inside and outside the cell. Such a mechanism would require that the crucial transporting conformational change take place only if an anion is bound to the transport site. Two alternative mechanisms can be described by sequential kinetics [37]. One obligatory exchange mechanism has two binding/transport sites that simultaneously have separate access to the two pools of ions on opposite sides of the membrane. These two sites must each be loaded with an anion from their respective sides before the transporting conformational process takes place to assure that there is obligatory exchange of the two ions. In the transport process the two sites switch orientation or the two ions exchange sites simultaneously so that they each become accessible to the opposite solutions. They then may dissociate and the two sites are free to complex with two other anions. Another mechanism with sequential kinetics is one in which the obligatory exchange element of the single site scheme above is relaxed so that the *empty* site (not just the loaded site) changes orientation from one side of the membrane to the other. This type of mechanism, commonly attributed to the red blood cell glucose transporter, has sequential kinetics but only a single binding/transport site. It is also referred to as the simple carrier mechanism. There is now a great deal of evidence that obligate anion exchange proceeds by a mechanism with a single binding site, accurately described by a ping-pong kinetic scheme. A review of the mathematical treatment, kinetics, asymmetry, recruitment, and inhibition properties of such a mechanism applied to transport via band 3 has recently been published [39]. Thus, another important characteristic that the ultimate structure of band 3 must identify is the single anion transport site.

There are many possible physical mechanisms that exhibit single site transport kinetics. Consequently, the single anion transport site might not be a single physical location (Figure 7). For example, it would be possible for the transporter to have two physical locations (such as two constellations of amino and carboxyl groups) that are arranged such that only one of them could have an

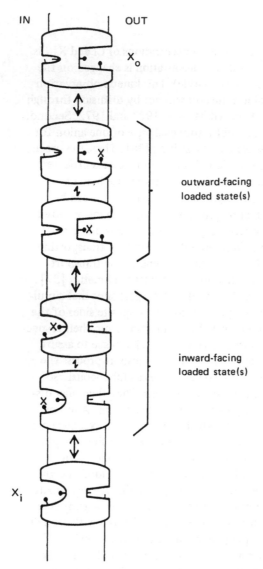

Figure 7 Ping-pong kinetics with several anion binding sites. The mathematics of anion exchange are that of a single site having alternate access to the two binding solutions. However, there may be a multiplicity of physical locations that have mutually exclusive occupancy, i.e., only one of the physical locations may be complexed with an anion at any given moment. Here there are two locations (↑) available to X_O at the top of the figure. Two outward-facing conformations loaded with the anion are shown. In one, the anion X is bound to the outermost superficial site. This is sometimes called a pre-binding site or a guid-

anion bound at any given time: a one-site-always-empty arrangement. Alternatively, two physical locations could be arranged such that only one location would be empty at any given time: a one-site-always-loaded arrangement. Either of these arrangements would behave mathematically as a single anion site mechanism in that the transporter would have only four classes of conformations: "empty" with access to the outside solution, "empty" with access to the inside solution, "loaded" with access to the outside solution, and "loaded" with access to the inside solution. In addition there may be occluded states not available to observation in the steady state. Such occluded states have been described in the Na/K pump (Chapter 16) where the transported ion is loaded but has access to neither solution, i.e., it is impossible for the ion to dissociate without the transporter first changing its conformation. In band 3 such occluded states have been postulated to explain the complex binding behavior of some fluorescent probes that ligate to the transport site [40]. Another important characteristic that any proposed model of the structure of band 3 must describe is the two major classes of conformations (one with an inward-facing binding site and the other with an outward-facing binding site) and how these sites shift back and forth *only* when an anion is complexed to the transport site.

D. Proton Interactions

Band 3 has significant titration reactions that both modulate the rate of exchange and provide the basis for another mode of transport, namely, anion-proton cotransport. For many years it has been appreciated that the pH dependence of anion transport was providing important clues as to the mechanism. This was particularly true when the opposite pH dependence of sulfate transport (activation by protons) and chloride transport (inhibition by protons) over the same pH range (6.2–8.5) raised insurmountable objections to the fixed charge

---

Figure 7 (continued)   ance site in the mouth of the channel to the deeper site. In the other, the anion X is bound to the deeper site which allows a subsequent conformational change, called translocation, to the inward-facing loaded states. Note that the outward binding sites do not necessarily disappear upon translocation of the ion to the inward-facing conformation but become occluded to other anions. The kinetics require as a first approximation that at most one of the four physical locations may be occupied at any time. In the lower half of the figure the anion X moves from deep site to superficial site and finally dissociates into the cytoplasm. To return the protein to the top conformation requires an intracellular anion to reverse the steps by following the sequence from the bottom of the figure to the top and thus complete the one-for-one obligatory exchange.

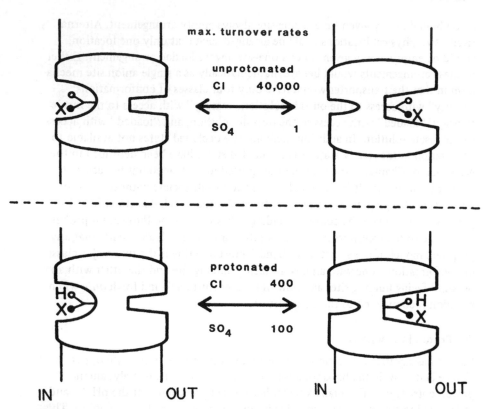

Figure 8 Cotransport of anions and protons. Protons inhibit chloride exchange, activate sulfate exchange, and are cotransported with both anions. This occurs through the titration of a single site, presumably a carboxyl group, shown here as an open circle on a stalk, with the anion binding/transport site shown as a solid circle. Shown at the top is the translocation step for the unprotonated band 3. The turnover rates are the maximum influxes (mmoles/(kg solids · min) for $X = Cl^-$ or $X = SO_4^=$ at $20°C$, $pH_{out} = 8.5$ in exchange for cytoplasmic chloride. Shown at the bottom is the translocation of the protonated band 3 with influxes at $20°C$, $pH_{out} = 3$, in exchange for cytoplasmic chloride. (The flux measurement is completed before $pH_{in} = 7.6$ changes more than 0.2 units.) Note that the chloride flux always exceeds the sulfate flux but protonation greatly reduces the difference in their relative maximum rates. Presumably H^+ reduces the negative charge on the exchanged complex making chloride behave more like a neutral species and making $SO_4^=$ behave more like a monovalent anion which the translocation reaction favors. The occluded forms of these sites on the unaccessible side of the protein are not drawn for simplicity.

model in 1972 [41]. The alternative titratable carrier model proposed a carrier that on protonation altered its selectivity in favor of divalent sulfate and against monovalent anions. However, this simplified view is no longer sufficient to explain all the existing data. We now know that protonation of a single critical site reduces the transport rate for chloride $\sim$100-fold and at the same time increases the transport rate of sulfate $\sim$100-fold (Figure 8) [13]. Inconsistent with the titratable carrier model is the fact that chloride transport is *always* favored over sulfate transport by at least a factor of 4 at low pH ($pH_O = 3$), although it favors chloride by a factor of 10^4 at high pH ($pH_O = 8$). These changes are primarily due to changes in the transport rate and not in the affinities, although external protonation enhances sulfate binding for chloride inhibition by a factor of 9 and slightly reduces Cl binding to the external facing binding site [42].

We now know that internal protons are mixed inhibitors of Cl transport [43]. A proton can bind to the inward-facing transporter whether the transport site is empty or is loaded with Cl. A similar relationship is believed to be true for the externally facing transporter. What is not proven is whether it is the same proton site that can be titrated on the two sides of the membrane and whether that site has alternate access to the two solutions in the same way the presumed single anion binding/transport site does. In 1976 M. Jennings first showed that protons were cotransported with sulfate by measuring sulfate flux and sulfate-activated net proton transport [44]. More recently, these studies have been expanded to measure Cl-H, NO_3-H, Br-H cotransport in red blood cells [45,46]. These experiments show that all of the anion-stimulated net proton flux is inhibitable by DNDS (a reversible stilbene derivative) and that this flux is most likely anion-proton cotransport, not anion-hydroxide exchange (which would also move proton equivalents in the same direction as the anion). These experiments are necessarily performed in CO_2-free cells because of the very rapid Cl-HCO_3 exchange, which also transports acid equivalents in the same direction as Cl. These anion-proton fluxes are small but may be an important component of net proton flux when an increased intracellular pCO_2 leads to increased internal H^+ and HCO_3^- concentrations and Cl-HCO_3 exchange is obligated to equilibrate HCO_3 and Cl at the expense of enhancing the proton gradient.

E. Self-Inhibition

At chloride concentrations up to physiological levels, anion exchange can be described by a simple ping-pong kinetic carrier scheme involving one anion binding site, but there are deviations from the Michaelis-Menten activation curve at elevated concentrations. That is, above a certain concentration exchange flux decreases with further increasing concentrations. This phenomenon is referred to as self-inhibition. It is more pronounced with bromide and iodide than with Cl. Dalmark [47], in his analysis of the self-inhibitory potency of the different

IN   OUT

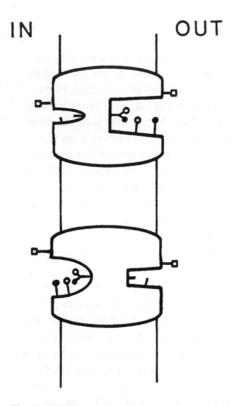

Figure 9 Mechanisms for self-inhibition. Three possible mechanisms for self-inhibition (at high substrate concentrations) are depicted in the figure. Solid circles on stalks are anion binding sites related to translocation of anions. Open circles on stalks are proton binding sites related to translocation of protons with anions. Open squares on the surfaces are anion binding sites called modifier sites that may alter the protein conformation and thus translocation rate without transfer of the bound anion. The empty stalks are anion binding sites on the occluded side of the membrane which may have low affinities for anions and thus may not be fully occluded or excluded from the ions in the adjacent solution. Historically the first mechanism is that represented by the open squares on the surface. Both internal and external modifier sites have been implied from kinetics. Self-inhibition by internal chloride is more significant than self-inhibition by external chloride at physiological concentrations, pH and temperature. The second mechanism is the possibility of double occupancy of the solid circles on stalks, the prebinding or guidance sites *and* the deep transport site at higher substrate concentrations. This double occupancy could slow or prevent translocation of the anion on the deep site. The third possibility proposed jointly by M. Jennings and M. Milanick is that the empty stalks are not occluded but only have low affinities for the anions in the conformation where the primary binding "site" is facing the opposite solution. When these trans-sites are occupied at high trans-substrate concentrations the translocation conformational change is impaired; or unimpaired but without either of the bound anions changing places. In that case the originally cis anion now finds itself on a low-affinity site and reverts to the solution on the cis side.

anions, ascribed it to the effect of a second, noncompetitive inhibitory anion binding site, which he referred to as a "modifier site." The location of this site on the anion transporter was not known at that time because those experiments were conducted under conditions in which Cl_O and Cl_i were varied simultaneously. By varying only extracellular Cl and bromide concentrations while keeping Cl_i or Br_i constant, Gunn and Fröhlich [36] provided evidence that there is a modifier site that inhibits by binding extracellular anions. Recent experiments by Knauf and Mann [48], however, indicate that the self-inhibitory effect of extracellular anions is considerably less than that of intracellular anions and that most of the self-inhibition that is observed under equilibrium exchange conditions is due to intracellular anions and thus an inward-facing modifier site. Altogether, one can state that the anion transport mechanism possesses inward-facing as well as outward-facing, anion-binding modifier sites (Figure 9). It remains unknown, however, on which conformational state of the anion transporter (inward-facing or outward-facing with respect to the anion transport site) these sites are available for anion binding.

The study by Knauf and Mann [48] also sheds light on the relationship between the anion modifier site and the binding site for an anion transport inhibitor, N-(4-azido-z-nitrophenyl)-z-aminoethylsulfonate, NAP-taurine. In previous studies (under equilibrium exchange conditions), Knauf et al. [49] demonstrated that extracellular NAP-taurine competed with Cl for a binding site that had an apparent Cl binding affinity of 150 mM. Because this value was larger than the $K_{1/2}$ of Cl exchange under these conditions and closer to the apparent affinity of the modifier site, they proposed that NAP-taurine and Cl competed for the modifier site. It is obvious that the extracellular NAP-taurine binding site and the internal self-inhibitory site cannot be the same.

Nevertheless there are still questions about the relationship between the extracellular self-inhibitory site and the noncompetitive NAP-taurine site (by its anionic nature, NAP-taurine can also bind competitively to the anion transport site, albeit probably at a considerably lower affinity; that there are two classes of binding sites can be shown by the nonlinearity of the Dixon plot [49,50]). Mutually exclusive binding of NAP-taurine and self-inhibiting Cl to a transport-inhibiting site might be difficult (but not impossible) to examine in transport experiments. Wieth and Bjerrum did, however, demonstrate a kinetic similarity between the transport inhibition by NAP-taurine and by self-inhibitory Cl [51]. They showed that self-inhibition only occurs when a titratable site with a pK of ~11 has bound an extracellular proton. In the same manner, flux inhibition by NAP-taurine exhibited the same dependence on an extracellular proton with the same pK. This means that the original conclusion by Knauf and co-workers [49] that NAP-taurine binds to a modifier site may still be partially correct. A full answer, however, will have to wait for further experimentation.

What is the role of the modifier site in the anion transport mechanism? As discussed by Knauf [10] and Wieth and Bjerrum [51], a close proximity be-

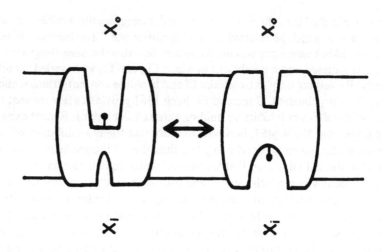

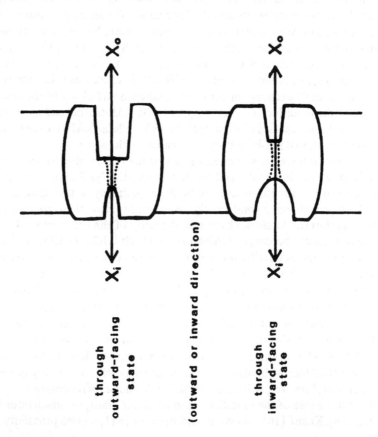

Figure 10 Conductive transport mechanisms. Two mechanisms for chloride conductance or net charge transfer are shown. Tunnelling is the movement of the anion through the channel-like elements of the protein without the conformational change associated with obligatory exchange. Because the protein can be either outward-facing or inward-facing there are in principle at least two modes of tunneling in each direction and they may occur at different rates (but both rates must be very much slower than exchange). Tunneling appears to be the dominant mechanism, and tunneling out through the outward-facing state appears easier than tunneling out through the inward-facing state. No measurements of tunneling have been reported. Slippage involves half of the usual transport of an exchanging anion (not shown here but in Fig. 6) followed by the reverse "translocation" conformational change but *without* an anion bound or being translocated. The result is a resetting of the usual transport mechanism without the obligatory return of an anion. Both slippage and tunneling involve band 3 or band 3– related proteins and thus would be inhibitable by the sulfonic stilbene derivatives.

tween transport and modifier sites could mean that the real purpose of the modifier site is to guide the anion to its transport site. This mechanism would be similar to the role of a prebinding site in an access channel that Tanford [52] proposed for self-inhibition, and it could possibly be useful in explaining the low apparent affinity obtained for the Cl_i activation of Cl net efflux by tunneling.

F. Conductive Transport

From the ping-pong kinetic behavior of chloride exchange one might suspect that the band 3-mediated chloride conductance (i.e., net transport) is a corollary of the carrier kinetics of anion transport. In terms of the schemes in Figure 10, anion net transport would be achieved if the transport site is empty during the return translocation step, i.e., if the outward-facing and inward-facing conformers can directly interconvert after a substrate anion has been transported across the membrane. This net transport mode has often been referred to as the "slippage" mode because one would expect, due to the much slower rate of net flux relative to exchange flux, that only occasionally the empty site would undergo the translocation reaction. However, when the notion of an internal "slip in the coupling" was introduced in the context of anion conductance [53], it could also have meant a different mechanism, namely the movement of the anion through the transport protein without the protein changing its conformational state. It is this latter mechanism that now appears to be the dominant mechanism of Cl net transport, and is called "tunneling" to indicate the channel-type behavior that it resembles [54].

In measurements of Cl net transport one has to ensure that net salt flux is not limited by the intrinsically low cation conductance. The erythrocyte membrane must be rendered cation-permeable so that every time an anion is translocated by band 3, this charge movement is compensated by a rapid and parallel movement of a cation so that the membrane potential remains constant. This is achieved by treating the red blood cells with either of the cationophores valinomycin or gramicidin. In fact, the cation gradient is fixed to impose a known membrane potential, thereby providing a known driving force for anion net flux.

In general, the total electrochemical potential difference that drives anion net flux is made up of the concentration gradient and the electrical potential difference. When the membrane potential is kept constant, one can determine the kinetics of anion net transport as a function of the substrate (anion) concentration. In such experiments, it was found that Cl net efflux does not decline to zero when external exchange sites are essentially saturated and that Cl net flux does not saturate at intracellular Cl concentrations that saturate the exchange process [54-57]. Both points argue against slippage as the mechanism for chloride net efflux. As a third point, net efflux by chloride, bromide and nitrate differ by as much as fourfold when they should be the same if the return of the empty site limits the net efflux. Therefore, tunneling is the major component of anion conductance, and the outward-facing conformation of band 3 permits at least a 10-fold higher rate of outward tunneling by Cl than the inward-facing conformation.

Recent experiments on sulfate net transport also support the conclusion of a very low rate of slippage [58]. Sulfate net efflux is much slower than Cl net efflux so that a possible contribution of slippage is more easily detectable. Using the noncompetitive inhibitor phloretin that inhibits the translocation reaction but not tunneling [59], it is possible to separate the contributions of slippage and tunneling to sulfate net flux and compare them with Cl net efflux. One can then compare the rates of net efflux by the different transport modes: at $20°C$, pH 7.0, and a membrane potential of -90 mV, Cl net efflux by tunneling has a rate of ~100 mmol $(kg\ Hb \cdot min)^{-1}$ through the outward-facing state, and <10 mmol $(kg\ Hb \cdot min)^{-1}$ through the inward-facing state. At the same time, the slippage rate, that is, the translocation rate of the unloaded transporter, is ~1 mmol $(kg\ Hb \cdot min)^{-1}$, and the rate of sulfate tunneling is ~3 mmol $(kg\ Hb \cdot min)^{-1}$.

The major question, naturally, concerns the mechanism of this hypothetical tunneling mode. So far tunneling has only been defined on kinetic grounds as a transport mechanism that permits the passage of an anion through the protein without the conformational change that is typical for carrier-type translocation. Thus, tunneling could be categorized kinetically as a channel-type mechanism. It is quite tempting to interpret this mechanistically as a clue about the physical nature of the anion transport pathway in the band 3 protein. This pathway could be envisioned as a very tight channel pathway characterized by a potential energy profile that possesses at least one major barrier [60]. Depending on the conformational state, this barrier is located either on the cytoplasmic side or the extracellular side of a central anion binding site, corresponding to the outward-facing or the inward-facing state, respectively (Figure 10). If the barrier were infinitely high, the only modes of transmembrane transport would be via the conformational change, i.e., carrier-type exchange and net transport by slippage (if the conformational change can occur when the binding site is empty). If, on the other hand, the major barrier can be crossed by the anion, this would constitute a channellike movement via the tunneling mode. At this point, it is not yet possible to correlate a physical structure belonging to the known band 3 peptide sequence with a potential energy barrier or the anion binding site, but further theoretical-structure and crystallographic efforts should in the future help in providing this correlation.

G. Cation Transport by Band 3

Sodium and Lithium in Bicarbonate Solutions

At first it may appear contradictory to discuss cation transport, specifically Na and Li^+, by the anion exchanger. But in physiological bicarbonate (HCO_3^-) solutions Na^+ an Li^+ can form the ion pairs $NaCO_3^-$ and $LiCO_3^-$. (Lithium is present in the blood at concentrations of 1-2 mM in the plasma of patients taking lithium for the treatment of manic–depressive disorders.) These anionic ion pairs are also transported by band 3 [61,62]. Their transport is inhibited by stilbenes and other transported anions. Although $NaCO_3^-$ and $LiCO_3^-$ are present in low concentrations, their transport is not trivial because the alternative pathways for Na^+ and Li^+ have even lower turnover rates. It is estimated that in plasma (Na^+ = 140 mM, HCO_3^- = 24 mM, pH 7.4, 37°C) $NaCO_3^-$ influx constitutes ~25-30% of the passive Na^+ influx. Of course, in the steady state this amount must also be pumped out by the Na/K pump. This cation leak pathway is discussed in more detail in Chapter 18.

In planning experiments this mechanism must be kept in mind. For example, removing red blood cells from a bicarbonate-buffered medium will necessarily decrease the Na^+ influx even if the Na^+ concentration in the new buffer matches that of the bicarbonate medium. Consequently, the rate of Na/K pump activity should decrease 25% because in the new steady state it will have to pump less Na out of the cell. Second, the steady state concentration of intracellular Na may measurably decrease compared with that in fresh cells. Third, there may be an ATP-sparing effect because ~40% of the total ATP consumption of red blood cells is required to operate the pump so that 10% less will be needed if the $NaCO_3^-$ leak is prevented. Another consideration is that if inhibitors of the anion exchange mechanism are added in vitro these secondary effects on the Na pump and ATP metabolism may also occur.

Lead in Bicarbonate Solutions

Recently the passive transport of lead (Pb) into human red blood cells by band 3 has been examined. It appears that >90% of the uptake is inhibited by stilbene inhibitors of anion exchange. Lead uptake is a linear function of Pb^{2+} and HCO_3 - concentrations and is inhibited by protons, suggesting that the neutral compound $PbCO_3$ may be the transported species [63].

H. Inhibitors of Band 3

In general, small inorganic anions are transported by band 3 and thus mutually compete for the limited number of transporters on each red blood cell. Thus Cl inhibits Br fluxes, sulfate inhibits phosphate and Cl fluxes, etc. In addition most organic anions are inhibitors of band 3. The major drugs that may be clinically important are salicylates [64], diuretics (furosemide–Lasix [65]), bumetanide

[66], thiazides [67], local anesthetics (benzocaine, tetracaine [68]), antifertility drugs (gossypol [69]), antidiabetic drugs (Tolrestat, [46]) and poisons such as dinitrophenol [70] and its many derivatives. For experimental use, the stilbenes, NAP-taurine, and phloretin have been the most valuable tools.

III. CLINICAL IMPLICATIONS

A. Failures of Band 3 Function

There are no known genetic abnormalities of band 3 function at this time. The great importance of band 3 in CO_2 transport to the lungs may mean that failure of band 3 function is incompatible with life or neonatal life. However, there is suggestive evidence that normal band 3 function may be marginal in certain conditions of stress.

B. Capillary Transit Time Problem

As described above, the time for 90% completion of band 3-mediated equilibration of HCO_3 between the erythrocyte cytoplasm and the plasma is $\sim$250 ms and this is the slowest and thus rate-limiting step in CO_2 transfer to the blood [71]. If red blood cells transit the capillaries in $<$250 ms then this equilibration will be incomplete and the maximum amount of CO_2 transport will not be achieved. It is possible to envision a situation like that in Figure 11, which graphs CO_2 removal and O_2 delivery as a function of blood flow to a specific region or organ. At low flow rates the removal of CO_2 and delivery of O_2 are proportional (their ratio being the respiratory quotient of the region or organ) to flow. The erythrocytes spend $>$250 ms on average in the capillary bed and the Jacobs-Stewart cycle reaches equilibrium. As flow increases in this linear range, either the transit time gets shorter or, as in most tissues, the number of open capillaries increases so that the transit time stays about the same. At very high flow rates the O_2 delivery remains nearly proportional to blood flow because O_2 unloading is very rapid compared with the loading of the plasma HCO_3^- space. (Some decrease may not occur because a proportional number of protons from H_2CO_3 will not be available because of the reduction in CO_2 removal.) The CO_2 removal rate at some flow rate, when erythrocyte transit times approach 250 ms, will continue to increase but at a lower slope. The greater blood flow will allow more dissolved CO_2 to be removed and more carbamino group transport, but these increases will not fully compensate for the reduction due to failure of $Cl-HCO_3$ exchange to be complete during transit of the erythrocyte. Of course, after the erythrocyte leaves the capillaries the exchange will continue and bicarbonate will equilibrate, but this will occur with a lower total amount of CO_2 than if the reaction had been completed when there was a continuing source of CO_2 as in the tissue capillaries. Perhaps equal in importance to the possible limitation in CO_2 removal is the possibility of O_2 "toxicity." Depend-

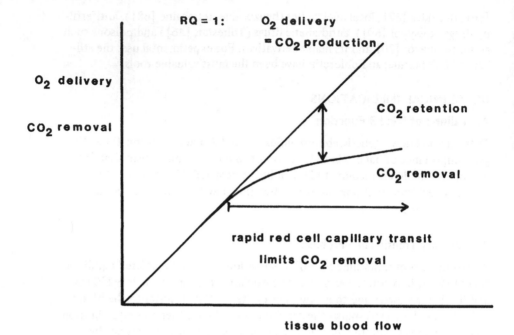

Figure 11 A graph of the hypothetical O_2 delivery, CO_2 production and removal (moles/(min · g tissue)) as a function of tissue blood flow (ml/(min · g tissue)). A theoretical gap can develop between CO_2 production and CO_2 removal at high rates of tissue blood flow if red cell transit time becomes less than CO_2–H_2CO_3–HCO_3^- equilibration time. The tissue respiratory quotient is assumed to be 1.0 which is approximately true at high metabolic rates. The CO_2 retention will raise tissue pCO_2 and cause tissue acidosis which may increase tissue blood flow which in turn will exacerbate the discrepancy between O_2 delivery and CO_2 removal. See text for discussion.

ing on whether O_2 tissue demands or tissue acidosis due to inadequate CO_2 removal are driving the increased flow, the tissue pO_2 may be low or increased. This theoretical gap between O_2 delivery and CO_2 removal could be of pathophysiological significance. When fluorocarbon solutions are used as "blood" substitutes to provide oxygen transport to tissues, the entire role of red blood cell carbonic anhydrase and Cl–HCO_3 exchange in CO_2 removal is neglected. These solutions only permit CO_2 transport as dissolved gas and thus have a much lower capacity than blood. Thus, there may be expected a gap between O_2 delivery and CO_2 removal which, in the long run, may be as damaging to tissues as hypoxia.

C. Possible Limitations On Cardiac Performance

Which tissues are at risk? Clearly those tissues with rapid red blood cell transit rates; or, put another way, those tissues with a high blood flow for the number of open capillaries. It is estimated that ~50% of the rabbit cardiac capillaries are open at rest [72] and 100% of rat cardiac capillaries are open at rest [73, 74], and that under hypoxia or high performance demands all of the capillaries open. Further increase of collaterals are not possible at this point so any fractional increase in coronary flow will result in an equal fractional decrease in erythrocyte transit time. There are no direct measurements of red blood cell capillary transit time except in the mouse cremasteric muscle [75], which has a transit time of 4.2 sec. It is estimated that in humans the erythrocyte transit time in alveolar capillaries is ~700 ms at rest and may decrease to half that value at high cardiac outputs even though there may be two- to threefold recruitment of collaterals during exercise. Recently, Brahm and colleagues reported that in anesthetized dogs right atrial infusion of DNDS to cause 99% inhibition of HCO_3^-/Cl^- exchange reduced the slope of the exhaled CO_2 curve by only 14%, and concluded that "anion exchange inhibition is tolerable and can be fully compensated" [76]. The heart and perhaps the lung aside there seems to be no other candidate tissues at risk given a normal functioning band 3 and hematocrit.

D. Salicylate Toxicity

Salicylate is one of several drugs that inhibit anion exchange on band 3. Patients with salicylate toxicity have inappropriate hyperpnea for their level of acidemia. Wieth and Brahm [64] have shown that these patients' plasma inhibits Cl exchange across normal red blood cells and that their red blood cells have reduced Cl exchange fluxes in plasmalike media. They postulate that the hyperpnea may not be a direct central action of salicylate but a reflecion of incomplete equilibration of the carbonic acid–bicarbonate system between the plasma and respiratory center. If local pCO_2 is increased because of decreased anion exchange either in red blood cells or central receptor cells themselves, ventilation may be stimulated at a rate inconsistent with the arterial pCO_2 and pH.

REFERENCES

1. Passow, H. (1986). Molecular aspects of band 3 protein-mediated anion transport across the red blood cell membrane. *Rev. Physiol. Biochem. Pharmacol.* 103:61.
2. Hamburger, H. J. (1892). Uber den Einfluss der Athmung auf die Permeabilität der Blutkörperchen. *Z. Biol.* 28:405.
3. Jacobs, M. H., and Stewart, D. R. (1942). The role of carbonic anhydrase in certain ionic exchanges involving the erythrocyte. *J. Gen. Physiol.* 25:539.

4. Rossi-Bernardi, L., Roughton, F. J. W., Pace, M., and Coven, E. (1972). The effects of organic phosphates on the binding of CO_2 to human haemoglobin and on CO_2 transport in the circulating blood. In *Oxygen Affinity of Hemoglobin and Red Cell Acid Base Status*, M. Rørth and P. Astrup (Eds.). Munksgaard, Copenhagen, p. 225.

5. Brahm, J. (1977). Temperature-dependent changes of chloride transport kinetics in human red cells. *J. Gen. Physiol.* 70:283.

6. Knauf, P. A., Fuhrmann, G. F., Rothstein, S., and Rothstein, A. (1977). The relationship between exchange and net anion flow across the human red blood cell membrane. *J. Gen. Physiol.* 69:363.

7. Lassen, U. V., Pape, L., and Vestergaard-Bogind, B. (1978). Chloride conductance of the Amphiuma red cell membrane. *J. Membr. Biol.* 39:27.

8. Hoffman, J. F., and Laris, P. C. (1974). Determination of membrane potentials in human and Amphiuma red blood cells by means of a fluorescent probe. *J. Physiol.* 239:519.

9. Freedman, J. C., and Hoffman, J. F. (1979). Ionic and osmotic equilibrium of human red blood cells treated with nystatin. *J. Gen. Physiol.* 74:157.

10. Knauf, P. A. (1979). Erythrocyte anion exchange and the band 3 protein: transport kinetics and molecular structure. *Curr. Top. Membr. Transp.* 12: 249.

11. Gunn, R. B. (1979). Transport of anions across red cell membranes. In *Membrane Transport in Biology*, G. Giebisch, D. C. Tosteson, and H. H. Ussing (Eds.). Springer-Verlag, Berlin, pp. 59-80.

12. Runyan, K. R., and Gunn, R. B. (1984). Phosphate-chloride exchange in human red blood cells: monovalent vs. divalent phosphate transport. *Biophys. J.* 45:18a.

13. Milanick, M. A., and Gunn, R. B. (1984). Proton-sulfate cotransport: external proton activation of sulfate influx into human red blood cells. *Am. J. Physiol.* 247:C247.

14. Gunn, R. B., Milanick, M. A., and Fröhlich, O. Phosphate and chloride binding to the external face of the anion transporter of human red blood cells. *Fed. Proc.* 39:1715.

15. Shoemaker, D. G., Bender, C. A., and Gunn, R. B. (1988). Sodium-phosphate cotransport: kinetics and role in membrane metabolism of human red blood cells. *J. Gen. Physiol.* 92:449.

16. Walker, J. J., Yan, T. S., and Quamme, G. A. (1987). Presence of multiple sodium-dependent phosphate transport processes in proximal brush-border membranes. *Am. J. Physiol.* 252:F226.

17. Parker, J. C., and Hoffman, J. F. (1967). The role of membrane phosphoglycerate kinase in the control of glycolytic rate by active cation transport in human red blood cells. *J. Gen. Physiol.* 50:893.

18. Proverbio, F., and Hoffman, J. F. (1977). Membrane compartmentalized ATP and its preferential use by the Na,K-ATPase of human red cell ghosts. *J. Gen. Physiol.* 69:605.

19. Mercer, R. W., and Dunham, P. B. (1981). Membrane-bound ATP fuels in the Na/K pump. Studies on membrane bound glycolytic enzymes on inside-out vesicles from human red cell membranes. *J. Gen. Physiol.* 78:547.

20. Kopito, R. R., and Lodish, H. F. (1985). Primary structure and transmembrane orientation of the murine anion exchange protein. *Nature* 316:234.
21. Mueckler, M., Caruso, C., Baldwin, S. A., Panico, M., Blench, I., Morris, H. R., Allard, W. J., Lienhard, G. E., and Lodish, H. F. (1985). Sequence and structure of a human glucose transporter. *Science* 229:941.
22. Deuticke, B. (1982). Monocarboxylate transport in erythrocytes. *J. Membr. Biol.* 70:89.
23. Jennings, M. L., and Adams-Lackey, M. (1982). A rabbit erythrocyte membrane protein associated with L-lactate transport. *J. Biol. Chem.* 257: 12866.
24. Halestrap, A. P. (1976). Transport of pyruvate and lactate into human erythrocytes. *Biochem. J.* 159:193.
25. Deuticke, B., Beyer, E., and Forst, B. (1982). Discrimination of three parallel pathways of lactate transport in the human erythrocyte membrane by inhibitors and kinetic properties. *Biochim. Biophys. Acta* 684:96.
26. Ellory, J. C., Jones, S. E. M., and Young, J. D. (1981). Glycine transport in human erythrocytes. *J. Physiol.* 320:403.
27. King, P. A., and Gunn, R. B. (1989). Na- and Cl-dependent glycine transport in human red blood cells and ghosts: a study of the binding of substrates to the outward-facing carrier. *J. Gen. Physiol.* 93: February.
28. King, P. A., and Gunn, R. B. (1989). Na-independent, DNDS-sensitive glycine transport by human red blood cells and ghosts: evidence for glycine and proton cotransport by band 3. Submitted.
29. Christensen, H. (1984). Organic ion transport during seven decades. The amino acids. *Biochim. Biophys. Acta* 779:255.
30. Young, J. D., Jones, S. E. M., and Ellory, J. C. (1980). Amino acid transport in human and in sheep erythrocytes. *Proc. R. Soc. Lond. [Biol.]* 209:355.
31. Ellory, J. C., Preston, R. L., Osotimehin, B., and Young, J. D. (1983). Transport of amino acids for glutathione biosynthesis in human and dog red blood cells. *Biomed. Biochim. Acta* 42:S48.
32. Milanick, M. A., and Gunn, R. B. (1981). The selectivity of the external anion exchange transport site of red blood cells. *Biophys. J.* 33:47a.
33. Wieth, J. O. (1979). Bicarbonate exchange through the human red cell membrane determined with [^{14}C] bicarbonate. *J. Physiol.* 294:521.
34. Falke, J. J., Pace, R. J., and Chan, S. L. (1984). Chloride binding to the anion transport binding sites of band 3. A ^{36}Cl NMR study. *J. Biol. Chem.* 259:6472.
35. Knauf, P. A., and Mann, N. A. (1984). Use of niflumic acid to determine the nature of the asymmetry of the human erythrocyte anion exchange system. *J. Gen. Physiol.* 83:703.
36. Gunn, R. B., and Fröhlich, O. (1979). Asymmetry in the mechanism for anion exchange in human red blood cell membranes. Evidence for reciprocating sites that react with one transported anion at a time. *J. Gen. Physiol.* 74:351.
37. Milanick, M., and Gunn, R. B. (1982). How the temperature dependence of the anion influx in human red blood cells is influenced by the tempera-

ture dependence of $K_{1/2}$ and V_{max}: Studies with Cl_{in}/SCN_{out} exchange. *Fed. Proc.* 41:975a.

38. Hunter, M. J. (1977). Human erythrocyte anion permeabilities measured under conditions of net charge transfer. *J. Physiol.* 268:35.

39. Fröhlich, O., and Gunn. R. B. (1986). Erythrocyte anion transport: the kinetics of a single-site obligatory exchange system. *Biochim. Biophys. Acta* 864:169.

40. Pimplikar, S. W., and Reithmeier, R. A. F. (1986). Two-stage binding of band 3, the anion exchange protein of human erythrocytes, to a matrix-bound inhibitor. *Biophys. J.* 49:140a.

41. Gunn, R. B. (1972). A titratable carrier model for both mono- and di-valent anion transport in human red blood cells. In *Oxygen Affinity of Hemoglobin and Acid-Base Status*, M. Rørth and P. Astrup (Eds.). Munksgaard, Copenhagen, pp. 823–827.

42. Milanick, M. A., and Gunn, R. B. (1982). Proton-sulfate cotransport. Mechanism of H^+ and sulfate addition to the chloride transporter of human red blood cells. *J. Gen. Physiol.* 79:87.

43. Milanick, M. A., and Gunn, R. B. (1986). Proton inhibition of chloride exchange: asynchrony of band 3 proton and anion transport sites? *Am. J. Physiol.* 250:C955.

44. Jennings, M. L. (1976). Proton fluxes associated with erythrocyte membrane anion exchange. *J. Membr. Biol.* 28:187.

45. Gunn, R. B. (1986). H^+ (or OH^-) fluxes activated by inorganic ions on the anion exchanger (band 3) of human red blood cells. *Biophys. J.* 49:579a.

46. Gunn, R. B., and Gunn, H. B. (1989). Inhibition of erythrocyte anion exchange by Tobrestal, an inhibitor of aldose reductase. Submitted.

47. Dalmark, M. (1976). Effects of halides and bicarbonate on chloride transport in human red blood cells. *J. Gen. Physiol.* 67:223.

48. Knauf, P. A., and Mann, N. A. (1986). Location of the chloride self-inhibitory site of the human erythrocyte anion exchange system. *Am. J. Physiol.* 251:C1.

49. Knauf, P. A., Ship, S., Breuer, W., McCulloch, L., and Rothstein, A. (1978). Asymmetry of the red cell anion exchange system: Different mechanisms of reversible inhibition by *N*-(4-azido-2-nitrophenyl)-2-aminoethylsulfonate (NAP-taurine) at the inside and outside of the membrane. *J. Gen. Physiol.* 72:607.

50. Fröhlich, O., and Gunn, R. B. (1987). Interactions of inhibitors on anion transporter of human erythrocyte. *Am. J. Physiol.* 252:C153.

51. Wieth, J. O., and Bjerrum, P. J. (1982). Titration of transport and modifier sites in the red cell anion transport system. *J. Gen. Physiol.* 79:253.

52. Tanford, C. (1985). Simple model can explain self-inhibition of red cell anion exchange. *Biophys. J.* 47:15.

53. Vestergaard-Bogind, B., and Lassen, U. V. (1974). Membrane potential of Amphiuma red cells: hyperpolarizing effect of phloretin. In *Comparative Biochemistry and Physiology of Transport*, L. Bolis, K. Bloch, S. Lurie, F. Lynen (Eds.). North-Holland, Amsterdam, pp. 346–353.

54. Fröhlich, O. F., Leibson, C., and Gunn, R. B. (1983). Chloride net efflux from intact erythrocytes under slippage conditions. Evidence for a positive charge on the anion binding/transport site. *J. Gen. Physiol.* 81:127.

55. Kaplan, J. H., Pring, M., and Passow, H. (1983). Band-3 protein-mediated anion conductance of the red cell membrane. Slippage vs ionic diffusion. *FEBS Lett.* 156:175.

56. Knauf, P. A., Law, F.-Y., and Marchant, P. J. (1983). Relationship of net chloride flow across the human erythrocyte membrane to the anion exchange mechanism. *J. Gen. Physiol.* 81:95.

57. Fröhlich, O. (1984). Relative contributions of the slippage and tunneling mechanisms to anion net efflux from human erythrocytes. *J. Gen. Physiol.* 84:877.

58. King, P. A., and Fröhlich, O. (1989). The effects of phoretin on sulfate exchange and net flux in human erythrocytes. Submitted.

59. Fröhlich, O., Bain, D. A., and Weimer, L. H. (1986). Effects of phloretin and DNDS on the chloride conductance (tunneling) in erythrocytes. *Biophys. J.* 49:141a.

60. Läuger, P. (1985). Channels with multiple conformational states: interrelation with carriers and pumps. *Curr. Top. Membr. Transp.* 21:309.

61. Funder, J., Tosteson, D. C., and Wieth, J. O. (1978). Effects of bicarbonate on lithium transport in human red cells. *J. Gen. Physiol.* 71:721.

62. Duhm, J., and Becker, B. F. (1978). Studies on Na^+-dependent Li^+ countertransport and bicarbonate-stimulated Li^+ transport in human erythrocytes. In *Cell Membrane Receptors for Drugs and Hormones: A Multidisciplinary Approach*, R. W. Straub and L. Bolis (Eds.). Raven Press, New York, pp. 281–299.

63. Simons, T. J. B. (1986). The role of anion transport in the passive movement of lead across the human red cell membrane. *J. Physiol.* 378:287.

64. Wieth, J. O., and Brahm, J. (1978). The inhibitory effect of salicylate on chloride and bicarbonate transport in human red cells. A hypothesis for the stimulatory effect of salicylate on the respiration. *Ugeskr. Laeger* 140:1859.

65. Brazy, P. C., and Gunn, R. B. (1976). Furosemide inhibition of chloride transport in human red blood cells. *J. Gen. Physiol.* 68:583.

66. Gunn, R. B. (1985). Bumetanide inhibition of anion exchange in human red blood cells. *Biophys. J.* 47:326a.

67. Gunn, R. B., Feldman, S. T., and Horton, J. M. (1977). Inhibition of chloride transport in human red cell by diuretics. *Fed. Proc.* 36:564.

68. Gunn, R. B., and Cooper, J. A. (1975). Effect of local anesthetics on chloride transport in erythrocytes. *J. Membrane Biol.* 25:311.

69. Haspel, H. C., Corin, R. E., and Sonnenburg, M. (1985). Effect of gossypol on erythrocyte membrane function: specific inhibition of inorganic anion exchange and interaction with band 3. *J. Pharmacol. Exp. Ther.* 234:575.

70. Gunn, R. B., and Tosteson, D. C. (1971). The effect of 2,4,6 trinitro-*m*-cresol on cation and anion transport in sheep red blood cells. *J. Gen. Physiol.* 57:593.

71. Brahm, J. (1986). The physiology of anion transport in red blood cell. *Prog. Hematol.* 14:1.
72. Weiss, H. R., and Conway, R. S. (1985). Morphometric study of the total and perfused arteriolar and capillary network of the rabbit left ventricle. *Cardiovasc. Res.* 19:343.
73. Steinhausen, M., Tillmanns, H., and Thederan, H. (1978). Microcirculation of the epimyocardial layer of the heart. *Pflügers Arch.* 378:9.
74. Vetterlein, F., Dal Ri, H., and Schmidt, G. (1982). Capillary density in rat myocardium during timed plasma staining. *Am. J. Physiol.* 242:H133.
75. Sarelius, I. H. (1986). Cell flow path influences transit time through striated muscle capillaries. *Am. J. Physiol.* 250:H899.
76. Brahm, J., Gronlund, J., Hlastala, M. P., Ohlsson, J., and Swenson, E. R. (1987). Effects of *in vivo* inhibition of red-cell anion exchange and Haldane effect on carbon dioxide output (V_{CO_2}) in the dog lung, *J. Physiol.* 390:245p.

20

The Kinetics and Thermodynamics of Glucose Transport in Human Erythrocytes

ALLAN G. LOWE *School of Biological Sciences, University of Manchester, Manchester, England*

ADRIAN R. WALMSLEY *Leicester University, Leicester, England*

I. INTRODUCTION

A large number of carrier systems catalyzing the transport of hydrophilic molecules across mammalian plasma membranes, the mitochondrial inner membrane, and bacterial cell membranes have been identified and, in many cases, the structure of the protein responsible has been deduced by sequencing of cDNA. For most of these carriers the kinetics of transport can be satisfactorily described by the conventional carrier model [1], albeit with provisions for cotransport of two species (e.g., the lactose carrier of *Escherichia coli* or the input of energy from the hydrolysis of adenosine triphosphate (ATP) (Na,K-ATPase). Unexpectedly, one of the most intensively studied transport systems, the glucose carrier of the human erythrocyte (for reviews see [2–6]), appears to have complex kinetic properties that cannot be reconciled with the conventional carrier model [6], even though its function in the cell seems to demand no complex regulatory mechanisms and its structure [7] seems typical of other carriers and membrane-spanning proteins. In this review our primary objective is to examine the reported data on the kinetics of glucose transport with a view to assessing whether the overall results of these studies can, despite the apparent kinetic anomalies, be considered consistent with the conventional carrier model. It turns out that most kinetic data for glucose transport based on initial rate measurements agree reasonably well with the conventional four-state carrier model (or the kinetically equivalent mobile carrier), and that the major reported anomalies have almost invariably arisen after use of the integrated rate equation for the analysis of kinetic data. It therefore seems unnecessary to invoke more complex mechanisms for glucose transport, such as the tetramer models [8,9], the lattice-pore model [10], the introverting pore [11], the paired asymmetric carrier model [12], or the allosteric pore model [13], until such time as there is better evidence for inconsistency with the simple carrier model. The most recently suggested simultaneous carrier model [5,14] is also both difficult to envisage in relation to the known structure of the transporter and inconsistent with some effects of inhibitors of glucose transport (see below). Once the reliability of the conventional four-state carrier model is accepted, detailed analysis of the thermodynamics of carrier-mediated glucose transport becomes possible [15] and, as will be discussed below, this combined with information about the specificity of substrate binding to the carrier [16,17] gives a useful insight into the molecular mechanism of the transport process.

II. THE METHODOLOGY USED IN STUDIES OF GLUCOSE TRANSPORT KINETICS

The wide range of experimental designs used, and kinetic parameters reported, in describing glucose transport in human erythrocytes make it inevitable that any assessment as to the sufficiency of a model in describing the transport system should be preceded by an evaluation of the relative reliabilities of the various techniques employed. The following section therefore attempts to outline the variety of techniques used in measurement of red blood cell glucose transport, and to highlight areas in which the experimental design, or method of analysis of the results, may lead to derivation of incorrect values for transport parameters.

A. Zero-trans Entry

At first glance, measurement of entry of glucose into glucose-free erythrocytes is a straightforward procedure requiring radiolabeled glucose and termination of transport with a suitable inhibitor, before removal of extracellular radiolabel and assay of radioactivity taken up into the cells. However, the rapid rate of glucose entry into the small intracellular volume makes measurements of the true initial rate of glucose entry very difficult. At concentrations of glucose below K_m the half-time for transport is on the order of 1 sec at 20°C, whereas at higher extracellular glucose intracellular glucose rapidly reaches concentrations at which there is an important backflux that can lead to underestimates of the rate of influx. These problems can be partly overcome by carrying out experiments near $0°C$, or by using rapid reaction techniques. Both a metronome [18] and automatically operated syringes [19] have been successfully applied to glucose influx measurements at $\sim20°C$, whereas Lowe and Walmsley [20] have described a quenched-flow apparatus capable of extending measurements to physiological temperatures.

The nature of the transport-quenching agent is also important in determining the accuracy of zero-trans entry measurements. Early studies employed mercuric chloride (usually in the presence of potassium iodide) to stop transport after incubations with D-[^{14}C]-glucose, but this inhibitor does not permit washing of the cells so that the error in the measured glucose uptake is increased by the substantial amount of isotope in the extracellular fluid collected with the cells after separation from the quenched transport medium by centrifugation. More recently, the more potent glucose transport inhibitor, phloretin, has become available (and economical to use) as a quenching agent. Phloretin-induced inhibition of glucose transport is virtually instantaneous [20] and, because red blood cells lose very little glucose when washed with phloretin–saline at $0°C$, washing of quenched cells makes it possible to decrease the extracellular counts to <1% of the potential cellular uptake of glucose at equilibrium.

The method by which red blood cells are washed before measurements of zero trans influx are made can also have an important effect on the results obtained. This is because, particularly at low temperatures, the presence of unlabeled glucose inside the cells substantially stimulates the entry of extracellular D-[^{14}C]-glucose [21]. For this reason red blood cells must be washed exhaustively (preferably with warm saline) to decrease intracellular glucose to micromolar levels.

An alternative means of monitoring zero trans entry of glucose is to use red blood cell ghosts after resealing with glucose oxidase inside [22]. This makes it possible to monitor glucose influx continuously by measuring glucose oxidase-dependent consumption of oxygen with an oxygen electrode, and also avoids any buildup of glucose concentration within the ghosts as transport proceeds. However, perfect "zero trans" conditions cannot be achieved because of the high K_m of glucose oxidase (24 mM), and in practice the intracellular glucose concentration is ~0.24 mM. This is much lower than K_{io}^{zt} (Table 1), but in practice the K_{oi}^{zt} found using the glucose oxidase method [22,23] are much higher than those found using labeled glucose (Table 1).

B. Zero-trans Exit

In zero-trans exit experiments red blood cells are preloaded with (normally) D-[^{14}C]-glucose, then suspended in glucose-free saline for varying times before inhibition of transport and measurement of radioactivity either remaining in the cells or appearing in the medium. Use of a large volume of saline avoids significant backflux, but when cells are counted the accuracy of initial rate meas-

Table 1 Reported and Predicted Values of K_m and V_{max} for Glucose Transport in Human Erythrocytes at Near 0° and 20°C

Type of experiment	Ref.	Method	°C	Measured V_{max} ($\mu M\ min^{-1}$)	Predicted V_{max} (four-state carrier model)		Measured K_m (mM)	Predicted K_m (four-state model)	
					(a) (+ = 0°C, * = 20°C)	(b)		(a)	(b)
Zero-trans exit	51	Initial rates	0	5.5	4.3+	7.0+	3.4	2.1+	3.8+
	19	Initial rates	0	4.3			1.6		
	27	Integrated rate equation	2	9.3			15		
	70	Integrated rate equation	2	7.3–14.6			40–122		
	37	Integrated rate equation/ light scattering	20	174			32.4		
	26	Integrated rate equation	20	138	154*	120*	25.4	5.1*	3.9*
	64	Integrated rate equation	20	212			22.9		
	65	Initial rate	20	112			7.4		
	70	Integrated rate equation	20	141–168			18–33		
	19	Initial rates	22	223			5.1		
	66	Initial rate	25	390			5.8		
Zero-trans entry	21	Initial rates	0	0.21	0.29+	0.32+	0.2	0.15+	0.17+
	19	Initial rates	0	0.33			0.15		
	64	Initial rate	20	21			1.6		
	19	Initial rates	20	52	59*	36*	1.6	1.6*	1.2*

Ref	Method							
Equilibrium exchange								
21	Initial rates	20	36			1.6		
22,23	Initial rates/glucoseoxidase	20	56b			7–9		
21	Initial rates	0	22	33+	27+	20	17+	14+
19	Initial rates	0	33			13		
27	Initial rates	2	30			25		
21	Initial rates	20	264			20		
67	Initial rates	20	260			38		
68	Initial rates	20	360			34		
69	Initial rates	20	357	353*	310*	32	11*	10*
31	Initial rates	20	300			14		
72	Initial rates	20	369			13		
19	Initial rates	20	353			17		
66	Initial rates	25	540			8.1		
Infinite-trans entry								
21	Initial rates	0	12.6	33+	27+	0.65	1.1+	0.63+
21	Initial rates	20	179	353b	310b	1.7–2.0	3.1b	2.6b
18	Initial rates	20	233			2.6b		
Infinite-trans exit								
57	Initial rate	0	27.5	33+	27+	8.7	15.7+	14+
Infinite-cis entry								
29	Integrated rate equation/light scattering	0.66	12	0.29+	0.32+	5.2	15.7+	14+

Table 1 (continued)

Type of experiment	Ref.	Method	°C	Measured V_{max} (μM min^{1})	Predicted V_{max}a (four-state carrier model) (a) (+ = 0°C, * = 20°C)	(b)	Measured K_m (mM)	Predicted K_ma (four-state model) (a)	(b)
Infinite-cis entry (continued)	37	Integrated rate equation/ light scattering	20	23	50b	36b	5.8	9.5b	8.6b
	42	Integrated rate equation	20	85			2.8		
Infinite-cis exit	27	Exit times	2	8.6	4.3+	7.0+	0.39	1.1+	0.63+
	32	Exit times/light scattering	20	66			1.7		
	67	Initial rates	20	104			1.8		
	41	Initial rates/analysis using glucose oxidase	20	210	154b	120b	1.9	3.1b	2.6b
	37	Integrated rate equation/ light scattering	20(?)	78			3.5		

aValues predicted from the conventional four-state carrier model (assuming rates of association with and dissociation from the carrier to be non-rate-limiting) were calculated (a) from rate constants c, d, g, and h (and their activation energies) and dissociation constants K_{so} and K_{si} according to Lowe and Walmsley (Table 2 and [19]) and (b) according to Wheeler [57] with $R_{12} = 3.09$, $R_{21} = 0.142$, $R_{ee} = 0.038$ and $K = 0.168$ at 0°C and $R_{12} = 0.0278$, $R_{21} = 0.083$, $R_{ee} = 0.0032$ and $K = 0.99$ at 20°C. Methods used to measure glucose transport employed radioisotope (unless stated otherwise) and in many cases initial rates of transport were derived from logarithmic plots of time courses at varying glucose concentrations.
bIn pink ghosts. Units = μmol min^{-1} (L packed ghosts)$^{-1}$.

urements is limited by the need to measure a ~10% decrease in cellular radioactivity. Measurement of the increase in radioactivity in the extracellular medium must also be made against a substantial starting background because cells must inevitably be used at considerably less than 100% hematocrit so that the radioactivity in the medium at "zero time" typically represents ~50% of the total radioactivity available to leave the cells. As in the case of influx measurements the rapid rate of zero-trans exit of glucose necessitates the use of rapid reaction techniques, including automated syringes [19], rapid (manual) filtration through a glass fiber membrane [24], and the rapid flow/membrane filtration method [25] for work at room temperature or above.

In view of the above difficulties, integrated forms of the Michaelis equation, which also take into account changes in red blood cell volume during glucose transport, have been used to derive values of K_m and V_{max} from nearly complete time courses of D-[^{14}C]-glucose exit from preloaded red blood cells [26, 27]. However, this approach introduces new uncertainties, which include the possibility that the α- and β-anomers of D-glucose have different kinetic properties and the difficulty of determining the final equilibrium position when glucose efflux is complete. Results of studies designed to investigate the way the transport system handles α- and β-D-glucose are conflicting because Faust [28] found a pronounced preferential transport of the β-anomer, whereas Carruthers and Melchior [29] state that the two anomers are transported identically. The latter conclusion is, however, based on data with very high standard deviations and appears to be an overstatement because the different stereochemistries of the α- and β-anomers must dictate some difference in their binding to the carrier, even though this could possibly be too small to significantly alter transport kinetics. Furthermore, Barnett et al. [17] found substantial differences in inhibitory effects of α- and β-fluoro-glucose on sorbose transport via the red blood cell glucose carrier.

The final point of equilibrium when efflux of D-glucose [^{14}C]-glucose is measured has an important effect on the calculated K_m for glucose efflux because measurements at the end of the time course (when the glucose concentration is near to K_m) have the greatest weight in determining the final estimate of K_m. Ideally, this point of equilibrium should be included as one of the parameter to be determined [30] rather than relying on expiration of ~10 transport half-times before measuring the equilibration point [26]. This problem is especially important for metabolized sugars such as glucose, because metabolites can account for a major proportion of the isotope content of the cell at equilibrium [31] and hence have a correspondingly large effect on the estimate for K_m.

C. Equilibrium Exchange

In equilibrium exchange experiments, glucose-containing red blood cells are suspended in saline containing an identical concentration of glucose and exchange of radiolabeled glucose (initially present only inside or outside the cells) with unlabeled glucose is measured with quenching of transport as for "zero trans" experiments. This type of experiment is uncomplicated by problems associated with cell filling or the transport properties of the anomers of glucose. Consequently, accurate measurements of initial rates of exchange transport are possible provided that the glucose concentration is sufficiently high. However, at low concentrations of glucose, equilibration of isotope is extremely rapid so that accurate measurements of K_m require either rapid reaction techniques [20,25], or low temperatures.

D. Infinite-Trans Entry and Exit

In infinite-trans uptake experiments, entry of radiolabeled glucose is measured into cells containing a high (relative to the transport K_m) concentration of unlabeled glucose. The maximum observable rates of transport are the same as those for equilibrium exchange, but at lower concentrations of radiolabeled glucose measurements of initial influx rates will tend to be underestimates because of the loss of glucose from the cells (leading to dilution of extracellular isotope) during the experiment. This limitation applies with even more force to infinite trans exit experiments because here the small volume of the cell leads to a more rapid dilution of intracellular isotope.

E. Net Flux Measurements

Infinite Cis Exit

In infinite cis exit experiments the net loss of glucose from red blood cells into saline containing lower (variable) concentrations of glucose is measured. In carrying out experiments of this design Sen and Widdas [32] exploited the very high water permeability of the red blood cell [33-35], which makes it possible to monitor glucose efflux by the changes in light scattering [36,37] associated with shrinkage of the red blood cells when osmotic activity associated with glucose is lost. Sen and Widdas [32] measured a parameter (the time taken for complete loss of glucose if this occurred at the initial rate) that was proportional to the initial rate of glucose efflux from glucose-loaded cells into media with varying external glucose concentration. However, integrated Michaelis equations, which take into account the effects of loss of cell water on the intracellular glucose concentration, have also been used to extract estimated of K_m and V_{max} from infinite cis exit experiments and, for instance, Carruthers and Melchior [38] obtained K_m and V_{max} from the final one-third of the time courses of glucose exit. The main advantage of this type of experiment is that volume changes can be easily and continuously monitored by measuring changes in turbidity (typically changes in absorbance at 610 nm). Some caution must be used in interpreting the results, however, because the work of Gary-Bobo and

Solomon [39] and Sidel and Solomon [33] has shown that changes in red blood cell volume in anisosmotic media are less than for an ideal osmometer, probably because the osmotic coefficient of hemoglobin varies with its concentration. Any glucose transport data determined by turbidometric or light scattering measurements therefore require confirmation by isotope studies.

An alternative method of carrying out infinite cis exit experiments is to monitor loss of radiolabeled glucose into saline contain a lower glucose concentration of the same specific radioactivity [40], or to analyze the extracellular glucose chemically [41]. These methods have seldom been used, probably because they are much more labor-intensive than measurements of turbidity, and because accuracy is limited by the high levels of starting levels of glucose (radioactivity) in both the cells and saline.

Infinite Cis Entry

Like infinite cis exit, infinite cis entry of glucose from saline containing a high glucose concentration into cells containing a lower (variable) glucose concentration has been measured both by monitoring light scattering [37] and movements of radiolabeled glucose [42]. Rapid filling of the cells makes initial rate measurements very difficult and so the integrated Michaelis equation has been used to extract K_m and V_{max} from both type of experiment.

F. Consistency of Results of Glucose Transport Experiments with the Conventional Four-state Carrier Model

Since the suggestion that a simple carrier was responsible for glucose transport [43], the steady-state equations describing the kinetics of the conventional four-state carrier model have been thoroughly evaluated in varying degrees of detail [43–53]. The analyses lead to the following equation for net flux of glucose between the extracellular medium (compartment 1) and the intracellular volume (compartment 2):

$$V_1 \gg_2 = \frac{[KG_1 + G_1 G_2]\ [C]}{K^2 R_{OO} + KR_{12}G_1 + KR_{21}G_2 + R_{ee}G_1 G_2}$$

where G_1 and G_2 are the concentrations of glucose in the two compartments, [C] is the concentration of glucose carrier (per liter cell water) and the resistance terms and constant, K, are related to the rate constants (a–h) in the carrier model (Figure 1) as follows:

$R_{12} = 1/e + 1/h + 1/c + d/ec$

$R_{21} = 1/b + 1/g + 1/d + c/bd$

$R_{ee} = 1/c + 1/d + c/bd + d/ec + 1/G + 1/e$

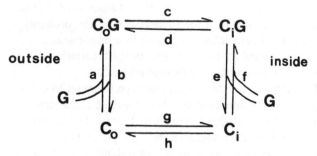

Figure 1 The conventional 4-state carrier model. C_o and C_i represent the outward- and inward-facing carrier conformations, G is glucose and a-h are the rate constants governing the various transitions.

$$R_{oo} = 1/g + 1/h$$

$$K = g/a + h/f + bg/ac$$

In this scheme the experimentally accessible maximum transport rates under conditions of zero trans entry, zero trans exit, and equilibrium exchange (V_{oi}^{zt}, V_{io}^{zt} and V^{ee}) are given by the products of the carrier concentration and the reciprocals of the resistance terms R_{12}, R_{21}, and R_{ee}, respectively, and knowledge of a further constant (K) is necessary to define the various Michaelis constants.

The complexity of the resultant expressions [6] makes extraction of individual rate constants from the experimentally measurable parameters an apparently hopeless goal, but fortunately simplification is possible because the nuclear magnetic resonance studies of Wang et al. [54] have demonstrated that the rate constants governing glucose association with and dissociation from the carrier are very much faster than those controlling reorientation of the carrier and carrier–glucose complexes. Furthermore, the rate-limiting nature of the reorientation steps is supported by the fact that (especially at low temperatures) the presence of glucose on the trans side of the red blood cell membrane substantially stimulates both entry and exit of isotopically labeled glucose [21,55, 56], a situation that would not pertain if rates of association/dissociation were slow compared with carrier reorientation steps.

Given the simplifying assumption that the rate constants for association and dissociation of the carrier/glucose complex are fast compared with the rates of carrier reorientation, the transport equations simplify so that the readily measurable V_{max} are given by

$$V_{oi}{}^{zt} = \frac{[C]}{(1/c + 1/h)}, \quad V_{io}{}^{zt} = \frac{[C]}{(1/d + 1/g)}, \quad \text{and } V^{ee} = \frac{[C]}{(1/c + 1/d)}$$

and the corresponding Michaelis constants by

$$K_{io}{}^{zt} = \frac{b\,(1 + (g/h))}{a\,(1 + (c/h))}, \quad K_{io}{}^{zt} = \frac{e\,(1+(h/g))}{f\,(1+(d/g))}, \quad \text{and } K^{ee} = \frac{b\,(1 + (g/h))}{a\,(1 + (c/d))}$$

where e/f (= K_{si}) and b/a (= K_{so}) are the true dissociation constants of the glucose complexes of the inward- and outward-facing carriers, respectively.

One of the difficulties in assessing whether the various experimental studies of glucose transport are consistent with this model is the possibility that spurious deviations from the model could arise from unrecognized problems arising from the very wide range of methods used in collecting the data. For this reason both Wheeler [57] and Lowe and Walmsley [19] have recently made new measurements of some of the kinetic parameters using methods intended to minimize any systematic errors between experiments. Derivations of K_m and V_{max} were all based on "initial rate" transport measurements to avoid the possibility of errors arising from use of the integrated Michaelis equation. Our own measurements showed that, despite the marked asymmetry of glucose transport kinetics, the thermodynamically based Haldane relationship, which demands that ratios of K_m/V_{max} must be equal under conditions of zero trans entry and exit, and equilibrium exchange, was obeyed reasonably satisfactorily at both 0°C and 20°C [19]. In addition, the degree of carrier asymmetry (as measured by the ratio, $V_{io}{}^{zt}:V_{oi}{}^{zt}$) decreased from ~10 to 3 between 0°C and 20°C, whereas the extent of trans-stimulation of glucose entry, as measured the ratio V^{ee} to $V_{oi}{}^{zt}$, also decreased from ~100 at 0°C to near unity at 37°C [19, 21].

Lowe and Walmsley [19] confirmed the finding of previous workers that Arrhenius plots of the temperature dependence of glucose transport in red blood cells [25,58,59] and red blood cell ghosts [60] are nonlinear, even when extrapolated V_{max} are used to avoid errors arising from variations in carrier saturation at different temperatures. Curved Arrhenius plots of processes taking place in membranes are sometimes ascribed to phase changes of the membrane lipids, but the plasticizing effect of cholesterol in the red blood cell membrane makes this an improbable explanation in this case. A simpler and more plausible explanation is that the rate constants that govern transport under the various conditions of measurement have different activation energies, so that the temperature dependence of the measured V_{max} are given by (e.g.)

$$V^{ee} = 1/(A_c e^{-Ec/RT} + A_d e^{-Ed/RT})$$

Table 2 Fitted Values of the Rate Constants Governing Glucose Transport and Their Activation Energies[a]

Parameters	Rate constant x[C] $(mmol \ 1^{-1}s^{-1})$ $(0°C)$	Rate constant $(s^{-1}) \ (0°C)$	Activation energy $Jmol^{-1}$
c	7.42 ± 3.29	1113 ± 494	31700 ± 5110
d	0.602 ± 0.0231	90.3 ± 3.47	88000 ± 6170
g	0.0809 ± 0.00654	12.1 ± 0.98	127000 ± 4780
h	0.00484 ± 0.000332	0.726 ± 0.498	173000 ± 3100

[a]Rate constants have been calculated assuming that the concentration of glucose carrier molecules [C] in human red blood cells is 6.67 μmol/L cell water.
Source: Ref. 19.

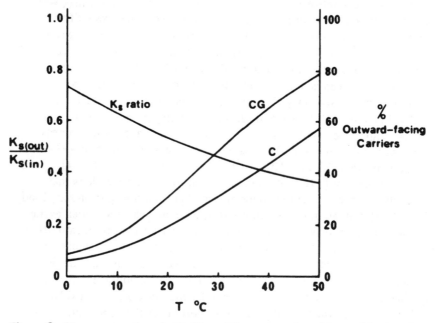

Figure 2 The asymmetry of affinities of the outward- and inward-facing glucose carriers (Ks(out)/Ks(in) = hc/dg = bf/ae) and the proportions of outward-facing carriers (C) and carrier-glucose complexities (CG) as a function of temperature. Values plotted are calculated from the fitted values of rate constants, c, d, g and h given in Table 1. From reference 19.

where A_c and A_d are preexponential terms and E_c and E_d are activation energies associated with rate constants c and d, respectively (Figure 1).

Given this analytical relationship between temperature and the expected transport rates under zero-trans entry, exit, and equilibrium conditions, it was possible to use a nonlinear regression procedure to find the best fits for the rate constants, c, d, g, and h governing carrier reorientation [19] (Table 2). This analysis also derived both the true dissociation constants of the carrier glucose complexes (from the equations shown above) and the extent of asymmetry in the affinities of the outward- and inward-facing complexes, with the ratios ($K_{si}:K_{so}$ ranging from ~0.8 at 0°C to 0.4 at 40°C (Figure 2), and K_{so} remaining fairly constant at ~10 mM throughout the temperature range. Another outcome of the analysis (Figure 2) [19] was that the transmembrane distributions of the unloaded carrier and carrier–substrate complexes are both asymmetric (as previously suggested by Baker et al. [61] and Widdas [62]) with the extent of this asymmetry decreasing with increasing temperature. The estimated rate constants for reorientation of the unloaded carrier [19] are also in reasonably good agreement with rate of carrier reorientation measured by Appleman and Lienhard [63] from changes in fluorescence during binding of 4,6-O-ethylidine-D-glucose to the purified glucose carrier.

Overall, the above analysis shows that the kinetics of glucose transport under conditions of zero-trans exit, entry, and equilibrium exchange entry are satisfactorily described by the conventional four-state carrier model throughout the temperature range 0–45°C. Wheeler's [57] recent reinvestigation came to a similar conclusion from measurements of initial rates of glucose transport under conditions of zero-trans efflux and infinite-trans efflux at 0°C. Furthermore, as indicated in Table 1, the values of K_m and V_{max} derived from or predicted by these recent studies are generally in reasonably good agreement with each other. Table 1 also summarizes the results of other measurements of the parameters governing red blood cell glucose transport and, in view of the apparent compatibility of the above studies [19,57] with the conventional carrier model, a careful evaluation of those results that appear discordant with the model seems appropriate.

G. Zero-trans Entry and Exit

Radioisotope studies of zero-trans entry [19,21,64] show a remarkable consistency among each other, and the discrepancy with the glucose oxidase technique [23] can probably be ascribed to the existence of significant concentrations of glucose inside the cell when this method is used. However, zero-trans exit studies are much more variable, and prominent among thost that are apparently inconsistent with the conventional carrier model are those of Karlish et al. [26], Baker and Naftalin [27], Challis et al. [64], and Naftalin et al. [70]. In all these, V_{max} for zero-trans exit of glucose are broadly in line with the four-state carrier model, but reported K_m are approximately five times higher than the values measured by Miller [65], Brahm [66], Lowe and Walmsley [19], and Wheeler [57]. Probably the most significant difference between these two groups of studies is that all those arriving at K_m too high to be consistent with the conventional carrier model used the integrated rate equation to analyze the full time course of glucose exit, whereas the remaining workers measured initial rates of radiolabeled glucose efflux. This raises the possibility that the factors mentioned in the previous section (differences in the kinetic parameters for α- and β-D-glucose, error in estimation of the final equilibrium position, cell heterogeneity) have combined in such a way as to cause overestimates in the value of $K_{io}{}^{zt}$ when the integrated rate equation is used. This possibility is further supported by thermodynamic considerations, which dictate that for a simple (passive) facilitated diffusion system the Haldane ratio must be obeyed, irrespective of the model used or the type and complexity of the mechanism of glucose transport. This condition is met more or less satisfactorily for the initial rate studies, whereas the higher measured $K_{io}{}^{zt}$ [26,27,64] imply either that glucose transport is being measured under nonequilibrium conditions arising from linkage of glucose transport to some source of energy, or that the two sets of measurements have been carried out with cells under different metabolic conditions leading to the presence of a modifier of transport in some but not all cases. There is no evidence that glucose transport is anything other than a passive non-energy-linked process, whereas differences in $K_{io}{}^{zt}$ determined by initial rate and integrated rate equation studies remain evident when cells of the same metabolic status are used. Furthermore, the kinetic simulations of Wheeler [57] have shown that complexing of intracellular glucose by hemoglobin or other intracellular constituents cannot account for the apparent kinetic anomalies of either zero-trans or infinite-trans exit, as suggested by Naftalin et al. [70]. It is therefore reasonable to deduce that $K_{io}{}^{zt}$ determined using the integrated rate equation are unreliable.

H. Infinite-cis Exit and Entry

Most of the results of infinite-cis entry and exit experiments can be considered compatible with the conventional carrier model because, although the measured K_m are generally somewhat lower than those predicted, the deviations are probably within the uncertainties expected from the precision of the experiments and any systematic errors in the different types of method employed. The results of two studies [29,42] do deviate from expected measurements by more than a factor of 2, but the statistical analysis of Foster and Jacquez [52] showed that there is a large (statistical) uncertainty in the K_m infinite-cis-entry reported by Hankin et al. [42], whereas the rather low K_m determined by Baker and Naftalin [29] for infinite-cis exit at 2°C was derived using the integrated rate equation and must therefore be regarded in a similar light to the zero-trans exit experiments discussed above. One of the experiments reported by Hankin et al. [42] was reassessed (Foster et al. [71]) as giving an even lower K_m for infinite-cis entry, but it is notable that other experiments detailed in the same article [42] gave results that are fully consistent with predictions of the conventional carrier model.

I. Infinite-trans Entry and Exit

Relatively few studies of glucose transport have been done under infinite-trans conditions but they are mostly in reasonable agreement with the conventional carrier model. For infinite-trans entry the measured K_m show fairly good agreement at both 0°C [21] and 20°C [18,21] even though the corresponding V_{max} are appreciably lower than those reported in equilibrium exchange studies. The infinite-trans exit study of Wheeler [57] agrees well with the model, and although Baker and Naftalin [27] reported a much lower infinite trans exit K_m for glucose it should be noted that the latter study was carried out with 50 mM galactose as the trans sugar (well below a saturating concentration) so that agreement with other glucose transport data could not be expected in this case.

J. Equilibrium Exchange

The results of equilibrium exchange studies are generally consistent with the conventional carrier model even though the measured K^{ee} tend to be somewhat higher than predicted, particularly in the earlier studies. This may be partly a result of the use of mercuric chloride rather than the more efficient phloretin as transport quenching agent, but the use of long-outdated transfusion blood is probably of importance because Weiser et al. [72] demonstrated significant decreases in K_m and V_{max} for equilibrium exchange glucose transport when erythrocytes were compared before and after being cold-stored for ~8 weeks. The possible role of ATP in this effect is discussed in a subsequent section.

K. Transport of Sugars Other than Glucose

Although the glucose carrier of the human red blood cell distinguishes between D- and L-glucose almost absolutely, a number of other sugars including D-galactose [73,74], D-mannose [75,76], D-xylose [76], D-arabinose [76], L-arabinose [76], L-sorbose [77], D-fructose [77], and 3-O-methyl-D-glucose [61] are also transported with kinetics that differ significantly from glucose with respect to both K_m and V_{max} under equilibrium exchange conditions. These differences in the kinetic parameters for different monosaccharides are predictable in terms of the conventional four-state carrier model, in that the intrinsic binding constants would be expected to differ both according to the different interactions of the sugars with aminoacid residues at the carrier binding site, and according to the values of rate constants c and d for each sugar. Differences between the "trans" stimulatory effects of glucose and other sugars such as galactose and mannose on glucose transport [40,78] can also be explained by differences in rate constants c and d for the different sugars; as can the phenomenon of "counter transport" [79] in which a gradient of one sugar can lead to temporary development of concentration gradient of a second sugar.

L. Kinetics of the Reconstituted Glucose Carrier

The solubilized, purified red blood cell glucose carrier has been reconstituted into lipid vesicles [80–86] and black lipid membranes [87], and shown to catalyze stereospecific glucose transport measured using both radioisotopes [80,81, 85] and turbidometrically [86]. Not all the carriers reconstituted into the vesicles are oriented in the same direction, but treatment with trypsin [88] can be used to degrade carriers with their globular domain exposed on the outward-facing surface of the lipid vesicles, and this leaves essentially all functional carriers oriented in same way as in red blood cells [88]. Using this type of preparation and initial rate isotopic measurements of glucose transport, Wheeler and Hinkle [88] found that transport kinetics resembled those in intact red blood cells in being asymmetric (with K_m for zero-trans efflux $\sim$ fourfold greater than for zero-trans influx, and $\sim$15-fold less than for equilibrium exchange at 25°C). These K_m were also lower than found in the intact red blood cell at 25°C and stimulation of [^{14}C]-D-glucose influx into carrier-containing vesicles by un-labeled glucose inside the vesicles was $\sim$30-fold, compared with $\sim$ fourfold in intact red blood cells. Thus, the reconstituted carrier at 25°C had kinetic properties resembling those of the carrier in the intact cell at near 0°C, and Wheeler and Hinkle [88] suggest that these differences could arise either from changes (perhaps induced by the solubilizing detergent) in the carrier during purification, or the different lipid environments in the reconstituted system and the red blood cell membrane.

In contrast to the above work, Carruthers and Melchior [86] concluded that the kinetics of the reconstituted carrier were symmetrical. This work depended on the differential effects of cytochalasin B in inhibiting glucose influx and efflux in a reconstituted system containing glucose carriers oriented in both directions, and is therefore less persuasive than the direct measurements of Wheeler and Hinkle [88]. Nevertheless, further work to resolve this discrepancy is desirable.

Carruthers and co-workers [86,89,90] have made extensive studies of the glucose carrier after reconstitution into lipid vesicles of varying phospholipid composition. These indicate that glucose transport is very sensitive to the type of phospholipid head-group, with (in dimyrystoyl phospholipids) maximum glucose transport rates increasing in the order phosphatidylcholine, phosphatidylglycerol, phosphatidic acid, phosphatidylserine, and with transport 100 times faster in dimyrystoyl phosphatidylserine than in dimyrystoyl phosphatidylcholine-liposomes [90]. The nature of the fatty acyl groups also affected glucose transport in reconstituted liposomes with (for phosphatidyl choline liposomes) both maximum rates and activation energies for glucose transport rates at high temperature fastest for the distearyl and dielaidoyl species, but fatty acyl groups were much less important than phospholipid head groups [90]. Effects of adding cholesterol to liposomes were somewhat variable, but in distearoyl phosphatidylcholine cholesterol did increase the rate of glucose transport below 20°C and decrease the rate above 20°C [89]. However, Tefft et al. [90] found little effect of melting of fatty acyl chains on the rate of glucose transport and conclude that membrane fluidity and "annular" lipids have only minor effects on glucose transport kinetics in reconstituted systems.

III. INHIBITION OF GLUCOSE TRANSPORT

A. Inhibitors as a Means for Testing the Four-state Carrier Model

Although the preceding analysis indicates that the kinetics of glucose transport are probably consistent with the conventional carrier model, it is conceivable that a linear model, in which the carrier simultaneously presents binding sites to both the inner and outward compartments (Carruthers [4,14]), could also possibly account for the measured kinetic parameters. No detailed appraisal of whether this possibility is consistent with the temperature dependence of glucose transport indicated in Table 1 has been carried out, but fortunately the effects of competitive inhibitors on the transport system provide a simple means of distinguishing between the conventional carrier model and linear (simultaneous) carrier models of the type suggested by Carruthers [4].

Studies of the inhibitory effects of cytochalasin B [91,92] and phloretin [92,93] have shown that both these compounds inhibit the equilibrium exchange of glucose with competitive kinetics, but have very different effects on zero-trans fluxes. In the case of cytochalasin B, zero-trans efflux of glucose is inhibited competitively and zero-trans influx noncompetively, whereas for phloretin zero-trans efflux is inhibited noncompetitively and zero-trans influx competitively. This result can be explained using the conventional carrier model if phloretin has a much higher affinity for the outside- than inside-facing carrier, whereas the reverse is true for cytochalasin B. The linear model could explain these differential inhibitory effects in a similar way, but the two models behave differently when both inhibitors are present. As shown by Krupka and Deves [93], the conventional carrier model demands that two such inhibitors, Io and Li, acting at the outside and inside of the membrane, respectively, should have combined effects such that

$$\frac{V}{V_{Io}L_i} - 1 = \frac{V}{V_{Io}} - 1 + \frac{V}{V_{Li}} - 1$$

where V is the rate of transport (zero-trans efflux or zero-trans influx) in the absence of inhibitor, V_{Io} and V_{Li} are the rate with the two inhibitors present singly, and $V_{Io}L_i$ is the rate with both inhibitors present. In contrast, any linear (simultaneous) mechanism demands that

$$\frac{V}{V_{Io}L_i} - 1 = \frac{V}{V_{Io}} - 1 + \frac{V}{V_{Li}} - 1 + \left(\frac{V}{V_{Io}} - 1\right)\left(\frac{V}{V_{Li}} - 1\right)$$

In fact, Krupka and Deves' measurements of the combined inhibitory effects of phloretin and cytochalasin B [93], and also both androstanedione and androstenedione and cytochalasin B [94], are consistent with the conventional carrier and not with the simultaneous, two-site carrier model, and studies of the effects of ligands such as 4,6-O-ethylidene-glucose and phenyl-β-glucopyranoside on cytochalasin B binding to the glucose carrier [95,96] are fully consistent with the findings of Krupka and Deves. However, in a recent article Helgerson and Carruthers [97] argued against these findings on the basis that a large outside:inside concentration gradient of glucose had little inhibitory effect on cytochalasin B binding to the glucose carrier in red blood cells. They deduced that inward- and outward-facing glucose binding sites must exist simultaneously on the carrier, but this conclusion is unjustified because, using the conventional carrier model, the glucose gradient would be expected to cause the carrier to be mainly in the inward-facing (cytochalasin B-binding) conformation, rather than outward-facing, as suggested [97]. Furthermore the finding [97] that inhibitory effects of maltose and 4,6-O-ethylidene glucose on cytochalasin B binding were biphasic in resealed red blood cell ghosts could be explained by slow entry of these ligands into the ghosts, rather than by the simultaneous existence of binding sites on both sides of the membrane.

B. Other Asymmetric Inhibitors

A number of inhibitors (both sugar analogues and other compounds) have been shown to have different apparent affinities for the inside- and outside-facing glucose carrier. In the case of nontransported species the relationship between the measured K_i and the dissociation constant of the inhibitor–carrier complex is given by

$$K_{Io} = \frac{\overline{K}_{Io}}{1 + g/h} \quad \text{and} \quad K_{Ii} = \frac{\overline{K}_{Ii}}{1 + h/g}$$

for competitive inhibition at the inside or outside facing carrier site or

$$K_{Io} = \frac{\widetilde{K}_{Io}}{1 + g/d} \quad \text{and} \quad K_{Ii} = \frac{\widetilde{K}_{Ii}}{1 + h/c}$$

for noncompetitive inhibition of glucose efflux (or influx) with the inhibitor
at the trans side (Deves and Krupka [98,99]). The true affinities of transported sugar analogues for the two carrier conformations are difficult to obtain
because the calculations require knowledge of the rate constants for movement
of each individual carrier-sugar analogue between inward- and outward-facing
conformations, as well as rate constants g and h. However, in the case of 4-6-O-
ethylidene-D-glucose and other glucose derivatives that are transported very
slowly if at all [60,102,103], the measured K_i at the outside of the cell is ~10
times lower than that at the inside and, because the carrier is oriented mainly
in the inward conformation (rate constant g greater than h) the true affinity for
the outside facing carrier can be calculated as being >50 times greater than that
of the inside-facing conformation (Table 3). 6-O-propyl-D-galactose, 6-O-benzyl-
D-galactose, and 6-O-pentyl-D-galactose also appear to have substantially higher
affinities for the outward, rather than the inward-facing carrier. For propyl-β-
D-glucopyranoside and benzyl-β-D-glucopyranoside the measured K_i at the inside of the cell is substantially less than at the outside. However, in these cases
the relative values of rate constants g and h will make the difference between the
true affinities for the inside- and outside-facing carriers less than is apparent
from the measured inhibition constants, and the true affinities are only moderately greater at the inside than the outside (Table 3). Maltose is sometimes said
to be specific for the outward-facing carrier [61], although it has been tested
only by addition to red blood cells into which is it not transported, and the
question of whether maltose is inhibitory when incorporated into resealed red
blood cell ghosts has not been answered. The bifunctional 1,3-bis(D-maltose)-
propylamine derivatives described by Holman and co-workers [104,105] also
come into the category of inhibitors, which are not transported, and therefore
inhibit only at the outside when added to intact red blood cells.

In the case of lipophilic, passively permeable inhibitors it is not possible to
measure "one-sided" K_i because of penetration of the inhibitor to both sides
of the membrane. However, by exploiting the very different affinities of D-
glucose and D-xylose for the glucose transport system, Krupka and Deves
succeeded in measuring the ratio $\widetilde{K}_{Io}/\widetilde{K}_{Ii}$ for a series of nontransported compounds including cytochalasin B, androstanedione, and corticosterone [100]
(Table 3). Given the relationships above and the values of rate constants c, d,
g, and h, and the inhibition constants, it is possible to calculate the relative
affinities of these inhibitors for the inside- and outside-facing carrier conformations from the relationship

Lowe and Walmsley

Table 3 Asymmetric Effects of Reversible Inhibitors of Sugar Transport in Human Red Boood Cells[a]

Inhibitor	°C	Ref	$\bar{K}_{Ii}$	$\tilde{K}_{Ii}$	K_{Ii}	$\bar{K}_{Io}$	$\tilde{K}_{Io}$	K_{Io}	$\tilde{K}_{Io}/\tilde{K}_{Ii}$	K_{Io}/K_{Ii}
6-O-propyl-D-glucose	25	103		90	80.4	17		4.15		0.052
6-O-propyl-D-galactose	25	103		>90	>80.4	17		4.15		<0.052
	15	103				24		3.5		<0.039
6-O-benzyl-D-galactose	25	103		6	5.4	1.2		0.29		0.055
	25	103				3		0.44		0.082
Propyl-β-D-glucopyranoside	25	103		9	8.0	30		>7.3		>0.91
	15	103				24		>3.5		>0.43b
Phenyl-β-D-glucopyranoside	25	103		0.5	0.45	6		1.47		3.3
	15	103			>24	>24		3.5		7.9b
Propyl-β-D-galactopyranoside	25	103		90	80.4	>30		>7.3		>0.9
4,6-O-ethylene-D-glucose	16	61, 110	135		93.1	11		1.7		0.018
	16	61					1.8	1.33		0.014
Methyl-2-3-di-O-methyl-α-D-glucopyranoside	16	61			114	290		45		0.39
1,2-Isopropylidene-D-glucofuranose	16	61	76		64	59		9.1		0.14
	25	100							0.9	0.77

Cytochalasin B	25	100	100	>86
Androstendione	25	100	100	>86
Androstanedione	25	100	40	34
5β-Androstane-3,17-dione	25	100	6.4	5.5
Deoxycorticosterone	25	100	4.5	3.9
Testosterone	25	100	2.4	2.1
Hydrocortisone	25	100	2.3	2.0
Cortisone	25	100	2.0	1.7
Corticosterone	25	100	1.1	0.95
Phloretin	25	100	0	0

aK_{Ii} are competitive inhibition constants derived from inhibition of equilibrium exchange exit [103] of radiolabeled D-glucose. K_{Io} are noncompetitive inhibition constants derived from inhibition of net entry of radiolabeled L-sorbose by intracellular inhibitors [61]. $\tilde{K}_{Io}$ are competitive inhibition constants derived from inhibition of zero-trans entry of radiolabeled L-sorbose or D-glucose [61] or equilibrium exchange entry of radiolabeled D-glucose [103] $\tilde{K}_{Io}$ are noncompetitive inhibition constants derived from inhibition of net exit of D-glucose by extracellular inhibitors in Sen-Widdas experiments [102]. Ratios of $\tilde{K}_{Io}$ (noncompetitive inhibition constants for extracellular inhibitors against net glucose exit) to $\tilde{K}_{Ii}$ (competitive inhibition constants for intracellular inhibitors against net xylose exit) are taken from Ref. 100. Dissociation constants for inhibitors (K_{Io} and K_{Ii}) were calculated as indicated in the text using values of c, d, g and h (at appropriate temperatures) from Ref. 19.

bThese ratios compare K_{Io} and K_{Ii} at different temperatures.

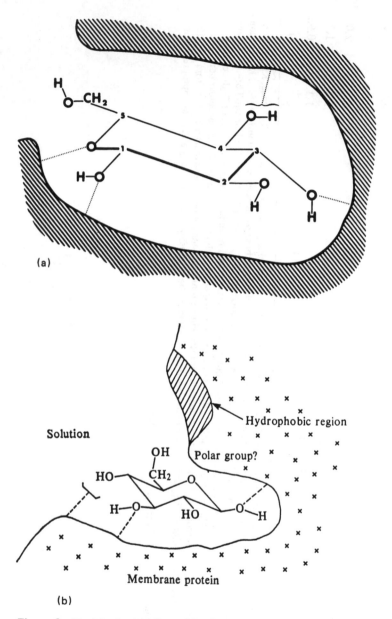

(a)

(b)

Figure 3 Models for binding of D-glucose to the human erythrocytes carrier according to (a) Kahlenberg & Dolansky [16] and (b) Barnett et al., [17].

$$K_{Io}/K_{Ii} = [\widetilde{K}_{Io}/\overline{K}_{Ii}]/[(1 + g/d)/(1 + h/g)]$$

(Table 3). This shows that most of the steroids have a strong preference for the inside-facing carrier binding site, but that cortisone and corticosterone bind preferentially to the outward-facing carrier site. Lacko et al. [101] have also measured the asymmetric inhibitory effects of steroids on red blood cell glucose transport. Other reversible inhibitors include tryptophan esters [106] and the phosphodiesterase inhibitors, theophylline and 3-isobutylmethylxanthine [62].

C. Inhibitors as Probes for Characterization of the Nature of Glucose Binding to the Carrier

In classical studies both Kahlenburg et al. [16,107] and Barnett et al. [17,103] investigated the specificity of the binding of different sugars to the glucose carrier. Barnett et al. measured the inhibitory effects of glucose and a range of (e.g.) fluoride-substituted sugar derivatives on the transport of L-sorbose into red blood cells. This study relied on the assumption that L-sorbose is in fact transported via the glucose carrier and that the measured K_i, although not true dissociation constants, give an index of the relative affinities of the inhibitory sugars. The results of the study led to the well-known model for glucose binding illustrated in Figure 3, in which the bound sugar is envisaged as being retained at the binding sites by hydrogen bonds from the carrier to the Cl-, C3-, and C4-hydroxyls, with the last group in a sterically unrestricted position in the outward-facing carrier site and some interaction between a hydrophobic region of the carrier with the C6-region of the sugar. Furthermore, the low affinity of 4,6-O-ethylidene glucose for the inward-facing form of the carrier is attributed to a conformational change that involves the carrier closing around the C4-hydroxyl and opening around the C1-hydroxyl [17].

Kahlenberg et al. [16,107] measured binding of D-([^{14}C)-glucose to glucose carriers in red blood cell ghosts directly, using a membrane filtration method. Extreme conditions (95% saturated ammonium sulphate) were required to increase the affinity of glucose sufficiently to make binding detectable, and for this reason the results seem to have attracted little attention. Nevertheless, the number of glucose binding sites per cell estimated from these binding studies (150,000–250,000 sites/cell, [16]) is very much in line with current estimates based on the number of cytochalasin B binding sites, and the model for glucose binding to the carrier derived from the inhibitory effects of other sugars on D-[^{14}C]-glucose binding [107] is remarkably similar to that of Barnett et al. [17] (Figure 3). Indeed, the only differences are that the C4-hydroxyl is in a relatively unrestricted position according to Barnett et al., and that Kahlenberg and Dolansky proposed a hydrogen bond between the carrier and the bridge-oxygen of glucose (tested using the thio-analogue). Barnett et al. did not test the latter possibility.

The picture of the glucose binding site developed from the above studies is also consistent with the inhibitory effects on glucose transport of the series of pyridine derivatives examined by Hershfield and Richards [109]. Although these molecules are chemically very different from sugars, they are sterically very similar to hexoses and appear to bind to the carrier in much the same was as glucose.

IV. IRREVERSIBLE INHIBITORS

Glucose transport in human red blood cells can be inhibited irreversibly by a number of reagents, the most extensively studied being FDNB,* which was first shown to inhibit glucose transport by Bowyer and Widdas [110]. Inactivation by FDNB is the result of alkylation of the carrier (probably reaction with amino or sulhydryl groups), but the rate of inactivation is highly sensitive to both temperature and the presence of compounds binding to the carrier including both transported sugars and reversible inhibitors. Krupka [111] made an extensive survey of the effects of different sugars on the rate of FDNB inactivation and concluded that transported sugars (including glucose, galactose, xylose, and also deoxyglucose [112]) accelerated inactivation, whereas nontransported sugars such as maltose and 4,6-O-ethylidene glucose [100,102,103] were protective. In addition, Krupka [114] found that a number of other reagents, notably, urethane and n-butanol, accelerated FDNB inactivation of glucose transport in effects correlated with their olive oil/water partition coefficients. To explain the effects of these diverse agents on the rate of inactivation, Krupka suggested that FDNB reacted with an intermediate state of the carrier–substrate complex that was normally formed during the transport process. Any transported sugar would be expected to promote the formation of this intermediate state, whereas hydrophobic agents such as urethane could stabilize the intermediate state because it presumably represented a complex between the carrier and a dehydrated sugar molecule. Similarly, the very high affinity of 2,3,4,5-tetrachloro-6-nitro-pyridine for the glucose carrier and the fact that 2-amino-5-chloropyridine potentiates FDNB inactivation [109] could be explained if these hydrophobic molecules form stable complexes with the intermediate state of the carrier. The existence of a FDNB-reactive intermediate state is also consistent with Jung's findings [112] that the rate of carrier inactivation in the absence of glucose was very slow indeed at 4°C (where the rate of zero-trans glucose transport is very slow) and that glucose both decreased the activation energy for FDNB inactivation of the carrier from 134 to 71 kJ mol^{-1} and modified the pH dependence of inactivation.

An alternative interpretation of the modifying effects of sugars on FDNB-inactivation has been put forward by Edwards [115] and Barnett et al. [103]. These workers have suggested that FDNB reacts with the inward-facing form

---

*1-fluoro-2,4-dinitro-benzene

of the carrier more rapidly than the outward-facing conformation. However, this interpretation seems inconsistent with the fact that cytochalasin B (which is generally accepted to react with and stabilize the inward-facing conformation) protects from FDNB inactivation, and that kinetic analysis [19] indicates that at low temperatures (where there is little reaction with FDNB unless glucose is present) the carrier is present predominantly in the inward-facing conformation in both the presence and absence of glucose.

In studies of the inactivation of red blood cell glucose transport by other irreversible inhibitors, Rampal and Jung [116] found that the inactivating effects of iodoacetamide, N-ethyl maleimide, and chloronitrobenzoxadiazole (in addition to FDNB) were all accelerated by glucose, and that glucose decreased the activation energies for the reactions in all cases. Rampal and Jung deduce that because glucose is an activator all the reactive groups of the carrier must be outside the glucose-binding site, and that the site of inactivation is probably in the hydrophobic domain (because maleimides of increasing hydrophobicity have increasing inactivation effects [116]). These results would also be consistent with the inhibitors reacting with the glucose-induced transition state of the carrier.

The glucose carrier is also irreversibly inactivated by tetrathionate [117], but in this case external application of maltose or addition of phloretin (which binds preferentially to the outward-facing carrier) to red blood cells accelerates inhibition, whereas cytochalasin B (which binds to the inward-facing carrier) slows down inactivation. These results can be interpreted in terms of an asymmetric carrier molecule, the outward-facing conformation of which reacts more rapidly with tetrathionate than the inward-facing conformation. In contrast to both FDNB and tetrationate, inactivation of glucose transport by 2,3-butanedione [118] is slowed down by transported sugars (e.g., glucose, deoxyglucose), maltose (which binds at the outside only), and 1,2-O-isopropylidene-D-glucose (which binds both inside and outside), but is slightly increased by cytochalasin B and androstanedione, which binds to the inward-facing carrier conformation [118]. Krupka suggests that effects of these modifying effects are best explained by proposing that inhibition by 2,3-butanedione depends on whether or not the site of reaction is obstructed by the modifying agent.

V. EFFECTS OF ATP

The change in the kinetics of red blood cell glucose transport that occurs during prolonged storage [57,79] has been attributed to changes in cellular ATP content. This is supported by the findings of Jacquez [119], Carruthers [14], and Hebert and Carruthers [120] that low-molecular-weight cytosolic constituents or depletion of ATP can modify glucose transport kinetics in red blood cells and red blood cell ghosts. Witters et al. [121] have also shown that the red blood cell glucose carrier can be phosphorylated by ATP, and that this phosphorylation is stimulated by phorbol esters, suggesting that there could be natural hormone-related variations in the extent of phosphorylation in vivo. However, although phosphorylation of the carrier is clearly a potential cause of the kinetic changes observed during storage of red blood cells, there is no positive evidence that this is the case and Gibbs et al. [122] have shown that insulin does not induce phosphorylation of the glucose carrier in cultured adipocytes.

Carruthers [14,123] and Hebert and Carruthers [120] have made extensive studies of the effects of ATP on both glucose transport kinetics and the fluorescence of the carrier protein. Carruthers found that the kinetics of glucose transport in (ATP-free) resealed ghosts and inside-out resealed vesicles (IOV) from ghosts were symmetrical, but that ATP converted the transport system of IOV to an asymmetric mechanism by lowering K_m and V_{max} for glucose efflux (measured by analysis of turbidometric data using the integrated rate equation). This effect of ATP was exerted with high affinity ($K_{0.5}$ about 40 μM) and was highly specific since GTP, UTP, ITP, AMP-CPP, AMPP-CP, AMP and ADP had no effect at 0.5 for 1 mM. In addition, V_{max} (but not K_m) for glucose efflux was decreased by Ca^{2+} in IOV that had previously been exposed to ATP (but not in freshly prepared IOV) [14].

In parallel studies Carruthers found that glucose and other sugars known to be transported or bound by the glucose carrier quenched the intrinsic fluorescence of both red blood cell ghosts and the purified glucose carrier [123]. Adenosine triphosphate also caused fluorescence quenching and increased the apparent affinity for glucose at a high-affinity quenching site. Furthermore, glucose modulated the fluorescence-quenching effects of cytochalasin B (which binds to the inward-facing carrier) and phloretin (which binds to the outward-facing carrier) with widely different K_i (29.4 mM and 2.1 mM, respectively). Carruthers [123] interpreted these findings, and also the biphasic nature of Eadie-Scatchard plots of glucose-induced fluorescence quenching of the purified carrier, as indicative of the simultaneous existence of low- and high-affinity glucose binding sites at opposite sides of the membrane and hence as providing supportive evidence for the simultaneous site model for glucose transport. Some caution is necessary before accepting this interpretation because the conventional carrier model also predicts different K_i for effects of glucose on cytochalasin B and phloretin fluorescence quenching, as a consequence of (a) the modest difference between the dissociation constants of the inward- and outward-facing carrier–glucose complexes, and (b) the fact that the operationally measured K_i is a function of both these dissociation constants and the asymmetry of carrier distribution across the membrane. Furthermore, assuming a one-site model, the raw data of Carruthers [123] indicate a $K_{0.5}$ for the quenching effect of glucose ~10-20 mM (similar to the K_s for glucose binding predicted by analysis of the conventional carrier model [19]), and it would be interesting to know whether the "goodness of fit" of Carruthers' data is significantly better to a two-site than a one-site binding model.

In work leading to conclusions very different from those discussed above, Wheeler [124] compared the effects of ATP on glucose transport in intact erythrocytes, resealed ghosts, and purified glucose carriers reconstituted into liposomes without the use of a detergent. The results were consistent with a stimulatory effect of ATP on glucose influx into both intact erythrocytes and resealed ghosts, but ATP had no significant effect on transport in the reconstituted carrier system. Wheeler [124] suggests that effects of ATP are likely to be due to nonspecific interactions of ATP in the erythrocyte membrane (cytoskeletal elements and changes in cell shape could be involved). This proposal leaves the high-affinity effect of ATP [123] without a functional role, but because the glucose carrier lacks the sequences of amino acids normally found in the ATP-binding site of enzymes, it is probably safest to regard specific effects of ATP on glucose carrier kinetics as unproven until more compelling evidence is generated.

VI. THERMODYNAMICS OF THE GLUCOSE CARRIER MECHANISM

Given that the kinetics of glucose transport in intact erythrocytes from fresh blood can be described by the conventional four-state carrier model, it has been possible to calculate the thermodynamic parameters associated with transitions of the carrier between inward- and outward-facing conformations, and also binding of glucose to the carrier, from the temperature dependence of the rate constants governing these transitions [15]. The initial kinetic analysis [19] yielded not only the individual rate constants but also the activation energies (Table 1) for the carrier reorientation processes, and so the enthalpy changes for reorientations of the free carrier and carrier-glucose complexes between inward- and outward-facing conformations could be obtained simply as the differences between the activation energies for rate constants g and h (free carrier) and c and d (carrier–glucose complexes). Standard free energies for these processes were obtained from the relationship

$$\Delta G° = RTln\,(K)$$

where K is the ratio g:h or c:d at a particular temperature. Entropies were also calculated from the relationship

$$\Delta G = \Delta H - T\Delta S$$

Table 4 Thermodynamic Parameters for the Transport System for Glucose in Human Red Blood Cells

Glucose carrier	ΔH^o kJ mol^{-1}	ΔS^o kJ^{-1} mol^{-1}	Basic Gibbs free energy change kJ mol^{-1}	
			0°C	37°C
Substrate binding to the:				
Outward-facing carrier	5.51 ± 3.39	58.1 ± 24.0	1.7	1.2
Inward-facing carrier	−4.95 ± 3.73	17.3 ± 21.9	2.3	3.3
Carrier reorientation from inside to outside:				
Unloaded carrier	46.0 ± 5.69	145 ± 21.0	6.4	1.05
Carrier-glucose complex	56.3 ± 8.01	185 ± 30.7	5.8	−1.05

[a]Standard enthalpies and entropies were calculated as described in the text. Basic Gibbs free energies [125] and the proportions of carrier states present were calculated for a glucose concentration of 5 mM.
Source: From Ref. 15, used with permission.

The results of this analysis (Table 4) indicate reorientations of both the free carrier and the carrier–glucose complex from inward- to outward-facing conformations are strongly endothermic processes associated with an increase in entropy. This characteristic leads to most carriers existing in the inward-facing conformation at and near 0°C, and it is this which largely accounts for the major asymmetry in the kinetics of transport at low temperatures. Baker et al. [61] came to a similar conclusion from consideration of the asymmetric effects of transport inhibitors. Increasing the temperature to the physiological range apparently leads to breakage of some of the bonds stabilizing the inward-facing conformation so that the inward- and outward-facing conformations of both the free carrier and its glucose complex are nearly equally populated [15]. We have recently obtained supportive evidence for this interpretation in "temperature-jump" experiments. A single half-turnover of the carrier was observed after warm red blood cells (38°C) were mixed with a larger volume of cold (0°C) saline containing D-[^{14}C]-glucose [126].

The nature of the forces holding the carrier in the inward-facing conformation at 0°C are uncertain, but there are at least two possibilities. The first is that the inward-facing conformations are more highly hydrated than the outward-facing forms, and the second that intracarrier bonds (possibly hydrogen bonds or salt bridges within the globular domain of the carrier) are broken when the carrier adopts the outward-facing conformation. In either case the transi-

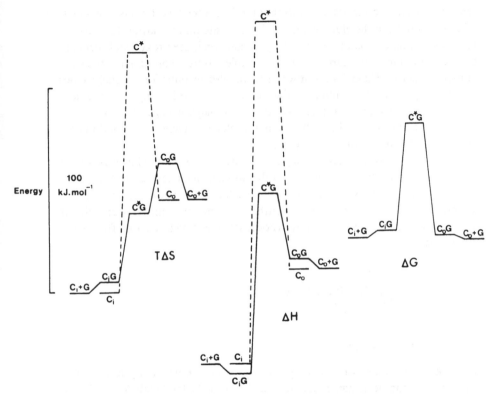

Figure 4 Gibbs free energy, enthalpy and entropy diagrams associated with transport of glucose via the glucose carrier at 37°C. Standard Gibbs free energies are shown except for glucose binding, for which basic Gibbs free energies (with glucose at 5 mM on both sides of the membrane) are given. Data were calculated from the dissociation constants of the glucose complexes of the inward- and outward facing carriers, and the rate constants and activation energies for carrier reorientation using transition state theory. C_i and C_o are the inward- and outward-facing carrier states, and G represents glucose.

tions from outward- to inward-facing conformations can be regarded as "entropy-driven."

Enthalpy changes for binding of glucose to the carrier can be obtained from van't Hoff plots of temperature dependences of the carrier-substrate binding constants [15]. In the case of the outward-facing carrier, glucose binding is almost temperature-independent with, if anything, a slightly negative binding enthalpy and a very small entropy of binding (calculated from the free energy of binding). At first glance this is surprising because simple chemical binding processes are normally exothermic. However, it seems likely that association of glucose with the carrier involves not only formation of this complex, but also dissociation of water from both the glucose and the binding site on the carrier. It is therefore possible that similar numbers of hydrogen bonds are broken and formed during formation of the carrier-glucose complex, thus accounting for the small enthalpy changes. Dissociation of water could also account for the slightly positive entropy change on binding. In the case of the inward-facing carrier, glucose binding is associated with a small exothermic enthalpy change and a small negative entropy change. Possibly binding of glucose involves dissociation of less water from the carrier in this case.

Changes in conformation of the glucose carrier between the inward- and outward-facing conformations can be regarded as equivalent to chemical reactions whose rate is limited by the requirement to attain sufficient energy to reach an intermediate transition state. The enthalpies, entropies, and free energies associated with attainment of such transition states can be calculated from the equations

$$\Delta H^* = E_A - RT$$

$$k = \frac{k_B Te^{\Delta S^*/R} e^{-\Delta H^*/RT}}{\hbar}$$

and $\Delta G^* = \Delta H^* - T\Delta S^*$

where k can be rate constant c, d, g or h, E_A is the corresponding activation energy, k_B is the Boltzman constant, T is the absolute temperature, $\hbar$ is Planck's constant, R is the gas constant, and ΔG^*, ΔS^*, and ΔH^* are the Gibbs free energy, entropy, and enthalpy changes associated with formation of the transition state. Applying this analysis to the kinetic data given in Table 1, and combining this with the thermodynamic data for glucose binding just discussed, leads to Figure 4 [15], which describes changes in entropy, enthalpy, and free energy during turnover of the carrier at 37°C. Naturally the exact energy levels in these diagrams are subject to considerable uncertainties, due to the errors in the rate constants used in the calculations, and the uncertainty as to the exact number of carriers per red blood cell. Nevertheless, the overall pattern of the diagram is likely to be correct and the features discussed below are of considerable interest in relation to the mechanism of transport.

Perhaps the most interesting feature of these diagrams (Figure 4) is that attainment of the transition state of the free carrier is associated with a very large increase in entropy, whereas there is much less change in entropy in going from either of the primary carrier-glucose complexes to the carrier-glucose transition state. This effect of glucose binding is also manifest in the enthalphy diagram because the activation energies for the free carrier conformations are much larger than for the glucose-carrier complexes. A possible interpretation of these changes is that formation of the transition state for the free carrier requires dis-

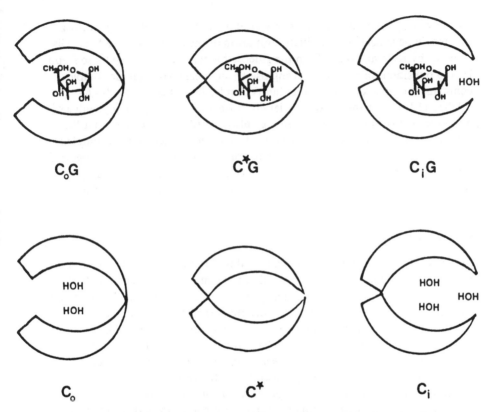

Figure 5 Diagrammatic representation of states of the carrier of carrier-glucose complexes based on the involvement of intermediate states (C* and C*G) whose formation is rate-limiting during transition between inward- and outward-facing conformations.

sociation of water from the glucose binding site with a consequent substantial increase in entropy (arising from the translational freedom of this water) and a change in entropy associated with the breakage of hydrogen bonds between the carrier and water and any additional changes in bonding within other parts of the carrier. The much lower changes in entropy and enthalpy associated with attainment of the transition state of the carrier-glucose complex can be explained if less dissociation of water is involved because some has already occurred during the initial binding of glucose before any conformational change of the carrier. Some evidence for the importance of water in the binding of glucose to the carrier has been provided by the finding of Kahlenberg et al. [107], who demonstrated that the presence of nearly saturated ammonium sulfate increased the affinity of the carrier for glucose sufficiently for binding to be detected by membrane filtration. This effect can be rationalized on the basis of the above suggestions because withdrawal of water by the high salt concentration would be expected to facilitate formation of the carrier-glucose Michaelis complex by removing water from the glucose binding site, and might also stabilize the transition state, thereby giving rise to an "occluded" glucose complex comparable to the "occluded" rubidium complex described for the Na,K-ATPase [127]. A change in the carrier conformation allowing strengthening of hydrogen bonds with bound glucose may also contribute to the fact that the enthalpies of formation of the transition state of the carrier-glucose complex are much smaller than those for the transition state of the free carrier (Figure 5). Figure 5 illustrates the possible role of the occluded carrier-glucose complex (C*G) as an intermediate between the inward- and outward-facing carrier-glucose complexes.

VII. CONCLUSION

The preceding kinetic and thermodynamic description of the glucose carrier of red blood cells depends on acceptance that the conventional carrier model adequately describes the kinetics and that glucose association with and dissociation from the carrier occur much more quickly than carrier reorientations (as indicated by nuclear magnetic resonance studies [54]. Bearing in mind these (possible) limitations, the conclusions reached give an important insight into the way in which biological membrane carriers achieve their ends. For all carriers the problem is one of increasing the partitioning of the carried molecule into the membrane phase, and the solution appears to be one of binding a dehydrated form of the carried species within the carrier protein. In the case of the glucose carrier this seems to be achieved in two stages in which (a) the carrier presents to the aqueous phase a hydrophilic pocket to which glucose can bind by exchanging hydrogen bonds with water for hydrogen bonds with the carrier binding site, and (b) the carrier adjusts its conformation to occlude the glucose, temporarily excluding it from the aqueous phases on the two sides. The transport process can then be completed when the carrier reverts to a conformation (inward- or outward-facing) in which glucose is again free to dissociate and enter the bulk aqueous phase. The intermediate (occluded) complex is difficult to detect in the case of the glucose carrier, but in the Na,K-ATPase [127] and Ca-ATPase [128] carrier conformations with occluded ions are stable under some conditions and can be detected relatively easily. It is also possible that occlusion occurs during transport via the red blood cell anion carrier, because the relatively hydrophobic iodide is transported much more slowly than more hydrophilic anions such as bicarbonate and chloride [24], and could form a stable occluded complex. Further studies of the nature and properties of occluded carrier-substrate complexes will almost certainly throw further light on the mechanisms of membrane transport processes.

REFERENCES

1. Vidaver, G. A. (1966). *J. Theor. Biol.* 10:301–336.
2. Naftalin, R. J., and Holman, G. D. (1977). In *Membrane Transport in Red Blood Cells*, J. C. Ellory and V. L. Lew (Eds.). Academic Press, New York.
3. Jones, M. M., and Nickson, J. K. (1981). *Biochim. Biophys. Acta* 650:1–20.
4. Carruthers, A. (1984). *Prog. Biophys. Mol. Biol.* 43:33–69.
5. Wheeler, T. J., and Hinckle, P. C. (1985). *Ann. Rev. Physiol.* 47:503–517.
6. Stein, W. D. (1986). *Transport and Diffusion across Cells Membranes.* Academic Press, New York.
7. Mueckler, M., Caruso, C., Baldwin, S. A., Panico, M., Blench, I., Morris, H. R., Allard, W. J., Lienhard, G. E., and Lodish, H. F. (1985). *Science* 229: 941–945.
8. Lieb, W. R., and Stein, W. D. (1970). *Biophys. J.* 10:585–609.
9. Eilam, Y. (1975). *Biochim. Biophys. Acta* 401:349–363.
10. Naftalin, R. J. (1970). *Biochim. Biophys. Acta* 211:65–78.
11. LeFevre, P. G. (1973). *J. Memb. Biol.* 11:1–9.
12. Ginsburg, H. (1978). *Biochim. Biophys. Acta* 506:119–135.
13. Holman, G. D. (1980). *Biochim. Biophys. Acta* 599:202–213.
14. Carruthers, A. (1986). *Biochemistry* 25:3592–3602.
15. Walmsley, A. R., and Lowe, A. G. (1987). *Biochim. Biophys. Acta* 901: 229–238.
16. Kahlenberg, A., and Dolansky, D. (1972). *Can. J. Biochem.* 50:638–643.
17. Barnett, J. E. G., Holman, G. D., and Munday, K. A. (1973). *Biochem. J.* 13:211–221.
18. Holman, G. D., Busza, A. L., Pierce, E. J., and Rees, W. D. *Biochim. Biophys. Acta* 649:503–514.
19. Lowe, A. G., and Walmsley, A. R. (1986). *Biochim. Biophys. Acta* 857: 146–154.
20. Lowe, A. G., and Walmsley, A. R. (1985). *Anal. Biochem.* 144:385–389.
21. Lacko, L., Wittle, B., and Kromphardt, H. (1972). *Eur. J. Biochem.* 25: 447–454.
22. Taverna, R. D., and Langdon, R. G. (1973). *Biochim. Biophys. Acta* 298: 412–421.
23. Krupka, R. M. (1985). *J. Membr. Biol.* 83:71–80.
24. Dalmark, M., and Wieth, J. O. (1972). *J. Physiol. (Lond.)* 224:583–610.
25. Brahm, J. (1977). *J. Gen. Physiol.* 70:283–306.
26. Karlish, S. J. D., Liew, W. R., Ram, D., and Stein, W. D. (1972). *Biochim. Biophys. Acta* 225:126–132.
27. Baker, G. F., and Naftalin, R. J. (1979). *Biochim. Biophys. Acta* 550:474–484.
28. Faust, R. G. (1960). *J. Cell Comp. Physiol.* 56:103–121.
29. Carruthers, A., and Melchior, D. L. (1985). *Biochemistry* 24:4244–4250.
30. Wharton, C. W. (1983). *Trans. Biochem. Soc.* 11:817–825.
31. Edwards, P. A. W. (1974). *Biochim. Biophys. Acta* 345:373–386.
32. Sen, A. K., and Widdas, W. F. (1962). *J. Physiol. (Lond.)* 160:392–403.

33. Sidel, V. W., and Solomon, A. K. (1957). *J. Gen. Physiol.* 41:243–257.
34. Paganelli, C. V., and Solomon, A. K. (1957). *J. Gen. Physiol.* 41:259–277.
35. Brahm, J. (1982). *J. Gen. Physiol.* 79:791–819.
36. Orskov, S. L. (1935). *Biochem. Zeit.* 269:241–249.
37. Widdas, W. F. (1954). *J. Physiol. (Lond.)* 125:163–180.
38. Carruthers, A., and Melchior, D. L. (1983). *Biochim. Biophys. Acta* 728: 254–266.
39. Gary-Bobo, C. M., and Solomon, A. K. (1968). *J. Gen. Physiol.* 52:825–853.
40. Miller, D. M. (1968). *Biophys. J.* 8:1329–1338.
41. Harris, E. J. (1964). *J. Physiol. (Lond.)* 173:344–353.
42. Hankin, B. L., Lieb, W. R., and Stein, W. D. (1972). *Biochim. Biophys. Acta* 288:114–126.
43. Widdas, W. F. (1952). *J. Physiol. (Lond.)* 118:23–39.
44. Regen, D. M., and Morgan, H. E. (1964). *Biochim. Biophys. Acta* 79:151–166.
45. Britton, H. G. (1964). *J. Physiol. (Lond.)* 170:1–20.
46. Britton, H. G. (1965). *J. Theor. Biol.* 10:28–52.
47. Geck, P. (1971). *Biochim. Biophys. Acta* 339:462–472.
48. Lieb, W. R., and Stein, W. D. (1971). *J. Theor. Biol.* 30:219–222.
49. Lieb, W. R., and Stein, W. D. (1974). *Biochim. Biophys. Acta* 373:178–196.
50. Eilam, Y., and Stein, W. D. (1974). *Methods Memb. Biol.* 11:282–354.
51. Regen, D. M., and Tarpley, H. L. (1974). *Biochim. Biophys. Acta* 339:218–233.
52. Foster, D. M., and Jacquez, J. A. (1976). *Biochim. Biophys. Acta* 436:210–216.
53. Lieb, W. R. (1982). *Red Cell Membranes. A Methodological Approach*, J. C. Ellroy and J. D. Young (Eds.). Academic Press, New York.
54. Wang, J-F, Falke, J. J., and Chan, S. I. (1986). *Proc. Natl. Acad. Sci. USA* 83:3277–3281.
55. Mawe, R. C., and Hempling, H. G. (1965). *J. Comp. Cell. Physiol.* 66:95–103.
56. Levine, M., Oxender, D. L., and Stein, W. D. (1965). *Biochim. Biophys. Acta* 109:151–163.
57. Wheeler, T. J. (1986). *Biochim. Biophys. Acta* 862:387–398.
58. Lacko, L., Wittke, B., and Geck, P. (1973). *J. Cell Physiol.* 82:213–218.
59. Bolis, L., Luly, P., Pethica, B. A., and Wilbrandt, W. (1970). *J. Memb. Biol.* 3:83–92.
60. Jung, C. Y. (1971). *Biochim. Biophys. Acta* 241:613–627.
61. Baker, G. F., Basketter, D. A., and Widdas, W. F. (1978). *J. Physiol. (Lond.)* 278:377–388.
62. Widdas, W. F. (1980). *Curr. Top. Bioenerget.* 14:165–162.
63. Appleman, J. R., and Lienhard, G. E. (1985). *J. Biol. Chem.* 260:4575–4578.
64. Challis, J. R. A., Taylor, L. P., and Holman, G. D. (1980). *Biochim. Biophys. Acta* 602:155–166.

65. Miller, D. M. (1971). *Biophys. J.* 11:915-923.
66. Brahm, J. (1983). *J. Physiol.* 339:339-354.
67. Miller, D. M. (1965). *Biophys. J.* 8:1339-1352.
68. Eilam, Y. (1975). *Biochim. Biophys. Acta* 401:364-369.
69. Eilam, Y., and Stein, W. D. (1972). *Biochim. Biophys. Acta* 266:161-173.
70. Naftalin, R. J., Smith, P. M., and Roselaar, S. E. (1985). *Biochim. Biophys. Acta* 820:235-249.
71. Foster, D. M., Jacquez, J. A., Lieb, W. R., and Stein, W. D. (1979). *Biochim. Biophys. Acta* 555:349-351.
72. Weiser, M. B., Razin, M., and Stein, W. D. (1983). *Biochim. Biophys. Acta* 727:379-388.
73. LeFevre, P. G. (1954). *Symp. Soc. Exp. Biol.* 8:118-136.
74. Ginsburg, H., and Stein, W. D. (1975). *Biochim. Biophys. Acta* 382:353-368.
75. Sen, A. K., and Widdas, W. F. (1962). *J. Physiol. (Lond.)* 160:404-416.
76. LeFevre, P. G. (1962). *Am. J. Physiol.* 203:286-290.
77. Miller, D. M. (1966). *Biochim. Biophys. Acta* 120:156-158.
78. Lacko, L., and Burger, M. (1963). *J. Biol. Chem.* 238:3478-3841.
79. Rosenberg, T., and Wilbrandt, W. (1957). *J. Gen. Physiol.* 41:289-296.
80. Kasahara, M., and Hinckle, P. C. (1976). *Proc. Natl. Acad. Sci. USA* 73:390-400.
81. Baldwin, S. A., Lienhard, G. E., and Baldwin, J. M. (1982). *Biochemistry* 21:3836-3842.
82. Baldwin, J. M., Lienhard, G. E., and Baldwin, S. A. (1980). *Biochim. Biophys. Acta* 599:699-714.
83. Baldwin, J. M., Gorga, J. C., and Lienhard, G. E. (1981). *J. Biol. Chem.* 256:3685-3689.
84. Goldin, S. M., and Rhoden, V. (1978). *J. Biol. Chem.* 253:2575-2583.
85. Kahlenberg, A., and Zala, C. A. (1977). *J. Supramol. Struct.* 7:287-300.
86. Carruthers, A., and Melchior, D. L. (1984). *Biochemistry* 23:6901-6911.
87. Phutrakul, S., and Jones, M. N. (1979). *Biochim. Biophys. Acta* 551:188-200.
88. Wheeler, T. J., and Hinkle, P. L. (1981). *J. Biol. Chem.* 256:8907-8914.
89. Connoly, T. J., Carruthers, A., and Melchior, D. L. (1985). *Biochemistry* 24:2865-2873.
90. Tefft, R. E., Carruthers, A., and Melchior, D. L. (1986). *Biochemistry* 25:3709-3718.
91. Deves, R., and Krupka, R. M. (1978). *Biochim. Biophys. Acta* 510:339-348.
92. Basketter, D. A., and Widdas, W. F. (1978). *J. Physiol. (Lond.)* 278:389-401.
93. Krupka, R. M., and Deves, R. (1981). *J. Biol. Chem.* 256:5410-5416.
94. Deves, R., and Krupka, R. M. (1980). *J. Biol. Chem.* 255:11870-11874.
95. Gorga, F. R., and Lienhard, G. E. (1981). *Biochemistry* 20:5108-5113.
96. Gorga, F. R., and Lienhard, G. E. (1982). *Biochemistry* 21:1905-1908.
97. Helgerson, A. L., and Carruthers, A. (1987). *J. Biol. Chem.* 262:5464-5475.

98. Deves, R., and Krupka, R. M. (1978). *Biochim. Biophys. Acta* 513:156-172.
99. Deves, R., and Krupka, R. M. (1979). *Biochim. Biophys. Acta* 556:533-547.
100. Krupka, R. M., and Deves, R. (1980). *Biochim. Biophys. Acta* 598:134-144.
101. Lacko, L., Wittke, B., and Geck, P. (1975). *J. Cell Physiol.* 86:673-680.
102. Baker, G. F., and Widdas, W. F. (1973). *J. Physiol. (Lond.)* 231:143-165.
103. Barnett, J. E. G., Holman, G. D., Chalkey, R. A., and Munday, K. A. (1975). *Biochem. J.* 145:417-429.
104. Midgley, P. J. N., Parker, B. A., and Holman, G. D. (1985). *Biochim. Biophys. Acta* 812:33-41.
105. Holman, G. D., Parkar, B. A., and Midgley, P. J. W. (1986). *Biochim. Biophys. Acta* 855:115-126.
106. Lacko, L., and Wittke, B. (1982). *Biochem. Pharmacol.* 31:1925-1929.
107. Kahlenberg, A., Urman, B., and Dolansky, D. (1971). *Biochemistry* 10:3154-3162.
108. Barnett, J. E. G., Holman, G. D., and Munday, K. A. (1973). *Biochem. J.* 135:537-541.
109. Hershfield, R., and Richards, F. M. (1976). *J. Biol. Chem.* 251:5141-5148.
110. Bowyer, F., and Widdas, W. F. (1958). *J. Physiol. (Lond.)* 141:219-232.
111. Krupka, R. M. (1971). *Biochemistry* 10:1143-1148.
112. Jung, C. Y. (1974). *J. Biol. Chem.* 249:3568-3573.
113. Baker, G. F., and Widdas, W. F. (1973). *J. Physiol. (Lond.)* 231:129-142.
114. Krupka, R. M. (1971). *Biochemistry* 10:1148-1153.
115. Edwards, P. A. W. (1973). *Biochim. Biophys. Acta* 307:415-418.
116. Rampal, A. L., and Jung, C. Y. (1987). *Biochim. Biophys. Acta* 896:287-294.
117. Krupka, R. M. (1985). *J. Membr. Biol.* 84:35-43.
118. Krupka, R. M., and Deves, R. (1980). *Biochim. Biophys. Acta* 598:127-133.
119. Jacquez, J. A. (1983). *Biochim. Biophys. Acta* 727:367-378.
120. Hebert, D. N., and Carruthers, A. (1986). *J. Biol. Chem.* 261:10093-10099.
121. Witters, L. A., Vater, C. A., and Lienhard, G. E. (1985). *Nature* 316:777-778.
122. Gibbs, E. M., Allard, W. J., and Lienhard, G. E. (1986). *J. Biol. Chem.* 261: 16597-16603.
123. Carruthers, A. (1987). *J. Biol. Chem.* 261:11028-11037.
124. Wheeler, T. J. (1986). *Biochim. Biophys. Acta* 859:180-188.
125. Hill, T. L. (1977). *Free Energy Transduction in Biology*. Academic Press, New York.
126. Lowe, A. G., and Walmsley, A. R. (1987). *Biochim. Biophys. Acta* 903:547-550.

127. Glynn, I. M., and Richards, D. E. (1982). *J. Physiol. (Lond.)* 330:17–43.
128. Inesi, G., Kurzmack, M., Kosh-Kosicka, D., Lewis, D., Scofano, H. C., and Vianna, M. (1982). *Z. Naturforsch [C]* 37:685–691.

21

Nucleoside Transport

WENDY P. GATI and ALAN R. P. PATERSON *University of Alberta, Edmonton, Alberta, Canada*

I. INTRODUCTION

Nucleoside fluxes across animal cell plasma membranes are mediated by nucleoside transport (NT)*systems of several types [1-5], including facilitated diffusion (equilibrative) systems of high or low sensitivity† to nitrobenzylthioinosine (NBMPR) [2,5,6-8], and concentrative, "secondary active" transport systems that are linked to fluxes of sodium ions [9-14]. In human erythrocytes, nucleoside permeation occurs through an equilibrative NT mechanism [15-17], which, in initiating anabolism of adenosine, plays an important role in maintaining cellular levels of adenosine triphosphate (ATP). Important components in the homeostasis of adenosine in blood plasma are (a) erythrocyte anabolism of adenosine, and (b) the uptake and formation of adenosine by the liver [18] and by the vascular endothelium [19].

Functional and structural properties of human erythrocytes have favored their use in NT studies. For example, physiological pyrimidine nucleosides

---

*Abbreviations: NT, nucleoside transport; NBMPR, nitrobenzylthioinosine (6-[(4-nitrobenzyl)thio]-9-β-D-ribofuranosylpurine); NBTGR, nitrobenzylthioguanosine (2-amino-6-[4-nitrobenzyl)thio]-9-β-D-ribofuranosylpurine); HNBMPR and HNBTGR, the 6-(2-hydroxy-5-nitrobenzyl)- derivatives of 6-thioinosine and 6-thioguanosine; pCMBS, *p*-chloromercuribenzenesulfonate; IOV, inside-out vesicle; ROV, right side-out vesicle; SDS, sodium dodecyl sulfate; PAGE, polyacrylamide gel electrophoresis; MAbs, monoclonal antibodies; IC50, inhibitor concentration at which a biological activity is reduced to 50% of values measured in the absence of the inhibitor.
†In the context of this report, NT systems of "high" and "low" NBMPR sensitivity are exemplified by those of human erythrocytes and cultured Walker 256 cells for which IC50 values for NBMPR inhibition of nucleoside influx of 24 n*M* (5-iododeoxyuridine [20]) and >1 μ*M* (adenosine [21]) have been reported.

are not metabolized in these cells [15], so that rates of uptake of such permeants represent membrane fluxes uncomplicated by subsequent anabolic steps.
Plasma membrane fractions free of cytoplasmic contaminants, readily obtainable
from human erythrocytes, are important starting material for the partial purification of NT proteins.

Although the equilibrative NT system of human erythrocytes has been studied in detail, this NT system is but one among several in animal cells and should
not be regarded as a general model at this early stage in our understanding of
NT systems.

II. THE NUCLEOSIDE TRANSPORT SYSTEM OF HUMAN
ERYTHROCYTES: A REVERSIBLE CARRIER WITH
LOW SUBSTRATE SPECIFICITY

In human erythrocytes, nucleoside fluxes are reversible, proceed down concentration gradients (and may therefore be termed equilibrative), and are mediated
by membrane elements of a single type. Associated with these NT elements are
specific "receptor" sites at which NBMPR (a potent inhibitor of equilibrative
NT) is bound with high affinity (K_D, ~1 nM). NBMPR occupancy of these sites
blocks transporter activity. Thus, in human erythrocytes, virtually complete occupancy of the NBMPR binding sites may be achieved with low (<100 nM)
NBMPR concentrations under equilibrium conditions, and the system is said
to be of high sensitivity to the inhibitor.

A. Kinetic Properties

The classical and comprehensive studies of Cabantchik and Ginsburg [16] on the
transport of uridine in stored human erythrocytes recognized that the kinetic
characteristics of uridine fluxes could be described by a "simple carrier" model*
[22], but asymmetries were evident, such as the observed differences between
the kinetic properties of efflux and influx processes under zero-trans† conditions, maximum velocities for the former being fourfold higher than for the latter. Asymmetry was also evident in the differences between the kinetic charac-

---

*In the simple carrier model, "simple" means that only single molecules of permeant are
 bound to the transport system (carrier), which exists in one of two conformations, each
 able to interact with permeant molecules at one membrane face or the other. Carrier "mobility" refers to the rates of change between these alternative conformations rather than to
 a physical translocation of a component of the transport system.
†In this convention [22], cis may refer to the solution at either membrane face and cross-
 membrane fluxes are understood to proceed in the cis to trans direction. The concentration terms (zero or infinite) specify experimental conditions in which the permeant concentration at a designated membrane face is (a) initially zero, whereas that at the other
 face is varied (e.g., zero-trans influx), or (b) limitingly high (infinite), whereas that at
 the other face is varied (e.g., infinite-cis efflux).

teristics of uridine fluxes under equilibrium exchange and zero-trans conditions. Kinetic analysis in terms of the simple carrier model was interpreted to mean that conformational changes were slower for the empty carrier than for the permeant-loaded carrier. It has become apparent since the Cabantchik and Ginsburg report [16] that the kinetic characteristics of the human erythrocyte NT mechanism change on storage of the cells under blood bank conditions. Jarvis et al. [23] and Plagemann and Wohlhueter [24] have shown that in fresh erythrocytes, the uridine transport mechanism has directional symmetry in that the kinetic constants for inward and outward zero-trans fluxes are similar. However, in the fresh cells, mobility of the empty carrier was lower than that of the loaded carrier, because maximum velocities of equilibrium exchange fluxes exceeded those of zero-trans fluxes and the trans-acceleration* phenomenon was shown. Jarvis et al. [23] and Plagemann and Wohlhueter [24] have demonstrated that storage of erythrocytes decreased maximum velocities for zero-trans influx of uridine and increased K_m values for zero-trans efflux of uridine. The storage-induced reduction in V_{max} values for zero-trans influx of uridine has been attributed to reduction in mobility of the unloaded carrier [23,24]. Also, V_{max} and K_m values for inward equilibrium exchange fluxes were reduced and increased, respectively, by storage. Thus, storage under blood bank conditions impairs both the mobility of the empty carrier and the directional symmetry of the human erythrocyte NT system.

B. Permeant Specificity

Trans-acceleration of nucleoside fluxes in erythrocytes has been used to determine if nonphysiological nucleosides were NT substrates [19,25-27], although Jarvis [28] has recently pointed out that the inability of a substance to induce trans-acceleration does not exclude the possibility that it is an NT substrate. Thus, although [^{3}H] 2-chloroadenosine was shown to be an NT substrate in human erythrocytes [29], the nonisotopic compound caused trans-inhibition of uridine efflux, a result explained by the relative V_{max} values for the influx of the permeants, that for 2-chloroadenosine being fourfold less than that for uridine.

In erythrocytic NT systems, (a) substrate specificity is broad, and (b) structural variations in the base moiety and at the 2′-position of 9-β-D-pentofuranosides are tolerated [19,25-27]; however, the presence of a 3′-hydroxyl group seems to be a determinant of permeant transportability [25,27]. It is consistent with this idea that the permeation of 3′-azido-3′-deoxythymidine in human erythrocytes appears to proceed by passive diffusion, rather than by a mediated pathway [30].

---

*The stimulation of a transporter-mediated flux by the presence of a transporter substrate at the trans membrane face is a consequence of the mobility of the empty nucleoside carrier being lower than that of the permeant-loaded carrier.

In some cultured neoplastic cells, as in human erythrocytes, NBMPR-sensitive NT systems of low permeant specificity have been identified. The remarkably broad specificity of the NT system of cultured RPMI 6410 human lymphoblastoid cells was evident in the protection of these cells against a variety of cytotoxic nucleoside analogs afforded by the presence of NBMPR [31]. Similarly, mutational deletion of NT activity in cultured S49 cells imparted resistance against a diverse group of cytotoxic nucleoside analogs [32], indicating that cellular uptake of the analogs was mediated by the deleted system. Low transporter substrate specificity was also evident in NT studies with cells of two cultured lines that possess equilibrative NT systems of low sensitivity to NBMPR, Walker 256 rat mammary carcinosarcoma cells [7], and Novikoff rat hepatoma cells [4,6].

Some cultured mammalian cells possess NT systems, which, unlike that of human erythrocytes, do not show trans-accelerative effects. Thus, in ATP-depleted HeLa cells and in thymidine kinase-deficient Novikoff cells, directional symmetry of NT processes was evident and kinetic parameters for zero-trans influx were similar to those for inward equilibrium exchange of nucleosides [33]. These characteristics signify similar mobilities of permeant-loaded and empty nucleoside carriers in these cells.

III. FLUXES OF NUCLEOSIDES ACROSS HUMAN ERYTHROCYTE MEMBRANES ARE RAPID

Nucleoside fluxes in human erythrocytes are rapid; for example, in fresh human erythrocytes at $25°C$, in medium containing $160\ \mu M$ uridine, a concentration close to the K_m value for zero-trans influx of uridine in these cells, the nucleoside equilibrated across the plasma membrane within 30 sec [12]. Accordingly, rapid sampling assays are required to obtain time courses of cellular uptake of nucleosides that define initial rates of uptake. Initial rates of permeant uptake measured in this way define rates of inward transport [1,33], providing corrections are made for nonmediated permeation. In a widely used method, replicate assay mixtures are employed, and brief intervals (a few seconds) of permeant exposure are initiated by completing cell-permeant mixtures and ended (a) by rapid centrifugal pelleting of cells under oil-layers (phthalate esters, silicone oils) [17,19,34], or (b) by addition of potent inhibitors of NT such as NBMPR [18,35,36] or dipyridamole [37], frequently followed by centrifugation of cells under oil layers ("inhibitor-oil stop" methodology) [6,23,38,39]. It may be noted that "inhibitor stop" methods are not of general applicability and their use to end intervals of nucleoside uptake depends on both the inhibitor and the sensitivity of the NT system under study to that inhibitor [40]. Thus, although it is an appropriate stopper for assays of uridine transport in human erythrocytes, NBMPR would not block uridine permeation in rat erythrocytes because a major component of uridine transport in rat erythrocytes is of low sensitivity to NBMPR [37]. Uridine transport in rat erythrocytes is sensitive to dipyridamole, which was used by Jarvis and Young [37] as a stopper in measurement of uridine fluxes in these cells. Rapid assays for measurement of NT rates are surveyed in [40].

Using "inhibitor stop" methods to end intervals of nucleoside uptake by suspended cells and conducting operations manually with timing by metronome signals, uptake intervals of 1 sec are close to the minimum practical limit. However, uptake intervals in the 50–500 msec range have been obtained by means of a quenched-flow procedure in which high concentrations of dilazep were employed to provide "instantaneous" NT blockade [39]. The latter technique allowed measurement of initial rates of adenosine influx into human erythrocytes at 37°C in studies that showed that both K_m and V_{max} values for adenosine influx were four to five times greater at 37°C than at 25°C [39]. Studies with human erythrocytes [41] and guinea pig erythrocytes [35], showed that both K_m and V_{max} values for the zero-trans influx of uridine increased with temperature and that temperature reduction impaired the mobility of the empty carrier more than that of the loaded carrier. Mobility differences between the two carrier states were exaggerated at lower temperatures [35,41].

IV. POTENT, HIGH-AFFINITY INHIBITORS HAVE BEEN VALUABLE PROBES OF NUCLEOSIDE TRANSPORT SYSTEMS IN ERYTHROCYTES

Various S^6- and N^6-nitrobenzyl-substituted 9-β-D-pentofuranosyl derivatives of 6-thiopurine and adenine have been identified as potent inhibitors of NT systems in human erythrocytes [42] and other cell types [43]. Of these, NBMPR has been most widely employed in studies of NT processes, of the cell biology of NT systems, and of membrane proteins involved in equilibrative NT processes. Congeners of NBMPR that are also potent NT inhibitors include nitrobenzylthioguanosine (NBTGR) and hydroxynitrobenzylthioguanosine (HNBTGR) [16,25, 44]. Concentrations of these substances that reduced the influx of 5-iododeoxyuridine into human erythrocytes by 50% (IC50 values) were of the same order of magnitude as that of NBMPR [20], and both NBTGR and HNBTGR inhibited binding of NBMPR at high-affinity sites on erythrocytes of several species [37, 45-47]. The pH dependence of the interaction of the 6-(2-hydroxy-5-nitrobenzyl)-derivative of 6-thioinosine (HNBMPR) with NBMPR binding sites on human erythrocytes indicated that the anionic form of this acidic NBMPR congener has low affinity for the binding site and that the undissociated form of HNBMPR, which has a higher binding site affinity than NBMPR, is the active NT inhibitor [48,49]. The same considerations likely apply to HNBTGR.

A. Do Inhibitor Binding Sites and Nucleoside Permeation Sites Have Common Regions?

In human erythrocytes, membrane sites that bind NBMPR with high affinity (K_D, ~1 nM) are present at an abundance of ~10^4 per cell [45,47,50], and are associated with a transmembrane polypeptide that has an M_r of ~56,000 [37, 85] (see Section VIII.A.). Inhibition of the equilibrium exchange of uridine was shown to be proportional to the fractional occupancy of NBMPR binding sites, suggesting that binding site densities reflected the abundance of functional transporters in these cells [45].

The abundance of NBMPR binding sites in erythrocyte membranes varies greatly from one species to another, with erythrocytes of some species being devoid of such sites [47,50]. Uridine translocation capacities, calculated as the V_{max}/B_{max} ratio (the maximum rate of uridine influx divided by the maximum number of NBMPR binding sites per cell), were found to be similar in erythrocytes from seven species, ranging from 123-190 molecules per carrier per sec [50]. This demonstration of a similar relationship between influx rates and cellular abundance of NBMPR binding sites in cells from several species supports the idea that NBMPR sites identify functional nucleoside transporters. Furthermore, in erythrocytes that constitutively lack NT activity (dog erythrocytes [50], nucleoside-impermeable* sheep erythrocytes [47]), NBMPR binding sites are not detectable. After a single-step mutagenization of cultured S49 mouse lymphoma cells, an adenosine-resistant clone (AE_1) was isolated in which the loss of both NT activity and NBMPR binding sites had occurred, suggesting that both properties were probably associated with a single protein [32,52]. Together, these findings suggest that in erythrocytes and in some cultured cells, NBMPR binding sites and permeation sites are associated, and that NBMPR sites identify transporter polypeptides. A physiological ligand for the NBMPR binding site has not been identified.

Although high-affinity binding sites for NBMPR are located on plasma membrane NT proteins, it is likely that the inhibitor enters erythrocytes by diffusion. Evidence for the diffusional entry of NBMPR into erythrocytes was found in studies that employed either intact erythrocytes or hemoglobin-free unsealed ghosts in measurement of NBMPR binding under equilibrium conditions [45,47]. In those studies, the nonspecific component of binding was considerably lower in ghosts than in intact cells. The lipophilicity of NBMPR, reflected in an octanol/water partition coefficient of 30 [53], may account for diffusion of this substance across the plasma membrane.

---

*Although mature erythrocytes from most sheep lack a functional NT system (nucleoside-impermeable type cells), in ~5% of sheep, erythrocytes possess this capability (nucleoside-permeable type cells). The expression of NT activity appears to be determined by two autosomal alleles, with the impermeable character being dominant [51].

Studies of the inhibition of uridine fluxes in nucleoside-permeable sheep erythrocytes have provided indirect evidence that in these cells NBMPR binding sites are located at the outer face of the plasma membrane [36]. In addition, studies of site-specific binding of NBMPR to unsealed ghosts, inside-out vesicles (IOVs) and right-side-out vesicles (ROVs) from pig erythrocytes suggested that NBMPR binding sites are located on the outside of the plasma membrane [54]. Jarvis et al. [36] observed that inhibition of zero-trans influx of uridine in nucleoside-permeable sheep erythrocytes was apparently competitive, whereas inhibition of zero-trans efflux was noncompetitive. The interpretation of this result in terms of an external location for the NBMPR binding site supposes competition between NBMPR and uridine at permeation sites, such as that expected in a transporter model in which sites for nucleoside permeation and NBMPR binding coincide or overlap* [3,36,55]. Support for this model was found in studies with nucleoside-permeable sheep erythrocytes that showed similarity between values for (a) the K_D of the NBMPR-site complex, and (b) the K_i for the apparently competitive inhibition by NBMPR of the zero-trans influx of uridine [36]. Similarities between K_m values for uridine equilibrium exchange in human erythrocytes [16,23] and K_i values for this permeant as an apparently competitive inhibitor of NBMPR binding to human erythrocyte membranes [36] further support this model. Although these observations are consistent with a transporter model in which permeation and NBMPR binding sites overlap,* they could also be accommodated in a transporter model in which these sites are associated, but physically distinct, and in which the mutual inhibition of permeant and inhibitor interactions is allosteric in nature.

Jarvis et al. [56] have explored the relationship between sites for nucleoside permeation and NBMPR binding in kinetic studies of [^{3}H] NBMPR association and dissociation with human erythrocyte membranes. The kinetics of NBMPR association with the high-affinity sites appeared to be those of a simple bimolecular reaction, and dissociation of [^{3}H] NBMPR from the [^{3}H] NBMPR-site complex was initially a first-order process in the presence of 5 nM NBMPR (which precluded significant rebinding of dissociated ^{3}H-ligand). The K_D (0.12 nM) for the NBMPR-site interaction derived from the observed rate constants fo association and dissociation was similar to that measured under equilibrium conditions (K_D, 0.13 nM). In this study, adenosine was found to be a competitive inhibitor (K_i, 0.1 mM) of NBMPR binding to erythrocyte membranes, as was uridine in earlier studies [36,57]; the K_i values for nucleoside inhibition of NBMPR binding in these instances are sufficiently similar to their K_m values for permeation [1,36]

---

*In this model [3,36,55], the transporter sites that bind nucleoside permeants and NBMPR are visualized as having common or "overlapping" regions; additional determinants for NBMPR binding are thought to be present at the latter's site.

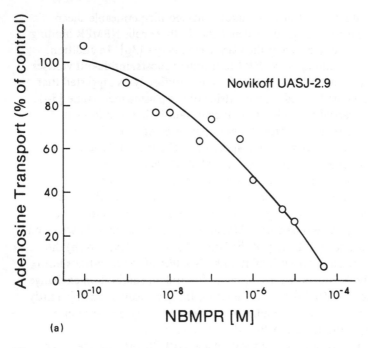

Figure 1   (a) *Low NBMPR sensitivity of NT in a cloned line of Novikoff UA*
cells that possess sites that bind NBMPR with high affinity. The effect of graded
concentrations of NBMPR on the influx of 10 μM adenosine. Novikoff UASJ-
2.9 cells, a twice-cloned line, were incubated with inhibitor-containing solutions
for 15 min at 22°C before exposure to isotopic permeant for 4-sec intervals.
Intervals of [³H] adenosine influx were ended by addition of non-isotopic permeant at a final concentration of 10 mM, followed by a rapid pelleting of the
cells under an oil layer. Adenosine influx rates are shown as percentages of the
adenosine influx rate in the absence of NBMPR (in the experiment shown, this
rate was 1.3 pmol/μl cell water/sec).

that nucleoside interactions at the permeation site would, indeed, take place
under these conditions. Although uridine, adenosine, and NBMPR enhance rates
of NBMPR dissociation from the ligand–site complex, and although dipyridamole slows this process, the very high concentrations required to achieve these
effects make it difficult to interpret the effects in terms of specific interactions

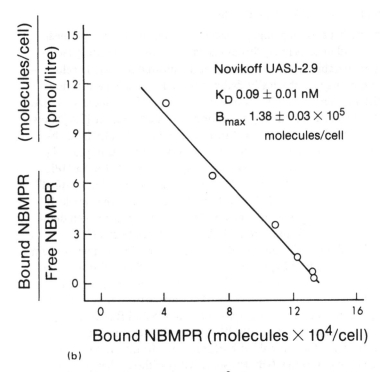

Figure 1 (continued) (b) *Binding of [³H] NBMPR to high-affinity sites on Novikoff UASJ-2.9 cells under equilibrium conditions.* Cells were incubated with graded concentrations of [³H] NBMPR for 30 min before separation of cells and free ligand by pelleting the cells under an oil layer, in order to determine the free NBMPR concentration in the medium and the cell content of site-bound [³H] NBMPR (difference between total pellet content of [³H] NBMPR and that acquired in the presence of 12.5 μM NBTGR). Illustrated is a mass law analysis (method of Scatchard) of data for the cell content of site-bound [³H] NBMPR in equilibrium with the measured concentration of free [³H] NBMPR. Similar data have been presented (2).

of the effectors at permeation or binding sites, as Jarvis has pointed out [56]. This comment also applies to similar attempts to explore permeation site-NBMPR site identities in cultured fibroblasts [58], and in human erythrocytes, CHO cells, and P388 cells [59], through perturbation of NBMPR-site dissociation kinetics.

B. Inhibitor Binding Sites in Nucleated Cells

Although a close, perhaps overlapping, physical relationship between permeation and inhibitor binding sites in erythrocytes has been shown [36,60] (see Section IV.A), considerable evidence argues that in cultured neoplastic cells, NBMPR binding sites may be physically distinct from nucleoside permeation sites. In three different lines of rat hepatoma cells (Novikoff UA, Morris 3924A, and Reuber H35), high-affinity binding sites for NBMPR are present [2,8]; however, NBMPR concentrations in excess of those required to saturate the high-affinity sites appear not to impair the inward transport of adenosine [2,8,61]. Thus, NT systems in these cells are of low sensitivity to NBMPR (for NBMPR inhibition of adenosine transport, IC50 $>$ 1 μM) and appear to be "uncoupled" from NBMPR binding sites. Figure 1 shows that a cloned line of the Novikoff UA hepatoma, UASJ-2.9, displays these properties, which have also been demonstrated in clones established from the 3294A and H35 cell lines [2].

In S49 mouse lymphoma cells, nucleosides are transported by a single system of high sensitivity to NBMPR [52], and from mutagenized S49 cell stocks the isolation of several clones with genetically altered NT properties has been reported [62,63]. In these mutant clones, NT systems differ from that in the wild type cells in (a) acquisition of an NT component of low NBMPR sensitivity, whereas total transport capacity and NBMPR binding site densities were unchanged [62], or (b) loss of about half of NBMPR high-affinity sites (and acquisition of an NT component of low NBMPR sensitivity) without change in total transport capacity [62,63]. These results suggest that NBMPR and nucleoside permeants interact at genetically distinct sites on membrane components.

V. NOT ALL ERYTHROCYTIC NUCLEOSIDE TRANSPORT SYSTEMS ARE OF HIGH SENSITIVITY TO NBMPR

Although nucleosides enter human erythrocytes by a single, mediated pathway of high sensitivity to NBMPR, only a portion of nucleoside permeation in rat erythrocytes is of high sensitivity to the inhibitor [37,64]. Jarvis and Young [37] have shown that in rat erythrocytes two NT systems contribute to uridine permeation, and these may be distinguished by their sensitivities to NBMPR and p-chloromercuribenzene sulfonate (pCMBS) (see Section VII). The two transporters also differ in their affinities for uridine, the NT system of low NBMPR sensitivity having a threefold lower affinity for this permeant than the system of high NBMPR sensitivity. The uridine transport component of low NBMPR sensitivity in rat erythrocytes accounts for ~80% of total influx at saturating concentrations of permeant [37]. In this study, the translocation capacity (turnover number) of transporters of high NBMPR sensitivity was ~ fivefold lower than an earlier estimate [50] and was not in the range of those earlier estimates for erythrocytes from several other species. This discrepancy might be related to the assumption made in the recent calculation [37] that NBMPR binding sites are associated only with transporters of high NBMPR sensitivity.*

In erythrocytes from normal mice, nucleoside permeation is mediated by a single NT system of high sensitivity to NBMPR; however, in parasitized erythrocytes from mice infected with the malarial parasite, *Plasmodium yoelii*, additional parasite-induced routes contribute to nucleoside uptake [65]. Adenosine and tubercidin (7-deazaadenosine, a cytotoxic analog of adenosine) enter erythrocytes harboring the parasites by the native NT system and by "new," induced pathways of low sensitivity to NBMPR. Inhibition of the zero-trans influx of adenosine by graded concentrations of NBMPR yielded a biphasic concentration-effect plot, whereas only a single component of adenosine influx with high NBMPR sensitivity was evident in uninfected erythrocytes [65]. Flux saturability studies (W. P. Gati, unpublished data) have indicated that a minor portion of the parasite-induced component of adenosine influx in *P. yoelii*-infected erythrocytes is transporter-mediated; the remainder of the induced permeability appears to be nonmediated, resembling the route described by others for the entry of other organic solutes into *P. falciparum*-infected erythrocytes [66,67].

VI. NONNUCLEOSIDE DRUGS ARE ALSO POTENT INHIBITORS OF NUCLEOSIDE PERMEATION

Dipyridamole and dilazep are among a group of *N*-heterocyclic agents originally recognized for vasodilatory activity that are now known to be potent inhibitors of the transport of adenosine and other nucleosides [7,20,43,55,68]. The vasoactivity and antiplatelet effects of these drugs may derive ultimately from inhibitory effects on cellular uptake of adenosine, an apparent means by which adenosine concentrations are reduced in the immediate vicinity of adenosine receptors [69,70].

Although inhibition by dipyridamole of NT in erythrocytes has been extensively reported [20,36,37,55,71,72,73], many concentration-effect studies did not take into account (a) depletion of dipyridamole from the incubation medium through cellular uptake (now recognized as substantial [55,73]), or (b) the time required for equilibrium to be established between bound and free forms of the inhibitor [74]. Jarvis and colleagues [36,55] have studied dipyridamole interactions with NT systems in erythrocytes of guinea pigs and sheep, cells in

---

*If it is assumed that (a) NBMPR binding sites represent the entire population of nucleoside transporters, and that (b) NBMPR occupancy of a fraction (80%) of these sites is without effect on transporter function, then these data yield a turnover number of 127 molecules per carrier per second for transporters of high NBMPR sensitivity, which is similar to the earlier value (140 molecules per carrier per second [50]) based on assumption (a). Assumption (b) is entirely speculative.

which nucleoside fluxes are slow (because of low numbers of permeation sites
[50]); NT assays in these cells require long intervals (minutes) of permeant uptake. In experiments that employed [^{3}H] dipyridamole to allow measurement
of concentrations of free dipyridamole (presumed to be in equilibrium with cell-
bound dipyridamole), Jarvis [55] found that dipyridamole was an apparently
competitive inhibitor (K_i, 1 nM) of zero-trans influx of uridine in guinea pig
erythrocytes, whereas inhibition of zero-trans efflux of uridine by dipyridamole
in this system was noncompetitive. Earlier studies of the inhibition of uridine
fluxes by dipyridamole in sheep erythrocytes led to similar results [36], and
in both instances it was concluded that the inhibitor interacted at sites on the
outer face of the membrane [55].

Jarvis has presented evidence that [^{3}H] dipyridamole interacts at the sites on
human erythrocyte membranes that bind NBMPR with high affinity [55]. Using human erythrocyte membranes rather than intact cells to minimize nonspecific cellular uptake of [^{3}H] dipyridamole, Jarvis demonstrated that the
ligand was bound by high-affinity sites of a single class (K_D, 0.65 nM) with
an abundance (29.7 pmol/mg protein) similar to that of NBMPR binding sites.
Adenosine and uridine, as well as NBMPR, NBTGR, and dilazep, inhibited site-
specific, high-affinity binding of [^{3}H] dipyridamole to the membranes. Similarly, dipyridamole inhibited the binding of NBMPR to human [75] and rat
[37] erythrocytes, and displaced NBMPR from sites on membranes from nucleoside-permeable sheep erythrocytes [47].

On the basis of the foregoing observations, Jarvis [55] suggested that in
erythrocytes with NT systems of high NBMPR sensitivity, dipyridamole and
NBMPR bind at overlapping sites on the outer face of the plasma membrane
that include the permeation site totally or partially. This interpretation does not
recognize the structural dissimilarities between uridine, NBMPR, dipyridamole
and dilazep, agents that in this model are thought to interact with sites that
overlap with or share the permeant binding site.

Woffendin and Plagemann [74] have also reported the presence of high-affinity binding sites for [^{3}H] dipyridamole (K_D, ~10 nM) on intact human erythrocytes at an abundance of ~5 × 10^5 sites per cell, which is severalfold higher
than that for NBMPR binding sites. This discrepancy may be related to the relatively high levels of low-affinity and nonspecific binding of dipyridamole found
in these experiments with intact cells. These studies demonstrated that dipyridamole binding was inhibited by nucleosides and other NT inhibitors (NBMPR,
dilazep, and lidoflazine), compounds that also displaced bound [^{3}H] dipyridamole. These authors concluded that binding sites for dipyridamole and NBMPR
overlap and appear to include the nucleoside permeation site [74].

For a number of cultured cell lines, concentration–effect relationships for
NT inhibition by NBMPR have indicated the presence of NT systems of both
high and low sensitivity to NBMPR. Two such components of transport have

been recognized in HeLa cells [68,74], L1210 cells [5,6], CHO cells [6], and in rat erythrocytes [37]. However, in concentration–effect studies of NT inhibition by dipyridamole with these cells [5,6,37,69,74], only single components of NT inhibition (complete) were apparent. Therefore, it is evident that dipyridamole does not distinguish between transporters of high and low sensitivity to NBMPR and that transporter binding sites for dipyridamole and NBMPR are not identical. The additional concept that dipyridamole and NBMPR compete at the same transporter site, as suggested by competitive inhibitory activity of each agent with respect to (a) binding of the other [55,68], and (b) NT inhibition [36,55], would appear to require that the ligand binding sites include as a shared or common region the transporter site involved in nucleoside permeation. An alternative interpretation of the competitive behavior of permeants and inhibitors (allosteric inhibition of inhibitor binding and NT, respectively) would not require common regions for sites of interaction of permeant or inhibitor molecules. Belt [7] has shown that in cells of the rat Walker 256 carcinosarcoma (which lack NBMPR binding sites), dipyridamole inhibited transport of uridine, thymidine, and adenosine; in the presence of 10 μM dipyridamole, transport rates were reduced by more than 50%. The low dipyridamole sensitivity of NT systems in rat tissues has been noted [37,76,77].

Because of high aqueous solubility and potent NT inhibitory activity, dilazep has been employed as a flux-stopping reagent in the measurement of adenosine influx in human erythrocytes in a quenched-flow procedure [39]. In human erythrocytes and in cultured cell lines that have large NT components of high NBMPR sensitivity, dilazep inhibited NT with IC50 values of 5–100 nM [20, 43,59]. The potent interaction of dilazep with NT systems of erythrocytes and cultured cells was also evident in inhibition of NBMPR binding by dilazep [37, 59,72]. Hexobendine, which is structurally similar to dilazep, inhibited both NT and binding of NBMPR in human erythrocytes and in some cultured cells with potencies only slightly lower than those of dilazep [69,72]. Lidoflazine, a vasodilator that inhibited adenosine influx in HeLa cells [68], inhibited binding of NBMPR to human erythrocytes with a K_i of 42 nM [72].

VII. NUCLEOSIDE TRANSPORT SYSTEMS OF HIGH AND LOW SENSITIVITY TO NBMPR ARE DIFFERENTIALLY SENSITIVE TO NONPENETRATING THIOL REAGENTS

pCMBS, an organomercurial thiol reagent that penetrates membranes poorly, inhibits site-specific binding of NBMPR and nucleoside influx by blocking sensitive thiol groups on NT polypeptides. The poor membrane penetrability of pCMBS has revealed asymmetries in the location of these thiol groups. In cells with NT systems of high NBMPR sensitivity (erythrocytes from fetal sheep [78] and humans [79]), treatment of intact cells with pCMBS did not affect NBMPR binding, whereas such treatment inhibited NBMPR binding to unsealed ghosts, and uridine protected against that inhibition, suggesting that the involved pCMBS-sensitive thiol groups were located on the cytoplasmic side of the membrane. In those studies, uridine influx into intact cells was not sensitive to pCMBS. In studies with vesicles prepared from human erythrocytes, Tse et al. [80] showed that NBMPR binding to ROVs was insensitive to pCMBS, whereas binding to IOVs was sensitive, and that pCMBS was a poor inhibitor of the NBTGR-sensitive, zero-trans influx of uridine into ROVs, whereas influx into IOVs was inhibited.

Jarvis and Young demonstrated that in intact rat erythrocytes, in which NT systems of both high and low sensitivity to NBMPR are present, pCMBS inhibited uridine influx through the system of low NBMPR sensitivity, but had little effect on the NT component of high NBMPR sensitivity [37]. It was also shown that pCMBS had no effect on NBMPR binding to intact rat erythrocytes, but did inhibit NBMPR binding to disrupted membranes. It was concluded that the transporter of low NBMPR sensitivity has a thiol group(s) accessible from the outer membrane surface, whereas in the NT system of high NBMPR sensitivity, a pCMBS-sensitive thio group(s) is accessible from the cytoplasmic face of the membrane. These studies indicated that pCMBS distinguishes between NT systems of high and low sensitivity to NBMPR, an observation made previously by Belt [5,7,80], who found that pCMBS inhibited NT systems of low NBMPR sensitivity in L1210 leukemia cells and Walker 256 carcinosarcoma cells, but did not inhibit the NT system of high NBMPR sensitivity in S49 lymphoma cells. These results suggested the presence in the NT systems of L1210 and Walker 256 cells of essential thiol groups accessible from the outer surface of the membrane. The differential effects of pCMBS appear to be evident only at temperatures that do not allow significant penetration of the thiol reagent into the membrane [7], because Plagemann and Wohlhueter, in studies at 37°C with L1210 and Walker 256 cells, did not detect such differences [6,81].

VIII. EXPLORATION OF THE MOLECULAR PROPERTIES OF ERYTHROCYTE NUCLEOSIDE TRANSPORT SYSTEMS WITH NUCLEOSIDE TRANSPORT INHIBITORS

The identity and molecular properties of polypeptide(s) responsible for nucleoside translocation and inhibitor binding in erythrocytes are not yet well defined, although NBMPR binding polypeptides have been identified. The assumption that the polypeptides involved in nucleoside translocation have NBMPR binding sites, or are closely associated with NBMPR binding polypeptides, derives from (a) proportionality between fractional inhibition of uridine transport and fractional occupancy of NBMPR binding sites in human erythrocytes [45], (b) proportionality between NBMPR binding site densities and nucleoside transport capacities of erythrocytes from several species [50], and (c) the coincident absence of inhibitor binding sites and transport capability in erythrocytes from particular species [47,50]. It has been established in human erythrocytes (a) that both NBMPR binding and NT activities are properties of erythrocytic band 4.5 polypeptides (nomenclature of Steck [82]) [83], and (b) that the NBMPR binding polypeptide traverses the membrane [84], and it is obvious that the NT polypeptides must also do so. However, it has not been established that both NT and NBMPR binding activities are properties of a single, transmembrane polypeptide in cultured cells.

Because the NBMPR binding polypeptide represents only ~0.1% of the membrane proteins in human erythrocytes [85], radiolabeled ligands with high affinity for the binding site have been of critical importance in attempts to identify and isolate the polypeptides. Jarvis and Young [86] reported solubilization of NBMPR binding proteins from human and nucleoside-permeable sheep erythrocytes in detergent (sodium cholate) solution, with recovery of binding activity after removal of the detergent by gel filtration. In a subsequent study, protein-depleted membranes from human erythrocytes were solubilized in Triton X-100 and NBMPR binding polypeptides therefrom were partially purified (13-fold) by passage through diethylaminoethyl (DEAE)-cellulose [87]. Copurification of the glucose transporter (D-glucose-sensitive, cytochalasin B binding activity) in the latter procedure signalled the still-unresolved difficulties in separation of the NBMPR binding and glucose transport (cytochalasin B binding) polypeptides. Both activities migrate in the band 4.5 region of sodium dodecyl sulfate (SDS) polyacrylamide gel electrophoretograms as broad bands (M_r, 45,000-66,000).

A.   The Use of Photoaffinity Probes in Molecular Studies of Nucleoside Transport Systems

The identification of NBMPR binding polypeptides on electrophoretograms followed the development of photoactivation procedures that result in covalent attachment of [3H]-labeled ligands to NBMPR binding site polypeptides. Young et al. [85] showed that when the ligand binding sites of human erythrocyte membranes were occupied by [3H]N^6-azidobenzyladenosine ([3H] ABA), a novel, photolabile ligand developed for that study, exposure to ultraviolet light resulted in the covalent attachment of [3H] ABA molecules, presumably at the binding sites [85]. The covalent linkage of the ligand was evident in the migration of [3H]-labeled material on SDS gel electrophoretograms as a band 4.5 polypeptide (M_r of 45,000-66,000). This study also showed that [3H] NBMPR was an effective photoaffinity ligand at the NBMPR binding site and that membrane polypeptides photolabeled with ABA and NBMPR were similar. As well,

Table 1 M_r Values of NBMPR Binding Polypeptides in Cell Types with Nucleoside Transport Systems of Low and High Sensitivity to NBMPR

Cell type	Sensitivity to NBMPR	$M_r \times 10^3$	Ref.
Erythrocyte, human	High	56	20, 37
Erythrocyte, rabbit	High	55	91, A. S. Clanachan, unpublished data
Erythrocyte, mouse	High	56	65
Erythrocyte, rat	High + low	62	37
Erythrocyte, pig	High	62–64	73, 88, 89, 91
S49 mouse lymphoma	High	50–54	8, 90
Novikoff UASJ-2.9 rat hepatoma	Low	72	W. P. Gati and A. R. P. Paterson, unpublished data
Morris 3924A-R1.7 rat hepatoma	Low	81–87	W. P. Gati and A. R. P. Paterson, unpublished data
Reuber H35-R1.3 rat hepatoma	Low	66	W. P. Gati and A. R. P. Paterson, unpublished data

the NBMPR binding site polypeptides of *intact* human erythrocytes have been covalently photolabelled with [^{3}H]NBMPR [88]. Woffendin and Plagemann have reported covalent photolabeling of band 4.5 proteins in human erythrocyte membranes using [^{3}H]dipyridamole [74].

Photoaffinity labeling with [^{3}H]NBMPR has allowed comparison of M_r values for NBMPR binding polypeptides from erythrocyte membranes of a variety of species (Table 1). Such polypeptides from rabbit and mouse erythrocytes are comparable in size to the human erythrocyte polypeptide, while those from rat and pig erythrocytes are significantly larger. NBMPR binding proteins have been identified also in plasma membrane-enriched fractions from rat and guinea pig liver and lung [76,77]. The rat liver polypeptides were larger (M_r, 60,000) than those from guinea pig liver (M_r, 55,000), although this difference could be abolished by enzymatic deglycosylation [60]. Although differences in M_r values of the rat and human erythrocyte NBMPR binding polypeptides were not apparent after enzymatic deglycosylation [37], M_r differences between pig and human erythrocyte polypeptides remained after treatment with glycosidases [89].

NBMPR binding polypeptides in crude plasma membrane fractions from S49 mouse lymphoma cells (which have a single NT system of high NBMPR sensitivity) migrate on SDS polyacrylamide gel electrophoresis (PAGE) with apparent M_r values similar to those of the human erythrocyte [90]. In cultured cells of the Novikoff UA rat hepatoma line, in which NT is of low sensitivity to NBMPR (despite the presence of high-affinity NBMPR binding sites), NBMPR binding polypeptides are large (M_r, 72,000-80,000) [8]. In two other cultured lines of rat hepatoma cells (Reuber H35, Morris 3924A), in which NT systems are insensitive to site-bound NBMPR [2,61], NBMPR binding polypeptides also have M_r values higher than those of human erythrocytes (Table 1). Thus, NBMPR binding polypeptides from several cell types in which the relation of NBMPR binding to NT systems is unclear have a lower SDS PAGE mobility than that of such polypeptides from other cell types.

Attempts to use a photolabile nucleoside transporter substrate, 8-azido-adenosine, to covalently label the nucleoside translocation polypeptide in human erythrocytes have yielded surprising results. Jarvis et al. [91] showed that [^{3}H] 8-azidoadenosine became covalently associated with band 4.5 polypeptides during exposure to UV light of membranes in [^{3}H] 8-azidoadenosine-containing medium. It was concluded that the glucose transporter polypeptide, known to be a band 4.5 constituent was labeled by [^{3}H] 8-azidoadenosine because of observations to the effect that covalent attachment of the nucleoside to membrane polypeptides was inhibited by the presence of nucleosides, D-glucose, and cytochalasin B, but not by the presence of NBMPR, NBTGR, or dilazep. It was also shown that nucleosides inhibited the covalent attachment of [^{3}H] cytochalasin B to band 4.5 polypeptides. These results suggest structural relationships between the nucleoside and glucose transporter polypeptides of human erythrocytes.

Radiation inactivation studies have indicated that the M_r of NT polypeptides in human erythrocytes is ~122,000, whether inactivation of NBMPR binding activity [92] or of NBMPR-sensitive uridine influx [93] was followed. A recent study has shown that the target sizes of NT and glucose transporter polypeptides determined by this technique are similar [94]. Comparison of these M_r values with those for inhibitor binding polypeptides determined by SDS PAGE suggest that both transporters may be present as dimers in the human erythrocyte plasma membrane.

B. Enzymatic Cleavage of NBMPR Binding Polypeptides

Proteolytic fragmentation of covalently-labeled NBMPR binding polypeptides from erythrocytes of different species has provided information about the membrane location of inhibitor binding proteins. Treatment of intact fetal sheep erythrocytes with trypsin under mild conditions did not affect either uridine transport or NBMPR binding, but treatment of disrupted erythrocyte mem-

branes in this way resulted in a considerable loss of NBMPR binding activity
[78]. These results suggested that a trypsin-sensitive site was located on the
cytoplasmic side of NT polypeptides in these cells. Janmohamed et al. [84]
showed that, after covalent labeling of intact human erythrocytes with [^{3}H]-
NBMPR, mild trypsin treatment did not release ^{3}H-labeled peptides, although
at higher enzyme:substrate ratios, or at low ionic strength, a radiolabeled tryptic
peptide (M_r, 22,500) was released from intact cells [84]. Tryptic digestion
under conditions for release of the latter peptide did not affect either uridine
uptake into the cells or NBMPR binding site abundance, although the affinity of
membrane sites for the inhibitor was approximately twofold lower. In the same
study, mild treatment of labeled, unsealed human erythrocyte ghosts with trypsin released two radiolabeled fragments with M_r values of 38,000 and 23,000
from the cytoplasmic side of the membrane. Thus, this study showed that at
either membrane face trypsin was able to effect partial digestion of the NBMPR
binding polypeptide in human erythrocytes, which therefore is a transmembrane protein.

Lacking glucose transport activity, erythrocytes from adult pigs depend on
inosine as a source of metabolic energy [95,96]. Because of the apparent absence of glucose transporter polypeptides, pig erythrocyte membranes have been
used as starting material in attempts to purify erythrocytic NBMPR binding
polypeptides, to avoid the problem of co-isolation of these two band 4.5 polypeptides by DEAE-cellulose chromatography. With this tactic, Kwong et al.
[89] achieved a 61-fold purification of NBMPR binding activity from pig
erythrocyte membranes by chromatography of an octylglucoside-solubilized
membrane extract on DEAE-cellulose. Woffendin and Plagemann [73] have confirmed that a band 4.5 polypeptide in pig erythrocyte ghosts may be photolabeled with [^{3}H]NBMPR. Structural differences between the human and pig
erythrocyte NBMPR binding proteins have been explored by digestion of [^{3}H]-
NBMPR-labeled proteins with endo-β-galactosidase and endoglycosidase F [89].
These studies established that both NBMPR binding proteins are glycoproteins
with N-linked oligosaccharides; however, treatment with endo-β-galactosidase
reduced the apparent molecular weight of the human protein, but did not affect
the pig protein. Endoglycosidase F treatment, which should remove virtually all
glycans that are N-linked to asparagine [97], yielded a human polypeptide with
a significantly lower M_r (44,000) than that of the pig polypeptide (57,000).
Good et al. have also demonstrated that treatment of the pig polypeptide with
crude endoglycosidase F reduced the apparent M_r from 64,000 to 57,000 [98].
These results suggest that the human and pig NBMPR binding proteins differ in
both carbohydrate and polypeptide structure.

C. Reconstitution of Nucleoside Transport Activity and Inhibitor Binding Activity

Reconstitution experiments have shown that band 4.5 polypeptides partly puri-fied from human erythrocytes by extraction with Triton X-100 and DEAE-cellulose chromatography imparted both NBMPR binding activity and NT ac-tivity when incorporated into proteoliposomes prepared from soybean phospho-lipids by a sonication technique [84]. The vesicles with reconstituted band 4.5 proteins displayed saturable uridine transport activity that was sensitive to NBTGR, dipyridamole, and dilazep, and was inhibited by adenosine and ino-sine. Turnover numbers for the reconstituted transporters were considerably lower than those of intact erythrocytes. Cytochalasin B-sensitive, stereospecific fluxes of glucose into vesicles with the reconstituted proteins were also demon-strated, indicating that transport activities for both glucose and nucleosides were present in the band 4.5 preparation and were reconstituted together.

D. Monoclonal Antibodies Against NBMPR Binding Proteins

Although NBMPR has enabled the identification of NT-associated polypeptides, the recent development of monoclonal antibodies (MAbs) against the NBMPR binding protein from pig erythrocytes [98] has provided new probes of the erythrocyte NT system. Good et al. [98] have developed MAbs that recognize detergent-solubilized [^{3}H]NBMPR-labeled polypeptides from pig erythrocytes. MAbs 11C4 and 3E3 do not recognize polypeptides from human, rabbit, or mouse erythrocytes, and do not recognize the cytochalasin B binding polypep-tides present in neonatal pig erythrocytes. These MAbs interact with external polypeptide epitopes of NBMPR binding proteins in intact pig erythrocytes or protein-depleted membranes, but do not inhibit NBMPR binding activity.

IX. CONCLUSION

During the past decade, exploration of NT systems has shown that not only are equilibrative NT systems of high NBMPR sensitivity, such as that of the human erythrocyte, widely distributed among animal cell types, but equilibrative and concentrative NT systems of low NBMPR sensitivity are also to be found in vari-ous cells and tissues, with some cells having NT systems of more than one type. NBMPR has been an exceedingly valuable probe of NT systems in studies of per-meation processes and of the cellular biology of NT systems. By photoaffinity labeling with [^{3}H]NBMPR, an NBMPR binding polypeptide, evidently the nu-cleoside transporter of human erythrocytes, has been identified, and with this step, exploration of the molecular biology of this system has begun. Challenges to the postulated identity of polypeptides involved in NBMPR binding and NT activity are found in the NT properties of various nucleated cells.

ACKNOWLEDGEMENT

We thank our colleagues Dr. Carol E. Cass and Dr. James D. Craik for reading
the manuscript and for helpful comments. We acknowledge support from the
Medical Research Council of Canada, the Alberta Cancer Board, and the National Cancer Institute of Canada during the preparation of this manuscript and
during the conduct of McEachern Laboratory research projects referenced there-
in. A. R. P. Paterson is a Senior Research Scientist of the National Cancer Institute of Canada.

NOTE ADDED IN PROOF

Since completion of this manuscript in late 1987, significant progress in the
identification and purification of NT polypeptides has been achieved. Craik
et al. [99] employed MAbs to show that glucose transporter and nucleoside
transporter proteins of neonatal pig erythrocytes are distinct entities and that
the glucose transporter is absent from erythrocytes of adult pigs. Kwong et al.
[100] reported purification of the NBMPR binding protein from human erythrocytes by immunoaffinity chromatography. The purified protein catalyzed
uridine uptake when reconstituted into phospholipid vesicles.

REFERENCES

1. Paterson, A. R. P., and Cass, C. E. (1986). Transport of nucleoside drugs in
   animal cells. In *Membrane Transport of Antineoplastic Agents, International
   Encyclopedia of Pharmacology and Therapeutics*, Section 118, I. D. Goldman (Ed.). Pergamon Press, Oxford, pp. 309–329.
2. Paterson, A. R. P., Jakobs, E. S., Ng, C. Y. C., Odegard, R. D., and Adjei,
   A. A. (1987). Nucleoside transport inhibition in vitro and in vivo. In *Topics
   and Perspectives in Adenosine Research*, E. Gerlach and B. F. Becker (Eds.).
   Springer-Verlag, Berlin and Heidelberg, pp. 89–101.
3. Young, J. D., and Jarvis, S. M. (1983). Nucleoside transport in animal cells.
   Biosci. Rep. 3:309–322.
4. Plagemann, P. G. W., and Wohlhueter, R. M. (1980). Permeation of nucleosides, nucleic acid bases, and nucleotides in animal cells. *Curr. Top. Membr.
   Transport* 14:225–330.
5. Belt, J. A. (1983). Heterogeneity of nucleoside transport in mammalian
   cells. Two types of transport activity in L1210 and other cultured neoplastic cells, *Mol. Pharmacol.* 24:479–484.
6. Plagemann, P. G. W., and Wohlhueter, R. M. (1984). Nucleoside transport
   in cultured mammalian cells. Multiple forms with different sensitivity to
   inhibition by nitrobenzylthioinosine or hypoxanthine. *Biochim. Biophys.
   Acta* 773:39–52.

7. Belt, J. A., and Noel, L. D. (1985). Nucleoside transport in Walker 256 rat carcinosarcoma and S49 mouse lymphoma cells. Differences in sensitivity to nitrobenzylthioinosine and thiol reagents. *Biochem. J.* 232:681–688.

8. Gati, W. P., Belt, J. A., Jakobs, E. S., Young, J. D., Jarvis, S. M., and Paterson, A. R. P. (1986). Photoaffinity labelling of a nitrobenzylthioinosine-binding polypeptide from cultured Novikoff hepatoma cells. *Biochem. J.* 236:665–670.

9. Ungemach, F. R., and Hegner, D. (1978). Uptake of thymidine into isolated rat hepatocytes. Evidence for two transport systems. *Hoppe-Seylers Z. Physiol. Chem.* 359:845–856.

10. Schwenk, M., Hegazy, E., and Lopez del Pino, V. (1984). Uridine uptake by isolated intestinal epithelial cells of guinea pig. *Biochim. Biophys. Acta* 805:370–374.

11. Spector, R., and Huntoon, S. (1984). Specificity and sodium dependence of the active nucleoside transport system in choroid plexus. *J. Neurochem.* 42:1048–1052.

12. Jakobs, E. S., and Paterson, A. R. P. (1986). Sodium-dependent, concentrative nucleoside transport in cultured intestinal epithelial cells. *Biochem. Biophys. Res. Commun.* 140:1028–1035.

13. Darnowski, J. W., Holdridge, C., and Handschumacher, R. E. (1987). Concentrative uridine transport by murine splenocytes: kinetics, substrate specificity and sodium dependency. *Cancer Res.* 47:2614–2619.

14. Dagnino, L., Bennett, L. L., and Paterson, A. R. P. (1987). Concentrative transport of nucleosides in L1210 mouse leukemia cells. *Proc. Am. Assoc. Cancer Res.* 28:15.

15. Oliver, J. M., and Paterson, A. R. P. (1971). Nucleoside transport. I. A mediated process in human erythrocytes. *Can. J. Biochem.* 49:262–270.

16. Cabantchik, Z. I., and Ginsburg, H. (1977). Transport of uridine in human red blood cells. Demonstration of a simple carrier-mediated process. *J. Gen. Physiol.* 69:75–96.

17. Plagemann, P. G. W., Wohlhueter, R. M., and Erbe, J. (1982). Nucleoside transport in human erythrocytes. A simple carrier with directional symmetry and differential mobility of loaded and empty carrier. *J. Biol. Chem.* 257:12069–12074.

18. Fox, I. H., and Kelly, W. N. (1978). The role of adenosine and 2'-deoxyadenosine in mammalian cells. *Ann. Rev. Biochem.* 47:655–686.

19. Gerlach, E., Becker, B. F., and Nees, S. (1987). Formation of adenosine by vascular endothelium: a homeostatic and antithrombogenic mechanism. In *Topics and Perspectives in Adenosine Research*, E. Gerlach and B. F. Becker (Eds.). Springer-Verlag, Berlin and Heidelberg, pp. 309–320.

20. Mahony, W. B., and Zimmerman, T. P. (1986). An assay for inhibitors of nucleoside transport based upon the use of 5-[^{125}I]iodo-2'-deoxyuridine as permeant. *Anal. Biochem.* 154:235–243.

21. Paterson, A. R. P., Jakobs, E. S., Harley, E. R., Cass, C. E., and Robins, M. J. (1983). Inhibitors of nucleoside transport as probes and drugs. In *Development of Target-Oriented Anticancer Drugs*, Y.-C. Cheng, B. Goz, and M. Minkoff (Eds.). Raven Press, New York, pp. 41–56.

22. Lieb, W. R. (1982). A kinetic approach to transport studies. In *Red Cell Membranes—A Methodological Approach*, J. C. Ellory and J. D. Young (Eds.). Academic Press, London, pp. 135–164.

23. Jarvis, S. M., Hammond, J. R., Paterson, A. R. P., and Clanachan, A. S. (1983). Nucleoside transport in human erythrocytes. A simple carrier with directional symmetry in fresh cells, but with directional asymmetry in cells from outdated blood. *Biochem. J.* 210:457–461.

24. Plagemann, P. G. W., and Wohlhueter, R. M. (1984). Kinetics of nucleoside transport in human erythrocytes. Alterations during blood preservation. *Biochim. Biophys. Acta* 778:176–184.

25. Cass, C. E., and Paterson, A. R. P. (1972). Mediated transport of nucleosides in human erythrocytes. Accelerative exchange diffusion of uridine and thymidine and specificity toward pyrimidine nucleosides as permeants. *J. Biol. Chem.* 247:3314–3320.

26. Cass, C. E., and Paterson, A. R. P. (1973). Mediated transport of nucleosides by human erythrocytes. Specificity toward purine nucleosides as permeants. *Biochim. Biophys. Acta* 291:734–746.

27. Gati, W. P., Misra, H. K., Knaus, E. E., and Wiebe, L. I. (1984). Structural modifications at the 2'- and 3'-positions of some pyrimidine nucleosides as determinants of their interaction with the mouse erythrocyte nucleoside transporter. *Biochem. Pharmacol.* 33:3325–3331.

28. Jarvis, S. M. (1986). *Trans*-stimulation and *trans*-inhibition of uridine efflux from human erythrocytes by permeant nucleosides. *Biochem. J.* 233:295–297.

29. Jarvis, S. M. (1985). 2-Chloroadenosine, a permeant for the nucleoside transporter. *Biochem. Pharmacol.* 34:3237–3241.

30. Zimmerman, T. P., Mahony, W. B., and Prus, K. L. (1987). 3'-Azido-3'-deoxythymidine. An unusual nucleoside analogue that permeates the membrane of human erythrocytes and lymphocytes by nonfacilitated diffusion. *J. Biol. Chem.* 262:5748–5754.

31. Paterson, A. R. P., Yang, S.-E., Lau, E. Y., and Cass, C. E. (1979). Low specificity of the nucleoside transport mechanism of RPMI 6410 cells. *Mol. Pharmacol.* 16:900–908.

32. Cohen, A., Ullman, B., and Martin, D. W. (1979). Characterization of a mutant mouse lymphoma cell with deficient transport of purine and pyrimidine nucleosides. *J. Biol. Chem.* 254:112–116.

33. Wohlhueter, R. M., and Plagemann, P. G. W. (1982). On the functional symmetry of nucleoside transport in mammalian cells. *Biochim. Biophys. Acta* 689:249–260.

34. Wohlhueter, R. M., Marz, R., Graff, J. C., and Plagemann, P. G. W. (1978). A rapid mixing technique to measure transport in suspended animal cells: application to nucleoside transport in rat hepatoma cells. *Methods Cell Biol.* 20:211–236.

35. Jarvis, S. M., and Martin, B. W. (1986). Effects of temperature on the transport of nucleosides in guinea pig erythrocytes. *Can. J. Physiol. Pharmacol.* 64:193–198.

36. Jarvis, S. M., McBride, D., and Young, J. D. (1982). Erythrocyte nucleoside transport: asymmetrical binding of nitrobenzylthioinosine to nucleoside permeation sites. *J. Physiol. (Lond.)* 324:31–46.
37. Jarvis, S. M., and Young, J. D. (1986). Nucleoside transport in rat erythrocytes: two components with differences in sensitivity to inhibition by nitrobenzylthioinosine and *p*-chloromercuriphenyl sulfonate. *J. Membr. Biol.* 93:1–10.
38. Pickard, M. A., and Paterson, A. R. P. (1972). Use of 4-nitrobenzylthioinosine in the measurement of rates of nucleoside transport in human erythrocytes. *Can. J. Biochem.* 50:839–840.
39. Paterson, A. R. P., Harley, E. R., and Cass, C. E. (1984). Inward fluxes of adenosine in erythrocytes and cultured cells measured by a quenched-flow method. *Biochem. J.* 224:1001–1008.
40. Paterson, A. R. P., Harley, E. R., and Cass, C. E. (1985). Measurement and inhibition of membrane transport of adenosine. In *Methods in Pharmacology*, Vol. 6, D. M. Paton (Ed.). Plenum Press, New York, pp. 165–180.
41. Plagemann, P. G. W., and Wohlhueter, R. M. (1984). Effect of temperature on kinetics and differential mobility of empty and loaded nucleoside transporter of human erythrocytes. *J. Biol. Chem.* 259:9024–9027.
42. Paul, B., Chen, M. F., and Paterson, A. R. P. (1975). Inhibitors of nucleoside transport: a structure-activity study using human erythrocytes. *J. Med. Chem.* 18:968–973.
43. Paterson, A. R. P., Jakobs, E. S., Harley, E. R., Fu, N.-W., Robins, M. J., and Cass, C. E. (1983). Inhibition of nucleoside transport. In *Regulatory Function of Adenosine*, R. M. Berne, T. W. Rall, and R. Rubio (Eds.). Martinus Nijhoff, The Hague, pp. 203–220.
44. Paterson, A. R. P., and Oliver, J. M. (1971). Nucleoside transport. II. Inhibition by *p*-nitrobenzylthioguanosine and related compounds. *Can. J. Biochem.* 49:271–274.
45. Cass, C. E., Gaudette, L. A., and Paterson, A. R. P. (1974). Mediated transport of nucleosides in human erythrocytes. Specific binding of the inhibitor nitrobenzylthioinosine to nucleoside transport sites in the erythrocyte membrane. *Biochim. Biophys. Acta* 345:1–10.
46. Pickard, M. A., Brown, R. R., Paul, B., and Paterson, A. R. P. (1973). Binding of the nucleoside transport inhibitor 4-nitrobenzylthioinosine to erythrocyte membranes. *Can. J. Biochem.* 51:666–672.
47. Jarvis, S. M., and Young, J. D. (1980). Nucleoside transport in human and sheep erythrocytes. Evidence that nitrobenzylthioinosine binds specifically to functional nucleoside-transport sites. *Biochem. J.* 190:377–383.
48. Gati, W. P., Wiebe, L. I., Knaus, E. E., and Paterson, A. R. P. (1987). [^{125}I] Iodohydroxynitrobenzylthioinosine: a new high-affinity nucleoside transporter probe. *Biochem. Cell Biol.* 65:467–473.
49. Cass, C. E., Gati, W. P., Odegard, R., and Paterson, A. R. P. (1985). The effect of pH on interaction of nitrobenzylthioinosine and hydroxynitrobenzylthioinosine with the nucleoside transporter of human erythrocyte membranes. *Mol. Pharmacol.* 27:662–665.

50. Jarvis, S. M., Hammond, J. R., Paterson, A. R. P., and Clanachan, A. S. (1982). Species differences in nucleoside transport. A study of uridine transport and nitrobenzylthioinosine binding by mammalian erythrocytes. *Biochem. J.* 208:83–88.

51. Young, J. D. (1978). Nucleoside transport in sheep erythrocytes: Genetically controlled transport variation and its influence on erythrocyte ATP concentrations. *J. Physiol. (Lond).* 277:325–339.

52. Cass, C. E., Kolassa, N., Uehara, Y., Dahlig-Harley, E., Harley, E. R., and Paterson, A. R. P. (1981). Absence of binding sites for the transport inhibitor nitrobenzylthioinosine on nucleoside transport-deficient mouse lymphoma cells. *Biochim. Biophys. Acta* 649:769–777.

53. Wohlhueter, R. M., Brown, W. E., and Plagemann, P. G. W. (1983). Kinetic and thermodynamic studies on nitrobenzylthioinosine binding to the nucleoside transporter of Chinese hamster ovary cells. *Biochim. Biophys. Acta* 731:168–176.

54. Agbanyo, F. R., Cass, C. E., and Paterson, A. R. P. (1988). External location of sites on pig erythrocyte membranes that bind nitrobenzylthioinosine. *Mol. Pharmacol.* 33:332–337.

55. Jarvis, S. M. (1986). Nitrobenzylthioinosine-sensitive nucleoside transport system: mechanism of inhibition by dipyridamole. *Mol. Pharmacol.* 30:659–665.

56. Jarvis, S. M., Janmohamed, S. N., and Young, J. D. (1983). Kinetics of nitrobenzylthioinosine binding to the human erythrocyte nucleoside transporter. *Biochem. J.* 216:661–667.

57. Cass, C. E., and Paterson, A. R. P. (1976). Nitrobenzylthioinosine binding sites in the erythrocyte membrane. *Biochim. Biophys. Acta* 419:285–294.

58. Koren, R., Cass, C. E., and Paterson, A. R. P. (1983). The kinetics of dissociation of the inhibitor of nucleoside transport, nitrobenzylthioinosine, from the high-affinity binding sites of cultured hamster cells. *Biochem. J.* 216:299–308.

59. Plagemann, P. G. W., and Kraupp, M. (1986). Inhibition of nucleoside and nucleobase transport and nitrobenzylthioinosine binding by dilazep and hexobendine. *Biochem. Pharmacol.* 35:2559–2567.

60. Jarvis, S. M., and Young, J. D. (1987). Photoaffinity labelling of nucleoside transporter polypeptides. *Pharmacol. Ther.* 32:339–359.

61. Ng, C. Y. C. (1986). Nucleoside transport systems of low sensitivity to nitrobenzylthioinosine in cultured rat hepatoma cells. Thesis, University of Alberta, Edmonton.

62. Aronow, B., Allen, K., Patrick, J., and Ullman, B. (1985). Altered nucleoside transporter in mammalian cells selected for resistance to the physiological effects of inhibitors of nucleoside transport. *J. Biol. Chem.* 260:6226–6233.

63. Cohen, A., Leung, C., and Thompson, E. (1985). Characterization of mouse lymphoma cells with altered nucleoside transport. *J. Cell. Physiol.* 123:431–434.

64. Plagemann, P. G. W., and Wohlhueter, R. M. (1985). Nitrobenzylthioinosine-sensitive and -resistant nucleoside transport in normal and transformed rat cells. *Biochim. Biophys. Acta* 816:387-395.
65. Gati, W. P., Stoyke, A. F.-W., Gero, A. M., and Paterson, A. R. P. (1987). Nucleoside permeation in mouse erythrocytes infected with *Plasmodium yoelli. Biochem. Biophys. Res. Commun.* 145:1134-1141.
66. Ginsburg, H., Kutner, S., Krugliak, M., and Cabantchik, Z. I. (1985). Characterization of permeation pathways appearing in the host membrane of *Plasmodium falciparum* infected red cells. *Mol. Biochem. Parasitol.* 14: 313-322.
67. Ginsburg, H., and Stein, W. D. (1987). Biophysical analysis of novel transport pathways induced in red blood cell membranes. *J. Membr. Biol.* 96: 1-10.
68. Paterson, A. R. P., Lau, E. Y., Dahlig, E., and Cass, C. E. (1980). A common basis for inhibition of nucleoside transport by dipyridamole and nitrobenzylthioinosine? *Mol. Pharmacol.* 18:40-44.
69. Berne, R. M., Gidday, J. M., Hill, R. E., Curnish, R. R., and Rubio, R. (1987). Adenosine in the local regulation of blood flow: some controversies. In *Topics and Perspectives in Adenosine Research*, E. Gerlach and B. F. Becker (Eds.). Springer-Verlag, Berlin and Heidelberg, pp. 395-405.
70. Dawicki, D. D., Agarwal, K. C., and Parks, Jr., R. E. (1985). Role of adenosine uptake and metabolism by blood cells in the antiplatelet actions of dipyridamole, dilazep and nitrobenzylthioinosine. *Biochem. Pharmacol.* 34: 3965-3972.
71. Jarvis, S. M., Young, J. D., Ansay, M., Archibald, A. L., Harkness, R. A., and Simmonds, R. J. (1980). Is inosine the physiological energy source of pig erythrocytes? *Biochim. Biophys. Acta* 597:183-188.
72. Hammond, J. R., and Clanachan, A. S. (1984). [3H]Nitrobenzylthioinosine binding to guinea pig CNS nucleoside transport system: a pharmacological characterization. *J. Neurochem.* 43:1582-1592.
73. Woffendin, C., and Plagemann, P. G. W. (1987). Nucleoside transporter of pig erythrocytes. Kinetic properties, isolation and reaction with nitrobenzylthioinosine and dipyridamole. *Biochim. Biophys. Acta* 903:18-30.
74. Woffendin, C., and Plagemann, P. G. W. (1987). Interaction of [3H]dipyridamole with the nucleoside transporters of human erythrocytes and cultured animal cells. *J. Membr. Biol.* 98:89-100.
75. Hammond, J. R., Paterson, A. R. P., and Clanachan, A. S. (1981). Benzodiazepine inhibition of site-specific binding of nitrobenzylthioinosine, an inhibitor of adenosine transport. *Life Sci.* 29:2207-2214.
76. Wu, J.-S. R., and Young, J. D. (1984). Photoaffinity labelling of nucleoside-transport proteins in plasma membranes isolated from rat and guinea-pig liver. *Biochem. J.* 220:499-506.
77. Shi, M. M., Wu, J.-S. R., Lee, C.-M., and Young, J. D. (1984). Nucleoside transport. Photoaffinity labelling of high-affinity nitrobenzylthioinosine binding sites in rat and guinea pig lung. *Biochem. Biophys. Res. Commun.* 118:594-600.

78. Jarvis, S. M., and Young, J. D. (1982). Nucleoside translocation in sheep reticulocytes and fetal erythrocytes: a proposed model for the nucleoside transporter. *J. Physiol. (Lond.)* 324:47–66.

79. Tse, C.-M., Wu, J.-S. R., and Young, J. D. (1985). Evidence for the asymmetrical binding of *p*-chloromercuriphenyl sulphonate to the human erythrocyte nucleoside transporter. *Biochim. Biophys. Acta* 818:316–324.

80. Belt, J. A. (1983). Nitrobenzylthioinosine-insensitive uridine transport in human lymphoblastoid and murine leukemia cells. *Biochem. Biophys. Res. Commun.* 110:417–423.

81. Plagemann, P. G. W., and Wohlhueter, R. M. (1984). Effect of sulfhydryl reagents on nucleoside transport in cultured mammalian cells. *Arch. Biochem. Biophys.* 233:489–500.

82. Steck, T. L. (1974). The organization of proteins in the human red blood cell membrane. A review. *J. Cell Biol.* 62:1–19.

83. Tse, C.-M., Belt, J. A., Jarvis, S. M., Paterson, A. R. P., Wu, J.-S., and Young, J. D. (1985). Reconstitution studies of the human erythrocyte nucleoside transporter. *J. Biol. Chem.* 260:3506–3511.

84. Janmohamed, S. N., Young, J. D., and Jarvis, S. M. (1985). Proteolytic cleavage of [^{3}H]nitrobenzylthioinosine-labelled nucleoside transporter in human erythrocytes. *Biochem. J.* 230:777–784.

85. Young, J. D., Jarvis, S. M., Robins, M. J., and Paterson, A. R. P. (1983). Photoaffinity labeling of the human erythrocyte nucleoside transporter by N^6-(*p*-azidobenzyl)adenosine and nitrobenzylthioinosine. Evidence that the transporter is a band 4.5 polypeptide. *J. Biol. Chem.* 258:2202–2208.

86. Jarvis, S. M., and Young, J. D. (1980). Solubilization of the nucleoside translocation system from human and nucleoside-permeable sheep erythrocytes. *FEBS Lett.* 117:33–36.

87. Jarvis, S. M., and Young, J. D. (1981). Extraction and partial purification of the nucleoside-transport system from human erythrocytes based on the assay of nitrobenzylthioinosine-binding activity. *Biochem. J.* 194:331–339.

88. Wu, J.-S. R., Kwong, F. Y. P., Jarvis, S. M., and Young, J. D. (1983). Identification of the erythrocyte nucleoside transporter as a band 4.5 polypeptide. Photoaffinity labeling studies using nitrobenzylthioinosine. *J. Biol. Chem.* 258:13745–13751.

89. Kwong, F. Y. P., Baldwin, S. A., Scudder, P. R., Jarvis, S. M., Choy, M. Y. M., and Young, J. D. (1986). Erythrocyte nucleoside and sugar transport. Endo-β-galactosidase and endoglycosidase-F digestion of partially purified human and pig transporter proteins. *Biochem. J.* 240:349–356.

90. Young, J. D., Jarvis, S. M., Belt, J. A., Gati, W. P., and Paterson, A. R. P. (1984). Identification of the nucleoside transporter in cultured mouse lymphoma cells. Photoaffinity labeling of plasma membrane-enriched fractions from nucleoside transport-competent (S49) and nucleoside transport-deficient (AE$_1$) cells with [^{3}H]nitrobenzylthioinosine. *J. Biol. Chem.* 259:8363–8365.

91. Jarvis, S. M., Young, J. D., Wu, J.-S. R., Belt, J. A., and Paterson, A. R. P. (1986). Photoaffinity labeling of the human erythrocyte glucose transporter with 8-azidoadenosine. *J. Biol. Chem.* 261:11077–11085.

92. Jarvis, S. M., Young, J. D., and Ellory, J. C. (1980). Nucleoside transport in human erythrocytes. Apparent molecular weight of the nitrobenzylthioinosine-binding complex estimated by radiation-inactivation analysis. *Biochem. J.* 190:373–376.

93. Jarvis, S. M., Fincham, D. A., Ellory, J. C., Paterson, A. R. P., and Young, J. D. (1984). Nucleoside transport in human erythrocytes. Nitrobenzylthioinosine binding and uridine transport activities have similar radiation target sizes. *Biochim. Biophys. Acta* 772:227–230.

94. Jarvis, S. M., Ellory, J. C., and Young, J. D. (1986). Radiation inactivation of the human erythrocyte nucleoside and glucose transporters. *Biochim. Biophys. Acta* 855:312–315.

95. Kim, H. D., and McManus, T. J. (1971). Studies on the energy metabolism of pig erythrocytes. I. The limiting role of membrane permeability in glycolysis. *Biochim. Biophys. Acta* 230:1–11.

96. Young, J. D., Paterson, A. R. P., and Henderson, J. F. (1985). Nucleoside transport and metabolism in erythrocytes from the Yucatan miniature pig. Evidence that inosine functions as an in vivo energy substrate. *Biochim. Biophys. Acta* 842:213–224.

97. Elder, J. H., and Alexander, S. (1982). Endo-β-N-acetylglucosaminidase F: Endoglycosidase from *Flavobacterium meningosepticum* that cleaves both high-mannose and complex glycoproteins. *Proc. Natl. Acad. Sci. USA* 79: 4540–4544.

98. Good, A. H., Craik, J. D., Jarvis, S. M., Kwong, F. Y. P., Young, J. D., Paterson, A. R. P., and Cass, C. E. (1987). Characterization of monoclonal antibodies that recognize band 4.5 polypeptides associated with nucleoside transport in pig erythrocytes. *Biochem. J.* 244:749–755.

99. Craik, J. D., Good, A. H., Gottschalk, R., Jarvis, S. M., Paterson, A. R. P., and Cass, C. E. (1988). Identification of glucose and nucleoside transport proteins in neonatal pig erythrocytes using monoclonal antibodies against band 4.5 polypeptides of adult human and pig erythrocytes. *Biochem. Cell Biol.* 66:839–852.

100. Kwong, F. Y. P., Davies, A., Tse, C. M., Young, J. D., Henderson, P. J. F., and Baldwin, S. A. (1988). Purification of the human erythrocyte nucleoside transporter by immunoaffinity chromatography. *Biochem. J.* 255: 243–249.

22

Regulated Transport
The Response of Ion Transport Pathways to Physiological Stimuli

MARK HAAS   *Yale University School of Medicine, New Haven, Connecticut*

I. INTRODUCTION

The regulation of specific pathways for the transport of ions across cell membranes plays a vital role in the normal function of many types of cells. Like many cell types, red blood cells from different animal species possess transport pathways that are regulated by physiological stimuli such as changes in cell volume and catecholamine hormones. Unfortunately, although the kinetic aspects of these ion transport pathways in red blood cells have in many cases been extensively studied, the cellular mechanisms involved in their regulation and the physiological consequences of such regulation are in general not as well understood in these cells as in many other cell types. Furthermore, it is not entirely clear whether such regulated pathways are actually needed in the mature red blood cells of certain species, as is discussed below for pathways activated by cell swelling or shrinkage.

Still, the study of regulated ion transport pathways in red blood cells remains important for a number of reasons. First, in several disease states involving red blood cells, including sickle cell disease, other hemoglobinopathies, and drug-induced hemolytic anemias, abnormalities in specific ion transport pathways have been found that may play an important role in the pathogenesis of the disease process. As will be discussed briefly in this chapter and in more detail in the chapter following the specific red blood cell transport pathways involved in these disease states appear to be among those affected by changes in cell volume and other physiological stimuli. In addition, red blood cells remain a readily accessible and relatively simple system for the study of ion transport, and much of our present understanding of both the general principles of ion transport across biological membranes and of specific transport pathways in many cell types has come from studies with red blood cells. It is almost certain that red blood cells will continue to be a useful model for the study of ion transport and its regulation, which can be applied to more complex cells and tissues.

In this chapter, I will review specific red blood cell ion transport pathways that respond to changes in cell volume and/or catecholamines, as well as what is known about the cellular mechanisms involved in the activation of transport by these stimuli. I will also briefly discuss other possible physiological effectors of ion transport in red blood cells, such as hypoxia and increased hydrostatic pressure, and draw parallels between ion transport pathways activated by the above stimuli and those affected in certain disease states that directly involve red blood cells. In reviewing the regulation of ion transport by changes in cell volume and by catecholamines, I will pay considerable attention to red blood cells from species other than human. In particular, studies of the nucleated erythrocytes of amphibians, birds, and fish have provided much of what is presently known about ion transport and its regulation in red blood cells.

II. ION TRANSPORT REGULATED BY CHANGES IN CELL VOLUME

Two classes of ion transport processes thought to be involved in the regulation of cell volume have been discussed in the literature: those involved in maintaining the cell at its normal, steady-state volume, and those activated in response to sudden perturbations in cell volume. The latter processes have received the most attention during the past decade, and will be the primary subject of this section. However, whether such processes actually play a role in cell volume regulation remains unclear, because mature erythrocytes in the circulation are, with one notable exception, unlikely to undergo volume changes of sufficient magnitude for these processes to perform their prescribed function. The exception is when the cells pass through the vessels of the renal medulla, encountering a markedly hypertonic environment that causes them to shrink. Here, some degree of volume regulation may be necessary to prevent a degree of shrinkage that would cause the hemoglobin to crystallize out of solution. However, most if not all pathways for ion transport that are activated by cell shrinkage transport at far too slow a rate to produce a significant increase in cell volume during the short transit time of red blood cells through the vasa recta. Instead, it appears that very rapid urea transport into the cells may serve the important role of limiting red blood cell shrinkage in the vasa recta [1]. There is, however, a definite need for those transport processes involved in maintaining the normal volume of red blood cells, and thus a brief discussion of such processes seems appropriate here.

Under physiological conditions, there is a driving force favoring a net flux of cations into the cells, due ultimately to the presence of impermeant, negatively charged molecules (e.g., hemoglobin, organic phosphates) within the cytoplasm. Unopposed, this would cause cells to swell, and possibly lyse, in their own plasma. However, in red blood cells of most species, the passive net influx of salt and water is counterbalanced by the action of the ATP-dependent Na/K pump, which transports three Na ions outward for each two K ions transported inward [2]. This is essentially the basis of the "pump-leak" hypothesis of Tosteson and Hoffman [3], who demonstrated in sheep red blood cells that the net cation uptake via diffusion is fully offset by net cation efflux via the Na/K pump. Of course, it is now known that all passive cation movements in red blood cells do not occur by simple diffusion, and thus the quantitative agreement between diffusional influx and pump efflux found in the sheep red blood cell [3], which under physiological conditions exhibits very little passive ion movement via specific mediated pathways, would not apply in red blood cells of many other species. For example, Duhm and Göbel [4] showed in human red blood cells that net salt efflux via (Na + K + Cl) cotransport appears to supplement the action of the Na/K pump in counterbalancing diffusional cation influx and maintaining normal cell volume. Still, the qualitative principle of the "pump-leak" hypothesis seems to apply to all red blood cells, even those of dogs, which completely lack Na/K pumps. In dog red blood cells, Na ions moving passively into the cells are exchanged for extracellular Ca ions via a Na/Ca exchange mechanism; the Ca is then extruded from the cells via an ATP-dependent Ca pump [5].

A. Pathways Activated by Cell Swelling

It has long been known that cell swelling increases potassium fluxes in red blood cells. Ponder and colleagues [6,7] and Davson [8] found that red blood cells of several species became more permeable to potassium when swollen, and Ponder [7] postulated that this potassium loss in hypotonic media was at least one reason why red blood cells fail to behave as perfect osmometers. In the last decade, it has been shown that the potassium loss from red blood cells incubated in hypotonic solutions occurs via specific, ouabain-insensitive pathways. In most species of red blood cells in which the kinetic mechanism of swelling-activated K transport has been elucidated, one of two types of pathways appears to mediate this transport: (K + Cl) cotransport or K/H exchange, with the former being more widespread. Other pathways for K transport that are activated by cell swelling have been described in other cell types; for example swelling of lymphocytes activates parallel K and Cl conductance pathways [9]. Red blood cells from most (although not all) species also have a K conductance pathway activated by increased intracellular ionized calcium ($[Ca^{2+}]_c$). However, to date activation of this pathway in a red blood cell by changes in cell volume or other physiological stimuli has not been demonstrated.

(K + Cl) Cotransport

This pathway is characterized by several features: (a) a saturable dependence on extracellular potassium ($[K]_o$), usually with low affinity ($K_{1/2}$ = 20-50 mM); (b) a strict requirement for chloride or bromide; and (c) inhibition by "loop" diuretics such as furosemide and bumetanide, although this usually requires high concentrations (0.1-1.0 mM) of these drugs. In fact, some workers (e.g., [10]) have somewhat misleadingly termed fluxes mediated by (K + Cl) cotransport "bumetanide-insensitive" based on minimal inhibition by 0.01 mM bumetanide, a concentration that fully inhibits (Na + K + Cl) cotransport [11,12]. For the purpose of this chapter, I will refer to K fluxes that satisfy the above three criteria as (K + Cl) cotransport, although this may be premature because in only a few cases (e.g., [13]) has such a transport process actually been shown to be electrically neutral, and even in these cases direct coupling of K and Cl fluxes has not been proven. (K + Cl) cotransport has been found to be stimulated by swelling of red blood cells from humans [14], sheep [15,16], dogs [17], fish [18,19], and birds [20,21], although the properties of these pathways do differ between red blood cells of different species with respect to the degree of "loop" diuretic sensitivity and affinity for $[K]_o$. In addition, DIDS, an inhibitor of the red blood cell anion exchange pathway, completely inhibits (K + Cl) cotransport in swollen fish red blood cells at 0.1 mM [18].

In high-Na, low-K hypotonic media, (K + Cl) cotransport mediates a net efflux of KCl (and osmotically obliged water) from the cells, resulting in a "volume regulatory decrease." Thus, swollen red blood cells from fish and birds, which exhibit relatively large ion movements via (K + Cl) cotransport, return to their normal, isotonic volume in 30-120 min [19-21]. However, in human red blood cells, the magnitude of (K + Cl) cotransport activated by cell swelling is very small. Furthermore, this response is transient and mainly involves only a subpopulation of the cells (the least dense, or youngest cells; see below). Thus, no detectable "volume regulatory decrease" is observed after hypotonic swelling of unfractionated human red cells [10,22].

K/H Exchange

In Amphiuma red blood cells, swelling activates a 1:1, electrically neutral K/H exchange [23,24]. In parallel with the red blood cell anion exchange pathway (band 3), which mediates a Cl/HCO_3 exchange, activation of the K/H exchange pathway by cell swelling results in a net KCl and water loss and a return of the cells to their isotonic volume ("volume regulatory decrease"). The K/H exchange pathway in swollen Amphiuma red blood cells is not inhibited by "loop" diuretics. DIDS partially inhibits the "volume regulatory decrease," although presumably by blocking the secondary Cl/HCO_3 exchange; the continued operation of the K/H exchanger then produces an increasing proton gradient, which makes this exchange less energetically favorable. The K/H exchange of Amphiuma red blood cells is also partially inhibited when Cl is replaced by NO_3 or SCN (although this inhibitory effect is less pronounced than with Na/H exchange; [25]). However, as is discussed below for Na/H exchange in shrunken cells, this effect appears not to be directly related to an anion transport site [as with (Na + K + Cl) and (K + Cl) cotransport], but rather to events involved in the activation of the transport system.

Other Swelling-Activated Transport Pathways

When flounder red blood cells are swollen in a low-K, high-Na hypotonic medium, the cells lose K (but not Na) and water and shrink back toward their isotonic volume [26]. However, in contrast to cells that accomplish this "volume regulatory decrease" via (K + Cl) cotransport or K/H exchange, the Cl content of flounder red cells does not decrease concurrently with the K content during this process [26]. Thus, some anion other than Cl must exit the cells to maintain electroneutrality of transport. The most likely candidates for such anions appear to be negatively charged amino acids as Fugelli [27] has shown that during "volume regulatory decrease" by flounder red blood cells the content of ninhydrin-positive substances in the cells decreases by 8–16%.

Poznansky and Solomon [28] reported that a 10% increase in human red blood cell volume increased both unidirectional Na and K effluxes by ∼10%, and decreased the influxes of these cations by ∼25%. However, subsequent studies have not confirmed these "linked" cation fluxes; cell swelling appears to increase both unidirectional K influx and efflux in human red blood cells [14, 22], and decrease passive Na efflux in these cells [29].

In dog red blood cells, swelling stimulates not only (K + Cl) cotransport but Na/Ca exchange as well [5,30]. As noted earlier, this Na/Ca exchange system appears to participate in the normal volume maintenance of these cells, which have Ca pumps but lack Na/K pumps. In contrast to the situation noted above with (K + Cl) cotransport and K/H exchange, Na/Ca exchange in swollen dog red blood cells is actually stimulated when Cl is replaced by NO_3, and even more when SCN is the predominant anion [30]. Since the relative conductances of these anions are $SCN > NO_3 > Cl$, the above finding is consistent with the Na/Ca exchange being electrogenic, and the rate of net cation transport via this exchange being limited by the conductance of the anion which must follow in the direction of net cation transport to maintain electroneutrality.

B. Pathways Activated by Cell Shrinkage

Ion transport pathways activated by cell shrinkage are present in red blood cells from a variety of species, although these pathways do not appear to be as widespread as those activated by cell swelling. As with the swelling-induced processes, however, there appears to be a central theme with two major variations. The central theme is Na (or more correctly NaCl and water) uptake by the cells, and the two major pathways effecting this uptake are (Na + K + Cl) cotransport and Na/H exchange. Both pathways are ouabain-insensitive, and transport Na down its gradient into the cells. In the case of (Na + K + Cl) cotransport, the Cl gradient also contributes largely to the driving force for net inward transport.

(Na + K + Cl) Cotransport

The (Na + K + Cl) is actually a generic term, since in actuality this pathway is electrically neutral, and transports equal amounts of cations and Cl. In duck red blood cells, a stoichiometry of 1Na:1K:2Cl has been demonstrated [31], and this is in agreement with stoichiometries found in most cells, including Ehrlich ascites tumor cells [32] and the epithelial cells of the thick ascending limb of Henle's loop of mammalian kidney [33]. However, in ferret red blood cells a Na:K influx stoichiometry of 2:1 has been reported [34]. Because this stoichiometry was obtained in tracer flux experiments with Na and K present on both sides of the membrane, it is possible that 2Na:1K is not the true net flux stoichiometry, because the (Na + K + Cl) cotransport pathway appears to mediate obligatory 1:1 Na/Na and K/K exchanges ("partial reactions") as well as net transport [35]. Because ferret red blood cells, which lack Na/K pumps, are high in Na, Na/Na exchange would be expected to be the predominant partial reaction in these cells [35], and the observed 2Na:1K stoichiometry could represent a combination of both net cotransport with 1Na:1K:2Cl stoichiometry and Na/Na (or NaCl/NaCl) exchange. However, it should be noted that in squid axon, Russell [36] has demonstrated a cotransport influx stoichiometry of 2Na:1K:3Cl in axons internally dialyzed with solutions containing low [Na], thus minimizing the possibility of significant Na/Na exchange. Therefore, 2Na:1K:3Cl is probably the true stoichiometry for (Na + K + Cl) cotransport in squid axon.

(Na + K + Cl) cotransport in all tissues is characterized by a high affinity for bumetanide. In turkey and duck red blood cells incubated in media containing physiological levels of Na, K, and Cl, (Na + K + Cl) cotransport is fully inhibited by 0.01 mM bumetanide [11,12,37], a concentration nearly two orders of magnitude lower than the concentration of furosemide required to fully inhibit this pathway [11,38]. Bumetanide inhibition of (Na + K + Cl) cotransport is, however, dependent on the ionic composition of the extracellular medium. If Na or K is completely removed from the medium, bumetanide becomes a far less potent inhibitor of (Na + K + Cl) cotransport efflux [39], consistent with the finding that the presence of both of these cations is required for maximal saturable binding of [3H]bumetanide to duck red blood cells [40] and dog kidney membranes [41]. Low levels of Cl are also required for saturable [3H]bumetanide binding [40,41], suggesting the stable inhibited conformation of the cotransporter is (Na + K + Cl + bumetanide). Higher concentrations of Cl inhibit [^{3}H]bumetanide binding [40,41], consistent with data from ion flux experiments in duck red blood cells, which indicate that Cl and bumetanide compete for a common site [12].

Although (Na + K + Cl) cotransport and its activation by cell shrinkage has been described in greatest detail in avian red blood cells, it is also present in red blood cells from several mammalian species. As noted above, ferret red blood cells have high rates of (Na + K + Cl) cotransport [34,42], and cell shrinkage increases this rate up to threefold [43]. Cell shrinkage also stimulates (Na + K + Cl) cotransport in rat red blood cells [44]; when shrunken the cotransport rate in these cells ($\sim$10-15 mmol Na or K/kg cell solid $\times$ hr) is much lower than in duck or ferret red blood cells (>100 mmol/kg cell solid $\times$ hr), but higher than in human red cells ($\sim$2 mmol/kg cell solid $\times$ hr; [45]). Furthermore, in human red blood cells there is no evidence for activation of (Na + K + Cl) cotransport by cell shrinkage (e.g., [29]).

Na/H Exchange

Cell shrinkage stimulates a 1:1, electrically neutral Na/H exchange in Amphiuma [23,24,46,47], dog [48], rabbit [49], and frog [50] red blood cells. This exchange can be reversibly inhibited by the diuretic agent amiloride with an IC$_{50}$ in the 1-10 μM range, and complete inhibition at 0.5-1.0 mM [46,48, 49]. As is the case with Na/H exchange in some epithelia (e.g., [51]), amiloride displays some competitive behavior with respect to Na [46]. In concert with rapid Cl/HCO$_3$ exchange via band 3, activation of Na/H exchange by shrinkage of Amphiuma red blood cells produces a net influx of NaCl and water that returns the cells to their isotonic volume within 1-2 hr [23]. DIDS and SITS, inhibitors of anion exchange, do not directly inhibit Na/H exchange, but do slow the volume recovery ("volume regulatory increase") of shrunken Amphiuma red cells [23,47], presumably because in the presence of these inhibitors Na/H exchange generates a proton gradient that makes net Na uptake via Na/H exchange less energetically favorable.

Na/H exchange in shrunken Amphiuma [25] and dog [48] red blood cells is substantially inhibited when intra- and extracellular Cl is replaced by the permeant anions NO$_3$ and SCN. Because this effect of these anion substitutions on Na/H exchange is also seen in DIDS-treated cells [25], it is unlikely that this effect is related to the Cl/HCO$_3$ exchange, which operates in parallel with Na/H exchange during "volume regulatory increase." Rather, the effect appears unrelated to the transport of these anions, based on the experiments of Parker [52] in which the Na/H exchange of dog red blood cells was fixed in the "on" conformation by treating shrunken cells with glutaraldehyde. The presence of Cl was required for fixation of the exchanger in the "on" conformation, however,

once fixed Na/H exchange was no longer inhibited when Cl was replaced by
SCN [52]. This suggests that the role of the anion may be in the regulation of
Na/H exchange, rather than in the transport process itself. More recent findings
that replacement of Cl by NO_3 or SCN leads to alkaline shifts in the intracellular
pH of Amphiuma red cells [53] are consistent with this idea (see below).

An important feature of Na/H exchange, first described in renal plasma membrane vesicles [54], is that intracellular protons modify the transport. Lithium,
which is also a substrate for Na/H exchange, has a similar effect [55]. Increasing
intracellular proton concentration (decreasing pH_c) activates Na/H exchange,
and in lymphocytes hypertonic cell shrinkage appears to lower the level of $[H]_c$
required to produce an activation of Na/H exchange [56]. As discussed below,
this is one possible mechanism by which cell shrinkage may activate Na/H exchange in red blood cells of various species. Another interesting phenomenon
possibly related to the regulation of Na/H exchange is the finding in shrunken
Amphiuma red blood cells that pretreatment of the cells with amiloride results
in a marked augmentation of Na/H exchange when the cells are washed free of
the drug and resuspended in an otherwise identical hypertonic medium [46].
This suggests the possibility that Na/H exchangers may become inserted into
the plasma membrane when the cells are shrunken, or that a substance that
activates (or relieves an inhibition of) Na/H exchange accumulates in the cytoplasm when the cells are shrunken.The finding that there is a several-minute
delay after initial cell shrinkage before Na/H exchange is maximally activated
[23,47] appears consistent with either of these hypotheses. Presumably, the
operation of the Na/H exchanger (with resultant changes in cell volume, pH_c,
and cell Na content) might then produce a feedback inhibition of the putative
activating process, thus accounting for the augmentation of transport seen when
cells are preincubated in hypertonic media under conditions where Na/H exchange cannot occur (e.g., with amiloride present).

Other Shrinkage-Induced Ion Transport Pathways

In human red blood cells, Poznansky and Solomon [28] reported that a 10%
cell shrinkage stimulated Na and K influxes 15-20%, and inhibited effluxes of
both of these cations by ~25%. Because the changes in Na and K fluxes in this
study are equivalent, the possibility of coupling of these cation fluxes arises.
However, ouabain was found to inhibit the increase in K influx, but not the increase in Na influx, in shrunken human red blood cells [57]. By contrast,
Adragna and Tosteson [29] found that shrinkage of human red blood cells increased ouabain-insensitive (but not furosemide-sensitive) Na efflux, and had
minimal effect on ouabain-insensitive K efflux. Shrinkage of human red blood
cells was also found to have no significant effect on ouabain-insensitive K influx in a recent study [14]. When flounder red blood cells are shrunken in a
hypertonic medium, the cells take up NaCl and swell back toward their isotonic

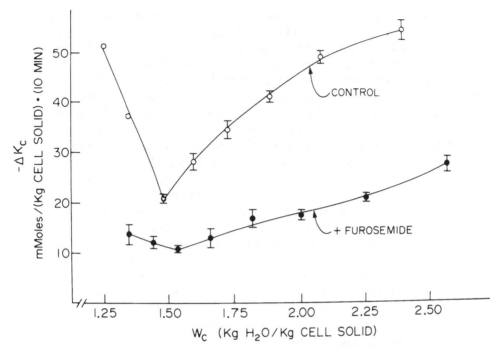

Figure 1 Effects of furosemide (1 mM) and cell volume (water content, W_C) on net potassium efflux from duck red cells into sodium- and potassium-free media. Cells were preincubated overnight at 2°C in a high-sodium medium to raise $[Na]_c$ to ~60 mM and thus augment outward (Na + K + Cl) cotransport. Test incubations were performed for 10 minutes at 3% hematocrit, 41°C in media containing 30 mM TMA-TES buffer (pH 7.4 at 41°C), 10 mM glucose, 0.1 mM ouabain, and sufficient TMA chloride to adjust the osmolality to between 170 and 380 mosmol/kg, in increments of 30 mosmol/kg. The cells' average water contents over the 10 minute incubation period are shown on the abscissa. Results shown are means of three separate determinations of potassium efflux ($-\Delta K_C$) in each medium, ± SEM (error bars).

volume [26]. Although additional details of this process are not known, Na/H exchange might be involved, because it is present in red blood cells of other fish species (see below).

C. Relationship Between Shrinkage- and Swelling-Induced Transport Pathways

Several pieces of indirect evidence suggest that (Na + K + Cl) cotransport and (K + Cl) cotransport may be related, perhaps as alternate modes of the same, highly complex transport pathway: (a) Both have a specific anion requirement satisfied only by Cl or Br (e.g., [58]). (b) Both are inhibited by "loop" diuretics, such as furosemide and bumetanide. Figure 1 illustrates an experiment in which K efflux from duck red blood cells was measured as a function of cell water content (1.5 kg $H_2 O$/kg cell solid = isotonic water content of these cells) by incubating them in Na- and K-free tetramethylammonium (TMA) chloride media of different osmolalities. Furosemide (1 mM) inhibited a significant fraction of the K efflux in cells of normal volume, as well as essentially all of the efflux stimulated by cell shrinkage ($W_c < 1.5$) and most of the K efflux stimulated by cell swelling ($W_c > 1.5$). However, other studies have shown that the IC_{50} for bumetanide inhibition of (Na + K + Cl) cotransport is generally 1-2 orders of magnitude lower than that for bumetanide inhibition of (K + Cl) cotransport [11,12,14]. (c) In human red blood cells, cell swelling and exposure to N-ethyl maleimide (NEM), conditions that stimulate (K + Cl) cotransport [29,59]; also see below), also reduce furosemide-sensitive Na fluxes (and thus apparently (Na + K + Cl) cotransport). These findings have led to speculation that swelling and NEM treatment might "uncouple" the (Na + K + Cl) cotransporter into a (K + Cl) and a presumably inactive (Na + Cl) form [59]. (d) Finally, exposure of swollen duck red blood cells to β-adrenergic catecholamines or cyclic AMP (cAMP) both stimulates (Na + K + Cl) cotransport and abolishes, or overrides, (K + Cl) cotransport [20]; also see below). However, although all of the above findings are consistent with the idea of a single pathway that under the influence of different physiological or chemical stimuli can mediate either (Na + K + Cl) or (K + Cl) cotransport, there remains the definite possibility that these two are in fact similar but separate pathways that happen to be oppositely affected by the above-mentioned stimuli.

More direct evidence exists linking the Na/H and K/H exchange processes in Amphiuma red blood cells. Normally, the swelling-induced K/H exchange in these cells is not inhibited by amiloride, whereas the shrinkage-activated Na/H exchange is amiloride-sensitive (see above). However, Cala [24] found that when cells were first shrunken in hypertonic media containing amiloride, then transferred to hypotonic media also containing amiloride, K/H exchange was blocked. Cells preincubated in amiloride-free hypertonic media or amiloride-containing isotonic media and then transferred to hypotonic media containing amiloride showed a normal K/H exchange response [24]. Thus, it appears that amiloride bound to the Na/H exchange pathway in shrunken cells could conceivably remain bound when the cells are subsequently swollen, and this volume-sensitive pathway converts to a K/H exchange mode.

D. Cellular Mechanisms Underlying Volume-Sensitive Ion Transport

Perhaps the most important general questions facing the field of cell volume regulation are: (a) How does the cell sense its volume? and (b) Once a change in

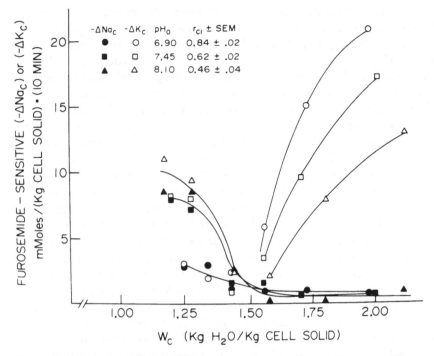

Figure 2 Effects of pH and cell volume on furosemide-sensitive net sodium and potassium effluxes from duck red cells into sodium- and potassium-free media. Fresh cells were washed three times in ice-cold, isotonic TMA chloride, then incubated for 10 minutes at 3% hematocrit, 41°C, in the presence and absence of 1 mM furosemide, in media containing 30 mM TMA-TES buffer, 10 mM glucose, 0.1 mM ouabain, and varying amounts of TMA chloride. The ranges of osmolalities over which the cells were incubated were (mosmol/kg); 240–440 (pH 6.90), 200–400 (pH 7.45), and 160–360 (pH 8.10), all in increments of 40 mosmol/kg. The cells' average water contents over the 10 minute incubation period (W_c) are plotted on the abscissa. Open symbols represent furosemide-sensitive net potassium efflux ($-\Delta K_c$), closed symbols represent furosemide-sensitive sodium efflux ($-\Delta Na_c$).

volume is sensed, what are the mechanisms by which the cell activates an appropriate ion transport response? From what we do know, the answers to these questions may be different in different cell types, and even in red blood cells from different species. In red blood cells, a change in cell volume results in a change in the concentration of fixed intracellular anions, A^- (e.g., hemoglobin, organic phosphates), and a resultant chloride shift into or out of the cell. This Cl shift is accompanied by a change in membrane potential, E_m (because E_m

approximates E_{Cl} in red blood cells), and a change in pH_c, through the action of the anion exchange pathway. This pathway maintains the ratio of $[Cl]_c/[Cl]_o$ (the Cl ratio, or r_{Cl}) equal to $[OH]_c/[OH]_o$, or $[H]_o/[H]_c$. Cell shrinkage increases $[A^-]_c$, causing a chloride shift out of the cells and a decrease in pH_c. Conversely, cell swelling increases pH_c. Because a change in red blood cell volume results in a concomitant change in E_m and pH_c, any one of these three could be the parameter used by the cell to sense its volume. In the duck red blood cell, volume itself appears to be sensed, based on experiments such as that shown in Figure 2. In this experiment, duck red blood cells were incubated in Na- and K-free TMA Cl media of varying osmolality and pH. The furosemide-sensitive effluxes of Na and K are plotted as a function of cell water content, W_c, which reflects cell volume. As discussed above, (K + Cl) cotransport (net efflux of K without a concomitant Na efflux) is stimulated in swollen cells ($W_c > 1.5$ kg H_2O/kg cell solid) and (Na + K + Cl) cotransport with Na:K stoichiometry of 1:1 is activated in shrunken cells [$W_c < 1.5$; note that (Na + K + Cl) cotransport is inhibited at acid pH]. However, regardless of medium pH, the cells show a transport minimum at $W_c = 1.5$ kg H_2O/kg cell solid, even though this corresponds to a different level of medium osmolality, r_{Cl}, E_m, and pH_c in each of the three cases. Thus, there appears to be a specific volume at which duck red blood cells sense they are neither swollen or shrunken. However, as discussed below, this apparent direct sensing of cell volume may not be the mechanism by which red blood cells from other species sense their volume. Furthermore, still other cellular mediators are likely to be involved in initiating the response to volume changes.

The Role of Cell pH

As noted above, intracellular protons play an important role in regulating Na/H exchange in red blood cells [48,49,55] as well as in lymphocytes [56] and epithelial cells [54]. Specifically, decreasing pH_c activates Na/H exchange. For example, in rabbit red blood cells of normal volume, the Na/H exchanger is inactive when cells are incubated in media of pH 7.4 (with pH_c thus being ~ 7.2). Lowering medium pH to 7.0 (and thus pH_c to ~ 6.8 via the action of the anion exchanger) activates Na/H exchange, and this exchange is maximally activated in media of pH ~ 6.5 [49]. Cell shrinkage not only causes a decrease in pH_c by the mechanism described above, but also changes the "set point" for stimulation of Na/H exchange by internal protons [48,49,55,56]. In shrunken rabbit red blood cells, Na/H exchange is easily detectable at physiological pH_c, and appears to be maximal at $pH_c \sim 6.7$-6.8 [49].

Cala [23] has likewise found that the magnitude of Na/H exchange in shrunken Amphiuma red blood cells is greater at acid than at alkaline pH, and found that Na/H exchange is even activated in hypotonic media if the pH is lowered to 6.65. More interestingly, when Amphiuma red blood cells were shrunken in media (pH 7.4) made hypertonic by the addition of the Na salt of the impermeant anion p-aminohippurate (PAH), the Na/H exchange response was markedly attenuated compared with the response in an all-Cl hypertonic medium [23]. In the PAH-containing medium, the impermeant PAH balanced the increase in fixed intracellular anion concentration resulting from cell shrinkage, so that no Cl shift or resultant decrease in pH_C occurred. This suggests that in Amphiuma red blood cells, the cell may sense its being shrunken by detecting a lowering of pH_C rather than a decrease in volume itself. It is not clear whether changes in pH_C play a similar role with regard to activation of K/H exchange in swollen Amphiuma red blood cells; however, this possibility is suggested by the finding that in cells swollen in a 183 mOsmol/kg medium, lowering external pH from 7.65 to 6.65 decreased K efflux by ~35% [23].

The Role of Divalent Cations

The role of calcium in the activation of Na/H and K/H exchanges in Amphiuma red blood cells has been extensively studied by Cala and co-workers [24,60]. Shrinkage of Amphiuma red blood cells results in up to a threefold decrease in the concentration of ionized free Ca in the cytoplasm ($[Ca^{2+}]_c$), whereas cell swelling causes a pronounced increase in $[Ca^{2+}]_c$ [60]. This latter increase in $[Ca^{2+}]_c$ appears to be involved in the activation of K/H exchange, because this transport process is also stimulated when cells are incubated in isotonic media containing calcium plus the ionophore A23187 [24]. In addition, cell swelling may lower the level of $[Ca^{2+}]_c$ required to activate K/H exchange [60]. Inhibitors of Ca–calmodulin-mediated processes, such as quinidine and phenothiazine compounds, inhibit K/H exchange in Amphiuma red blood cells, whether the stimulus for this exchange is cell swelling or Ca plus A23187 [60]. Furthermore, when shrunken cells are exposed to phenothiazines, the stimulation of Na/H exchange is augmented, suggesting that the decrease in $[Ca^{2+}]_c$ accompanying cell shrinkage may play a role in the activation of Na/H exchange [60].

Although a decrease in $[Ca^{2+}]_c$ may activate Na/H exchange in Amphiuma red blood cells, an increase in $[Ca^{2+}]_c$ activates amiloride-sensitive Na influx and efflux in human red blood cells [61]; these fluxes may represent Na/H exchange, which has been reported in human red blood cells [62]. However, it is not known if this pathway, which mediates only very small fluxes under physiological conditions, is affected by cell volume. Na/H exchange activated by shrinkage of gallbladder epithelial cells is inhibited by quinidine and phenothiazines [63]; these effects may be related to events involving microfilaments (see below).

Increased levels of $[Ca^{2+}]_c$ produced by incubation of cells with A23187 and Ca were reported to markedly inhibit (Na + K + Cl) cotransport in human red blood cells [64], and increased levels of both extra- and intracellular Mg were found to inhibit bumetanide-sensitive and Cl-dependent Na and K fluxes in these cells [65]. In ferret red blood cells, high levels of $[Ca^{2+}]_c$ inhibit (Na + K + Cl) cotransport, although a threefold increase over the physiological level of $[Ca^{2+}]_c$ has little effect on ion fluxes via this pathway [66]. High levels of $[Mg^{2+}]_c$ also inhibit (Na + K + Cl) cotransport in ferret red blood cells; however, increasing $[Mg^{2+}]_c$ in the physiological range (0.1-1.0 mM) increases cotransport activity in these cells [66]. Again, the relationship of these effects of divalent cations to effects of changes in cell volume is not known. However, the apparent stimulatory effect of increasing $[Mg^{2+}]_c$ in the physiological range on (Na + K + Cl) cotransport could perhaps explain the stimulatory effect of hypoxia on this pathway in avian red blood cells (see below).

The Role of the Membrane Skeleton

Red blood cells, and many other types of cells as well, possess a network of proteins associated with the cytoplasmic face of the plasma membrane, termed the cytoskeleton or membrane skeleton. The proteins comprising this network have been extensively characterized in the human red blood cell, and are the subject of several chapters in this volume. As discussed in Chapter 4, the membrane skeleton of human red blood cells plays a role in maintaining the normal shape and deformability of the cell, and defects in one or more of the proteins comprising the membrane skeleton can result in hemolytic anemias. Because of its association with the plasma membrane (including certain integral membrane proteins, such as band 3), and because it is a dynamic structure (i.e., specific cytoskeletal proteins such as spectrin and ankyrin associate with each other, and this association is influenced by their interaction with the cytoplasmic domain of band 3; [67]), the membrane skeleton would appear to be an ideal means for sensing cell volume. Unfortunately, studies directly showing a role for the membrane skeleton in the activation of volume-sensitive ion transport pathways have not yet been performed. This may be largely because experimental manipulation of the red blood cell membrane skeleton must be done in ghost or vesicle preparations, and demonstration of most volume-sensitive transport processes in red blood cell ghosts or vesicles has not yet been accomplished. Furthermore, the process of preparing ghosts (which involves hypotonic lysis) may itself activate certain swelling-induced transport processes; human red blood cell ghosts were found to have approximately threefold higher levels of furosemide-sensitive K influx than intact cells, and this additional K transport appears to represent (K + Cl) cotransport [68].

One study in Necturus gallbladder epithelium has addressed the possible role of microfilaments in the activation of volume-sensitive ion transport pathways. Foskett and Spring [63] found that the bumetanide-sensitive loss of K and Cl from swollen cells was inhibited when the cells were treated with cytochalasin B to disrupt microfilaments. Colchicine, which disrupts microtubules, was without effect. Thus, actin microfilaments may be involved in sensing a volume increase in gallbladder epithelial cells and/or transmitting this information to ultimately activate a bumetanide-sensitive K transport pathway [possible (K + Cl) cotransport].

The Role of Membrane Sulfhydryl Groups

It has been recognized for more than two decades that agents that interact with membrane sulfhydryl groups alter the cation permeability of red blood cells (e.g., [69]). However, only more recently has it been found that these agents affect specific transport pathways, which in some cases appear to be the same ones that are affected by changes in cell volume. In LK sheep (for review, see [70]) and human [59,71,72] red blood cells, NEM, an alkylating agent that forms stable addition products with sulfhydryl groups, stimulates a (K + Cl) cotransport that appears kinetically identical to that stimulated by swelling of these cells. NEM-stimulated (K + Cl) cotransport in LK sheep red blood cells is also inhibited by treatment of the cells with anti-L_1 [16,73], an antibody that inhibits swelling-induced (K + Cl) cotransport in these cells to an equivalent extent [16]. In addition, as noted above, both swelling and NEM treatment of human red blood cells have been reported to inhibit (Na + K + Cl) cotransport [29,59]. Thus, it appears that membrane sulfhydryl groups may play a role in the regulation of certain ion transport processes by changes in cell volume, and that the effect of swelling could be mimicked (to a maximal extent) in cells of normal volume by directly modifying these sulfhydryl groups with NEM. In support of this idea is the finding that the effects of cell swelling and NEM treatment on (K + Cl) cotransport in sheep red blood cells are not additive; Cl-dependent K efflux in NEM-treated cells was found to be maximal and essentially volume-dependent [16]. In addition, depletion of intracellular divalent cations by incubation in a medium containing A23187 plus EGTA amplifies swelling-induced (K + Cl) cotransport in sheep red blood cells (to approximately the magnitude seen in NEM-treated cells), but does not increase (K + Cl) cotransport in NEM-treated cells [74].

Whether membrane sulfhydryl groups are involved in the sensing of cell volume, the activation of ion transport, or directly in mediating the transport process itself is not known, although some insight into this may be derived from studies using sulfhydryl cross-linking agents. Parker [75] found that treating swollen dog red blood cells with diamide, a thiol oxidizer, locks the swelling-induced Na/Ca exchange in these cells in an "on" conformation. Because this action of diamide is influenced by Na and Ca concentrations, it is likely that the sulfhydryl groups involved may be closely related to the transporter itself, and not simply involved in the sensing of cell volume. As noted above, shrinkage-induced Na/H exchange in dog red blood cells can be locked "on" or "off," respectively, if shrunken or swollen cells are treated with glutaraldehyde [52] or with mono- or bifunctional maleimides [75]. Because the ability of glutaraldehyde to lock "on" the Na/H exchanger is unaffected by the presence of the transport inhibitor amiloride, but is blocked when Cl is replaced by SCN ([52; as discussed above, the inhibitory effect of replacing Cl with SCN or NO_3 on Na/H exchange appears to be related to the activation of the pathway, and not to the transport itself), it appears that sulfhydryl groups involved in this effect are probably not part of the transporter itself, but rather are related to the sensing of cell volume and/or subsequent events which occur before transport activation.

III. CATECHOLAMINE-STIMULATED ION TRANSPORT

An effect of catecholamine hormones on potassium transport in pigeon and frog red blood cells was first noted >20 years ago [76]. Later studies have shown that catecholamines exert marked effects on ion transport in red blood cells of birds, fish, and amphibians through interaction with β-adrenergic receptors, and that the effects of these hormones on ion transport occur via a cAMP-dependent mechanism. These studies have also shown that two major transport pathways are stimulated by β-adrenergic agonists: (Na + K + Cl) cotransport in avian red blood cells, and Na/H exchange in fish and amphibian red blood cells. In stimulating Na/H exchange, catecholamines appear to play a role in maintaining red blood cell volume in vivo. In addition, other ion transport pathways are influenced directly or indirectly by catecholamines, most notably the swelling-induced (K + Cl) cotransport pathway in both avian and fish red blood cells.

A. Avian Red Blood Cells

Addition of norepinephrine, other β-adrenergic catecholamines, or cAMP (or one of its analogs) to duck [39] or turkey [37] red blood cells stimulates a cotransport of (Na + K + Cl) with kinetic properties identical to the process activated by shrinkage of these cells (see above). Kregenow [77] initially proposed a volume-regulatory function for this catecholamine-stimulated transport, based on the finding that in the presence of elevated $[K]_o$, addition of norepinephrine to duck red cells caused them to swell to a new, higher steady-state volume. However, it was later found that in these cells, under physiological conditions, the chemical potential gradients of the transported ions are balanced so that no significant net transport results from activation of (Na + K + Cl) cotransport [20,38,39].

Still, catecholamines do appear to influence duck red blood cell volume in vivo, as first noted by Riddick and co-workers [78]. They found that when cells are incubated in their own plasma, they maintain a constant volume, but on addition of propranolol (a β-adrenergic antagonist) they lose K and Cl (but not Na) and shrink to a new, lower steady-state (LSS) volume. This fresh-LSS transition

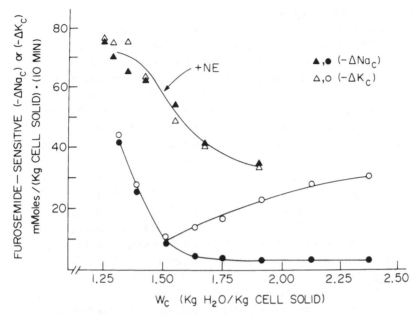

Figure 3 Effects of cell volume and norepinephrine (NE) on furosemide-sensitive net sodium and potassium effluxes from duck red cells into sodium- and potassium-free media. In this experiment, cells were preincubated to raise $[Na]_c$ to ~60 mM, though qualitatively similar results are obtained with cells of normal $[Na]_c$. Test incubations were performed for 10 minutes at 3% hematocrit, 41°C in media containing 30 mM TMA-TES (pH 7.4 at 41°C), 10 mM glucose, 0.1 mM ouabain, and sufficient TMA chloride to adjust the osmolality to between 200 and 410 mosmol/kg, in increments of 30 mosmol/kg. All test incubations were done in the presence and absence of 10^{-6} M NE and 1 mM furosemide. The cells' average water contents (W_c) over the 10 minute incubation period are plotted on the abscissa. Open symbols represent furosemide-sensitive net potassium efflux ($-\Delta K_c$), closed symbols represent furosemide-sensitive net sodium efflux ($-\Delta Na_c$). Reproduced from *The Journal of General Physiology*, 1985, volume 85, pp. 649–667, by copyright permission of The Rockefeller University Press.

also occurs if freshly drawn cells are incubated in an isotonic, catecholamine-free synthetic medium with an ionic composition similar to that of plasma, but not in an identical synthetic medium containing norepinephrine [78]. More recently, it was shown that the fresh-LSS transition of duck red blood cells involves the same (K + Cl) cotransport pathway that is stimulated when cells are hypotonically swollen; indeed, freshly drawn cells ($W_c \cong 1.6$ kg H_2O/kg cell solid) appear to behave as if they are mildly swollen [20]. Furthermore, addition of norepinephrine to hypotonically swollen cell blocks (K + Cl) cotransport and the resultant "volume regulatory decrease" just as it blocks the fresh LSS transition. The ionic basis for this effect is shown in Figure 3. In this experiment, duck red blood cells (with $[Na]_c$ raised to 60 mM to maximize outward (Na + K + Cl) cotransport; see legend) were incubated in Na- and K-free TMA Cl media of varying osmolalities, with or without 10^{-6} M norepinephrine (NE) and 1 mM furosemide. As in Figure 2, the furosemide-sensitive net effluxes of Na and K are plotted as a function of cell water content. In the absence of NE, swollen cells (W_C > 1.5 kg H_2O/kg cell solid) lose K and not Na (K + Cl cotransport), whereas shrunken cells (W_C < 1.5) show (Na + K + Cl) cotransport with Na:K stoichiometry of 1:1. By contrast, with NE present only (Na + K + Cl) cotransport is observed at all cell volumes, and the (K + Cl) cotransport seen in swollen cells in the absence of the catecholamine is abolished, or overridden, in its presence [20] (Figure 3). The finding that cell swelling causes a decrease in the level of norepinephrine-stimulated (Na + K + Cl) cotransport (Figure 3) is almost certainly not due entirely to lowering of cellular Na, K, and Cl concentrations (though this may contribute somewhat to this decrease), nor is it related to a swelling-induced inhibition of adenylate cyclase, because we observe essentially identical results when 8-Br-cAMP (1 mM, plus a phosphodiesterase inhibitor) is used in lieu of norepinephrine [20]. This decrease in cotransport could be related to other cellular events involved in catecholamine activation of cotransport (see below), or possibly to a direct inhibitory effect of swelling on this pathway (as has been noted for human red blood cell (Na + K + Cl) cotransport; [29]).

B. Fish and Amphibian Red Blood Cells

Exposure of trout red cells to β-adrenergic agonists causes them to swell, and pH_C to transiently rise [79]. More recent studies [80-82] have shown that these effects are due to the stimulation of a 1:1 Na/H exchange, coupled with the action of the anion exchanger. This Na/H exchanger shares two important properties with the Na/H exchangers activated by shrinkage of dog and Amphiuma red blood cells and lymphocytes: it is amiloride-sensitive [80] and it is stimulated by internal protons [83]. Catecholamine-treated trout red blood cells also develop increased Cl-dependent K fluxes that are inhibited by 1 mM furosemide or bumetanide and appear to represent (K + Cl) cotransport [19,84]. However, this (K + Cl) cotransport does not appear to be directly stimulated by catecholamines, but rather appears secondary to cell swelling resulting from Na/H exchange; it is not observed when β-adrenergic agonists are added to trout red blood cells in the presence of amiloride [85], and it is markedly reduced when external Na is replaced by choline [84]. This conclusion is also consistent with the finding that hypotonic swelling induces (K + Cl) cotransport in red blood cells from trout [19] and other fish species [18]; in each case (K + Cl) cotransport in fish red blood cells was found to be inhibited not only by "loop" diuretics but also by the stilbene inhibitors of band 3 mediated anion exchange, DIDS and SITS [18,19,84].

As is the case with duck red blood cells, catecholamines appear to influence the in vivo volume of trout red blood cells. Thus, when trout red blood cells are washed free of plasma and incubated in a synthetic isotonic medium similar to plasma in ionic composition, the cells shrink if this medium is free of catecholamines, but not if it contains norepinephrine [86]. Although (K + Cl) cotransport may contribute to this shrinkage, the prevention of shrinkage by catecholamines appears only to be related to stimulation of Na/H exchange (and thus NaCl uptake to counterbalance the KCl loss). Unlike the case with duck red blood cells, the swelling-induced (K + Cl) cotransport pathway of trout red blood cells appears to operate normally in the presence of catecholamines of cAMP [84].

The Na/H exchange pathway of frog red blood cells is activated by β-adrenergic agonists, cAMP, and prostaglandin E_1, as well as by cell shrinkage [50,87]. Although little is known about the kinetics of this Na/H exchange (it is amiloride-sensitive; [50]), biochemical studies have been performed with these cells as well as with avian red blood cells examining the role of membrane protein phosphorylation in the activation of transport by catecholamines (see below).

C. Mammalian Red Blood Cells

Human red blood cells lack β-adrenergic receptors coupled to adenylate cyclase [88]; however, human red blood cell membranes do show cAMP-dependent protein kinase activity [89]. Garay [64] found that exposure of human red blood cells to cAMP in the presence of a phosphodiesterase inhibitor significantly lowered furosemide-sensitive Na and K effluxes in these cells. Similarly, Mercer and Hoffman [43] reported that exposure of ferret red blood cells to 8-Br-cAMP (1 mM) inhibited (Na + K + Cl) cotransport in these cells by 16%.

D. Cellular Events in the Activation of Ion Transport by Catecholamines

(Na + K + Cl) cotransport systems in several different cell types (e.g., [32,34, 36,43,90]) including avian red blood cells [91] require intracellular ATP, as do some (K + Cl) cotransport sytems (e.g., [70]). This ATP does not appear to be required as energy to drive transport, because net transport via these pathways is determined only by the gradients of the cotransported ions (see above). Indeed, Geck et al. [32] found that ATP hydrolysis does not obligatorily accompany (Na + K + Cl) cotransport in Ehrlich ascites cells. Thus, ATP may play a role in the regulation of these (and possibly other) transport pathways. One possible such need for ATP could be in protein phosphorylation, and this has been examined in some detail in turkey and frog red blood cells.

In turkey red blood cells, Alper and co-workers [92] found that exposure of cells to the β-adrenergic agonist isoproterenol caused an incorporation of ^{32}P (from the terminal phosphate of ^{32}P-labeled ATP) into a 230,000-dalton membrane protein, termed goblin. The phosphorylation of goblin correlated fairly well with the activation of (Na + K + Cl) cotransport with respect to dose-response for isoproterenol and initial time course. An even better correlation was obtained between the isoproterenol dose–response and time course of activation for cotransport and these same parameters for the phosphorylation of a 37,000-dalton peptide obtained through proteolytic cleavage of goblin by *Staphylococcus aureus* V8 protease [92]. These findings suggest that phosphorylation of goblin at a site or sites on the 37,000-dalton peptide may be involved in catecholamine activation of (Na + K + Cl) cotransport in avian red blood cells. However, a cascade involving goblin phosphorylation is not the only means via which (Na + K + Cl) cotransport can be activated in these cells, because cell shrinkage and hypoxia (both activators of (Na + K + Cl) cotransport in avian red blood cells) do not increase ^{32}P incorporation into goblin or its 37,000-dalton fragment [92], nor do these stimuli increase cellular cAMP content [93].

Rudolph and Greengard [87] reported that incubation of frog red blood cells in the presence of isoproterenol or prostaglandin E_1, plus a phosphodiesterase inhibitor, results in the phorphorylation of a 240,000-dalton intrinsic membrane protein. This phosphorylation showed a similar isoproterenol dose–response and initial time course as the activation of Na influx (Na/H exchange). However, while the level of Na influx began to decline 30 min after initial isoproterenol exposure, the level of phosphorylation of this 240,000-dalton protein did not. Palfrey et al. [50] reported that incubation of frog red blood cells with isoproterenol or cAMP, plus a phosphodiesterase inhibitor, resulted in the phosphorylation of a 180,000-dalton protein having some homology with goblin by peptide mapping. As with goblin, phosphorylation of this 180,000-dalton protein from frog red blood cell membranes was not found to be stimulated by cell shrinkage, although Na/H exchange in these cells is [50].

Thus, the final common pathway for the activation of (Na + K + Cl) cotransport in avian red blood cells and Na/H exchange in amphibian red blood cells remains unknown. Also not known is whether catecholamines and cell shrinkage increase transport rates by increasing the number of transporters in the membrane, the turnover number of the transporters, or both. As noted above in discussing shrinkage-activated Na/H exchange in Amphiuma red blood cells, the delay period observed between initial shrinkage and maximal activation of Na/H exchange, as well as the augmentation of the Na/H exchange response by preincubation of the cells in amiloride-containing hypertonic media [46,47], suggest the possibility that additional transporters may become inserted into the plasma membrane when the cells are shrunken. Perhaps binding studies using a radiolabeled amiloride analog may help resolve this issue. [^{3}H] bumetanide binding studies have been done with duck red blood cells, and it was found that both cell shrinkage and norepinephrine exposure increase the number of high-affinity [3H] bumetanide binding sites by approximately the same extent that these stimuli increase (Na + K + Cl) cotransport rates [40]. This might suggest that these stimuli promote insertion of cotransporters into the plasma membrane; however, bumetanide appears to be sufficiently lipophilic [41] that it would be expected to pass through the cell membrane and bind to putative cotransporters contained in cytoplasmic vesicles within the time required to achieve equilibrium binding of [3H] bumetanide to duck red blood cells [40]. Thus, cell shrinkage and catecholamines could promote [3H] bumetanide binding to duck red blood cells by simply altering the conformation (to an active form that binds bumetanide with high affinity) of cotransporters already within the plasma membrane.

E. Possible Physiological Function of Catecholamine-Stimulated Ion Transport

The fact that activation of avian red blood cell (Na + K + Cl) cotransport under physiological ionic conditions does not produce any significant net salt or water movement [20] certainly argues against this process being involved in cell volume regulation, as was initially suggested [77]. One possible alternative role that has been proposed for this pathway in avian red blood cells [20,25] is in extrarenal K regulation. Schmidt and McManus [38] found that net (Na + K + Cl) cotransport in duck red blood cells responds linearly to changes in $[K]_0$ in the range from ~2.5–7.5 mM. Thus, during transient hypokalemia, (Na + K + Cl) cotransport efflux could mobilize potassium from red blood cells. During hyperkalemia, excess K could be taken up by the red blood cells, remaining there until renal or other mechansms adjusted plasma K levels back to normal. Although such an uptake would produce cell swelling, the excess cellular K would not be lost back to the plasma via (K + Cl) cotransport because plasma catecholamines abolish this swelling-induced pathway in duck red blood cells [20].

In fish red blood cells, catecholamine-activated Na/H exchange does influence cell volume, but perhaps more importantly increases pH_c [78], thus raising the oxygen affinity of hemoglobin. This could be of benefit to the animals during strenuous exercise or other conditions of hypoxic stress (during which plasma catecholamine levels might also be expected to rise), because under these conditions oxygen loading in the gills is likely to be the major factor determining oxygen availability to the tissues [78,94].

IV. OTHER PHYSIOLOGICAL EFFECTORS OF ION TRANSPORT

Incubation of duck red blood cells in a 100% nitrogen atmosphere increases both ouabain-insensitive Na and K fluxes across the cell membranes [95], and further studies suggest the major process involved is (Na + K + Cl) cotransport. Whether this phenomenon is limited to avian red blood cells is not known; however, in a recent study furosemide-sensitive Na and K movements in human red blood cells were not found to be increased when the cells were incubated in a nitrogen atmosphere [96]. An intriguing aspect of the apparent stimulatory effect of anoxia on (Na + K + Cl) cotransport in duck red blood cells is that when cells are first incubated in carbon monoxide, then transferred to 100% N_2, the activation of cotransport is prevented. Thus, it appears that hemoglobin must be in the deoxy conformation for stimulation of cotransport by anoxia to occur. Beyond this, the mechanism(s) involved in this stimulation of (Na + K + Cl) cotransport by anoxia are not known. Anoxia does not increase cellular cAMP levels or stimulate phosphorylation of the 37,000-dalton fragment of goblin [92]. One interesting speculation involves divalent cations, because $[Ca^{2+}]_c$ and $[Mg^{2+}]_c$ would be expected to increase with the conversion of oxy- to deoxyhemoglobin. This is because deoxyhemoglobin binds inositol pentaphosphate (IP_5, the major organic phosphate compound of mature avian erythrocytes, which lack 2,3 DPG; [97]), whereas when hemoglobin is in the oxy conformation IP_5 would instead chelate divalent cations. In this regard, glucose transport in avian red blood cells, which is stimulated by hypoxia [98], is also stimulated when cells are exposed to media containing A23187 and Ca (and presumably Mg as well; [99]). In addition, as noted above, Flatman [66] found that increasing $[Mg^{2+}]_c$ in the physiological range increased (Na + K + Cl) cotransport in ferret red blood cells.

Hall and co-workers [100] have found that very high hydrostatic pressures (200–400 atm) increase ouabain-insensitive, Cl-dependent K fluxes in human red blood cells. Because these fluxes are not inhibited by 0.05 mM bumetanide and not accompanied by equivalent Na fluxes (although ouabain-insensitive Na influx is increased somewhat at 400 atm), it is likely that these K fluxes represent (K + Cl) cotransport. Furthermore, the effect of hydrostatic pressure on K transport is augmented by hypotonic cell swelling [101], which as noted above also stimulates (K + Cl) cotransport in human red blood cells [14]. Finally, the Cl-dependent K transport stimulated by both swelling and increased hydrostatic pressure is markedly accentuated in the youngest or least dense population of cells [102].

V. (K + Cl) COTRANSPORT IN RED BLOOD CELL DISORDERS

Although this will be discussed in much more detail in Chapter 23, it is worth noting here that the (K + Cl) cotransport pathway, which is activated in human red blood cells by cell swelling, hydrostatic pressure, and NEM treatment (see above) also appears to be activated in red blood cells from patients with sickle cell (Hb SS) disease [103,104] and homozygous hemoglobin C (Hb CC) disease

[10,105]. (K + Cl) cotransport may also be responsible for cell shrinkage in those drug-induced hemolytic anemias associated with oxidative damage to the cell membrane and Heinz body formation [106]. In Hb SS [103,104] and CC [10] red blood cells, Cl-dependent K efflux is stimulated by hypotonic cell swelling and by acid pH (which also causes swelling, although (K + Cl) cotransport is activated in Hb CC cells in a 410 mosmol/kg medium of pH 7.0; [10]). Because (K + Cl) cotransport is accentuated in the least dense population of human red blood cells [102], it might be expected that the increase in this pathway noted in Hb SS and CC red blood cells is related to the increased number of young cells in the blood of patients with these disorders. However, although the least dense Hb SS cells do have the highest level of (K + Cl) cotransport, SS cells of intermediate density (middle one-third on a density gradient) and even (although to a far lesser extent) the densest Hb SS cells exhibit levels of K efflux higher than in even the least dense (top one-third on a density gradient) normal red blood cells [103]. In addition, sickle cells also exhibit higher levels of NEM-stimulated (K + Cl) cotransport than do normal (Hb AA) cells [104].

Why Hb SS and CC red blood cells have elevated levels of (K + Cl) cotransport is not clear, although this could indeed play a role in the shortened life of these cells in the circulation. Loss of KCl and osmotically obliged water from these cells would decrease their deformability [107], thus promoting their removal from the circulation (e.g., by the spleen). Cell dehydration would also increase the concentration of hemoglobin in the cytoplasm, which would have deleterious effects in sickle cell disease. Activation of (K + Cl) cotransport with resultant cell shrinkage may also be involved in the pathogenesis of certain drug-induced hemolytic anemias, particularly those associated with oxidant stress and Heinz body formation. Orringer [106] found that exposure of normal human red blood cells to acetylphenylhydrazine (APH), an oxidant compound that promotes Heinz body formation, activates a Cl-dependent K efflux, which is sodium-independent and inhibited by 1 mM furosemide. Because membrane-associated Heinz bodies (denatured hemoglobin) are present in APH-treated cells, the findings of Orringer [106] suggest that association of hemoglobin with the cell membrane may be involved in activating (K + Cl) cotransport. Because hemoglobins C [108] and S [109] bind to the red blood cell membrane with a higher affinity than does hemoglobin A, these findings in APH-treated cells also offer an interesting basis for speculation as to the mechanism underlying the increased levels of (K + Cl) cotransport in Hb SS and CC red blood cells.

VI. SUMMARY

A number of physiological stimuli, most notably catecholamine hormones and alterations of cell volume, greatly influence the activities of specific ion transport pathways in red blood cells from many different species. Several of these regulated transport pathways, including Na/H exchange, (Na + K + Cl) cotransport, and (K + Cl) cotransport, are also present in a variety of multicellular organs and tissues where they serve a number of vital and well-defined physiological functions. Just as red blood cells have in many cases proven to be an ideal tissue for the kinetic characterization of ion transport pathways, they may likewise prove to be highly valuable for further study of the complex molecular events involved in the regulation of these pathways. Such studies should not only provide a basis for understanding regulation of ion transport in more complex cells and tissues, but they may also help us better understand the physiological roles served by specific regulated ion transport pathways in red blood cells themselves, and the pathophysiology of disease processes in which these transport pathways function abnormally.

ACKNOWLEDGMENTS

The experiments shown in Figures 1–3 were performed in the laboratory of Dr. T. J. McManus, Department of Physiology, Duke University Medical Center, Durham, North Carolina, and were done in collaboration with Dr. McManus. I thank Dr. McManus and Dr. John Parker for their helpful comments. The author is a John A. and George L. Hartford Fellow.

NOTATION

APH, acetylphenylhydrazine; ATP, adenosine triphosphate; $[B]_c$, intracellular concentration of B (mmol/liter cell H_2O); $[B]_o$, extracellular concentration of B (mM); $[Ca^{2+}]_c$, intracellular concentration of free ionized calcium; cAMP, cyclic adenosine $3',5'$-monophosphate; DIDS, 4,4'-diisothiocyanostilbene-2,2'-disulfonic acid; 2,3 DPG, 2,3 diphosphoglycerate; E_B, equilibrium (Nernst) potential of ion B; E_m, membrane potential; EGTA, ethylene glycol bis-N,N,N', N'-tetraacetic acid; Hb, hemoglobin; IC_{50}, concentration of inhibitor producing half-maximal inhibition; IP_5, inositol pentaphosphate; $K_{1/2}$, concentration of ion producing half-maximal transport rate; $[Mg^{2+}]_c$, intracellular concentration of free ionized magnesium; NE, norepinephrine; NEM, N-ethyl maleimide; PAH, para-aminohippurate; pH_c, intracellular pH; r_{Cl}, chloride ratio; $[Cl]_c/[Cl]_o$; SITS, 4-acetamido-4'-isothiocyanostilbene-2,2'-disulfonic acid; TMA, tetramethylammonium; W_c, cell water content (kg H_2O/kg cell solid).

REFERENCES

1. Macey, R. I. (1984). Am. J. Physiol. 246:C195–C203.
2. Post, R. L., and Jolly, P. C. (1957). Biochim. Biophys. Acta 25:118–128.

3. Tosteson, D. C., and Hoffman, J. F. (1960). *J. Gen. Physiol.* 44:169-194.
4. Duhm, J., and Göbel, B. O. (1984). *J. Membr. Biol.* 77:243-254.
5. Parker, J. C., Gitelman, H. J., Glosson, P. S., and Leonard, D. L. (1975). *J. Gen. Physiol.* 65:84-96.
6. Ponder, E., and Robinson, E. J.(1934). *Biochem. J.* 28:1940-1943.
7. Ponder, E. (1948). *Hemolysis and Related Phenomena.* Grune & Stratton, New York.
8. Davson, H. J. (1937). *Cell. Comp. Physiol.* 10:247-264.
9. Grinstein, S., Rothstein, A., Sarkadi, B., and Gelfand, E. W. (1984). *Am. J. Physiol.* 246:C204-C215.
10. Brugnara, C., Kopin, A. S., Bunn, H. F., and Tosteson, D. C. (1985). *J. Clin. Invest.* 75:1608-1617.
11. Palfrey, H. C., Feit, P. W., and Greengard, P. (1980). *Am. J. Physiol.* 238:C139-C148.
12. Haas, M., and McManus, T. J. (1983). *Am. J. Physiol.* 245:C235-C240.
13. McManus, T. J., Haas, M., Starke, L. C., and Lytle, C. Y. (1985). *Ann. N.Y. Acad. Sci.* 456:183-186.
14. Kaji, D. (1986). *J. Gen. Physiol.* 88:719-738.
15. Dunham, P. B., and Ellory, J. C. (1981). *J. Physiol. (Lond.)* 318:511-530.
16. Lauf, P. K. (1984). *J. Membr. Biol.* 82:167-178.
17. Parker, J. C. (1981). *J. Gen. Physiol.* 78:141-150.
18. Lauf, P. K. (1982). *J. Comp. Physiol.* 146:9-16.
19. Bourne, P. K., and Cossins, A. R. (1984). *J. Physiol. (Lond.)* 347:361-375.
20. Haas, M., and McManus, T. J. (1985). *J. Gen. Physiol.* 85:649-667.
21. Kregenow, F. M. (1981). *Ann. Rev. Physiol.* 43:493-505.
22. Berkowitz, L. R., and Orringer, E. P. (1987). *Am. J. Physiol.* 252:C300-C306.
23. Cala, P. M. (1980). *J. Gen. Physiol.* 76:683-708.
24. Cala, P. M. (1983). *J. Gen. Physiol.* 82:761-784.
25. Cala, P. M. (1983). *Mol. Physiol.* 4:33-52.
26. Cala, P. M. (1977). *J. Gen. Physiol.* 69:537-552.
27. Fugelli, K. (1967). *Comp. Physiol. Biochem.* 22:253-260.
28. Poznansky, M., and Solomon, A. K. (1972). *J. Membr. Biol.* 10:259-266.
29. Adragna, N. C., and Tosteson, D. C. (1984). *J. Membr. Biol.* 78:43-52.
30. Parker, J. C. (1983). *Am. J. Physiol.* 244:C318-C323.
31. Haas, M., Schmidt, W. F. III, and McManus, T. J. (1982). *J. Gen. Physiol.* 80:125-147.
32. Geck, P., Pietrzyk, C., Burckhardt, B. C., Pfeiffer, B., and Heinz, E. (1980). *Biochim. Biophys. Acta* 600:432-447.
33. Greger, R., and Schlatter, E. (1981). *Pflugers Arch.* 392:92-94.
34. Hall, A. C., and Ellory, J. C. (1985). *J. Membr. Biol.* 85:205-213.
35. Lytle, C., Haas, M., and McManus, T. J. (1986). *Fed. Proc.* 45:548.
36. Russell, J. M. (1983). *J. Gen. Physiol.* 81:909-925.
37. Ueberschar, S., and Bakker-Grunwald, T. (1983). *Biochim. Biophys. Acta* 731:243-250.
38. Schmidt, W. F. III, and McManus, T. J. (1977). *J. Gen. Physiol.* 70:59-79.

39. Haas, M., and McManus, T. J. (1982). *Biophys. J.* 37:214a.
40. Haas, M., and Forbush, B. III (1986). *J. Biol. Chem.* 261:8434–8441.
41. Forbush, B. III, and Palfrey, H. C. (1983). *J. Biol. Chem.* 258:11787–11792.
42. Flatman, P. W. (1983). *J. Physiol. (Lond.)* 341:545–557.
43. Mercer, R. W., and Hoffman, J. F. (1985). *Biophys. J.* 47:157a.
44. Duhm, J., and Göbel, B. O. (1984). *Am. J. Physiol.* 246:C20–C29.
45. Brugnara, C., Canessa, M., Cusi, D., and Tosteson, D. C. (1986). *J. Gen. Physiol.* 87:91–112.
46. Siebens, A. W., and Kregenow, F. M. (1986). *J. Gen. Physiol.* 86:527–564.
47. Kregenow, F. M., Caryk, T., and Siebens, A. W. (1986). *J. Gen. Physiol.* 86:565–584.
48. Parker, J. C. (1983). *Am. J. Physiol.* 244:C324–C330.
49. Jennings, M. L., Douglas, S. M., and McAndrew, P. E. (1986). *Am. J. Physiol.* 251:C32–C40.
50. Palfrey, H. C., Stapleton, A., Alper, S. L., and Greengard, P. (1980). *J. Gen. Physiol.* 76:25a.
51. Benos, D. J. (1982). *Am. J. Physiol.* 242:C131–C145.
52. Parker, J. C. (1984). *J. Gen. Physiol.* 84:789–803.
53. Adorante, J. S., and Cala, P. M. (1986). *J. Gen. Physiol.* 88:10a.
54. Aronson, P. S., Nee, J., and Suhm, M. A. (1982). *Nature* 299:161–163.
55. Parker, J. C. (1986). *J. Gen. Physiol.* 87:189–200.
56. Grinstein, S., Rothstein, A., and Cohen, S. (1985). *J. Gen. Physiol.* 85:765–787.
57. Poznansky, M., and Solomon, A. K. (1972). *Biochim. Biophys. Acta* 274:111–118.
58. Kregenow, F. M., and Caryk, T. (1979). *Physiologist* 22:73.
59. Lauf, P. K., Adragna, N. C., and Garay, R. P. (1984). *Am. J. Physiol.* 246:C385–C390.
60. Cala, P. M., Mandel, L. J., and Murphy, E. (1986). *Am. J. Physiol.* 250:C423–C429.
61. Escobales, N., and Canessa, M. (1985). *J. Biol. Chem.* 260:11914–11923.
62. Milanick, M. A., Dissing, S. D., and Hoffman, J. F. (1985). *Biophys. J.* 47:490a.
63. Foskett, J. K., and Spring, K. R. (1985). *Am. J. Physiol.* 248:C27–C36.
64. Garay, R. P. (1982). *Biochim. Biophys. Acta* 688:786–792.
65. Ellory, J. C., Flatman, P. W., and Stewart, G. W. (1983). *J. Physiol. (Lond.)* 340:1–17.
66. Flatman, P. W. (1988). *J. Physiol. (Lond.)* 397:471–487.
67. Cianci, C. D., Giorgi, M., and Morrow, J. S. (1988). *J. Cell. Biochem.* 37:301–315.
68. Dunham, P. B., and Logue, P. J. (1986). *Am. J. Physiol.* 250:C578–C583.
69. Jacob, H. S., and Jandl, J. L. (1962). *J. Clin. Invest.* 41:779–792.
70. Lauf, P. K. (1985). *J. Membr. Biol.* 88:1–13.
71. Wiater, L. A., and Dunham, P. B. (1983). *Am. J. Physiol.* 245:C348–C356.
72. Kaji, D., and Kahn, T. (1985). *Am. J. Physiol.* 249:C490–C496.

73. Logue, P., Anderson, C., Kanik, C., Farquharson, B., and Dunham, P. (1983). *J. Gen. Physiol.* 81:861–885.
74. Lauf, P. K. (1985). *Am. J. Physiol.* 249:C271–C278.
75. Parker, J. C. (1986). *J. Gen. Physiol.* 88:45a–46a.
76. Ørskov, S. L. (1956). *Acta Physiol. Scand.* 37:299–306.
77. Kregenow, F. M. (1973). *J. Gen. Physiol.* 61:509–527.
78. Riddick, D. H., Kregenow, F. M., and Orloff, J. (1971). *J. Gen. Physiol.* 57: 752–766.
79. Nikinmaa, M. (1983). *J. Comp. Physiol.* 152:67–72.
80. Baroin, A., Garcia-Romeu, F., LaMarre, T., and Motais, R. (1984). *J. Physiol. (Lond.)* 356:21–31.
81. Cossins, A. R., and Richardson, P. A. (1985). *J. Exp. Biol.* 118:229–246.
82. Borgese, F., Garcia-Romeu, F., and Motais, R. (1986). *J. Gen. Physiol.* 87: 551–566.
83. Borgese, F., Garcia-Romeu, F., and Motais, R. (1987). *J. Physiol. (Lond.)* 382:145–157.
84. Borgese, F., Garcia-Romeu, F., and Motais, R. (1987). *J. Physiol. (Lond.)* 382:123–144.
85. Baroin, A., Garcia-Romeu, F., LaMarre, T., and Motais, R. (1984). *J. Physiol. (Lond.)* 350:137–157.
86. Bourne, P. K., and Cossins, A. R. (1983). *J. Exp. Biol.* 101:93–104.
87. Rudolph, S. A., and Greengard, P. (1980). *J. Biol. Chem.* 255:8534–8540.
88. Rasmussen, H., Lake, W., and Allen, J. E. (1975). *Biochim. Biophys. Acta* 411:63–73.
89. Rubin, C. S., Erlichman, J., and Rosen, O. M. (1972). *J. Biol. Chem.* 247: 6135–6139.
90. Dagher, G., Brugnara, C., and Canessa, M. (1985). *J. Membr. Biol.* 86:145–155.
91. Ueberschar, S., and Bakker-Grunwald, T. (1985). *Biochim. Biophys. Acta* 818:260–266.
92. Alper, S. L., Beam, K. G., and Greengard, P. (1980). *J. Biol. Chem.* 255: 4864–4871.
93. Kregenow, F. M., Robbie, E. D., and Orloff, J. (1976). *Am. J. Physiol.* 231:306–312.
94. Nikinmaa, M. (1982). *Mol. Physiol.* 2:287–297.
95. McManus, T. J., and Allen, D. W. (1968). In *Metabolism and Membrane Permeability of Erythrocytes and Thrombocytes*, E. Deutsch, E. Gerlach, and K. Moser (Eds.). Georg Thieme Verlag, Stuttgart, pp. 428–430.
96. Berkowitz, L. R., and Orringer, E. P. (1985). *Am. J. Physiol.* 249:C208–C214.
97. Borgese, T. A., and Lampert, L. M. (1975). *Biochem. Biophys. Res. Comm.* 65:822–827.
98. Wood, R. E., and Morgan, H. E. (1969). *J. Biol. Chem.* 244:1451–1460.
99. Bihler, I., and Sawh, P. W. (1979). *Fed. Proc.* 38:698.
100. Hall, A. C., Ellory, J. C., and Klein, R. A. (1982). *J. Membr. Biol.* 68: 47–56.

101. Ellory, J. C., and Hall, A. C. (1985). *J. Physiol. (Lond.)* 362:15P.
102. Hall, A. C., and Ellory, J. C. (1986). *Biochim. Biophys. Acta* 858:317–320.
103. Brugnara, C., Bunn, H. F., and Tosteson, D. C. (1986). *Science* 232:388–390.
104. Canessa, M., Spalvins, A., and Nagel, R. L. (1986). *FEBS Lett.* 200:197–202.
105. Berkowitz, L. R., and Orringer, E. P. (1985). *J. Gen. Physiol.* 86:40a–41a.
106. Orringer, E. P. (1984). *Am. J. Hematol.* 16:355–366.
107. Gulley, M. L., Ross, D. W., Feo, C., and Orringer, E. P. (1982). *Am. J. Hematol.* 13:283–291.
108. Reiss, G. H., Ranney, H. M., and Shaklai, N. (1982). *J. Clin. Invest.* 70:946–952.
109. Shaklai, N., Sharma, V. S., and Ranney, H. M. (1981). *Proc. Natl. Acad. Sci. USA* 78:65–68.

23

Ion Transport in Red Blood Cell Disorders

LEE R. BERKOWITZ and ANNE M. GRIFFIN   *University of North Carolina at Chapel Hill School of Medicine, Chapel Hill, North Carolina*

I. INTRODUCTION

The packaging of hemoglobin in the human erythrocyte has advantages that include stabilization of the hemoglobin molecule, modification of its oxygen dissociation, and protection from oxidation [1]. There are consequences, however, of keeping hemoglobin within a cell membrane. The osmotic gradient created by separating hemoglobin from the plasma drives water into the cell, and the electrical gradient established by the "fixed anions" of hemoglobin creates Donnan forces that also lead to an increase in cell water [2]. The red blood cell must compensate for these hemoglobin-induced gains in cell water or hemolysis will occur. This is done by creating a disequilibrium of the permeant cations Na and K [3]. A great number of studies have defined ion transport processes important in establishing and maintaining this cation disequilibrium in red cells.

In this chapter we review reports of abnormalities of such ion transport processes in three abnormal human red blood cells, the sickle erythrocyte, the hemoglobin CC erythrocyte, and the stomatocyte. Although the literature contains numerous reports of altered ion transport in many pathologic erythrocytes, we have focused on these particular red blood cells because there is some evidence to suggest that abnormalities of ion transport in these cells contribute to their shortened life span. We also review attempts to link abnormal transport to altered membrane structure.

II. THE SICKLE ERYTHROCYTE

The sickle erythrocyte was described in 1910 [4]. Almost 50 years later, a defect in this abnormally shaped red blood cell was found in its hemoglobin, where Glu is replaced by Val at the sixth position from the amino terminus of the β-globin chain [5]. Due to this amino acid substitution, the abnormal hemoglobin, called hemoglobin S, polymerizes into a gel when deoxygenated [6]. Polymerization is thought to be the key pathogenetic step in sickle cell anemia because polymerized hemoglobin S results in a poorly deformable red blood cell susceptible to hemolysis and trapping in the microcirculation [7]. Hemoglobin S is inherited in an autosomal dominant fashion with homozygosity necessary for the manifestations of sickle cell anemia. The heterozygous state, affecting ~10% of black Americans, is virtually asymptomatic [8].

A number of functional and structural abnormalities also have been detected in the sickle erythrocyte membrane. These abnormalities are presumed to be acquired as a consequence of the hemoglobin abnormality. They are not thought to represent coexistent genetic defects in the membrane of the sickle red blood cell.

A. Functional Abnormalities

In 1952, Tosteson et al. were the first to detect functional changes in the membrane of the sickle erythrocyte. Seeking to explain elevated levels of venous K in a small number of patients, experiments were designed to investigate the possibility that the sickle erythrocyte was the source of the elevated venous K. Sickle erythrocytes showed a net gain of Na and loss of K when deoxygenated [9]. The Na gain and K loss were of equal magnitude and both were reversible with reoxygenation. Additional experiments demonstrated that the Na gain and K loss were not seen when normal cells were deoxygenated or when sickling was inhibited by preexposing sickle cells to carbon monoxide before deoxygenation. Sickling-induced ion movements were also found not to depend on the presence of plasma, and could not be explained by reticulocytosis [10,11].

Subsequent work has sought to characterize further the Na and K movements and to identify cellular factors that influence them. These fluxes do not occur via the Na-K pump as they are found in sickle cells treated with ouabain or in cells with decreased levels of adenosine triphosphate (ATP) [12,13,14]. Rather, the initiating event for Na and K fluxes is likely mechanical stress on the membrane by bundles of hemoglobin S polymer because cation fluxes are inhibited when polymer growth is prevented by artificially increasing mean corpuscular hemoglobin concentration [15]. The membrane site for these fluxes could be the anion transport protein band 3, because both Na and K fluxes are inhibited by the 4,4'-diisothiocyanostilbene-2,2' disulfonate [16]. Another possibility is that the membrane sites for the Na and K movements are different. This possibility stems from the observation that sulfhydryl cross-linking has been shown to inhibit deoxygenation-induced K movements without affecting deoxygenation-induced Na movements [17]. Biological factors shown to influence Na and K fluxes are α globin gene number and intracellular levels of hemoglobin F. Sickle

Table 1 Postulated Pathogeneses of
Dehydration in Sickle Erythrocytes

Deoxygenation-induced cation movements

Calcium-dependent K pathway

Dehydrative effect of the Na/K pump

Volume/pH-dependent K pathway

Heinz body pathway

---

erythrocytes with a decreased number of α globin genes or increased levels of hemoglobin F have decreased Na and K fluxes when deoxygenated [18,19].

A second focus of research, stemming from the experiments of Tosteson et al. [9], has sought to determine whether altered ion transport in sickle erythrocytes causes cell dehydration (Table 1). Interest in this possibility began with the in vitro observation that a slight increase in hemoglobin S concentration greatly enhances the tendency toward gelation, and this interest was heighted with the demonstration that subpopulations of sickle erythrocytes, particularly irreversibly sickled cells, have extensive loss of cell K and cell water [20,21,22]. It was then reasoned that the cause of cell dehydration in sickle erythrocytes could be deoxygenation-induced ion movements. If so, a link between abnormal ion transport and the pathogenesis of sickling would exist as cell dehydration would increase the concentration of hemoglobin and thereby enhance gelation. Furthermore, therapies for sickle cell anemia could be devised that would prevent cell dehydration. Results of experiments regarding this line of reasoning have created controversy as evidence has been accumulated both supporting and disputing the possibility that acute deoxygenation causes cell dehydration in sickle erythrocytes.

Direct measurements of cell ion and water content in sickle cells have failed to demonstrate loss of cell water when cells are deoxygenated acutely. Tosteson et al. did not find a loss of cell water when cells were deoxygenated for 60 min, and Glader and Nathan reported no significant loss of cell water after a 4-hr deoxygenation period [9,12]. These studies have been criticized in that control experiments in the Tosteson work did not show a gain in cell water, an expected finding with deoxygenation, and cell water in the work of Glader and Nathan was expressed as mean cell volume, which may be inaccurate for sickle cells [23, 24]. However, other studies in which cell water was directly measured also find that sickle cells deoxygenated for 1 hr have no loss of cell water and actually behave like controls, gaining cell water with deoxygenation [13,25].

Support for the idea that acute deoxygenation of sickle cells causes cell dehydration comes from experiments in which cell water was measured indirectly.

Masys et al. used radiolabeled albumin as an indicator of the plasma space and reported that albumin concentration decreased when sickle erythrocytes were deoxygenated, suggesting that cell water loss was responsible for the albumin dilution [26]. Kaul et al., employing a continuous density gradient system, found that sickle erythrocytes became more dense with deoxygenation in contrast to normal erythrocytes, which became less dense with deoxygenation [27]. Additional reports have also demonstrated increasing cell density when sickle cells are deoxygenated [28,29].

Why there is a discrepancy in findings between experiments measuring cell water directly versus indirectly is not clear. One possibility is that indirect measurements of cell water detect smaller losses of cell water than direct measurements do. Another possibility, however, is that indirect measurements may be influenced by a number of factors, only one of which is cell water, so that changes in these measurements are not a reflection of a loss of cell water but a change in another parameter.

That experiments cannot unequivocally link acute deoxygenation with loss of cell water has led to further investigations seeking to explain how sickle cells become dehydrated (Table 1). One line of research has focused on calcium-induced loss of cell water. This is based on the observations that sickle erythrocytes have elevated total calcium concentrations and that an increase in ionized calcium in erythrocytes can initiate calcium-dependent K loss and cell dehydration [30,31]. The acceptance of this line of reasoning has fluctuated with time. Initial reports finding some evidence for calcium-dependent K loss in sickle erythrocytes generated much enthusiasm [32,33]. However, this enthusiasm diminished when subsequent reports did not directly demonstrate sustained calcium-dependent K loss in these cells [34]. The finding that most calcium in sickle cells is contained in vesicles was then put forth as an explanation for a lack of sustained calcium-dependent K loss despite increased total cell calcium [35]. This explanation gained further credibility when measurements of ionized calcium in sickle erythrocytes, done with nuclear magnetic resonance, showed no elevation compared with normal controls regardless of whether sickle cells were deoxygenated or oxygenated [36]. Recently, the possibility of a transient increase in ionized calcium with subsequent brief periods of K loss in sickle erythrocytes has been raised based on reports from two groups. One group has shown that substitution of intracellular chloride with the more permeant anion thiocyanate allows identification of brief periods of calcium-dependent K loss in sickle erythrocytes. A second group has found that under shear conditions there is a calcium-dependent decrease in cell deformability when sickle cells are reoxygenated, which is eliminated if calcium is chelated or if Na is replaced by K in the incubation buffer [37,38].

Another explanation for dehydration in sickle erythrocytes is increased activity of the Na-K pump. Studies in normal erythrocytes have shown that increases in intracellular Na produce increased Na-K pump activity that is de-hydrative due to the 3 Na/2 K stoichiometry of the pump [39,40]. Because there is a gain in intracellular Na when sickle cells are deoxygenated, the possibility of a pump response identical to that found in normal erythrocytes exists. In sup-port of this idea, two groups have found that ouabain inhibits increases in cell density when sickle cells are deoxygenated for 2–18 hr [14,29]. However, the findings of Ohnishi et al. do not support this because no dense cells were found after a 16-hr deoxygenation in the absence of ouabain [41]. Lew and Bookchin have also raised objections to this idea, based on theoretical calculations of the red cell response to high intracellular Na [42].

Brugnara et al. have provided yet another explanation for dehydration in sickle erythrocytes, by showing that sickle cells have a volume- and pH-sensitive K efflux pathway, activated by cell swelling or a decrease in intracellular pH. This pathway is present in ouabain-treated cells and also has been found in hemoglobin CC erythrocytes and in young normal red blood cells [43,44].

A final possibility accounting for dehydration in sickle erythrocytes is in-creased K loss due to denatured hemoglobin, Heinz bodies. That denatured hemoglobin produces loss of cell K was originally shown in normal cells treated with the oxidizing agent acetylphenylhydrazine [45]. Subsequent work with normal cells found that this K loss is ouabain-resistant, dependent on the presence of intracellular chloride, and is inhibited by loop diuretics. It is con-ceivable that such a pathway exists in the sickle erythrocyte because the mem-brane of the sickle erythrocyte has excess denatured hemoglobin, presumably due to increased oxidant stress. Experimental data have not been published sup-porting or negating that K loss in sickle erythrocytes could be due to a Heinz-body–like pathway.

B. Structural Abnormalities

Structural abnormalities in the sickle erythrocyte membrane (Table 2) are pre-sumed to be secondary either to increased membrane-bound denatured hemo-globin S, in the form of hemichromes, or to repeated membrane stretching from hypoxia-induced polymerization of hemoglobin S. The presence of denatured hemoglobin S in the membrane is thought to initiate increased production of oxidants. Hebbel and co-workers have shown that sickle erythrocytes generate increased amounts of superoxide, peroxide, and hydroxyl radicals [46]. The additional findings that levels of oxidant production were found to correlate significantly with measurements of membrane-bound hemichrome and that normal erythrocyte membranes induced to acquire increased hemichrome gen-erated increased amounts of hydroxyl radical substantiated the idea that oxi-dized iron in hemichromes is the source of oxidant stress. End results of this oxidant stress may include membrane intramolecular thiol oxidation, lipid per-oxidation, and increased static rigidity [47,48].

Table 2 Structural Abnormalities in the
Sickle Erythrocyte Membrane

Intramolecular thiol oxidation
Lipid peroxidation
Phospholipid reorientation
Increased adherence to endothelium
Defective spectrin binding site on ankyrin
Membrane vesiculation

The event of sickling has been shown to change phospholipid orientation in
the membrane of the sickle erythrocyte. Lubin et al. have found that with
sickling, phosphatidyl choline is translocated from its normal position in the
outer leaflet of the membrane to the inner leaflet, and phosphatidyl ethanol-
amine and phosphatidyl serine are translocated from the inner leaflet to the
outer leaflet. These changes were reversible with reoxygenation and were not observed in deoxygenated normal red blood cells [49]. Recently it has been appreciated that ATP depletion in sickle red blood cells leads to similar phospholipid orientation changes independent of deoxygenation and that a consequence
of this phospholipid alteration is increased adherence of sickle erythrocytes to
macrophages [50,51].

Other detected changes in the sickle erythrocyte membrane include increased
adherence to vascular endothelial cells, decreased binding of spectrin to inside-
out vesicles made from sickle red blood cell membranes, and membrane vesiculation. Whether these changes are due to membrane-bound hemichrome, repeated
sickling, or a combination of both is not known. Hebbel and co-workers were
the first to demonstrate abnormal adherence of sickle erythrocytes to cultured
vascular endothelial cells [52]. To demonstrate this property, ^{51}Cr-labeled
sickle red blood cells were incubated with cultured endothelial cells, washed,
and plates were counted for residual activity. After repeated washing, one to ten
sickle erythrocytes remained adherent to each endothelial cell, whereas repeat
experiments with normal erythrocytes found no adherence. Adherence was not
influenced by deoxygenation but was increased when the most dense subpopulations of sickle erythrocytes were tested. Enzymatic removal of sialic acid was
found to decrease adherence. This work has been confirmed and extended by
other investigators, showing that external divalent cations and plasma proteins
are necessary for adherence [53]. Adherence also may be influenced by infection because increased adherence has been demonstrated to virus-infected endothelial cells [54]. Platt et al. have shown that there is a defect in the sickle
erythrocyte cytoskeleton at the spectrin binding site on ankyrin [55]. Because

the ankyrin binding site of spectrin from sickle erythrocytes was normal and purified sickle erythrocyte ankyrin bound spectrin normally, it was hypothesized that sickle hemoglobin or cytoskeletal damage led to the decreased binding to ankyrin in situ on the inside-out vesicle. Wagner et al. have demonstrated that a variety of abnormal red blood cells, including the sickle erythrocyte, spontaneously lose membrane as vesicles that contain band 3, band 4.1, and glycophorin A. Vesicles also were found to have abnormal phospholipid orientation [56].

A summary of all data regarding functional and structural properties of the sickle erythrocyte membrane substantiates the idea that abnormalities are acquired throughout the life span of the cell. As yet, though, none of the abnormalities detected has been unequivocally linked to the pathogenesis of hemolysis and/or vaso-occlusion in sickle cell anemia. Further investigation should improve our understanding of how hemoglobin S alters the cell membrane and may lead to therapeutic options in the treatment of sickle cell anemia.

III. THE HEMOGLOBIN CC ERYTHROCYTE

Human erythrocytes containing only hemoglobin C, hemoglobin CC erythrocytes, have altered cation and water content. Several reports have shown that intracellular Na is slightly increased, intracellular K is substantially decreased, and cell water is lower in hemoglobin CC cells compared with normal erythrocytes [57-59]. Because the reduced K and water content of hemoglobin CC cells are thought to be the major factors responsible for poor cell deformability and subsequent shortened cell survival, characterization of ion transport abnormalities in these erythrocytes may be integral to understanding the pathogenesis of hemoglobin CC disease [60]. In addition, hemoglobin CC cells have a single genetic defect in the structure of hemoglobin ($\alpha_2 \beta_2$ 6 Glu → Lys), so that the possibility that hemoglobin C itself contributes to abnormalities of ion transport must be considered [61].

The first investigations of ion transport in hemoglobin CC cells were done by Murphy. In a series of papers, the reduced total ion and water content of hemoglobin CC cells were reported, plus the observations that hemoglobin CC cells have reduced intracellular pH and reduced intracellular chloride content [57,62]. More recently, Brugnara et al. reinvestigated ion transport in these cells, finding that hemoglobin CC cells have a normal intracellular chloride content and an increase in unidirectional K efflux [63]. The increased K efflux was found to be insensitive to ouabain and calcium, but was stimulated by cell swelling or reduced intracellular pH. The findings of Brugnara et al. have been confirmed and extended. It has been shown that there are likely two separate ouabain-insensitive K efflux pathways in hemoglobin CC cells because the increased K efflux seen in cells at original volume is not affected by anion substitution or loop diuretics, whereas the K efflux stimulated by cell swelling is inhibited by loop diuretics or by replacing intracellular chloride with another permeant anion [59,64]. It has also been demonstrated that sickle erythrocytes and normal young erythrocytes have a volume-sensitive K efflux similar to that described in hemoglobin CC cells, which raises the possibility that the volume-sensitive K efflux in hemoglobin CC cells is not unique but rather a reflection of the young age of the cell population [43,44]. Interpretation of these data does not define the specific cause of K loss in hemoglobin CC cells, although comparison of these findings to other studies improves out understanding of volume regulation in hemoglobin CC cells. The transport characteristics of K leak stimulated by cell swelling in hemoglobin CC cells are similar to those described in swollen avian erythrocytes, suggesting that there may be a common cell-volume-sensing apparatus in these cells. That the increased K efflux in hemoglobin CC cells at original volume is not reduced by anion substitution, loop diuretics, or calcium suggests that this K efflux is not the result of sulfhydryl injury or opening of the Gardos channel [31,65].

The inability of physiologic experiments to define a known pathway of K loss in hemoglobin CC cells raises the possibility that hemoglobin C itself is involved in stimulating K efflux. Some experiments of the structural properties of hemoglobin C may shed light on a connection between the altered hemoglobin and reduced cation content in these cells. Hemoglobin C has decreased solubility compared with hemoglobin A so that crystals of hemoglobin C have been observed in the red blood cells of homozygous patients, particularly those patients who have undergone splenectomy [66,67]. Hemoglobin C has also been demonstrated to have increased affinity for the erythrocyte membrane at a band 3 binding site [68]. These data raise the possibility that hemoglobin C, either in crystal form or solubilized, directly damages the erythrocyte membrane, producing loss of ions and cell water. Another possibility is that hemoglobin C crystals contribute to poor deformability of these cells with subsequent membrane damage in the microcirculation.

IV. THE STOMATOCYTE

A human erythrocyte is called a stomatocyte if, on examination on a blood smear, the expected central pallor of the red blood cell is replaced by a slit or single concavity [69]. Stomatocytes have been found in vivo in a small number of cases of familial hemolytic anemia in patients from Europe and North America. Up to 10% of aborigines in Melanesia also have stomatocytes in their blood, but these stomatocytes exhibit minimal or no reduction in red blood cell survival [70,71]. In vitro, normal erythrocytes can be transformed into stomatocytes by lowering pH in a buffer or by exposure to cationic detergents or phenothiazines [69,72].

Studies of ion transport in stomatocytes have consistently shown that ouabain-insensitive fluxes of both Na and K are increased. Initial reports indicated that the increase in ouabain-insensitive Na flux far exceeded the increase in ouabain-insensitive K flux, so that net cation and cell water content were increased (hydrocytes) compared with normal erythrocytes [73]. This ouabain-insensitive Na flux has been investigated further with conflicting results. The increased Na flux in red blood cells from one patient had a linear dependence on external Na and was not affected by K or furosemide, suggesting that the increase represented an increase in ground permeability for Na rather than increased Na:K cotransport [74]. In another study of red blood cells from five related patients, the increased Na influx was largely furosemide-sensitive and K fluxes were insensitive to furosemide. This was interpreted as representing imbalanced furosemide-sensitive Na:K:2Cl cotransport [75]. Regardless of whether the increased Na flux in these stomatocytes is due to Na:K:2Cl cotransport or an increase in ground permeability for Na, there is likely a defect in the cell membrane responsible for the increased Na flux because the Na flux can be substantially reduced and cell volume restored to normal when stomatocytes are exposed to the bifunctional imidoester dimethyl adipimidate [76]. The specific structural defect has not been identified. In most instances, gel electrophoresis of stomatocytic membranes shows no differences compared with normal red blood cell membranes, although two patients have been reported with a deficiency in band 7 and the presence of an abnormal protein has been found in red blood cell membranes from one other patient [77,78].

Ion transport studies also have detected another variant of stomatocytes, the xerocyte. Red blood cells from these patients have increases in ouabain-insensitive fluxes of Na and K, but in contrast to hydrocytes, K flux is greater in magnitude than Na flux, resulting in reduced total cation content and cell water. In the original description of xerocytosis, a mother and her son were studied [79]. Ouabain-insensitive K flux was three times that of normal controls and ouabain-insensitive Na flux was increased to 1.6 times controls. The Na-K pump rate was twice normal but unable to compensate for the K leak. These xerocytes demonstrated osmotic resistance, and on density gradients young and old cells alike were shown to have abnormal cation contents.

Studies done on the red blood cells of another patient with xerocytosis demonstrated an increased shear sensitivity compared with that of normal red blood cells [80]. This was attributed to the dehydrated state of the patient's cells because if xerocytes were rehydrated artificially, their shear sensitivity became like that of normal red blood cells. Conversely, artificially dehydrated normal red blood cells demonstrated an increase in their shear sensitivity. These data have been confirmed in two other reports [81,82].

It is presumed that the primary defect in the xeroxyte is located in the cell membrane. This presumption is based on the inability to detect an abnormality in hemoglobin or metabolism in xerocytes from a few patients, although studies of the components of the xerocyte membrane have found no abnormalities that could not be explained by the young age of the cell population [79,83–85]. How the putative membrane defect leads to increased cation permeability is also not understood. One possibility is the presence of cation channels in the membrane. Another possibility is age-dependent loss of cell membrane, based on the observation of accelerated membrane fragmentation in metabolically depleted xerocytes [86].

There is some evidence to suggest a secondary defect in the xerocyte due to high internal sodium. Based on the work of Clark et al., showing that normal red blood cells made to have increased internal Na suffer loss of total ion content and water by increased activity of the Na-K pump, it has been shown that cation loss and dehydration in xerocytes are inhibited by ouabain [14,40]. The supposition is that the Na-K pump, stimulated by high internal Na in the xerocyte, exacerbates dehydration due to the 3Na-2K stoichiometry.

ACKNOWLEDGMENTS

The authors thank Sharon Pigott for expert secretarial assistance. This work was supported in part by National Institutes of Health Grant HL-30467.

REFERENCES

1. Darling, R. C., and Roughton, F. J. W. (1942). Effect of methemoglobin on equilibrium between oxygen and hemoglobin. *Am. J. Physiol.* 137:56–68.
2. Hoffman, E. K. (1977). Control of cell volume. In *Transport of Ions and Water in Animals*, B. L. Gupta, R. B. Moreton, J. L. Oschman, and B. J. Wall (Eds.). Academic Press, New York, pp. 285–332.
3. Hoffman, J. F. (1986). Active transport of Na^+ and K^+ by red blood cells. In *Physiology of Membrane Disorders*, T. E. Andreoli, J. F. Hoffman, D. D. Fanestil, and S. G. Schultz (Eds.). Plenum, New York, pp. 221–234.
4. Herrick, J. B. (1910). Peculiar elongated and sickle-shaped red blood corpuscles in a case of severe anemia. *Arch. Intern. Med.* 6:517–521.
5. Kan, Y. W., and Dozy, A. M. (1980). Evolution of the hemoglobin S and C genes in world populations. *Science* 209:388–391.
6. Dean, J., and Schecter, A. N. (1978). Sickle cell anemia: molecular and cellular bases of therapeutic approaches. *N. Engl. J. Med.* 299:752–763.
7. Noguchi, C. T., and Schecter, A. N. (1981). The intracellular polymerization of sickle hemoglobin and its relevance to sickle cell disease. *Blood* 58:1057–1068.
8. Heller, P., Best, W. R., Nelson, R. B., and Becktel, J. (1979). Clinical implications of sickle cell trait and glucose-6-phosphate dehydrogenase deficiency in hospitalized black male patients. *N. Engl. J. Med.* 300:1001–1005.
9. Tosteson, D. C., Shea, E., and Darling, R. C. (1952). Potassium and sodium of red blood cells in sickle cell anemia. *J. Clin. Invest.* 31:4061–411.
10. Tosteson, D. C., Carlsen, E., and Dunham, E. T. (1955). The effects of sickling on ion transport. *J. Gen. Physiol.* 39:31–53.

11. Tosteson, D. C. (1955). The effects of sickling on ion transport. *J. Gen. Physiol.* 39:55-67.
12. Glader, B. E., and Nathan, D. G. (1978). Cation permeability alterations during sickling: Relationship to cation composition and cellular hydration of irreversibly sickled cells. *Blood* 51:983-989.
13. Berkowitz, L. R., and Orringer, E. P. (1985). Passive sodium and potassium movements in sickle erythrocytes. *Am. J. Physiol.* C208-C214.
14. Joiner, C. H., Platt, O. S., and Lux, S. E. IV. (1986). Cation depletion by the sodium pump in red cells with pathologic cation leaks. *J. Clin. Invest.* 78:1487-1496.
15. Mohandas, N., Rossi, M. E., and Clark, M. R. (1986). Association between morphologic distortion of sickle cells and deoxygenation-induced cation permeability increase. *Blood* 68:450-454.
16. Joiner, C. H., and Dew, A. (1986). Deoxy cation fluxes in sickle cells are inhibited by DIDS. *Fed. Proc.* 45:446a.
17. Wall, S. N., and Berkowitz, L. R. (1987). The effect of N,N'-p-phenylenedimaliemide (PMD) on deoxygenation-induced K loss in sickle erythrocytes. *Am. J. Med. Sci.* 294:105-109.
18. Embury, S. K., Becker, K., and Glader, B. E. (1985). Monovalent cation changes in sickle erythrocytes: A direct reflection of β-globin gene number. *J. Lab. Clin. Med.* 106:75-79.
19. Fabry, M. E., Buchanan, I. D., and Nagel, R. L. (1984). Red cell density and rate of potassium loss are affected by hemoglobin F levels in sickle cell anemia. *Clin. Res.* 32:551a.
20. Hofrichter, J., Ross, P. D., and Eaton, W. A. (1974). Kinetics and mechanism of deoxyhemoglobin S gelation: A new approach to understanding sickle cell disease. *Proc. Natl. Acad. Sci. USA* 71:4864-4868.
21. Coletta, M., Hofrichter, J., Ferrone, F. A., and Eaton, W. A. (1982). Kinetics of haemoglobin polymerization in single red cells. *Nature* 300:194-197.
22. Clark, M. R., Mohandas, N., Embury, S. H., and Lubin, B. H. (1982). A simple laboratory alternative to irreversibly sickled cell (ISC) counts. *Blood* 60:659-662.
23. Van Slyke, D. D., Wu, H., and McLean, F. C. (1923). Studies of gas and electrolyte equilibria in the blood. *J. Biol. Chem.* 53:765-849.
24. Kamentsky, L. A. (1980). Objective measures of information from blood cells. *Blood Cells* 6:121-140.
25. Joiner, C. H. (1985). Physiologic characteristics of sodium and potassium fluxes stimulated by deoxygenation of sickle red cells. *Pediatr. Res.* 19:263a.
26. Masys, D. R., Bromberg, P. A., and Balcerzak, S. P. (1974). Red cells shrink during sickling. *Blood* 44:885-889.
27. Kaul, D. K., Fabry, M. E., Windish, P., Baez, S., and Nagel, R. L. (1983). Erythrocytes in sickle cell anemia are heterogeneous in their rheological and hemodynamic characteristics. *J. Clin. Invest.* 72:22-31.

28. Izumo, H., Williams, M., Rosa, R., Flier, J., and Epstein, F. H. (1985). Na/K ATPase and ion fluxes in sickle cell anemia. Implications for therapy. *Clin. Res.* 33:343a.
29. Izumo, H., Lear, S., Williams, M., Rosa, R., and Epstein, F. H. (1988). Sodium-potassium pump, ion fluxes, and cellular dehydration in sickle cell anemia. *J. Clin. Invest.* 79:1621–1628.
30. Eaton, J. W., Skelton, T. D., Swofford, H. S., Kolpin, C. E., and Jacob, H. S. (1973). Elevated erythrocyte calcium in sickle cell disease. *Nature* 246:105–106.
31. Gardos, G. (1959). The role of calcium in the potassium permeability of human erythrocytes. *Acta Physiol. Acad. Sci. Hung.* 15:121–125.
32. Steinberg, M. H., Eaton, J. W., Berger, E., Coleman, M. B., and Oelshlegel, F. G. (1978). Erythrocyte calcium abnormalities and the clinical severity of sickling disorders. *Br. J. Haematol.* 40:533–539.
33. Berkowitz, L. R., and Orringer, E. P. (1981). The effect of cetiedil, an in vitro antisickling agent, on erythrocyte membrane cation permeability. *J. Clin. Invest.* 68:1215–1220.
34. Lew, V. L., and Bookchin, R. M. (1980). A Ca^{2+}-refractory state of the Ca sensitive K^+ permeability mechanism in sickle cell anaemia red cells. *Biochim. Biophys. Acta* 602:196–200.
35. Lew, V. L., Hockaday, A., Sepulveda, M. I., Somlyo, A. P., Somlyo, A. V., Ortiz, O. E., and Bookchin, R. M. (1985). Compartmentalization of sickle-cell calcium in endocytic inside-out vesicles. *Nature* 315:586–589.
36. Murphy, E., Berkowitz, L. R., Orringer, E., Levy, S. A., Gabel, S. A., and London, R. E. (1987). Cytosolic free calcium levels in sickle red blood cells. *Blood* 69:1469–1474.
37. Bookchin, R. M., Ortiz, O. E., and Lew, V. L. (1985). Cell dehydration from sickling-induced calcium-sensitive K^+-channel activation. *Blood* 66:56a.
38. Orringer, E. P., Thomas, R. P., and Shipp, G. W. (1986). Role of calcium in the sickling phenomenon: an ektacytometric analysis. *Clin. Res.* 34:661a.
39. Garay, R. P., and Garrahan, P. J. (1973). The interaction of sodium and potassium with the sodium pump in red cells. *J. Physiol.* 231:297–325.
40. Clark, M. R., Guatelli, J. C., White, A. T., and Shohet, S. B.(1981). Study on the dehydrating effect of the red cell Na^+/K^+ pump in nystatin-treated cells with varying Na^+ and water contents. *Biochim. Biophys. Acta* 646:422–432.
41. Ohnishi, S. T., Horiuchi, K. Y., and Horiuchi, K. (1986). The mechanism of in vitro formation of irreversibly sickled cells and modes of action of its inhibitors. *Biochim. Biophys. Acta* 886:119–129.
42. Lew, V. L., and Bookchin, R. M. (1986). Volume, pH, and ion-content regulation in human red cells: Analysis of transient behavior with an integrated model. *J. Membr. Biol.* 92:57–74.
43. Brugnara, C., Bunn, H. F., and Tosteson, D. C. (1986). Regulation of erythrocyte cation and water content in sickle cell anemia. *Science* 232:388–390.

44. Brugnara, C., and Tosteson, D. C. (1987). Cell volume, K transport and cell density in human erythrocytes. *Am. J. Physiol.* 252:C269–C276.
45. Orringer, E. P. (1984). A further characterization of the selective K movements observed in human red blood cells following acetylphenylhydrazine exposure. *Am. J. Hematol.* 16:335–366.
46. Hebbel, R. P., Eaton, J. W., Balasingam, M., and Steinberg, M. H. (1982). Spontaneous oxygen radical generation by sickle erythrocytes. *J. Clin. Invest.* 70:1253–1259.
47. Rank, B. H., Carlsson, J., and Hebbel, R. P. (1985). Abnormal redox status of membrane-protein thiols in sickle erythrocytes. *J. Clin. Invest.* 75:1531–1537.
48. Evans, E., and Mohandas, N. (1986). Membrane-associated sickle hemoglobin: A major determinant of sickle erythrocyte rigidity. *Blood* 68:61a.
49. Lubin, B., Bastacky, J., Roelofsen, B., and Van Deenen, L. L. M. (1981). Abnormalities in membrane phospholipid organization in sickled erythrocytes. *J. Clin. Invest.* 67:1643–1649.
50. Middlekoop, E., Bevers, E. M., Lubin, B. H., Op den Kamp, J. A. F., and Roelofsen, B. (1986). The effect of ATP on translocation of (amino) phospholipids in sickle erythrocytes. *Blood* 68:63a.
51. Schwartz, R. S., Tanaka, Y., Fidler, I. J., Chiu, D., Lubin, B., and Schroit, A. J. (1985). Increased adherence of sickled and phosphatidylserine-enriched human erythrocytes to cultured human peripheral blood monocytes. *J. Clin. Invest.* 75:1965–1972.
52. Hebbel, R. P., Yamada, O., Moldow, C. F., Jacob, H. S., White, J. G., and Eaton, J. W. (1980). Abnormal adherence of sickle erythrocytes to cultured vascular endothelium. *J. Clin. Invest.* 65:154–160.
53. Mohandas, N., and Evans, E. (1985). Sickle erythrocyte adherence to vascular endothelium. *J. Clin. Invest.* 76:1605–1612.
54. Hebbel, R. P., Visser, M. R., Vercellotti, G. M., and Jacob, H. S. (1986). Enhanced adhesion of sickle rbc to virus-infected endothelium: Novel mechanism linking vassoocclusive crisis and systemic infection. *Blood* 68:62a.
55. Platt, O. S., Falcone, J. F., and Lux, S. E. (1985). Molecular defect in the sickle erythrocyte skeleton. *J. Clin. Invest.* 75:266–271.
56. Wagner, G., Chiu, D., Yee, M., Claster, S., Wang, W., and Lubin, B. (1984). Red cell vesiculation—a common pathophysiologic event. *Blood* 64:32a.
57. Murphy, J. R. (1968). Hemoglobin CC disease: rheologic properties of erythrocytes and abnormalities in cell water. *J. Clin. Invest.* 47:1483–1495.
58. Brugnara, C., Kopin, A. S., Bunn, H. F., and Tosteson, D. C. (1984). Electrolyte composition and transport in hemoglobin CC red cells. *Clin. Res.* 32:551A.
59. Berkowitz, L. R., and Orringer, E. P. (1987). Cell volume regulation in hemoglobin CC and AA erythrocytes. *Am. J. Physiol.* 252:C300–C306.
60. Bunn, H. F., and Forget, B. G. (1986). Human hemoglobin variants. In *Hemoglobin: Molecular, Genetic, and Clinical Aspects*, H. F. Bunn and B. G. Forget (Eds.). W. B. Saunders, Philadelphia, pp. 421–425.

61. Hunt, J. A., and Ingram, V. M. (1960). Abnormal human haemoglobins. IV. The chemical difference between normal human haemoglobin and haemoglobin C. *Biochim. Biophys. Acta* 42:409–421.
62. Murphy, J. R. (1976). Hemoglobin CC erythrocytes: Decreased intracellular pH and decreased O_2 affinity-anemia. *Semin. Hematol.* 13:177–180.
63. Brugnara, C., Kopin, A. S., Bunn, H. F., and Tosteson, D. C. (1985). Regulation of cation content and cell volume in hemoglobin erythrocytes from patients with homozygous hemoglobin C disease. *J. Clin. Invest.* 75:1608–1617.
64. Berkowitz, L. R., and Orringer, E. P. (1985). Chloride-dependent K transport mediates cell volume regulation in hemoglobin CC red blood cells. *J. Gen. Physiol.* 86:40a.
65. Lauf, P. K. (1985). $K^+:Cl^-$ cotransport: sulfhydryls, divalent cations, and the mechanism of volume activation in a red cell. *J. Membr. Biol.* 88:1–13.
66. Diggs, L. W., Kraus, A. P., Morrison, D. B., and Rudnicki, R. P. T. (1954). Intraerythrocytic crystals in a white patient with hemoglobin C in the absence of other types of hemoglobin. *Blood* 9:1172–1184.
67. Fabry, M. E., Kaul, D. K., Raventos, C., Baez, S., Rieder, R., and Nagel, R. L. (1981). Some aspects of the pathophysiology of homozygous Hb CC erythrocytes. *J. Clin. Invest.* 67:1284–1291.
68. Reiss, C., Ranney, H. M., and Shaklai, N. (1982). Association of hemoglobin C with erythrocyte ghosts. *J. Clin. Invest.* 70:946–952.
69. Weed, R. I., and Bessis, M. (1973). The discocyte-stomatocyte equilibrium of normal and pathologic red cells. *Blood* 41:471–475.
70. Dutcher, P. O., Segal, G. B., Feig, S. A., Miller, D. R., and Klemperer, M. R. (1975). Cation transport and its altered regulation in human stomatocytic erythrocytes. *Pediatr. Res.* 9:924–927.
71. Lux, S. E. (1983). Disorders of the red cell membrane skeleton: hereditary spherocytosis and hereditary elliptocytosis. In *The Metabolic Basis of Inherited Disease*, J. B. Stanbury, J. B. Wyngaarden, D. S. Frederickson, J. L. Goldstein, and M. S. Brown (Eds.). McGraw-Hill, New York, pp. 1572–1605.
72. Deuticke, B. (1968). Transformation and restoration of biconcave shape of human erythrocytes induced by amphiphilic agents and changes of ionic environment. *Biochim. Biophys. Acta* 163:494–500.
73. Miller, D. R., Rickles, F. R., Lichtman, M. A., LaCelle, P. L., Bates, J., and Weed, R. I. (1971). A new variant of hereditary anemia with stomatocytosis and erythrocyte cation abnormality. *Blood* 38:184–204.
74. Wiley, J. S. (1977). Genetic abnormalities of cation transport in the human erythrocyte. In *Membrane Transport in Red Cells*, J. C. Ellory and V. L. Lew (Eds.). Academic Press, New York, pp. 337–361.
75. Chailley, B., Feo, C., Garay, R., Dagher, G., Bruckdorfer, R., Fischer, S., Piau, J. P., and Delaunay, J. (1981). Evidence fo imbalanced furosemide-sensitive Na+K cotransport in hereditary stomatocytosis. *Scand. J. Haematol.* 27:365–373.

76. Mentzer, W. C., Lubin, B. H., and Emmons, S. (1974). Correction of the permeability defect in hereditary stomatocytosis by dimethyl adipimidate. *N. Engl. J. Med.* 294:1200–1204.
77. Lande, W. M., Thiemann, P. V. W., and Mentzer, W. C. (1982). Missing band 7 membrane protein in two patients with high Na, low K erythrocytes. *J. Clin. Invest.* 70:1273–1280.
78. Bienzle, V., Niethammer, D., Keeberg, V., Ungefahr, K., Konhe, K., and Kleihauer, E. (1975). Congenital stomatocytosis and chronic hemolytic anemia. *Scand. J. Haematol.* 15:339–346.
79. Glader, B. E., Fortier, N., Albala, M. M., and Nathan, D. G. (1974). Congenital hemolytic anemia associated with dehydrated erythrocytes and increased potassium loss. *N. Engl. J. Med.* 291:491–496.
80. Platt, O. S., Lux, S. E., and Nathan, D. G. (1981). Exercise-induced hemolysis in xerocytosis. *J. Clin. Invest.* 68:631–638.
81. Clark, M. R., Mohandas, N., Caggiano, V., and Shohet, S. B. (1978). Effects of abnormal cation transport on deformability of dessicytes. *J. Supramol. Struct.* 8:521–532.
82. Nathan, D. G., and Shohet, S. B. (1970). Erythrocyte ion transport defects and hemolytic anemia: "hydrocytosis" and "dessicytosis." *Semin. Hematol.* 7:381–408.
83. Fairbanks, G. (1980). The red cell membrane in normal and abnormal status. In *Red Blood Cell and Lens Metabolism*, S. K. Srivistava (Ed.). Elsevier, Amsterdam, pp. 191–212.
84. Sauberman, N., Fortier, N. L., Fairbanks, G., O'Connor, R. J., and Snyder, L. M. (1979). Red cell membrane in hemolytic disease: Studies of variables affecting electrophoretic analysis. *Biochim. Biophys. Acta* 556:292–313.
85. Sauberman, N., Fairbanks, G., Lutz, H. U., Fortier, N. L., and Snyder, L. M. (1981). Altered red blood cell surface area in hereditary xerocytosis. *Clin. Chem. Acta* 114:149–161.
86. Snyder, L. M., Lutz, H. U., Sauberman, N., Jacobs, J., and Fortier, N. L. (1978). Fragmentation and myelin formation in hereditary xerocytosis and other hemolytic anemias. *Blood* 52:750–761.

24

Partial Deficiencies of Erythrocyte Spectrin in Hereditary Spherocytosis

PETER AGRE   *The Johns Hopkins University School of Medicine, Baltimore, Maryland*

I. A CORPUSCULAR DEFECT

Hereditary spherocytosis is a congenital hemolytic anemia that has been a long-standing puzzle to hematologists and geneticists (see recent reviews [1-3]). It is a common disorder affecting ~1 in 5000 individuals of Northern European ancestry [4]. The genetic defect is incompletely defined but results in erythrocytes that are abnormally fragile and become increasingly spherical rather than retaining the normal biconcave disk shape.

Spherocytosis has been recognized for more than 100 years, and a report that appeared in the European literature in 1871 captured the salient features of the disease [5]: the patients were anemic, had splenomegaly, experienced periods of jaundice, had similarly affected relatives, and had abnormally shaped erythrocytes (drawn as microcytic spherocytes in the original lithograph). The therapeutic value of splenectomy was subsequently discovered to reduce the pace of hemolysis [6], because the abnormal erythrocytes survive much longer in the circulation after removal of the spleen. Despite advances in the understanding of the disease, splenectomy remains the treatment of choice.

Spherocytes were noted to have increased sensitivity to lysis in hypotonic saline, and this led to the development of the osmotic fragility test [7], a standard clinical test used to establish the diagnosis. Due to a reduction in surface area relative to volume, spherocytes are unable to swell as far as normal erythrocytes and consistently rupture at hypotonic saline concentrations slightly above the saline concentration at which normal erythrocytes will rupture (at ~0.45% NaCl instead of 0.4%). Cross-transfusion studies unequivocally demonstrated that the fundamental defect resides within the spherocytic erythrocytes themselves, because their survival in the circulation was reduced when transfused into normal individuals, whereas normal erythrocytes survived normally when transfused into spherocytosis patients [9].

Several workers astutely concentrated on the erythrocyte membrane as the likely site for the underlying defect. However, the investigations were confounded by a large number of observations that were subsequently recognized to be secondary to a more fundamental (and undiscovered) molecular defect. One important observation is the consistent partial reduction of membrane phospholipid noted in spherocytosis membranes [10,11]. The reduction is especially curious because it appears to be corrected by splenectomy.

Additional confusion arose because of the clinical heterogeneity of the disorder. Although spherocytosis is usually inherited as an autosomal dominant mutation, the number of affected relatives was found to be smaller than expected [12]. It was estimated that as many as 25% of cases were born to apparently normal parents, and speculation existed that these may be new mutations or the result of a nondominant form of inheritance. Although the anemia is generally noted to be mild, the clinical severity of the illness is somewhat variable, although affected family members frequently express anemias of comparable severity. Many patients are diagnosed only because a relative was known to be affected. Relatively few patients require multiple transfusions, and most patients undergo elective splenectomy as older children or adolescents, after which blood counts nearly always rise to normal. Although certain families are more seriously affected, only rare patients experience prolonged, severe anemia.

II. SPHEROCYTOSIS IN MICE

Workers at the Jackson Laboratories in Bar Harbor, Maine, identified an important group of mutations in mice that permitted the first clear elucidation of a molecular basis for inherited spherocytosis. These mice were first noted to be jaundiced, and subsequent evaluations indicated that their erythrocytes are unusually fragile and spherocytic (see review [13]). There are several differences, however, between the forms of spherocytosis affecting humans and these mice. All of the mouse mutations are inherited recessively, and the simple and compound heterozygotes are apparently normal. The mice are affected by an extremely severe anemia with many of the homozygotes dying in utero. Furthermore, unlike the small increase in osmotic fragility seen in human spherocytosis, the spherocytic mouse blood is inordinately sensitive to osmotic lysis, indicating that there is a large reduction in membrane surface area relative to volume. When analyzed by gel electrophoresis, membranes from mice homozygous for the sph/ sph mutation were found to be strikingly deficient in spectrin, the long filamentous protein that is the principal structural protein of the membrane associ-

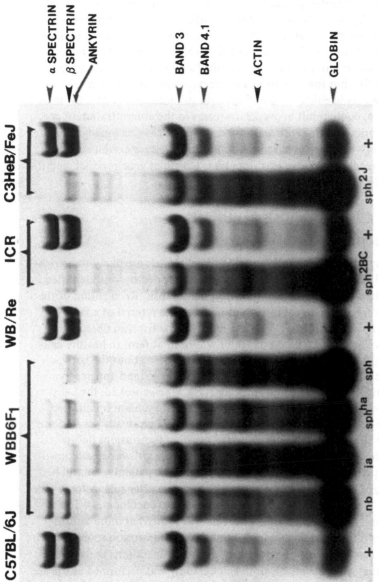

Figure 1 Coomassie blue stained polyacrylamide electrophoresis gel of membrane proteins from normal and mutant mouse erythrocytes and reticulocytes. The animals were homozygous for the following mutations as listed at the figure bottom: +/+ = normal; nb/nb = "normoblastic" due to unstable ankyrin variant; ja/ja = "jaundiced" due to absent beta spectrin synthesis; sph^{ha}/sph^{ha} = "spherocytosis locus, hemolytic anemia" due to unstable alpha spectrin; sph/sph = "spherocytosis locus" due to absent alpha spectrin synthesis. Reprinted from *Cell* with permission [16].

ated cytoskeleton [14]. Membranes from the other mouse mutants were also found to be partially or totally deficient in spectrin [15], and distinct abnormalities in membrane biogenesis were recently identified in reticulocytes and circulating normoblasts obtained from each of the mouse mutants [16]. Spectrin consists of two similar but nonidentical subunits (α and β spectrin) that are individually unstable but together form a stable heterodimer coil. It would be expected that defects in either the α or β spectrin subunits or ankyrin, the membrane binding site, could result in overall decreases in the concentration of spectrin on the membrane. Indeed, this was found to be the case, as shown in Figure 1. While spectrin-like molecules have been found in all tissues examined, it is certain that the spectrin found in erythrocytes is a distinct gene product, and the mutant mice only expressed a defect in their erythrocytes.

III. RECESSIVE SPHEROCYTOSIS IN HUMANS

A. Clinical Presentation

Numerous examples in biology indicate that animal models are likely to be relevant to human disease, and a deficiency of erythrocyte spectrin was identified in a severe, recessively inherited form of human spherocytosis. Recognition of the human form of recessive spherocytosis resulted from the referral of a family to the University of North Carolina [17]. This family came from an obscure part of coastal North Carolina. An infant daughter was born at term to healthy caucasian parents in their mid 20s. Nothing unusual was noted about the baby, although the maternal grandmother insisted she was too pale (and apparently grandmothers are rarely wrong about such matters). When checked, the baby was found to severely anemic (hematocrit 12-15%), and evaluation indicated that the defect was intracorpuscular and seemed to be an unusually severe form of spherocytosis (Figure 2). The child received periodic transfusions before undergoing splenectomy at the age of 1½ years. The little girl was greatly improved by the operation, but unlike typical spherocytosis patients, she still had a significant degree of incompletely compensated anemia after splenectomy (the hematocrit only rose to 28-30% with 5-10% circulating reticulocytes). The couple subsequently had their second child. She was somewhat more seriously affected than her older sister and underwent exchange transfusion soon after birth because of severe jaundice. She required regular transfusions thereafter to maintain her hematocrit above 12%, until splenectomy was performed at the age of 6 months. Both girls have done well subsequent to their splenectomies, and although remaining mildly anemic, they lead normal lives.

B. Parental Consanguinity

The birth of two severely affected children to apparently normal parents was puzzling to the physicians caring for the children, but the possibility that the

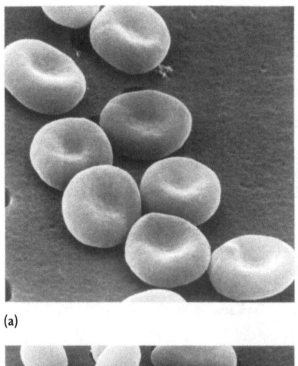

(a)

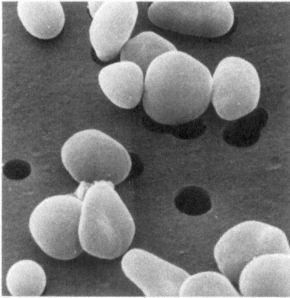

(b)

Figure 2 Scanning electron micrograph of erythrocytes. The normal appearing cells are from the mother (a), and from her daughter, a homozygote for recessive spherocytosis (b). Reprinted from the *New England Journal of Medicine* with permission [17].

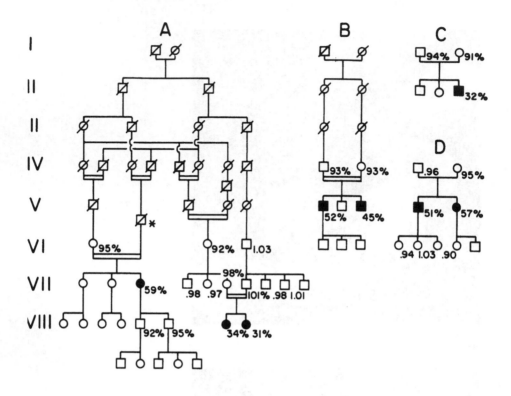

(a)

(b)

Figure 3 Selected genealogies of six families with spherocytosis: (a) nondominant inheritance, (b) dominant heritance. Symbols indicate that family members were asymptomatic with normal clinical laboratory data (□, ○); deceased, with no clinical laboratory data available (⌀, ⌀); deceased, with normal clinical laboratory data and no history of anemia (⌀*); nondominant spherocytosis (■, ●); or had dominant spherocytosis (◨, ◑). The symbols also indicate the percentage of normal spectrin content as determined by radioimmunoassay (e.g., ○ 98%), the normalized spectrin: band 3 ratio (e.g., □ .98), or the occurrence of a consanguineous marriage (══). Reprinted from the *New England Journal of Medicine* with permission [16].

first child was a new mutation was ruled out by the birth of the second. A detailed geneological search had been made [18], and both parents were descendants of the family that originally settled in the region at the time of the Revolutionary War. The parents were linked by at least three loops of consanguinity, being simultaneously fourth cousins, fifth cousins, and fifth cousins once removed (Figure 3, family A). Despite the serious anemias of the two children, there was no history of anemia in any of the aunts or uncles or in any close relatives, and it became increasingly clear that this was an example of human spherocytosis of recessive inheritance.

Given the size of the kindred and the region being relatively remote, it was considered likely that other patients would be found. Indeed, there was a notable degree of profligacy in one ancestor who by geneology was considered likely to have carried the mutation. The individual lived from 1826–1893 and fathered 17 children by his two wives and was still remembered 80 years after his death as a "strong-minded, hard-headed woman chaser" about whom additional information is apparently contained in court records [19]. A distant cousin was subsequently discovered to have undergone splenectomy as a youngster (Figure 3, family A, generation VII). Now in her 50s, the distant cousin described a moderately severe anemia during her childhood that prompted splenectomy at the age of 10 years. This patient had never required frequent transfusions, and she was unaware of any residual anemia in herself, or any history of anemia in her father who died of a stroke in his late 70s, in her 80-year-old mother, or in either of her two sons of several grandchildren. However, both of her parents were descended from the same ancestors as the two little girls (see Figure 3) and were related by multiple loops of consanguinity, being at the same time first cousins once removed, second cousins once removed, and third cousins once removed (the last by two different pathways—Figure 3).

C. Spectrin Quantitation

When erythrocyte membranes were prepared from these individuals and analyzed with Coomassie Blue-stained sodium dodecylsulfate polyacrylamide electrophoresis gels, it was found that the two little girls and their distant cousin were the only ones who had an apparent deficiency of spectrin, and the membranes of all other relatives appeared normal. In addition to the gross deficiency of spectrin, it was noted that there were smaller reductions in ankyrin, protein 4.1, and other smaller proteins (detectable in Figure 4, right lane). Nevertheless, the relative decrease of spectrin was most dramatic, and it was considered likely that inherited defects in spectrin were responsible for the anemia. Additional studies demonstrated that the relative reduction in spectrin was not the result of in vitro proteolytic degradation, because western blots using antispectrin antibodies failed to demonstrate increased proteolytic degradation of the patient preparations. Also, using binding studies, it was shown that the patient's mem-

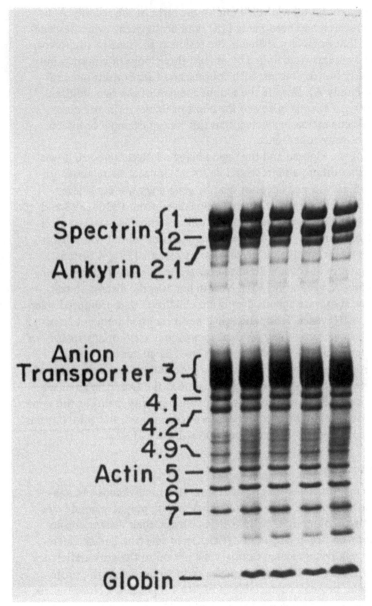

Figure 4 Coomassie Blue-stained sodium dodecylsulfate polyacrylamide electrophoresis gel of membrane proteins from a normal individual (far left lane), and four patients with a range of spectrin deficiencies (2nd lane from left = small deficience, middle lanes = larger deficiencies, and far right lane = most deficient). Reprinted from *Nature* with permission [20].

branes bound normal spectrin normally, and the patient spectrin bound to normal membrane sites normally. Furthermore, unlike pyropoikilocytosis and certain forms of elliptocytosis, the recessive spherocytosis spectrin was >90% in the tetramer form on the membrane. Thus, the spectrin appeared quantitatively reduced but functionally normal.

The relative degree of spectrin deficiency can be estimated from electrophoresis gels when the intensity of the dye in the spectrin region is compared with that of band 3, the anion transporter that is an index of membrane surface area ("spectrin:band 3 ratio"). Measurement of the actual spectrin deficiency (rather than the relative deficiency) required development of a radioimmunoassay ("spectrin-RIA") to determine the number of copies of spectrin per cell. Pure spectrin prepared from a normal individual is ^{125}I-labeled, and an immunoadsorbent is prepared by coating protein A-bearing staphylococci with antispectrin antibodies. Competition for sites on the immunoadsorbent is measured between ^{125}I-labeled spectrin and unlabeled spectrin in detergent lysates prepared from a known number of washed erythrocytes. Displacement by spectrin in the patient lysates is compared with normal control lysate [20]. Using the spectrin-RIA, it was shown that the two little girls with the extremely severe anemia had spectrin contents at 31 and 34% of normal whereas their less severely affected distant cousin had 59% of the normal spectrin level (Figure 3). It is assumed from the pedigree that all three are homozygotes for the same mutation. Although the level of expression is similar between the sisters, it is quite different in the distant cousin, indicating that other factors may regulate expression.

D. Additional Families

More than a dozen additional spherocytosis families have subsequently been identified with moderately to severely affected children born to clinically normal parents. One of these additional families was found to have parental consanguinity, but no consanguity could be established in the other families, although it was frequently impossible to obtain information more than one or two generations into the past (Figure 3). In each case it was found that the patients' erythrocytes were significantly deficient in spectrin (ranging from 30–74% of normal), whereas erythrocytes from the other relatives were normal. Because two affected siblings were found in several of the families, all cannot be new dominant mutations. Indeed, none of the children born to any of these nondominant patients has any significant hematological abnormality.

E. Presumed Heterozygotes

When the parents of the nondominantly affected patients were examined for subtle erythrocyte defects, minor abnormalities were consistently found (Figure 5). Fourteen mothers and fathers were examined with the spectrin-RIA, and the

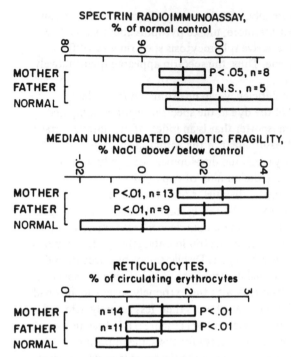

Figure 5 Subtle erythrocyte abnormalities in the parents of patients with non-dominantly inherited spherocytosis. The vertical line = mean and the horizontal bar = plus/minus one standard deviation. Reprinted from the *New England Journal of Medicine* with permission [26].

mean spectrin level was 95% of normal. The osmotic fragility was compared between blood samples drawn from 22 of the parents and simultaneously from normal controls and processed identically. The parents invariably had a slight increase in sensitivity to lysis (~0.02% NaCl above control). The percentage of circulating erythrocytes was measured in 25 of the parents, and the mean of 1.6% was somewhat above the normal of 1.0%. Although suggestive of a very small deficiency in surface membrane and a subtle decrease in erythrocyte survival, none of these parameters is sufficiently convincing to absolutely establish that any individual is a heterozygote for a recessive mutation. Nevertheless, taken as a group, it seems very likely that many of these parents, if not most, are indeed silent carriers for a mutation that affects the homozygote in a dramatic way.

IV. DOMINANT SPHEROCYTOSIS IN HUMANS

Approximately 75% of spherocytosis cases are inherited as classical dominant mutations, and these patients are usually affected with only a mild to moderately severe anemia. It was noted on electrophoresis gels that erythrocyte membranes from these patients were very slightly deficient in spectrin. Multiple members of nine families with dominantly inherited spherocytosis were examined with the spectrin-RIA. All were found to be partially deficient in spectrin, with spectrin contents ranging from 63–81% of the normal level. The levels of spectrin in affected family members in different generations were reduced very similarly (see families E and F, Figure 3). Likewise, the spectrin level did not change when measured before and after splenectomy in the same individuals. It seems likely that the partial reduction in spectrin is a common feature of dominantly inherited spherocytosis. It had previously been reported that a subset of patients with the dominant form of spherocytosis inherited a spectrin mutant with a reduced affinity for association with protein 4.1 [21,22]. One such patient was investigated for spectrin content with an improved gel electrophoresis method, and her membranes were reported to be partially deficient in spectrin [3].

V. CORRELATIONS WITH SPECTRIN DEFICIENCY

A. Clinical Severity

The deficiency of spectrin is likely to be specific for hereditary spherocytosis and hereditary pyropoikilocytosis, a rare disorder with erythrocyte fragmentation and increased osmotic fragility resulting from inheritance of variants in the N-terminal region of α spectrin [23–25]. No deficiency of spectrin was found when patients with a large variety of other erythrocyte abnormalities were examined including: thalassemia trait, iron deficiency, macrocytic anemias, traumatic splenectomies, elliptocytosis, and spherocytosis resulting from warm-reacting autoantibodies [20,26].

When 33 individuals with spherocytosis of the different inheritance forms and clinical severities were examined with the spectrin-RIA, all were found to be deficient in spectrin with the spectrin levels ranging from 30–80% of normal spectrin [26]. Furthermore, there is a notable relationship between the quantity of spectrin and the increased sensitivity of the erythrocytes to lysis in hypotonic saline (Figure 6, panel A).

An earlier publication indicated that the increased sensitivity to osmotic lysis correlated approximately with the reduction in red blood cell survival in a small number of patients with hereditary spherocytosis [27]. A relationship was noted between the quantity of spectrin and factors reflecting the clinical severity of the anemia (Figure 6, panels B, C, and D). Patients with >70% of the normal spectrin level generally had no evidence of hemolysis persisting after splenec-

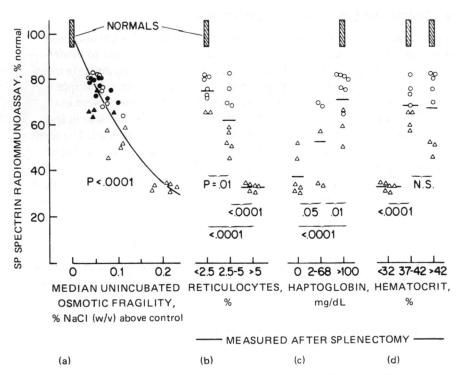

Figure 6 Correlation of spectrin level with osmotic fragility (a) and parameters splenectomy clinical status after splenectomy (b, c, and d). Reprinted from the *New England Journal of Medicine* with permission [26].

tomy, whereas patients with 40–70% of the normal spectrin content generally had some evidence of persistent postsplenectomy hemolysis (either elevated reticulocyte count, reduced haptoglobin level, or both). Only patients with < 40% of the normal spectrin content were found to remain anemic after splenectomy with hematocrits stabilizing in the range of 25–31%. In addition, it was found that the clinical course before splenectomy reflected the erythrocyte spectrin level. All nine patients with <55% of the normal spectrin level received multiple transfusions before splenectomy, and seven of these patients underwent splenectomy before 2 years of age. Only 3 of the 15 patients with >70% of the normal spectrin level underwent splenectomy before the age of 12 years, and only two received more than a single transfusion before splenectomy. Spectrin levels reflected the type of inheritance with most of the dominant patients having levels >65% of normal and most of the nondominant patients <65%.

B. Physical Measurements

If partial deficiencies of spectrin are responsible for the membrane instability, it would be expected that measurements of physical membrane properties would reflect the relative degrees of spectrin deficiency. It was fortunate that investigators with expertise in measuring membrane physical properties had developed methods for evaluating erythrocyte membranes at same time this cohort of spherocytosis patients with defined levels of spectrin deficiency was identified.

The ektacytometer was used to evaluate membrane deformability over a gradient of saline concentrations at a constant shear stress, and a remarkable correlation was noted between the deformability index and the erythrocyte spectrin content [28]. Membrane viscoelastic properties can be measured for single cells by using micropipet analyses. The fractional reduction of membrane shear modulus was compared with erythrocyte membrane content (estimated by spectrin-RIA) and membrane spectrin density (determined spectrin:band 3 ratios on electrophoresis gels). Although correlations were noted for both methods of quantitating spectrin, the correlation with spectrin density indicated that shear elasticity is directly proportional to membrane surface density of spectrin (see Chapter 15). The rate of lateral redistribution of fluorescently labeled erythrocyte membrane surface markers can be measured after photobleaching with a highly focused laser. It was found that the rates of lateral diffusion of band 3 and glycophorin were increased in direct proportion to the quantity of spectrin (see Chapter 13). Thus, three distinct physical properties of the erythrocyte membrane were all found to reflect the quantity of spectrin. Although these observations do not unequivocally prove that all physical factors leading to increased hemolysis in spherocytosis are directly caused by the deficiency in spectrin, it seems very likely that at least several of these factors are related.

VI. GENETIC DEFECTS RESULTING IN SPECTRIN DEFICIENCY

The genetic explanations for the recessive spherocytosis mutations in mice were clearly established by the Jackson Laboratory group to include reduced synthesis or reduced stability of either the α spectrin, β spectrin, or ankyrin [16]. Thus, defects may occur in multiple sites with similar phenotypic results. Recent work by two laboratories at Yale University [29] indicate that most of the parents of the nondominantly affected spherocytosis patients are themselves heterozygotes for a spectrin variant that is abnormal in the α II domain, as demonstrated with two-dimensional gel analysis of partial tryptic digests. The severely affected children all appear to be homozygotes for this variant (see Chapter 4). One family that did not express this spectrin variant (Figure 3, family B) and was shown to have inherited the defect in linkage with a common polymorphism near the N-terminus or in the promoter regions of the α spectrin gene (see Chap-

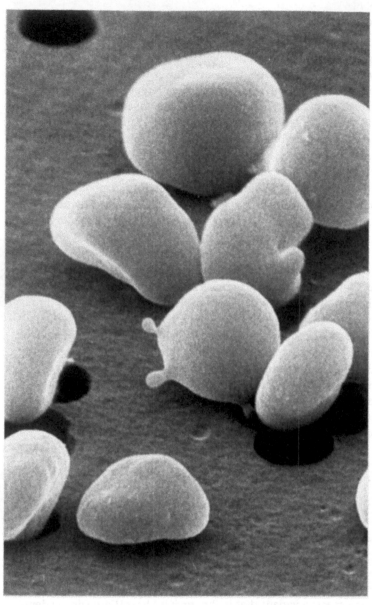

(a)

Figure 7 (a) Scanning electron micrograph of erythrocytes from a patient with recessive spherocytosis and severe anemia, spectrin level 31% of normal. Note points of membrane vesiculation. (b) Diagram representing a hypothesis of how spectrin deficiency leads to loss of surface membrane.

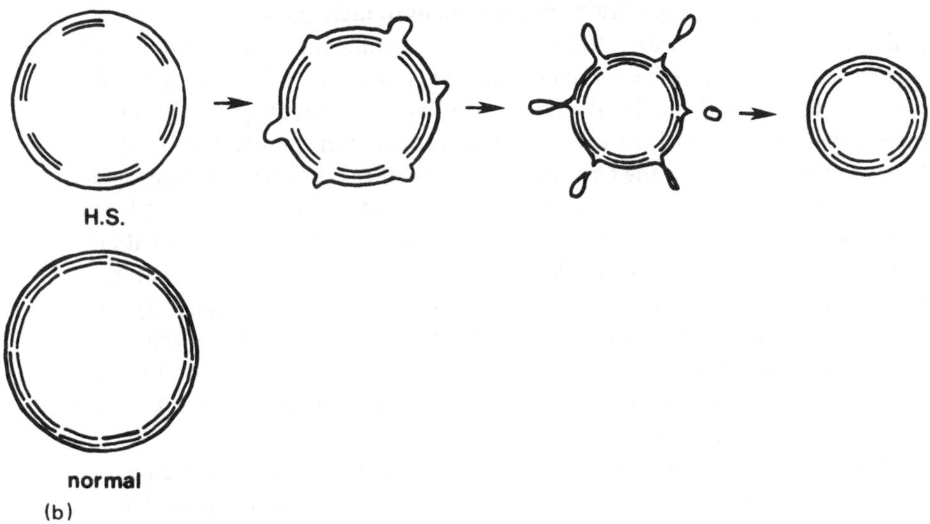

H.S.

normal

(b)

Figure 7 (continued)

ter 5). It therefore appears likely that spectrin deficiency in most families with nondominantly and recessively inherited spherocytosis have mutations in the coding or noncoding regions of the α spectrin gene.

Unlike the experience with nondominant spherocytosis, no spectrin variants were detected when members of large kindreds with dominant spherocytosis were investigated. Neither the two-dimensional analysis of tryptic peptides nor the α spectrin gene polymorphisms segregated in linkage with the clinical defect [30]. Recent studies of one very large kindred with dominant spherocytosis established that the anemia was inherited in linkage with ankyrin, when assessed by restriction fragment polymorphism analysis.

VII. PATHOGENESIS OF SPHEROCYTOSIS MEMBRANES

Several observations fit together, providing a potential explanation for the membrane pathology of hereditary spherocytosis. Although the genetic and clinical heterogeneity of the disease indicate that no single lesion is likely to be responsible for all forms of the disease, all forms characteristically bear certain common features. The patients have an inherited hemolytic anemia; spherocytes are present in the peripheral blood; the patients experience clinical improvement after splenectomy. Numerous studies over the past several decades have consistently identified a deficiency of erythrocyte membrane surface area. The intriguing finding that membrane lipid is reduced ~15% in spherocytes but rises to normal after splenectomy must be taken as an extremely important clue to the pathogenesis of the disorder [10,11]. However, normal individuals who undergo splenectomy have a rise in their membrane lipids to levels ~15% still higher. We have also found small fractional reductions of band 3 levels in spherocytosis membranes with a partial correction toward normal after splenectomy [20]. Characteristic changes in the osmotic fragility profiles are also noted after splenectomy indicating that the most spherocytic population of cells (the "tail") is no longer found. It is not doubted that splenectomy is beneficial to the patient because it permits continued circulation of abnormal erythrocytes. However, these observations also indicate that the lack of a spleen probably partially corrects the relative deficiency in membrane surface area, because the spleen plays an essential role in the continued loss of sections of membrane and leads to the increasingly spherical cell shape. However, splenectomy does not correct the primary underlying defects in the disorder.

It is our hypothesis that partial deficiency of spectrin is the fundamental underlying membrane defect in most forms of hereditary spherocytosis. The deficiency of spectrin probably leads to reduced membrane stability with subsequent loss of unsupported areas of the plasma membrane, including lipid bilayer and a portion of the untethered fraction of band 3 and other integral membrane proteins. Indeed, high-resolution scanning electron micrographs of erythrocytes from the most seriously affected spherocytosis patients after splenectomy demonstrate a continued loss of surface by membrane budding (Figure 7), a phenomenon that likely occurs much more rapidly in the spleen.

The deficiency of spectrin may result from reduced spectrin biosynthesis or increased spectrin degradation. Our experience with density gradient separated erythrocytes indicates that the densest, intermediate, and least dense cells (corresponding roughly to the oldest, younger, and youngest) are all deficient in spectrin, and increased degradation of spectrin is not seen with western blots. The observations with the spherocytic mice indicate that defects in the synthesis or stability of spectrin or ankyrin result in overall spectrin deficiencies. The recent findings by the groups at Yale strongly support the concept that the inherited defect is within the coding or noncoding portions of the α spectrin gene in patients with nondominantly inherited spherocytosis [30]. This is consistent with a model of membrane biogenesis depicted in Figure 8, which is based on the work of the group at California Institute of Technology [32]; see also Chapter 3). It has been demonstrated in chick embryo erythroid cells that α spectrin is synthesized in a threefold excess of β spectrin. Therefore the concentration of β is limiting for α assembly. In such a setting, one would expect that a defect in synthesis or assembly of an α spectrin variant would be a silent trait in a heterozygote, because sufficient α spectrin is synthesized off the normal gene to hybridize with all of the available β spectrin. A homozygote for the α defect could be very seriously affected, although the level would not neces-

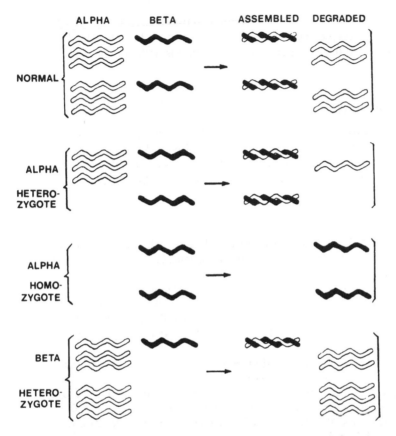

Figure 8 Diagrams representing spectrin assembly for a normal individual and models for speculative defects in spectrin synthesis: a heterozygote for reduced alpha spectrin synthesis, a homozygote for reduced alpha spectrin synthesis, and a heterozygote for reduced beta spectrin synthesis. Note that the relative excess of alpha spectrin may mask the alpha heterozygote since beta spectrin is limiting.

sarily be zero because some α synthesis might occur. The unhybridized β spectrin would be in excess and degraded, resulting in an overall spectrin deficiency. Indeed, as noted above, all but one of the nondominant/recessive spherocytosis families clearly have variants in the coding or noncoding regions of the α spectrin. Families with dominantly inherited spherocytosis would be expected to have a defect in the stability of a subpopulation of spectrin within the membrane skeleton or a defect in the synthesis of a factor limiting the assembly of spectrin onto the membrane, such as β spectrin or ankyrin. The answer to this problem is still forthcoming. No spectrin degradation has been identified, and it has been shown that inheritance of dominant spherocytosis does not segregate with known α spectrin polymorphisms [30]. It remains to be determined if the anemia will be linked to ankyrin in additional kindreds with dominant spherocytosis [31].

ACKNOWLEDGMENTS

The enthusiastic participation of our patients, their families, and referring physicians were essential factors in these studies. This work supported by NIH grant R01 HL33991, March of Dimes Basil O'Connor Award 5-490, and an Established Investigator Award from the American Heart Association.

REFERENCES

1. Dacie, J. V. (1980). The life span of the red blood cell and circumstances of its premature death. In *Blood, Pure and Eloquent*, M. M. Wintrobe (Ed.). McGraw-Hill, New York, 211–256.
2. Lux, S. E. (1983). Disorders of the red cell membrane skeleton: hereditary spherocytosis and hereditary elliptocytosis. In *The Metabolic Basis of Inherited Disease*, J. B. Stanbury, J. B. Wyngaarden, D. S. Fredrickson, J. L. Goldstein, and M. S. Brown (Eds.). McGraw-Hill, New York, pp. 1573–1605.
3. Becker, P. S., and Lux, S. E. (1985). Hereditary spherocytosis and related disorders. *Clin. Haematol.* 14:15–43.
4. Morton, N. E., MacKinney, A. A., Kosower, N., Schilling, R. F., and Gray, M. P. (1962). Genetics of spherocytosis. *Am. J. Hum. Genet.* 14:170.
5. Vanlair, V. F., and Masius, J. B. (1871). De la microcythemie. *Bull. R. Acad. Med. Belg.* 5:515.
6. Michaeli, F. (1911). Unmittelbare Effecte der Splenektomie bei einem Fall von erworbenem haemolytischen Splenomegalischen Ikterus Typus Hayem-Widal (Splenohaemolytischer Ikterus), *Wien. Klin. Wochenschr.* 24:1269.
7. Chauffard, M. A. (1907). Pathogene de l'ictere congenital de l'adulte. *Sem. Med. (Paris)* 27:25.
8. Castle, W. B., and Daland, G. A. (1937). Susceptibility of erythrocytes to hypotonic hemolysis as a function of discoidal form. *Am. J. Physiol.* 120: 371–383.
9. Emerson, C. P. Jr., Shen, S. C., Ham, T. H., Fleming, E. M., and Castle, W. B. (1956). Studies on the destruction of red blood cells. *AMA Arch. Intern. Med.* 97:1–38.
10. Reed, C. F., and Swisher, S. N. (1966). Erythrocyte lipid loss in hereditary spherocytosis. *J. Clin. Invest.* 45:777–781.
11. Cooper, R. A., and Jandl, J. H. (1969). The role of membrane lipids in the survival of red cells in hereditary spherocytosis. *J. Clin. Invest.* 48:736–744.
12. Young, L. E., Izzo, M. J., and Platzer, R. G. (1951). Hereditary spherocytosis. I. Clinical, hematologic and genetic features in twenty-eight cases, with

particular reference to the osmotic and mechanical fragility of incubated erythrocytes. *Blood* 6:1073.

13. Bernstein, S. E. (1980). Inherited hemolytic disease in mice: a review and undate. *Lab. Anim. Sci.* 30:197–205.
14. Greenquist, A. C., Shohet, S. B., and Bernstein, S. E. (1978). Marked reduction of spectrin in hereditary spherocytosis in the common house mouse. *Blood* 51:1149–1155.
15. Lux, S. E., Pease, B., Tomaselli, M. B., John, K. M., and Bernstein, S. E. (1979). Hemolytic anemias associated with deficient or dysfunctional spectrin. In *Normal and Abnormal Red Cell Membranes*, S. E. Lux, V. T. Marchesi, and C. F. Fox (Eds.). Alan R. Liss, New York, pp. 463–469.
16. Bodine, D. M. IV, Birkenmeier, C. S., and Barker, J. E. (1984). Spectrin deficient inherited hemolytic anemias in the mouse: characterization by spectrin synthesis and mRNA activity in reticulocytes. *Cell* 37:721–729.
17. Agre, P., Orringer, E. P., and Bennett, V. (1982). Deficient red-cell spectrin in severe, recessively inherited spherocytosis. *N. Engl. J. Med.* 306:1155–1161.
18. Prince, B. O. (1962). *Let's Meet the Gores*. Columbia, South Carolina.
19. Recollections and Records of Bug Hill Township, Bug Hill, North Carolina, 1977.
20. Agre, P., Casella, J. F., Zinkham, W. H., McMillan, C., and Bennett, V. (1985). Partial deficiency of erythrocyte spectrin in hereditary spherocytosis. *Nature* 314:380–383.
21. Good man, S. R., Shiffer, K. A., Casoria, L. A., and Eyster, M. E. (1982). Identification of the molecular defect in the erythrocyte membrane skeleton of some kindreds with hereditary spherocytosis. *Blood* 60:772–784.
22. Wolfe, L. C., John, K. M., Falcone, J. C., Byrne, A. M., and Lux, S. E. (1982). A genetic defect in the binding of protein 4.1 to spectrin in a kindred with hereditary spherocytosis. *N. Engl. J. Med.* 307:1367–1374.
23. Palek, J., Liu, S.-C., Liu, P. Y., Prchal, J., and Castleberry, R. P. (1981). Altered assembly of spectrin in red cell membranes in hereditary pyropoikilocytosis. *Blood* 57:130–139.
24. Knowles, W. J., Morrow, J. S., Specher, D. W., Zarkowsky, H. S., Mohandas, N., Mentzer, W. C., Shohet, S. B., and Marchesi, V. T. (1983). Molecular and functional changes in spectrin from patients with hereditary pyropoikilocytosis. *J. Clin. Invest.* 71:1867–1877.
25. Coetzer, T. L., and Palek, J. (1986). Partial spectrin deficiency in hereditary pyropoikilocytosis. *Blood* 67:919–924.
26. Agre, P., Asimos, A., Casella, J. F., and McMillan, C. (1986). Inheritance pattern and clinical response to splenectomy as a reflection of erythrocyte spectrin deficiency in hereditary spherocytosis. *N. Engl. J. Med.* 315:1579–1583.
27. Wiley, J. S. (1970). Red cell survival studies in hereditary spherocytosis. *J. Clin. Invest.* 49:666–672.
28. Chasis, J. A., Agre, P., and Mohandas, N. (1988). Decreased membrane mechanical stability. *J. Clin. Invest.* 82:617–623.

29. Winkelmann, J. C., Marchesi, S. L., Watkins, P., Linnenbach, A. J., Agre, P., and Forget, B. G. (1986). Recessive hereditary spherocytosis is associated with an abnormal alpha spectrin subunit. *Clin. Res.* 34:474A.

30. Winkelman, J., Marchesi, S., Gillespie, F., Agre, P., and Forget, B. (1986). Dominant hereditary spherocytosis is not closely linked to the gene for alpha spectrin in three kindreds. *Blood* 68 (suppl.) 68:47A.

31. Costa, F. F., Lux, S. E., Agre, P., Watkins, P., John, K., and Forget, B. (1988). Dominant hereditary spherocytosis is linked to the gene for the erythrocyte membrane protein ankyrin. *Blood* (Suppl.) 72:64A.

32. Moon, R. T., and Lazarides, E. (1983). Beta spectrin limits alpha spectrin assembly on membranes following synthesis in a chicken erythroid cell lysate. *Nature* 305:62-65.

Index

9 780367 451141